TABLE OF INTEGRALS

Basic Forms

1. $\int u^n\,du = \dfrac{1}{n+1} u^{n+1} + C, \; n \neq -1$

2. $\int \dfrac{du}{u} = \ln|u| + C$

3. $\int e^u\,du = e^u + C$

4. $\int a^u\,du = \dfrac{1}{\ln a} a^u + C$

5. $\int \sin u\,du = -\cos u + C$

6. $\int \cos u\,du = \sin u + C$

7. $\int \sec^2 u\,du = \tan u + C$

8. $\int \csc^2 u\,du = -\cot u + C$

9. $\int \sec u \tan u\,du = \sec u + C$

10. $\int \csc u \cot u\,du = -\csc u + C$

11. $\int \tan u\,du = \ln|\sec u| + C$

12. $\int \cot u\,du = \ln|\sin u| + C$

13. $\int \sec u\,du = \ln|\sec u + \tan u| + C$

14. $\int \csc u\,du = \ln|\csc u - \cot u| + C$

15. $\int \dfrac{du}{\sqrt{a^2 - u^2}} = \sin^{-1} \dfrac{u}{a} + C$

16. $\int \dfrac{du}{a^2 + u^2} = \dfrac{1}{a} \tan^{-1} \dfrac{u}{a} + C$

17. $\int \dfrac{du}{u\sqrt{u^2 - a^2}} = \dfrac{1}{a} \sec^{-1} \dfrac{u}{a} + C$

Rational Forms

18. $\int \dfrac{du}{a+bu} = \dfrac{1}{b} \ln|a+bu| + C$

19. $\int \dfrac{u\,du}{a+bu} = \dfrac{1}{b^2}(a+bu - a\ln|a+bu|) + C$

20. $\int \dfrac{u\,du}{(a+bu)^2} = \dfrac{a}{b^2(a+bu)} + \dfrac{1}{b^2} \ln|a+bu| + C$

21. $\int \dfrac{u^2\,du}{(a+bu)^2} = \dfrac{1}{b^3}\left(a+bu - \dfrac{a^2}{a+bu} - 2a\ln|a+bu|\right) + C$

22. $\int \dfrac{du}{u(a+bu)^2} = \dfrac{1}{a(a+bu)} - \dfrac{1}{a^2} \ln\left|\dfrac{a+bu}{u}\right| + C$

23. $\int \dfrac{du}{u^2(a+bu)} = -\dfrac{1}{au} + \dfrac{b}{a^2} \ln\left|\dfrac{a+bu}{u}\right| + C$

24. $\int u(a+bu)^n\,du = \dfrac{(a+bu)^{n+1}}{b^2}\left(\dfrac{a+bu}{n+2} - \dfrac{a}{n+1}\right) + C, \; n \neq -1, -2$

25. $\int \dfrac{du}{a^2 - u^2} = \dfrac{1}{2a} \ln\left|\dfrac{u+a}{u-a}\right| + C$

26. $\int \dfrac{du}{(a+bu)(c+du)} = \dfrac{1}{ad-bc} \ln\left|\dfrac{c+du}{a+bu}\right| + C, \; ad-bc \neq 0$

27. $\int \dfrac{u\,du}{(a+bu)(c+du)} = \dfrac{1}{ad-bc}\left(\dfrac{a}{b} \ln|a+bu| - \dfrac{c}{d} \ln|c+du|\right) + C, \; ad-bc \neq 0$

Forms Containing $\sqrt{a+bu}$

28. $\int u\sqrt{a+bu}\,du = \dfrac{2}{15b^2}(3bu - 2a)(a+bu)^{3/2} + C$

29. $\int u^n\sqrt{a+bu}\,du = \dfrac{2}{b(2n+3)}\left[u^n(a+bu)^{3/2} - na\int u^{n-1}\sqrt{a+bu}\,du\right]$

30. $\int \dfrac{u\,du}{\sqrt{a+bu}} = \dfrac{2}{3b^2}(bu-2a)\sqrt{a+bu} + C$

31. $\int \dfrac{u^2\,du}{\sqrt{a+bu}} = \dfrac{2}{15b^3}(8a^2 + 3b^2u^2 - 4abu)\sqrt{a+bu} + C$

32. $\int \dfrac{u^n\,du}{\sqrt{a+bu}} = \dfrac{2u^n\sqrt{a+bu}}{b(2n+1)} - \dfrac{2na}{b(2n+1)} \int \dfrac{u^{n-1}\,du}{\sqrt{a+bu}}$

33. $\int \dfrac{du}{u\sqrt{a+bu}} = \begin{cases} \dfrac{1}{\sqrt{a}} \ln\left|\dfrac{\sqrt{a+bu} - \sqrt{a}}{\sqrt{a+bu} + \sqrt{a}}\right| + C, & a > 0 \\ \dfrac{2}{\sqrt{|a|}} \tan^{-1} \sqrt{\dfrac{a+bu}{|a|}} + C, & a < 0 \end{cases}$

34. $\int \dfrac{du}{u^n\sqrt{a+bu}} = -\dfrac{\sqrt{a+bu}}{a(n-1)u^{n-1}} - \dfrac{b(2n-3)}{2a(n-1)} \int \dfrac{du}{u^{n-1}\sqrt{a+bu}}$

35. $\int \dfrac{\sqrt{a+bu}}{u}\,du = 2\sqrt{a+bu} + a\int \dfrac{du}{u\sqrt{a+bu}}$

36. $\int \dfrac{\sqrt{a+bu}}{u^2}\,du = -\dfrac{\sqrt{a+bu}}{u} + \dfrac{b}{2} \int \dfrac{du}{u\sqrt{a+bu}}$

Forms Containing $\sqrt{u^2 \pm a^2}$

37. $\int \sqrt{u^2 \pm a^2}\,du = \dfrac{u}{2}\sqrt{u^2 \pm a^2} \pm \dfrac{a^2}{2} \ln\left|u + \sqrt{u^2 \pm a^2}\right| + C$

38. $\int u\sqrt{u^2 \pm a^2}\,du = \dfrac{1}{3}(u^2 \pm a^2)^{3/2} + C$

39. $\int u^2\sqrt{u^2 \pm a^2}\,du = \dfrac{u}{8}(2u^2 \pm a^2)\sqrt{u^2 \pm a^2} - \dfrac{a^4}{8} \ln\left|u + \sqrt{u^2 \pm a^2}\right| + C$

40. $\int \dfrac{\sqrt{u^2 + a^2}}{u}\,du = \sqrt{u^2 + a^2} - a\ln\left|\dfrac{a + \sqrt{u^2 + a^2}}{u}\right| + C$

2

MULTIVARIABLE CALCULUS

E D I T I O N

2

MULTIVARIABLE CALCULUS

Leonard I. Holder
Gettysburg College

James DeFranza
St. Lawrence University

Jay M. Pasachoff
Williams College

Brooks/Cole Publishing Company
Pacific Grove, California

To

*Jean, Bill, and Steve
Regan, Sara, and David
Naomi, Eloise, and Deborah*

for their love and support

Brooks/Cole Publishing Company
A Division of Wadsworth, Inc.

Chapter 10 © 1994, 1988; Chapters 11–17 © 1995, 1988 by Wadsworth, Inc., Belmont, California 94002. All rights reserved. No part of this book may be reproduced, stored in a retrieval system, or transcribed, in any form or by any means—electronic, mechanical, photocopying, recording, or otherwise—without the prior written permission of the publisher, Brooks/Cole Publishing Company, Pacific Grove, California 93950, a division of Wadsworth, Inc.

Printed in the United States of America

10 9 8 7 6 5 4 3 2 1

Library of Congress Cataloging in Publication Data
Holder, Leonard Irvin, [date]–
   [Calculus. Chapter 10–17]
    Multivariable calculus / Leonard I. Holder, James DeFranza, Jay M.
  Pasachoff. — 2nd ed.
    p.   cm.
    Excerpt from: Calculus. 2nd ed. 1994.
    Sequel to: Single variable calculus. 2nd ed. c1994.
    Includes index.
    ISBN 0-534-24912-4
    1. Calculus.   I. DeFranza, James, [date]– . II. Pasachoff, Jay
M.   III. Title.
[QA303.H57825   1994]
515—dc20   94-10328
                                                                                    CIP

Sponsoring Editor: *Jeremy Hayhurst*
Editorial Assistant: *Elizabeth Barelli Rammel*
Production Coordinator: *Marlene Thom*
Production Assistant: *Tessa A. McGlasson*
Manuscript Editor: *Janet Tilden*
Interior and Cover Design: *Sharon L. Kinghan*
Cover Photo: *Orrery, Adler Planetarium; Satellite, European Space Agency/Photo Researchers, Inc.*
Art Coordinator: *Lisa Torri*
Interior Illustration: *MacArt Design; Tech-Graphics*
Photo Coordinator: *Robert Western*
Photo Researchers: *Diana Mara Henry; Joan Meyers*
Typesetting: *Electronic Technical Publishing*
Cover Printing: *Phoenix Color Corporation*
Printing and Binding: *Arcata Graphics/Hawkins*

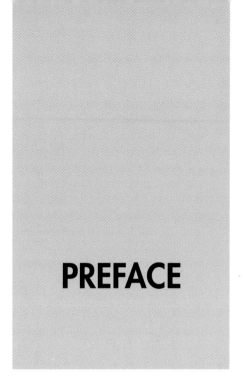

PREFACE

This book is a sequel to our text *Single Variable Calculus, Second Edition*, published by Brooks/Cole in 1994. It is equivalent to the last eight chapters of our combined text *Calculus, Second Edition*. Chapter 10, on infinite series, is repeated here in order to provide flexibility as to the semester, or quarter, in which this topic is covered. Chapters 11–17 cover the topics usually found in multivariable calculus texts: vectors in two and three dimensions, vector-valued functions, functions of more than one variable, multivariable differential calculus, multiple integrals, vector field theory, and differential equations.

We have revised this edition with special attention to making the material clearer and more concrete as well as more readable and interesting, while retaining the appropriate level of rigor. The organization, the explanations and examples, the exercises, and the text layout have all been substantially improved. We have also added several new features to the text that we believe will enhance students' perception and understanding of the subject.

NEW FEATURES

- In addition to an ample number of drill exercises, applied problems, and more theoretical exercises, we included many problems calling for the use of graphing calculators or computer algebra systems. These latter problems are marked with a special icon so that instructors can easily skip over or emphasize them, depending on local preferences and resources. The number of examples and figures has more than doubled.
- Selected applied problems have been added to the end of each chapter in special exercise sets called Applying Calculus. Some of these problems will be appropriate for class projects and, at the request of some reviewers, answers to these exercises are not in the text. Solutions are provided in the Complete Solutions Manual, Volumes 1 and 2, for instructors.

- Preceding each chapter is a "Personal View" interview or essay "about calculus." People interviewed here include representatives of statistics, history, economics, physics, biology, engineering, and computer science, as well as mathematicians. The purposes are severalfold. We hope to give a glimpse of the diversity of opinion and varied personalities of scientists and mathematicians, to neutralize some myths about mathematics, and to show that mathematics is not a static discipline. We also want students to understand that calculus is widely useful in different professions, and that teaching and mathematical research are not without controversy. We hope these "personal views" will provide a little extra motivation and perhaps even inspiration for those students who ask "where will I ever use this stuff?" We hope more students can be led to perceive that the course has utility and significance beyond the immediate grade they achieve.
- Each chapter concludes with a Concept Quiz. Answers to these quizzes are provided in the Study Guides, but students should be encouraged to attempt these quizzes and verify the answers on their own.
- Computer algebra subsections have been provided, illustrating the use of Maple and Mathematica. We want students to know what today's technology can do, even while stressing that they must learn the fundamentals of calculus rather than merely learning how to use the technology. Along this line, we have given some examples of how to follow certain topics using Maple and Mathematica. These examples can be read as prose, given the way Maple and Mathematica work, or can be worked on computers. These subsections are on a screened background so that they can be easily identified. Instructors not wishing their students to use them can teach the course without them without affecting the flow of the text.
- The use of color screening in the text makes theorems and other important results stand out. Color has also been used more effectively in figures to bring out important features. Color photographs are used in various places throughout the text to help in the understanding of concepts and add to the visual appeal of the book.

PEDAGOGICAL CONSIDERATIONS

Most students learn best by proceeding from the concrete to the more abstract, and our consistent approach is to discuss new concepts intuitively before stating formal definitions or theorems. Our style is conversational, speaking directly to the student. We have tried to explain the mathematics in simple language. We do not "talk down" to students and we do not avoid rigorous argument where it is needed. (A number of the more technical proofs are given in the Supplementary Appendices.)

We begin each chapter with a summary of the chapter goals so that students will have some idea of where the material is leading. We try to motivate new ideas via concrete problems and often ask students to make their own conjectures before we state the results. We hope that, with the right tone in the exposition, many students can be persuaded to read the text before they come to class. We even dare to hope that more of them will truly enjoy calculus and decide to major in mathematics or in a mathematically related field.

ACKNOWLEDGMENTS

First, we give our sincere thanks to the contributors of the Personal Views: Constance McMillan Elson, Allan Cormack, Joseph Newhouse, David Lieberman, N. Scott Urquhart, Alfred L. Goldberg, and Frank McGrath. The efforts of supplement authors and accuracy checkers Raymond Southworth, Ben Brown, John Banks, David Royster, Marian Hoyle, and Terri Bittner are also gratefully acknowledged. Special thanks are due to Raymond Southworth, who labored long and hard to ensure accurate answers and solutions to the exercises. Any remaining errors are our responsibility.

Jay Pasachoff thanks his daughters, Eloise and Deborah, for inspiring him through *their* studies of calculus to work on this book. He further thanks Deborah Pasachoff for her comments on the manuscript and proofs.

In addition, we thank the following reviewers and colleagues, who have also contributed a great deal to the text: Donna Bailey, Northeast Missouri State University; Nancy Blachman, Variable Symbols, Inc.; Barbara Bohannon, Hofstra University; Richard Bonnano, Deerfield Academy; Stephen Brady, Wichita State University; Fred Brauer, University of Wisconsin–Madison; Jon Breitenbucher, Ohio State University; Chris Caldwell, University of Tennessee–Martin; Mervin Childers; Roger Cooper, Canada College; Branko Curgus, Western Washington University; Deborah Frantz, Kutztown University; Anthony Ferzola, University of Scranton; Dante Giarusso, St. Lawrence University; James Hart, Middle Tennessee State University; Nabil Husni, Palm Beach Community College–North Campus; Nathaniel Martin, University of Virginia; Montie Monzingo, Southern Methodist University; Lance Nielsen, University of Nebraska–Omaha; John Randolph, West Virginia University; Janice Rech, University of Nebraska–Omaha; Robert Stanton, St. John's University; M. J. Still, Palm Beach Community College–North Campus; Richard Tucker, North Carolina A&T State University; Andrei Verona, California State University–Los Angeles; and Roger Waggoner, University of Southwestern Louisiana. We thank Joel Cohen of Rockefeller University for comments regarding the logistic equation.

We also thank the indexer, Nancy Kutner, for her expert work.

Thanks also to the Brooks/Cole staff: Elizabeth Barelli Rammel, editorial assistant, who coordinated volumes of handwritten, verbal, and electronic communication and resolved numerous details essential to the book's development; Audra Silverie, assistant editor; Sharon Kinghan, senior designer; Lisa Torri, senior art coordinator; Margaret Parks, promotions manager; Patrick Farrant, marketing manager; and Bob Western, photo coordinator. Special thanks go to Marlene Thom, production coordinator, for the superb job she did guiding the book through the complex all-electronic production process. She did so with skill, and somehow maintained her good humor, despite what was often a demanding schedule. It was a delight to work with her. Finally, we wish to thank our editor, Jeremy Hayhurst, for his support and guidance. His ideas, his vision, and his encouragement were critical in making this book possible.

Leonard I. Holder
James DeFranza
Jay M. Pasachoff

PROLOGUE FOR STUDENTS

COMPUTERS, CALCULATORS, AND CALCULUS

During your course you will undoubtedly encounter some form of technology. Both graphing calculators and computer software can be valuable aids in studying calculus, if they are used properly; however, they are not substitutes for learning the basic mathematical concepts. In this text we discuss ways in which calculators and currently available computer software can be used to best advantage in learning calculus. We are trying to give you a general sense of where these systems are most helpful and appropriate, even if you don't have a computer or a calculator at hand. We have included discussion of software in the text and graphing calculators are discussed in the Supplementary Appendices.

With the software programs and mathematical languages known as *computer algebra systems* (abbreviated "CAS"), almost all calculus operations and computations can be done quickly and efficiently. The best-known of these systems are Maple, Mathematica, and DERIVE. The term *computer algebra system* is slightly misleading, because although these systems do algebra, they do much more, and they are forcing a reevaluation of what it is that we emphasize when teaching mathematics. CAS are designed to do symbolic manipulation, solve numerical problems, and graph functions, and they can be programmed as well. Other software programs, called *spreadsheets*, are optimized for numerical calculations. They are widely used in the business world and can be useful in calculus too, especially in work with numerical methods. Spreadsheets also have the ability to graph functions. Some widely used spreadsheets are Microsoft's Excel, Lotus's 1-2-3, and Novell's Quattro Pro.

The Graphing Calculator program distributed free with Power Macintosh computers, which use the PowerPC™ chip, graphs in two and three dimensions, carries out algebraic simplifications, does numeric evaluations, and performs simple differentiations. It does not include integration. Thus, although the program does not have as much capability as Maple, Mathematica, or DERIVE, it will be useful for graphing three-dimensional surfaces in multivariable calculus. Choosing "derivative" from a menu or "differentiate" from a keyboard

that you select to have on the screen allows you to find the derivative of the function you type in. This Graphing Calculator program can be a helpful aid as you work problems to understand calculus.

Do not let yourself be intimidated by this type of technology. It is a servant in the course, not a master. No one is encouraging you to compete with or to try to outcompute an algebra system. It's important to understand that in many cases the software is using the same algorithm (correct mathematical procedure) that you use when you solve the problem "by hand," but in other cases it is using an algorithm that is only practical to use within the software. You will sometimes have to ask yourself how the computer or calculator is getting its results, and occasionally you will have to use the theory of calculus to verify that your results are correct. Technology changes rapidly, and there will undoubtedly be newer devices or releases of these programs available when you take this course. It is important to remember that the mathematics you are learning has not changed, and the problem-solving skills that you develop in this course will never be obsolete.

The text includes "computer algebra subsections" on a yellow screened background at the end of certain sections of the text. These subsections give guidance on carrying out calculus operations using one of the three common CAS. Not all the details are shown, but you should be able to follow the instructions as given. Don't panic if the commands don't work as indicated on your version of the software. You may have a different release that uses different commands. Also, your computer may have some other CAS, in which case you can adapt the instructions to your particular system. You may need to consult the program documentation carefully.

Basic scientific calculators have many uses in calculus just as they do in precalculus courses. The more advanced programmable calculators, especially those with graphing capabilities, are especially well suited to calculus. These advanced calculators share many features with computers. They have smaller viewing screens and do numerical calculations rather than the more general symbolic operations carried out by CAS. Although the calculators are more limited, they are more compact and portable and are cheaper than software and a computer. However, if you already have a computer, consider getting a CAS program. They rarely cost more than a good graphing calculator and will take you much further than a calculator will. Other advantages of the CAS are that you can program and update them easily as well as use them with a word processor.

Of course, it may not be necessary for you to choose between a computer and a calculator. If you have access to both, you will sometimes want to use one and sometimes the other; in most cases, we do not specify which device or program to use. We do not show instructions for using calculators in the text. (See the Supplementary Appendices for graphing calculator examples.) There are specially marked exercises for graphing calculators and CAS where technology is all but required, but we have deliberately not marked all the exercises where graphing calculators or CAS could be used.

APPROPRIATE USE OF TECHNOLOGY

What should you learn to do by hand and what should you have a computer or calculator do for you? This is analogous to asking whether grade-school students should learn fundamentals of multiplication and multiplication tables, or simply learn to push keys on calculators. Most mathematicians would argue that it is always better to understand the principles before you push the buttons. But perhaps there is more in common here than is first apparent. Using a multiplication or log table, memorized or not, and pushing buttons on a calculator both involve the use of an algorithm for carrying out a mathematical operation. In a sense, both you and the calculating device are simply employing an algorithm. The key is in understanding what the algorithm means and why it works. As your experience and confidence increase, you'll develop a sense of when the calculator or software will be most helpful. In many cases you will find that it is calculus itself, not the technology, that is the most useful tool.

In using technology to find derivatives, for example, it is essential that you first learn what a derivative is and what information it conveys. Only then is it meaningful to calculate derivatives using software, and only for more complicated derivatives (the simple ones can be found more easily and even faster by hand). If you rely solely on software to find all derivatives you encounter, you will have learned nothing about derivatives. Also, if you place too heavy a reliance on a computer or a calculator, you might draw incorrect conclusions from the output unless you have some idea of what to expect. When plotting a graph, for example, if you don't chose the x-scale and y-scale properly, you may not see all the important features of the graph. Without a knowledge of calculus, you would not know what the important features are and could easily come to incorrect conclusions. With a basic understanding of calculus, you can have total confidence in the correctness of your results.

Calculators and computers with CAS or spreadsheets are powerful devices and can be real assets in a calculus course, but they do not eliminate the need to learn the fundamental mathematical concepts. If your instructor prefers to overlook the CAS sections or the exercises marked for calculators or computers, you will not be disadvantaged in the course. We have tried to show how to take advantage of the availability of these devices, but the course is not dependent on them. If your instructor permits and encourages it, take advantage of what technology has to offer, but use it to learn your calculus. Once you truly understand what a derivative is, your instructor is unlikely to criticize you for using software to calculate one. If you don't understand what a derivative is, no degree of proficiency with software will compensate for this deficiency.

SOME SUGGESTIONS FOR STUDYING CALCULUS

Although a calculus course provides challenging goals, your success will have a great deal to do with how you study from your lecture notes and from this text. First, you will understand far more if you read the relevant section of the text before your lecture and before you attempt to work the exercises you are assigned. Don't treat this text as a novel. You need to read actively, with a pencil and paper at hand. Try to fill in any missing details you encounter in the algebraic steps. Second, do your assignments when they are assigned. Letting homework accumulate is a prescription for disaster in this or any other mathematics course. To learn mathematics, you must do mathematics. Watching your instructor do it is not enough. Don't give up too quickly if you have difficulty with a problem. If your instructor permits it, you might find it helpful to study with some of your classmates. Group study has proved to be more effective for some students than individual study.

As a student of calculus, you have much to look forward to as this rich and engaging subject unfolds before you. Read the chapter-opening Personal Views, not to learn calculus, but to learn about calculus and mathematics and their influence on the modern world. Recognize that calculus contains powerful ideas that are very general and applicable to a huge variety of problems, whatever your chosen discipline. Use whatever technology you have at your disposal to explore and investigate the mathematical ideas you encounter. This course should help you think about mathematics in a more logical fashion and to understand that definitions are not theorems, that theorems require proofs, and that abstract mathematical arguments are of great importance in efficiently describing the phenomena we see in the "real world." It is impossible to say what calculus is and to explain what all it can do in a brief introduction, because that, of course, is what the rest of the book is about. We hope that these preliminary remarks have served to whet your appetite and that you are ready now for the "main course." If you give calculus the attention it deserves, you will not regret the effort devoted to it in your future courses and career.

CONTENTS

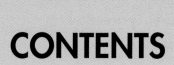

A Personal View   Calculus, Fractals, & Mathematical Intuition **Benoit Mandelbrot**   *717*

CHAPTER 10   INFINITE SERIES   719

- 10.1 **Approximation of Functions by Polynomials; Taylor's Theorem** 721
- 10.2 **Sequences** 736
- 10.3 **Infinite Series of Constants** 754
- 10.4 **Series of Positive Terms; The Integral Test and Comparison Tests** 769
- 10.5 **Series of Positive Terms; The Ratio Test and the Root Test** 777
- 10.6 **Series with Terms of Variable Signs; Absolute Convergence** 781
- 10.7 **Power Series** 789
- 10.8 **Differentiation and Integration of Power Series** 794
- 10.9 **Taylor Series** 802
- 10.10 **Fourier Series** 813

Chapter 10 Review Exercises 821
Chapter 10 Concept Quiz 823
Applying Calculus 824

A Personal View   *Vectors and Multivariable Calculus* **Constance McMillan Elson** *827*

CHAPTER 11   VECTORS IN TWO AND THREE DIMENSIONS   829

- 11.1 **Vectors in the Plane** 830
- 11.2 **The Dot Product** 837
- 11.3 **Vectors in Space** 846
- 11.4 **The Cross Product** 853
- 11.5 **Lines in Space** 864
- 11.6 **Planes** 870

Chapter 11 Review Exercises 881
Chapter 11 Concept Quiz 882
Applying Calculus 883

A Personal View   *Computer Tomography and Calculus* **Allan Cormack** *885*

CHAPTER 12   VECTOR-VALUED FUNCTIONS   887

- 12.1 **Vector-Valued Functions** 888
- 12.2 **The Calculus of Vector Functions** 893
- 12.3 **Arc Length** 902
- 12.4 **Unit Tangent and Normal Vectors; Curvature** 907
- 12.5 **Motion Along a Curve** 917
- 12.6 **Kepler's Laws** 928

Chapter 12 Review Exercises 932
Chapter 12 Concept Quiz 933
Applying Calculus 934

A Personal View   *Health-care Economics and Calculus* **Joseph Newhouse** *937*

CHAPTER 13   FUNCTIONS OF MORE THAN ONE VARIABLE   939

- 13.1 **Functions of Two and Three Variables** 940
- 13.2 **Sketching Surfaces** 955
- 13.3 **Limits and Continuity** 967
- 13.4 **Partial Derivatives** 977

Chapter 13 Review Exercises 992
Chapter 13 Concept Quiz 994
Applying Calculus 995

A Personal View   Signal Processing and Calculus **David Lieberman**   999

CHAPTER 14   MULTIVARIABLE DIFFERENTIAL CALCULUS   1001

- 14.1 **Differentiability** 1001
- 14.2 **Chain Rules** 1012
- 14.3 **Directional Derivatives** 1023
- 14.4 **Tangent Planes and Normal Lines** 1032
- 14.5 **Extreme Values** 1037
- 14.6 **Constrained Extremum Problems** 1051

  Chapter 14 Review Exercises   1060
  Chapter 14 Concept Quiz   1062
  Applying Calculus   1063

A Personal View   Multivariable Calculus in Biological Analysis **N. Scott Urquhart**   1065

CHAPTER 15   MULTIPLE INTEGRALS   1067

- 15.1 **Double Integrals** 1068
- 15.2 **Evaluating Double Integrals by Iterated Integrals** 1073
- 15.3 **Double Integrals in Polar Coordinates** 1085
- 15.4 **Moments, Centroids, and Centers of Mass Using Double Integration** 1092
- 15.5 **Surface Area** 1103
- 15.6 **Triple Integrals** 1110
- 15.7 **Triple Integrals in Cylindrical and Spherical Coordinates** 1117
- 15.8 **Changing Variables in Multiple Integrals** 1126

  Chapter 15 Review Exercises   1137
  Chapter 15 Concept Quiz   1139
  Applying Calculus   1140

A Personal View   Medicine, Cell Biology, and Calculus **Alfred L. Goldberg**   1143

CHAPTER 16   VECTOR FIELD THEORY   1145

- 16.1 **Vector Fields** 1146
- 16.2 **Line Integrals and Work** 1149
- 16.3 **Gradient Fields and Path Independence** 1161
- 16.4 **Green's Theorem** 1173
- 16.5 **Surface Integrals** 1184

- 16.6 **The Divergence Theorem** 1192
- 16.7 **Stokes's Theorem** 1199

    Chapter 16 Review Exercises 1207
    Chapter 16 Concept Quiz 1209
    Applying Calculus 1210

A Personal View *Calculus and Large Software and Hardware Systems*
Frank McGrath *1213*

CHAPTER 17 DIFFERENTIAL EQUATIONS 1215

- 17.1 **Basic Concepts; General Solutions and Particular Solutions** 1216
- 17.2 **First-Order Separable and First-Order Homogeneous Differential Equations** 1224
- 17.3 **First-Order Exact and First-Order Linear Differential Equations** 1232
- 17.4 **Applications of First-Order Differential Equations** 1239
- 17.5 **Second-Order Linear Differential Equations with Constant Coefficients: The Homogeneous Case** 1251
- 17.6 **Second-Order Linear Differential Equations with Constant Coefficients: The Nonhomogeneous Case** 1259
- 17.7 **The Vibrating Spring** 1268
- 17.8 **Series Solutions** 1275

    Chapter 17 Review Exercises 1280
    Chapter 17 Concept Quiz 1281
    Applying Calculus 1283

Answers 1286

Photo Credits 1317

Index 1318

MULTIVARIABLE CALCULUS

A PERSONAL VIEW

Benoit B. Mandelbrot

Calculus is required as the first real mathematics course taught in most colleges. Is that still a reasonable requirement in view of the mathematics that is most applicable today?

It remains perfectly reasonable for many, or even most, students, but I would prefer to see the coexistence of several distinct introductory courses. At Yale, I work actively at promoting one that is based on fractal geometry. Calculus can be taught so as to leave little room for subjective grading, so it is tempting to use it as a kind of filter to select students among otherwise qualified candidates for professional schools. This may increase the number of jobs for math teachers, but mathematics as a profession is left the poorer. On the other hand, mathematicians find it hard to claim that their field must be supported simply because it is a noble and great adventure of humanity, like great music and drama. This is because music and drama are widely understood and appreciated in the general public, whereas mathematics is even less understood than the most extreme avant-garde in art.

Do fractals have a place in the teaching of calculus?

Yes, I think they have a very important place. The popular appeal of fractals shows, in my opinion, that almost everybody has some interest for some part of mathematics. Even when the overall presentation of calculus is very traditional and basic, fractals can help with a critically important lesson early on, when derivatives and tangents are introduced. My claim is that to understand tangents, it is best to start with curves that have no tangents, such as fractal curves. To the contrary, around 1977, an eminent international mathematical education committee recommended that teachers not confuse students by explaining that a curve may fail to have a tangent at any point. The only exceptions they allowed were those courses reserved for future professional mathematicians. Obviously, I think that this recommendation was badly timed and terribly out of touch. First of all, it happened to come out the same year my book *Fractals, Form, Chance and Dimension* was published. This book treated fractal geometry as a mathematical subspecialty. It effectively destroyed the well-entrenched belief that nondifferentiability was an abstract nonphysical notion, a kind of mathematical pathology. On the contrary, it is a very useful and broad notion in describing nature. In fact, the very same features that had been called "mathematical pathologies" are the very stuff of nature. Secondly, this committee seemed to be convinced that before starting calculus students' minds are blank slates. I take issue with this perception. I am convinced that every person, at a very early age, has a kind of vague but valid intuition based on their everyday experience that denies the physical existence of tangent lines or planes. Even a child knows that one cannot draw a tangent line to a coastline or a tangent plane to the bark of an oak tree. Even those who have never heard of fractals are at ease with the basic ideas of fractal geometry; yet textbooks throughout history have argued as if those intuitive ideas simply did not exist. Today, many high school students are exposed to fractal geometry through popular science books. They already know examples of theoretical curves without tangents. Therefore, a calculus course would benefit from including some form of the following argument. "Let's talk about something that you probably already know, either intuitively or from reading or class work. Most curves you either see or can draw are best viewed as having no tangents. On the other hand, a circle has a tangent and so do other simple abstract shapes. You may feel that this is a difficult notion; but practice makes it intuitive." Calculus can serve to model natural phenomena as well as abstract phenomena. It is indispensable in science as well as higher mathematics, and if you pursue calculus far enough, you'll find it extraordinarily beautiful.

Your life has been very interesting and unconventional. How do you think your experiences influenced your decision to become a mathematician?

One of my uncles was a successful mathematician. I could have followed directly in his footsteps, but despite this family connection, my path into mathematics was convoluted. My schooling was always chaotic and thoroughly disrupted by the Second World War. My mother was a doctor and terrified of epidemics, so she made me skip early school grades and I studied privately with another of my uncles. He despised rote learning, including the alphabet and multiplication tables. This definitely made an impression; I'm still not good with either. Instead, he encouraged me to read and train my memory. Mostly we played chess. My father was a map nut, and I could read maps as early as I can remember. I've since thought that maybe the constant daily practice of playing chess and reading maps in my childhood helped develop my geometric intuition, which became my most important scientific skill.

Before World War II started we moved from Poland to Paris, then to the country outside Paris. From mid–1942 to the end of 1943, the German occupation of France was tightening, and my family decided that it was not safe for me to attend school. Periods of intense danger and fear—several times I narrowly escaped being executed or deported—were separated by periods when not much was happening. I had plenty of time for study on my own. Finally I landed in a large lycée (high school) that included two postgraduate grades designed to prepare good students for Ecole Normale and Ecole Polytechnique. You had to pass a series of notoriously tough "killer exams." Many students, some of whom later became great scientists, had to repeat the second grade. During the second week after I joined this postgraduate high school in Lyons, I found out that I had a talent I had never known or suspected.

Your geometric intuition?

Yes. Our math professor would describe a problem that he enjoyed, always stating it in terms of algebra or analytic geometry. But I didn't seem to hear the analytic problem; instead I heard a geometric problem. I soon realized that during my self-schooling, I had become intimately familiar with a large "zoo" of geometric shapes that I could instantly recognize and call upon, even when they were dressed up in analytic clothes.

The five months I spent in Lyons were among the most important of my life. I hardly ever left the school grounds because I couldn't afford to buy a meal elsewhere and we feared and avoided the Nazi boss in Lyons (I later found out his name was Klaus Barbie). I had a burning desire to do well. I worked at a rate that I could not have sustained indefinitely, learning the basics for the exams and polishing my geometric skills. But becoming an algebra whiz was not my goal. Anyhow, I was supposed to take the exams "for the practice." To my amazement and that of nearly everyone else, not only did I pass, I finished a very close second in all of France.

Did these exams contain calculus topics?

Yes, I remember a conversation with a professor who complained about a problem involving a triple integral that he couldn't solve in the time allowed on the exam. I told him that I had recognized that the integral was simply the volume of the sphere in certain coordinates suggested by the geometry. After I had explained my reasoning he went away repeating "of course, of course."

So you were mathematically precocious because of your gift?

Yes. My university schooling continued and I became certified in several disciplines but only because of my geometric gift. But I missed the main goal of certification, which is to train students for a profession by providing guidance and role models. This never happened to me, in any profession. Unlike most mathematicians I did my best known work later in life. I didn't devote substantial time to mathematical research until I was 40 and I discovered the Mandelbrot Set in my mid-fifties.

Your schooling may have adversely affected your early academic career, but it obviously gave you other opportunities.

There are many reasons why my particular story isn't a good model for academic success. Even though established fields find it hard to deal with diversity in training, diversity is just as indispensable to science as it is to society. There are those who believe that a given discipline at a given instant should have a single "best way" of training new workers in that field, and many believe that this best way is determined by some kind of inevitable tide of history. Were this true, those who depart from the one true way would deserve to be shunned by the mainstream. That is indeed how things happen, but this is an absolutely deplorable situation. It may not be corrected, but it must be mitigated. To mitigate it, society must intervene to allow at least a few exceptions to survive.

Could you say very briefly how you came to the ideas of fractal geometry?

It would be nice to say that fractal geometry was born in some kind of "Eureka moment," like when Archimedes ran from his overflowing bathtub. It would also be nice to respond by identifying the source in some early application, like records of prices, errors in telephone lines, or the shape of coastlines. But this would be misleading. Fractal geometry didn't start with an overreaching idea centered on an already formulated problem, much less a problem already recognized as significant enough to justify the effort needed to answer it. Let me point out that *problem solving* and *problem setting* seem to involve different talents. In the popular view—and in the restricted views of many mature disciplines—science is viewed as problem solving. I regret this, perhaps because my talent lies mainly in problem setting.

The study of fractals proceeded from bottom to top. It went from a mess of disregarded and disconnected odds and ends, on to ill-organized parts, and then toward increasing organization. Of the disconnected odds and ends with which I started, some were theoretical. In that sense, fractal geometry has incorporated many geometric shapes that were known long before I was born, but remained scattered with no common idea behind them. This is also why they were given no name, other than "mathematical monsters." My present name for them is *protofractals*. They were part of the zoo of shapes that played a central role in my early career. A second source of disconnected odds and ends were miscellaneous facts about nature. Most people don't register facts that do not yet fit into any established theory, but I was blessed with a sponge-like memory and could recall without conscious effort all these odd shapes and all these odd facts. I also read about mathematics a lot. On many occasions I was reading about some new mathematical shape and thought instantly and spontaneously of some real empirical phenomenon that somehow had the same "taste." On other occasions, I went from fact to formula. Each time, I dropped whatever I had been doing and followed up this new link between form and substance. So the "flavor" of my scientific activity has been oscillating back and forth. It was, at different times, dominated either by the search for more facts—to be identified in the literature or discovered in the laboratory, or by the search for new mathematics—again, to be identified in the literature or developed from scratch. My skills can best be described as "having a nose" for the subject.

Benoit B. Mandelbrot is best known for his *Fractal Geometry of Nature* (1982). "The father of fractal geometry," Professor Mandelbrot continues to work in this new field, with its concrete applications, and its growing impact on art and on high-school and college science teaching. Professor Mandelbrot worked for 35 years at the IBM T. J. Watson Research Center, where he is now IBM Fellow Emeritus. His contributions, first viewed as isolated and peripheral, are now widely recognized as central to the new ideas concerning chaos and complexity. Latest among his numerous awards is the 1993 Wolf Prize for Physics.

CHAPTER

INFINITE SERIES

Leonhard Euler (1707–1783) began his research in mathematics and physics at age 18. He joined the two Bernoulli brothers in St. Petersburg, Russia, in 1725 at the newly organized St. Petersburg Academy of Sciences and stayed for 14 years.

Euler's mathematical work ranged from the foundations of mathematical physics to the theory of lunar and planetary motion to differential geometry of surfaces. He devoted much attention to the problem of logarithms of negative numbers, the equation of a vibrating string, and problems in optics.

Euler investigated many functions and extensively studied power series. He discovered, for example, that $\frac{\pi}{2} - \frac{x}{2} = \sin x + \sin \frac{2x}{2} + \sin \frac{3x}{3} + \ldots$, a type of Fourier series.

Euler provided many notations we use today: the symbol e for the base of natural logarithms, the use of f and of parentheses for a function, the modern signs for trigonometric functions, the Greek Sigma (Σ) for sum, and the letter i for the square root of -1. The use of the letters a, b, and c for the sides of a triangle is also due to Euler. One of the most prolific mathematicians in history, Euler published more than 500 books and articles.

In 1736 the great Swiss mathematician Leonhard Euler (pronounced "oiler") discovered the remarkable formula

$$1 + \frac{1}{2^2} + \frac{1}{3^2} + \frac{1}{4^2} + \cdots = \frac{\pi^2}{6} \tag{10.1}$$

The sum on the left-hand side of the equation is an example of what is called an *infinite series* (or simply a *series*). The three dots mean that we are to continue adding terms indefinitely. The question immediately arises as to how we can find the sum of infinitely many terms. We will give a formal definition soon, but for now you can think of the sum as the value that is *approached* (if such a value exists) as we add more and more terms. The discovery of Formula 10.1 is a fascinating story in the history of mathematics.

We will not try to show how Euler found the sum $\pi^2/6$ for the series in Equation 10.1 but will indicate another way of obtaining the result in Exercise 8 of Exercise Set 10.10. To illustrate the idea of the sum of an infinite series, we consider the simpler series

$$1 + \frac{1}{2} + \frac{1}{4} + \frac{1}{8} + \frac{1}{16} + \cdots \tag{10.2}$$

in which each term after the first is one-half the preceding one. By starting with the first term, and listing the finite sums obtained by successively adding on one more term, we get the following:

1	$1\frac{1}{2}$	$1\frac{3}{4}$	$1\frac{7}{8}$	$1\frac{15}{16}$
first term	sum of first two terms	sum of first three terms	sum of first four terms	sum of first five terms

This list of finite sums is called the *sequence of partial sums* of the series. It appears that as we add more and more terms, the finite sums come arbitrarily close to 2. For this reason we say that the *infinite* sum *is* 2, and write

$$1 + \frac{1}{2} + \frac{1}{4} + \frac{1}{8} + \frac{1}{16} + \cdots = 2$$

We say that the series *converges* to 2.

Euler was one of the most prolific writers of mathematics in history.

Not all series converge. For example, the sum of the series $1+2+3+4+\cdots$ clearly becomes arbitrarily large as we add on more and more terms. As another example, the series $1-1+1-1+1-\cdots$ does not approach any one value as we add more and more terms, since the finite sums alternate between 1 and 0, depending on where we stop. When a series fails to converge to a unique finite number, we say that it *diverges*. So both of the series $1+2+3+4+\cdots$ and $1-1+1-1+\cdots$ diverge.

As one final example, consider the series

$$1 + \frac{1}{2} + \frac{1}{3} + \frac{1}{4} + \cdots$$

composed of the reciprocals of consecutive positive integers. Does this series converge? If so, can you find its sum? Or does it diverge? You should make a conjecture by considering the sums of the first 10, 20, or even 100 terms using a CAS, spreadsheet, or calculator. We will return to this series in Section 10.2 and answer the questions raised here.

Infinite series can also be made up of variable terms. The simplest ones involve integral powers of a variable. An example is

$$1 + x + x^2 + x^3 + \cdots \tag{10.3}$$

We call such a series a *power series* in x, since each term is a power of x. We would expect the sum, if it exists, to depend on x; that is, the sum is a function of x. So, for the series 10.3, we can define a function f by

$$f(x) = 1 + x + x^2 + x^3 + \cdots$$

In general, a power series will converge for some values of x and diverge for other values (although it may converge for all values). For example, if we set $x = \frac{1}{2}$ in the series 10.3, we get the series 10.2, which converges to 2, as we have seen; but if we set $x = 1$, we get $1 + 1 + 1 + \cdots$, which clearly diverges.

If we consider only a finite number of terms of the series 10.3, we have a polynomial of the form

$$1 + x + x^2 + \cdots + x^n \quad (n \geq 1)$$

which is an approximation to the infinite series and consequently to the function determined by the series. As we take polynomials of higher and higher degree (that is, as $n \to \infty$), the approximations become better and better, and in the limit we obtain the exact value of the function (assuming x is restricted to those values for which the series converges). Suppose now, instead of beginning with a series such as 10.3, we begin with a function $f(x)$. An important question is whether we can find an nth degree polynomial approximating $f(x)$ in such a way that as $n \to \infty$, the resulting "infinite polynomial" converges to $f(x)$ exactly.

For example, we will show in the next section that

$$e^x \approx 1 + x + \frac{x^2}{2!} + \frac{x^3}{3!} + \frac{x^4}{4!} + \cdots + \frac{x^n}{n!} \tag{10.4}$$

and that the approximation becomes better and better as we let n increase. So we might expect that we can let $n \to \infty$ and obtain the exact equality

$$e^x = 1 + x + \frac{x^2}{2!} + \frac{x^3}{3!} + \frac{x^4}{4!} + \cdots \tag{10.5}$$

Passing from the finite sum in Equation 10.4 to the infinite sum in Equation 10.5 and concluding that this latter sum converges to e^x requires justification. In this case the result is valid, as we will see later.

Our procedure in this chapter will be to begin with approximation of functions by polynomials. Next, we will discuss sequences and their limits, since, as we have indicated, the sum of an infinite series is determined from its sequence of partial sums. Then we will consider the general question of when an infinite series converges or diverges. In particular, we will study various convergence criteria for series of constants. Afterward, we will study power series and their properties. Then we will return to the crucial question of whether we can extend finite polynomial approximations of a function to an infinite power series that represents the function. We conclude the chapter by showing another way to represent functions by infinite series, one that involves sines and cosines instead of powers of x. This last kind of series, known as a *Fourier series,* is especially important in physics.

10.1 APPROXIMATION OF FUNCTIONS BY POLYNOMIALS; TAYLOR'S THEOREM

Polynomial functions are in many ways the simplest functions of all. First, they are the easiest to evaluate, requiring only addition and multiplication (since raising to positive integral powers is just repeated multiplication). Second, they possess derivatives of all orders, which are easily computed, for all real numbers. Third, they can be integrated with ease over any interval. In contrast, nonpolynomial functions can be difficult or impossible to evaluate, differentiate, or integrate. Even the exponential function $f(x) = e^x$ does not lend itself to easy evaluation. For example, try finding $e^{0.1}$ without the aid of a calculator, a computer, or tables. Similarly, $\ln 2$, $\sin 0.5$, and $\sqrt[3]{9.2}$ cannot readily be approximated as decimal quantities. We encounter the same problem in evaluating definite integrals. Although we can sometimes readily get the "answer," as in

$$\int_{\pi/6}^{\pi/2} \cot x \, dx = \ln(\sin x)\Big]_{\pi/6}^{\pi/2} = \ln\left(\sin\frac{\pi}{2}\right) - \ln\left(\sin\frac{\pi}{6}\right)$$

$$= \ln 1 - \ln\frac{1}{2} = \ln 2$$

we are again confronted with the difficulty of giving this result as a decimal quantity. The situation is even worse with an integral such as

$$\int_0^1 e^{-x^2} dx$$

where no elementary antiderivative of the integrand exists. We could use Simpson's Rule (see Section 5.5) to approximate the integral in this case. If, however, we could find a polynomial approximation to e^{-x^2}, we could easily perform the integration on the polynomial, and the result would be an approximation to the given integral. We have seen that the function $f(x) = e^{-x^2}$ is especially important in probability theory.

Of course, calculators, computers, and tables *are* available, and with these a very large number of functions can be evaluated to a high degree of accuracy.

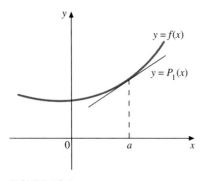

FIGURE 10.1
$P_1(x) = f(a) + f'(a)(x - a)$ is the linear approximation to $f(x)$ near $x = a$.

It is instructive, however, to learn how calculators and computers might arrive at the answers and how tables are constructed. Also, it is important to be able to determine just how accurate the approximations are. We will see answers to these questions in the procedures we develop in this section.

Linear and Quadratic Approximation

In Section 3.9 we discussed linear approximation of a function f near a point $x = a$. We found that if $f'(a)$ exists, then for x near a,

$$f(x) \approx f(a) + f'(a)(x - a)$$

If we let

$$P_1(x) = f(a) + f'(a)(x - a) \tag{10.6}$$

then P_1 is the first-degree polynomial that best approximates f near a. The graph of P_1 is just the tangent line to the curve $y = f(x)$ at the point for which $x = a$. (See Figure 10.1.) The graph of P_1 has two things in common with the graph of f: both graphs go through the point $(a, f(a))$, and they both have the slope $f'(a)$ at that point.

If we also know the concavity of the curve $y = f(x)$ at $x = a$, it would make sense to approximate f with a polynomial that not only goes through the point $(a, f(a))$ and has slope $f'(a)$ there, but also has the same concavity as f at $x = a$. (See Figure 10.2.) That is, we want a quadratic polynomial $P_2(x)$ that satisfies the following three conditions:

$$P_2(a) = f(a) \quad \text{Same value at } x = a$$
$$P_2'(a) = f'(a) \quad \text{Same slope at } x = a$$
$$P_2''(a) = f''(a) \quad \text{Same concavity at } x = a$$

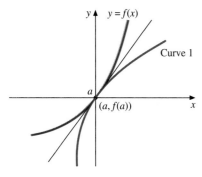

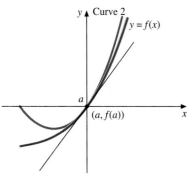

FIGURE 10.2
Curves 1 and 2 both go through the point $(a, f(a))$ with slope $f'(a)$, but Curve 2 better approximates the graph of f near that point, because its concavity agrees with that of f.

(We use the subscript 2 to indicate that P_2 is a polynomial of degree 2.) Let us write such a polynomial as

$$P_2(x) = c_0 + c_1 x + c_2 x^2$$

and determine what values the coefficients c_0, c_1, and c_2 must have so that our three conditions are satisfied. By differentiating $P_2(x)$ and substituting $x = a$ in $P_2(x)$, $P_2'(x)$, and $P_2''(x)$, our three conditions can be written as

$$P_2(a) = c_0 + c_1 a + c_2 a^2 = f(a)$$
$$P_2'(a) = c_1 + 2c_2 a = f'(a)$$
$$P_2''(a) = 2c_2 = f''(a)$$

From the third equation, we see that

$$c_2 = \frac{f''(a)}{2}$$

Thus, on substituting this value of c_2 in the second equation, we have

$$c_1 = f'(a) - 2\left(\frac{f''(a)}{2}\right)a = f'(a) - af''(a)$$

and from the first equation,

$$c_0 = f(a) - \frac{f''(a)}{2}a^2 - [f'(a) - af''(a)]a$$

$$= f(a) - af'(a) + \frac{a^2}{2}f''(a)$$

Our polynomial can now be written in the form

$$P_2(x) = \overbrace{f(a) - af'(a) + \frac{a^2}{2}f''(a)}^{c_0} + \overbrace{[f'(a) - af''(a)]}^{c_1} x + \overbrace{\frac{f''(a)}{2}}^{c_2} x^2$$

By rearranging the right-hand side and factoring out the derivatives $f'(a)$ and $f''(a)$, we obtain

$$P_2(x) = f(a) + f'(a)(x - a) + \frac{f''(a)}{2}(x^2 - 2ax + a^2)$$

or, finally,

$$P_2(x) = f(a) + f'(a)(x - a) + \frac{f''(a)}{2}(x - a)^2 \qquad (10.7)$$

EXAMPLE 10.1 Find the linear and quadratic polynomial approximations $P_1(x)$ and $P_2(x)$ for the function $f(x) = e^x$ near $x = 0$. Approximate $e^{0.1}$ using both $P_1(0.1)$ and $P_2(0.1)$.

Solution Taking $a = 0$ in Equations 10.6 and 10.7, we get

$$P_1(x) = f(0) + f'(0)x$$

and

$$P_2(x) = f(0) + f'(0)x + \frac{f''(0)}{2}x^2$$

Since $f(x) = e^x$, both $f'(x)$ and $f''(x)$ also equal e^x. So $f(0) = 1$, $f'(0) = 1$, and $f''(0) = 1$. We therefore have

$$P_1(x) = 1 + x \qquad \text{and} \qquad P_2(x) = 1 + x + \frac{x^2}{2}$$

In Figure 10.3 we show the graph of $y = e^x$ along with the graphs of $y = P_1(x)$ and $y = P_2(x)$.

When $x = 0.1$, we have

$$P_1(0.1) = 1.1 \qquad \text{and} \qquad P_2(0.1) = 1.105$$

Using a calculator, we can find that

$$e^{0.1} \approx 1.10517092$$

to eight places. So $P_2(0.1)$ approximates $e^{0.1}$ to three places of accuracy. ∎

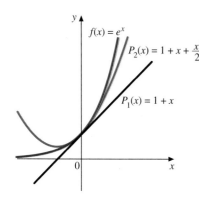

FIGURE 10.3

Taylor Polynomials of Degree n

Suppose now that f possesses derivatives up through the nth order at $x = a$. Our objective now is to find the polynomial of the nth degree, $P_n(x)$, whose first n derivatives at $x = a$ agree with the corresponding first n derivatives of

f at $x = a$ and whose value agrees with f at $x = a$. We know the result for $n = 1$ and $n = 2$. The pattern of the polynomials in these two cases suggests that we write $P_n(x)$ in the form

$$P_n(x) = c_0 + c_1(x - a) + c_2(x - a)^2 + \cdots + c_n(x - a)^n$$

Computing the successive derivatives, we obtain

$$P_n'(x) = c_1 + 2c_2(x - a) + 3c_3(x - a)^2 + \cdots + nc_n(x - a)^{n-1}$$
$$P_n''(x) = \quad\quad 2c_2 \quad\quad + 3 \cdot 2c_3(x - a) + \cdots + n(n - 1)c_n(x - a)^{n-2}$$
$$P_n'''(x) = \quad\quad\quad\quad\quad 3 \cdot 2c_3 \quad\quad + \cdots + n(n - 1)(n - 2)c_n(x - a)^{n-3}$$
$$\vdots$$
$$P_n^{(n)}(x) = \quad\quad\quad\quad\quad\quad\quad\quad\quad\quad n!\, c_n$$

Setting $x = a$ and equating $P_n(a)$ with $f(a)$ and $P_n^{(k)}(a)$ with $f^{(k)}(a)$ for $k = 1, \ldots, n$, gives

$$P_n(a) = c_0 \quad = f(a)$$
$$P_n'(a) = c_1 \quad = f'(a)$$
$$P_n''(a) = 2c_2 \quad = f''(a), \quad \text{so } c_2 = \frac{f''(a)}{2!}$$
$$P_n'''(a) = 3 \cdot 2c_3 = f'''(a), \quad \text{so } c_3 = \frac{f'''(a)}{3!}$$
$$\vdots$$
$$P_n^{(n)}(a) = n!\, c_n \quad = f^{(n)}(a), \quad \text{so } c_n = \frac{f^{(n)}(a)}{n!}$$

We can write the kth coefficient c_k as

$$c_k = \frac{f^{(k)}(a)}{k!}, \quad k = 0, 1, 2, \ldots, n$$

where $f^{(0)}(a)$ means $f(a)$. The polynomial $P_n(x)$ with these coefficients is named after the English mathematician Brook Taylor (1685–1731).

Definition 10.1
The nth Taylor Polynomial

Let f possess derivatives up through the nth order at $x = a$. The polynomial

$$P_n(x) = f(a) + f'(a)(x - a) + \frac{f''(a)}{2!}(x - a)^2 + \cdots + \frac{f^{(n)}(a)}{n!}(x - a)^n \quad (10.8)$$

is called the **nth Taylor polynomial for f about $x = a$**.

REMARK
■ As our derivation of $P_n(x)$ shows, $P_n(x)$ is the polynomial of degree n in powers of $x - a$ that agrees with $f(x)$ at $x = a$ and whose first n derivatives agree with the corresponding derivatives of $f(x)$ at $x = a$.

EXAMPLE 10.2 Find the Taylor polynomials $P_3(x)$, $P_4(x)$, and $P_5(x)$ about $x = 0$ for the function $f(x) = e^x$. Discuss the approximations to $e^{0.1}$ and $e^{1.0}$ by each of these polynomials.

Solution Recall that in Example 10.1 we found $P_1(x)$ and $P_2(x)$ for this same function. To find $P_3(x)$, $P_4(x)$, and $P_5(x)$ we use Equation 10.8 with $a = 0$. Since all derivatives of $f(x) = e^x$ also equal e^x, their common value at $x = 0$ is 1. So we have

$$P_3(x) = 1 + x + \frac{x^2}{2!} + \frac{x^3}{3!}$$

$$P_4(x) = 1 + x + \frac{x^2}{2!} + \frac{x^3}{3!} + \frac{x^4}{4!}$$

$$P_5(x) = 1 + x + \frac{x^2}{2!} + \frac{x^3}{3!} + \frac{x^4}{4!} + \frac{x^5}{5!}$$

where $k! = 1 \cdot 2 \cdot 3 \cdot \cdots \cdot k$. The computer-generated graphs in Figure 10.4 show that P_3, P_4, and P_5 approximate f more and more closely as the degree of the Taylor polynomial increases. For x near 0, the polynomial graphs are almost indistinguishable from the graph of f. As we move farther from 0 the approximation is not as good. To confirm this observation, we compute values of each of the polynomials at $x = 0.1$ and at $x = 1.0$.

$$P_3(0.1) \approx 1.105167 \qquad P_3(1.0) = 2.6167$$
$$P_4(0.1) \approx 1.1051708 \qquad P_4(1.0) = 2.7083$$
$$P_5(0.1) \approx 1.105170917 \qquad P_5(1.0) = 2.71667$$

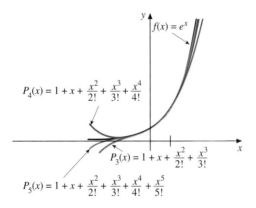

FIGURE 10.4

As we found with a calculator in Example 10.1, $f(0.1) = e^{0.1} \approx 1.10517092$ to eight places. Also, $f(1.0) = e^{1.0} \approx 2.71828183$ to eight places. We see, then, that for $x = 0.1$ each of the polynomials gives an answer close to the true value. In fact, $P_3(0.1)$, if rounded, is correct to five places, $P_4(0.1)$ is correct to six places, and $P_5(0.1)$ is correct to eight places.

In contrast, $P_3(1.0)$ is not even correct to one decimal place, $P_4(1.0)$ is correct to one place only, and $P_5(1.0)$ is correct only to two places. If we wanted more accuracy at $x = 1$, we would have to use a higher-degree Taylor polynomial. Also, to approximate e^x for values of x near $x = 1$, it would be better to use Taylor polynomials about $x = 1$ than about $x = 0$. ■

As this example shows, the farther x is from the point $x = a$ (in our case $a = 0$), the less accurate is the Taylor polynomial approximation $P_n(x)$ for any fixed n. Also, for a fixed n we get better and better approximations by $P_n(x)$ as

n increases. Although these observations are valid for the function $f(x) = e^x$, it is not clear that they are true for other functions. We will have more to say about errors in approximating functions by their Taylor polynomials shortly. First, let us look at another example.

EXAMPLE 10.3 Find the Taylor polynomial $P_n(x)$ about $x = 0$ for the function $f(x) = \cos x$.

Solution To apply Formula 10.8, we need to calculate successive derivatives of $f(x) = \cos x$ and evaluate them at $x = 0$.

$$\begin{aligned}
f(x) &= \cos x & f(0) &= 1 \\
f'(x) &= -\sin x & f'(0) &= 0 \\
f''(x) &= -\cos x & f''(0) &= -1 \\
f'''(x) &= \sin x & f'''(0) &= 0 \\
f^{(4)}(x) &= \cos x & f^{(4)}(0) &= 1 \\
&\vdots & &\vdots
\end{aligned}$$

The pattern is now clear: $f^{(n)}(0) = \pm 1$ when n is even, and $f^{(n)}(0) = 0$ when n is odd. Let $n = 2m$, where $m = 0, 1, 2, 3, \ldots$. Then n is even. Since the signs of $f^{(2m)}(0)$ alternate from $+1$ to -1, we can write

$$f^{(2m)}(0) = (-1)^m, \qquad m = 0, 1, 2, \ldots$$

Thus, by Formula 10.8,

$$P_{2m}(x) = 1 - \frac{x^2}{2!} + \frac{x^4}{4!} - \frac{x^6}{6!} + \cdots + \frac{(-1)^m x^{2m}}{(2m)!} = \sum_{k=0}^{m} \frac{(-1)^k x^{2k}}{(2k)!}$$

Note that since $f^{(2m+1)}(0) = 0$, it follows that $P_{2m+1}(x) = P_{2m}(x)$ for $f(x) = \cos x$.

In Figure 10.5 we show a computer-generated graph of $f(x)$ along with $P_0(x)$, $P_2(x)$, $P_4(x)$, $P_6(x)$, $P_8(x)$, and $P_{10}(x)$. ∎

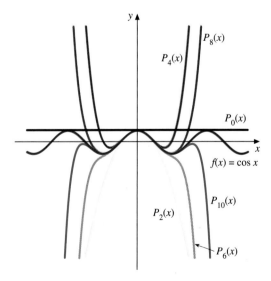

FIGURE 10.5

The Error in the Taylor Polynomial Approximation

We know that the nth Taylor polynomial $P_n(x)$ and its first n derivatives agree with $f(x)$ at $x = a$, and that for x near a, $P_n(x) \approx f(x)$. But just how good is this approximation? The following theorem provides the answer. We give its proof in Supplementary Appendix 5.

THEOREM 10.1

Taylor's Formula with Remainder

Let f have derivatives up through the $(n+1)$st order in an open interval I centered at $x = a$. Then for each x in I there is a number c between a and x such that

$$f(x) = f(a) + f'(a)(x-a) + \frac{f''(a)}{2!}(x-a)^2 + \cdots + \frac{f^{(n)}(a)}{n!}(x-a)^n + R_n(x) \tag{10.9}$$

where

$$R_n(x) = \frac{f^{(n+1)}(c)}{(n+1)!}(x-a)^{n+1} \tag{10.10}$$

REMARKS

- Formula 10.9 (Taylor's Formula) can be written more briefly as $f(x) = P_n(x) + R_n(x)$.
- $R_n(x)$ is called the **Lagrange form of the remainder**, after Joseph Louis Lagrange (1736–1813), one of the greatest mathematicians of the eighteenth century. The term *remainder* is used here to mean what is left when the approximation $P_n(x)$ is subtracted from $f(x)$. It is the error in approximating $f(x)$ by $P_n(x)$.
- The number c as stated in the theorem is dependent on x. It is between a and x, whether x is greater than a or less than a.
- The interval I may be the entire real line, $(-\infty, \infty)$, or it may be finite, in which case it will be of the form $I = (a - r, a + r)$ for some $r > 0$. The size of the interval is determined by the point closest to a for which $f^{(n+1)}(x)$ fails to exist. If this point is r units from a, then $I = (a - r, a + r)$. If $f^{(n+1)}(x)$ exists everywhere, then $I = (-\infty, \infty)$.
- Note that the remainder term $R_n(x)$ follows the same pattern as the terms in $P_n(x)$ *except* that the derivative is evaluated at the intermediate value c between a and x, rather than at a.

When $n = 0$ in Taylor's Formula and we replace x with b, where b is in the interval I and $b \neq a$, the result is

$$f(b) = f(a) + f'(c)(b-a)$$

or equivalently,

$$f'(c) = \frac{f(b) - f(a)}{b - a}$$

where c is between a and b. But this formula is just the statement of the conclusion in the Mean-Value Theorem (Theorem 4.9). For this reason Theorem 10.1 is sometimes called the *Extended Mean-Value Theorem*. (We should note that the Mean-Value Theorem is used in the proof of Theorem 10.1, so we cannot use this theorem as an independent way of getting the Mean-Value Theorem.)

EXAMPLE 10.4 Find $P_n(x)$ and $R_n(x)$ about $x = 0$ for $f(x) = e^x$. Use the result to find a bound on the error in estimating $e^{0.1}$ using $P_3(0.1)$.

Solution Note first that $f(x) = e^x$ possesses derivatives of all orders everywhere, all equal to e^x, so that in this case Theorem 10.1 is valid on $(-\infty, \infty)$. Taking $a = 0$ in Theorem 10.1, we have

$$P_n(x) = 1 + x + \frac{x^2}{2!} + \frac{x^3}{3!} + \cdots + \frac{x^n}{n!}$$

and

$$R_n(x) = \frac{f^{(n+1)}(c)}{(n+1)!} x^{n+1} = \frac{e^c}{(n+1)!} x^{n+1}$$

where c is between 0 and x. Thus, the error in using $P_3(0.1)$ to estimate $e^{0.1}$ is

$$R_3(0.1) = \frac{e^c}{4!}(0.1)^4, \qquad 0 < c < 0.1$$

Although we do not know c, we can get an upper bound on e^c by the fact that

$$e^c < e^{0.1} < e^1 < 3$$

This is a very generous upper bound, since $e^{0.1}$ is much less than e, but since we presumably do not know $e^{0.1}$, we can take something larger, with a value that we do know. Thus,

$$R_3(0.1) < \frac{3}{4!}(0.1)^4 = \frac{3}{4 \cdot 3 \cdot 2 \cdot 1}(0.0001) = 0.0000125$$

Even using the generous upper bound on e^c, we see that the error is small, resulting in accuracy to at least four decimal places. Recall that in Example 10.2 we found that $P_3(0.1) \approx 1.10517$, which is correct to five decimal places. ∎

EXAMPLE 10.5 Find Taylor's Formula with remainder for arbitrary n for the function $f(x) = \ln x$ about $x = 1$, and state the interval in which it is valid.

Solution In abbreviated form, Taylor's Formula is $f(x) = P_n(x) + R_n(x)$. For P_n we need the first n derivatives of f, and for R_n we need $f^{(n+1)}$. So we begin by calculating the successive derivatives:

$$\begin{aligned}
f(x) &= \ln x & f(1) &= 0 \\
f'(x) &= \tfrac{1}{x} = x^{-1} & f'(1) &= 1 \\
f''(x) &= -x^{-2} & f''(1) &= -1 \\
f'''(x) &= 2x^{-3} & f'''(1) &= 2 \\
f^{(4)}(x) &= -3 \cdot 2x^{-4} = -3!\, x^{-4} & f^{(4)}(1) &= -3! \\
f^{(5)}(x) &= 4 \cdot 3!\, x^{-5} = 4!\, x^{-5} & f^{(5)}(1) &= 4! \\
&\vdots & &\vdots \\
f^{(n)}(x) &= (-1)^{n-1}(n-1)!\, x^{-n} & f^{(n)}(1) &= (-1)^{n-1}(n-1)!
\end{aligned}$$

For $R_n(x)$ we need $f^{(n+1)}(x)$ evaluated at $x = c$:
$$f^{(n+1)}(x) = (-1)^n n! \, x^{-(n+1)} \qquad f^{(n+1)}(c) = (-1)^n n! \, c^{-(n+1)}$$

Observe carefully how we handled the coefficients of the successive derivatives. First, they alternate in sign—they are positive when the order is odd and negative when the order is even. The factor $(-1)^{n-1}$ in the nth derivative accomplishes the alternation of the signs. Second, we did not multiply out the factors as they accumulated but rather indicated the product as a factorial. Doing so enabled us to recognize the general pattern. We also used the fact that $k! = k(k-1)!$ for $k \geq 1$. You should convince yourself of this equality.

Since
$$f^{(n+1)}(x) = \frac{(-1)^n n!}{x^{n+1}}$$

is not defined when $x = 0$, the largest interval about $x = 1$ for which Taylor's Formula is valid is $(0, 2)$. (Later we will see that the formula remains valid at the right endpoint $x = 2$, but for now we will consider only the open interval.)

Now we can find $P_n(x)$ and $R_n(x)$. For $x \in (0, 2)$,

$$\begin{aligned}f(x) &= P_n(x) + R_n(x) \\ &= f(1) + f'(1)(x-1) + \frac{f''(1)}{2!}(x-1)^2 + \frac{f'''(1)}{3!}(x-1)^3 \\ &\quad + \frac{f^{(4)}(1)}{4!}(x-1)^4 + \cdots + \frac{f^{(n)}(1)}{n!}(x-1)^n + \frac{f^{(n+1)}(c)}{(n+1)!}(x-1)^{n+1} \\ &= 0 + 1(x-1) + \frac{-1}{2!}(x-1)^2 + \frac{2}{3!}(x-1)^3 + \frac{-3!}{4!}(x-1)^4 + \cdots \\ &\quad + \frac{(-1)^{n-1}(n-1)!}{n!}(x-1)^n + \frac{(-1)^n n!}{(n+1)! \, c^{n+1}}(x-1)^{n+1}\end{aligned}$$

or
$$\begin{aligned}f(x) &= (x-1) - \frac{(x-1)^2}{2} + \frac{(x-1)^3}{3} - \frac{(x-1)^4}{4} + \cdots \\ &\quad + \frac{(-1)^{n-1}(x-1)^n}{n} + R_n(x)\end{aligned}$$

where
$$R_n(x) = \frac{(-1)^n (x-1)^{n+1}}{(n+1)c^{n+1}}, \qquad c \text{ between } 1 \text{ and } x$$

In Figure 10.6 we show the graphs of $f(x) = \ln x$, along with the graphs of $P_n(x)$ for $n = 2, 5,$ and 10. Notice how all the graphs of $P_n(x)$ differ significantly from that of $\ln x$ for $x > 2$. ∎

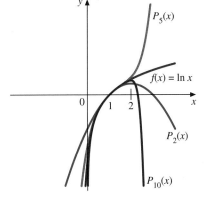

FIGURE 10.6

EXAMPLE 10.6 Using the results of Example 10.5,

(a) find $P_5(1.2)$ and an upper bound on the error in using $P_5(1.2)$ to approximate $\ln 1.2$, and

(b) find the smallest integer n that will guarantee that $P_n(1.2)$ estimates $\ln 1.2$ with eight-decimal-place accuracy.

Solution

(a) From Example 10.5,
$$P_5(x) = (x-1) - \frac{(x-1)^2}{2} + \frac{(x-1)^3}{3} - \frac{(x-1)^4}{4} + \frac{(x-1)^5}{5}$$

and
$$R_5(x) = \frac{(-1)^5(x-1)^6}{6c^6}, \qquad c \text{ between } 1 \text{ and } x$$

So
$$P_5(1.2) = 0.2 - \frac{(0.2)^2}{2} + \frac{(0.2)^3}{3} - \frac{(0.2)^4}{4} + \frac{(0.2)^5}{5}$$
$$\approx 0.182331$$

The error is given by
$$|R_5(1.2)| = \left|\frac{(-1)^5(0.2)^6}{6c^6}\right| = \frac{0.000064}{6c^6}$$

Since $1 < c < 1.2$, we get a larger fraction by replacing c by 1. Thus,
$$|R_5(1.2)| \leq \frac{0.000064}{6} \approx 0.000011$$

So the answer is correct to at least four decimal places.

(b) From Example 10.5,
$$R_n(x) = \frac{(-1)^n(x-1)^{n+1}}{(n+1)c^{n+1}}, \qquad c \text{ between } 1 \text{ and } x$$

With $x = 1.2$, an upper bound on $|R_n(1.2)|$ can be obtained by replacing c with 1. Then $c^{n+1} = 1^{n+1} = 1$, so we have
$$|R_n(1.2)| \leq \frac{(0.2)^{n+1}}{n+1}$$

To obtain eight decimal places of accuracy, we find the smallest integer n such that
$$\frac{(0.2)^{n+1}}{n+1} \leq 5 \times 10^{-9}$$

By trial and error, using a calculator, a CAS, or a spreadsheet, we find that $n + 1 = 11$, or $n = 10$, is the smallest value that works. So $P_{10}(1.2)$ estimates $\ln 1.2$ to at least eight decimal places of accuracy. ∎

EXAMPLE 10.7 Find an upper bound on the error in using the nth Taylor polynomial $P_n(x)$ about $x = 0$ to approximate $f(x) = \cos x$. Use $P_6(x)$ to approximate $\cos 10°$ and estimate the error.

Solution In Example 10.3 we found that
$$P_{2n}(x) = 1 - \frac{x^2}{2!} + \frac{x^4}{4!} + \cdots + \frac{(-1)^n}{(2n)!}x^{2n}$$

and that $P_{2n+1}(x) = P_{2n}(x)$. The remainder term in Taylor's Formula (Equation 10.9) can therefore be written
$$R_{2n}(x) = R_{2n+1}(x) = \frac{f^{(2n+2)}(c)}{(2n+2)!}x^{2n+2}$$

where c is between 0 and x. Since an even-order derivative of $\cos x$ is either $\cos x$ or $-\cos x$, we therefore have
$$|R_{2n}(x)| = \frac{|\cos c|}{(2n+2)!}|x|^{2n+2} \leq \frac{|x|^{2n+2}}{(2n+2)!}$$

since $|\cos c| \leq 1$. Thus, the error in using $P_{2n}(x)$ to approximate $\cos x$ for any real number x is at most $|x|^{2n+2}/(2n+2)!$.

To use $P_6(x)$ to approximate $\cos 10°$, we first express $10°$ in radians:

$$10° = 10°\left(\frac{\pi \text{ radians}}{180°}\right) = \frac{\pi}{18} \text{ radians}$$

Then we have

$$P_6\left(\frac{\pi}{18}\right) = 1 - \frac{1}{2!}\left(\frac{\pi}{18}\right)^2 + \frac{1}{4!}\left(\frac{\pi}{18}\right)^4 - \frac{1}{6!}\left(\frac{\pi}{18}\right)^6 \approx 0.9848077530$$

The upper bound we found above on the error is

$$\left|R_6\left(\frac{\pi}{18}\right)\right| \leq \frac{\left(\frac{\pi}{18}\right)^8}{8!} \approx 2.135 \times 10^{-11}$$

So our answer is correct to at least ten decimal places. ∎

A Preview of Taylor Series

In Example 10.4 we found the nth Taylor polynomial about $x = 0$ for $f(x) = e^x$ to be

$$P_n(x) = 1 + x + \frac{x^2}{2!} + \frac{x^3}{3!} + \cdots + \frac{x^n}{n!}$$

So we have the approximation

$$e^x \approx 1 + x + \frac{x^2}{2!} + \frac{x^3}{3!} + \cdots + \frac{x^n}{n!}$$

and the approximation gets better and better as n increases. Later we will show that the approximation is valid for all real values of x. A natural question to ask is whether we can let n "go to infinity" and get the exact value of e^x. That is, can we say that

$$e^x = 1 + x + \frac{x^2}{2!} + \frac{x^3}{3!} + \frac{x^4}{4!} + \cdots \tag{10.11}$$

for all real values of x? Recall from this chapter's introduction that we call a series such as this one a *power series in x*. If we set $x = 1$, for example, our question is whether it is valid to write

$$e = 1 + 1 + \frac{1}{2!} + \frac{1}{3!} + \frac{1}{4!} + \cdots$$

We are immediately confronted with another question: What do we mean by the sum of infinitely many numbers? Clearly, we must have some new definition of addition in such a case. We will give a definition in the next section and will explore its consequences in the sections that follow. We will find that Equation 10.11 is valid for all real values of x.

For now, we will say that the expression on the right-hand side of Equation 10.11 is the **Taylor Series of e^x about $x = 0$** and that, under the meaning yet to be given, the series **converges** to e^x for all real x. More generally, by letting $n \to \infty$ in Taylor's Formula with Remainder, Equation 10.9, we obtain the **Taylor Series for the function f about $x = a$**:

$$f(a) + f'(a)(x-a) + \frac{f''(a)}{2!}(x-a)^2 + \frac{f'''(a)}{3!}(x-a)^3 + \cdots$$

As you might guess, the condition under which this series converges to $f(x)$ is that the error in the approximation of $f(x)$ by its nth Taylor polynomial gets smaller and smaller as $n \to \infty$; that is,

$$\lim_{n \to \infty} R_n(x) = 0$$

We will return to this important question of the representation of a function by its Taylor series after our discussion of infinite series in general.

COMPUTING TAYLOR POLYNOMIALS USING A COMPUTER ALGEBRA SYSTEM

Taylor's Theorem presents a method for using polynomials to approximate functions that cannot be explicitly evaluated. The theorem also provides a useful means of analyzing the error in such approximations. The degree of accuracy in using Taylor polynomials to approximate functions generally increases as the degree of the polynomial increases. The higher the degree of the polynomial desired, the greater the amount of computation necessary. For this reason, a computer algebra system becomes a valuable aid.

CAS 36

Use Taylor polynomials to approximate the function $f(x) = e^{-x^2}$ close to 0.

We will begin by using Maple, Mathematica, and DERIVE to compute the Taylor polynomial of degree 8 for f near $x = 0$.

Maple:

f := x–>exp(–x^2);
taylor(f(x),x=0,9);

The output is:

$$1 - x^2 + \frac{1}{2}x^4 - \frac{1}{6}x^6 + \frac{1}{24}x^8 + O(x^9)$$

Notice the big-O term on the end of the output from Maple. This indicates that all higher degree terms have an exponent of at least 9. Because of this form of the output in Maple, if the Taylor polynomial of degree 8 is desired, the last parameter in the **taylor** command should be specified as 9.

Mathematica:

f[x_] = Exp[–x^2]
Normal[Series[f[x],{x,0,8}]]

DERIVE:

(At the ☐ symbol, go to the next step.)
a (author) ☐ exp(–x^2) ☐ c (calculus) ☐ t (taylor) ☐ [choose expression] ☐ x (variable) ☐ 8 (degree) ☐ 0 (point) ☐ s (simplify)

Notice that Mathematica and DERIVE do not include a big-O term in the output. In Mathematica, the Normal command removes this big-O term.

To see the relative accuracy of the approximation near 0 a plot would be helpful. Maple does not directly plot the result of the command taylor(f(x),x=0,9);. We could enter a plot and retype the 8th degree polynomial as the function to plot. We can, however, avoid the problem by using a more general command in Maple, called mtaylor. To access this command, it must be read from a Maple library via readlib('mtaylor'):

Section 10.1 Approximation of Functions by Polynomials; Taylor's Theorem   733

Maple:

readlib('mtaylor'):
P := mtaylor(f(x),[x],9);

The output is:

$$P := 1 - x^2 + \frac{1}{2}x^4 - \frac{1}{6}x^6 + \frac{1}{24}x^8$$

plot({f(x),P},x=-2..2);

We see from Figure 10.1.1 that near 0 the approximation looks very good.

Mathematica:

P[x_] = Normal[Series[f[x], {x,0,8}]]
Plot[{f[x],P[x]},{x,-2,2}]

DERIVE:

(At the □ symbol, go to the next step.)
[choose exp(-x^2)] □ p (plot window)
□ p (plot) □ a (algebra) □ [choose Taylor polynomial] □ p (plot window)
□ p (plot)

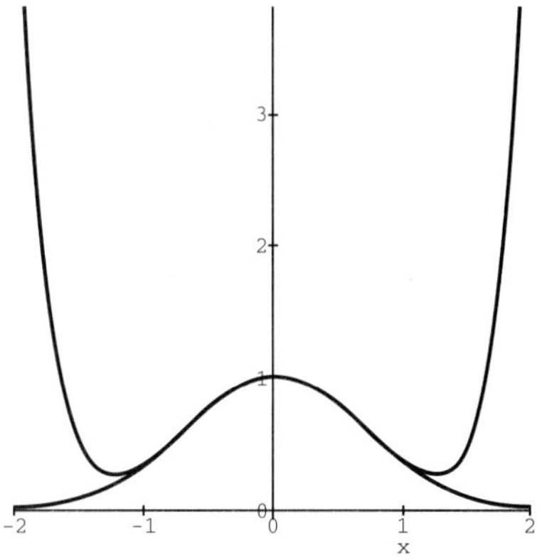

FIGURE 10.1.1

To see the relative accuracy obtained in approximating f by Taylor polynomials of different degrees, we could repeat what we have already done with different polynomials and compare the pictures obtained. This would be tedious. However, the computer algebra systems allow us to make this kind of comparison very quickly by using a loop structure.

Maple:

for n from 1 by 1 to 10 do
 P := mtaylor(f(x),[x],n);
od;

The output gives Taylor polynomials of degrees 0 to 9.

$$P := 1$$
$$P := 1$$
$$P := 1 - x^2$$
$$P := 1 - x^2$$
$$P := 1 - x^2 + \frac{1}{2}x^4$$
$$P := 1 - x^2 + \frac{1}{2}x^4$$
$$P := 1 - x^2 + \frac{1}{2}x^4 - \frac{1}{6}x^6$$
$$P := 1 - x^2 + \frac{1}{2}x^4 - \frac{1}{6}x^6$$
$$P := 1 - x^2 + \frac{1}{2}x^4 - \frac{1}{6}x^6 + \frac{1}{24}x^8$$
$$P := 1 - x^2 + \frac{1}{2}x^4 - \frac{1}{6}x^6 + \frac{1}{24}x^8$$

Notice the 1st and 2nd, 3rd and 4th, 5th and 6th, etc., Taylor polynomials are identical.

To get a series of plots in Maple, enter:

for n from 1 by 1 to 10 do
P[n] := mtaylor(f(x),[x],n);
plot({f(x),P[n]},x=-2..2);
od;

To plot f and the Taylor polynomials for f about $x = 0$, of orders 2, 4, 6, 8, and 10 on the same axes, use the Maple command:

plot({f(x),P[2],P[4],P[6],P[8],P[10]}, x=-5..5,y=-5..5);

Mathematica:

Do[p[x_,k] = Normal[Series[f[x],{x,0,k}]],
{k,0,8}]
Do[Plot[{f[x],p[x,k]},
{x,-2,2}],{k,0,8}]

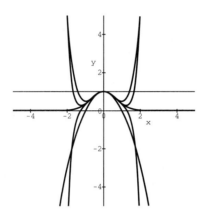

FIGURE 10.1.2
$f(x)$, $P_2(x)$, $P_4(x)$, $P_6(x)$, $P_8(x)$, and $P_{10}(x)$

The graphical analysis above gives an intuitive feel for where the Taylor polynomial is a good approximation to the function. For more precise numerical information, the error term should be analyzed.

Exercise Set 10.1

In Exercises 1–6, find the Taylor polynomial of the specified degree for f about $x = 0$.

1. $f(x) = e^{-x}$; $n = 5$
2. $f(x) = \sin x$; $n = 7$
3. $f(x) = 1/(x+1)$; $n = 6$
4. $f(x) = \sqrt{x+1}$; $n = 5$
5. $f(x) = \cosh x$; $n = 8$
6. $f(x) = \tan x$; $n = 5$

In Exercises 7–12, find the Taylor polynomial of degree n for f about the specified point a.

7. $f(x) = 1/x$; $a = 1$
8. $f(x) = e^{x-2}$; $a = 2$
9. $f(x) = \sin x$; $a = 0$
10. $f(x) = 1/(1-x)^2$; $a = -1$
11. $f(x) = \ln(2-x)$; $a = 1$
12. $f(x) = \sinh x$; $a = 0$

In Exercises 13–18, find $R_n(x)$ for the functions specified in Exercises 7–12 and give the interval about which $f(x) = P_n(x) + R_n(x)$.

In Exercises 19–32, estimate the specified function value using P_n for the given value of n and an appropriate value of a. Find an upper bound on the error, and determine the number of decimal places of accuracy. (When appropriate, you may use the results of preceding exercises or examples.)

19. e^{-1}; $n = 5$
20. $\sqrt{1.2}$; $n = 5$
21. $\sin 1.5$; $n = 7$
22. $\ln 1.5$; $n = 6$
23. $\cos 2$; $n = 8$
24. $\sinh 2$; $n = 7$
25. $\sqrt{5}$; $n = 5$ (*Hint:* Take $a = 4$.)
26. $\ln 4/5$; $n = 8$
27. e^3; $n = 12$
28. $\sqrt{e}$; $n = 4$
29. $\cosh 1$; $n = 6$
30. $\sin 32°$; $n = 5$ (*Hint:* Change to radians and use $a = \pi/6$.)
31. $\cos 58°$; $n = 4$ (*Hint:* Take $a = \pi/3$.)
32. $\sqrt[3]{7}$; $n = 4$

33. Use the result of Example 10.7 to find the degree of the Taylor polynomial for $\cos x$ about $x = 0$ that approximates $\cos 72°$ to six decimal places of accuracy. Find this approximation and compare it with the value given by a calculator.

34. Find $\sin 80°$ correct to five decimal places of accuracy using the appropriate Taylor polynomial about $x = 0$. (See Exercises 9 and 15.)

35. Find $\sinh 2$ to five decimal places of accuracy using the appropriate Taylor polynomial about $x = 0$. (See Exercises 12 and 18.)

36. Find $P_n(x)$ and $R_n(x)$ for $f(x) = \sqrt{x}$ about $x = 4$. In what interval is the result valid? Use the result to find $\sqrt{3}$ correct to four decimal places.

37. Let $f(x) = (1+x)^\alpha$, where α is real.
(a) Find $P_n(x)$ and $R_n(x)$ for f about $x = 0$.
(b) Show that if $\alpha = m$, where m is a positive integer, then $f(x) = P_m(x)$ for all x.
(c) If α is not a positive integer, show that $f(x) = P_n(x) + R_n(x)$ for $-1 < x < 1$.
(d) Use the result of part (a) to estimate $\sqrt[3]{1.5}$, taking $n = 4$. Find an upper bound on the error.

38. (a) Show that if $f(x)$ is a polynomial of degree n, then the Taylor polynomial $P_n(x)$ for f about $x = a$ is identical to $f(x)$.

(b) Express the polynomial
$$f(x) = 2x^5 - 3x^4 + 7x^3 - 8x^2 + 2$$
in powers of $x - 1$.

39. For $-1 < x < 1$, let
$$f(x) = \ln \frac{1+x}{1-x}$$
(a) Show that if u is any positive number, there is an x in $(-1, 1)$ such that $f(x) = \ln u$.
(b) Find $P_n(x)$ and $R_n(x)$ for f about $x = 0$. (*Hint:* Rewrite $f(x)$ using properties of logarithms before taking derivatives.)
(c) Use parts (a) and (b) to find $\ln 2$ correct to three decimal places.

40. (a) In Exercise 39(a), let $u = (N+1)/N$, where N is a positive integer. What is x so that $f(x) = \ln u$?
(b) Use the result of part (a), together with Exercise 39, to show that

$$\ln(N+1) = \ln N + 2\sum_{k=1}^{n} \frac{1}{(2k-1)(2N+1)^{2k-1}} + R_{2n}\left(\frac{1}{2N+1}\right)$$

and that

$$\left| R_{2n}\left(\frac{1}{2N+1}\right) \right| \leq \frac{2}{(2n+1)(2N)^{2n+1}}$$

(Note that $R_{2n-1}(x) = R_{2n}(x)$, since the Taylor polynomial consists of odd powers only.)
(c) Find $\ln 3$ correct to three decimal places, making use of part (b) and Exercise 39(c).

41. Let $f(x) = \ln(1+x)$.
(a) Find the nth Taylor polynomial $P_n(x)$ for f about $x = 0$.
(b) Use a CAS or graphing calculator to obtain the graphs of $f(x)$, $P_5(x)$, $P_{10}(x)$, and $P_{20}(x)$ on the same set of axes.
(c) Describe the behavior of the graphs obtained in part (b). Pay particular attention to values of x in the interval $(-1, 1)$ as compared with values outside this interval.
(d) Conjecture, based on your graphs, the values of x for which the remainder term $R_n(x)$ approaches 0 as $n \to \infty$.

42. Use the remainder term $R_n(x)$ in Taylor's Formula to determine the values of x for which the approximation

$$\sin x \approx x - \frac{x^3}{6}$$

is accurate to within 0.001 unit.

43. Repeat Exercise 42 with
$$\cos x \approx 1 - \frac{x^2}{2}$$

44. Repeat Exercise 42 with
$$\sqrt{1+x^2} \approx 1 + \frac{x^2}{2}$$

45. Letting $f(x) = \sin x$,
(a) find the nth Taylor polynomial for f about $x = 0$;
(b) generate graphs of f and P_n on the same axes for $n = 1, 3, 5, 10$;
(c) use the graphs to estimate on which interval $P_3(x)$ approximates $f(x)$ to within about 0.01 unit. (Zoom in.) Repeat for $P_5(x)$ and for $P_{10}(x)$.

46. Repeat Exercise 45 with $f(x) = e^{-x}$.

47. Repeat Exercise 45 with $f(x) = (1-x)^{1/3}$.

48. Repeat Exercise 47 using the nth Taylor polynomial about $x = 2$.

49. Repeat Exercise 45 with $f(x) = \sqrt{x+1}$, first using $P_n(x)$ about $x = 0$, and then using $P_n(x)$ about $x = 3$.

50. Let $f(x) = e^x$. Find a value of n so that $P(0.1)$ is within 10^{-6} units of $e^{0.1}$.

10.2 SEQUENCES

In the introduction to this chapter we indicated how the sum of an infinite *series* can be defined. The key lies in computing what we called the *sequence of partial sums* and determining if these partial sums approach some finite value as more and more terms are added. To illustrate this relationship between a series and its sequence of partial sums, we used the series

$$1 + \frac{1}{2} + \frac{1}{4} + \frac{1}{8} + \frac{1}{16} + \cdots \tag{10.12}$$

Its sequence of partial sums is

$$1, \ 1\frac{1}{2}, \ 1\frac{3}{4}, \ 1\frac{7}{8}, \ 1\frac{15}{16}, \ldots \tag{10.13}$$

which comes arbitrarily close to 2 as we add more and more terms. Thus, we concluded that the sum of series 10.12 is 2. Recall that we obtain the sequence 10.13 by writing, in order, the first term of the series 10.12, the sum of the first two terms, the sum of the first three terms, and so on.

Note carefully that in a series, such as 10.12, the terms are added together. In contrast, in a sequence, such as 10.13, the members (also called terms) are separated by commas. Though the words *sequence* and *series* are often used interchangeably in nonmathematical usage, in mathematics we are careful to distinguish between them.

We will explore more fully the relationship between a series and its sequence of partial sums in the next section. In this section we investigate general properties of sequences. Later, we will apply these properties to sequences of partial sums of a series. Since the convergence or divergence of a series is of central importance, one of our main goals is to define the notion of the *limit* of a sequence and to learn some methods of finding this limit when it exists.

Let us be more precise about the meaning of a sequence. One way of defining a sequence is to say it is an ordered collection of elements, called *terms*, formed according to some rule that gives a unique term for each positive integer. Since a rule that assigns a unique element to each positive integer is a function having the positive integers as its domain, we can state the definition more succinctly as follows:

> **Definition 10.2**
> **A Sequence**
>
> A **sequence** is a function whose domain is the set of all positive integers.

REMARK
■ According to the definition, a sequence is a function, but it is customary to refer to the ordered list of function values (the range) as the sequence itself.

For most of the sequences we will be dealing with, the terms will be real numbers, but they could be other mathematical entities. Note that by our definition, there will always be infinitely many terms, one for each positive integer (although the terms may not be distinct—there can be repetitions). So when we speak of a sequence, we will always mean an *infinite sequence*. In Definition 10.2, if the domain is replaced by the set of the first n positive integers, the resulting function is called a *finite sequence*.

Notation

It is customary, in giving the rule for a sequence, to use subscript notation, such as a_n, rather than the usual functional notation $f(n)$. For example, if we write

$$a_n = \frac{1}{n}$$

then by substituting, in turn, $n = 1, 2, 3, \ldots$, we get the sequence

$$\frac{1}{1}, \frac{1}{2}, \frac{1}{3}, \frac{1}{4}, \ldots \qquad (10.14)$$

We refer to a_n as the **nth term**, or **general term**, of the sequence. Giving a formula for the nth term effectively defines the entire sequence.

When a formula for the nth term a_n of a sequence is known, we can designate the sequence with braces around a_n, such as $\{a_n\}$ or, more explicitly, $\{a_n\}_{n=1}^{\infty}$. The latter notation makes it explicit that n takes on the positive integers starting with 1. If the simple notation $\{a_n\}$ is used (as will usually be the case), we will understand it to mean the same as $\{a_n\}_{n=1}^{\infty}$. (There may, however, be times when we want to start with $n = 0$ or some other positive integer n_0 with $n_0 \neq 1$. If so, we will indicate the limits.) Using this notation, we can designate the sequence 10.14 by $\{1/n\}$.

In the next example we show several sequences using this notation.

EXAMPLE 10.8 Show the first five terms of each of the following sequences, indicating its continuation by three dots.

(a) $\{2n - 1\}$ (b) $\{(-1)^{n-1}\}$ (c) $\left\{\dfrac{n}{n+1}\right\}$

(d) $\left\{\dfrac{2^n}{n!}\right\}$ (e) $\{\sin n\}$

Solution

(a) $1, 3, 5, 7, 9, \ldots$ (the odd positive integers)
(b) $(-1)^0, (-1)^1, (-1)^2, (-1)^3, (-1)^4, \ldots$, or on simplifying, $1, -1, 1, -1, 1, \ldots$
(c) $\dfrac{1}{2}, \dfrac{2}{3}, \dfrac{3}{4}, \dfrac{4}{5}, \dfrac{5}{6}, \ldots$
(d) $\dfrac{2^1}{1!}, \dfrac{2^2}{2!}, \dfrac{2^3}{3!}, \dfrac{2^4}{4!}, \dfrac{2^5}{5!}, \ldots.$ Since $1! = 1$, $2! = 2 \cdot 1 = 2$, $3! = 3 \cdot 2 \cdot 1 = 6$, $4! = 4 \cdot 3 \cdot 2 \cdot 1 = 24$, and $5! = 5 \cdot 4 \cdot 3 \cdot 2 \cdot 1 = 120$, we can rewrite the sequence as

$$\dfrac{2}{1}, \dfrac{4}{2}, \dfrac{8}{6}, \dfrac{16}{24}, \dfrac{32}{120}, \ldots$$

which becomes, after reducing the fractions,

$$2, 2, \dfrac{4}{3}, \dfrac{2}{3}, \dfrac{4}{15}, \ldots$$

(e) $\sin 1, \sin 2, \sin 3, \sin 4, \sin 5, \ldots.$ Using a calculator (in radian mode), we get (rounding to four decimal places) approximately $0.8415, 0.9093, 0.1411, -0.7568, -0.9589, \ldots.$ ∎

If the pattern of the terms is clear, so that we can infer the formula for the nth term, it is permissible to designate the sequence by merely showing a few of its terms. For example, the sequence 10.14 could be given simply by writing

$$1, \dfrac{1}{2}, \dfrac{1}{3}, \dfrac{1}{4}, \ldots$$

We could also show the sequence in parts (a), (b), and (c) of Example 10.8 in this way. However, showing the sequence in part (d) in its simplified form,

$$2, 2, \dfrac{4}{3}, \dfrac{2}{3}, \dfrac{4}{15}, \ldots$$

is *not* in itself a satisfactory way of designating the sequence, since it is not clear from this list alone how to form succeeding terms.

EXAMPLE 10.9 Assume that each of the following sequences continues in the pattern of the terms shown. Write the general term.

(a) $\dfrac{1}{2}, -\dfrac{1}{4}, \dfrac{1}{6}, -\dfrac{1}{8}, \ldots$ (b) $\dfrac{1}{2}, \dfrac{3}{4}, \dfrac{5}{8}, \dfrac{7}{16}, \dfrac{9}{32}, \ldots$

Solution

(a) The alternating signs can be indicated by means of the factor $(-1)^{n-1}$, as we saw in Example 10.8(b). Each denominator is twice the number of the term. So we can write

$$a_n = \dfrac{(-1)^{n-1}}{2n}$$

(b) The numerators are the odd positive integers, which are indicated by the factor $2n - 1$, as we saw in Example 10.8(a). The denominators appear to be powers of 2, with the nth term denominator equal to 2^n. So we can write

$$a_n = \frac{2n-1}{2^n}$$

∎

It is not always possible to give an explicit formula for the nth term. For example, consider the sequence $\{a_n\}$ where a_n is the nth prime number. Recall that a prime number is a positive integer greater than 1 whose only positive divisors are 1 and itself. Several terms of this sequence are

$$2, 3, 5, 7, 11, 13, 17, 19, 23, \ldots$$

It is known that there are infinitely many primes, but there is no known formula for the nth prime.

As another example, let a_n denote the digit in the nth decimal place of the decimal representation of π. The first few terms are

$$1, 4, 1, 5, 9, 2, 6, 5, \ldots$$

Incidentally, π has been calculated to more than two billion places! Again, no formula has been found for the nth decimal digit.

Recursive Definitions

Sometimes sequences are defined *recursively*. For example, suppose we are given that $a_1 = \sqrt{2}$ and that for all $n \geq 2$,

$$a_n = \sqrt{2 + a_{n-1}}$$

This formula giving a_n in terms of a_{n-1} is called a **recursion formula**. Since we know that $a_1 = 1$, we can find a_2 by the recursion formula as follows:

$$a_2 = \sqrt{2 + a_1} = \sqrt{2 + \sqrt{2}}$$

Now we use the recursion formula again to get a_3:

$$a_3 = \sqrt{2 + a_2} = \sqrt{2 + \sqrt{2 + \sqrt{2}}}$$

and again to get a_4:

$$a_4 = \sqrt{2 + a_3} = \sqrt{2 + \sqrt{2 + \sqrt{2 + \sqrt{2}}}}$$

and so on. We could approximate each term by using a calculator, but the pattern would no longer be evident.

Another example of a recursive definition is the sequence of approximations by Newton's Method to a root of an equation $f(x) = 0$ (see Section 4.5). After the initial approximation x_0 is made, we find subsequent approximations by the recursion formula:

$$x_{n+1} = x_n - \frac{f(x_n)}{f'(x_n)} \qquad (n = 0, 1, 2, \ldots)$$

Note that in this case we begin with $n = 0$.

The Fibonacci sequence is named after the Italian businessman and mathematician Leonardo Fibonacci (who lived about 1170–1250). He posed the following problem:

> How many pairs of rabbits can be produced from a single pair in a year if every month each pair begets a new pair that from the second month onward becomes productive?

If we assume the rabbits in the initial pair are newborn (so that they produce the first new pair after two months), then the number of pairs present in the nth month is f_n, the nth term of the Fibonacci sequence.

Perhaps the best-known sequence defined recursively is the **Fibonacci sequence** (see the note in the margin), defined by

$$f_1 = 1, \quad f_2 = 1, \quad \text{and} \quad f_n = f_{n-1} + f_{n-2} \quad \text{for } n \geq 2$$

In this case two initial terms are given, and the recursion formula gives the nth term as the sum of the two preceding terms. The first few terms are

$$1, 1, 2, 3, 5, 8, 13, \ldots$$

The numbers f_n are called *Fibonacci numbers*, and they occur in nature in various ways. For example, if you count the number of spirals on a pine cone or a pineapple, you will get a Fibonacci number. The same is true for the number of spirals formed by the seeds of a sunflower. Also, the number of petals on a daisy is normally equal to a Fibonacci number.

Graphical Representation of Sequences

There are two useful ways of showing sequences graphically. In the first we use a two-dimensional coordinate system just as with other functions. We plot the points (x, y), where $x = n$ and $y = a_n$ for $n = 1, 2, 3, \ldots$. The graph consists of isolated points that we do not connect. In the second method we show the numbers a_n, where $n = 1, 2, 3, \ldots$, as points on a number line. We illustrate both methods in the next example.

EXAMPLE 10.10 Show each of the following sequences graphically by two methods.

(a) $\left\{ \dfrac{1}{n} \right\}$ (b) $\left\{ \dfrac{n}{n+1} \right\}$

Solution

(a) By the first method we plot the points $(1, 1), (2, \frac{1}{2}), (3, \frac{1}{3}), (4, \frac{1}{4}), \ldots$. We show the result in Figure 10.7(a). By the second method, we show the terms

$$1, \frac{1}{2}, \frac{1}{3}, \frac{1}{4}, \ldots$$

as points on a number line in Figure 10.7(b).

(b) In the first method we plot the points $(1, \frac{1}{2}), (2, \frac{2}{3}), (3, \frac{3}{4}), (4, \frac{4}{5}), \ldots$, as shown in Figure 10.8(a) at the top of page 741. In the second method we show the points

$$\frac{1}{2}, \frac{2}{3}, \frac{3}{4}, \frac{4}{5}, \ldots$$

as points on a number line in Figure 10.8(b). ∎

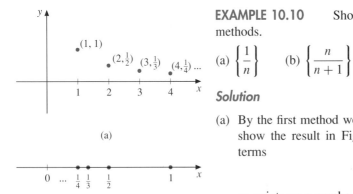

FIGURE 10.7

Limits of Sequences

In Figure 10.7(a) it appears that if we were to plot more and more points, they would come arbitrarily close to the x-axis, that is, to the line $y = 0$. Similarly, in Figure 10.7(b) the points seem to be coming arbitrarily close to 0. In fact, we can see from the nth term

$$a_n = \frac{1}{n}$$

that as $n \to \infty$, a_n comes arbitrarily close to 0.

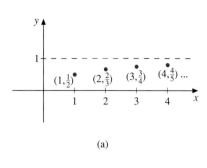

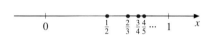

(b)

FIGURE 10.8

In Figure 10.8(a) the points seem to be getting arbitrarily close to the horizontal line $y = 1$, and in Figure 10.8(b) the points on the number line seem to be getting close to 1. If we consider the nth term in this case,

$$a_n = \frac{n}{n+1}$$

we can see intuitively that for large n, a_n is almost 1, since the numerator and denominator differ by only one unit. For example, with $n = 1,000$,

$$a_n = \frac{1000}{1001}$$

which is only slightly less than 1.

When the terms of a sequence come arbitrarily close to some number L as $n \to \infty$, we say that the sequence **converges** and that its **limit** is L. From our discussion in Example 10.10, we strongly suspect that both sequences $\{1/n\}$ and $\{n/(n+1)\}$ converge, the first to the limit 0 and the second to the limit 1. In the case of the second sequence, we can use the "dominant term" approach, just as we did with rational functions of a continuous variable in Chapters 1 and 2. We reason that for large n, both numerator and denominator are approximately the same as their terms of highest degree. In this case, then, we could say that

$$\frac{n}{n+1} \approx \frac{n}{n} = 1$$

for large n.

Another way to see this same result is to divide numerator and denominator by n:

$$\frac{n}{n+1} = \frac{1}{1 + \frac{1}{n}}$$

As $n \to \infty$, $1/n \to 0$, so we conclude that the limit is 1.

The precise definition of the limit of a sequence is given below.

Definition 10.3
The Limit of a Sequence

We say that the **limit** of the sequence $\{a_n\}$ is L and write

$$\lim_{n \to \infty} a_n = L$$

provided that corresponding to each positive number ε, there is a positive integer N such that

$$|a_n - L| < \varepsilon$$

for all $n > N$. If the limit of $\{a_n\}$ is L, we say that the sequence **converges** to L. If the limit of $\{a_n\}$ does not exist, we say that the sequence **diverges**.

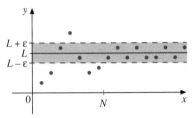

REMARKS

- Since ε can be any positive number, by saying that $|a_n - L| < \varepsilon$ for all $n > N$, we are saying that the term a_n of the sequence will be *arbitrarily* close to L (closer than ε, however small ε may be) if we go far enough out in the sequence (beyond the Nth term).
- A sequence can have at most one limit, since the terms cannot simultaneously all be arbitrarily close to two different values.
- Definition 10.3 is the same as Definition 2.3 for the limit of a function $f(x)$ as $x \to \infty$, except that the continuous variable x is replaced by the integer variable n.

y
L + ε
L
L − ε
0
N
x
For n > N, the points (n, a_n) lie within an ε-band about L.
(a)
a_1 a_2 a_3 ... a_N L − ε L L + ε x
For n > N, a_n is the interval (L − ε, L + ε)
(b)

FIGURE 10.9

In Figure 10.9 we illustrate Definition 10.3 graphically by both methods.

EXAMPLE 10.11 Use Definition 10.3 to show that

$$\lim_{n\to\infty} \frac{n}{n+1} = 1$$

Solution Consider the difference $|a_n - L|$, where $a_n = n/(n+1)$ and $L = 1$:

$$\left|\frac{n}{n+1} - 1\right| = \left|\frac{n-n-1}{n+1}\right| = \left|\frac{-1}{n+1}\right| = \frac{1}{n+1}$$

Now let ε denote any positive number. We want to see how large to make n so that $1/(n+1) < \varepsilon$. Since

$$\frac{1}{n+1} < \frac{1}{n}$$

we can simplify our choice by making $1/n < \varepsilon$. Solving this inequality for n, we get $n > 1/\varepsilon$. Thus, if we let N denote any integer larger than $1/\varepsilon$, we will have, for all $n > N$,

$$\left|\frac{n}{n+1} - 1\right| < \varepsilon$$

So by Definition 10.3,

$$\lim_{n\to\infty} \frac{n}{n+1} = 1$$

■

EXAMPLE 10.12 Show that each of the following sequences is divergent.
(a) $\{(-1)^{n-1}\}$ (b) $\{2n-1\}$ (c) $\{\sin n\}$

y
(1, 1) (3, 1) (5, 1) (7, 1)
1
0
1 2 3 4 5 6 7
x
−1
(2, −1) (4, −1) (6, −1)

FIGURE 10.10
Graph of $\{(-1)^{n-1}\}$

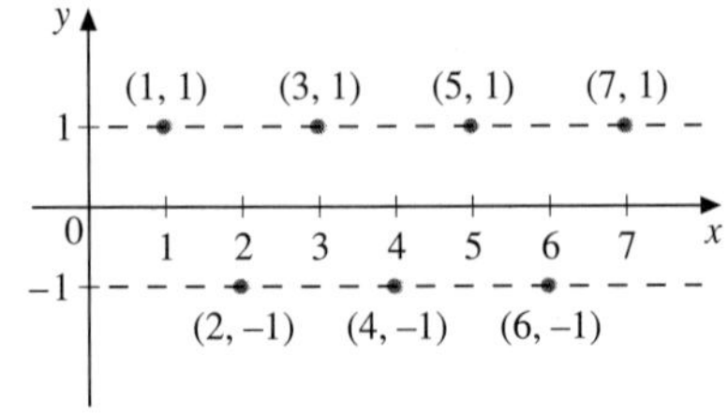

Solution

(a) The sequence is $1, -1, 1, -1, 1, \ldots$. There is no *single* number L that all terms come arbitrarily close to, no matter how far out in the sequence we go. More formally, if we choose $\varepsilon = \frac{1}{2}$, say, then regardless of the value of L, and regardless of how large N is, there are infinitely many terms a_n with $n > N$ such that $|a_n - L| \geq \frac{1}{2}$, contrary to the requirement on L given in Definition 10.3. The graph of $\{(-1)^{n-1}\}$ in Figure 10.10 clearly shows that the terms do not approach a limit.

(b) The sequence is $1, 3, 5, 7, 9, \ldots$, whose terms become arbitrarily large as $n \to \infty$, so the sequence diverges. We show its graph in Figure 10.11.

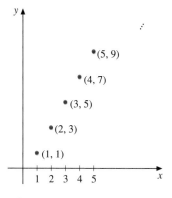

FIGURE 10.11
Graph of $\{2n - 1\}$

(c) In Figure 10.12 we show by means of the dashed line the graph of $f(x) = \sin x$. The points $(n, \sin n)$ lie on this graph. It is evident that as $n \to \infty$ these points do not come arbitrarily close to any horizontal line $y = L$. That is, $\lim_{n \to \infty} \sin n$ does not exist, so the sequence $\{\sin n\}$ diverges. ∎

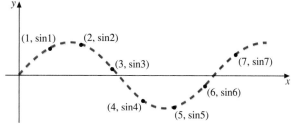

FIGURE 10.12
Graph of $\{\sin n\}$

EXAMPLE 10.13 Use Definition 10.3 to prove that if $|r| < 1$, then $\lim_{n \to \infty} r^n = 0$.

Solution If $|r| < 1$, then its reciprocal $1/|r|$ is greater than 1, so we can write

$$\frac{1}{|r|} = 1 + p$$

for some positive number p. By the Binomial Theorem (see Appendix 4),

$$(1 + p)^n = 1 + np + \frac{n(n-1)}{2!}p^2 + \cdots + p^n > 1 + np > np$$

Thus, inverting, we get

$$|r|^n = \frac{1}{(1+p)^n} < \frac{1}{np}$$

If ε denotes any positive integer, then we can make $1/np < \varepsilon$ by choosing $n > 1/\varepsilon p$. So if we choose N any integer greater than $1/\varepsilon p$, we have, for all $n > N$,

$$|r^n - 0| = |r^n| = |r|^n < \frac{1}{np} < \varepsilon$$

By Definition 10.3, it follows that

$$\lim_{n \to \infty} r^n = 0 \qquad \blacksquare$$

n	$(0.9999)^n$
10	0.9990
100	0.9900
1000	0.9048
10,000	0.3679
100,000	0.000045

Even if we choose r only slightly less than 1—say, $r = 0.9999$—it is still true that r^n approaches 0 as $n \to \infty$. The table in the margin gives pretty convincing evidence for $r = 0.9999$.

If $r = 1$, then $r^n = 1$ for all n, and the sequence $\{r^n\}$ is the *constant* sequence

$$1, 1, 1, \ldots$$

whose limit is clearly 1. (After all, $|a_n - 1| = |1 - 1| = 0$, which is less than ε for any positive ε and for all n.)

If $r = -1$, the sequence $\{r^n\}$ is

$$-1, 1, -1, 1, -1, \ldots$$

which diverges, similar to the sequence in Example 10.12(a).

If $|r| > 1$, then, as we showed in Example 10.13, $|r| = 1 + p$ for some $p > 0$, and $|r|^n = (1 + p)^n > np$. Since $p > 0$, $np \to \infty$ as $n \to \infty$. Thus, $\{r^n\}$ diverges. For $r > 1$, r^n becomes arbitrarily large as $n \to \infty$. For $r < -1$, r^n oscillates in sign but becomes large in absolute value.

We now summarize our findings about the sequence $\{r^n\}$. (See also Figure 10.13.)

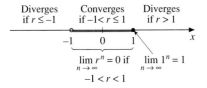

FIGURE 10.13
Convergence of $\{r^n\}$

> **The Sequence $\{r^n\}$**
>
> (a) If $|r| < 1$, $\lim_{n\to\infty} r^n = 0$.
>
> (b) If $r = 1$, $\lim_{n\to\infty} r^n = 1$.
>
> (c) For all other values of r, the sequence diverges.

Infinite Limits

As we saw with the sequence $\{2n - 1\}$ in Example 10.12(b) and again with the sequence $\{r^n\}$ when $r > 1$, some sequences diverge because their terms become arbitrarily large as $n \to \infty$. We indicate this type of divergence of a sequence $\{a_n\}$ by writing

$$\lim_{n\to\infty} a_n = \infty$$

Similarly, if the terms a_n are negative (at least for all sufficiently large n) but become arbitrarily large in absolute value as $n \to \infty$, we write

$$\lim_{n\to\infty} a_n = -\infty$$

In the first case we may say that "a_n approaches infinity" and in the second that "a_n approaches minus infinity," but in neither case does the sequence $\{a_n\}$ converge. We are simply saying it diverges in a particular way.

REMARKS
- A sequence can diverge without approaching ∞ or $-\infty$, as we saw in Example 10.12, parts (a) and (c).
- When we write $\lim_{n\to\infty} a_n = L$, we will always mean that L is finite.

Limit Properties of Sequences

The limit properties for functions given in Section 2.2 also hold true for sequences. After all, sequences are particular kinds of functions. We state these properties below in terms of sequences.

Limit Properties for Sequences

If $\lim\limits_{n\to\infty} a_n$ and $\lim\limits_{n\to\infty} b_n$ both exist, then the following properties hold true:

1. $\lim\limits_{n\to\infty} ca_n = c\left(\lim\limits_{n\to\infty} a_n\right)$ for any constant c
2. $\lim\limits_{n\to\infty} (a_n + b_n) = \lim\limits_{n\to\infty} a_n + \lim\limits_{n\to\infty} b_n$
3. $\lim\limits_{n\to\infty} (a_n - b_n) = \lim\limits_{n\to\infty} a_n - \lim\limits_{n\to\infty} b_n$
4. $\lim\limits_{n\to\infty} a_n b_n = \left(\lim\limits_{n\to\infty} a_n\right)\left(\lim\limits_{n\to\infty} b_n\right)$
5. $\lim\limits_{n\to\infty} \dfrac{a_n}{b_n} = \dfrac{\lim\limits_{n\to\infty} a_n}{\lim\limits_{n\to\infty} b_n}$ if $b_n \neq 0$ for all n and $\lim\limits_{n\to\infty} b_n \neq 0$

The next two theorems are often helpful in finding limits of sequences. We omit the proofs.

THEOREM 10.2

If $\lim\limits_{n\to\infty} a_n = L$, and f is a function whose domain includes L and a_n for $n \geq N$, and if f is continuous at $x = L$, then
$$\lim_{n\to\infty} f(a_n) = f(L)$$

In particular, since $f(x) = x^k$, for k a positive integer, is continuous for all x, we have
$$\lim_{n\to\infty} (a_n)^k = L^k$$
provided the sequence $\{a_n\}$ converges to L. Similarly,
$$\lim_{n\to\infty} \sqrt[k]{a_n} = \sqrt[k]{L}$$
provided $a_n > 0$ and $L > 0$ for even-ordered kth roots.

THEOREM 10.3

Let $\{a_n\}$ be a sequence and f a function such that
$$f(n) = a_n, \quad n = 1, 2, 3, \ldots$$
If
$$\lim_{x\to\infty} f(x) = L$$
then also
$$\lim_{n\to\infty} a_n = L$$

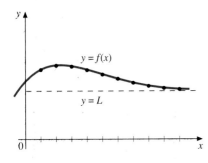

FIGURE 10.14
If $\lim\limits_{x \to \infty} f(x) = L$, then $\lim\limits_{n \to \infty} a_n = L$, where $a_n = f(n)$.

You will be asked to give a proof of Theorem 10.3 in Exercise 48 of Exercise Set 10.2, but the result is obvious intuitively by looking at the graph in Figure 10.14. The points (n, a_n) lie on the graph of $y = f(x)$, since $a_n = f(n)$. So if the graph of f is asymptotic to the line $y = L$, the a_n values come arbitrarily close to L.

REMARK
■ The converse of Theorem 10.3 is not true, as can be seen from the sequence $\{\sin \pi n\}$, which is constantly 0 and hence is convergent, but the corresponding function $f(x) = \sin \pi x$ does not approach a limit as $x \to \infty$. (See Figure 10.15.) So in this case $\lim\limits_{n \to \infty} f(n) = 0$, but $\lim\limits_{x \to \infty} f(x)$ does not exist.

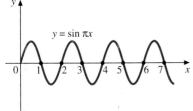

FIGURE 10.15
The sequence $\{\sin \pi n\} = \{0\}$, but $\lim\limits_{x \to \infty} \sin \pi x$ does not exist.

In the next example we illustrate how we can make use of Theorem 10.3 to find limits of sequences.

EXAMPLE 10.14 Find the limit of each of the following sequences.
(a) $\left\{\dfrac{\ln n}{n}\right\}$ (b) $\{\sqrt[n]{n}\}$

Solution
(a) Let
$$f(x) = \frac{\ln x}{x}$$
As $x \to \infty$, the function is indeterminant of the type ∞/∞. Applying L'Hôpital's Rule, we get
$$\lim_{x \to \infty} \frac{\ln x}{x} = \lim_{x \to \infty} \frac{1/x}{1} = 0$$
Since $f(n) = (\ln n)/n$, we conclude by Theorem 10.3 that
$$\lim_{n \to \infty} \frac{\ln n}{n} = 0$$

(b) Let $f(x) = x^{1/x}$ for $x > 0$. To find the limit, write $y = x^{1/x}$ and take the natural logarithm of both sides:
$$\ln y = \frac{1}{x} \ln x = \frac{\ln x}{x}$$
As we have seen in part (a), $(\ln x)/x \to 0$ as $x \to \infty$. Thus, since $\ln y \to 0$, it follows that $y \to e^0 = 1$. By Theorem 10.3 we conclude that
$$\lim_{n \to \infty} n^{1/n} = \lim_{n \to \infty} \sqrt[n]{n} = 1 \qquad ■$$

The Squeeze Theorem (Theorem 2.3) also holds true for sequences. We restate it here in terms of sequences.

THEOREM 10.4

The Squeeze Theorem for Sequences

If $\lim_{n \to \infty} a_n = L$ and $\lim_{n \to \infty} b_n = L$ and if for all sufficiently large n the inequality
$$a_n \leq c_n \leq b_n$$
holds true, then $\lim_{n \to \infty} c_n = L$ also.

EXAMPLE 10.15 Find the limit of the sequence

$$\left\{\frac{n \sin n}{1 + n^2}\right\}$$

or show that it diverges.

Solution Since $|\sin n| \leq 1$ for all n, we have

$$\left|\frac{n \sin n}{1 + n^2}\right| \leq \frac{n}{1 + n^2} < \frac{n}{n^2} = \frac{1}{n}$$

Thus,

$$-\frac{1}{n} < \frac{n \sin n}{1 + n^2} < \frac{1}{n}$$

for all positive integers n. Since both $\{1/n\}$ and $\{-1/n\}$ converge to 0, it follows by the Squeeze Theorem that $\{n \sin n/(1 + n^2)\}$ converges to 0 also. That is,

$$\lim_{n \to \infty} \frac{n \sin n}{1 + n^2} = 0$$

∎

Monotonicity and Boundedness

Sequences whose terms either always increase or always decrease (or else remain the same) play an especially important role in the study of infinite series. We use the following terminology to describe such sequences.

Definition 10.4
Monotone Sequences

A sequence $\{a_n\}$ is said to be

(a) **increasing** if $a_n \leq a_{n+1}$, or
(b) **decreasing** if $a_n \geq a_{n+1}$

for all positive integers n. A sequence that is either always increasing or always decreasing is said to be **monotone**.

REMARKS
- Frequently we will say *monotone increasing* or *monotone decreasing* to emphasize the monotonicity.
- The terms *nondecreasing* and *nonincreasing* are sometimes used where we have used increasing and decreasing, respectively.

If it is necessary to distinguish between a sequence for which $a_n < a_{n+1}$ for all n and one for which $a_n \leq a_{n+1}$, we will refer to the former as *strictly increasing*. By our definition, each of the following is increasing:

$$1, 2, 3, 4, 5, \ldots$$

$$1, 2, 2, 3, 4, 4, 5, \ldots$$

Only the first one is strictly increasing. Similarly, we can say a sequence in which $a_n > a_{n+1}$ is *strictly* decreasing.

Definition 10.5
Bounded Sequences

A sequence $\{a_n\}$ is said to be **bounded** if there is some positive constant M such that

$$|a_n| \leq M$$

for all positive integers n.

The sequences $\{(-1)^{n-1}\}$ and $\{\sin n\}$ considered earlier are bounded, since in each case $|a_n| \leq 1$. On the other hand, the sequence $\{2n - 1\}$ is unbounded.

Alone, neither monotonicity nor boundedness guarantees convergence. The sequence $\{(-1)^{n-1}\}$ is bounded but not convergent (also not monotone). The sequence $\{2n - 1\}$ is monotone but not convergent (also not bounded). We will shortly prove the important result that if a sequence is *both* monotone and bounded, then it must be convergent.

Before proving this result we need several more notions related to boundedness. A set S of real numbers is said to be **bounded above** if there is a real number M such that all elements of S are less than or equal to M. The number M is then said to be an **upper bound** of the set S. Similarly, S is **bounded below** if there is a real number m such that all elements of S are greater than or equal to m, in which case m is a **lower bound** of S. Clearly, if a sequence is bounded both above and below, then it is bounded, and the converse is also true. A fundamental property of the real-number system, called the *completeness property*, is that every nonempty set of real numbers that is bounded above has a **least upper bound**. To say L is the least upper bound of S means not only that L is an upper bound of S, but also that no number less than L is an upper bound of S. Similarly, if S is bounded below, it has a **greatest lower bound**. (These concepts are discussed further in Appendix 1.)

We are ready now for the main theorem on monotone bounded sequences.

THEOREM 10.5

Monotone Bounded Sequence Theorem

If $\{a_n\}$ is a sequence of real numbers that is both monotone and bounded, then it converges.

Proof We will prove the theorem for a monotone increasing sequence only. The proof for the decreasing case is similar.

Let $\{a_n\}$ be monotone increasing and bounded. The terms of the sequence constitute a set of real numbers that is bounded above. By the completeness property of real numbers just discussed, there is a least upper bound, say L, for this set. That is, $a_n \leq L$ for all n, where L is the smallest number having this property.

Let ε denote any positive number. Then, since $L - \varepsilon < L$, the number $L - \varepsilon$ is *not* an upper bound to the sequence $\{a_n\}$. (Remember that L is the *least* upper bound, so $L - \varepsilon$ cannot be an upper bound.) Thus, there is some member

Section 10.2 Sequences 749

FIGURE 10.16

of the sequence, say a_N, such that $a_N > L - \varepsilon$. Since the sequence is increasing, it follows that $a_n \geq a_N$ for all $n > N$ (see Figure 10.16). Thus, for all $n > N$,

$$L - \varepsilon < a_N \leq a_n \leq L < L + \varepsilon$$

In particular, for $n > N$,

$$L - \varepsilon < a_n < L + \varepsilon$$

So, subtracting L from each member of this inequality,

$$-\varepsilon < a_n - L < \varepsilon$$

or equivalently,

$$|a_n - L| < \varepsilon \qquad \text{if } n > N$$

By Definition 10.3, we see that

$$\lim_{n \to \infty} a_n = L \qquad \blacksquare$$

Note that we have proved not only that the sequence converges but also that its limit is the least upper bound of the points in the sequence. (In many cases, however, finding this least upper bound is difficult.)

In applying Theorem 10.5 it is sufficient that the sequence be monotone *from some point onward*, since convergence or divergence depends only on the behavior of the sequence as n gets arbitrarily large. For example, the sequence

$$20, \quad 33, \quad -75, \quad 1, \quad \frac{1}{2}, \quad \frac{1}{4}, \quad \frac{1}{8}, \quad \frac{1}{16}, \ldots$$

converges since it is monotone decreasing starting with the fourth term and is bounded. In fact, in this case we see that the limit is 0, which is the greatest lower bound of the terms starting with the fourth term.

Showing Monotonicity

To apply Theorem 10.5, we need to know that the sequence is monotone. Sometimes monotonicity is obvious. For example, the sequence $\{1/n\}$ is clearly decreasing. Some sequences, however, we must analyze further to determine if they are monotone. The next example illustrates three ways to test for monotonicity.

EXAMPLE 10.16 Show that each of the following sequences is monotone.

(a) $\left\{\dfrac{2n+3}{n}\right\}$ (b) $\left\{\dfrac{n}{\sqrt{1+n^2}}\right\}$ (c) $\left\{\dfrac{n!}{n^n}\right\}$

Solution

(a) The first technique to test for monotonicity is to look at the difference $a_{n+1} - a_n$ between successive terms. If this difference is always greater than or equal to 0, the sequence is increasing; if it is always less than or equal to 0, the sequence is decreasing. For the sequence $\{(2n+3)/n\}$, we have

$$a_{n+1} - a_n = \frac{2(n+1)+3}{n+1} - \frac{2n+3}{n} = \frac{(2n+5)n - (2n+3)(n+1)}{n(n+1)}$$

$$= \frac{2n^2 + 5n - (2n^2 + 5n + 3)}{n(n+1)} = \frac{-3}{n(n+1)} < 0$$

Thus, for all n, $a_{n+1} - a_n < 0$, so $a_{n+1} < a_n$, and we conclude that the sequence is (strictly) decreasing.

(b) As a second technique for testing for monotonicity, we consider a function f for which $f(n) = a_n$. If we can show monotonicity of f on the interval $[1, \infty)$, then we can conclude that $\{a_n\}$ is also monotone for $n \geq 1$. If f is differentiable, we can test it for monotonicity by considering $f'(x)$.
For the sequence $\{a_n\} = \{n/\sqrt{1+n^2}\}$, the natural choice for f is

$$f(x) = \frac{x}{\sqrt{1+x^2}}$$

Taking its derivative, we have

$$f'(x) = \frac{\sqrt{1+x^2} - x(\tfrac{1}{2})(1+x^2)^{-1/2}(2x)}{1+x^2}$$

$$= \frac{1+x^2-x^2}{(1+x^2)^{3/2}} = \frac{1}{(1+x^2)^{3/2}} > 0$$

(You should supply the missing algebra here.) Since $f'(x) > 0$ for all x, we know by Theorem 4.1 that f is an increasing function. Thus, since $a_n = f(n)$, we see that the sequence $\{a_n\}$ is also increasing.

(c) In the third approach to testing for monotonicity, we consider the ratio a_{n+1}/a_n of a given term (after the first) to the preceding term. If this ratio is always greater than or equal to 1, we conclude that $a_{n+1} \geq a_n$, and the sequence is increasing. If the ratio is less than or equal to 1, then $a_{n+1} \leq a_n$, and the sequence is decreasing. (In order to apply this technique, we must have $a_n > 0$ for all n.)
With $a_n = n!/n^n$, we have

$$\frac{a_{n+1}}{n} = \frac{\dfrac{(n+1)!}{(n+1)^{n+1}}}{\dfrac{n!}{n^n}} = \frac{(n+1)!}{(n+1)^{n+1}} \cdot \frac{n^n}{n!} = \frac{(n+1)!}{n!} \cdot \frac{n^n}{(n+1)^{n+1}}$$

$$= (n+1) \cdot \frac{n^n}{(n+1)^{n+1}} \qquad \text{Since } (n+1)! = (n+1)n!$$

$$= \frac{n^n}{(n+1)^n} < \frac{n^n}{n^n} = 1 \qquad \text{Since } (n+1)^n > n^n$$

Thus,

$$\frac{a_{n+1}}{a_n} < 1 \qquad \text{for all } n$$

Hence, $a_{n+1} < a_n$, and we conclude that $\{a_n\}$ is a decreasing sequence. ■

Let us summarize the three ways of showing monotonicity illustrated in the preceding example.

> **Tests for Monotonicity**
>
> 1. Calculate the difference $a_{n+1} - a_n$.
>
>   If $\begin{cases} a_{n+1} - a_n \geq 0 & \text{for all } n, \text{ then } \{a_n\} \text{ is increasing;} \\ a_{n+1} - a_n \leq 0 & \text{for all } n, \text{ then } \{a_n\} \text{ is decreasing.} \end{cases}$
>
> 2. Let $f(x)$ be such that $f(n) = a_n$. Calculate $f'(x)$ if it exists.
>
>   If $\begin{cases} f'(x) \geq 0 & \text{on } [1, \infty), \text{ then } \{a_n\} \text{ is increasing;} \\ f'(x) \leq 0 & \text{on } [1, \infty), \text{ then } \{a_n\} \text{ is decreasing.} \end{cases}$
>
> 3. If $a_n > 0$ for all n, calculate the ratio a_{n+1}/a_n.
>
>   If $\begin{cases} \dfrac{a_{n+1}}{a_n} \geq 1 & \text{for all } n, \text{ then } \{a_n\} \text{ is increasing;} \\ \dfrac{a_{n+1}}{a_n} \leq 1 & \text{for all } n, \text{ then } \{a_n\} \text{ is decreasing.} \end{cases}$

Again, for purposes of applying Theorem 10.5, it is sufficient that the sequence be monotone from some point onward, say, for $n \geq n_0$. Thus, in applying Tests 1 and 3 listed above, we can modify "for all n" by requiring only that the conditions be true "for all $n \geq n_0$." For Test 2, $f'(x)$ need be either positive or negative on $[a, \infty)$ for some $a > 0$.

If a sequence of nonnegative terms is found to be decreasing, then it must converge, since it is bounded (bounded below by 0 and above by its first term). Thus, we can conclude that the sequences in parts (a) and (c) of Example 10.16 both converge. In fact, for part (a) we can find the limit directly:

$$\lim_{n \to \infty} \frac{2n + 3}{n} = \lim_{n \to \infty} \left(2 + \frac{3}{n}\right) = 2$$

The limit in part (c) is 0, as you will be asked to show in the exercises (Exercise 50 of this section). In part (b) we showed that the sequence is increasing. It is bounded above by 1, since

$$\frac{n}{\sqrt{1 + n^2}} < \frac{n}{\sqrt{n^2}} = \frac{n}{n} = 1$$

(Of course, it is also bounded below by its first term, since subsequent terms are larger.) Thus, the sequence converges. The limit is actually 1, as we see from the dominant term approach:

$$\frac{n}{\sqrt{1 + n^2}} \approx \frac{n}{\sqrt{n^2}} = \frac{n}{n} = 1 \quad \text{for large } n$$

So

$$\lim_{n \to \infty} \frac{n}{\sqrt{1 + n^2}} = 1$$

Exercise Set 10.2

In Exercises 1–3, write out the first five terms of the sequence $\{a_n\}$.

1. (a) $a_n = \dfrac{(-1)^{n-1} n}{2n - 1}$ (b) $a_n = \dfrac{2^n}{n!}$

2. (a) $a_n = \dfrac{2 \cdot 4 \cdot 6 \cdots (2n)}{1 \cdot 3 \cdot 5 \cdots (2n - 1)}$

   (b) $a_n = \dfrac{\sin[(2n + 1)\pi/2]}{n(n + 1)}$

3. (a) $a_1 = 1$, $a_n = -n a_{n-1}$ if $n \geq 2$
   (b) $a_1 = 1$, $a_2 = 2$, $a_n = \dfrac{a_{n-1}}{a_{n-2}}$ if $n \geq 3$

In Exercises 4–6, determine a formula for the nth term of the sequence.

4. (a) $1, -\dfrac{1}{3}, \dfrac{1}{5}, -\dfrac{1}{7}, \dfrac{1}{9}, \ldots$

   (b) $\dfrac{1}{2}, \dfrac{2}{5}, \dfrac{3}{10}, \dfrac{4}{17}, \dfrac{5}{26}, \ldots$

5. (a) $\dfrac{1}{2}, \dfrac{3}{4}, \dfrac{7}{8}, \dfrac{15}{16}, \dfrac{31}{32}, \ldots$

   (b) $\dfrac{2}{5}, -\dfrac{4}{7}, \dfrac{6}{9}, -\dfrac{8}{11}, \dfrac{10}{13}, \ldots$

6. (a) $\dfrac{1}{7}, -\dfrac{3}{10}, \dfrac{5}{13}, -\dfrac{7}{16}, \dfrac{9}{19}, \ldots$

   (b) $-1, 3, \dfrac{5}{3}, \dfrac{7}{5}, \dfrac{9}{7}, \ldots$

In Exercises 7–30, find $\lim_{n \to \infty} a_n$ or show that the sequence diverges.

7. $a_n = \dfrac{n^2 - 2n}{3n^2 + n - 1}$

8. $a_n = \dfrac{n + 1}{2n^2 - n + 1}$

9. $a_n = \dfrac{2n - n^3}{(n + 1)(n + 3)}$

10. $a_n = \dfrac{n}{\sqrt{1 + n^2}}$

11. $a_n = \dfrac{\cos n\pi}{\sqrt{n}}$

12. $a_n = \dfrac{1 + \sin n}{\ln(n + 1)}$

13. $a_n = \dfrac{1 - (-1)^n}{2}$

14. $a_n = n^3 e^{-n}$

15. $a_n = \tanh n$

16. $a_n = \dfrac{\ln n}{\sqrt{n}}$

17. $a_n = (-0.999)^n$

18. $a_n = \dfrac{1 - \cosh n}{\sinh n}$

19. $a_n = \ln\left(1 + \dfrac{1}{n}\right)$

20. $a_n = \dfrac{\ln n}{\ln(\ln n)}$

21. $a_n = \dfrac{2^n}{e^{n-1}}$

22. $a_n = \dfrac{(-1)^{n-1}(2 \sin n + \cos n)}{n \sec n}$

23. $a_n = \sin n + \cos n$

24. $a_n = (1.00001)^n$

25. $a_n = \dfrac{n + 1}{n}\left(\dfrac{2}{3}\right)^n$

26. $a_n = \dfrac{(-1)^n e^{2n}}{1 + 9^n}$

27. $a_n = \dfrac{n \ln n}{1 + n^2}$

28. $a_n = \dfrac{\sqrt{1 + n^3}}{\sqrt[3]{1 + n^4}}$

29. $a_n = n^2 \cdot 2^{-n}$

30. $a_n = n \sin \dfrac{1}{n}$

In Exercises 31–40, show that the sequence $\{a_n\}$ is monotone for $n \geq n_0$. If $n_0 \neq 1$, give the smallest value of n_0.

31. $a_n = \dfrac{2n - 1}{n + 1}$

32. $a_n = \dfrac{n}{2n - 3}$

33. $a_n = n^2 e^{-n}$

34. $a_n = \dfrac{(\ln n)^2}{n}$

35. $a_n = \dfrac{2^n}{n!}$

36. $a_n = 2^{1/n}$

37. $a_n = \dfrac{n!}{1 \cdot 3 \cdot 5 \cdots (2n - 1)}$

38. $a_n = \dfrac{1 \cdot 3 \cdot 5 \cdots (2n - 1)}{2^n \cdot n!}$

39. $a_n = \ln n - \ln(n + 1)$

40. $a_n = \tan^{-1} n$

In Exercises 41–46, prove that the sequence $\{a_n\}$ converges.

41. $a_n = \dfrac{e^n}{n!}$

42. $a_n = \dfrac{2^n n!}{(2n)!}$

43. $a_n = n^{1/n}$

44. $a_n = \dfrac{1 \cdot 3 \cdot 5 \cdots (2n - 1)}{2 \cdot 4 \cdot 6 \cdots 2n}$

45. $a_1 = 1$, $a_n = 1 + \dfrac{a_{n-1}}{2}$ for $n \geq 2$

46. $a_1 = 2$, $a_n = \sqrt{a_{n-1}}$ for $n \geq 2$

47. Prove that a sequence can have at most one limit.

48. Prove Theorem 10.3.

49. Prove that $\lim_{n\to\infty} a_n = 0$ if and only if $\lim_{n\to\infty} |a_n| = 0$.

50. Prove that $\lim_{n\to\infty}(n!/n^n) = 0$. (*Hint:* Write
$$a_n = \frac{n!}{n^n} = \frac{1 \cdot 2 \cdot 3 \cdots n}{n \cdot n \cdot n \cdots n}$$
and show that $a_n < 1/n$.)

51. Prove that if
$$a_n = \frac{1}{n+1} + \frac{1}{n+2} + \cdots + \frac{1}{2n}$$
then $\{a_n\}$ is convergent. *Hint:* Use Theorem 10.4. To show boundedness, observe that
$$a_n \leq n\left(\frac{1}{n+1}\right)$$

52. Let $a_1 = \sqrt{3}$ and for $n \geq 2$, $a_n = \sqrt{3a_{n-1}}$. Prove that $\{a_n\}$ converges, and find its limit. (*Hint:* Use mathematical induction (see Supplementary Appendix 4) to prove that $a_{n+1} > a_n$ and $a_n < 3$ for all n.)

53. For the Fibonacci sequence $\{f_n\}$, define
$$a_n = \frac{f_{n+1}}{f_n}, \qquad n = 1, 2, 3, \ldots$$
Use a spreadsheet, CAS, or calculator to find the first twenty terms of the sequence $\{a_n\}$. Assuming that $\lim_{n\to\infty} a_n = L$, find the exact value of L. (*Hint:* Show that $a_n = 1 + (1/a_{n-1})$ for $n \geq 2$ and take the limit as $n \to \infty$ to get $L = 1 + (1/L)$. Now solve for L.)

Note: The number L is the *golden mean*, defined as shown in the figure.

The golden mean L is the mean proportional between 1 and $1 - L$:
$$\frac{1}{L} = \frac{L}{1-L}$$

54. Use a spreadsheet, CAS, or calculator to do the following:
(a) Find the first twenty terms of the Fibonacci sequence $\{f_n\}$.
(b) Find the first twenty terms of the "Fibonacci-like" sequence $\{g_n\}$ in which $g_1 = 1$, $g_2 = 3$, and $g_n = g_{n-1} + g_{n-2}$ for $n \geq 2$.
(c) Find the ratio g_n/f_n for $n = 1, 2, \ldots, 20$ for f_n and g_n in parts (a) and (b). Conjecture the limit
$$\lim_{n\to\infty} \frac{g_n}{f_n}$$

55. (a) Let x_1 be given and $x_{n+1} = \sqrt{x_n}$ for $n \geq 1$. Use a spreadsheet, CAS, or calculator to find the first twenty terms of the sequence $\{x_n\}$ for four different values of x_1. Conjecture the limit in each case. Prove your conjecture. Does the answer depend on x_1?
(b) Repeat part (a) if $x_{n+1} = x_n^2 - 2$.

56. (1) Choose an arbitrary integer $A > 0$.
(2) If $A = 1$, stop.
(3) If A is even, replace A with $A/2$ and go to step 2.
(4) If A is odd, replace A with $3A + 1$ and go to step 2.

Let $A = 3$, $A = 34$, and $A = 75$, and in each case compute the terms generated. Is it true that the larger the value of A, the more steps are required before the algorithm stops?

57. Let $a_1 = 2$ and $a_n = \sqrt{2 + a_{n-1}}$ for $n \geq 2$.
(a) Plot several points of the sequence $\{a_n\}$ on a number line. Does the sequence appear to be bounded above? If so, what appears to be an upper bound?
(b) Use mathematical induction to prove that $a_n \leq 2$ for all n.
(c) Show that $a_{n+1}^2 - a_n^2 = (2 - a_n)(1 + a_n)$. Use this result to show that $\{a_n\}$ is monotone increasing.
(d) Let $\lim_{n\to\infty} a_n = L$. By taking the limit of both sides of the equation $a_n = \sqrt{2 + a_{n-1}}$ as $n \to \infty$, find the value of L. (*Hint:* Note that $a_{n-1} \to L$ as $n \to \infty$.)

58. (a) Is it possible for a term of a convergent sequence to exceed the limit of the sequence? Is it possible for some terms to exceed and some terms to be less than the limit? Is it possible for some terms to equal the limit? Use examples in your answers to these questions.
(b) If the terms of a sequence alternate in sign, can the sequence converge? If you think it is possible, give an example.

59. In each of the following give an example, whenever possible, satisfying the given condition. If it is not possible, explain why.
(a) A sequence with limit 12 and no term equal to 12
(b) An increasing sequence that is not bounded
(c) A bounded sequence that does not converge
(d) A bounded, convergent sequence that is not monotone
(e) An increasing sequence that does not converge
(f) A convergent sequence with infinitely many terms equal to 7, infinitely many terms not equal to 7, that converges to 7
(g) A convergent sequence with infinitely many terms equal to 7 and infinitely many terms equal to 8
(h) A convergent sequence that is not bounded
(i) Two sequences with the same limit but with no common terms
(j) A sequence with limit 2 that has infinitely many negative terms

10.3 INFINITE SERIES OF CONSTANTS

In the introduction to this chapter we discussed the question of how to define the sum of infinitely many terms. Now we want to explore this question in more detail.

Consider first the familiar decimal representation of the fraction 1/3. By repeated division, we get

$$\frac{1}{3} = 0.3333\ldots$$

We can write the repeating decimal on the right as the infinite sum

$$0.3 + 0.03 + 0.003 + 0.0003 + \cdots \qquad (10.15)$$

We know, then, that the sum of this infinite series is 1/3.

As a second example, consider again the infinite series

$$1 + \frac{1}{2} + \frac{1}{4} + \frac{1}{8} + \cdots \qquad (10.16)$$

in which each term after the first is half of the preceding term. We proceed as we did in the introduction to the chapter by recording its **partial sums**—that is, the successive finite sums:

$$\begin{aligned}
\text{First term} &= 1 \\
\text{Sum of first two terms} &= 1\frac{1}{2} \\
\text{Sum of first three terms} &= 1\frac{3}{4} \\
\text{Sum of first four terms} &= 1\frac{7}{8} \\
&\vdots
\end{aligned}$$

FIGURE 10.17

In Figure 10.17 we show these partial sums as points on a number line. Notice that each successive point after the first is halfway between the previous point and 2. Clearly, as we add more and more terms, the partial sums come arbitrarily close to 2, so we say that the sum of the series is 2 and write

$$1 + \frac{1}{2} + \frac{1}{4} + \frac{1}{8} + \cdots = 2$$

Let us now generalize what we have illustrated with the two particular infinite series 10.15 and 10.16. Suppose that $a_1, a_2, a_3, \ldots$ are numbers (not necessarily all distinct). The sum

$$a_1 + a_2 + a_3 + \cdots \qquad (10.17)$$

is called an **infinite series**, or simply a **series**. Writing the three dots at the end is important as an indication that the series continues indefinitely. For brevity, we will often write series 10.17 in summation notation as

$$\sum_{n=1}^{\infty} a_n$$

The particular letter used as the index of summation is not important, since the index is replaced by $1, 2, 3, \ldots$ when we write out the series. For example, we could write $\sum_{k=1}^{\infty} a_k$ for the series 10.17. Sometimes we will omit the range of the index and simply write $\sum a_n$, with the understanding that the index goes from 1 to ∞. There will be times when we want the index to begin with 0. Note, for example, that

$$\sum_{n=0}^{\infty} \frac{1}{2^n} \quad \text{and} \quad \sum_{n=1}^{\infty} \frac{1}{2^{n-1}}$$

are two ways of representing the same series. (Try writing the first few terms in each case.)

Just as we did with the series 10.16, we list the partial sums of the series 10.17:

$$\begin{aligned}
\text{First partial sum} &= a_1 \\
\text{Second partial sum} &= a_1 + a_2 \\
\text{Third partial sum} &= a_1 + a_2 + a_3 \\
&\vdots \\
n\text{th partial sum} &= a_1 + a_2 + a_3 + \cdots + a_n \\
&\vdots
\end{aligned}$$

The question is: Do these partial sums approach more and more closely some limiting value as we let n increase indefinitely? If so, this limiting value is what we define as the sum of the series 10.17. In order to state this definition more succinctly, we introduce the notation

$$S_n = a_1 + a_2 + a_3 + \cdots + a_n$$

That is, $S_1 = a_1$, $S_2 = a_1 + a_2$, $S_3 = a_1 + a_2 + a_3$, and so on. For each positive integer n, S_n is the **nth partial sum** of the series 10.17. These partial sums, in order, $S_1, S_2, S_3, \ldots$, constitute a sequence $\{S_n\}$, called the **sequence of partial sums** of the series.

Note carefully that when we list the terms of a sequence, we separate them with commas, whereas the terms of a series are added together. For example, the *series* 10.16 is

$$1 + \frac{1}{2} + \frac{1}{4} + \frac{1}{8} + \cdots$$

and its *sequence* of partial sums is

$$1, 1\frac{1}{2}, 1\frac{3}{4}, 1\frac{7}{8}, \ldots$$

For this example, we can find a simple formula for S_n, namely,

$$S_n = 2 - \frac{1}{2^{n-1}}$$

since S_1 is 1 less than 2, S_2 is $\frac{1}{2}$ less than 2, S_3 is $\frac{1}{4}$ less than 2, and so on. As $n \to \infty$, we get the limit

$$\lim_{n \to \infty} S_n = 2$$

since $2 - (1/2^{n-1})$ can be made arbitrarily close to 2 by taking n sufficiently large. This result confirms again that series 10.16 converges to 2.

We now give the formal definition of the sum of an infinite series.

> **Definition 10.6**
> **The Sum of an Infinite Series**
>
> The infinite series
> $$\sum_{n=1}^{\infty} a_n = a_1 + a_2 + a_3 + \cdots$$
> is said to **converge to the sum S** provided
> $$\lim_{n \to \infty} S_n = S$$
> where $S_n = a_1 + a_2 + \cdots + a_n$. In this case we write $\sum_{n=1}^{\infty} a_n = S$. If $\lim_{n \to \infty} S_n$ does not exist, we say that the series **diverges**.

Briefly, we can write

$$\sum_{n=1}^{\infty} a_n = \lim_{n \to \infty} (a_1 + a_2 + \cdots + a_n)$$

provided the limit on the right exists (that is, the limit is finite).

Recall from Section 10.2 that the precise meaning of

$$\lim_{n \to \infty} S_n = S$$

is that, given any positive number ε, there exists a positive number N such that for all $n > N$, $|S_n - S| < \varepsilon$.

In most cases it is not feasible to apply Definition 10.6 directly to determine whether a given series converges and if so, the value of its sum. In the next two sections we will develop means of testing for convergence or divergence without having to apply the definition directly. There are, however, certain series for which the definition can be used. The most important of these is called a *geometric series*, which we now describe.

Geometric Series

A **geometric series** is an infinite series of the form

$$\sum_{n=1}^{\infty} ar^{n-1} = a + ar + ar^2 + ar^3 + \cdots \qquad (10.18)$$

in which a and r are fixed real numbers with $a \neq 0$. The number a is its first term, and the number r is called the **common ratio**. Note that the ratio of any term (after the first) to the preceding term is r if $r \neq 0$ (for example, $ar^3/ar^2 = r$), hence the name common ratio. Geometric series are particularly easy to work with not only because we can completely describe when they converge and diverge, but also because when they converge, we can easily compute the sum. To verify these facts, we investigate the sequence $\{S_n\}$ of partial sums, where

$$S_n = a + ar + ar^2 + \cdots + ar^{n-1}$$

First, suppose $r = 1$. Then

$$S_n = a + a + a + \cdots + a = na$$

Clearly, $\lim_{n \to \infty} S_n$ does not exist as a finite number in this case, since na will be arbitrarily large in absolute value if n is sufficiently large. That is, $\lim_{n \to \infty} na = \infty$ if $a > 0$ and $\lim_{n \to \infty} na = -\infty$ if $a < 0$. Thus, when $r = 1$, the geometric series 10.18 diverges.

If $r \neq 1$, we can write S_n in what is called a *closed form* by the following trick. We note that if we multiply S_n by r, we get many terms just like those in S_n. If we subtract rS_n from S_n, we get

$$S_n - rS_n = (a + ar + ar^2 + ar^3 + \cdots + ar^{n-1}) - (ar + ar^2 + ar^3 + \cdots + ar^n)$$

All the terms except a and ar^n cancel, so we have

$$S_n - rS_n = a - ar^n$$

or, equivalently,

$$S_n(1 - r) = a(1 - r^n)$$

Since with $r \neq 1$, the factor $1 - r$ is nonzero, we can solve for S_n:

$$S_n = \frac{a(1 - r^n)}{1 - r}$$

We know from Example 10.13 and the discussion following it that when $|r| < 1$,

$$\lim_{n \to \infty} r^n = 0$$

and when $r = -1$ or $|r| > 1$,

$$\lim_{n \to \infty} r^n \text{ does not exist}$$

Thus, when $|r| < 1$,

$$\lim_{n \to \infty} S_n = \lim_{n \to \infty} \frac{a(1 - r^n)}{1 - r}$$

$$= \frac{a}{1 - r}$$

and the geometric series converges to $a/(1 - r)$. If $|r| \geq 1$, $\{S_n\}$ does not converge, so the geometric series diverges.

Because of the importance of geometric series, we state our findings as a theorem.

THEOREM 10.6

If $|r| < 1$, the geometric series $\sum_{n=1}^{\infty} ar^{n-1}$ converges to the sum $a/(1-r)$. That is,

$$a + ar + ar^2 + ar^3 + \cdots = \frac{a}{1 - r} \qquad \text{if } |r| < 1 \qquad (10.19)$$

If $|r| \geq 1$, the series diverges.

Both of the series 10.15 and 10.16 are geometric. For 10.15 we see from the series

$$0.3 + 0.03 + 0.003 + \cdots$$

that $a = 0.3$ and $r = 0.1$. Since $|r| = 0.1 < 1$, we have, by Equation 10.19,

$$0.3 + 0.03 + 0.003 + \cdots = \frac{0.3}{1 - 0.1} = \frac{0.3}{0.9} = \frac{3}{9} = \frac{1}{3}$$

confirming what we already knew.

For the series 10.16,

$$1 + \frac{1}{2} + \frac{1}{4} + \frac{1}{8} + \cdots$$

$a = 1$ and $r = \frac{1}{2}$. Again, by Equation 10.19, the sum is

$$\frac{a}{1 - r} = \frac{1}{1 - \frac{1}{2}} = \frac{1}{\frac{1}{2}} = 2$$

The next four examples further illustrate Theorem 10.6.

EXAMPLE 10.17 Determine whether each of the following series is convergent or divergent. If convergent, find the sum.

(a) $1 - \dfrac{2}{3} + \dfrac{4}{9} - \dfrac{8}{27} + \cdots$

(b) $3 + 4 + \dfrac{16}{3} + \dfrac{64}{9} + \cdots$

Solution

(a) The series is geometric, with $a = 1$ and $r = -2/3$. One way to determine r is to divide any term after the first by the one before it. Another way is to ask by what factor you would multiply a given term to get the next term. In either case, be sure to check that the ratio (or the multiplicative factor) remains the same for all pairs of consecutive terms. Confirming that $r = -2/3$, we see that its absolute value is less than 1, so by Equation 10.19 the series converges to

$$\frac{a}{1 - r} = \frac{1}{1 - \left(-\frac{2}{3}\right)} = \frac{1}{\frac{5}{3}} = \frac{3}{5}$$

We can therefore write

$$1 - \frac{2}{3} + \frac{4}{9} - \frac{8}{27} + \cdots = \frac{3}{5}$$

(b) The series is geometric, with $a = 3$ and $r = 4/3$. Since $r > 1$, the series diverges. ∎

EXAMPLE 10.18 Show that each of the following series is geometric and determine whether it is convergent or divergent. If convergent, find the sum.

(a) $\displaystyle\sum_{n=0}^{\infty} e^{-n}$ (b) $\displaystyle\sum_{k=1}^{\infty} \frac{(-1)^{k-1}}{2^k}$

Solution

(a) When a series is written in summation notation, it is sometimes helpful to write out the first few terms. In this case, we have

$$\sum_{n=0}^{\infty} e^{-n} = 1 + e^{-1} + e^{-2} + e^{-3} + \cdots = 1 + \frac{1}{e} + \frac{1}{e^2} + \frac{1}{e^3} + \cdots$$

Each term is obtained by multiplying the preceding one by $1/e$. So $r = 1/e$, and the series is geometric. Since $e \approx 2.7$, $|1/e| < 1$. The sum of the series is therefore

$$\frac{a}{1-r} = \frac{1}{1 - \frac{1}{e}} = \frac{e}{e-1}$$

(b) By writing out the first few terms,

$$\sum_{k=1}^{\infty} \frac{(-1)^{k-1}}{2^k} = \frac{1}{2} - \frac{1}{4} + \frac{1}{8} - \frac{1}{16} + \cdots$$

we see that the series is geometric, with $a = \frac{1}{2}$ and $r = -\frac{1}{2}$. Since $|-\frac{1}{2}| < 1$, the sum is

$$\frac{a}{1-r} = \frac{\frac{1}{2}}{1-(-\frac{1}{2})} = \frac{\frac{1}{2}}{\frac{3}{2}} = \frac{1}{3}$$ ∎

EXAMPLE 10.19 Use a geometric series to express the repeating decimal $1.272727\ldots$ as the ratio m/n of two positive integers.

Solution First, write the repeating decimal as the infinite series

$$1 + (0.27 + 0.0027 + 0.000027 + \cdots)$$

The series in parentheses is geometric, with $a = 0.27$ and $r = 0.01$. Thus, its sum is

$$\frac{a}{1-r} = \frac{0.27}{1 - 0.01} = \frac{0.27}{0.99} = \frac{27}{99} = \frac{3}{11}$$

The entire sum is therefore $1 + \frac{3}{11}$. That is,

$$1.272727\ldots = 1 + \frac{3}{11} = \frac{14}{11}$$ ∎

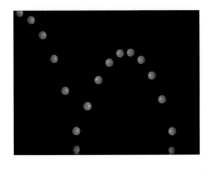

EXAMPLE 10.20 A ball is dropped from a height of 50 cm, and on each bounce it goes three-fourths as high as before. Approximate the total distance traveled by the ball in coming to rest.

Solution Although the ball moves only along a vertical line, we can get a better picture by showing the path as in Figure 10.18 (on page 760). First it drops 50 cm, then it rises $37\frac{1}{2}$ cm and falls the same distance. Thereafter it rises and falls three-fourths of the up-and-down total for the previous bounce. To make the first motion analogous to the others, we can pretend it started from the ground, rising 50 cm and then falling 50 cm. We will subtract this imaginary initial 50-cm rise after getting the total. With this understanding, the total distance covered is

$$\left[100 + \frac{3}{4}(100) + \left(\frac{3}{4}\right)^2 (100) + \left(\frac{3}{4}\right)^3 (100) + \cdots \right] - 50$$

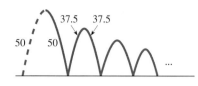

FIGURE 10.18

The expression in brackets is a geometric series, with $a = 100$ and $r = \frac{3}{4}$. So its sum is

$$\frac{a}{1-r} = \frac{100}{1-\frac{3}{4}} = 400$$

Thus, on subtracting the initial imaginary 50-cm rise, we find the total distance the ball covers to be 350 cm. ∎

Telescoping Series

The next example illustrates another type of series in which the limit of the nth partial sum can be found. For reasons that will become apparent, the series is called a **telescoping series**.

EXAMPLE 10.21 Show that the series

$$\sum_{k=1}^{\infty} \frac{1}{k^2 + k}$$

converges, and find its sum.

Solution Using partial fractions, we can rewrite the general term in the form

$$\frac{1}{k^2 + k} = \frac{1}{k(k+1)} = \frac{1}{k} - \frac{1}{k+1}$$

Thus, the nth partial sum of the series is

$$S_n = \sum_{k=1}^{n} \left(\frac{1}{k} - \frac{1}{k+1} \right)$$

$$= \left(\frac{1}{1} - \frac{1}{2}\right) + \left(\frac{1}{2} - \frac{1}{3}\right) + \left(\frac{1}{3} - \frac{1}{4}\right) + \cdots + \left(\frac{1}{n} - \frac{1}{n+1}\right)$$

By regrouping the parentheses, we see that all terms except the first and last cancel out (this expression "telescopes," as in an old type of telescope that is collapsible), and so

$$S_n = 1 - \frac{1}{n+1}$$

As $n \to \infty$, $S_n \to 1$. This calculation shows that the original series converges, since the sequence $\{S_n\}$ converges, and also that its sum is 1. That is,

$$\sum_{k=1}^{\infty} \frac{1}{k^2 + k} = 1$$

∎

A Necessary Condition for Convergence

Another very useful result can be obtained from Definition 10.6, which we state as a theorem.

THEOREM 10.7 If the series $\sum_{n=1}^{\infty} a_n$ converges, then $\lim_{n\to\infty} a_n = 0$.

Proof Let S_n be the nth partial sum of $\sum_{n=1}^{\infty} a_n$; that is,
$$S_n = a_1 + a_2 + a_3 + \cdots + a_n$$
Then, if $n > 1$, we also have
$$S_{n-1} = a_1 + a_2 + a_3 + \cdots + a_{n-1}$$
On subtracting S_{n-1} from S_n, all terms drop out except a_n.
$$a_n = S_n - S_{n-1}$$
Now, since the series converges, we know that $\lim_{n\to\infty} S_n$ exists. Call its value S. But as $n \to \infty$, we also have $(n-1) \to \infty$. So both $\lim_{n\to\infty} S_n = S$ and $\lim_{n\to\infty} S_{n-1} = S$. Thus,
$$\lim_{n\to\infty} a_n = \lim_{n\to\infty} (S_n - S_{n-1})$$
$$= \lim_{n\to\infty} S_n - \lim_{n\to\infty} S_{n-1}$$
$$= S - S = 0 \qquad \blacksquare$$

REMARK
■ Here we have made use of the limit property from Section 10.2 that says if $\lim_{n\to\infty} a_n$ and $\lim_{n\to\infty} b_n$ both exist, then $\lim_{n\to\infty} (a_n - b_n)$ exists and equals $\lim_{n\to\infty} a_n - \lim_{n\to\infty} b_n$.

An equivalent way of stating Theorem 10.7 (a logical equivalent called its *contrapositive*) is the following:

The nth-Term Test for Divergence
If $\lim_{n\to\infty} a_n \neq 0$, then $\sum_{n=1}^{\infty} a_n$ diverges.

Thus, if for a given series we can show that $\lim_{n\to\infty} a_n \neq 0$, we can conclude that the series $\sum a_n$ definitely diverges. For example, the series
$$\sum_{n=1}^{\infty} \frac{n}{2n+1}$$
diverges, since
$$\lim_{n\to\infty} a_n = \lim_{n\to\infty} \frac{n}{2n+1} = \frac{1}{2} \neq 0$$

If you find merely that $\lim_{n\to\infty} a_n = 0$, you cannot draw any definite conclusion regarding the convergence or divergence of $\sum a_n$ except that the series at least has a chance to converge. It is a common mistake to assume that the converse of Theorem 10.7 is true; that is, that $\lim_{n\to\infty} a_n = 0$ implies the convergence of

$\sum a_n$. That this converse is *not true* can be seen by finding a divergent series whose terms approach 0, as we do now.

The Harmonic Series

The best-known example of a divergent series whose nth term tends to 0 is the so-called **harmonic series**—namely,

$$\sum_{n=1}^{\infty} \frac{1}{n} = 1 + \frac{1}{2} + \frac{1}{3} + \cdots + \frac{1}{n} + \cdots$$

Clearly, $\lim_{n \to \infty} 1/n = 0$, yet the series diverges, as the following analysis of certain of its partial sums shows:

$$S_1 = 1$$

$$S_2 = 1 + \frac{1}{2}$$

$$S_4 = 1 + \frac{1}{2} + \frac{1}{3} + \frac{1}{4} > 1 + \frac{1}{2} + \left(\frac{1}{4} + \frac{1}{4}\right) = 1 + \frac{1}{2} + \frac{1}{2}$$

$$S_8 = 1 + \frac{1}{2} + \frac{1}{3} + \frac{1}{4} + \frac{1}{5} + \frac{1}{6} + \frac{1}{7} + \frac{1}{8}$$

$$> 1 + \frac{1}{2} + \left(\frac{1}{4} + \frac{1}{4}\right) + \left(\frac{1}{8} + \frac{1}{8} + \frac{1}{8} + \frac{1}{8}\right) = 1 + \frac{1}{2} + \frac{1}{2} + \frac{1}{2}$$

Notice that the relationship between these partial sums and the number of times $\frac{1}{2}$ is added is as follows:

$$S_1 = S_{2^0} = 1 + 0\left(\frac{1}{2}\right)$$

$$S_2 = S_{2^1} > 1 + 1\left(\frac{1}{2}\right)$$

$$S_4 = S_{2^2} > 1 + 2\left(\frac{1}{2}\right)$$

$$S_8 = S_{2^3} > 1 + 3\left(\frac{1}{2}\right)$$

$$\vdots \qquad \vdots$$

In general, we have

$$S_{2^n} \geq 1 + n\left(\frac{1}{2}\right), \qquad n = 0, 1, 2, \ldots$$

(Mathematical induction can be used to give a formal proof.) Thus, $\lim_{n \to \infty} S_{2^n} = \infty$ and it follows that $\lim_{n \to \infty} S_n$ does not exist. Thus, the series diverges. Because we will have frequent occasions to refer to this result, we set it off for emphasis.

Section 10.3 Infinite Series of Constants 763

The Harmonic Series

The harmonic series

$$\sum_{n=1}^{\infty} \frac{1}{n} = 1 + \frac{1}{2} + \frac{1}{3} + \cdots + \frac{1}{n} + \cdots$$

diverges.

Since for the harmonic series $\lim_{n \to \infty} S_n = \infty$, we sometimes write $\sum_{n=1}^{\infty} 1/n = \infty$. It is interesting to note that the sum of the first billion terms of the harmonic series is only about 21. That is,

$$S_{10^9} = 1 + \frac{1}{2} + \frac{1}{3} + \cdots + \frac{1}{10^9} \approx 21$$

and each succeeding term is *very* small. Yet if we continue adding terms indefinitely, the cumulative effect is to cause the sum to become infinite! Sometimes infinite processes defy one's intuition.

Properties of Convergent Series

The next theorem is a direct consequence of Definition 10.6. We call for a proof in Exercise 38 of Exercise Set 10.3.

THEOREM 10.8

If $\sum_{n=1}^{\infty} a_n$ and $\sum_{n=1}^{\infty} b_n$ are convergent series, and if c is any real number, then each of the series $\sum_{n=1}^{\infty} ca_n$, $\sum_{n=1}^{\infty} (a_n + b_n)$, and $\sum_{n=1}^{\infty} (a_n - b_n)$ converges, and

1. $\displaystyle\sum_{n=1}^{\infty} ca_n = c \sum_{n=1}^{\infty} a_n$

2. $\displaystyle\sum_{n=1}^{\infty} (a_n + b_n) = \sum_{n=1}^{\infty} a_n + \sum_{n=1}^{\infty} b_n$

3. $\displaystyle\sum_{n=1}^{\infty} (a_n - b_n) = \sum_{n=1}^{\infty} a_n - \sum_{n=1}^{\infty} b_n$

EXAMPLE 10.22 Show that the series

$$\sum_{n=1}^{\infty} \left(\frac{1}{2^{n-1}} + \frac{1}{n^2 + n} \right)$$

converges, and find its sum.

Solution The series

$$\sum_{n=1}^{\infty} \frac{1}{2^{n-1}}$$

is the geometric series with $a = 1$ and $r = \frac{1}{2}$, whose sum is

$$\frac{a}{1-r} = \frac{1}{1-\frac{1}{2}} = 2$$

The series

$$\sum_{n=1}^{\infty} \frac{1}{n^2 + n}$$

is the telescoping series we considered in Example 10.21, where we found that it converges to 1. It follows by Property 2 of Theorem 10.8 that

$$\sum_{n=1}^{\infty} \left(\frac{1}{2^{n-1}} + \frac{1}{n^2 + n} \right) = 2 + 1 = 3 \qquad \blacksquare$$

If $\sum a_n$ converges and $\sum b_n$ diverges, then we can see from Theorem 10.8 that $\sum (a_n + b_n)$ and $\sum (a_n - b_n)$ both diverge. For suppose $\sum (a_n + b_n)$ converges. Then we can write $b_n = (a_n + b_n) - a_n$. Property 3 of Theorem 10.8 then gives

$$\sum b_n = \sum [(a_n + b_n) - a_n] = \sum (a_n + b_n) - \sum a_n$$

since by our assumption both series on the right converge. But we are given that $\sum b_n$ diverges. So $\sum (a_n + b_n)$ cannot converge. We can prove that $\sum (a_n - b_n)$ diverges in a similar way. If $\sum a_n$ and $\sum b_n$ both diverge, no definite conclusion can be drawn about convergence or divergence of $\sum (a_n + b_n)$ or $\sum (a_n - b_n)$. They may converge or diverge, depending on the particular series in question. (See Exercise 39 in Exercise Set 10.3.)

Another consequence of Theorem 10.8 is that if $\sum a_n$ diverges, so does $\sum c a_n$ for any $c \neq 0$. This result follows from the fact that we can write

$$a_n = \frac{1}{c}(ca_n)$$

So by Property 1 of Theorem 10.8 (with c replaced by $1/c$ and a_n replaced by ca_n), if $\sum ca_n$ converges, it would follow that $\sum a_n = \sum 1/c(ca_n)$ also would converge, contrary to what we are given.

We summarize these properties as a corollary to Theorem 10.8.

COROLLARY 10.8

1. If $\sum a_n$ converges and $\sum b_n$ diverges, then both $\sum (a_n + b_n)$ and $\sum (a_n - b_n)$ diverge.
2. If $\sum a_n$ diverges and $c \neq 0$, then $\sum ca_n$ diverges.

We mention one other property that will sometimes be useful.

The convergence or divergence of an infinite series is unaltered if we delete or add any finite number of terms.

Suppose, for example, we delete some or all of the first N terms of the series $\sum a_n$. For convenience of notation, let us replace each term we deleted with 0 and call the new series $\sum b_n$. Then some or all of the terms $b_1, b_2, \ldots, b_N$ equal 0, but for $n > N$, $b_n = a_n$. Let $\{A_n\}$ denote the sequence of partial sums of the series $\sum a_n$ and $\{B_n\}$ denote the sequence of partial sums of the series $\sum b_n$. Then for all $n > N$,

$$A_n - A_N = a_{N+1} + a_{N+2} + \cdots + a_n = b_{N+1} + b_{N+2} + \cdots + b_n = B_n - B_N$$

Since A_N and B_N are fixed constants (each being the sum of the first N terms of a series, where N is fixed), we see that if either A_n or B_n approaches a finite limit as $n \to \infty$, so does the other. It follows that $\sum a_n$ and $\sum b_n$ either both converge or both diverge. A similar argument can be given if a finite number of terms are added.

Note carefully that while altering a finite number of terms does not alter convergence or divergence, it *does* alter the sum when the series is convergent.

Fractals

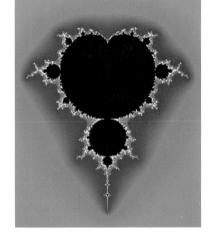

The Mandelbrot Set

In 1904 the Swedish mathematician Helge von Koch gave an example of a curve without a tangent anywhere. To his contemporaries this was a "pathological" example, whereas now mathematicians are finding that such examples occur frequently in pure and applied mathematics. They are called **fractals**, geometric figures in which a set pattern repeats on an ever-decreasing scale. The Polish-born mathematician Benoit Mandelbrot, while working in this country at IBM during the late 1970s, did pioneering work in the theory of fractals. His findings have played a key role in the rapidly expanding field of fractal geometry. To construct the Koch curve, we start with the unit interval, remove the middle third and replace it with two pieces of equal length, forming the sides of an equilateral triangle. Then we repeat the process over and over. The first three stages of the construction are shown in Figure 10.19. A good approximation is given in Figure 10.20. The Koch curve is the "limit" of the sets shown. This limiting curve can never be drawn, existing only as a mathematical concept. In Figure 10.21, we can see that magnification of one-third of the edge of the curve by a factor of 3 results in the entire curve. If this property is preserved at every stage of any iterative construction, the curve is called self-similar. Fractals possess this property of self-similarity.

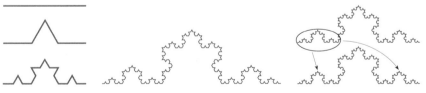

FIGURES 10.19–10.21

The construction of the **Koch island** is similar to the construction of the Koch curve described above. In the case of the island, we start with an equilateral triangle (Figure 10.22(a)). Now we remove the middle third of each of the three sides and replace each with two pieces of equal length, resulting in the star-shaped region in Figure 10.22(b). Now we repeat the process over and over. The next two approximations to the Koch island are given in Figure 10.23.

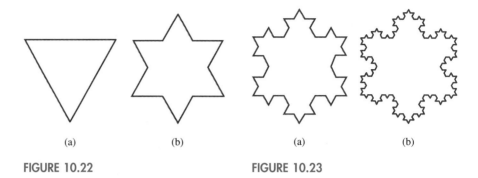

(a) (b) (a) (b)
FIGURE 10.22   FIGURE 10.23

In Exercise 49 of Exercise Set 10.3 we will ask you to show that the perimeter of the Koch island is infinite and that its area is finite.

INFINITE SERIES USING COMPUTER ALGEBRA SYSTEMS

Computer algebra systems have commands for finding partial sums of infinite series and, in some cases, the sum of a convergent series.

CAS 37

Find several partial sums for the series

$$\sum_{n=1}^{\infty} \frac{1}{n^2}$$

If possible, determine the sum of the series.

Maple:

sum(1/k^2, k=1..4);
$$\frac{205}{144}$$
evalf(sum(1/k^2, k=1..100));
  1.634983901
evalf(sum(1/k^2,
  k=1..1000));
  1.643934568

Mathematica:

Sum[1/n^2,{n,1,4}]
$$\frac{205}{144}$$
N[Sum[1/n^2,{n,1,100}],10]
  1.6349839
N[Sum[1/n^2,{n,1,1000}],10]
  1.643934567

DERIVE:

(At the □ symbol, go to the next step.)
a (author) □ 1/k^2 □ c (calculus) □ s (sum) □ [choose expression] □ k (variable) □ 1 (lower limit) □ 4 (upper limit)

Output: $\frac{205}{144}$

Another approach:
a (author) □ 1/k^2 □ c (calculus) □ s (sum) □ [choose expression] □ k (variable) □ 1 (lower limit) □ n (upper limit) □ m (manage) □ s (substitute) □ k (variable) □ 100 (value for *n*) □ x (approximate) □ [choose expression]
Output: 1.63498

Maple can take an example like this one step further. In fact, the sum of this infinite series can be computed.

In Maple enter:

sum(1/k^2,k=1..infinity);

The output is the constant given by Euler, $\pi^2/6$.

Maple can also return formal infinite sums by using the Sum command (note the capital S). In Maple, sum and Sum are different commands. The command sum computes a sum, and Sum returns the formal sum in summation notation.

For example, in Maple enter:

Sum(1/k^2,k=1..infinity);

The output is:

$$\sum_{k=1}^{\infty} \frac{1}{k^2}$$

As one final example, in Maple enter:

sum(k^2, k=1..n);

The output is:

$$\frac{1}{3}(n+1)^3 - \frac{1}{2}(n+1)^2 + \frac{1}{6}n + \frac{1}{6}$$

If the output is factored, we see the formula given in Equation 5.5.

factor(");

The output is:

$$\frac{1}{6}n(n+1)(2n+1)$$

Exercise Set 10.3

In Exercises 1–4, write out several terms of the series and give the first four partial sums.

1. $\sum_{k=1}^{\infty} \frac{k}{k^2+1}$

2. $\sum_{n=1}^{\infty} \frac{(-1)^{n-1}}{\sqrt{2n-1}}$

3. $\sum_{m=1}^{\infty} \frac{\ln(m+1)}{m!}$

4. $\sum_{n=0}^{\infty} \frac{(-1)^n(n+1)}{2^n}$

In Exercises 5–8, write the series using summation notation. Also, give the nth partial sum.

5. $1 - \frac{1}{3} + \frac{1}{9} - \frac{1}{27} + \frac{1}{81} - \cdots$

6. $1 - \frac{1}{2!} + \frac{1}{4!} - \frac{1}{6!} + \frac{1}{8!} - \cdots$

7. $\frac{1}{2\ln 2} + \frac{1}{3\ln 3} + \frac{1}{4\ln 4} + \cdots$

8. $1 - \frac{2}{3^2} + \frac{3}{5^2} - \frac{4}{7^2} + \frac{5}{9^2} - \cdots$

In Exercises 9–12, the nth partial sum, S_n, of a series $\sum a_n$ is given. Determine whether the series converges and, if it does, give its sum.

9. $S_n = \frac{n}{n+1}$

10. $S_n = 1 - (-1)^n$

11. $S_n = 2 - \frac{\ln(n+1)}{n+1}$

12. $S_n = \frac{n^3 - 2n + 1}{\sqrt{n^4 + n^2 + 4}}$

In Exercises 13–20, show that the given series is geometric and determine whether it converges. If it converges, find its sum.

13. $2 - 1 + \dfrac{1}{2} - \dfrac{1}{4} + \dfrac{1}{8} - \cdots$

14. $\displaystyle\sum_{k=1}^{\infty} \dfrac{2}{3^{k-1}}$

15. $\displaystyle\sum_{n=0}^{\infty} 3 \cdot 2^{-n}$

16. $\dfrac{1}{\ln 3} + \dfrac{1}{(\ln 3)^2} + \dfrac{1}{(\ln 3)^3} + \cdots$

17. $\dfrac{2}{e} - \dfrac{4}{e^2} + \dfrac{8}{e^3} - \dfrac{16}{e^4} + \cdots$

18. $\displaystyle\sum_{n=1}^{\infty} \left(\dfrac{5}{4}\right)^n$

19. $\displaystyle\sum_{n=0}^{\infty} (0.99)^n$

20. $\displaystyle\sum_{k=1}^{\infty} \dfrac{(-1)^{k-1} 3^k}{4^{k-1}}$

In Exercises 21–24, express the repeating decimal in the form m/n, in lowest terms, where m and n are integers, making use of geometric series.

21. $0.151515\ldots$

22. $2.181818\ldots$

23. $0.148148148\ldots$

24. $1.135135135\ldots$

In Exercises 25–28, show that the series is telescoping, and find its sum.

25. $\displaystyle\sum_{k=1}^{\infty} \dfrac{1}{(k+1)(k+2)}$

26. $\displaystyle\sum_{n=1}^{\infty} \dfrac{2}{n(n+2)}$

27. $\displaystyle\sum_{n=1}^{\infty} \dfrac{2}{4n^2 - 1}$

28. $\displaystyle\sum_{k=1}^{\infty} \dfrac{1}{k^2 + 4k + 3}$

In Exercises 29–32, show that the series converges, and find its sum.

29. $\displaystyle\sum_{n=1}^{\infty} \left(\dfrac{1}{2^n} + \dfrac{2}{3^n}\right)$

30. $\displaystyle\sum_{k=1}^{\infty} \left[\left(\dfrac{2}{3}\right)^k - \left(\dfrac{3}{4}\right)^{k-1}\right]$

31. $\displaystyle\sum_{n=1}^{\infty} \left(\dfrac{1}{n^2 + n} - \dfrac{1}{3^{n-1}}\right)$

32. $\displaystyle\sum_{n=1}^{\infty} \left[\dfrac{(-1)^{n-1} 5}{2^n} + \dfrac{2}{n^2 + 2n}\right]$

33. A ball is dropped from a height of 10 m, and on each successive bounce it rises two-thirds as high as on the preceding bounce. Find the total distance the ball travels.

34. Determine whether the series $\displaystyle\sum_{k=100}^{\infty} 1/k$ converges or diverges. Justify your answer.

In Exercises 35 and 36, show that each series diverges.

35. (a) $\displaystyle\sum_{n=1}^{\infty} \dfrac{n}{100n + 1}$ (b) $\displaystyle\sum_{n=2}^{\infty} \dfrac{n}{(\ln n)^2}$

36. (a) $\displaystyle\sum_{n=1}^{\infty} \dfrac{2n^2 - 3n + 4}{3n^2 + n + 5}$

   (b) $\displaystyle\sum_{n=1}^{\infty} \dfrac{(-1)^{n-1} n}{\sqrt{1 + n^2}}$

37. Indicate which of the following statements are true and which are false.
   (a) If $\sum a_n$ diverges, then $\lim_{n \to \infty} a_n = 0$.
   (b) If $\lim_{n \to \infty} a_n = 0$, then $\sum a_n$ converges.
   (c) If $a_1 + a_2 + \cdots + a_n = 1/n$, then $\sum a_n$ converges to 0.
   (d) If $\sum (a_n + b_n)$ converges, so do $\sum a_n$ and $\sum b_n$.
   (e) If $c \neq 0$ and $\sum c a_n$ converges, then $\sum a_n$ also converges.

38. Prove Theorem 10.8, making use of the properties of limits of sequences from Section 10.2.

39. (a) Prove that if $\sum a_n$ converges and $\sum b_n$ diverges, then $\sum (a_n - b_n)$ diverges.
   (b) Give examples to show that if $\sum a_n$ and $\sum b_n$ both diverge, then $\sum (a_n + b_n)$ and $\sum (a_n - b_n)$ may converge or may diverge.

40. Show that the series

$$\sum_{n=1}^{\infty} \ln \dfrac{n}{n+1}$$

diverges. (*Hint:* Show that it is a telescoping series and find S_n.)

41. Prove that if $a_n \geq 0$ and $a_1 + a_2 + a_3 + \cdots + a_n \leq k$ for all n, where k is a constant, then $\sum a_n$ converges.

In Exercises 42 and 43, show that the series is geometric. Find the values of x for which the series converges, and give the sum as a function of x.

42. (a) $\displaystyle\sum_{n=0}^{\infty} \left(\dfrac{x}{2}\right)^n$ (b) $\displaystyle\sum_{n=0}^{\infty} \dfrac{(x-1)^n}{3^{n+1}}$

43. (a) $\sum_{n=0}^{\infty} \dfrac{(-1)^n 3^n}{(x+2)^{n+1}}$ (b) $\sum_{n=0}^{\infty} (\ln x)^n$

44. A pendulum 1 m long is released from a position in which its angle with the vertical is 60°. On each swing after the first, it reaches a maximum angle with the vertical that is 0.9 times as large as the angle reached on the previous swing. Find the total distance covered by the bob of the pendulum in coming to rest.

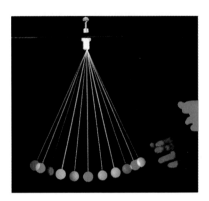

45. **Regrouping of terms.** Consider the divergent series $\sum_{n=1}^{\infty}(-1)^{n+1}$. If we group the terms by pairs starting with the first two terms, we get

$$(1-1)+(1-1)+(1-1)+\cdots = 0+0+0+\cdots = 0$$

If, instead, we group by pairs starting with the second and third terms, we can write the result as

$$1-(1-1)-(1-1)-(1-1)-\cdots = 1-0-0-0-\cdots = 1$$

Prove that this phenomenon cannot occur in a convergent series. That is, prove that the terms of a convergent series can be regrouped, preserving order, in any way and the resulting series has the same sum as the original one.

46. Use a CAS or a spreadsheet to approximate the value of $\sum_{n=1}^{\infty} 1/n^2$ by the partial sum S_{100}. Compare your answer with the known result $\pi^2/6$.

47. Use a CAS or a spreadsheet to approximate S_{100} for the harmonic series $\sum_{n=1}^{\infty} 1/n$.

48. Find the smallest integer n for which the partial sum S_n of the series $\sum_{n=0}^{\infty} 1/3^n$ agrees with the exact sum to 10 decimal places.

49. For the Koch island, let the initial equilateral triangle have sides each of length a.
    (a) Find formulas for the number of sides, the length of each side, and the perimeter of the nth approximation to the Koch island. Denote these by s_n, l_n, and P_n, respectively.
    (b) Show that the perimeter of the Koch island is infinite. That is,

    $$\lim_{n \to \infty} P_n = \infty$$

    (c) Express the area of the Koch island as an infinite geometric series and show that the area equals

    $$\left(\dfrac{2\sqrt{3}}{5}\right)a^2$$

    (*Hint:* The initial triangle has area $(\sqrt{3}/4)a^2$, and three triangles are added at the second stage, each with area $(\sqrt{3}/4)(a/3)^2$. So the combined area at the second stage is $(\sqrt{3}/4)a^2 + 3(\sqrt{3}/4)(a/3)^2$. Continue in this manner.)

10.4 SERIES OF POSITIVE TERMS; THE INTEGRAL TEST AND COMPARISON TESTS

Given an infinite series of constants, we usually want to answer two primary questions: (1) Does the series converge? and (2) If it does converge, what is its sum? The second question generally is much harder to answer than the first. (Geometric series and telescoping series are exceptions.) However, if we know that a series converges, we can at least approximate its sum by using a partial sum S_n for sufficiently large n. So it is very useful just to be able to answer the first question. In this section and the next we give certain tests for convergence that are applicable when the terms of the series are all positive or zero.

The basis for our tests is the following theorem.

THEOREM 10.9 If $a_n \geq 0$ for all n, then $\sum a_n$ converges if and only if its sequence of partial sums is bounded.

Proof Let $\{S_n\}$ be the sequence of partial sums. Then, since $S_1 = a_1$, $S_2 = a_1 + a_2$, $S_3 = a_1 + a_2 + a_3, \ldots$, and all of the a_n's are nonnegative, we see that $S_1 \leq S_2 \leq S_3 \leq \cdots$. That is, $\{S_n\}$ is a monotone increasing sequence. If $\{S_n\}$ is bounded, we know from Theorem 10.5 that the sequence converges. Thus, the series $\sum a_n$ converges.

Suppose now we are given that $\sum a_n$ converges. Since $\{S_n\}$ is an increasing sequence, it is either bounded or it diverges to ∞. But if $\lim_{n \to \infty} S_n = \infty$, the series would diverge, contrary to our hypothesis. Thus $\{S_n\}$ must be bounded. Our proof is therefore complete. ∎

The Integral Test

As our first application of Theorem 10.9, we prove a test for convergence of a series based on our knowledge of improper integrals (see Section 8.7).

THEOREM 10.10 **The Integral Test**

Let f be a continuous, positive, monotone decreasing function for $x \geq 1$, and let $a_n = f(n)$ for $n = 1, 2, 3, \ldots$. Then $\sum_{n=1}^{\infty} a_n$ is convergent if and only if the improper integral $\int_1^{\infty} f(x)\,dx$ is convergent. That is:

(a) If $\int_1^{\infty} f(x)\,dx$ converges, then $\sum_{n=1}^{\infty} a_n$ converges.
(b) If $\int_1^{\infty} f(x)\,dx$ diverges, then $\sum_{n=1}^{\infty} a_n$ diverges.

Proof Suppose first that $\int_1^{\infty} f(x)\,dx$ converges. As Figure 10.24(a) shows, if we partition the interval $[1, n]$ at the integer points, the sum of the areas of the inscribed rectangles is

$$f(2) \cdot 1 + f(3) \cdot 1 + f(4) \cdot 1 + \cdots + f(n) \cdot 1$$

Since $a_n = f(n)$, the sum can be written as

$$a_2 + a_3 + a_4 + \cdots + a_n = S_n - a_1$$

where S_n is the nth partial sum of the series. The areas of the inscribed rectangles cannot exceed the area under the graph of f. So we have

$$S_n - a_1 \leq \int_1^n f(x)\,dx \leq \int_1^{\infty} f(x)\,dx$$

Thus,

$$S_n \leq a_1 + \int_1^{\infty} f(x)\,dx$$

By our hypothesis, the integral on the right converges. Therefore, the sequence $\{S_n\}$ is bounded. We also know that $a_n \geq 0$, since $a_n = f(n)$ and f is a positive function. It now follows from Theorem 10.9 that $\sum a_n$ converges.

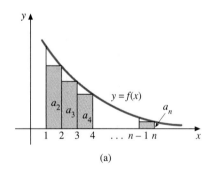

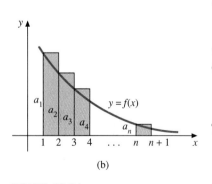

FIGURE 10.24

Suppose now that $\int_1^\infty f(x)dx$ diverges. From Figure 10.24(b), by partitioning the interval $[1, n+1]$ and using circumscribed rectangles, we see that

$$S_n \geq \int_1^{n+1} f(x)dx$$

As $n \to \infty$, the integral on the right becomes arbitrarily large (otherwise $\int_1^\infty f(x)dx$ would be finite). Thus, $\lim_{n\to\infty} S_n = \infty$, and $\sum a_n$ diverges. ∎

REMARK
■ Since deleting or adding any finite number of terms of a series does not affect its convergence or divergence, it is sufficient that the conditions of the Integral Test hold for $x \geq m$, where m is some positive integer.

EXAMPLE 10.23 Test each of the following series for convergence or divergence. (a) $\sum_{n=1}^{\infty} \frac{1}{1+n^2}$ (b) $\sum_{n=2}^{\infty} \frac{1}{n \ln n}$

Solution

(a) Let $f(x) = 1/(1 + x^2)$ (see Figure 10.25). Then f satisfies the hypotheses of Theorem 10.10. So we consider the improper integral

$$\int_1^\infty \frac{dx}{1+x^2} = \lim_{t\to\infty} \int_1^t \frac{dx}{1+x^2} = \lim_{t\to\infty} \left[\tan^{-1} t - \tan^{-1} 1\right] = \frac{\pi}{2} - \frac{\pi}{4} = \frac{\pi}{4}$$

Since the integral converges, we conclude from Theorem 10.10 that the series

$$\sum_{n=1}^{\infty} \frac{1}{1+n^2}$$

also converges.

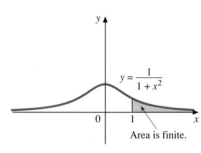

FIGURE 10.25

Note that while we know the value of the improper integral is $\pi/4$, we cannot conclude that the series converges to this value. All we know is that the series does converge.

(b) The function $f(x) = 1/(x \ln x)$ (see Figure 10.26) satisfies the hypotheses of Theorem 10.10 for $x \geq 2$. So we consider

$$\int_2^\infty \frac{1}{x \ln x} dx = \lim_{t\to\infty} \int_2^t \frac{1}{x \ln x} dx$$

Since $d(\ln x) = (1/x)dx$, the integrand is in the form du/u with $u = \ln x$. So an antiderivative is $\ln |u| = \ln |\ln x|$. Thus,

$$\lim_{t\to\infty} \int_2^t \frac{1}{x \ln x} dx = \lim_{t\to\infty} [\ln |\ln t| - \ln |\ln 2|] = \infty$$

since $\ln t \to \infty$, and hence, $\ln |\ln t| \to \infty$ as $t \to \infty$. Thus, by Theorem 10.10, the series

$$\sum_{n=2}^{\infty} \frac{1}{n \ln n}$$

diverges. ∎

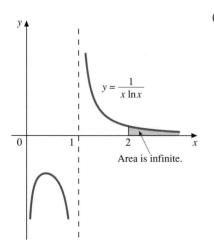

FIGURE 10.26

EXAMPLE 10.24 Determine all values of p for which the series $\sum 1/n^p$ converges.

Solution If $p > 0$, let $f(x) = 1/x^p$. Then f is a positive, continuous, monotone decreasing function for $x \geq 1$. Furthermore, $f(n) = 1/n^p$. So the conditions of Theorem 10.10 are met. We found in Example 8.28 that

$$\int_1^\infty \frac{1}{x^p} dx$$

converges if $p > 1$ and diverges if $p \leq 1$. Thus, by Theorem 10.10, $\sum 1/n^p$ also converges if $p > 1$ and diverges if $p \leq 1$.

If $p \leq 0$, write $p = -\alpha$, where $\alpha \geq 0$. Then $1/n^p = n^\alpha \geq 1$. That is, each term of the series is at least 1, so the partial sums become arbitrarily large, and the series diverges.

The series in the preceding example is referred to as the **p-series**. We give special prominence to what we have found.

The p-Series

The p-series

$$\sum_{n=1}^\infty \frac{1}{n^p}$$

converges if $p > 1$ and diverges if $p \leq 1$.

Note that by taking $p = 1$, we see again that the harmonic series diverges. Also, we confirm the fact that the series $\sum_{n=1}^\infty 1/n^2$ converges. This series is the one Euler proved converges to $\pi^2/6$. In fact, Euler also proved that $\sum_{n=1}^\infty 1/n^4$ converges to $\pi^4/90$, and he found a general formula for the sum of $\sum_{n=1}^\infty 1/n^{2k}$, but he was unable to find the sum of the reciprocals of odd powers (greater than or equal to 3) of the positive integers. In particular, he did not find the sum of the series

$$\sum_{n=1}^\infty \frac{1}{n^3}$$

Nor has anyone else ever been able to find this sum. It is one of the famous unsolved problems in mathematics. You will become famous overnight if you find the sum.

Comparison Tests

We can frequently determine whether a series converges or diverges by comparing its terms with those of some known convergent series or known divergent series. There are two ways to make such comparisons. We state the first way in the following theorem.

THEOREM 10.11

The Comparison Test

Let $\sum a_n$ and $\sum b_n$ be series of nonnegative terms, with $a_n \leq b_n$ for all n.

(a) If $\sum b_n$ converges, $\sum a_n$ converges.
(b) If $\sum a_n$ diverges, $\sum b_n$ diverges.

In other words, if each term of a positive-term series is smaller than the corresponding term of a known convergent series, the original series also converges, and if each term is larger than the corresponding term of a known divergent series of positive terms, the original series also diverges.

Proof Let A_n and B_n denote the nth partial sums of $\sum a_n$ and $\sum b_n$, respectively. Suppose first that $\sum b_n$ converges. Then, by Theorem 10.9, $B_n \leq M$ for some positive constant M. Since $a_n \leq b_n$, it follows that

$$A_n = a_1 + a_2 + \cdots + a_n \leq b_1 + b_2 + \cdots + b_n = B_n$$

So $A_n \leq B_n \leq M$. Using Theorem 10.9 again, we see that $\sum a_n$ converges, since $\{A_n\}$ is bounded.

Suppose now that $\sum a_n$ diverges. Then the sequence $\{A_n\}$ is unbounded (if it were bounded, $\sum a_n$ would be convergent). Since $B_n \geq A_n$, it follows that $\{B_n\}$ also is unbounded. Thus, by Theorem 10.9, $\sum b_n$ diverges. ∎

REMARKS

■ It is sufficient that $a_n \leq b_n$ from some n onward, say for $n \geq N$, since convergence or divergence is unaffected by neglecting any finite number of terms.

■ When a_n and b_n are nonnegative and $a_n \leq b_n$ for all $n \geq N$, we say that the series $\sum b_n$ *dominates* the series $\sum a_n$. Equivalently, we may say that $\sum a_n$ is *dominated by* $\sum b_n$. Theorem 10.11 can therefore be rephrased by saying that for series of nonnegative terms, when a series to be tested is dominated by a known convergent series, the series converges. Similarly, when a series to be tested dominates a known divergent series, the series diverges.

If $a_n \leq b_n$ and $\sum a_n$ converges, we can draw no conclusion from the Comparison Test about convergence or divergence of $\sum b_n$. Similarly, if $\sum b_n$ diverges, we can draw no conclusion about $\sum a_n$. For example,

$$\frac{1}{n^2} \leq \frac{1}{n}$$

but $\sum 1/n^2$ converges and $\sum 1/n$ diverges. Figure 10.27 shows the situation graphically. In the figure, C represents a known convergent series and D a known divergent series. Any series with terms less than those of C converges, and any series with terms greater than those of D diverges. No conclusion can be drawn about series between C and D, however.

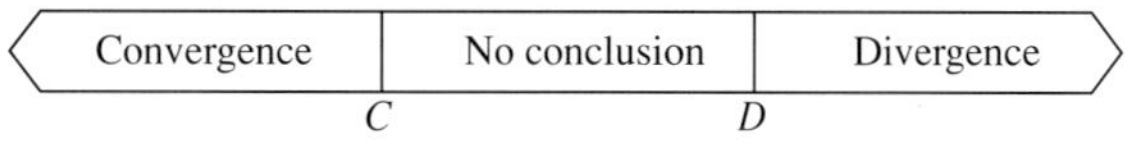

Convergence	No conclusion	Divergence
C		D

FIGURE 10.27

EXAMPLE 10.25 Test each of the following series for convergence or divergence.

(a) $\sum_{n=1}^{\infty} \dfrac{1}{\sqrt{n^3+3}}$ (b) $\sum_{n=1}^{\infty} \dfrac{1}{2n-1}$

Solution

(a) By ignoring the 3 under the radical in the denominator, we see that the general term behaves like $1/n^{3/2}$. But $\sum 1/n^{3/2}$ is a convergent *p*-series, and we might suspect, by the Comparison Test, that the given series also converges. Indeed, since for $n \geq 1$,

$$\frac{1}{\sqrt{n^3+3}} < \frac{1}{\sqrt{n^3}} = \frac{1}{n^{3/2}}$$

it follows by the Comparison Test that

$$\sum_{n=1}^{\infty} \frac{1}{\sqrt{n^3+3}}$$

also converges.

(b) Here we might suspect that the terms $1/(2n-1)$ behave similarly to $1/n$. In fact,

$$\frac{1}{2n-1} > \frac{1}{2n} \quad \text{for } n \geq 1$$

since a smaller denominator results in a larger fraction. Now, we know that the harmonic series $\sum 1/n$ diverges. It follows that

$$\sum \frac{1}{2n}$$

also diverges (see Corollary 10.8). Thus, by the Comparison Test,

$$\sum \frac{1}{2n-1}$$

diverges. ∎

The next test is a variation on the Comparison Test that often is easier to apply than the Comparison Test itself. In it we consider the *ratio* of the terms of a series we are testing to those of a series that we know to be either convergent or divergent. If this ratio approaches some positive (finite) limit as $n \to \infty$, we can conclude that the two series are similar with regard to convergence or divergence. That is, both series converge or else both series diverge.

THEOREM 10.12

The Limit Comparison Test

If $\sum a_n$ and $\sum b_n$ are series of positive terms and

$$\lim_{n \to \infty} \frac{a_n}{b_n} = L$$

where $0 < L < \infty$, then $\sum a_n$ and $\sum b_n$ both converge or else both diverge.

Section 10.4 Series of Positive Terms; The Integral Test and Comparison Tests   775

Proof   Since a_n/b_n approaches L as a limit, it follows that a_n/b_n will lie in the interval $(L/2, 3L/2)$ for all sufficiently large n, say for $n \geq N$. Thus, for $n \geq N$, since $a_n/b_n < 3L/2$, we have

$$a_n < \frac{3L}{2} b_n$$

If we know that $\sum b_n$ converges, then so does $\sum (3L/2) b_n$ (Theorem 10.8). Thus, by the Comparison Test, $\sum a_n$ also converges.

Since for $n \geq N$, the ratio a_n/b_n is in the interval $(L/2, 3L/2)$, we also have $a_n/b_n > L/2$, or

$$a_n > \frac{L}{2} b_n$$

when $n \geq N$. Thus, if we know that $\sum b_n$ diverges, so does $\sum (L/2) b_n$, by Corollary 10.8. Thus, again by the Comparison Test, $\sum a_n$ diverges. ∎

REMARK
■ When the conditions of Theorem 10.12 are met (i.e., $\lim_{n \to \infty} a_n/b_n = L$, with $0 < L < \infty$), we say that a_n and b_n are *of the same order of magnitude* as $n \to \infty$. You can often tell the order of magnitude of the nth term of a series by neglecting all but the highest powers in the numerator and denominator. You can also ignore any coefficients. (See also Exercise 50 in Exercise Set 10.4.)

EXAMPLE 10.26   Test each of the following for convergence or divergence.

(a) $\sum_{n=1}^{\infty} \frac{n^2 - n}{2n^3 + 3n - 4}$ (b) $\sum_{n=1}^{\infty} \frac{1}{\sqrt{3n^4 - 2n}}$

Solution   Observe that in each case it would be difficult to apply the Integral Test. Also, finding a suitable series for comparison and showing that the appropriate inequality in Theorem 10.11 is satisfied would be difficult. So we try the Limit Comparison Test in each case.

(a) Neglecting the coefficients and all but the highest powers, we see that

$$\frac{n^2 - n}{2n^3 + 3n - 4} \sim \frac{n^2}{n^3} = \frac{1}{n}$$

where we are using "∼" to mean "is of the same order of magnitude." To confirm that the orders of magnitude are the same, we take the limit of the quotient:

$$\lim_{n \to \infty} \left[\frac{n^2 - n}{2n^3 + 3n - 4} \div \frac{1}{n} \right] = \lim_{n \to \infty} \frac{n^3 - n^2}{2n^3 + 3n - 4} = \frac{1}{2}$$

Since we know that the harmonic series $\sum 1/n$ diverges, we conclude by Theorem 10.12 that the given series also diverges.

(b) Again using the dominant-term approach, we have

$$\frac{1}{\sqrt{3n^4 - 2n}} \sim \frac{1}{\sqrt{n^4}} = \frac{1}{n^2}$$

To confirm that the orders of magnitude are the same, we divide and take the limit as $n \to \infty$:

$$\lim_{n \to \infty} \left[\frac{1}{\sqrt{3n^4 - 2n}} \div \frac{1}{n^2} \right] = \lim_{n \to \infty} \frac{n^2}{\sqrt{3n^4 - 2n}}$$

$$= \lim_{n \to \infty} \sqrt{\frac{n^4}{3n^4 - 2n}} = \frac{1}{\sqrt{3}}$$

We know that the p-series $\sum 1/n^2$ converges (since $p = 2 > 1$), so by Theorem 10.12,

$$\sum_{n=1}^{\infty} \frac{1}{\sqrt{3n^4 - 2n}}$$

also converges. ∎

Exercise Set 10.4

In Exercises 1–10, use the Integral Test to determine convergence or divergence.

1. $\sum_{n=1}^{\infty} \frac{1}{\sqrt{2n-1}}$
2. $\sum_{n=2}^{\infty} \frac{1}{n(\ln n)^2}$
3. $\sum_{n=1}^{\infty} \frac{1}{n^2+4}$
4. $\sum_{n=1}^{\infty} n e^{-n^2}$
5. $\sum_{n=3}^{\infty} \frac{1}{n\sqrt{\ln n}}$
6. $\sum_{n=1}^{\infty} \frac{n}{(n^2+1)^{3/2}}$
7. $\sum_{n=1}^{\infty} \frac{1}{3n+2}$
8. $\sum_{n=1}^{\infty} \frac{\ln n}{n}$
9. $\sum_{n=1}^{\infty} \frac{n}{n^2+1}$
10. $\sum_{n=1}^{\infty} n e^{-n}$

In Exercises 11–20, use the Comparison Test to determine convergence or divergence.

11. $\sum_{n=1}^{\infty} \frac{1}{n^2+1}$
12. $\sum_{n=1}^{\infty} \frac{1}{\sqrt{n^3+2n}}$
13. $\sum_{n=1}^{\infty} \frac{2}{2n-1}$
14. $\sum_{n=1}^{\infty} \frac{n+1}{n^2}$
15. $\sum_{n=1}^{\infty} \sqrt{\frac{n+1}{n}}$
16. $\sum_{n=1}^{\infty} \frac{n}{2n^3+1}$
17. $\sum_{n=1}^{\infty} \frac{2}{3^n+1}$
18. $\sum_{n=1}^{\infty} \frac{3^n}{2^n-1}$
19. $\sum_{n=1}^{\infty} \frac{1+\sin^2 n}{n\sqrt{n}}$
20. $\sum_{n=1}^{\infty} \frac{n+1}{\sqrt{2n^3-1}}$

In Exercises 21–30, use the Limit Comparison Test to determine convergence or divergence.

21. $\sum_{n=1}^{\infty} \frac{2n-1}{3n^2+4n-2}$
22. $\sum_{n=1}^{\infty} \frac{3n}{n^4-2}$
23. $\sum_{n=1}^{\infty} \frac{\sqrt{2n+3}}{n^3}$
24. $\sum_{n=1}^{\infty} \frac{n}{\sqrt{n^3-2n+4}}$
25. $\sum_{n=1}^{\infty} \frac{1+2n}{(n^2+1)^{3/2}}$
26. $\sum_{n=1}^{\infty} \frac{n^2-2}{\sqrt{3n^5+1}}$
27. $\sum_{n=1}^{\infty} \frac{2n^3-3n^2+4}{5n^4+2n^3-1}$
28. $\sum_{n=1}^{\infty} \sqrt{\frac{n}{n^5+2}}$
29. $\sum_{n=1}^{\infty} \frac{\sqrt{n^2+3n}}{n^3+1}$
30. $\sum_{n=2}^{\infty} \frac{1}{\sqrt{n^3-n}}$

In Exercises 31–48, test for convergence or divergence by any appropriate means.

31. $\sum_{n=1}^{\infty} \frac{n \cos^2 n}{1+n^3}$
32. $\sum_{k=2}^{\infty} \frac{\sqrt{\ln k}}{k}$
33. $\sum_{n=1}^{\infty} \frac{1}{n \cdot 2^n}$
34. $\sum_{i=1}^{\infty} \frac{2i^2}{3i^2-2}$

35. $\sum_{n=0}^{\infty} \dfrac{\sqrt{n+1}}{n^2+1}$

36. $\sum_{j=0}^{\infty} \dfrac{j+1}{j^3+2}$

37. $\sum_{m=0}^{\infty} \dfrac{3^m}{4^m+3}$

38. $\sum_{n=0}^{\infty} \dfrac{1}{\cosh n}$

39. $\sum_{k=0}^{\infty} \dfrac{e^k}{1+e^{2k}}$

40. $\sum_{k=2}^{\infty} \dfrac{2k^2-3k+1}{k^5+2k^2+1}$

41. $\sum_{n=1}^{\infty} \dfrac{3n-2}{(n+1)(n+2)}$

42. $\sum_{m=1}^{\infty} \dfrac{m \sec^2 m}{1+m^2}$

43. $\sum_{n=1}^{\infty} \dfrac{1+\cos^2 n}{n^{3/2}}$

44. $\sum_{k=1}^{\infty} \dfrac{k}{\sqrt{100k^2+1}}$

45. $\sum_{n=3}^{\infty} \dfrac{1}{n(\ln n)[(\ln(\ln n)]}$

46. $\sum_{n=1}^{\infty} \dfrac{\sinh n}{\cosh 2n}$

47. $\sum_{n=1}^{\infty} \dfrac{n \ln n}{1+n^3}$

48. $\sum_{n=1}^{\infty} \dfrac{\sqrt{n+1}-\sqrt{n}}{n}$

49. Determine all values of p for which the series
$\sum_{n=2}^{\infty} \dfrac{1}{n(\ln n)^p}$ converges.

50. Prove the following extensions of the Limit Comparison Test for series $\sum a_n$ and $\sum b_n$ of positive terms:
   (a) If $\lim_{n \to \infty} a_n/b_n = 0$ and $\sum b_n$ converges, then $\sum a_n$ also converges.
   (b) If $\lim_{n \to \infty} a_n/b_n = \infty$ and $\sum b_n$ diverges, then $\sum a_n$ also diverges.

51. Use the results of Exercise 50 to test each of the following series for convergence or divergence.
   (a) $\sum_{n=2}^{\infty} \dfrac{1}{\ln n}$
   (b) $\sum_{n=2}^{\infty} \dfrac{\ln n}{n}$
   (c) $\sum_{n=2}^{\infty} \dfrac{\ln n}{n^2}$

52. Prove that if $a_n \geq 0$ and $\sum a_n$ converges, then $\sum a_n^2$ also converges.

53. Prove that if $a_n \geq 0$ and $\lim_{n \to \infty} na_n$ exists and is positive, then $\sum a_n$ diverges.

54. Prove that if $\sum_{k=1}^{\infty} a_k$ converges, then $\lim_{n \to \infty} \sum_{k=n}^{\infty} a_k = 0$.

55. Use a CAS or a spreadsheet to find the first 100 terms of each of the series in Example 10.26. In part (a) compare your result with the first 100 terms of the harmonic series, and in part (b) compare your result with the first 100 terms of the series $\sum 1/n^2$. Interpret your findings.

10.5 SERIES OF POSITIVE TERMS; THE RATIO TEST AND THE ROOT TEST

The next test is particularly well suited to series whose nth terms involve powers or products, especially factorials. In it, we consider the ratio of each term (after the first) to the preceding one.

THEOREM 10.13

The Ratio Test

Let $\sum a_n$ be a series of positive terms such that
$$\lim_{n \to \infty} \dfrac{a_{n+1}}{a_n} = L$$
(a) If $L < 1$, then $\sum a_n$ converges.
(b) If $L > 1$, then $\sum a_n$ diverges.
(c) If $L = 1$, then the test is inconclusive.

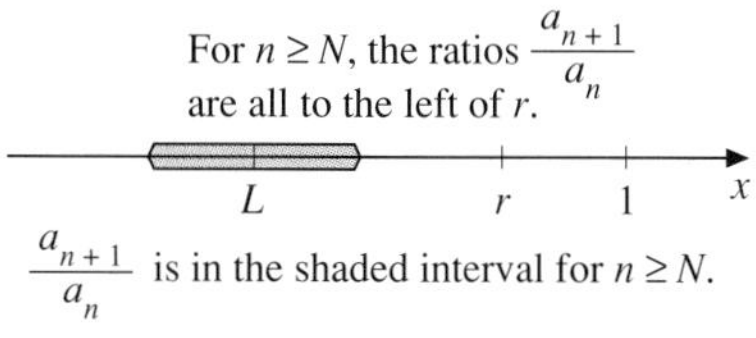

For $n \geq N$, the ratios $\dfrac{a_{n+1}}{a_n}$ are all to the left of r.

$\dfrac{a_{n+1}}{a_n}$ is in the shaded interval for $n \geq N$.

FIGURE 10.28

Proof Suppose first that $L < 1$. Let r denote any number such that $L < r < 1$. Since the ratios a_{n+1}/a_n eventually come arbitrarily close to L, they all lie to the left of r for n sufficiently large, say for $n \geq N$. (See Figure 10.28.)

Then we have

$$\frac{a_{n+1}}{a_n} < r \qquad \text{for all } n \geq N$$

or equivalently, $a_{n+1} < r a_n$ for all $n \geq N$. So

$$a_{N+1} < r a_N$$
$$a_{N+2} < r a_{N+1} < r^2 a_N$$
$$a_{N+3} < r a_{N+2} < r^3 a_N$$
$$\vdots$$

Thus, the terms of the series $\sum_{n=N+1}^{\infty} a_n$ are dominated by the terms of the series $\sum_{n=1}^{\infty} a_N r^n$. This latter series is a convergent geometric series, since the common ratio r is positive and less than 1. It follows by the Comparison Test that $\sum_{n=N+1}^{\infty} a_n$ converges. When we add the first N terms, the resulting series is also convergent; that is, the entire series $\sum_{n=1}^{\infty} a_n$ also converges.

If $L > 1$, then for all sufficiently large n,

$$\frac{a_{n+1}}{a_n} > 1, \qquad \text{or equivalently,} \qquad a_{n+1} > a_n$$

so the terms increase and thus cannot approach 0. Thus, $\sum a_n$ diverges (the nth term test for divergence).

Finally, the series $\sum 1/n$ and $\sum 1/n^2$ both satisfy

$$\lim_{n \to \infty} \frac{a_{n+1}}{a_n} = 1$$

(see Exercise 27 in Exercise Set 10.5), yet the first series diverges and the second converges. So when $L = 1$, the Ratio Test does not give us any conclusive information, and some other test is required. ∎

In applying the Ratio Test to show convergence, it is not enough to show $a_{n+1}/a_n < 1$ for all n. For example, consider the harmonic series

$$\sum \frac{1}{n} = 1 + \frac{1}{2} + \frac{1}{3} + \frac{1}{4} + \cdots$$

which we know to be divergent. Yet

$$\frac{a_{n+1}}{a_n} = \frac{1}{n+1} \div \frac{1}{n} = \frac{n}{n+1}$$

which is less than 1 for all n. It is the *limit* of this ratio as $n \to \infty$ that must be less than 1 to ensure convergence. In the case of the harmonic series, the limit of the ratio is 1. So the Ratio Test is not applicable.

EXAMPLE 10.27 Test each of the following series for convergence.

(a) $\displaystyle\sum_{n=1}^{\infty} \frac{2^n}{n!}$ (b) $\displaystyle\sum_{n=1}^{\infty} \frac{3^{n-1}}{n^2 \cdot 2^n}$

Solution Since products and powers are involved, the Ratio Test appears to be the test to use.

(a) $\displaystyle\lim_{n \to \infty} \frac{a_{n+1}}{a_n} = \lim_{n \to \infty} \frac{2^{n+1}}{(n+1)!} \cdot \frac{n!}{2^n} = \lim_{n \to \infty} \frac{2^{n+1}}{2^n} \cdot \frac{n!}{(n+1)!} = \lim_{n \to \infty} \frac{2}{n+1} = 0$

Note that instead of writing the quotient of a_{n+1} over a_n, we inverted the denominator and multiplied. Note also that $(n+1)! = (n+1)n!$, so we canceled the $n!$ that appeared in both numerator and denominator. Since $L = 0$ and $0 < 1$, the series converges.

(b)
$$\lim_{n \to \infty} \frac{a_{n+1}}{a_n} = \lim_{n \to \infty} \frac{3^n}{(n+1)^2 \, 2^{n+1}} \cdot \frac{n^2 2^n}{3^{n-1}}$$
$$= \lim_{n \to \infty} \frac{n^2}{(n+1)^2} \cdot \frac{3^n}{3^{n-1}} \cdot \frac{2^n}{2^{n+1}}$$
$$= \lim_{n \to \infty} \left(\frac{n}{n+1}\right)^2 \cdot \frac{3}{2} = \frac{3}{2}$$

Since $L = \frac{3}{2} > 1$, the series diverges. ∎

EXAMPLE 10.28 Test the following series for convergence or divergence.
$$\sum_{n=1}^{\infty} \frac{1 \cdot 3 \cdot 5 \cdot \ldots \cdot (2n-1)}{3^n \cdot n!}$$

Solution Before applying the Ratio Test, we simplify the quotient a_{n+1}/a_n. The numerator of the nth term is the product of the first n odd positive integers, so the numerator of the $(n+1)$st term is the product of the first $n+1$ odd positive integers, namely,
$$1 \cdot 3 \cdot 5 \cdot \ldots \cdot [2(n+1) - 1] = 1 \cdot 3 \cdot 5 \cdot \ldots \cdot (2n+1)$$
This product has all of the factors of the numerator of the nth term, together with the additional factor $2n+1$. Thus,
$$\frac{a_{n+1}}{a_n} = \frac{1 \cdot 3 \cdot 5 \cdot \ldots \cdot (2n+1)}{3^{n+1}(n+1)!} \cdot \frac{3^n n!}{1 \cdot 3 \cdot 5 \cdot \ldots \cdot (2n-1)} = \frac{2n+1}{3(n+1)}$$
Now we take the limit:
$$\lim_{n \to \infty} \frac{a_{n+1}}{a_n} = \lim_{n \to \infty} \frac{2n+1}{3n+3} = \frac{2}{3}$$
Since $\frac{2}{3} < 1$, we conclude that the series converges. ∎

We conclude the tests of positive term series with the **Root Test**. Its proof is similar to that of the Ratio Test, and we leave it for the exercises (Exercise 29 in Exercise Set 10.5).

THEOREM 10.14

The Root Test

Let $\sum a_n$ be a series of nonnegative terms such that
$$\lim_{n \to \infty} \sqrt[n]{a_n} = L$$

(a) If $L < 1$, then $\sum a_n$ converges.
(b) If $L > 1$, then $\sum a_n$ diverges.
(c) If $L = 1$, then the test is inconclusive.

The Root Test is more powerful than the Ratio Test in the sense that whenever the Ratio Test gives a definite conclusion concerning convergence or divergence, the same will be true of the Root Test. But there are series for which the Ratio Test is inconclusive and the Root Test gives definite information. (See Exercise 31 in Exercise Set 10.5.) However, the Ratio Test is usually easier to apply.

EXAMPLE 10.29 Test the series

$$\sum_{n=1}^{\infty} \left(\frac{n}{2n+1}\right)^n$$

Solution Here, the Root Test seems appropriate, since the general term involves the nth power. When we take the nth root, we get a simple expression whose limit we can find:

$$\lim_{n \to \infty} \sqrt[n]{a_n} = \lim_{n \to \infty} \frac{n}{2n+1} = \frac{1}{2} < 1$$

So the series converges. ■

REMARK
■ In applying the Root Test you may sometimes need the following limits:

$$\lim_{n \to \infty} \sqrt[n]{a} = 1 \quad \text{for } a > 0$$

$$\lim_{n \to \infty} \sqrt[n]{n} = 1$$

We proved the second of these limits in Example 10.14. You will be asked to prove the first in Exercise 28 of this section. You can also convince yourself that the results are reasonable with a CAS or a spreadsheet (Exercise 32 of this section).

You may have wondered why there are so many tests for convergence (there are still more, but we have given the main ones). The answer is that no single test works on all positive-term series. Even when more than one test *could* be used, one may be much easier to apply than the others. With a little practice you will probably develop a feel for which test to apply in a given situation.

Exercise Set 10.5

In Exercises 1–26, use the Ratio Test or Root Test to determine convergence or divergence.

1. $\sum_{n=1}^{\infty} ne^{-n}$

2. $\sum_{n=1}^{\infty} \frac{n^2 \cdot 2^n}{3^{n-1}}$

3. $\sum_{n=1}^{\infty} \frac{n!}{10^n}$

4. $\sum_{n=1}^{\infty} \left(\frac{n-1}{2n+1}\right)^n$

5. $\sum_{n=1}^{\infty} \frac{2^{3n}}{3^{2n}}$

6. $\sum_{n=1}^{\infty} \frac{n^2}{2^n}$

7. $\sum_{n=1}^{\infty} \frac{2^{n^2}}{n!}$

8. $\sum_{n=1}^{\infty} n^3 3^{-n}$

9. $\sum_{n=1}^{\infty} \frac{n}{\sqrt{e^n}}$

10. $\sum_{k=2}^{\infty} \left(\frac{4k^2+3}{3k^2-4}\right)^k$

11. $\sum_{m=1}^{\infty} \dfrac{(m+1)^{2m}}{3^{m^2}}$

12. $\sum_{k=1}^{\infty} k^2 \left(\dfrac{2}{3}\right)^k$

13. $\sum_{n=0}^{\infty} \dfrac{(100)^{2n+1}}{(2n+1)!}$

14. $\sum_{n=1}^{\infty} \dfrac{(n+1)^{n/2}}{2 \cdot 2^{n+1}}$

15. $\sum_{n=1}^{\infty} \dfrac{\sqrt{3^n}}{2^n}$

16. $\sum_{n=1}^{\infty} \dfrac{2^{3n+1}}{e^{2n}}$

17. $\sum_{n=1}^{\infty} \dfrac{n^n}{2^{n^2}}$

18. $\sum_{n=1}^{\infty} \dfrac{4^n}{n \cdot 3^n}$

19. $\sum_{n=1}^{\infty} \dfrac{2^n \cdot n!}{(2n)!}$

20. $\sum_{n=1}^{\infty} \dfrac{(n!)^2}{(2n)!}$

21. $\sum_{n=1}^{\infty} \dfrac{n!\,(2n+1)!}{(2n+3)!}$

22. $\sum_{n=1}^{\infty} \dfrac{n^n}{n!}$

23. $\sum_{n=1}^{\infty} \dfrac{n!}{1 \cdot 3 \cdot 5 \cdot \cdots \cdot (2n-1)}$

24. $\sum_{n=1}^{\infty} \dfrac{n!}{2 \cdot 4 \cdot 6 \cdot \cdots \cdot (2n)}$

25. $\sum_{n=1}^{\infty} \dfrac{1 \cdot 4 \cdot 7 \cdot \cdots \cdot (3n-2)}{3 \cdot 5 \cdot 7 \cdot \cdots \cdot (2n+1)}$

26. $\sum_{n=1}^{\infty} \dfrac{2^n \cdot n!}{1 \cdot 3 \cdot 5 \cdot \cdots \cdot (2n-1)}$

27. Prove that
$$\lim_{n \to \infty} \dfrac{a_{n+1}}{a_n} = 1$$
where (a) $a_n = 1/n$ and (b) $a_n = 1/n^2$. What conclusion can you draw?

28. Prove that
$$\lim_{n \to \infty} a^{1/n} = 1 \qquad \text{for } a > 0$$
(Hint: Let $y = a^{1/x}$. Take logs and use L'Hôpital's Rule. Then apply Theorem 10.2.)

29. Prove the Root Test (Theorem 10.14).

30. Prove the following version of the Ratio Test for the positive-term series $\sum a_n$.
    (a) If there is a number r such that $0 < r < 1$ for which
    $$\dfrac{a_{n+1}}{a_n} \leq r \qquad \text{for all } n \geq N$$
    where N is some positive integer, then $\sum a_n$ converges.
    (b) If
    $$\dfrac{a_{n+1}}{a_n} \geq 1 \qquad \text{for all } n \geq N$$
    where N is some positive integer, then $\sum a_n$ diverges.

31. Consider the series
$$\sum_{n=1}^{\infty} 2^{(-1)^n - n} = \dfrac{1}{2^2} + \dfrac{1}{2} + \dfrac{1}{2^4} + \dfrac{1}{2^3} + \dfrac{1}{2^6} + \dfrac{1}{2^5} + \cdots$$
    (a) Use the Root Test to show that the series converges.
    (b) Show that
    $$\dfrac{a_{n+1}}{a_n} = 2^{2(-1)^{n+1} - 1}$$
    and $\lim_{n \to \infty} (a_{n+1}/a_n)$ does not exist, so that the Ratio Test does not apply.

32. Use a CAS or a spreadsheet to find
    (a) $\sqrt[n]{2}$ for $n = 1$ to $n = 20$
    (b) $\sqrt[n]{100}$ for $n = 1$ to $n = 20$
    What do you observe in each case?

33. Use a CAS or a spreadsheet to find $\sqrt[n]{n}$ for $n = 1$ to $n = 20$. What do you observe?

10.6 SERIES WITH TERMS OF VARIABLE SIGNS; ABSOLUTE CONVERGENCE

In general, when a series consists of both positive and negative terms, tests for convergence are more complex than those for series consisting only of positive terms. There is one important test, however, that is relatively simple, and that is applicable to series in which the terms are alternately positive and negative. Such series are called **alternating series**. Though alternating series in general could begin with either a positive or a negative first term, it is sufficient to consider such series in which the first term is positive, writing, for example,

$$\sum_{n=1}^{\infty}(-1)^{n-1}a_n = a_1 - a_2 + a_3 - a_4 + \cdots$$

where for all n, $a_n > 0$. If the first term were negative, we could write

$$-a_1 + a_2 - a_3 + a_4 - \cdots = -(a_1 - a_2 + a_3 - a_4 + \cdots)$$

and determine convergence or divergence by studying the series in parentheses.

An example of an alternating series is

$$\sum_{n=1}^{\infty}\frac{(-1)^{n-1}}{n} = 1 - \frac{1}{2} + \frac{1}{3} - \frac{1}{4} + \cdots$$

This series is called the **alternating harmonic series**. Though the harmonic series itself, with all terms positive, diverges, we will soon see (Example 10.30) that the alternating harmonic series converges.

The following test shows that if the absolute values of the terms of an alternating series decrease monotonically and approach 0 as $n \to \infty$, then the series converges.

THEOREM 10.15

Alternating Series Test

If $a_n > 0$ for all n and the following two conditions are satisfied for all n,

(i) $a_{n+1} \leq a_n$

(ii) $\lim_{n \to \infty} a_n = 0$

then the alternating series

$$\sum_{n=1}^{\infty}(-1)^{n-1}a_n$$

converges.

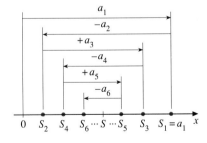

FIGURE 10.29

Proof The idea of the proof is indicated in Figure 10.29. We start with the first partial sum S_1, which is just a_1. For the purpose of this illustration, let us suppose that the strict inequality $a_{n+1} < a_n$ holds true. Then $S_2 = a_1 - a_2$ is to the left of S_1, and $S_3 = a_1 - a_2 + a_3 = S_2 + a_3$ is to the right of S_2. Also, since $a_3 < a_2$, S_3 is less than S_1. Similarly, $S_4 = S_3 - a_4$ is to the left of S_3 but to the right of S_2, and so on. We see that the even-ordered partial sums $S_2, S_4, S_6, \ldots$ increase and are all to the left of the odd-ordered partial sums $S_1, S_3, S_5, \ldots$, which decrease. It seems reasonable to suppose that both even-ordered and odd-ordered partial sums converge to the same limit S, from which we would conclude that S is the sum of the alternating series. We now proceed to the proof.

First, consider the partial sums of even order:

$$S_2 = a_1 - a_2$$

$$S_4 = (a_1 - a_2) + (a_3 - a_4)$$

$$\vdots$$

$$S_{2n} = (a_1 - a_2) + (a_3 - a_4) + \cdots + (a_{2n-1} - a_{2n})$$

By condition (i), the pairs in parentheses are all nonnegative. So $\{S_{2n}\}$ is a sequence of nonnegative terms that is monotone increasing. Furthermore, this sequence is bounded above, since we can also write S_{2n} in the form

$$S_{2n} = a_1 - (a_2 - a_3) - (a_4 - a_5) - \cdots - (a_{2n-2} - a_{2n-1}) - a_{2n}$$

The pairs in parentheses are all positive or zero, and a_{2n} is positive. So $S_{2n} \leq a_1$. As we saw in the proof of Theorem 10.5, such a bounded, monotone increasing sequence must converge. Let the limit be S. That is,

$$\lim_{n \to \infty} S_{2n} = S$$

Now let us consider an odd-ordered partial sum. Except for S_1, we can write a typical odd-ordered partial sum as

$$S_{2n+1} = S_{2n} + a_{2n+1}$$

For example,

$$S_3 = (a_1 - a_2) + a_3 = S_2 + a_3$$
$$S_5 = (a_1 - a_2 + a_3 - a_4) + a_5 = S_4 + a_5$$

and so on. By condition (ii), $\lim_{n \to \infty} a_{2n+1} = 0$. So we have

$$\lim_{n \to \infty} S_{2n+1} = \lim_{n \to \infty} S_{2n} + \lim_{n \to \infty} a_{2n+1} = S + 0 = S$$

That is, both even- and odd-ordered partial sums approach the same limit S. Thus,

$$\lim_{n \to \infty} S_n = S$$

So the series $\sum_{n=1}^{\infty} (-1)^{n-1} a_n$ converges to S. ∎

REMARK

■ As with our other tests for convergence, it is sufficient that the conditions of Theorem 10.15 hold true from some point onward. That is, the series converges if the signs alternate and $a_{n+1} \leq a_n$ from some n onward (not necessarily starting with $n = 1$), provided $\lim_{n \to \infty} a_n = 0$.

Error in Using S_n to Approximate S for Alternating Series

When a series satisfies the hypotheses of the Alternating Series Test, there is an easy way of getting an upper bound on the error when using any given partial sum, S_n, to approximate the sum S. The following corollary shows how.

COROLLARY 10.15

Under the conditions of Theorem 10.15,

$$|S_n - S| \leq a_{n+1}$$

That is, the error in approximating the sum S by the nth partial sum S_n does not exceed the absolute value of the $(n+1)$st term of the series.

Proof   In the proof of Theorem 10.15 we showed that the even-ordered partial sums were bounded above by a_1, the first term of the series. It follows that $\lim_{n\to\infty} S_{2n} = S$ must also not exceed a_1. Figure 10.29 clearly shows the fact that $S \leq a_1$.

Thus, for any alternating series whose terms in absolute value decrease monotonically and approach 0 as $n \to \infty$, the sum does not exceed the first term of the series (which we assumed to be positive). But

$$|S_n - S| = a_{n+1} - a_{n+2} + a_{n+3} - a_{n+4} + \cdots$$

is itself an alternating series whose terms in absolute value decrease monotonically and have 0 as a limit. By what we have just shown, the sum of the alternating series on the right does not exceed its first term, a_{n+1}. That is,

$$|S_n - S| \leq a_{n+1} \qquad \blacksquare$$

EXAMPLE 10.30   **The Alternating Harmonic Series**   Show that the alternating harmonic series

$$\sum_{n=1}^{\infty} \frac{(-1)^{n-1}}{n} = 1 - \frac{1}{2} + \frac{1}{3} - \frac{1}{4} + \cdots$$

converges. Determine an upper bound on the error in estimating the sum using S_{100}.

Solution   The conditions of Theorem 10.15 are clearly satisfied. That is, the series is an alternating series in which (i) $1/(n+1) < 1/n$, and (ii) $\lim_{n\to\infty} 1/n = 0$. The series therefore converges. Call its sum S. Then, by Corollary 10.15,

$$|S_{100} - S| \leq \frac{1}{101} \approx 0.009$$

So if we take the first 100 terms of the series to estimate the sum, we can be assured of accuracy to only one decimal place. This series converges very slowly. $\blacksquare$

EXAMPLE 10.31   Determine the convergence or divergence of each of the following series: (a) $\sum_{n=1}^{\infty} (-1)^{n-1} \ln\left(\frac{n+1}{n}\right)$ (b) $\sum_{n=1}^{\infty} (-1)^{n-1} \frac{n}{2n+1}$

Solution

(a) To apply Theorem 10.15, we must show both that $a_{n+1} \leq a_n$ and that $a_n \to 0$ as $n \to \infty$. To show that the terms $a_n = \ln((n+1)/n)$ decrease, we will use Method 2 of the tests for monotonicity given in Section 10.2. Replacing n with x in the formula for a_n and differentiating, we have

$$\frac{d}{dx} \ln\left(\frac{x+1}{x}\right) = \frac{d}{dx}[\ln(x+1) - \ln x]$$

$$= \frac{1}{x+1} - \frac{1}{x} = \frac{x - (x+1)}{x(x+1)} = \frac{-1}{x(x+1)}$$

Since the derivative is negative for all positive values of x, the function $\ln((x+1)/x)$ decreases. It therefore follows that for $n \geq 1$

$$\ln\left(\frac{n+2}{n+1}\right) \leq \ln\left(\frac{n+1}{n}\right)$$

That is, $a_{n+1} \leq a_n$ for all $n \geq 1$.

To see if $a_n \to 0$ as $n \to \infty$, we calculate the limit

$$\lim_{n \to \infty} \ln\left(\frac{n+1}{n}\right) = \ln\left[\lim_{n \to \infty} \left(\frac{n+1}{n}\right)\right] = \ln 1 = 0$$

(Note that we used the continuity of $\ln x$ here.)

Both conditions of the Alternating Series Test are met, so the series converges.

(b) Since

$$\lim_{n \to \infty} \frac{n}{2n+1} = \frac{1}{2} \neq 0$$

condition (ii) of the Alternating Series Test is not satisfied. We know, in fact, by the nth-Term Test for Divergence given in Section 10.3 that the series diverges. ∎

Absolute Convergence

Some series with variable signs have the property that when the signs are changed to be all positive, the resulting series converges. In this case we say that the original series is *absolutely convergent*. We can state the definition more succinctly as follows.

Definition 10.7
Absolute Convergence

The series $\sum a_n$ is said to be **absolutely convergent** if $\sum |a_n|$ converges.

REMARKS
- Here we are allowing a_n to be either positive, negative, or zero, and we are not explicitly showing the signs, as we did with alternating series. No restriction is placed on the signs of the terms. They need not alternate but may do so.
- If $a_n \geq 0$ for all n, then absolute convergence is equivalent to ordinary convergence. The concept of absolute convergence is useful only when the original series has variable signs (which may or may not alternate).

EXAMPLE 10.32 Test each of the following for absolute convergence.

(a) $\sum_{n=1}^{\infty} \frac{(-1)^{n-1}}{n}$ (b) $\sum_{n=1}^{\infty} \frac{(-1)^{n-1}}{n^2}$

Solution

(a) This series is the alternating harmonic series, which we have shown to be convergent, using the Alternating Series Test. However, when we take absolute values, we get the harmonic series $\sum 1/n$ itself, which we know to be divergent. So the alternating harmonic series is not absolutely convergent.

(b) Taking absolute values, we get the *p*-series $\sum 1/n^2$, with $p = 2$. Since $p > 1$, the series converges. Thus, the original series converges absolutely. ■

Conditional Convergence

The alternating harmonic series in part (a) of Example 10.32 converges, but it is not absolutely convergent. This type of convergence is called *conditional convergence*.

Definition 10.8
Conditional Convergence

If $\sum a_n$ converges but $\sum |a_n|$ diverges, then $\sum a_n$ is said to be **conditionally convergent**.

Thus, of the two series in Example 10.32, the first is conditionally convergent and the second is absolutely convergent.

We now know that a series may be convergent without being absolutely convergent (in which case it is conditionally convergent). However, if a series is absolutely convergent, then it must also be convergent, as the next theorem shows.

THEOREM 10.16

If $\sum a_n$ is absolutely convergent, then it is convergent.

Proof Since $|a_n|$ is either a_n or $-a_n$, it follows that

$$0 \leq a_n + |a_n| \leq 2|a_n|$$

(For example, if $a_n = -2$, then $|a_n| = 2$, and $0 = -2 + 2 < 2(2)$. Similarly, if $a_n = 4$, then $|a_n| = 4$, and $0 < 4 + 4 = 2(4)$.) Thus, since $\sum |a_n|$ (and hence also $\sum 2|a_n|$) converges, it follows by the Comparison Test that $\sum (a_n + |a_n|)$ converges. Finally, $a_n = (a_n + |a_n|) - |a_n|$, so by Theorem 10.8, part 3,

$$\sum a_n = \sum (a_n + |a_n|) - \sum |a_n|$$

Since each series on the right converges, we conclude that $\sum a_n$ converges. ■

The next theorem, which we state without proof, shows that absolute convergence is a much stronger condition than conditional convergence. It deals with *rearrangements* of terms—that is, altering the order in which the terms occur. For example,

$$1 + \frac{1}{3} - \frac{1}{2} + \frac{1}{5} - \frac{1}{4} + \frac{1}{7} - \frac{1}{6} + \cdots$$

is a rearrangement of the alternating harmonic series.

Section 10.6 Series with Terms of Variable Signs; Absolute Convergence

THEOREM 10.17

> If $\sum a_n$ is absolutely convergent, then every rearrangement of it converges to the same sum. If $\sum a_n$ is conditionally convergent, then if S is any real number, there exists a rearrangement of the series that converges to S. Also, there are rearrangements that diverge to ∞ or to $-\infty$.

(See Exercises 39–43 in Exercise Set 10.6 for the ideas involved in the proof.)

REMARK

■ According to this theorem, no amount of rearranging of terms disturbs the convergence or the value of the sum of an absolutely convergent series. On the other hand, a conditionally convergent series is delicately balanced. Any change in the order may cause it to diverge or to converge to another sum. In fact, by a suitable rearrangement, the series can be made to converge to any sum we want. (See Exercise 43 in Exercise Set 10.6.) The sum is therefore *conditional* on the particular way the terms are written.

A Strategy for Testing Series with Variable Signs

When you are confronted with a series of variable signs, a good way to proceed is to test first to see that $a_n \to 0$ (so that the series has a chance to converge). If $a_n \to 0$, then test for absolute convergence, using one of the tests for positive-term series. If the given series is absolutely convergent, then we know by Theorem 10.16 that it is convergent. If it is not absolutely convergent and the signs alternate, then the Alternating Series Test can be applied. If the signs do not alternate, there are more delicate tests that can be used, but we will not study them here.

Exercise Set 10.6

In Exercises 1–10, use the Alternating Series Test to determine convergence or divergence.

1. $\sum_{n=1}^{\infty} \frac{(-1)^{n-1}}{\sqrt{n}}$

2. $\sum_{n=1}^{\infty} \frac{(-1)^{n-1}}{\ln(n+1)}$

3. $\sum_{n=1}^{\infty} \frac{(-1)^{n-1} \ln n}{n}$

4. $\sum_{n=1}^{\infty} \frac{(-1)^n n}{\sqrt{1+n^2}}$

5. $\sum_{n=2}^{\infty} \frac{(-1)^n}{n \ln n}$

6. $\sum_{n=1}^{\infty} (-1)^n \ln\left(1 + \frac{1}{n}\right)$

7. $\sum_{n=1}^{\infty} \frac{(-1)^n n}{100n + 1}$

8. $\sum_{n=1}^{\infty} \frac{\cos n\pi}{n}$

9. $\sum_{n=1}^{\infty} \frac{(-1)^n (n+1)}{n^2 - 2}$

10. $\sum_{n=1}^{\infty} (-1)^n \ln \frac{2n+3}{2n+1}$

In Exercises 11–14, show that the given series converges, and determine an upper bound on the error in using S_n to approximate the sum, for the specified value of n.

11. $\sum_{k=1}^{\infty} \frac{(-1)^{k-1}}{k^3}$; $n = 9$

12. $\sum_{k=1}^{\infty} \frac{(-1)^{k-1}}{k!}$; $n = 7$

13. $\sum_{k=1}^{\infty} \frac{(-1)^{k-1}}{k^k}$; $n = 5$

14. $\sum_{k=1}^{\infty} \frac{(-1)^k \sqrt{k}}{k+1}$; $n = 9999$

In Exercises 15–18, show that the series converges, and determine the smallest value of n for which the error in using S_n to approximate the sum does not exceed 0.005.

15. $\sum_{k=1}^{\infty} \dfrac{(-1)^k}{\sqrt{k+1}}$

16. $\sum_{k=2}^{\infty} \dfrac{(-1)^k}{k(\ln k)^2}$

17. $\sum_{k=1}^{\infty} \dfrac{(-1)^k(k+1)}{2k^2 - 3}$

18. $\sum_{k=1}^{\infty} \dfrac{(-1)^{k-1}(\ln k)^3}{k^2}$

19. Find an estimate, with an error of at most 0.0005, of the sum of the series

$$\sum_{n=1}^{\infty} \dfrac{(-1)^{n-1}(0.7)^n}{n}$$

In Exercises 20–31, test for absolute convergence, conditional convergence, or divergence.

20. $\sum_{n=1}^{\infty} \dfrac{(-1)^{n-1}}{\sqrt{n^3+1}}$

21. $\sum_{n=1}^{\infty} \dfrac{(-1)^{n-1}\sqrt{n+1}}{n}$

22. $\sum_{n=2}^{\infty} \dfrac{(-1)^n}{n \ln \sqrt{n}}$

23. $\sum_{n=1}^{\infty} \dfrac{(-1)^{n-1} n}{2^n}$

24. $\sum_{n=2}^{\infty} \dfrac{(-1)^n}{\ln n}$

25. $\sum_{n=0}^{\infty} (-1)^n \dfrac{e^n}{e^{2n}+1}$

26. $\sum_{n=1}^{\infty} \dfrac{(-1)^{n-1} n}{\sqrt{2n^2+3}}$

27. $\sum_{n=1}^{\infty} \dfrac{(-1)^{n-1}}{n(n+2)}$

28. $\sum_{n=1}^{\infty} \dfrac{\cos n}{n^2}$

29. $\sum_{n=1}^{\infty} \dfrac{n \sin n}{1 + n^3}$

30. $\sum_{n=0}^{\infty} n e^{-n} \cos n$

31. $\sum_{n=0}^{\infty} \dfrac{(1-\pi)^n}{2^{n+1}}$

32. Determine the fallacy in the following "proof" that $2 = 1$. Let S denote the sum of the alternating harmonic series:

$$S = 1 - \dfrac{1}{2} + \dfrac{1}{3} - \dfrac{1}{4} + \dfrac{1}{5} - \dfrac{1}{6} + \dfrac{1}{7} - \dfrac{1}{8} \cdots$$

Then, on multiplying both sides by 2, we get

$$2S = 2 - 1 + \dfrac{2}{3} - \dfrac{1}{2} + \dfrac{2}{5} - \dfrac{1}{3} + \dfrac{2}{7} - \dfrac{1}{4} + \cdots$$

$$= 1 - \dfrac{1}{2} + \left(\dfrac{2}{3} - \dfrac{1}{3}\right) - \dfrac{1}{4} + \left(\dfrac{2}{5} - \dfrac{1}{5}\right) - \dfrac{1}{6} + \cdots$$

$$= 1 - \dfrac{1}{2} + \dfrac{1}{3} - \dfrac{1}{4} + \dfrac{1}{5} - \dfrac{1}{6} + \cdots = S$$

So $2S = S$, and since $S \neq 0$, $2 = 1$.

33. Let

$$a_n = \dfrac{2^n n!}{3 \cdot 5 \cdot 7 \cdot \, \cdots \, \cdot (2n+1)}$$

Prove that $\sum_{n=0}^{\infty}(-1)^n a_n$ is conditionally convergent by carrying out the following steps: Show that a_n can be written in the form

$$a_n = \dfrac{2 \cdot 4 \cdot 6 \cdot \, \cdots \, \cdot (2n)}{3 \cdot 5 \cdot 7 \cdot \, \cdots \, \cdot (2n+1)}$$

To show the series is not absolutely convergent, write

$$a_n = \dfrac{2}{1} \cdot \dfrac{4}{3} \cdot \dfrac{6}{5} \cdots \dfrac{2n}{2n-1} \cdot \dfrac{1}{2n+1}$$

Explain how it follows that

$$a_n > \dfrac{1}{2n+1}$$

To show $\{a_n\}$ is monotone decreasing, consider a_{n+1}/a_n. To show that $a_n \to 0$, verify the following:

$$a_n < \dfrac{3 \cdot 5 \cdot 7 \cdot \, \cdots \, \cdot (2n+1)}{4 \cdot 6 \cdot 8 \cdot \, \cdots \, \cdot (2n+2)} = \dfrac{1}{a_n(n+1)}$$

so

$$a_n^2 < \dfrac{1}{n+1}, \quad \text{or equivalently,} \quad a_n < \dfrac{1}{\sqrt{n+1}}$$

Now show that $a_n \to 0$ as $n \to \infty$.

34. Prove that

$$\sum_{n=1}^{\infty}(-1)^{n-1} \dfrac{1 \cdot 3 \cdot 5 \cdot \, \cdots \, \cdot (2n-1)}{2 \cdot 4 \cdot 6 \cdot \, \cdots \, \cdot (2n)}$$

is conditionally convergent. (*Hint:* Use steps similar to those in Exercise 33.)

35. Under the conditions of Theorem 10.15 show that the partial sums S_{2n+1} of odd order form a monotone decreasing, bounded sequence.

36. It can be shown (see Exercise 43 in Exercise Set 10.3) that in any convergent series the associative law holds unrestrictedly; that is, we may group terms by introducing parentheses (but not rearranging the order of the terms). Use this fact to show that

$$\sum_{n=1}^{\infty} \dfrac{1}{(2n-1)2n} = \dfrac{1}{1 \cdot 2} + \dfrac{1}{3 \cdot 4} + \dfrac{1}{5 \cdot 6} + \cdots$$

converges to the same sum as the alternating harmonic series. (*Hint:* In the alternating harmonic series, group terms by pairs.)

37. Let $\sum a_n$ be conditionally convergent and write
$$p_n = \begin{cases} a_n & \text{if } a_n \geq 0 \\ 0 & \text{if } a_n < 0 \end{cases} \quad \text{and} \quad q_n = \begin{cases} 0 & \text{if } a_n \geq 0 \\ -a_n & \text{if } a_n < 0 \end{cases}$$
   (a) Show that $\sum a_n = \sum (p_n - q_n)$ and $\sum |a_n| = \sum (p_n + q_n)$.
   (b) Show that $\sum p_n$ and $\sum q_n$ both diverge.

38. Show that in an absolutely convergent series the series $\sum p_n$ and $\sum q_n$ both converge, where p_n and q_n are defined as in Exercise 37. (*Hint:* Use part (a) of Exercise 37.)

 39. Let p_n and q_n be defined as in Exercise 37 for the alternating harmonic series. Use a CAS or a spreadsheet to do the following. Add terms of $\sum p_n$ until the sum first exceeds 2. Then add one or more terms of $\sum (-q_n)$ until the sum is less than 2. Then continue with terms of $\sum p_n$ until the sum again exceeds 2. Continue in this way until you have added a total of 50 nonzero terms. If this process were continued, what could you conclude about the convergence of this rearrangement of the alternating harmonic series? Justify your conclusion.

40. Use the idea of Exercise 39 with the roles of $\sum p_n$ and $\sum (-q_n)$ reversed in order to obtain a rearrangement of the alternating harmonic series with partial sums that can be made arbitrarily close to -1.

41. Rearrange the alternating harmonic series so that the resulting series diverges.

42. Rearrange the alternating harmonic series so that the resulting series converges to 4.

43. Use the result of Exercise 37 to explain how a suitable rearrangement of any conditionally convergent series can be made to converge to any sum S we choose.

10.7 POWER SERIES

In Section 10.1 we indicated that we would be considering "infinite polynomials" that result from letting $n \to \infty$ in the Taylor polynomial representation of a function. Such infinite series differ from those we have been studying in the previous two sections in that the terms involve powers of a variable x, or more generally of $(x - a)$. In fact, our motivation for studying infinite series of constants was that such series arise when we substitute a particular value of x into a series involving variables.

In the next section we will return to the question of representing a function by its extended Taylor polynomial. In this section we consider properties of series of powers of x, or of $(x - a)$. Our emphasis will be on the series itself, rather than on what function may have given rise to the series.

Definition 10.9
Power Series

A series of the form
$$\sum_{n=0}^{\infty} a_n x^n = a_0 + a_1 x + a_2 x^2 + \cdots \qquad (10.20)$$
is called a **power series in x**, and a series of the form
$$\sum_{n=0}^{\infty} a_n (x-a)^n = a_0 + a_1(x-a) + a_2(x-a)^2 + \cdots \qquad (10.21)$$
is called a **power series in $(x - a)$**.

REMARK
■ If we let $x = 0$ on the left-hand side of Equation 10.20, the first term of the summation is $a_0(0)^0$, which is not defined. We will adopt the convention, however, that the first term in this case is a_0, in agreement with the right-hand side. Similar remarks apply to the summation in Equation 10.21 when $x = a$ and $n = 0$.

The power series 10.20 is clearly the special case of 10.21, in which $a = 0$. Although we could develop our theory for the general case of power series in $(x - a)$, we will concentrate instead primarily on the simpler case of power series in x. All our results can be extended in obvious ways to power series in $(x - a)$.

An appropriate question to ask for a power series is not "Does the series converge?" but rather "For what values of x does the series converge?" This change arises because convergence is defined in terms of series of constants, and a power series becomes a constant series only when x is given a value. All power series in x converge in a trivial way for $x = 0$, as can be seen by putting $x = 0$ in Equation 10.20. Similarly, the series in Equation 10.21 converges for $x = a$. Some series converge for all real values of x, and others converge on an interval and diverge outside that interval. We will see examples of each of these types shortly. The nature of the set of x values where convergence occurs can be determined from the following theorem.

THEOREM 10.18 If the power series $\sum a_n x^n$ converges at $x_0 \neq 0$, then it converges absolutely for all x such that $|x| < |x_0|$. If the series diverges at $x_1 \neq 0$, then it diverges for all x such that $|x| > |x_1|$.

Simply stated, if a power series in x converges for any nonzero value of x, it converges absolutely for all x that are smaller in absolute value. If it diverges for a given value of x, it diverges for all x that are greater in absolute value.

REMARK
■ For simplicity, we will sometimes drop the range of the index n (from 0 to ∞) in $\sum_{n=0}^{\infty} a_n x^n$ and write simply $\sum a_n x^n$.

Proof Since $\sum a_n x_0^n$ converges, we know by Theorem 10.7 that $\lim_{n \to \infty} a_n x_0^n = 0$. So for n sufficiently large, say $n > N$, $|a_n x_0^n| < 1$. Now let x be any number for which $|x| < |x_0|$, and denote the ratio $|x/x_0|$ by r, so that $0 \leq r < 1$. Then we have for $n > N$

$$|a_n x^n| = |a_n x_0^n| \cdot \left|\frac{x}{x_0}\right|^n = |a_n x_0^n| \cdot r^n < r^n$$

Since $\sum r^n$ is a convergent geometric series, it follows by the Comparison Test that $\sum |a_n x^n|$ converges; that is, $\sum a_n x^n$ converges absolutely.

For the second part, if $\sum a_n x_1^n$ diverges and $|x| > |x_1|$, the series could not converge at x. If it did, since $|x_1| < |x|$, it would converge absolutely at x_1

(and thus would converge), by what we have just proved. But this conclusion contradicts the given fact that the series is assumed to diverge at x_1. So $\sum a_n x^n$ diverges when $|x| > |x_1|$. ∎

Radius of Convergence and Interval of Convergence

As a consequence of Theorem 10.18 the following result can be proved. (See Exercise 40 in Exercise Set 10.7.)

THEOREM 10.19

For the power series $\sum a_n x^n$, one and only one of the following cases holds true:

1. The series converges only for $x = 0$.
2. The series converges absolutely for all x.
3. There is a positive number R such that the series converges absolutely for $|x| < R$ and diverges for $|x| > R$.

The number R in Case 3 is called the **radius of convergence** of the series. For convenience, if Case 1 holds, we agree to call the radius of convergence 0, and if Case 2 holds, we say the radius of convergence is ∞. For Case 3, in which $0 < R < \infty$, nothing is said about convergence at $x = R$ or at $x = -R$. The series may or may not converge at these points, and if it does converge, the convergence may be absolute or conditional. These endpoints of the interval $(-R, R)$ must be individually tested using means studied in the previous three sections. Depending on whether the series converges or diverges at these endpoint values, the series will converge in an interval of one of the types $(-R, R)$, $[-R, R]$, $[-R, R)$, or $(-R, R]$, and it will diverge elsewhere. Do you see that if only one endpoint is included, the convergence is necessarily conditional there? The appropriate interval for a given series is called its **interval of convergence**. When $R = 0$, the interval of convergence degenerates to the single point $x = 0$, and if $R = \infty$, it is the entire real line $(-\infty, \infty)$.

Theorem 10.19 as well as the notions of radius of convergence and interval of convergence extend, with obvious modifications, to power series of the form $\sum a_n (x - a)^n$. If the radius of convergence is R, for example, where $0 < R < \infty$, the interval of convergence is of the form $(a - R, a + R)$, or this interval together with one or both of its endpoints.

Using the Ratio Test to Find the Radius of Convergence

When $\lim_{n \to \infty} |a_{n+1}/a_n|$ exists, the radius of convergence can be found using the Ratio Test. We illustrate this procedure in the next three examples.

EXAMPLE 10.33   Find the interval of convergence of the series

$$\sum_{n=0}^{\infty} \frac{x^n}{2n+1}$$

Solution We use the Ratio Test on the series of absolute values. (Remember that the Ratio Test is applicable only to series of positive terms, and since x can be either positive or negative, we must consider absolute values.)

$$\lim_{n \to \infty} \left| \frac{x^{n+1}}{2(n+1)+1} \cdot \frac{2n+1}{x^n} \right| = \lim_{n \to \infty} \frac{2n+1}{2n+3} |x| = |x|$$

(Note that for this limit x is held fixed and $n \to \infty$.) The series therefore converges absolutely when $|x| < 1$ and diverges when $|x| > 1$. The radius of convergence is then $R = 1$. Now we must test the endpoint values $x = \pm 1$. We test them by substituting in the original series:

$$x = -1: \quad \sum_{n=0}^{\infty} \frac{(-1)^n}{2n+1} = 1 - \frac{1}{3} + \frac{1}{5} - \frac{1}{7} + \cdots$$

This series is an alternating series, and $1/(2n+1)$ decreases monotonically to 0. Thus, the series converges.

$$x = 1: \quad \sum_{n=0}^{\infty} \frac{1}{2n+1} = 1 + \frac{1}{3} + \frac{1}{5} + \frac{1}{7} + \cdots$$

The term $1/(2n+1)$ appears to be of the same order of magnitude as $1/n$, so we can try the Limit Comparison Test:

$$\lim_{n \to \infty} \frac{1}{2n+1} \div \frac{1}{n} = \lim_{n \to \infty} \frac{n}{2n+1} = \frac{1}{2}$$

This limit confirms that our series and the harmonic series $\sum 1/n$ behave in the same way. Thus, at $x = 1$ our series diverges.

The complete interval of convergence of the original series is therefore $-1 \leq x < 1$. Note that the convergence is conditional at $x = -1$. ∎

EXAMPLE 10.34 Find the interval of convergence of the series

$$\sum_{n=0}^{\infty} \frac{(-1)^n x^{2n}}{(2n)!}$$

Solution We will again apply the Ratio Test to the series of absolute values. Since $|(-1)^n| = 1$ for any n, we can ignore the factor $(-1)^n$. We then have

$$\lim_{n \to \infty} \left| \frac{x^{2(n+1)}}{[2(n+1)]!} \div \frac{x^{2n}}{(2n)!} \right| = \lim_{n \to \infty} \left| \frac{x^{2n+2}}{(2n+2)!} \cdot \frac{(2n)!}{x^{2n}} \right|$$

$$= \lim_{n \to \infty} \frac{x^2}{(2n+2)(2n+1)} = 0$$

for all fixed values of x. Since $0 < 1$, we conclude that the series converges absolutely for all real x; that is, $R = \infty$ and the interval of convergence is $(-\infty, \infty)$. ∎

EXAMPLE 10.35 Find the interval of convergence of the series $\sum_{n=0}^{\infty} n! \, x^n$.

Solution To apply the Ratio Test we consider the following limit for $x \neq 0$,

$$\lim_{n \to \infty} \left| \frac{(n+1)! \, x^{n+1}}{n! \, x^n} \right| = \lim_{n \to \infty} (n+1)|x|$$

Thus, if $x \neq 0$, the limit is ∞, and so by the Ratio Test the series diverges. If $x = 0$, only the first term of the series is different from 0, so the series is finite, hence convergent. Thus, $R = 0$ and the interval of convergence degenerates to the single point $x = 0$. ∎

We will now see how the same technique can be applied to power series in $(x - a)$ to find the interval of convergence.

EXAMPLE 10.36   Find the interval of convergence of the series

$$\sum_{n=0}^{\infty} \frac{(-1)^n (x-2)^n}{(n+1)^2 \cdot 3^n}$$

Solution   Consider the limit

$$\lim_{n \to \infty} \left| \frac{(x-2)^{n+1}}{(n+2)^2 3^{n+1}} \cdot \frac{(n+1)^2 3^n}{(x-2)^n} \right| = \lim_{n \to \infty} \frac{1}{3} \left(\frac{n+1}{n+2} \right)^2 |x-2| = \frac{|x-2|}{3}$$

Thus, by the Ratio Test the series converges absolutely if $|x - 2|/3 < 1$ or, equivalently, if $|x - 2| < 3$, and it diverges if $|x - 2| > 3$. Now we test the values $(x - 2) = \pm 3$, which correspond to the endpoint values $x = 5$ and $x = -1$.

$$(x - 2) = 3: \quad \sum_{n=0}^{\infty} \frac{(-1)^n 3^n}{(n+1)^2 \cdot 3^n} = \sum_{n=0}^{\infty} \frac{(-1)^n}{(n+1)^2}$$

and this series converges absolutely, since the series of absolute values is of the same order of magnitude as a p-series with $p = 2$.

$$(x - 2) = -3: \quad \sum_{n=0}^{\infty} \frac{(-1)^n (-3)^n}{(n+1)^2 \cdot 3^n} = \sum_{n=0}^{\infty} \frac{(-1)^n \cdot (-1)^n \cdot 3^n}{(n+1)^2 \cdot 3^n}$$

$$= \sum_{n=0}^{\infty} \frac{1}{(n+1)^2}$$

and again we have a convergent p-series. So the complete interval of convergence is defined by $|x - 2| \leq 3$, and the convergence is absolute for all values of x satisfying this inequality. We can determine the interval of convergence as follows:

$$|x - 2| \leq 3$$
$$-3 \leq x - 2 \leq 3$$
$$-1 \leq x \leq 5$$

Thus, the interval of convergence is $[-1, 5]$. Notice that the center of the interval is $x = 2$ and the radius is $R = 3$. ∎

Exercise Set 10.7

In Exercises 1–37, find the interval of convergence.

1. $\sum_{n=0}^{\infty} x^n$

2. $\sum_{n=1}^{\infty} \frac{x^n}{n}$

3. $\sum_{n=0}^{\infty} \frac{x^n}{n!}$

4. $\sum_{n=0}^{\infty} \frac{(-1)^n n x^n}{n+1}$

5. $\sum_{n=0}^{\infty} \frac{(-1)^n x^n}{1+n^2}$

6. $\sum_{n=0}^{\infty} \frac{x^{2n+1}}{(2n+1)!}$

7. $\displaystyle\sum_{n=0}^{\infty} \frac{(x-1)^n}{\sqrt{n^2+1}}$

8. $\displaystyle\sum_{n=0}^{\infty} \frac{(x+1)^n}{2^n}$

9. $\displaystyle\sum_{n=0}^{\infty} \frac{(x+2)^n}{3^n(n+1)}$

10. $\displaystyle\sum_{n=0}^{\infty} \frac{(-1)^n 2^{n+1} x^n}{3^n}$

11. $\displaystyle\sum_{n=2}^{\infty} \frac{(-1)^n x^{n-2}}{n \ln n}$

12. $\displaystyle\sum_{n=2}^{\infty} \frac{\ln n}{n} x^{n-2}$

13. $\displaystyle\sum_{n=1}^{\infty} \frac{2^n x^{n-1}}{n^2}$

14. $\displaystyle\sum_{n=0}^{\infty} (-1)^n \sqrt{\frac{n+1}{n+2}} x^n$

15. $\displaystyle\sum_{n=0}^{\infty} \frac{n+1}{n^2+1} x^n$

16. $\displaystyle\sum_{n=0}^{\infty} \frac{n!(x-1)^n}{2^n}$

17. $\displaystyle\sum_{n=0}^{\infty} \frac{n!}{(2n)!} x^n$

18. $\displaystyle\sum_{n=0}^{\infty} n e^{-n} x^n$

19. $\displaystyle\sum_{n=1}^{\infty} \frac{(-1)^{n-1} 2^n (x-3)^{n-1}}{n^2}$

20. $\displaystyle\sum_{n=0}^{\infty} \frac{(-1)^n \sqrt{n}}{n+1} (x+2)^n$

21. $\displaystyle\sum_{n=2}^{\infty} \left(\frac{3}{4}\right)^{n-2} (x+1)^n$

22. $\displaystyle\sum_{n=2}^{\infty} \frac{(-2x)^{n-2}}{\ln n}$

23. $\displaystyle\sum_{n=0}^{\infty} \frac{n(-3x)^n}{2n+1}$

24. $\displaystyle\sum_{n=0}^{\infty} \frac{(-1)^n n^4 x^n}{e^n}$

25. $\displaystyle\sum_{n=0}^{\infty} \frac{\sqrt{n}(x-5)^n}{1+n^2}$

26. $\displaystyle\sum_{n=0}^{\infty} \frac{n^2-1}{n^2+1} x^{2n}$

27. $\displaystyle\sum_{n=0}^{\infty} \frac{(-1)^n x^{2n+1}}{(2n+1)!}$

28. $\displaystyle\sum_{n=1}^{\infty} \frac{(2x-1)^n}{n\sqrt{n+1}}$

29. $\displaystyle\sum_{n=0}^{\infty} \frac{(3x-2)^n}{(n+1)^{2/3}}$

30. $\displaystyle\sum_{n=0}^{\infty} \frac{(1-x)^n}{2^{n+1}}$

31. $\displaystyle\sum_{n=0}^{\infty} \frac{n!(x+4)^{2n}}{3^n}$

32. $\displaystyle\sum_{n=0}^{\infty} \frac{n! x^n}{1 \cdot 3 \cdot 5 \cdot \cdots \cdot (2n+1)}$ (See Exercise 33 in Exercise Set 10.6.)

33. $\displaystyle\sum_{n=1}^{\infty} \frac{1 \cdot 3 \cdot 5 \cdot \cdots \cdot (2n-1)}{2 \cdot 4 \cdot 6 \cdot \cdots \cdot (2n)} x^{n-1}$ (See Exercise 34 in Exercise Set 10.6.)

34. $\displaystyle\sum_{n=1}^{\infty} \frac{n^n x^n}{2^{n+1}}$

35. $\displaystyle\sum_{n=0}^{\infty} \frac{(n!)^2}{(2n)!} x^{2n}$

36. $\displaystyle\sum_{n=0}^{\infty} \frac{2^n+n}{3^n+2} x^n$

37. $\displaystyle\sum_{n=1}^{\infty} \frac{x^n}{(\sqrt{n})^n}$

38. If $\sum a_n x^n$ has radius of convergence R, where $0 < R < \infty$, prove that the radius of convergence of $\sum a_n x^{2n}$ is $\sqrt{R}$.

39. Find the radius of convergence of the series $\sum_{n=0}^{\infty} \binom{\alpha}{n} x^n$, where α is a fixed real number and
$$\binom{\alpha}{n} = \frac{\alpha(\alpha-1)(\alpha-2)\cdots(\alpha-n+1)}{n!}$$

40. Prove Case 3 of Theorem 10.19 as follows.
   (a) Let R denote the least upper bound of the set of x values for which $\sum a_n x^n$ is absolutely convergent. If $x_0 \in (-R, R)$, choose x_1 such that $|x_0| < |x_1| \leq R$, and explain why $\sum a_n x_1^n$ is absolutely convergent. Based on the definition of R, why can such an x_1 be chosen? (*Hint:* Otherwise $|x_0|$ would be an upper bound, smaller than R, of the set of points where $\sum a_n x^n$ is absolutely convergent.)
   (b) Now use Theorem 10.18. How do you conclude that $\sum a_n x^n$ converges absolutely for *all* x in $(-R, R)$?
   (c) To prove divergence for $|x| > R$, assume to the contrary that $\sum a_n x^n$ converges for some x such that $|x| > R$, and arrive at a contradiction.

10.8 DIFFERENTIATION AND INTEGRATION OF POWER SERIES

Within its interval of convergence, a power series defines a function. So it is appropriate to write, for example,
$$f(x) = \sum_{n=0}^{\infty} a_n x^n$$
where x is in the interval I of convergence. For any x_0 in I, $f(x_0)$ is equal to the sum of the convergent series of constants, $\sum a_n x_0^n$. It is appropriate to

ask, then, what properties such a function has. For example, we may want to know about continuity, differentiability, or integrability. The following rather remarkable theorem answers these questions. We do not give its proof, which is quite technical. (A proof can be found in most textbooks on advanced calculus.) The theorem says that a power series can essentially be treated as a polynomial within its interval of convergence.

THEOREM 10.20

Let $\sum a_n x^n$ have nonzero radius of convergence R, and for $-R < x < R$ write
$$f(x) = \sum_{n=0}^{\infty} a_n x^n$$
Then

1. f is continuous on the interval $(-R, R)$.
2. f is differentiable on $(-R, R)$ and
$$f'(x) = \sum_{n=0}^{\infty} \frac{d}{dx}(a_n x^n) = \sum_{n=1}^{\infty} n a_n x^{n-1}$$
   The series on the right also has radius of convergence R.
3. f is integrable over any interval $[a, b]$ contained in $(-R, R)$, and
$$\int_a^b f(x)dx = \sum_{n=0}^{\infty} \int_a^b a_n x^n dx$$
   Furthermore, f has an antiderivative in $(-R, R)$ given by
$$\int f(x)dx = \sum_{n=0}^{\infty} \int a_n x^n dx = \sum_{n=0}^{\infty} \frac{a_n x^{n+1}}{n+1} + C$$
   The series on the right also has radius of convergence R.

REMARKS

■ Property 2 says that a power series may be differentiated term by term, and the resulting power series converges to the derivative of the function and has the same radius of convergence as the original series. Since the differentiated series is itself a power series with radius of convergence R, it too can be differentiated term by term to give $f''(x)$. Continuing in this way gives the very powerful result that within its interval of convergence (excluding endpoints) *a power series has derivatives of all orders.*

■ We can paraphrase Properties 2 and 3 by saying that a power series can be differentiated or integrated term by term within $(-R, R)$. If the integral is a definite integral over $[a, b]$, then we must have $-R < a < b < R$. If it is an indefinite integral (an antiderivative), the integration is valid for any x in $(-R, R)$.

■ None of the properties can be assumed to be true at the endpoints $x = \pm R$, even if the original series converges at one or both of these points. Actually, it can be proved that continuity *does* extend to an endpoint if the series converges there. So a power series is continuous on its entire interval of convergence.

■ The theorem extends in an obvious way to $f(x) = \sum a_n (x-a)^n$.

The following examples illustrate some of the consequences of Theorem 10.20. In the examples we will make repeated use of the geometric series

$$a + ar + ar^2 + \cdots = \frac{a}{1-r}, \quad |r| < 1$$

By setting $r = x$ and $a = 1$ and reversing the sides of the equation, we obtain the following result.

$$\frac{1}{1-x} = 1 + x + x^2 + x^3 + \cdots = \sum_{n=0}^{\infty} x^n, \quad |x| < 1 \quad (10.22)$$

EXAMPLE 10.37 Find the sum of the series

$$\sum_{n=1}^{\infty} nx^n = x + 2x^2 + 3x^3 + 4x^4 + \cdots$$

and state the domain of validity. Use the result to find the sum of the series $\sum_{n=1}^{\infty} n(1/2)^n$.

Solution On factoring out an x we get

$$x(1 + 2x + 3x^2 + 4x^3 + \cdots)$$

and we observe that the series in parentheses is the derivative of

$$x + x^2 + x^3 + x^4 + \cdots$$

Factoring out an x, we can write this series as $x(1 + x + x^2 + x^3 + \cdots)$. The series in parentheses is the geometric series in Equation 10.22. So if $|x| < 1$,

$$x + x^2 + x^3 + x^4 + \cdots = x(1 + x + x^2 + x^3 + \cdots) = x\left(\frac{1}{1-x}\right) = \frac{x}{1-x}$$

Thus, by Property 2 of Theorem 10.20, we have

$$\sum_{n=1}^{\infty} nx^n = x \frac{d}{dx}\left(x + x^2 + x^3 + x^4 + \cdots\right)$$

$$= x\left(\frac{d}{dx} \frac{x}{1-x}\right)$$

$$= \frac{x}{(1-x)^2}$$

and this result is valid for $|x| < 1$. In particular, for $x = 1/2$, we have

$$\sum_{n=1}^{\infty} n\left(\frac{1}{2}\right)^n = \frac{\frac{1}{2}}{\left(1 - \frac{1}{2}\right)^2} = 2$$

∎

EXAMPLE 10.38 Find a power series whose sum is $\ln(1+x)$. Use the result to find a series that converges to $\ln(1/2)$.

Solution We begin with the fact that
$$\ln(1+x) = \int \frac{dx}{1+x}$$

If in Equation 10.22 we replace x with $-x$, we obtain
$$\frac{1}{1+x} = \sum_{n=0}^{\infty} (-1)^n x^n, \qquad |x| < 1$$

Thus, by Theorem 10.20,
$$\int \frac{dx}{1+x} = \int \sum_{n=0}^{\infty} (-1)^n x^n \, dx$$
$$= \sum_{n=0}^{\infty} \int (-1)^n x^n \, dx$$

We therefore have
$$\ln(1+x) = \sum_{n=0}^{\infty} \frac{(-1)^n x^{n+1}}{n+1} + C, \qquad |x| < 1$$

To find C, we can substitute $x = 0$:
$$\ln 1 = 0 + C$$

Thus,
$$C = 0$$

So we have
$$\ln(1+x) = \sum_{n=0}^{\infty} \frac{(-1)^n x^{n+1}}{n+1} = x - \frac{x^2}{2} + \frac{x^3}{3} - \frac{x^4}{4} + \cdots$$

and this result is valid for $|x| < 1$. In particular, setting $x = -1/2$, we have
$$\ln\left(\frac{1}{2}\right) = -\frac{1}{2} - \frac{1}{2 \cdot 2^2} - \frac{1}{3 \cdot 2^3} - \frac{1}{4 \cdot 2^4} - \cdots$$

since $\ln(1/2) = -\ln 2$, we can also conclude that
$$\ln 2 = \frac{1}{2} + \frac{1}{2 \cdot 2^2} + \frac{1}{3 \cdot 2^3} + \frac{1}{4 \cdot 2^4} + \cdots$$
$$= \sum_{n=1}^{\infty} \frac{1}{n \cdot 2^n} \qquad \blacksquare$$

Note that when $x = 1$, the series for $\ln(1+x)$ that we obtained in Example 10.38 is convergent (verify), and as stated in the third remark after Theorem 10.20, the continuity of the series thus extends to include the endpoint $x = 1$. We can therefore say that
$$\lim_{x \to 1} \ln(1+x) = \lim_{x \to 1} \sum_{n=0}^{\infty} \frac{(-1)^n x^{n+1}}{n+1}$$

and since the natural logarithm function also is continuous, we get the following result.

> **The Sum of the Alternating Harmonic Series**
>
> $$\ln 2 = 1 - \frac{1}{2} + \frac{1}{3} - \frac{1}{4} + \cdots$$

So we now know that the sum of the alternating harmonic series is $\ln 2$.

EXAMPLE 10.39 Find a power series representation of $\tan^{-1} x$ valid near $x = 0$.

Solution We use the idea of the previous example. Since

$$\tan^{-1} x = \int \frac{1}{1+x^2} dx$$

we can solve the problem provided we can determine a power series representation of $f(x) = 1/(1+x^2)$. All we need to do is to replace x with $-x^2$ in Equation 10.22, giving

$$\frac{1}{1+x^2} = 1 - x^2 + x^4 - x^6 + \cdots = \sum_{n=0}^{\infty} (-1)^n x^{2n}$$

where $x^2 < 1$, or equivalently, $|x| < 1$. So

$$\int \frac{1}{1+x^2} dx = \int \sum_{n=0}^{\infty} (-1)^n x^{2n} dx$$

$$\tan^{-1} x = \sum_{n=0}^{\infty} \int (-1)^n x^{2n} dx \quad \text{By Theorem 10.20}$$

$$= \sum_{n=0}^{\infty} \frac{(-1)^n x^{2n+1}}{2n+1} + C$$

For $x = 0$ we get $\tan^{-1} 0 = 0 + C$. Thus, $C = 0$. So

$$\tan^{-1} x = \sum_{n=0}^{\infty} \frac{(-1)^n x^{2n+1}}{2n+1}$$

$$= x - \frac{x^3}{3} + \frac{x^5}{5} - \frac{x^7}{7} + \cdots$$

valid so long as we have $|x| < 1$.

We can readily see that the series on the right converges at both endpoints. (Check this.) So, as we have indicated, the series represents a continuous function on the closed interval $[-1, 1]$. Since $\tan^{-1} x$ is also continuous at $x = \pm 1$, it follows as in the preceding example that

$$\tan^{-1} 1 = 1 - \frac{1}{3} + \frac{1}{5} - \frac{1}{7} + \cdots$$

and

$$\tan^{-1}(-1) = -1 + \frac{1}{3} - \frac{1}{5} + \frac{1}{7} - \cdots$$

Using the first of these equations we find that

$$\frac{\pi}{4} = 1 - \frac{1}{3} + \frac{1}{5} - \frac{1}{7} + \cdots$$

or
$$\pi = 4\left[1 - \frac{1}{3} + \frac{1}{5} - \frac{1}{7} + \cdots\right]$$

We can view this result in two ways: first, it gives a means of calculating π to any degree of accuracy (although it is not efficient, since the series converges very slowly), and second, it is a formula for the sum of the series on the right. ∎

In the next example we give one way to verify that the nth Taylor polynomial for e^x about $x = 0$ can be extended to an infinite power series (the Taylor series) that converges to e^x.

EXAMPLE 10.40 Show that the series
$$\sum_{n=0}^{\infty} \frac{x^n}{n!} = 1 + x + \frac{x^2}{2!} + \frac{x^3}{3!} + \cdots + \frac{x^n}{n!} + \cdots$$
converges to e^x for all real x.

Solution Let
$$f(x) = \sum_{n=0}^{\infty} \frac{x^n}{n!}$$

We first show that this series converges for all real x. Then we will show that $f'(x) = f(x)$, suggesting the possibility that $f(x) = e^x$. Applying the Ratio Test, we have
$$\lim_{n \to \infty} \left| \frac{x^{n+1}}{(n+1)!} \cdot \frac{n!}{x^n} \right| = \lim_{n \to \infty} \frac{|x|}{n+1} = 0$$

Since this limit is always less than 1, the given series converges for all values of x. Its derivative is, by Theorem 10.20,
$$f'(x) = \sum_{n=1}^{\infty} \frac{nx^{n-1}}{n!} = \sum_{n=1}^{\infty} \frac{x^{n-1}}{(n-1)!} \quad \text{Since } n!/n = (n-1)!$$
$$= 1 + x + \frac{x^2}{2!} + \frac{x^3}{3!} + \cdots = f(x)$$

That is, $f'(x) = f(x)$ for all values of x. We know that e^x is a function with the property that its derivative equals itself. To show that our $f(x)$ and e^x are identical, consider the derivative
$$\frac{d}{dx}\left(\frac{f(x)}{e^x}\right) = \frac{f(x) \cdot e^x - e^x \cdot f'(x)}{e^{2x}}$$

Since $f'(x) = f(x)$, the numerator is 0 and it follows that
$$\frac{d}{dx}\left(\frac{f(x)}{e^x}\right) = 0$$

Thus,
$$\frac{f(x)}{e^x} = C$$

for some constant C. Setting $x = 0$, we get $C = 1$, since both $f(0)$ and e^0 equal 1. Thus, $f(x) = e^x$. ∎

A Power Series for e^x

$$e^x = \sum_{n=0}^{\infty} \frac{x^n}{n!} = 1 + x + \frac{x^2}{2!} + \frac{x^3}{3!} + \cdots + \frac{x^n}{n!} + \cdots \quad (10.23)$$

valid for $-\infty < x < \infty$.

A consequence of the convergence of the series in Equation 10.23 is that the nth term goes to 0 as $n \to \infty$, and this result, too, is one we will need later.

A Special Limit

$$\lim_{n \to \infty} \frac{x^n}{n!} = 0 \quad \text{for all real } x \quad (10.24)$$

Exercise Set 10.8

In Exercises 1–6, differentiate the given series term by term, and determine in what interval the resulting series is the derivative of the function defined by the given series.

1. $1 + x + \dfrac{x^2}{2} + \dfrac{x^3}{3} + \cdots$

2. $x - \dfrac{x^3}{3} + \dfrac{x^5}{5} - \dfrac{x^7}{7} + \cdots$

3. $\displaystyle\sum_{n=0}^{\infty} \frac{(-1)^n x^{2n}}{(2n)!}$

4. $\displaystyle\sum_{n=0}^{\infty} \frac{(-1)^n x^{2n+1}}{(2n+1)!}$

5. $\displaystyle\sum_{n=0}^{\infty} \frac{x^n}{2^{n+1}}$

6. $\displaystyle\sum_{n=1}^{\infty} \frac{x^n}{n^2 + n}$

In Exercises 7–12, find the indicated antiderivatives and give the domain of validity.

7. $\displaystyle\int \sum_{n=0}^{\infty} \frac{x^n}{n+1} \, dx$

8. $\displaystyle\int \sum_{n=1}^{\infty} (-1)^{n-1} n x^n \, dx$

9. $\displaystyle\int \sum_{n=0}^{\infty} \frac{(-1)^n x^{2n}}{(2n)!} \, dx$

10. $\displaystyle\int \sum_{n=1}^{\infty} \frac{x^{2n}}{1 \cdot 3 \cdot 5 \cdots (2n-1)} \, dx$

11. $\displaystyle\int \sum_{n=0}^{\infty} \frac{(-1)^n (2n+1) x^{2n}}{2^n} \, dx$

12. $\displaystyle\int \sum_{n=1}^{\infty} \frac{(-1)^{n-1} x^{2n-1}}{(2n-1)!} \, dx$

In Exercises 13–18, show that term-by-term integration is valid over the given interval, and carry out the integration. Where possible, express the answer in closed form.

13. $\displaystyle\int_0^{1/2} \sum_{n=0}^{\infty} (n+1) x^n \, dx$

14. $\displaystyle\int_{-1}^{1} \sum_{n=0}^{\infty} \frac{x^n}{2^n} \, dx$

15. $\displaystyle\int_1^2 \sum_{n=0}^{\infty} \frac{x^n}{n!} \, dx$

16. $\int_{-1}^{2} \sum_{n=0}^{\infty} \frac{(-1)^n x^n}{3^{n+1}} dx$

17. $\int_{-1/2}^{1/2} \sum_{n=0}^{\infty} \frac{(-1)^n x^{2n}}{2n+1} dx$

18. $\int_{-1}^{3} \sum_{n=0}^{\infty} \frac{(-1)^n x^{2n}}{(2n)!} dx$

In Exercises 19–31, use differentiation or integration of an appropriate geometric series to find a series representation in powers of x of the given function, and give its radius of convergence.

19. $\dfrac{1}{(1+x)^2}$

20. $\dfrac{2}{(2-x)^2}$

21. $\ln(1-x)$

22. $\tanh^{-1} x$

23. $\dfrac{2x}{(1-x^2)^2}$

24. $\dfrac{2x}{(1-2x)^2}$

25. $\ln\sqrt{1+x}$

26. $\ln\dfrac{1+x}{1-x}$

27. $\tan^{-1}\dfrac{x}{2}$

28. $6\left(\dfrac{x}{3-2x}\right)^2$

29. $\dfrac{1}{(1-x)^3}$

30. $\ln(3+2x)$

31. $\dfrac{x^2}{(1-x)^2}$

32. Show, by integrating the series in Equation 10.23, that $\int e^x dx = e^x + C$.

33. Let $f(x) = e^{2x}$. In Equation 10.23 replace x with $2x$ to find a series representation of f. Use this result to calculate $\int_0^1 f(x)dx$. What conclusion can you draw concerning the sum of the following series?

$$1 + \frac{2}{2!} + \frac{2^2}{3!} + \frac{2^3}{4!} + \cdots + \frac{2^n}{(n+1)!} + \cdots$$

34. Find the sum of the series $\sum_{n=1}^{\infty}(-1)^{n-1}nx^n$ for $|x| < 1$. (*Hint:* Differentiate an appropriate geometric series and then multiply by x.)

35. Find the sum of the series $\sum_{n=0}^{\infty} n^2 x^n$. (*Hint:* Write $n^2 = n + n(n-1)$ and use the idea of Example 10.37.)

36. Find the sum of the series

$$\sum_{n=1}^{\infty} \frac{n}{2^n}$$

(*Hint:* See Example 10.37.)

37. Find the sum of the series

$$\sum_{n=1}^{\infty} \frac{n^2}{3^n}$$

(*Hint:* See Exercise 35.)

38. Let $f(x) = xe^x$. Make use of Equation 10.23 and $f'(1)$ to find the sum of the series

$$\sum_{n=0}^{\infty} \frac{n+1}{n!}$$

39. Make use of Equation 10.23 to find a series for $(e^x - 1)/x$. By differentiating the result, show that

$$\frac{1}{2!} + \frac{2}{3!} + \frac{3}{4!} + \cdots = 1$$

40. (a) Use partial fractions and geometric series to find a power series in x for the function

$$f(x) = \frac{1}{2+x-x^2}$$

and give its radius of convergence.
(b) Using part (a), find a power series representation of

$$\frac{2x-1}{(2+x-x^2)^2}$$

What is the radius of convergence?

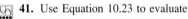
41. Use Equation 10.23 to evaluate

$$\int_0^1 e^{-x^2} dx$$

correct to eight decimal places.

42. Define f and g on $(-\infty, \infty)$ by

$$f(x) = \sum_{n=0}^{\infty} \frac{(-1)^n x^{2n}}{(2n)!} \qquad g(x) = \sum_{n=0}^{\infty} \frac{(-1)^n x^{2n+1}}{(2n+1)!}$$

Show the following:
(a) $f''(x) + f(x) = 0$ and $g''(x) + g(x) = 0$
(b) $f(0) = 1$, $f'(0) = 0$, $g(0) = 0$, and $g'(0) = 1$
(c) $f'(x) = -g(x)$ and $g'(x) = f(x)$
Do the results in part (b) and part (c) remind you of functions you have seen before? If so, what are they?

43. Use a CAS, a spreadsheet, or a calculator to verify the plausibility of Equation 10.24 with the following values of x and n.
(a) $x = 10$, $n = 50$ (b) $x = -20$, $n = 100$
(c) $x = 100$, $n = 1000$

10.9 TAYLOR SERIES

Recall (from Section 10.1) Taylor's Formula, which states that if f has derivatives up through the $(n+1)$st order in an open interval I centered at $x = a$, then

$$f(x) = P_n(x) + R_n(x) \qquad (10.25)$$

for all x in I. Here $P_n(x)$ is the Taylor polynomial

$$P_n(x) = f(a) + f'(a)(x-a) + \frac{f''(a)}{2!}(x-a)^2 + \cdots + \frac{f^{(n)}(a)}{n!}(x-a)^n \qquad (10.26)$$

and $R_n(x)$ is the remainder term, given by

$$R_n(x) = \frac{f^{(n+1)}(c)}{(n+1)!}(x-a)^{n+1} \qquad (10.27)$$

where c is some number between a and x.

We want to consider now what happens when we let $n \to \infty$. First of all, we must require that f have derivatives of *all* orders in the interval I. By letting $n \to \infty$ in Equation 10.26 we get the infinite series shown below.

Taylor Series for f about $x = a$

$$\sum_{n=0}^{\infty} \frac{f^{(n)}(a)}{n!}(x-a)^n = f(a) + f'(a)(x-a) + \frac{f''(a)}{2!}(x-a)^2$$
$$+ \frac{f'''(a)}{3!}(x-a)^3 + \cdots \qquad (10.28)$$

This series is called the **Taylor series for f about $x = a$**. In particular, if $a = 0$, we get the **Taylor series for f about $x = 0$**:

Taylor Series for f about $x = 0$: Maclaurin Series

$$\sum_{n=0}^{\infty} \frac{f^{(n)}(0)}{n!}x^n = f(0) + f'(0)x + \frac{f''(0)}{2!}x^2 + \frac{f'''(0)}{3!}x^3 + \cdots \qquad (10.29)$$

The Taylor series for f about $x = 0$ in Equation 10.29 is also called the **Maclaurin series for f**, after the Scottish mathematician Colin Maclaurin (1698–1746).

The crucial question is whether the Taylor series for a function f converges to that function in some interval. If so, then we say that f is *represented* by its Taylor series in the interval where the convergence is valid. The answer to this question of representation hinges on the remainder term $R_n(x)$ given by Equation 10.27. Note that the Taylor polynomial $P_n(x)$ is the nth partial sum (starting with $n = 0$) of the Taylor series in Equation 10.28. By Equation 10.25, we have

$$|f(x) - P_n(x)| = |R_n(x)|$$

It follows that $P_n(x)$ will come arbitrarily close to $f(x)$ if and only if $|R_n(x)|$ comes arbitrarily close to 0 as $n \to \infty$. That is, $\lim_{n\to\infty} P_n(x) = f(x)$ if and only if $\lim_{n\to\infty} R_n(x) = 0$. We summarize our findings as a theorem.

THEOREM 10.21

Taylor's Theorem

Let f have derivatives of all orders in an open interval I centered at $x = a$. Then the Taylor series for f about $x = a$ converges to $f(x)$ for x in I if and only if for all x in I,

$$\lim_{n\to\infty} R_n(x) = 0$$

where $R_n(x)$ is the remainder term in Taylor's Formula, given in Equation 10.27.

EXAMPLE 10.41 Show that the Taylor series for $f(x) = e^x$ about $x = 0$ converges to e^x for all real x.

Solution In Example 10.4 we found that the nth Taylor polynomial for e^x is

$$P_n(x) = 1 + x + \frac{x^2}{2!} + \frac{x^3}{3!} + \cdots + \frac{x^n}{n!}$$

and that

$$R_n(x) = \frac{e^c}{(n+1)!} x^{n+1}$$

where c is between 0 and x. If $0 < c < x$, then $e^c < e^x$ since $f(x) = e^x$ is an increasing function. Thus, for $x > 0$,

$$|R_n(x)| \le \frac{e^x}{(n+1)!} x^{n+1}$$

From the special limit given in Equation 10.24 we know that $x^{n+1}/(n+1)! \to 0$ as $n \to \infty$. So for fixed $x > 0$, e^x is constant and it follows that

$$\lim_{n\to\infty} R_n(x) = 0$$

If $x < c < 0$, then $e^c < e^0 = 1$. Thus,

$$|R_n(x)| \le \left| \frac{x^{n+1}}{(n+1)!} \right|$$

Again, by the special limit in Equation 10.24 the right-hand side goes to 0 as $n \to \infty$. Thus, for all $x < 0$,

$$\lim_{n\to\infty} R_n(x) = 0$$

By Theorem 10.21, we conclude that the Taylor series for e^x about $x = 0$ converges to e^x for all real x. That is,

$$e^x = 1 + x + \frac{x^2}{2!} + \frac{x^3}{3!} + \cdots = \sum_{n=0}^{\infty} \frac{x^n}{n!}$$

In the preceding section (see Example 10.40) we arrived at the result of Example 10.41 in a different way. We began with the power series $\sum x^n/n!$ and proved that its radius of convergence was infinite. Then we proved that the function represented by the series was, in fact, identical to e^x. As you might suspect, it is no accident that we got the same series for e^x by the two different approaches. The next theorem explains why.

THEOREM 10.22

> If f can be represented by a power series in $x - a$ in an open interval I centered at $x = a$, then that power series is the Taylor series for f about $x = a$.

Proof For simplicity, we prove the theorem for $a = 0$. The proof for $a \neq 0$ is similar. Suppose, then, that $f(x)$ is represented by the power series

$$f(x) = a_0 + a_1 x + a_2 x^2 + a_3 x^3 + a_4 x^4 + a_5 x^5 + \cdots$$

for all x in an interval I centered at $x = 0$. Then we have, on successive differentiation,

$$f'(x) = a_1 + 2a_2 x + 3a_3 x^2 + 4a_4 x^3 + 5a_5 x^4 + \cdots$$
$$f''(x) = 2a_2 + (3 \cdot 2)a_3 x + (4 \cdot 3)a_4 x^2 + (5 \cdot 4)a_5 x^3 + \cdots$$
$$f'''(x) = (3 \cdot 2)a_3 + (4 \cdot 3 \cdot 2)a_4 x + (5 \cdot 4 \cdot 3)a_5 x^2 + \cdots$$
$$f^{(4)}(x) = (4 \cdot 3 \cdot 2)a_4 + (5 \cdot 4 \cdot 3 \cdot 2)a_5 x + \cdots$$
$$\vdots$$

On setting $x = 0$ in each equation, we get

$$f(0) = a_0, \quad f'(0) = a_1, \quad f''(0) = 2a_2, \quad f'''(0) = 3!\, a_3,$$

$$f^{(4)}(0) = 4!\, a_4, \quad \ldots$$

and in general,

$$f^{(n)}(0) = n!\, a_n$$

If we solve for a_n, we get

$$a_n = \frac{f^{(n)}(0)}{n!}, \quad n = 0, 1, 2, \ldots$$

Thus, the coefficients in the given power series representation of f are precisely those of the Taylor series for f about $x = 0$. The given series therefore is the Taylor series. ∎

An important consequence of Theorem 10.22 is that by whatever valid means we arrive at a power series representation of a function (such as by differentiation or integration of a geometric series, for example), the series will be the Taylor series. In particular, then, the series for $\ln(1+x)$ and for $\tan^{-1} x$ that we found in Examples 10.37 and 10.38 are in fact the Taylor series about $x = 0$ for these functions.

It is important to emphasize in Theorem 10.22 the hypothesis that f can be represented by a power series. That is, we begin with the assumption that there is a power series in $x - a$ that *does converge to* f throughout some interval centered at $x = a$. Then we can conclude that f has derivatives of all orders at a, and that the given power series is, in fact, the Taylor series for f about $x = a$. If, on the other hand, we knew only that f had derivatives of all orders at $x = a$, we would know that it has a Taylor series about $x = a$, *but we would not know that f is represented by its series*. That is, we would not know that the series converges to $f(x)$. Theorem 10.21 gives the necessary and sufficient condition for the function to be represented by its Taylor series, namely, that $\lim_{n \to \infty} R_n(x) = 0$. In Exercise 29 of Exercise Set 10.9 we will ask you to show that the function f defined by

$$f(x) = \begin{cases} e^{-1/x^2} & \text{if } x \neq 0 \\ 0 & \text{if } x = 0 \end{cases}$$

has a Taylor series about $x = 0$ but that, except for $x = 0$, the series does not converge to $f(x)$.

In the next example we show two ways of finding the Taylor series about $x = 0$ of $f(x) = (1 - x)^{-2}$.

EXAMPLE 10.42 Find the Maclaurin series for $f(x) = 1/(1 - x)^2$ by two methods. (*Note:* Recall that the Maclaurin series is the Taylor series about $x = 0$.)

Solution We first use the direct method to calculate the coefficients in the Taylor series about $x = 0$. We calculate the successive derivatives of f and evaluate them at $x = 0$.

$$f(x) = \frac{1}{(1-x)^2} = (1-x)^{-2} \qquad\qquad f(0) = 1!$$
$$f'(x) = -2(1-x)^{-3}(-1) = 2!(1-x)^{-3} \qquad\qquad f'(0) = 2!$$
$$f''(x) = -3 \cdot 2(1-x)^{-4}(-1) = 3!\,(1-x)^{-4} \qquad\qquad f''(0) = 3!$$
$$f'''(x) = -4 \cdot 3!\,(1-x)^{-5}(-1) = 4!\,(1-x)^{-5} \qquad\qquad f'''(0) = 4!$$
$$\vdots \qquad\qquad\qquad\qquad \vdots$$
$$f^{(n)}(x) = (n+1)!\,(1-x)^{-(n+2)} \qquad\qquad f^{(n)}(0) = (n+1)!$$

The general term of the Maclaurin series in Equation 10.29 is therefore

$$\frac{f^{(n)}(0)}{n!} x^n = \frac{(n+1)!}{n!} x^n = (n+1)x^n$$

So the Maclaurin series for f is

$$\sum_{n=0}^{\infty} (n+1)x^n = 1 + 2x + 3x^2 + 4x^3 + \cdots$$

Note that we have not shown that this series represents f in any interval. To do so, we would have to show that the remainder term in Taylor's Formula, $R_n(x)$, approaches 0 as $n \to \infty$. Rather than show this limit, we defer the question of whether the series represents f to our second method, where the representation is an automatic consequence of the method.

For the second method, we note that

$$\frac{1}{(1-x)^2} = \frac{d}{dx}\left(\frac{1}{1-x}\right)$$

From Equation 10.22 we know that

$$\frac{1}{1-x} = 1 + x + x^2 + x^3 + \cdots = \sum_{n=0}^{\infty} x^n, \qquad |x| < 1$$

By Theorem 10.20, therefore,

$$\frac{1}{(1-x)^2} = \frac{d}{dx}(1 + x + x^2 + x^3 + \cdots)$$

$$= 1 + 2x + 3x^2 + \cdots = \sum_{n=0}^{\infty} (n+1)x^n, \qquad |x| < 1$$

This result is the same as we found by the direct method, but we actually have the added information that the series does represent the function in the interval $-1 < x < 1$. ∎

EXAMPLE 10.43

(a) Find the Maclaurin series for $f(x) = \cos x$.
(b) Use the result of part (a) to approximate the integral

$$\int_0^1 \frac{\sin^2 x}{x^2} dx$$

correct to three decimal places.

Solution

(a) In Example 10.3 we found the nth Taylor polynomial for $\cos x$ about $x = 0$ to be

$$P_n(x) = 1 - \frac{x^2}{2!} + \frac{x^4}{4!} - \frac{x^6}{6!} + \cdots + \frac{(-1)^n}{(2n)!} x^{2n}$$

and in Example 10.7, we found the remainder term

$$R_{2n}(x) = \frac{(-1)^{n+1} \cos c}{(2n+2)!} x^{2n+2}, \qquad c \text{ between } 0 \text{ and } x$$

To show where the Taylor series converges to $\cos x$, we need to find the values of x for which

$$\lim_{n \to \infty} R_{2n}(x) = 0$$

For any real value of x,

$$|R_{2n}(x)| = \frac{|\cos c|}{(2n+2)!} |x|^{2n+2} \le \frac{|x|^{2n+2}}{(2n+2)!}$$

since $|\cos c| \le 1$. By the special limit in Equation 10.24, the last expression approaches 0 as $n \to \infty$. Therefore, $\lim_{n \to \infty} R_{2n}(x) = 0$ for all real x. By Theorem 10.21 it follows that

$$\cos x = 1 - \frac{x^2}{2!} + \frac{x^4}{4!} - \frac{x^6}{6!} + \cdots = \sum_{n=0}^{\infty} \frac{(-1)^n}{(2n)!} x^{2n}$$

for all real x.

(b) To use the result of part (a) to evaluate the integral, we need to relate $\sin^2 x$ to the first power of the cosine. Such a relationship is provided by the trigonometric identity

$$\sin^2 x = \frac{1}{2}(1 - \cos 2x) \quad \text{See Supplementary Appendix 2.}$$

Now we replace x with $2x$ in the result of part (a) to get

$$\sin^2 x = \frac{1}{2}\left[1 - \left(1 - \frac{(2x)^2}{2!} + \frac{(2x)^4}{4!} - \frac{(2x)^6}{6!} + \cdots\right)\right]$$

$$= \frac{2x^2}{2!} - \frac{2^3 x^4}{4!} + \frac{2^5 x^6}{6!} - \cdots, \quad -\infty < x < \infty$$

Dividing by x^2 gives

$$\frac{\sin^2 x}{x^2} = 1 - \frac{2^3 x^2}{4!} + \frac{2^5 x^4}{6!} - \cdots \quad (x \neq 0)$$

Thus, by Theorem 10.20,

$$\int_0^1 \frac{\sin^2 x}{x^2} dx = \int_0^1 \left(1 - \frac{2^3 x^2}{4!} + \frac{2^5 x^4}{6!} - \frac{2^7 x^6}{8!} + \cdots\right) dx$$

$$= \left. x - \frac{(2x)^3}{3 \cdot 4!} + \frac{(2x)^5}{5 \cdot 6!} - \frac{(2x)^7}{7 \cdot 8!} + \cdots \right]_0^1$$

$$= 1 - \frac{2^3}{3 \cdot 4!} + \frac{2^5}{5 \cdot 6!} - \frac{2^7}{7 \cdot 8!} + \cdots$$

Using the first three terms, we get

$$\int_0^1 \frac{\sin^2 x}{x^2} dx \approx 0.898$$

Since the series alternates, we know from Corollary 10.15 that the error does not exceed the fourth term:

$$\text{Error} \leq \frac{2^7}{7 \cdot 8!} \approx 0.000454$$

So our approximation is correct to three places. ∎

REMARK ─────────────────────────────────────

■ Although $(\sin^2 x)/x^2$ is not defined when $x = 0$, it has a removable discontinuity there. In fact, we see from its series representation that its limit is 1 as $x \to 0$.

───

Binomial Series

We wish to consider finally an important series known as the **binomial series**, which is the Maclaurin series for $f(x) = (1+x)^\alpha$ for an arbitrary real number α. If $\alpha = 0$, then $f(x) = 1$. If α is a positive integer, say $\alpha = n$, we know from the **Binomial Formula** that

$$(1+x)^\alpha = (1+x)^n = 1 + nx + \frac{n(n-1)}{2!}x^2 + \frac{n(n-1)(n-2)}{3!}x^3 + \cdots + x^n \quad (10.30)$$

which is a finite series—that is, a polynomial—so there is no question of convergence. For all other values of α we proceed as usual to find the Maclaurin series:

$$f(x) = (1+x)^\alpha \qquad\qquad f(0) = 1$$
$$f'(x) = \alpha(1+x)^{\alpha-1} \qquad\qquad f'(0) = \alpha$$
$$f''(x) = \alpha(\alpha-1)(1+x)^{\alpha-2} \qquad\qquad f''(0) = \alpha(\alpha-1)$$
$$f'''(x) = \alpha(\alpha-1)(\alpha-2)(1+x)^{\alpha-3} \qquad\qquad f'''(0) = \alpha(\alpha-1)(\alpha-2)$$
$$\vdots \qquad\qquad\qquad\qquad \vdots$$
$$f^{(n)}(x) = \alpha(\alpha-1)\cdots(\alpha-n+1)(1+x)^{\alpha-n} \qquad f^{(n)}(0) = \alpha(\alpha-1)\cdots(\alpha-n+1)$$

So the binomial series is

$$1 + \alpha x + \frac{\alpha(\alpha-1)}{2!}x^2 + \frac{\alpha(\alpha-1)(\alpha-2)}{3!}x^3 + \cdots$$
$$+ \frac{\alpha(\alpha-1)(\alpha-2)\cdots(\alpha-n+1)}{n!}x^n + \cdots \qquad (10.31)$$

Note that if $\alpha = n$, then the series terminates with the term of nth degree since the factor $(\alpha - n)$ appears in all successive terms and is 0 when $\alpha = n$. Thus, series 10.31 reduces to the finite series 10.30, that is, to the Binomial Formula. It is convenient to introduce the notation

$$\binom{\alpha}{n} = \frac{\alpha(\alpha-1)(\alpha-2)\cdots(\alpha-n+1)}{n!}$$

with $\binom{\alpha}{0}$ defined as 1. Then we can write the series in the compact form

$$\sum_{n=0}^{\infty} \binom{\alpha}{n} x^n$$

It is not difficult to show that the radius of convergence for this series is $R = 1$ (see Exercise 27 in Exercise Set 10.9), so that the series converges in $(-1, 1)$, with only convergence at the endpoints remaining in question. This convergence in itself, however, does not guarantee that the function represented by the series in this interval is what we hope it will be—namely, $(1+x)^\alpha$. We can show that the series does converge to $(1+x)^\alpha$ by proving that $\lim_{n\to\infty} R_n(x) = 0$. The details are tedious, however. We simply state the result in the following theorem.

THEOREM 10.23

If α is any real number other than a nonnegative integer, the binomial series

$$\sum_{n=0}^{\infty} \binom{\alpha}{n} x^n = 1 + \alpha x + \frac{\alpha(\alpha-1)}{2!}x^2 + \frac{\alpha(\alpha-1)(\alpha-2)}{3!}x^3 + \cdots$$

converges to $(1+x)^\alpha$ in the interval $(-1, 1)$ and diverges if $|x| > 1$. If α is a nonnegative integer, the series is finite and represents $(1+x)^\alpha$ for all real values of x.

REMARK

■ When endpoint values are taken into consideration, it can be shown that the series represents the function precisely in the following intervals, depending on the size of α:

$$\begin{cases} -1 \leq x \leq 1 & \text{if } \alpha > 0 \\ -1 < x < 1 & \text{if } \alpha \leq -1 \\ -1 < x \leq 1 & \text{if } -1 < \alpha < 0 \end{cases}$$

EXAMPLE 10.44 Find the Maclaurin series for $f(x) = 1/\sqrt{1-x^2}$.

Solution Since $f(x) = (1-x^2)^{-1/2}$, we can use the binomial series in which $\alpha = -\frac{1}{2}$ and x is replaced by $-x^2$. So we have

$$\frac{1}{\sqrt{1-x^2}} = \sum_{n=0}^{\infty} \binom{-\frac{1}{2}}{n}(-x^2)^n = \sum_{n=0}^{\infty}(-1)^n \binom{-\frac{1}{2}}{n} x^{2n}$$

$$= 1 - \left(-\frac{1}{2}\right)x^2 + \frac{\left(-\frac{1}{2}\right)\left(-\frac{3}{2}\right)}{2!}x^4$$

$$- \frac{\left(-\frac{1}{2}\right)\left(-\frac{3}{2}\right)\left(-\frac{5}{2}\right)}{3!}x^6 + \cdots$$

$$= 1 + \frac{1}{2}x^2 + \frac{1 \cdot 3}{2^2 \cdot 2!}x^4 + \frac{1 \cdot 3 \cdot 5}{2^3 \cdot 3!}x^6 + \cdots$$

$$= 1 + \sum_{n=1}^{\infty} \frac{1 \cdot 3 \cdot 5 \cdot \cdots \cdot (2n-1)}{2^n \cdot n!} x^{2n}$$

The series converges to the given function for $|-x^2| < 1$ or, equivalently, $-1 < x < 1$. ∎

EXAMPLE 10.45 Find the Maclaurin series for $\sin^{-1} x$ and give its radius of convergence.

Solution Since

$$\sin^{-1} x = \int \frac{dx}{\sqrt{1-x^2}} + C$$

we can integrate the series obtained in the preceding example to get

$$\sin^{-1} x = \int \left[1 + \sum_{n=1}^{\infty} \frac{1 \cdot 3 \cdot 5 \cdot \cdots \cdot (2n-1)}{2^n \cdot n!} x^{2n}\right] dx + C$$

$$= x + \sum_{n=1}^{\infty} \int \frac{1 \cdot 3 \cdot 5 \cdot \cdots \cdot (2n-1)}{2^n \cdot n!} x^{2n} dx + C$$

$$= x + \sum_{n=1}^{\infty} \frac{1 \cdot 3 \cdot 5 \cdot \cdots \cdot (2n-1)}{2^n \cdot n!} \cdot \frac{x^{2n+1}}{2n+1} + C$$

Putting $x = 0$ on each side, we see that $C = 0$. So we can write

$$\sin^{-1} x = x + \frac{1}{2} \cdot \frac{x^3}{3} + \frac{1 \cdot 3}{2^2 \cdot 2!} \cdot \frac{x^5}{5} + \frac{1 \cdot 3 \cdot 5}{2^3 \cdot 3!} \cdot \frac{x^7}{7} + \cdots$$

Since the radius of convergence of the series for $(1-x^2)^{-1/2}$ is $R = 1$, it follows from Theorem 10.20 that the series for $\sin^{-1} x$ has the same radius of convergence. ∎

Summary of Some Important Maclaurin Series

For reference, we list some of the more frequently used Maclaurin series. The first three, in particular, are especially useful to know.

Frequently Used Maclaurin Series

$$e^x = \sum_{n=0}^{\infty} \frac{x^n}{n!} = 1 + x + \frac{x^2}{2!} + \frac{x^3}{3!} + \cdots, \quad -\infty < x < \infty$$

$$\sin x = \sum_{n=0}^{\infty} \frac{(-1)^n x^{2n+1}}{(2n+1)!} = x - \frac{x^3}{3!} + \frac{x^5}{5!} - \frac{x^7}{7!} + \cdots, \quad -\infty < x < \infty$$

$$\cos x = \sum_{n=0}^{\infty} \frac{(-1)^n x^{2n}}{(2n)!} = 1 - \frac{x^2}{2!} + \frac{x^4}{4!} - \frac{x^6}{6!} + \cdots, \quad -\infty < x < \infty$$

$$\ln(1+x) = \sum_{n=1}^{\infty} \frac{(-1)^{n-1} x^n}{n} = x - \frac{x^2}{2} + \frac{x^3}{3} - \frac{x^4}{4} + \cdots, \quad -1 < x \leq 1$$

$$\tan^{-1} x = \sum_{n=1}^{\infty} \frac{(-1)^{n-1} x^{2n-1}}{2n-1} = x - \frac{x^3}{3} + \frac{x^5}{5} - \frac{x^7}{7} + \cdots, \quad -1 \leq x \leq 1$$

$$\sinh x = \sum_{n=0}^{\infty} \frac{x^{2n+1}}{(2n+1)!} = x + \frac{x^3}{3!} + \frac{x^5}{5!} + \frac{x^7}{7!} + \cdots, \quad -\infty < x < \infty$$

$$\cosh x = \sum_{n=0}^{\infty} \frac{x^{2n}}{(2n)!} = 1 + \frac{x^2}{2!} + \frac{x^4}{4!} + \frac{x^6}{6!} + \cdots, \quad -\infty < x < \infty$$

$$(1+x)^\alpha = \sum_{n=0}^{\infty} \binom{\alpha}{n} x^n = 1 + \binom{\alpha}{1} x + \binom{\alpha}{2} x^2 + \binom{\alpha}{3} x^3 + \cdots, \quad -1 < x < 1,$$

$$\text{where } \binom{\alpha}{k} = \frac{\alpha(\alpha-1) \cdots (\alpha-k+1)}{k!}$$

COMPUTATIONS WITH POWER SERIES AND COMPUTER ALGEBRA SYSTEMS

When a function can be approximated via the Taylor series representation, then the function can be treated essentially as an infinite polynomial. In particular, differentiation and integration can be performed on the series representation term by term.

CAS 38

Approximate $\int_0^1 e^{-x^2}\, dx$.

We will use the representation

$$e^{-x^2} = \sum_{n=1}^{\infty} (-1)^n \frac{x^{2n}}{n!}$$

and to approximate the integral, we replace the integrand with a 10th degree Taylor polynomial, which is a partial sum of the series representation for the function.

Maple:

f := x->exp(-x^2);
readlib(`mtaylor`):   Access mtaylor
P := mtaylor(f(x),[x],12);

Output:

$$P := 1 - x^2 + \frac{1}{2}x^4 - \frac{1}{6}x^6 + \frac{1}{24}x^8 - \frac{1}{120}x^{10}$$

int(P,x);

Output:

$$x - \frac{1}{3}x^3 + \frac{1}{10}x^5 - \frac{1}{42}x^7 + \frac{1}{216}x^9 - \frac{1}{1320}x^{11}$$

int(P,x=0..1); $\frac{31049}{41580}$

evalf(");   .7467291967

Mathematica:

f[x_] = Exp[-x^2]
P[x_] = Normal[Series[f[x], {x,0,10}]]
Integrate[P[x],x]
Integrate[P[x],{x,0,1}]

DERIVE:

(At the □ symbol, go to the next step.)
a (author) □ exp(−x^2) □ c (calculus) □ t (taylor) □ [choose expression] □ x (variable) □ 10 (degree) □ 0 (point) □ s (simplify) □ [choose expression] □ c (calculus) □ i (integrate) □ [choose expression] □ x (variable) □ enter (no limits) □ s (simplify) □ [choose expression] □ c (calculus) □ i (integrate) □ [choose expression] □ x (variable) □ 0 (lower limit) □ 1 (upper limit) □ x (approximate) □ [choose expression]

Finally, we compare this result with that obtained from Simpson's Rule.

Maple:

with(student):
evalf(value(simpson(f(x), x=0..1,100)));

Output: .7468241331

Exercise Set 10.9

In Exercises 1–18, find the Maclaurin series for the given function and determine its interval of convergence.

1. $\sin x$
2. $\sinh x$
3. $\cosh x$
4. $\ln(1-x)$
5. $\tanh^{-1} x$
6. $\sqrt{1-x}$ (Do not test endpoints.)
7. $\sin^2 x$ (*Hint:* $\cos 2x = 1 - 2\sin^2 x$.)
8. $\cos^2 x$ (*Hint:* $\cos 2x = 2\cos^2 x - 1$.)
9. $\dfrac{1}{(1-x)^3}$
10. $\sin x \cos x$ (*Hint:* $\sin 2x = 2\sin x \cos x$.)
11. 2^x
12. xe^{-x}
13. $\dfrac{1-\cos x}{x^2}$ (defined as its limiting value at 0)
14. $\sin x^2$
15. $\sqrt[3]{8-x}$ (Do not test endpoints.)
16. $\dfrac{\sin x}{x}$ (defined at 0 to be its limiting value as $x \to 0$)
17. $\cos \sqrt{x}$
18. $x \ln \sqrt{1+x}$

In Exercises 19–24, find the Taylor series for the given function about $x = a$ for the specified value of a, and determine the interval of convergence of the series.

19. e^x; $a = 1$
20. $\cos x$; $a = \dfrac{\pi}{3}$
21. $\dfrac{1}{x}$; $a = -1$
22. $\sqrt{x+3}$; $a = 1$
23. $\sin \dfrac{\pi x}{6}$; $a = 1$
24. $1/(x-1)^2$; $a = 2$

25. For each of the following, show that the Maclaurin series of the given function represents the function for all values of x.
   (a) $\sin x$ (b) $\sinh x$ (c) $\cosh x$

26. Find the Taylor series about $x = 1$ for $f(x) = \ln x$ by replacing x with $x - 1$ in the Maclaurin series for $\ln(1+x)$. In what interval does the series represent $\ln x$?

27. Show that the radius of convergence of the binomial series is 1.

28. Prove that if $f^{(n)}(x)$ exists for all n and x and there exists a constant M such that $|f^{(n)}(x)| \leq M$ for all n and x, then the Taylor series of f about $x = a$ exists and converges to $f(x)$ everywhere.

29. For $x \neq 0$, define $f(x) = e^{-1/x^2}$ and let $f(0) = 0$. Prove that the Maclaurin series for f exists and converges everywhere but that it represents f only at $x = 0$.

In Exercises 30–33, approximate the integrals, making use of three terms of the Maclaurin series of the integrand. Estimate the error.

30. $\displaystyle\int_{-1}^{1} \dfrac{1-\cos x}{x^2}\,dx$
31. $\displaystyle\int_{0}^{0.5} \sqrt{1+x^3}\,dx$
32. $\displaystyle\int_{0}^{0.4} \dfrac{\tan^{-1} x}{x}\,dx$
33. $\displaystyle\int_{-0.2}^{0} e^{-x^2}\,dx$

34. Find the first three nonzero terms of the Maclaurin series for $\tan x$.

35. Find the first three nonzero terms of the Maclaurin series for $\sec^2 x$. (*Hint:* Use the results of Exercise 34.)

36. Use a CAS to find the first eight terms of the Maclaurin series for
   (a) $\tan x$;   (b) $\sec^2 x$.
   (Compare Exercises 34 and 35.)

37. The **error function** is defined by
$$\mathrm{erf}(x) = \dfrac{2}{\sqrt{\pi}} \int_0^x e^{-t^2}\,dt, \qquad -\infty < x < \infty$$
Find the Maclaurin series for erf(x), and use it to approximate erf(1) correct to three decimal places.

38. Approximate the integral
$$\int_0^2 xe^x\,dx$$

   (a) using Simpson's Rule with 10 subintervals;
   (b) using the first 10 nonzero terms of the Taylor series about $x = 0$.
   Compare the results.

39. (a) Use a CAS to obtain the graph of
$$f(x) = |\sin^{-1}(\sin x)|$$
Explain the appearance of the graph.
(b) Define the function
$$g(x) = \sum_{n=0}^{\infty} \left(\frac{3}{4}\right)^n f(4^n x)$$
where f is the function of part (a). Approximate the graph of $g(x)$ by several partial sums of the series.
Note: $g(x)$ is a nowhere-differentiable function first given by Karl Weierstrass (1815–1897).

10.10 FOURIER SERIES

The representation of functions by Taylor series is of fundamental importance. Yet there are many functions that cannot be expanded in Taylor series. For one thing, only functions that are infinitely differentiable have such expansions. Another type of series, whose terms consist of sines and cosines, can be used to represent a much broader class of functions. Series of this type are called Fourier (pronounced "foor-ee-yay") series, after the French mathematician and physicist Jean Baptiste Joseph Fourier (1768–1830), who studied these series in connection with his investigation of heat conduction.

Fourier series are especially important in studying all sorts of wave phenomena, such as light waves, sound waves, radio waves, and signals in electronic instruments. A concept closely related to a Fourier series, called a Fourier integral, has been used in such diverse ways as analyzing the sunspot cycle and investigating X rays in computer-assisted tomography (CAT) scanners.

We will give only a brief introduction here to Fourier series. There is a vast literature on this subject, and research related to Fourier series and Fourier integrals is ongoing. In this discussion we can only scratch the surface.

Let us begin by looking at the series

$$\cos x - \frac{\cos 3x}{3} + \frac{\cos 5x}{5} - \cdots = \sum_{k=1}^{\infty} \frac{(-1)^{k-1} \cos(2k-1)x}{2k-1} \quad (10.32)$$

This series is an example of a Fourier series involving cosine terms only. (Others involve only sine terms or combine sines and cosines.) In Figure 10.30 we show graphs of the first three partial sums,

$$S_1(x) = \cos x, \quad S_2(x) = \cos x - \frac{\cos 3x}{3}, \quad S_3(x) = \cos x - \frac{\cos 3x}{3} + \frac{\cos 5x}{5}$$

Notice how adding the second term modifies the first by adding more "ripples" with smaller amplitude. The tops and bottoms seem to be flattening out. Adding the third term (getting $S_3(x)$) further flattens the tops and bottoms and adds "ripples" of still smaller amplitude. To see that this trend continues, we show the graph of the first ten terms,

$$S_{10}(x) = \sum_{k=1}^{10} \frac{(-1)^{k-1} \cos(2k-1)x}{2k-1}$$

in Figure 10.31. It appears that the partial sums are converging to the "square wave" shown in Figure 10.32. In Exercise 9 of Exercise Set 10.10 we will ask you to verify that series 10.32 is the Fourier series for the function whose graph is given in Figure 10.32.

814 Chapter 10 Infinite Series

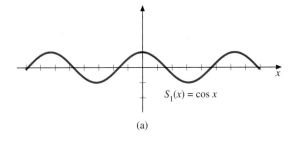
(a)

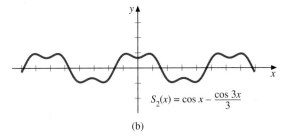
(b)

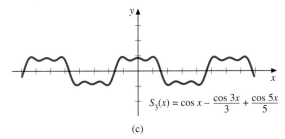
(c)

FIGURE 10.30

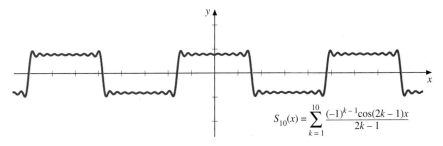

FIGURE 10.31

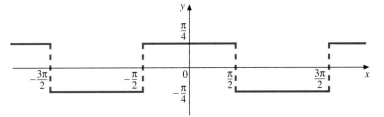

FIGURE 10.32
The square wave

$$f(x) = \begin{cases} \frac{\pi}{4} & \text{if } -\frac{\pi}{2} < x < \frac{\pi}{2} \\ -\frac{\pi}{4} & \text{if } \frac{\pi}{2} < x < \frac{3\pi}{2} \end{cases}$$

$f(x + 2\pi) = f(x)$.

In our example above, we began with a series and conjectured the function to which it converged. Now we want to reverse this procedure. We will begin with a function and attempt to find a series of sines and cosines converging to the function. This process will lead us to the definition of a Fourier series.

Deriving the Fourier Series

We begin with a function f that is integrable on the interval $[-\pi, \pi]$. (Later, we will indicate how we can use a more general interval of the form $[-L, L]$ for any $L > 0$.) At present we place no other restriction on f. It may not even be continuous. Let us assume for the moment that $f(x)$ can be represented by the following series of sines and cosines:

$$f(x) = \frac{a_0}{2} + \sum_{n=1}^{\infty}(a_n \cos nx + b_n \sin nx) \tag{10.33}$$

We will see later why it is convenient to write the constant term as $a_0/2$ rather than a_0. (By doing so, it will turn out that the formula we develop for a_n when $n \geq 1$ will be true for $n = 0$ as well.)

To find the coefficients in Equation 10.33, we will need the following results, which can be obtained using the integration techniques of Section 8.3 (see Exercise 42 in Exercise Set 8.3).

For m and n positive integers,

$$\int_{-\pi}^{\pi} \cos mx \cos nx \, dx = \begin{cases} 0 & \text{if } m \neq n \\ \pi & \text{if } m = n \end{cases} \tag{10.34}$$

$$\int_{-\pi}^{\pi} \sin mx \sin nx \, dx = \begin{cases} 0 & \text{if } m \neq n \\ \pi & \text{if } m = n \end{cases} \tag{10.35}$$

$$\int_{-\pi}^{\pi} \sin mx \cos nx \, dx = 0 \tag{10.36}$$

By direct integration, you can also show that

$$\int_{-\pi}^{\pi} \cos nx \, dx = \int_{-\pi}^{\pi} \sin nx \, dx = 0 \tag{10.37}$$

Let us return now to Equation 10.33 and integrate both sides from $-\pi$ to π. We will assume at this stage that term-by-term integration on the right-hand side is valid.

$$\int_{-\pi}^{\pi} f(x) \, dx = \int_{-\pi}^{\pi} \frac{a_0}{2} \, dx + \sum_{n=1}^{\infty} \left[\int_{-\pi}^{\pi} a_n \cos nx \, dx + \int_{-\pi}^{\pi} b_n \sin nx \, dx \right]$$

By Equation 10.37 the integrals in the summation both equal 0 for all n. Thus, we get

$$\int_{-\pi}^{\pi} f(x) \, dx = \frac{a_0}{2}(2\pi)$$

Solving for a_0 gives

$$a_0 = \frac{1}{\pi} \int_{-\pi}^{\pi} f(x) \, dx \tag{10.38}$$

We now have found the value of a_0. To find a_m for $m \geq 1$, we multiply both sides of Equation 10.33 by $\cos mx$, where m is a fixed, but unspecified, positive integer. Then, we again integrate over the interval $[-\pi, \pi]$:

$$\int_{-\pi}^{\pi} f(x) \cos mx \, dx = \frac{a_0}{2} \int_{-\pi}^{\pi} \cos mx \, dx + \sum_{n=1}^{\infty} \left[a_n \int_{-\pi}^{\pi} \cos mx \cos nx \, dx \right.$$
$$\left. + b_n \int_{-\pi}^{\pi} \cos mx \sin nx \, dx \right]$$

By Equation 10.37 the first integral on the right-hand side is 0 if $m \geq 1$. By Equation 10.34 the only nonzero term in the summation is the term for which $n = m$. Thus,

$$\int_{-\pi}^{\pi} f(x) \cos mx \, dx = a_m \int_{-\pi}^{\pi} \cos^2 mx \, dx$$

By Equation 10.34 the integral on the right equals π. So, on solving for a_m, we have

$$a_m = \frac{1}{\pi} \int_{-\pi}^{\pi} f(x) \cos mx \, dx, \quad m = 1, 2, \ldots \tag{10.39}$$

Notice that if we let $m = 0$ in this equation, the result agrees with the value of a_0 in Equation 10.38.

In order to find b_m, we proceed in a similar way, multiplying both sides of Equation 10.33 by $\sin mx$ and integrating. The result is (see Exercise 3 in Exercise Set 10.10)

$$b_m = \frac{1}{\pi} \int_{-\pi}^{\pi} f(x) \sin mx \, dx, \quad m = 1, 2, \ldots \tag{10.40}$$

REMARK
■ We used the letter m as a subscript in deriving Equations 10.39 and 10.40, since n was used as the index of summation. Now that we have the formulas, we can replace m with n.

Although we have made a number of assumptions in deriving the formulas for the coefficients, the formulas themselves have meaning provided only that $f(x)$ is integrable on $[-\pi, \pi]$. We therefore can make the following definition.

Definition 10.10
The Fourier Series for f

Let f be integrable on $[-\pi, \pi]$. Then the series

$$\frac{a_0}{2} + \sum_{n=1}^{\infty} (a_n \cos nx + b_n \sin nx)$$

where

$$a_n = \frac{1}{\pi} \int_{-\pi}^{\pi} f(x) \cos nx \, dx, \quad n = 0, 1, 2, \ldots$$

and

$$b_n = \frac{1}{\pi} \int_{-\pi}^{\pi} f(x) \sin nx \, dx, \quad n = 1, 2, 3, \ldots$$

is called the **Fourier series for f** on $[-\pi, \pi]$.

EXAMPLE 10.46 Find the Fourier series for the function f defined by $f(x) = x$ on $[-\pi, \pi]$.

Solution By Equation 10.39,

$$a_n = \frac{1}{\pi} \int_{-\pi}^{\pi} x \cos nx \, dx$$

Since $x \cos nx$ is the product of an odd function and an even function, the result is odd. Thus, as we showed in Section 5.4, the integral over a symmetric interval is 0. That is,

$$a_n = 0 \quad \text{for } n = 0, 1, 2, \ldots$$

Also, by Equation 10.40 and by what we showed in Section 5.4,

$$b_n = \frac{1}{\pi} \int_{-\pi}^{\pi} x \sin nx \, dx = \frac{2}{\pi} \int_{0}^{\pi} x \sin nx \, dx$$

since $x \sin nx$ is an even function (the product of two odd functions is even). Now we integrate by parts, with $u = x$ and $dv = \sin nx$.

$$b_n = \frac{2}{\pi} \left[-\frac{x \cos nx}{n} \Big|_0^\pi + \frac{1}{n} \int_0^\pi \cos nx \, dx \right]$$

$$= \frac{2}{\pi} \left[-\frac{\pi \cos n\pi}{n} \right] + \frac{2}{\pi n^2} \left[\sin nx \right]_0^\pi$$

$$= -\frac{2}{n} \cos n\pi$$

Since $\cos n\pi = 1$ when n is even and $\cos n\pi = -1$ when n is odd, we can write $\cos n\pi = (-1)^n$. Thus,

$$b_n = \frac{2(-1)^{n+1}}{n} \quad n = 1, 2, 3, \ldots$$

By Definition 10.10 the Fourier series for f is

$$2 \left[\sin x - \frac{\sin 2x}{2} + \frac{\sin 3x}{3} - \cdots \right] \quad (10.41)$$

We will graph some of the partial sums shortly. ∎

A Convergence Theorem

The question of convergence of the Fourier series for a function is a deep one and is better left to more advanced courses. We will state below one set of criteria that will ensure that the series converges to the function, at least for most values of x. First, though, observe that since $\sin nx$ and $\cos nx$ have period 2π for all values of n, if the Fourier series converges in the interval $[-\pi, \pi]$, then it also converges outside that interval in a periodic manner. Thus, if $f(x)$ is to be represented by its Fourier series for all x, it must be periodic, of period 2π. That is, $f(x)$ must satisfy

$$f(x + 2\pi) = f(x)$$

If f is not periodic, its Fourier series still may converge to $f(x)$ on $[-\pi, \pi]$. Outside that interval it would then converge to the *periodic extension* of f.

The following theorem was first stated by the German mathematician P. G. Lejeune Dirichlet (1805–1859).

THEOREM 10.24

Dirichlet's Theorem

Let f be bounded, piecewise-continuous, and have a finite number of maxima and minima on the interval $-\pi < x \leq \pi$. For x outside this interval let f be defined periodically by $f(x + 2\pi) = f(x)$. Then the Fourier series for f converges to $f(x)$ for all points of continuity of f. At each point of discontinuity the series converges to the average of the right-hand and left-hand limits of f at that point.

REMARK

■ A function is piecewise continuous on an interval if it has at most finitely many points of discontinuity, and each such discontinuity is a simple (jump) discontinuity. (See Section 2.3.)

Graphing the Partial Sums

Let us illustrate the manner in which the partial sums of a Fourier series converge by looking again at some graphs. We will use series 10.41 for the function $f(x) = x$. In Figure 10.33 we show the graph of f on $(-\pi, \pi]$ and extended periodically outside that interval. Because of the appearance of its graph, this function is often referred to as a "sawtooth" function.

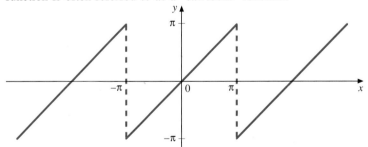

FIGURE 10.33
The "sawtooth" function $f(x) = x$ on $(-\pi, \pi]$, $f(x + 2\pi) = f(x)$

The hypotheses of Theorem 10.24 are satisfied for our function in Figure 10.33, so its Fourier series converges in the manner specified by the theorem. The points of discontinuity are the integral multiples of π. At each such point the average of the right-hand and left-hand limits of f is 0. If we redefine $f(x)$ to be 0 at these points, then we can write

$$f(x) = 2\left[\sin x - \frac{\sin 2x}{2} + \frac{\sin 3x}{3} - \cdots\right] = 2\sum_{k=1}^{\infty} \frac{(-1)^{k+1} \sin kx}{k}$$

In Figure 10.34 we show the graphs of the three partial sums $S_3(x)$, $S_6(x)$, and $S_9(x)$. Notice how they increasingly approach the sawtooth shape of the graph of f as n increases. Also, observe that each partial sum graph crosses the x-axis (the value of the partial sum is 0) at each integral multiple of π.

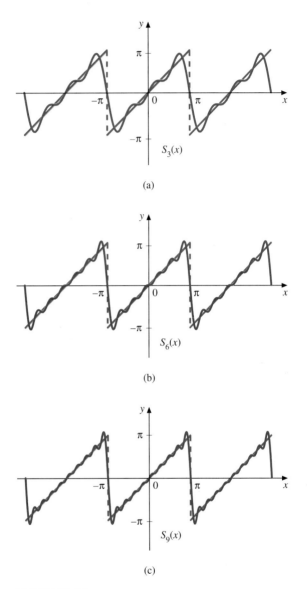

FIGURE 10.34

Harmonics

In the Fourier series for a function f,

$$\frac{a_0}{2} + \sum_{n=1}^{\infty}(a_n \cos nx + b_n \sin nx)$$

the constant term $a_0/2$ is known as *the fundamental*. In Exercise 1 of this section you will be asked to show that it is the average value of f over $[-\pi, \pi]$. The term $a_1 \cos x + b_1 \sin x$ is called the *first harmonic*, the term $a_2 \cos 2x + b_2 \sin 2x$ the *second harmonic*, and so on. When we listen to music, we are hearing the sum of many harmonics. The human ear and brain make a wonderful Fourier analyzer of these harmonics.

Other Intervals

By means of a horizontal stretching or shrinking of the type we studied in Section 1.2, it is possible to obtain the Fourier series for a function f defined on $[-L, L]$ for an arbitrary positive number L instead of on $[-\pi, \pi]$. The result, which we ask you to show in Exercise 16 of Exercise Set 10.10, is

$$f(x) = \frac{a_0}{2} + \sum_{n=1}^{\infty} \left(a_n \cos \frac{n\pi x}{L} + b_n \sin \frac{n\pi x}{L} \right) \quad (10.42)$$

where

$$a_n = \frac{1}{L} \int_{-L}^{L} f(x) \cos \frac{n\pi x}{L} \, dx \quad (10.43)$$

and

$$b_n = \frac{1}{L} \int_{-L}^{L} f(x) \sin \frac{n\pi x}{L} \, dx \quad (10.44)$$

Exercise Set 10.10

1. Show that the constant term $a_0/2$ in the Fourier series for a function f is the average value of f on $[-\pi, \pi]$. (See Definition 5.3.)

2. (a) Show that if f is even, then $b_n = 0$ for $n = 1, 2, 3, \ldots$, and the Fourier series for f consists of cosine terms only.
   (b) Show that if f is odd, then $a_n = 0$ for $n = 0, 1, 2, \ldots$, and the Fourier series for f consists of sine terms only.

3. Carry out the details of deriving Formula 10.40 for b_m.

In Exercises 4–7, find the Fourier series for f.

4. $f(x) = |x|$ on $[-\pi, \pi]$

5. $f(x) = \begin{cases} 0 & \text{if } -\pi < x < 0 \\ x & \text{if } 0 \leq x \leq \pi \end{cases}$ and $f(x + 2\pi) = f(x)$

6. $f(x) = x^2$ on $[-\pi, \pi]$

7. $f(x) = \cos 2x$ on $[-\pi, \pi]$ (How could you have determined the answer without calculation?)

8. By substituting $x = \pi$ in the Fourier series for $f(x) = x^2$ (Exercise 6), obtain the sum

$$\sum_{n=1}^{\infty} \frac{1}{n^2} = \frac{\pi^2}{6}$$

(Recall that this sum was referred to in the introduction to this chapter, and it was first discovered by Euler. He used a completely different approach from the one used here.)

9. Show that the series 10.32 is the Fourier series for the square wave function shown in Figure 10.32.

10. Find the Fourier series for $f(x) = \cos^2 x$ for $-\pi \leq x \leq \pi$.

In Exercises 11–15, use a CAS to graph the partial sums $S_n(x)$ for the specified series and the specified values of n.

11. The series of Example 10.46; $n = 2, 5, 7, 10$

12. The series of Exercise 4; $n = 2, 5, 8, 11$

13. The series of Exercise 5; $n = 1, 3, 5, 7$

14. The series of Exercise 6; $n = 2, 4, 6, 8$

15. The series of Exercise 10; $n = 2, 5, 8, 11$

16. Derive Equations 10.42, 10.43, and 10.44 as follows:
    (a) Let f satisfy the conditions given in Theorem 10.24 on the interval $[-L, L]$. Make the change of variables $u = \pi x/L$, and let
    $$g(u) = f\left(\frac{Lu}{\pi}\right)$$
    (b) Show that $g(u)$ is defined for $-\pi \leq u \leq \pi$.
    (c) Write the Fourier series for $g(u)$ and the formulas for a_n and b_n in terms of u.
    (d) In the results of part (c) replace u with $\pi x/L$. Simplify the formulas for a_n and b_n.

Chapter 10 Review Exercises

1. Find the Taylor polynomial of degree 5 about $x = 0$ for $f(x) = \ln(1 - x)$, and determine how accurately it approximates $f(x)$ for $|x| \leq 0.2$.

2. Compute $\cos 62°$ with six decimal places of accuracy using Taylor's Formula with an appropriate value of n and a.

3. Find the Taylor polynomial of degree 5 about $x = \pi/4$ for $f(x) = \tan x$.

4. Find the nth Taylor polynomial for $f(x) = \cos^2 x$ about $x = 0$. What is $R_n(x)$ for this function? (Hint: Use a trigonometric identity.)

5. Find the nth Taylor polynomial about $x = \pi/4$ for $f(x) = \sin x - \cos x$. (Hint: Show that $f(x) = \sqrt{2}\sin(x - \pi/4)$.)

6. Use a cubic polynomial to approximate $\sqrt[3]{25}$. Estimate the error. (Hint: $\sqrt[3]{25} = (27 - 2)^{1/3} = (27)^{1/3}(1 - \frac{2}{27})^{1/3} = 3(1 - \frac{2}{27})^{1/3}$.)

7. Find $P_3(x)$ and $R_3(x)$ about $x = 1$ for $f(x) = \tan^{-1} x$.

8. Show that the Taylor polynomial
$$1 - \frac{x^2}{2!} + \frac{x^4}{4!} - \frac{x^6}{6!} + \frac{x^8}{8!}$$
for $\cos x$ gives accuracy to at least seven decimal places for x in $[-\pi/4, \pi/4]$.

9. Let $P_n(x)$ be the nth Taylor polynomial about $x = 0$ for $f(x) = e^x$. Use a calculator to make a table showing each of the following for $x = \pm 0.25$, $x = \pm 0.50$, $x = \pm 0.75$, $x = \pm 1$: e^x, $P_1(x)$, $P_2(x)$, $P_3(x)$, and $P_4(x)$.

10. Write out Taylor's Formula with Remainder with $a = 0$ for $f(x) = xe^{-x}$.

In Exercises 11 and 12, find $\lim_{n \to \infty} a_n$, or show the limit fails to exist.

11. (a) $a_n = \dfrac{2n^3 - 3n + 1}{n^3 + 2n + 3}$ (b) $a_n = \dfrac{\sqrt{n + 10}}{n + 2}$

12. (a) $a_n = \dfrac{\sqrt{n^5 + 1}}{10n^2 + 3n}$ (b) $a_n = \dfrac{n \cos n\pi}{n + 1}$

In Exercises 13 and 14, show that each sequence is monotone. Then determine whether it converges by investigating boundedness.

13. (a) $\{e^{1-(1/n)}\}$ (b) $\left\{\dfrac{2n - 3}{n + 3}\right\}$

14. (a) $\left\{\dfrac{e^{2n} n!}{(2n - 1)!}\right\}$
    (b) $\{n \ln(n + 1) - (n + 1) \ln n\}$

15. Prove that if $\lim_{n \to \infty} a_n = L$, then $\lim_{n \to \infty} a_n^2 = L^2$. Is the converse also true? Prove or disprove.

16. Let
$$S_n = \frac{1}{2n + 1} + \frac{1}{2n + 2} + \frac{1}{2n + 3} + \cdots + \frac{1}{3n}$$
Prove that $\{S_n\}$ is a convergent sequence.

17. A sequence $\{S_n\}$ is defined recursively by $S_1 = 1$, and for $n \geq 2$, $S_n = S_{n-1} + 2/(n^2 + n)$. Find an explicit formula for S_n. Then show that $\{S_n\}$ converges, and find its limit.

18. What can you conclude about convergence or divergence of the series $\sum_{k=1}^{\infty} a_k$ under the following conditions?
    (a) $a_k = \dfrac{2k^2 - 1}{k^2 + 1}$
    (b) $a_1 + a_2 + \cdots + a_n = \dfrac{2n^2 - 1}{n^2 + 1}$
    Give reasons.

19. Find the sum of each of the following series:
(a) $\sum_{n=0}^{\infty} \frac{(-1)^n 2^{n-1}}{3^n}$
(b) $\sum_{k=2}^{\infty} \frac{1}{k^2 - 1}$

20. Use geometric series to express each of the following repeating decimals as the ratio of two integers:
(a) $1.297297297\ldots$
(b) $3.2454545\ldots$

In Exercises 21–44, test for convergence or divergence. If the series has variable signs, test also for absolute convergence.

21. $\sum_{n=1}^{\infty} \frac{n}{\sqrt{n^3 + 2}}$

22. $\sum_{n=1}^{\infty} \frac{\sin n + \cos n}{n\sqrt{n+1}}$

23. $\sum_{n=0}^{\infty} \frac{e^{-n}}{1 + e^{-n}}$

24. $\sum_{n=2}^{\infty} \frac{\cos n\pi}{\ln n}$

25. $\sum_{n=1}^{\infty} \frac{\tanh n}{n}$

26. $\sum_{n=1}^{\infty} n e^{-2n}$

27. $\sum_{k=1}^{\infty} \frac{\ln k}{k^2}$

28. $\sum_{n=1}^{\infty} \left(\frac{n}{3n+4}\right)^n$

29. $\sum_{k=1}^{\infty} \frac{(-1)^{k-1} \ln k}{\sqrt{k}}$

30. $\sum_{k=1}^{\infty} \frac{3k^2 + 2k - 1}{k^5 + 2}$

31. $\sum_{k=2}^{\infty} \frac{\sin k}{k(\ln k)^2}$

32. $\sum_{n=1}^{\infty} \frac{n - 1}{(n+1)^2}$

33. $\sum_{n=0}^{\infty} \frac{2^n n!}{(2n)!}$

34. $\sum_{k=0}^{\infty} (-1)^k k^2 e^{-k}$

35. $\sum_{k=1}^{\infty} \frac{k^k}{3^k k!}$

36. $\sum_{n=2}^{\infty} \frac{\ln n}{n^3 + 4}$

37. $\sum_{n=2}^{\infty} (-1)^n \ln\left(\frac{n+1}{n-1}\right)$

38. $\sum_{k=0}^{\infty} \frac{\sin(k + \frac{1}{2})\pi}{k + \frac{1}{2}}$

39. $\sum_{n=1}^{\infty} \frac{n^{2n}}{(n^3 + 1)^n}$

40. $\sum_{n=1}^{\infty} \left(\frac{1}{n} - \frac{1}{n+2}\right)$

41. $\sum_{n=1}^{\infty} \frac{2 + \cos n}{\sqrt{2n^2 - 1}}$

42. $\sum_{n=0}^{\infty} \frac{\text{sech}^2 n}{1 + \tanh n}$

43. $\sum_{n=0}^{\infty} (-1)^n (\sqrt{n+1} - \sqrt{n})$

44. $\sum_{n=0}^{\infty} \frac{(-1)^n (n-1)}{2n+1}$

In Exercises 45–52, find the interval of convergence.

45. $\sum_{n=0}^{\infty} \frac{2^n x^n}{3^{n+1}}$

46. $\sum_{n=0}^{\infty} \frac{(-1)^n n x^n}{2n+1}$

47. $\sum_{k=0}^{\infty} \frac{(x-2)^k}{\sqrt{2k+1}}$

48. $\sum_{k=0}^{\infty} \frac{(-1)^k (x+1)^{2k}}{3^k}$

49. $\sum_{k=1}^{\infty} \frac{k x^k}{\ln(k+1)}$

50. $\sum_{n=1}^{\infty} \frac{(nx)^n}{n! e^n}$ (Do not check endpoints.)

51. $\sum_{n=0}^{\infty} \frac{(-1)^n (2x - 1)^n n!}{1 \cdot 3 \cdot 5 \cdot \ldots \cdot (2n+1)}$ (Hint: In testing endpoints, see Exercise 33 in Exercise Set 10.6.)

52. $\sum_{n=1}^{\infty} \frac{\cosh n}{n^2} (x - 1)^n$

In Exercises 53–56, use integration or differentiation of an appropriate series to find the Maclaurin series of f. Give the radius of convergence.

53. $f(x) = \ln(1 - x^2)$

54. $f(x) = x/(2+x)^2$

55. $f(x) = \sinh^{-1} x$

56. $f(x) = \frac{1}{a} \tan^{-1} \frac{x}{a}$

57. Use infinite series to verify the following limits:
(a) $\lim_{x \to 0} \frac{\sin x}{x} = 1$
(b) $\lim_{x \to 0} \frac{1 - \cos x}{x^2} = \frac{1}{2}$
(c) $\lim_{x \to 0} \frac{e^x - 1}{x} = 1$

In Exercises 58–60, find the Taylor series about $x = a$.

58. $f(x) = 1/(2 - x);\ a = 1$

59. $f(x) = \ln\sqrt{x + 2};\ a = -1$

60. $f(x) = \sin x;\ a = \frac{\pi}{4}$. Show that the series represents the function for all values of x.

[CAS] In Exercises 61 and 62, approximate the integral using the first five nonzero terms of the Maclaurin series of the integrand. Give an upper bound on the error.

61. $\int_0^{1/2} \frac{1 - e^{-x}}{x} dx$

62. $\int_0^{1/2} \frac{dx}{\sqrt{1 + x^3}}$

63. Find the Taylor series for $f(x) = \cos^2 x - \sin^2 x$ about $x = \pi/4$, and show that it converges to f for all x. (*Hint:* First use a trigonometric identity.)

64. Find the Fourier series for each of the functions
(a) $f(x) = \begin{cases} 0 & \text{if } -\pi < x < 0 \\ 1 & \text{if } 0 \leq x \leq \pi \end{cases}$
(b) $f(x) = |\sin x|$ on $[-\pi, \pi]$

Chapter 10 Concept Quiz

1. Define each of the following:
   (a) the limit of a sequence $\{a_n\}$
   (b) the sum of a series $\sum_{n=1}^{\infty} a_n$
   (c) a geometric series
   (d) an absolutely convergent series
   (e) the nth Taylor polynomial for a function f about $x = a$
   (f) the Fourier series for a function f on $[-\pi, \pi]$

2. State each of the following:
   (a) the nth-Term Divergence Test
   (b) the Integral Test
   (c) the Comparison Test
   (d) the Limit Comparison Test
   (e) the Ratio Test
   (f) the Alternating Series Test
   (g) Taylor's Formula with Remainder
   (h) a necessary and sufficient condition for the Taylor series for a function f about $x = a$ to converge to $f(x)$
   (i) conditions that guarantee the Fourier series for a function will converge to the function

3. (a) Suppose that $a_n \leq a_{n+1} \leq M$ for all n, where M is a constant. What can you conclude about convergence or divergence of the sequence $\{a_n\}$, and why?
   (b) Suppose the power series $\sum a_n x^n$ converges when $x = -1$ and diverges when $x = 2$. State everything you can about where else the series converges and where else it diverges.
   (c) Suppose that $a_n \geq 0$, $b_n \geq 0$, and $c_n \geq 0$ for all n, and that $\sum a_n$ converges and $\sum b_n$ diverges. State everything you can about convergence or divergence of $\sum c_n$ if for all n:
   (i) $c_n \leq a_n$
   (ii) $a_n \leq c_n \leq b_n$
   (iii) $c_n \geq b_n$
   (d) Let $f^{(n)}(x)$ exist for all n for $-R < x < R$, and suppose $\sum_{n=0}^{\infty} a_n x^n$ converges to $f(x)$ for all x in $(-R, R)$. What can you conclude about a_n?
   (e) Explain under what conditions on a and b the equation

   $$\int_a^b \left(\sum_{n=0}^{\infty} a_n x^n \right) dx = \sum_{n=0}^{\infty} \int_a^b a_n x^n \, dx$$

   is valid.

4. Fill in the blanks.
   (a) The series $\sum_{n=1}^{\infty} a_n$ converges to the sum S if and only if its _____ converges to S.
   (b) The p-series $\sum_{n=1}^{\infty} 1/n^p$ converges if _____ and diverges if _____.
   (c) The geometric series $\sum_{n=1}^{\infty} ar^{n-1}$ converges to _____ if _____ and diverges if _____.
   (d) A sequence is a function whose domain is _____.
   (e) If the power series $\sum_{n=0}^{\infty} a_n x^n$ has positive radius of convergence R and its sum is $f(x)$, then $\sum_{n=1}^{\infty} n a_n x^{n-1}$ converges to _____ for each x in the interval _____.

5. Indicate which of the following statements are true and which are false.
   (a) If $\lim_{n \to \infty} a_n = 0$, then $\sum_{n=1}^{\infty} a_n$ converges.
   (b) If $\sum_{n=1}^{\infty} a_n$ diverges, then $\lim_{n \to \infty} a_n \neq 0$.
   (c) If $\lim_{n \to \infty} a_n \neq 0$, then $\sum_{n=1}^{\infty} a_n$ does not converge.
   (d) If $\sum_{n=1}^{\infty} a_n$ is a series of nonnegative terms, and $a_1 + a_2 + a_3 + \cdots + a_n \leq 10$ for all n, then $\sum_{n=1}^{\infty} a_n$ converges.
   (e) If $\{a_n\}$ is a monotone decreasing sequence of positive numbers, then the sequence converges.
   (f) If $\sum_{n=1}^{\infty} a_n$ is a series of positive terms and $a_{n+1}/a_n < 1$ for all n, then $\sum_{n=1}^{\infty} a_n$ converges.

APPLYING CALCULUS

1. When money is spent on goods and services, those that receive money also spend some of it. The people receiving some of the twice-spent money will spend some of that, and so on. Economists call this chain reaction the *multiplier effect*. In a hypothetical isolated community, the local government begins the process by spending D. Suppose that each recipient of spent money spends $100\,c\%$ and saves $100\,s\%$ of the money that he or she receives. The values c and s are called the *marginal propensity to consume* and the *marginal propensity to save*, and, of course, $c + s = 1$.
   (a) Let S be the total spending that has been generated after n transactions. Determine an equation for S.
   (b) Show that $\lim_{n \to \infty} S = kD$, where $k = 1/s$. The number k is called the *multiplier*.
   What is the multiplier if the marginal propensity to consume is 0.8?
   Note: The Federal Government uses this principle to justify deficit spending. Banks use this principle to justify lending out a large percentage of the money that they receive in deposits.

2. A certain ball has the property that each time it falls from height h onto a hard level surface, it rebounds to a height rh, where $0 < r < 1$. Suppose the ball is dropped from an initial height of H meters.
   (a) Assuming that the ball continues to bounce indefinitely, determine the total distance that it travels.
   (b) Calculate the total time that the ball is traveling. (Hint: If the ball is dropped from a height h, then the distance s it travels in t seconds is $s = 4.9t^2$).
   (c) Suppose that each time the ball strikes the surface with velocity v, it rebounds with velocity kv, where $0 < k < 1$. How long will it take for the ball to come to rest?

3. Let $S(x)$ and $C(x)$ be defined by the power series
$$S(x) = \sum_{k=0}^{\infty} \frac{(-1)^k (\omega x)^{2k+1}}{(2k+1)!}$$
$$C(x) = \sum_{k=0}^{\infty} \frac{(-1)^k (\omega x)^{2k}}{(2k)!}$$
where ω is a constant.
   (a) Determine the domain of S and of C.
   (b) Show that
$$\int_0^x C(t)\,dt = \frac{1}{\omega} S(x) \quad \text{and} \quad C'(x) = -\omega S(x)$$
   (c) Show that S and C are solutions of the differential equation
$$y'' + \omega^2 y = 0$$
satisfying the initial conditions: $y(0) = 0$, $y'(0) = 1$ and $y(0) = 1$, $y'(0) = 0$, respectively.
   (d) Show that $S^2(x) + C^2(x) = 1$. (Hint: Put $U(x) = S^2(x) + C^2(x)$ and calculate U'.)

4. Consider the initial value problem
$$y' - 2y = e^x; \qquad y(0) = 0$$

Assume that the problem has a solution $y = y(x)$ that can be expressed as a power series expansion

$$y(x) = \sum_{n=0}^{\infty} a_n x^n = a_0 + a_1 x + a_2 x^2 \ldots$$

with radius of convergence $R > 0$.

(a) Express y' as a power series and calculate the power series expansion for $y' - 2y$.

(b) Since two power series are equal if and only if the coefficients of the corresponding powers of x are equal, equate the power series found in part (a) to the MacLaurin series for e^x and solve for the coefficients. The result will be a recursion formula that gives a_{n+1} in terms of a_n, $n = 0, 1, 2, \ldots$.

(c) Show that the initial condition $y(0) = 0$ implies $a_0 = 0$ and determine the solution of the initial value problem.

(d) Show that the power series found in (c) is equal to $y(x) = e^{2x} - e^x$.

5. What if you watch a TV showing a TV camera watching it?

(a) The "nested" sequence of squares shown in the margin is formed as follows: join the midpoints of the sides of the outermost square S_1 to form the square S_2; join the midpoints of the sides of S_2 to form S_3; and so on.
Determine the sum of the areas of all the squares if the first square has area A.

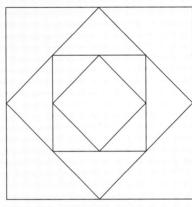

(b) The "nested" sequence of circles and squares shown in the margin is formed as follows: Inscribe a square in the circle of radius r; inscribe a circle in the square; inscribe a square in the circle; and so on.
Determine the area of the shaded region.

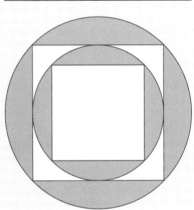

6. The curve defined parametrically by the equations

$$x = a \cos t, \qquad y = b \sin t, \qquad 0 < t < 2\pi$$

where a and b are positive numbers, is an ellipse whose center is at the origin and whose axes are on the coordinate axes. Assuming that $a > b$, the major axis is on the x-axis. The arc length of this ellipse is

$$L = \int_0^{2\pi} \sqrt{a^2 \sin^2 t + b^2 \cos^2 t}\, dt \qquad (*)$$

(a) Show that the formula $(*)$ can be simplified to

$$L = 4a \int_0^{\pi/2} \sqrt{1 - e^2 \cos^2 t}\, dt$$

where e is the eccentricity of the ellipse. This form is an elliptic integral; the integrand does not have an elementary antiderivative.

(b) Use the binomial series for $\sqrt{1-x}$ to show that

$$L = 4a \int_0^{\pi/2} \left[1 - \frac{1}{2} e^2 \cos^2 t - \frac{1}{8} e^4 \cos^4 t - \frac{1}{16} e^6 \cos^6 t - \ldots \right] dt$$

(c) Show that the arc length of the ellipse is given by

$$L = 2\pi a \left[1 - \frac{1}{4} e^2 - \frac{3}{64} e^4 - \frac{5}{256} e^6 - \ldots \right]$$

by evaluating the integral in part (b). Hint: Integrate termwise using the identity

$$\int_0^{\pi/2} \cos^{2n} t\, dt = \frac{1 \cdot 3 \cdot 5 \cdots (2n-1)}{2 \cdot 4 \cdot 6 \cdots (2n)} \cdot \frac{\pi}{2}$$

See Exercise 49 in Exercise Set 8.1. This formula gives the arc length of an ellipse in terms of its semimajor axis and its eccentricity. Note that this formula reduces to $2\pi a$, the circumference of a circle of radius a when $e = 0$.

(d) The orbit of the moon about the Earth is (almost) a perfect ellipse with the Earth at one focus. Assume that $a = 400,000$ km and $e = 0.055$. Use the first three terms of the series in (c) to determine the length of the moon's path around the earth.

7. In the accompanying figure is a region bounded by two circles of radius 1 that are tangent to one another and by a straight line which is tangent in both circles. A sequence of smaller circles, each having the largest possible radius, is inscribed in the region as shown. Show geometrically that the lengths of the diameters of these smaller circles are the terms of a convergent series whose sum is one. Then show that this series is

$$\sum_{n=1}^{\infty} \frac{1}{n(n+1)}.$$

8. Construct a sequence of circles with radii $r_1, r_2, r_3, \ldots,$ as follows. Inscribe a circle of radius $r_1 = 1$ in an equilateral triangle. Inscribe a circle of radius r_2, passing through the vertices of the triangle, in a square. Inscribe a circle of radius r_3, passing through the vertices of the square, in a regular pentagon. Continue in this way: inscribe a circle of radius r_{n-1}, passing through the vertices of a regular polygon of n sides, in a regular polygon of $n+1$ sides.

(a) Sketch a diagram of what is happening.

(b) Show that $r_2 = r_1 \sec(\pi/3)$ and $r_3 = r_2 \sec(\pi/4)$. Find a general formula for r_n.

(c) Show that

$$\ln r_n = \ln r_1 + \ln \sec \frac{\pi}{3} + \ln \sec \frac{\pi}{4} + \cdots + \ln \sec \frac{\pi}{n+1}$$

(d) Does r_n have a limit as n tends to infinity? (Hint: Use the Limit Comparison Test with $\sum [\ln \sec(\pi/n)]$ and $\sum (1/n^2)$.

9. An indicator of future population growth is the fertility rate, which is the average number of children a woman has during her childbearing years. If this rate exceeds the theoretical replacement level of 2 (it is slightly greater than 2 in the more developed countries and even higher in the less developed countries), the size of the population will continue to increase. One factor that affects a country's fertility rate is the type of family considered desirable. In each of the following assume that the probability of bearing male and female children is the same.

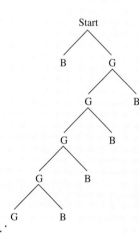

(a) Suppose each family will have children until they have a son. What is the fertility rate for the culture? (See the tree diagram.)

(b) Suppose each family will have children until they have at least one male and one female. Draw a tree diagram showing the possible outcomes. What is the fertility rate for this culture?

(c) Suppose each family has children until they have at least two males. Draw a tree diagram showing the possible outcomes. What is the fertility rate for this culture?

(d) A more general case is to suppose each family has children until they have at least m males and n females. Draw a tree diagram showing the possible outcomes. What is the fertility rate for this culture?

A PERSONAL VIEW

Constance McMillan Elson

What's the basic diference between multivariable calculus and vector calculus?

Some people use the terms interchangeably, but one could argue that multivariable calculus is the study of points, called vectors, in higher-dimensional spaces (dimension 2, 3, ..., n) and the study of functions whose inputs or outputs are vectors. Multivariable calculus deals with differentiating such functions and integrating them over flat regions in Euclidean space. It has applications in almost any area of human inquiry that is carried out quantitatively.

Vector calculus can include multivariable calculus, but it specifically refers to the extension of this calculus to curved spaces. Its principal mathematical objects are vector fields (functions whose inputs and outputs are vectors), special types of derivatives called the curl and divergence of these vector fields, and integrals of vector fields over curves and surfaces. In particular, vector calculus includes the study of Green's, Stokes's, and Gauss's Theorems, which are higher-dimensional analogs of the Fundamental Theorem of Calculus and relate various types of integrals. Vector calculus is the mathematical tool of greatest significance in the study of the basic forces of nature: electricity, magnetism, the nuclear forces, and gravity.

Are there different representations of the term vector?

There are three common ways to represent vectors: as algebraic objects called n-tuples, as points in n-dimensional Euclidean space, and, in the two- and three-dimensional cases, as arrows. Each representation has its advantages in specific situations, and it is important to be able to move conceptually among all three representations. Since an n-tuple is just an ordered collection of n real numbers, it is particularly easy to define the basic algebra of vectors in terms of n-tuples and to perform computations in this setting.

Equating vectors with points in space allows us to explore the geometry of multidimensional spaces, even in spaces of dimension 4 or higher. We do this by using vectors to describe the geometry of spaces we can visualize, i.e., regions in two- and three-dimensional Euclidean space. Then we use vector algebra to extend this to higher dimensions.

Finally, vectors can be interpreted as quantities having magnitude and direction; in $\mathbb{R}^2$ and $\mathbb{R}^3$ this means that we can represent them as arrows. These arows can be shifted at will (preserving magnitude and direction) without being altered as a vector; this property is one of the main reasons that vectors were developed and are used in physics.

The concepts of motion and force are fundamental in understanding the natural world. Since they are characterized by direction and magnitude, they are intrinsically vector quantities. When an object moves under the influence of two or more forces, its motion is related to the vector sum of the various forces. For instance, a canoe being paddled straight across a moving river will be carried downstream by the current. The actual course of the canoe is the vector sum of its still-water velocity and the velocity of the current. Similarly, an airplane flight path requires a pilot to set her course in an ocean of air. All the fancy navigational aids available to pilots and ship captains depend on the basic principal of vector addition.

Juggling the different representations of a vector can be tricky. It is important to be aware of what representation we are actually using at any given moment, but the interplay of these three points of view provides rich results.

Do vectors play a role in disciplines other than physics and engineering?

Definitely. Vectors can be used to organize information in any quantitative area. An investment officer of a bank or corporation, for instance, might manage a number of different stock portfolios, each of which can be represented as a vector. The components of the vector are the number of shares of the different stocks in the portfolio. If the prices of the stocks are summarized in a vector **P**, the current value of the portfolio represented by the vector **X** can be computed using the dot product from vector algebra. One can track the value of the portfolio over time by simply updating the price vector and recomputing the dot product. Modern spreadsheet software depends heavily on this kind of vectorization of information.

Vectors have important applications in social sciences. You could use vectors to try to find connections between several different socioeconomic variables. Often statistical techniques are used to test the strength of the various relationships between variables. These multivariate statistical methods are based on the algebra and geometry of vectors.

For example, say an archeologist is excavating a site and finds a huge number of different pottery fragments all jumbled together. He might conjecture that at least two different cultures have occupied the site and hopes to demonstrate this by means of the pottery fragments (shards). By assigning three numbers to each fragment, giving its thickness, granularity, and color, the archaeologist maps each shard into a vector. If these vectors are then interpreted as points in space and are found to separate into two (or more) distinct clusters, the archaeologist has found evidence supporting his conjecture. If there were several clusters of vectors, the geometric relationship between clusters might indicate a migration of skills or materials from one culture to another.

Are you concerned that students sometimes view calculus as just a set of rules for differentiating and integrating functions?

Describing calculus as a set of rules for differentiation and integration is about as informative as saying that a rock concert is loud. Both descriptions miss the essence of the event. It's better to develop the idea that a certain mathematical process is repeated indefinitely, with some controlled change taking place, and the resulting numerical or algebraic values converge to a limiting value. This idea underlies all of calculus. What gives calculus its practical utility is that this involved procedure can usually be reduced to certain computational rules for differentiation and integration. However, what gives calculus its power is that the process itself can be applied in very general circumstances. To make full use of this power requires a deep understanding of the idea of limit. For most students this kind of understanding is achieved only in an advanced analysis course, and so to many students in engineering and the sciences, limits seem abstruse and irrelevant because "you can get the answers without them." Using graphics and visualization, students can develop an intuitive and basic grasp of the idea of a limit in several dimensions, together with some algebraic results for computing limits. This gives a foundation that allows a student to understand other concepts in vector calculus quite rigorously. This is important because the interesting problems that lead to new results require thinking about things from first principles, rather than just applying an existing integration formula.

How well do most students make the transition from derivatives to partial derivatives and eventually to partial differential equations?

For functions of several variables, students usually understand partial derivatives very well, but they are often foggy about what a derivative is in this context. This is because a partial derivative is a number that represents a slope in a particular direction; thus it seems similar to the derivative that students are already familiar with from one-variable calculus. One way to visualize a partial derivative is to use the analogy between a function of several variables and a machine with several input hoppers and one output hopper. What happens to the output when we tweak one of the inputs by changing the quantity that we add to that particular hopper? Since that is the only input we are changing, the partial derivative with respect to that variable contains the answer to that question. It tells us the instantaneous rate at which the function output changes as one input variable changes.

A derivative of a multivariable function is a more abstract entity, but it's not hard to understand why it is the correct generalization of the single-variable derivative. We define the derivative at a point as a vector, or as a matrix, which makes a certain difference quotient go to zero in a controlled way. The derivative is composed of partial derivatives that makes it easy to compute, but the derivative is much richer than any single partial derivative and it generates a wealth of information about the function. If the derivative is not zero, we can find the rate of change of the function for any combination of changes in the inputs, the maximum rate of change of the function at that point, the direction (i.e., the combination of inputs) that produces the maximum rate of change, and the direction that produces no change in the function. Functions of several variables are mathematically rich objects and their derivatives reflect this.

A partial differential equation is an equation involving one or more partial derivatives and solving it involves finding a function whose partials satisfy the equation. For instance, three of the most famous partial differential equations—Laplace's equation, the heat equation, and the wave equation—involve only second partials. These equations are elegantly simple to write down, but efforts to understand their solutions have led to the development of an amazingly large part of modern mathematics. The problems that a student encounters may take a few hours to solve or at most, for a project, a few weeks. These three equations have engaged some of the best mathematical minds for the past two centuries and they're still yielding interesting results. The creation of mathematics is a continuing endeavor.

How do you expect students' attitudes and understanding of mathematics to develop while taking this course?

Ideally I hope three modes of learning will happen simultaneously. First, a student should develop proficiency in three-dimensional thinking and see vector functions as natural extensions of scalar functions. The goal is to make a student as comfortable and intuitively capable of solving problems in several dimensions as in one dimension. Second, the multidimensional calculus should reinforce and enrich a student's understanding of one-dimensional calculus. I try to stress the differences in moving from one to several variables, while revealing the unity of the concepts underlying the calculus. Good notation can facilitate this, but it is even more productive to state explicitly that the concepts of function, limit, derivative, and integral are at the core of calculus, and to explore in what way they are independent of dimension. The third mode is less tangible, and often neglected, but is the hallmark of a first-rate course. It's really about mathematical thinking. A student should progress from the perception that "mathematics is just a way to solve problems" to a true appreciation of the power of mathematics to find similarity and unity in seemingly different objects. Multivariable calculus has a dual geometric and algebraic approach that develops an appreciation for the uses and power of mathematical abstraction, without a heavy emphasis on the idea of "proof." I think multivariable calculus is the best possible foundation for a subsequent study of modern mathematics. It provides wonderful motivation for the study of linear algebra, differential geometry, topology, integration theory, and partial differential equations. And for a student in any other field that uses mathematics, the subject provides an opportunity to demonstrate how mathematics builds on known structures to explore unknown areas.

Constance McMillan Elson is professor of mathematics at Ithaca College. She has conducted mathematical research at NASA; M.I.T.; University of Wisconsin, Madison; and the National Center for Atmospheric Research. Professor Elson has a strong interest in teaching and promotes the use of computer algebra systems and other interactive media, particularly in undergraduate courses beyond calculus. She received her B.S. in mathematics from Stanford University and her Ph.D. in mathematics from the University of California at San Diego.

CHAPTER 11

VECTORS IN TWO AND THREE DIMENSIONS

FIGURE 11.1
A vector

Distance is a scalar quantity, but a quantity with both magnitude (distance) *and* direction is a vector quantity.

Certain physical quantities have both *magnitude* and *direction*; examples are force, velocity, acceleration, and displacement of a moving particle. A convenient way to represent such quantities is with a directed line segment, such as the one in Figure 11.1. The length of the segment, to some scale, represents the magnitude of the quantity in question, and the direction is indicated by the inclination of the segment and by the arrowhead. Such a directed line segment is called a **vector**. For example, the vector in Figure 11.1 might represent a wind velocity of 20 mph blowing in a northeasterly direction. The length would then be taken as 20 units, to some convenient scale. Any quantity that has both magnitude and direction can be represented in this way and is therefore called a *vector quantity*.

Vector quantities are different from measures of area, mass, time, and distance, which can be adequately described by a single number. These are called *scalar quantities* (since they are measured according to some scale), and the numbers used to measure them are called **scalars**. For our purposes, then, scalars are just real numbers.

The subject now called vector analysis was developed in the latter part of the nineteenth century by the American physicist and mathematician Josiah Willard Gibbs (1839–1903) and the English engineer Oliver Heaviside (1850–192?) working independently. Many of the ideas came earlier, however, especially from the Irish mathematician William Rowan Hamilton (1805–1865) in his work on *quaternions*, and the Scottish physicist James Clerk Maxwell (1831–18?), who used some of Hamilton's ideas in his study of electromagnetic field theory. So the subject is strongly grounded in the physical sciences and engineering and is still an important tool in these fields. The applications have expanded greatly now, even to economics and some of the other social sciences. Moreover, vectors have contributed significantly to the continuing development of mathematics itself.

In early work with vectors, their geometric properties were dominant. As we shall see, though, the advantages of vectors can be realized fully only when their algebraic properties are used in conjunction with their geometric ones. In the next section we explore the geometric nature of vectors in the plane and then formulate vectors and their properties in algebraic terms.

11.1 VECTORS IN THE PLANE

FIGURE 11.2
Equivalent vectors

FIGURE 11.3

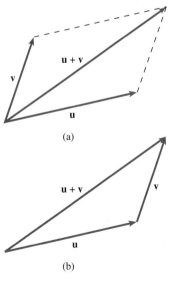

FIGURE 11.4

Geometric Vectors

Typically, boldface letters such as **u**, **v**, and **w** are used to designate vectors. Throughout this section we consider vectors that lie in a plane, although most of the results have natural extensions to three (or more) dimensions. Two vectors **u** and **v** are said to be **equivalent** if they have the same magnitude and direction, and in this case we write **u** = **v**. Three equivalent vectors are illustrated in Figure 11.2. We do not distinguish between equivalent vectors, so that in effect we can shift a vector from one location to another as long as its original magnitude and direction are retained. Because of this freedom of movement, we say that we are working within a system of *free* vectors.

Suppose, as in Figure 11.3, a vector extends from a point P to a point Q. When we wish to emphasize this fact we use the notation $\overrightarrow{PQ}$ to designate the vector. The point P is called the **initial point** and Q the **terminal point**. Sometimes we also use "tail" and "tip" instead of initial point and terminal point, respectively.

Vector Addition

Two nonparallel vectors are **added** according to the **parallelogram law**, illustrated in Figure 11.4(a). The vectors are drawn with a common initial point, and a parallelogram is constructed with **u** and **v** as adjacent sides. The vector **u** + **v** is then defined as the vector along the diagonal from the common initial point to the opposite vertex. An alternative method is to place the initial point of **v** at the terminal point of **u**. Then **u** + **v** is the vector shown in Figure 11.4(b), drawn from the initial point of **u** to the terminal point of **v**. You should convince yourself that the triangle in part (b) is just the lower half of the parallelogram in part (a). This second method is sometimes called the "tail to tip" method of adding. If **u** and **v** are parallel vectors, then the parallelogram of part (a) is degenerate. The tail to tip method still works, however.

From our definition of addition it is easy to see that

$$\mathbf{u} + \mathbf{v} = \mathbf{v} + \mathbf{u}$$

In other words, vector addition is commutative. It is also associative; that is,

$$\mathbf{u} + (\mathbf{v} + \mathbf{w}) = (\mathbf{u} + \mathbf{v}) + \mathbf{w}$$

You will be asked in the exercises to give a geometric argument for this property.

Vector addition is consistent with observed results. For example, if **u** and **v** represent forces acting on an object, then the net effect is **u** + **v**; that is, the two individual forces **u** and **v** could be replaced by the force **u** + **v**, and the effect would be the same. In this case we call **u** + **v** the **resultant** of **u** and **v**. Similarly, if **u** is a vector representing the indicated velocity of an airplane and **v** is the wind velocity vector, then the true velocity of the airplane relative to the ground is **u** + **v**.

It is convenient to introduce the notion of the **zero vector**, denoted by **0**, with a magnitude of 0 and assigned no direction. We may think of the zero vector as a single point. If **v** is a nonzero vector, then −**v** is the vector that has the same length as **v** but with a direction opposite to that of **v**. We now define

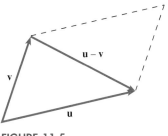

FIGURE 11.5

subtraction of vectors by

$$\mathbf{u} - \mathbf{v} = \mathbf{u} + (-\mathbf{v})$$

Thus, $\mathbf{u} - \mathbf{v}$ is the vector that when added to $\mathbf{v}$ gives $\mathbf{u}$. This definition is illustrated in Figure 11.5. Notice that when $\mathbf{u}$ and $\mathbf{v}$ are drawn with the same initial point, $\mathbf{u} - \mathbf{v}$ is the vector *from the tip of* $\mathbf{v}$ *to the tip of* $\mathbf{u}$. Notice also that when we construct the parallelogram with $\mathbf{u}$ and $\mathbf{v}$ as adjacent sides, $\mathbf{u} - \mathbf{v}$ is directed along the diagonal from the tip of $\mathbf{v}$ to the tip of $\mathbf{u}$, in contrast to $\mathbf{u} + \mathbf{v}$, which is directed along the other diagonal. From this definition we see that, as we would expect,

$$\mathbf{v} - \mathbf{v} = \mathbf{0}$$

Scalar Multiplication

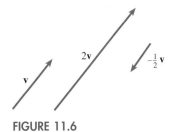

FIGURE 11.6

Vectors can be multiplied by scalars as follows. If $k > 0$ and $\mathbf{v}$ is a nonzero vector, then $k\mathbf{v}$ is a vector that has the same direction as $\mathbf{v}$ and magnitude k times the magnitude of $\mathbf{v}$. If $k < 0$, then $k\mathbf{v}$ has direction opposite to that of $\mathbf{v}$ and magnitude $|k|$ times the magnitude of $\mathbf{v}$. If $k = 0$, we define $k\mathbf{v}$ as the zero vector, and if $\mathbf{v} = \mathbf{0}$, then $k\mathbf{v} = \mathbf{0}$ for all scalars k. In Figure 11.6, we depict a vector $\mathbf{v}$, along with the vectors $2\mathbf{v}$ and $-\frac{1}{2}\mathbf{v}$.

Algebraic Vectors

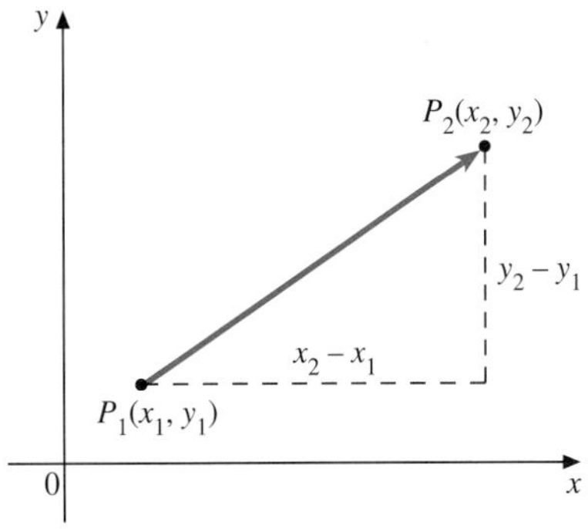

We can gain further insight into the properties of vectors by introducing a rectangular coordinate system. Suppose that a vector $\mathbf{v} = \overrightarrow{P_1 P_2}$, where the coordinates of P_1 and P_2 are (x_1, y_1) and (x_2, y_2), respectively. As shown in Figure 11.7, the horizontal displacement from P_1 to P_2 is $x_2 - x_1$ and the vertical displacement is $y_2 - y_1$. We call $x_2 - x_1$ the **horizontal component** (or x component) and $y_2 - y_1$ the **vertical component** (or y component) of $\mathbf{v}$. For example, if the coordinates of P_1 are $(3, 2)$ and those of P_2 are $(7, 5)$, then the horizontal component of $\mathbf{v}$ is $7 - 3 = 4$ and the vertical component is $5 - 2 = 3$. So every vector has a unique pair of components. Conversely, if we are given a pair of components, then these uniquely determine the collection of equivalent vectors that have these components. For example, given the x component 3 and y component 2, we can determine all vectors that have these components. The simplest of these is the one with initial point at the origin and terminal point at $(3, 2)$. Since we do not distinguish between equivalent vectors, we can in effect say that a vector is uniquely determined by its components.

This identification of a vector with its components enables us to look at vectors in a new way. We use the symbol $\langle a, b \rangle$ to indicate a vector with x component a and y component b, and we refer to this ordered pair of numbers as a vector. When we wish to distinguish between vectors as directed line segments and vectors as ordered pairs, we say the former is a *geometric vector* and the latter an *algebraic vector*. By the preceding discussion, given a geometric vector $\mathbf{v}$, we can determine the corresponding algebraic vector $\langle a, b \rangle$ and conversely. Because of this correspondence we write $\mathbf{v} = \langle a, b \rangle$. Any geometric vector corresponding to $\langle a, b \rangle$ is called a **geometric representative** of $\langle a, b \rangle$. The simplest geometric representative of a vector $\langle a, b \rangle$ is the vector from the origin to the point (a, b). We call this geometric representative the **position vector** of $\langle a, b \rangle$. (See Figure 11.8.)

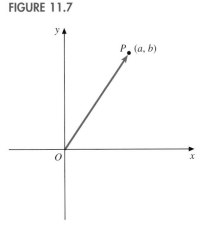

FIGURE 11.7

FIGURE 11.8
Position vector of P is $\overrightarrow{OP}$, O the origin $(0, 0)$

832   Chapter 11   Vectors in Two and Three Dimensions

In summary, we have the following:

> 1. If $P_1 = (x_1, y_1)$ and $P_2 = (x_2, y_2)$, then $\overrightarrow{P_1 P_2} = \langle x_2 - x_1, y_2 - y_1 \rangle$.
> 2. If $\mathbf{v} = \langle a, b \rangle$ and the initial point of $\mathbf{v}$ is $P_1 = (x_1, y_1)$, then the terminal point is $P_2 = (x_1 + a, y_1 + b)$. In particular, if P_1 is the origin, then $P_2 = (a, b)$, and $\overrightarrow{P_1 P_2}$ is the position vector of P_2.

EXAMPLE 11.1

(a) Express the vector $\overrightarrow{AB}$ in algebraic form, where $A = (-1, 2)$ and $B = (3, -4)$. Draw the geometric vector.

(b) Draw the geometric representative of the vector $\langle -2, 3 \rangle$ whose initial point is $(4, -1)$. What is its terminal point?

Solution

(a) $\overrightarrow{AB} = \langle 3 - (-1), -4 - 2 \rangle = \langle 4, -6 \rangle$
We show the vector geometrically in Figure 11.9.

(b) Beginning at $(4, -1)$, we go 2 units to the left and 3 units up, giving the terminal point $(2, 2)$, as shown in Figure 11.10. ∎

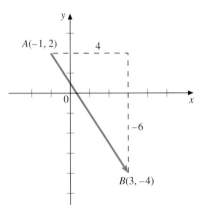

FIGURE 11.9

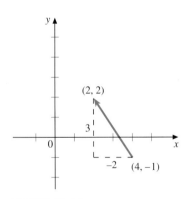

FIGURE 11.10

Properties of Vectors

We can now state properties of vectors in terms of their algebraic representations. First, we give a definition of addition and multiplication by a scalar. You will be asked in the exercises to show that these are consistent with the corresponding geometric definitions.

Definition 11.1
Two-Dimensional Vector Space

For ordered pairs $\langle a, b \rangle$ and $\langle c, d \rangle$ of real numbers,

1. $\langle a, b \rangle = \langle c, d \rangle$ if and only if $a = c$ and $b = d$
2. $\langle a, b \rangle + \langle c, d \rangle = \langle a + c, b + d \rangle$
3. $k\langle a, b \rangle = \langle ka, kb \rangle$ for any scalar k

The set of all such ordered pairs of real numbers with the definitions of equality, addition, and multiplication by a scalar given by Equations 1, 2, and 3 is called a **vector space of dimension two**, and each ordered pair in this set is called a **two-dimensional vector**.

THEOREM 11.1

If $\mathbf{u} = \langle a, b \rangle$, $\mathbf{v} = \langle c, d \rangle$, and $\mathbf{w} = \langle e, f \rangle$ are arbitrary vectors, then

1. $\mathbf{u} + \mathbf{v} = \mathbf{v} + \mathbf{u}$
2. $\mathbf{u} + (\mathbf{v} + \mathbf{w}) = (\mathbf{u} + \mathbf{v}) + \mathbf{w}$

and for any scalars k and l,

3. $k(\mathbf{u} + \mathbf{v}) = k\mathbf{u} + k\mathbf{v}$
4. $(k + l)\mathbf{u} = k\mathbf{u} + l\mathbf{u}$
5. $k(l\mathbf{u}) = (kl)\mathbf{u}$

The proof will be called for in the exercises.

Definition 11.2
The Zero Vector, the Negative of a Vector, and Subtraction

1. The element $\langle 0, 0 \rangle$ is called the **zero vector** and is denoted by **0**.
2. If $\mathbf{u} = \langle a, b \rangle$ is any vector, then $-\mathbf{u} = \langle -a, -b \rangle$.
3. If $\mathbf{u} = \langle a, b \rangle$ and $\mathbf{v} = \langle c, d \rangle$ are arbitrary vectors, then

$$\mathbf{u} - \mathbf{v} = \mathbf{u} + (-\mathbf{v})$$

THEOREM 11.2

For any vector $\mathbf{u} = \langle a, b \rangle$,

1. $\mathbf{u} + \mathbf{0} = \mathbf{u}$
2. $\mathbf{u} + (-\mathbf{u}) = \mathbf{0}$
3. $1\mathbf{u} = \mathbf{u}$
4. $(-1)\mathbf{u} = -\mathbf{u}$
5. $0\mathbf{u} = \mathbf{0}$
6. $k\mathbf{0} = \mathbf{0}$ for all scalars k.

We again call for the proof in the exercises.

REMARK

■ Because the addition of **0** to a vector leaves that vector unchanged, **0** is the *additive identity*. Also, because $-\mathbf{u}$ added to $\mathbf{u}$ gives **0**, $-\mathbf{u}$ is the *additive inverse* of $\mathbf{u}$.

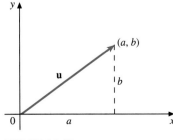

FIGURE 11.11
Length of $\mathbf{u} = \sqrt{a^2 + b^2}$

The Magnitude of a Vector

For a vector $\mathbf{u} = \langle a, b \rangle$, its geometric counterpart can be represented by the position vector of the point (a, b), as in Figure 11.11. The length of this geometric vector, by the Pythagorean Theorem, is $\sqrt{a^2 + b^2}$, which leads to the following definition.

Definition 11.3
The Magnitude of a Vector

Let $\mathbf{u} = \langle a, b \rangle$ be any vector. The **magnitude** (or *length*) of $\mathbf{u}$, denoted by $|\mathbf{u}|$, is defined by

$$|\mathbf{u}| = \sqrt{a^2 + b^2} \qquad (11.1)$$

REMARK

■ It is reasonable to use the same symbol to designate the magnitude of a vector as the one used to denote the absolute value of a real number. If x is a real number, then $|x|$ can be interpreted geometrically as the distance from 0 to x on a number line. Similarly, when a vector $\mathbf{u}$ is interpreted geometrically with initial point at the origin, $|\mathbf{u}|$ is the distance from the origin to the terminal point of $\mathbf{u}$.

The components of a vector determine its magnitude, as Definition 11.3 shows. They also determine the *direction* of the vector. For if $\mathbf{u} = \langle a, b \rangle$, the direction of $\mathbf{u}$ is the same as that of the position vector shown in Figure 11.11, directed from $(0, 0)$ toward the point (a, b).

THEOREM 11.3

Let $\mathbf{u} = \langle a, b \rangle$ and $\mathbf{v} = \langle c, d \rangle$ be any two vectors, and let k be any scalar. Then

1. $|\mathbf{u}| \geq 0$ and $|\mathbf{u}| = 0$ if and only if $\mathbf{u} = \mathbf{0}$
2. $|-\mathbf{u}| = |\mathbf{u}|$
3. $|k\mathbf{u}| = |k|\,|\mathbf{u}|$
4. $|\mathbf{u} + \mathbf{v}| \leq |\mathbf{u}| + |\mathbf{v}|$ Triangle Inequality

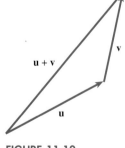

FIGURE 11.12

Proofs of Properties 1, 2, and 3 are called for in the exercises. We will give an algebraic proof of Property 4 in the next section, but its validity is evident geometrically, as an examination of Figure 11.12 shows. The inequality simply reflects the fact that the length of one side of a triangle is less than the sum of the lengths of the other two sides (which explains the name "triangle inequality"). This inequality also reflects the fact that the shortest distance between two points is a straight line. You should think about the circumstances in which equality occurs.

Unit Vectors and Basis Vectors

A **unit vector** is a vector with magnitude equal to 1. For example, $\langle 1, 0 \rangle$, $\langle 0, 1 \rangle$, $\langle \frac{3}{5}, \frac{4}{5} \rangle$, and $\langle \frac{\sqrt{2}}{2}, -\frac{\sqrt{2}}{2} \rangle$ are unit vectors. For any nonzero vector $\mathbf{v}$, the vector

$$\left(\frac{1}{|\mathbf{v}|}\right)\mathbf{v} = \frac{\mathbf{v}}{|\mathbf{v}|}$$

is a unit vector in the same direction as **v**. The fact that its length is 1 can be seen by

$$\left|\frac{\mathbf{v}}{|\mathbf{v}|}\right| = \left(\frac{1}{|\mathbf{v}|}\right)|\mathbf{v}| = 1$$

In words, to make a nonzero vector **v** into a unit vector, divide **v** by its own length.

EXAMPLE 11.2 Find a unit vector in the direction from $P_1(3, 2)$ toward $P_2(5, -2)$.

Solution Let $\mathbf{v} = \overrightarrow{P_1 P_2}$. Then $\mathbf{v} = \langle 5-3, -2-2 \rangle = \langle 2, -4 \rangle$. So the desired unit vector is

$$\frac{\mathbf{v}}{|\mathbf{v}|} = \frac{\langle 2, -4 \rangle}{\sqrt{20}} = \frac{\langle 2, -4 \rangle}{2\sqrt{5}} = \left\langle \frac{1}{\sqrt{5}}, -\frac{2}{\sqrt{5}} \right\rangle \qquad \blacksquare$$

If **v** is a nonzero vector, we know that $\mathbf{v}/|\mathbf{v}|$ is a unit vector, so for any scalar k, $k\mathbf{v}/|\mathbf{v}|$ is a vector of magnitude $|k|$ that has the same direction as **v** when $k > 0$ and the opposite direction when $k < 0$. This fact enables us to find a vector of a specified length in the direction of a given vector or in the opposite direction.

EXAMPLE 11.3 Find a vector in the same direction as $\langle -1, 2 \rangle$ that has magnitude 10.

Solution Let $\mathbf{v} = \langle -1, 2 \rangle$. The desired vector is

$$10 \frac{\mathbf{v}}{|\mathbf{v}|} = 10 \left\langle -\frac{1}{\sqrt{5}}, \frac{2}{\sqrt{5}} \right\rangle = \langle -2\sqrt{5}, 4\sqrt{5} \rangle \qquad \blacksquare$$

The two unit vectors $\langle 1, 0 \rangle$ and $\langle 0, 1 \rangle$ are of particular importance. They are given the special names

$$\mathbf{i} = \langle 1, 0 \rangle \qquad \text{and} \qquad \mathbf{j} = \langle 0, 1 \rangle$$

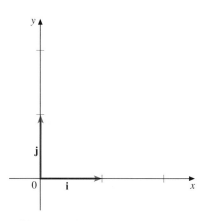

FIGURE 11.13

So **i** is a unit vector in the positive x direction and **j** is a unit vector in the positive y direction. (See Figure 11.13.)

Suppose $\mathbf{u} = \langle a, b \rangle$ is an arbitrary vector. Then we can write

$$\mathbf{u} = \langle a, 0 \rangle + \langle 0, b \rangle = a\langle 1, 0 \rangle + b\langle 0, 1 \rangle$$

or

$$\langle a, b \rangle = a\mathbf{i} + b\mathbf{j} \qquad (11.2)$$

The expression $a\mathbf{i} + b\mathbf{j}$ is called a **linear combination** of **i** and **j**. So every two-dimensional vector is uniquely expressible as a linear combination of **i** and **j**. Because of this property the vectors **i** and **j** constitute what is called a **basis** for two-dimensional vector space. As Exercise 49 in Exercise Set 11.1 shows, any two nonzero vectors that are not parallel also form a basis for this vector space, but **i** and **j** provide the simplest basis.

EXAMPLE 11.4 Let $P_1 = (7, -4)$ and $P_2 = (-3, 1)$. Express $\overrightarrow{P_1 P_2}$ as a linear combination of **i** and **j**.

Solution $\overrightarrow{P_1 P_2} = \langle -3 - 7, 1 + 4 \rangle = \langle -10, 5 \rangle = -10\mathbf{i} + 5\mathbf{j} \qquad \blacksquare$

Because of Equation 11.2, we can use the representation $a\mathbf{i} + b\mathbf{j}$ as an alternative to $\langle a, b \rangle$. In the future both representations will be used. Thus, for example, we may speak of the vector $2\mathbf{i} - 3\mathbf{j}$, which we will understand to mean the same thing as the vector $\langle 2, -3 \rangle$. Using this alternative representation, we have, in particular,

$$|a\mathbf{i} + b\mathbf{j}| = \sqrt{a^2 + b^2}$$

Parallel Vectors

In keeping with the geometric relationship between $\mathbf{u}$ and $k\mathbf{u}$, we have the following definition.

Definition 11.4
Parallel Vectors

Two nonzero vectors $\mathbf{u}$ and $\mathbf{v}$ are said to be **parallel** if there exists a nonzero scalar k such that $\mathbf{v} = k\mathbf{u}$. We also say that $\mathbf{0}$ is parallel to every vector.

Exercise Set 11.1

Exercises 1–12 refer to the vectors $\mathbf{u} = \langle -2, 3 \rangle$, $\mathbf{v} = \langle 4, 2 \rangle$, and $\mathbf{w} = \langle -1, -2 \rangle$. In each case, give the result as an algebraic vector. Also give a geometric construction illustrating the given operations.

1. $\mathbf{u} + \mathbf{v}$
2. $\mathbf{v} + \mathbf{w}$
3. $\mathbf{u} + \mathbf{w}$
4. $\mathbf{u} + \mathbf{v} + \mathbf{w}$
5. $\mathbf{u} - \mathbf{v}$
6. $\mathbf{v} - \mathbf{w}$
7. $\mathbf{w} - \mathbf{u}$
8. $2\mathbf{u} + \frac{1}{2}\mathbf{v}$
9. $\mathbf{u} - 2\mathbf{w}$
10. $\mathbf{u} + \frac{3}{2}\mathbf{v} - \mathbf{w}$
11. $-2\mathbf{u} + \mathbf{v} - 3\mathbf{w}$
12. $2\mathbf{u} - \frac{1}{2}\mathbf{v} + 3\mathbf{w}$

In Exercises 13–16, find the algebraic vector corresponding to $\overrightarrow{P_1 P_2}$.

13. $P_1 = (3, 4)$, $P_2 = (-1, 2)$
14. $P_1 = (-4, -2)$, $P_2 = (3, -1)$
15. $P_1 = (0, 4)$, $P_2 = (-3, 0)$
16. $P_1 = (7, -3)$, $P_2 = (-1, -8)$

In Exercises 17–20, draw the vector $\overrightarrow{P_1 P_2}$ that corresponds to the given algebraic vector and the given initial point P_1. Determine the coordinates of P_2.

17. $\langle 3, -2 \rangle$; $P_1 = (0, 0)$
18. $\langle -2, 4 \rangle$; $P_1 = (1, 2)$
19. $\langle 0, 3 \rangle$; $P_1 = (-2, -3)$
20. $\langle -3, -4 \rangle$; $P_1 = (4, 2)$

In Exercises 21–26, find $|\mathbf{v}|$.

21. $\mathbf{v} = \langle 3, 4 \rangle$
22. $\mathbf{v} = \langle -8, 6 \rangle$
23. $\mathbf{v} = 8\mathbf{i} + 15\mathbf{j}$
24. $\mathbf{v} = -12\mathbf{i} - 5\mathbf{j}$
25. $\mathbf{v} = 2\mathbf{i} + \mathbf{j}$
26. $\mathbf{v} = 4\mathbf{i} - 6\mathbf{j}$

27. If $\mathbf{u} = -3\mathbf{i} + \mathbf{j}$ and $\mathbf{v} = 2\mathbf{i} - 3\mathbf{j}$, find each of the following:
    (a) $|\mathbf{u} + \mathbf{v}|$
    (b) $|\mathbf{u} - \mathbf{v}|$
    (c) $|2\mathbf{u} + 3\mathbf{v}|$
    (d) $|3\mathbf{u} - 2\mathbf{v}|$

In Exercises 28–31, find a unit vector in the direction of $\mathbf{v}$.

28. $\mathbf{v} = 3\mathbf{i} - 4\mathbf{j}$
29. $\mathbf{v} = 5\mathbf{i} + 12\mathbf{j}$
30. $\mathbf{v} = \langle -4, 8 \rangle$
31. $\mathbf{v} = \langle -2, -3 \rangle$

In Exercises 32–37, find a vector $\mathbf{w}$ that is in the direction of the given vector $\mathbf{v}$, with the specified magnitude.

32. $\mathbf{v} = \langle -4, 3 \rangle$; $|\mathbf{w}| = 10$
33. $\mathbf{v} = \langle 2, -4 \rangle$; $|\mathbf{w}| = 10$
34. $\mathbf{v} = \mathbf{i} + \mathbf{j}$; $|\mathbf{w}| = 2$
35. $\mathbf{v} = 6\mathbf{i} + 8\mathbf{j}$; $|\mathbf{w}| = 4$
36. $\mathbf{v} = 7\mathbf{i} - 24\mathbf{j}$; $|\mathbf{w}| = 5$
37. $\mathbf{v} = 3\mathbf{i} - 6\mathbf{j}$; $|\mathbf{w}| = 15$

38. In the accompanying figure, $\mathbf{F}_1$ and $\mathbf{F}_2$ are forces acting on the object as shown. If $|\mathbf{F}_1| = 80$ lb, $|\mathbf{F}_2| = 60$ lb, $\alpha = 25°$, and $\beta = 115°$, find the magnitude and direction of the resultant both geometrically and by using trigonometry. (*Hint:* Use the Pythagorean Theorem.)

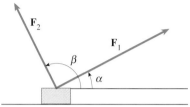

39. Repeat Exercise 38 if $|\mathbf{F}_1| = 20$ N, $|\mathbf{F}_2| = 30$ N, $\alpha = 10°$, and $\beta = 70°$. (*Hint:* For the trigonometric solution use the law of cosines.)

40. In air navigation, direction is given by the angle measured clockwise from north (this angle is called the *heading*). If an airplane is flying at an indicated heading of 120° at a speed of 300 kph, and a wind of 50 kph is blowing from 210°, find the actual speed and direction of the airplane (relative to the ground). Do this geometrically and also by using trigonometry.

41. Repeat Exercise 40 for an airplane flying at an indicated heading of 230° at 260 kph with a 60 kph wind blowing from 110°.

42. Show by a geometric argument that vector addition is associative; that is, $\mathbf{u} + (\mathbf{v} + \mathbf{w}) = (\mathbf{u} + \mathbf{v}) + \mathbf{w}$.

43. Show that Definition 11.1 is consistent with the corresponding geometric definitions of equivalence of vectors, addition of vectors, and multiplication of a vector by a scalar.

44. Prove Theorem 11.1.

45. Prove Theorem 11.2.

46. Prove Parts 1, 2, and 3 of Theorem 11.3.

47. Let $\mathbf{u} = \langle 1, 2 \rangle$, $\mathbf{v} = \langle -2, 1 \rangle$, and $\mathbf{w} = \langle 4, 3 \rangle$. Find scalars a and b such that $\mathbf{w} = a\mathbf{u} + b\mathbf{v}$.

48. Prove that if $\mathbf{u} = \langle u_1, u_2 \rangle$ and $\mathbf{v} = \langle v_1, v_2 \rangle$ are nonzero vectors, then they are parallel if and only if $u_1 v_2 - u_2 v_1 = 0$.

49. Prove that if $\mathbf{u} = \langle u_1, u_2 \rangle$ and $\mathbf{v} = \langle v_1, v_2 \rangle$ are nonzero vectors with $\mathbf{u} \neq k\mathbf{v}$ for all scalars k, and $\mathbf{w} = \langle w_1, w_2 \rangle$ is any other vector, then there exist scalars a and b such that $\mathbf{w} = a\mathbf{u} + b\mathbf{v}$.

50. Prove that if $\mathbf{u} = k\mathbf{v}$, with $k \geq 0$, then

$$|\mathbf{u} + \mathbf{v}| = |\mathbf{u}| + |\mathbf{v}|$$

Give a geometric argument to show that if $\mathbf{u}$ and $\mathbf{v}$ are not related in this way, $|\mathbf{u} + \mathbf{v}| < |\mathbf{u}| + |\mathbf{v}|$.

51. Use vectors to show that the diagonals of a parallelogram bisect each other.

52. Let P_1, P_2, P_3, and P_4 be any four points in the plane. Show both geometrically and algebraically that $\overrightarrow{P_1 P_2} + \overrightarrow{P_2 P_3} + \overrightarrow{P_3 P_4} + \overrightarrow{P_4 P_1} = \mathbf{0}$.

53. Use vectors to show that the line segments joining consecutive midpoints of the sides of an arbitrary quadrilateral form a parallelogram.

54. Three forces of magnitude $|\mathbf{F}_1| = 30$ lb, $|\mathbf{F}_2| = 45$ lb, and $|\mathbf{F}_3| = 56$ lb are acting on an object as shown in the figure. Find the magnitude of the resultant and its angle from $\mathbf{F}_1$ both geometrically and by using trigonometry. (See the hint for Exercise 39.)

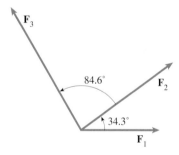

55. A pilot is to fly from A to B, 350 km due north of A, and then return to A. There is a wind blowing from 310° at 55 kph. If the average air speed of the plane is 210 kph, find the heading (see Exercise 40) the pilot should take on each part of the trip. What will be the total flying time?

11.2 THE DOT PRODUCT

In the previous section we considered addition and subtraction of vectors and multiplication by a scalar, but we have not yet considered the product of two vectors. We introduce one type of product now and later consider a second type, applicable to three-dimensional vectors only.

Definition 11.5
The Dot Product

Let $\mathbf{u} = \langle u_1, u_2 \rangle$ and $\mathbf{v} = \langle v_1, v_2 \rangle$ be any two vectors. Then the **dot product** of $\mathbf{u}$ and $\mathbf{v}$, written $\mathbf{u} \cdot \mathbf{v}$, is defined as

$$\mathbf{u} \cdot \mathbf{v} = u_1 v_1 + u_2 v_2 \tag{11.3}$$

REMARKS

- Observe that the dot product of two vectors is not a vector but is a scalar. For this reason the dot product is sometimes called the **scalar product**.
- If $\mathbf{u}$ and $\mathbf{v}$ are written in the form $\mathbf{u} = u_1 \mathbf{i} + u_2 \mathbf{j}$ and $\mathbf{v} = v_1 \mathbf{i} + v_2 \mathbf{j}$, then we have

$$\mathbf{u} \cdot \mathbf{v} = (u_1 \mathbf{i} + u_2 \mathbf{j}) \cdot (v_1 \mathbf{i} + v_2 \mathbf{j}) = u_1 v_1 + u_2 v_2$$

EXAMPLE 11.5   Let $\mathbf{u} = \langle 3, -2 \rangle$ and $\mathbf{v} = \langle -4, -5 \rangle$. Find $\mathbf{u} \cdot \mathbf{v}$.

Solution

$$\mathbf{u} \cdot \mathbf{v} = 3(-4) + (-2)(-5) = -12 + 10 = -2$$

Using the alternative way of writing $\mathbf{u}$ and $\mathbf{v}$ in terms of the basis vectors $\mathbf{i}$ and $\mathbf{j}$, we can also write

$$\mathbf{u} \cdot \mathbf{v} = (3\mathbf{i} - 2\mathbf{j}) \cdot (-4\mathbf{i} - 5\mathbf{j}) = -12 + 10 = -2 \qquad \blacksquare$$

Properties of the Dot Product

The dot product of vectors shares several properties with products of real numbers, as the following theorem shows.

THEOREM 11.4

If $\mathbf{u}$, $\mathbf{v}$, and $\mathbf{w}$ are vectors and k is a scalar, then

1. $\mathbf{u} \cdot \mathbf{v} = \mathbf{v} \cdot \mathbf{u}$ Commutative law
2. $\mathbf{u} \cdot (\mathbf{v} + \mathbf{w}) = \mathbf{u} \cdot \mathbf{v} + \mathbf{u} \cdot \mathbf{w}$ Distributive law
3. $k(\mathbf{u} \cdot \mathbf{v}) = (k\mathbf{u}) \cdot \mathbf{v} = \mathbf{u} \cdot (k\mathbf{v})$
4. $\mathbf{u} \cdot \mathbf{0} = 0$
5. $\mathbf{u} \cdot \mathbf{u} = |\mathbf{u}|^2$

Proof   We will verify Property 2 and leave the other properties for the exercises. Let $\mathbf{u} = \langle u_1, u_2 \rangle$, $\mathbf{v} = \langle v_1, v_2 \rangle$, and $\mathbf{w} = \langle w_1, w_2 \rangle$. We verify Property 2 by calculating the value of each side independently and showing that we get the same result. For the left-hand side, we have

$$\begin{aligned}
\mathbf{u} \cdot (\mathbf{v} + \mathbf{w}) &= \langle u_1, u_2 \rangle \cdot [\langle v_1, v_2 \rangle + \langle w_1, w_2 \rangle] \\
&= \langle u_1, u_2 \rangle \cdot \langle v_1 + w_1, v_2 + w_2 \rangle && \text{By Definition 11.1} \\
&= u_1(v_1 + w_1) + u_2(v_2 + w_2) && \text{By Definition 11.5} \\
&= (u_1 v_1 + u_1 w_1) + (u_2 v_2 + u_2 w_2) && \text{By the distributive law of real numbers}
\end{aligned}$$

For the right-hand side, we have

$$\mathbf{u} \cdot \mathbf{v} + \mathbf{u} \cdot \mathbf{w} = \langle u_1, u_2 \rangle \cdot \langle v_1, v_2 \rangle + \langle u_1, u_2 \rangle \cdot \langle w_1, w_2 \rangle$$
$$= (u_1 v_1 + u_2 v_2) + (u_1 w_1 + u_2 w_2) \quad \text{By Definition 11.5}$$
$$= (u_1 v_1 + u_1 w_1) + (u_2 v_2 + u_2 w_2) \quad \text{By commutativity and associativity of real numbers}$$

The equality of the results proves Property 2. ∎

The Angle Between Two Vectors

An important property of the dot product has to do with the angle between two vectors, which we now define.

Definition 11.6
Angle Between Vectors

The **angle** between two nonzero vectors $\mathbf{u}$ and $\mathbf{v}$ is the smallest positive angle between geometric representatives of $\mathbf{u}$ and $\mathbf{v}$ that have the same initial point.

If we denote the angle between $\mathbf{u}$ and $\mathbf{v}$ as θ, it follows that $0 \leq \theta \leq \pi$. The angle is 0 if $\mathbf{u}$ and $\mathbf{v}$ are in the same direction, and it is π if they are in opposite directions. Figure 11.14 illustrates various possibilities for θ.

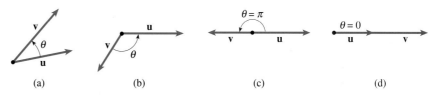

FIGURE 11.14

THEOREM 11.5

If θ is the angle between nonzero vectors $\mathbf{u}$ and $\mathbf{v}$, then

$$\mathbf{u} \cdot \mathbf{v} = |\mathbf{u}||\mathbf{v}| \cos \theta \tag{11.4}$$

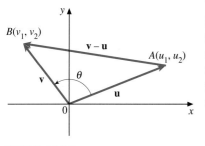

FIGURE 11.15

Proof Assume first that $\mathbf{u}$ and $\mathbf{v}$ are not parallel, and choose geometric representatives of these vectors such that each has its initial point at the origin, as in Figure 11.15. If $\mathbf{u}$ and $\mathbf{v}$ have components given by $\langle u_1, u_2 \rangle$ and $\langle v_1, v_2 \rangle$, respectively, then it follows that the terminal point A of $\mathbf{u}$ is (u_1, u_2) and the terminal point B of $\mathbf{v}$ is (v_1, v_2). The vector $\overrightarrow{AB} = \mathbf{v} - \mathbf{u}$, and so by the Law of Cosines (see the margin note on next page) we have

$$|\mathbf{v} - \mathbf{u}|^2 = |\mathbf{u}|^2 + |\mathbf{v}|^2 - 2|\mathbf{u}||\mathbf{v}| \cos \theta$$

or,

$$(v_1 - u_1)^2 + (v_2 - u_2)^2 = u_1^2 + u_2^2 + v_1^2 + v_2^2 - 2|\mathbf{u}||\mathbf{v}| \cos \theta$$

The Law of Cosines states that in a triangle with sides a, b, and c, if C is the angle opposite side c, then

$$c^2 = a^2 + b^2 - 2ab\cos C$$

After expanding and collecting terms, we get

$$|\mathbf{u}|\,|\mathbf{v}|\cos\theta = u_1v_1 + u_2v_2$$
$$= \mathbf{u}\cdot\mathbf{v}$$

If $\mathbf{u}$ and $\mathbf{v}$ are parallel, then $\mathbf{v} = k\mathbf{u}$, and $\theta = 0$ or $\theta = \pi$, according to whether $k > 0$ or $k < 0$.

For $k > 0$, we have $\cos\theta = \cos 0 = 1$, and

$$|\mathbf{u}||\mathbf{v}|\cos\theta = |\mathbf{u}||k\mathbf{u}|(1) = k|\mathbf{u}|^2 = k(\mathbf{u}\cdot\mathbf{u}) = \mathbf{u}\cdot(k\mathbf{u}) = \mathbf{u}\cdot\mathbf{v}$$

so the result is true in this case. For $k < 0$, $\cos\theta = \cos\pi = -1$, and

$$|\mathbf{u}||\mathbf{v}|\cos\theta = |\mathbf{u}||k\mathbf{u}|(-1) = -|k||\mathbf{u}|^2 = k(\mathbf{u}\cdot\mathbf{u}) = \mathbf{u}\cdot(k\mathbf{u}) = \mathbf{u}\cdot\mathbf{v}$$

Here we used the fact that since $k < 0$, $-|k| = -(-k) = k$. The result is therefore true in all cases. ∎

From Equation 11.4, if $\mathbf{u}$ and $\mathbf{v}$ are nonzero, we have

$$\cos\theta = \frac{\mathbf{u}\cdot\mathbf{v}}{|\mathbf{u}||\mathbf{v}|} \tag{11.5}$$

In the next example, we use Equation 11.5 to find the angle between two vectors.

EXAMPLE 11.6 Find the angle between the vectors $\mathbf{u} = \langle 3, -4\rangle$ and $\mathbf{v} = \langle 1, 7\rangle$. Draw the vectors with initial point at the origin.

Solution By Equation 11.5,

$$\cos\theta = \frac{\mathbf{u}\cdot\mathbf{v}}{|\mathbf{u}|\,|\mathbf{v}|} = \frac{3(1) + (-4)(7)}{\sqrt{9+16}\sqrt{1+49}} = \frac{-25}{\sqrt{25}\sqrt{50}} = \frac{-25}{5(5\sqrt{2})} = -\frac{1}{\sqrt{2}}$$

So $\theta = \frac{3\pi}{4}$. We show the vectors in Figure 11.16. ∎

EXAMPLE 11.7 Find the angle between $2\mathbf{i} - 4\mathbf{j}$ and $\mathbf{i} + \mathbf{j}$.

Solution

$$\cos\theta = \frac{(2\mathbf{i}-4\mathbf{j})\cdot(\mathbf{i}+\mathbf{j})}{|2\mathbf{i}-4\mathbf{j}|\,|\mathbf{i}+\mathbf{j}|} = \frac{2-4}{\sqrt{20}\sqrt{2}} = \frac{-2}{2\sqrt{10}} = \frac{-1}{\sqrt{10}}$$

Using a calculator, we find $\theta \approx 1.65$ radians. ∎

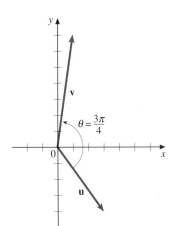

FIGURE 11.16

The following corollary is an immediate consequence of Equation 11.5.

COROLLARY 11.5a

The Cauchy-Schwarz Inequality

For any two vectors $\mathbf{u}$ and $\mathbf{v}$,

$$|\mathbf{u}\cdot\mathbf{v}| \leq |\mathbf{u}|\,|\mathbf{v}| \tag{11.6}$$

Proof The result is trivial if either $\mathbf{u}$ or $\mathbf{v}$ is the zero vector, since both sides of Equation 11.6 are 0. For nonzero vectors $\mathbf{u}$ and $\mathbf{v}$, with angle θ between them, we have, by Equation 11.5,

$$\frac{|\mathbf{u}\cdot\mathbf{v}|}{|\mathbf{u}|\,|\mathbf{v}|} = |\cos\theta| \leq 1$$

Thus, $|\mathbf{u} \cdot \mathbf{v}| \leq |\mathbf{u}| \, |\mathbf{v}|$. ∎

This corollary enables us to give an algebraic proof of the Triangle Inequality, as follows.

$$\begin{aligned}
|\mathbf{u} + \mathbf{v}|^2 &= (\mathbf{u} + \mathbf{v}) \cdot (\mathbf{u} + \mathbf{v}) \\
&= \mathbf{u} \cdot \mathbf{u} + 2\mathbf{u} \cdot \mathbf{v} + \mathbf{v} \cdot \mathbf{v} = |\mathbf{u}|^2 + 2\mathbf{u} \cdot \mathbf{v} + |\mathbf{v}|^2 \\
&\leq |\mathbf{u}|^2 + 2|\mathbf{u} \cdot \mathbf{v}| + |\mathbf{v}|^2 \leq |\mathbf{u}|^2 + 2|\mathbf{u}| \, |\mathbf{v}| + |\mathbf{v}|^2 \quad \text{By the Cauchy-Schwarz Inequality} \\
&= (|\mathbf{u}| + |\mathbf{v}|)^2
\end{aligned}$$

Now we take square roots to get the desired result:

$$|\mathbf{u} + \mathbf{v}| \leq |\mathbf{u}| + |\mathbf{v}|$$

Orthogonal Vectors

If the angle between two nonzero vectors is $\frac{\pi}{2}$, the vectors are said to be **orthogonal**. So geometric representatives of orthogonal vectors are perpendicular to each other. Since $\cos \frac{\pi}{2} = 0$, it follows from Theorem 11.5 that if $\mathbf{u}$ and $\mathbf{v}$ are orthogonal, then $\mathbf{u} \cdot \mathbf{v} = 0$. Conversely, if $\mathbf{u} \cdot \mathbf{v} = 0$, then $\cos \theta = 0$ by Equation 11.5, and so $\theta = \frac{\pi}{2}$. Thus, $\mathbf{u}$ and $\mathbf{v}$ are orthogonal. If either $\mathbf{u}$ or $\mathbf{v}$ is the zero vector, then $\mathbf{u} \cdot \mathbf{v} = 0$, and it is convenient in this case, too, to call $\mathbf{u}$ and $\mathbf{v}$ orthogonal; that is, we agree to say that $\mathbf{0}$ is orthogonal to every vector. We therefore have the following additional corollary to Theorem 11.5.

COROLLARY 11.5b   Two vectors $\mathbf{u}$ and $\mathbf{v}$ are orthogonal if and only if $\mathbf{u} \cdot \mathbf{v} = 0$.

EXAMPLE 11.8   Show that the vectors $\mathbf{u} = \langle 6, -4 \rangle$ and $\mathbf{v} = \langle -2, -3 \rangle$ are orthogonal.

Solution   Since $\mathbf{u} \cdot \mathbf{v} = 6(-2) + (-4)(-3) = -12 + 12 = 0$, by Corollary 11.5b, $\mathbf{u}$ and $\mathbf{v}$ are orthogonal. We show the vectors $\mathbf{u}$ and $\mathbf{v}$ as position vectors of the points $(6, -4)$ and $(-2, -3)$ in Figure 11.17. ∎

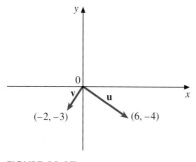

FIGURE 11.17

EXAMPLE 11.9   Find x so that the vectors $\langle x, 2 \rangle$ and $\langle 1 - x, 3 \rangle$ are orthogonal.

Solution   Let $\mathbf{u} = \langle x, 2 \rangle$ and $\mathbf{v} = \langle 1 - x, 3 \rangle$. Then

$$\begin{aligned}
\mathbf{u} \cdot \mathbf{v} &= \langle x, 2 \rangle \cdot \langle 1 - x, 3 \rangle = x - x^2 + 6 = 6 + x - x^2 \\
&= (3 - x)(2 + x)
\end{aligned}$$

and so $\mathbf{u} \cdot \mathbf{v} = 0$ if $x = 3$ or $x = -2$. Either value of x causes $\mathbf{u}$ and $\mathbf{v}$ to be orthogonal. For $x = 3$ we get $\mathbf{u} = \langle 3, 2 \rangle$ and $\mathbf{v} = \langle -2, 3 \rangle$, and for $x = -2$ we get $\mathbf{u} = \langle -2, 2 \rangle$ and $\mathbf{v} = \langle 3, 3 \rangle$. ∎

The Component of a Vector Along Another Vector

When we write $\mathbf{u} = a\mathbf{i} + b\mathbf{j}$, the numbers a and b are the components of $\mathbf{u}$ in the directions of $\mathbf{i}$ (horizontal) and $\mathbf{j}$ (vertical), respectively. (See Figure 11.18.) Sometimes it is useful to find the component of a vector in a direction other than horizontal and vertical. To understand what we mean, let $\mathbf{u}$ be any vector and suppose we want to find the displacement of $\mathbf{u}$ in the direction of some nonzero vector $\mathbf{v}$, as we show in Figure 11.19. We use geometric representatives $\overrightarrow{OP}$ and $\overrightarrow{OQ}$ of $\mathbf{u}$ and $\mathbf{v}$, respectively, and designate by P' the foot of the perpendicular from P to the line joining O and Q. Let θ be the angle between $\mathbf{u}$ and $\mathbf{v}$. Then we define the **component of u along v**, designated $\text{Comp}_{\mathbf{v}} \mathbf{u}$, by

$$\text{Comp}_{\mathbf{v}} \mathbf{u} = |\mathbf{u}| \cos \theta \tag{11.7}$$

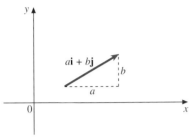

FIGURE 11.18

If $0 \leq \theta \leq \frac{\pi}{2}$, then $\text{Comp}_{\mathbf{v}} \mathbf{u} \geq 0$, as in Figure 11.19(a), whereas if $\frac{\pi}{2} < \theta \leq \pi$, $\text{Comp}_{\mathbf{v}} \mathbf{u} < 0$, as in Figure 11.19(b). If $\mathbf{u} \neq \mathbf{0}$, by Equation 11.5, we have $\cos \theta = \frac{\mathbf{u} \cdot \mathbf{v}}{|\mathbf{u}||\mathbf{v}|}$, so that Equation 11.7 becomes

$$\text{Comp}_{\mathbf{v}} \mathbf{u} = \frac{\mathbf{u} \cdot \mathbf{v}}{|\mathbf{v}|} \tag{11.8}$$

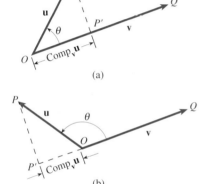

FIGURE 11.19

This result is valid also for $\mathbf{u} = \mathbf{0}$. It is easy to show (see Exercise 22 in Exercise Set 11.2) that if $\mathbf{u} = a\mathbf{i} + b\mathbf{j}$ and $\mathbf{v}$ is directed along the positive x-axis, then $\text{Comp}_{\mathbf{v}} \mathbf{u} = a$. Similarly, if $\mathbf{v}$ is directed along the positive y-axis, then $\text{Comp}_{\mathbf{v}} \mathbf{u} = b$. So our definition generalizes horizontal and vertical components.

Work

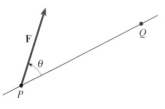

FIGURE 11.20

As an application of this concept, consider the work done by a constant force $\mathbf{F}$ in moving a particle along a straight line from P to Q. If $\mathbf{F}$ acts in the direction of $\overrightarrow{PQ}$, then since work equals force times distance, $\text{work} = |\mathbf{F}||\overrightarrow{PQ}|$. Suppose, however, that $\mathbf{F}$ acts at some fixed angle θ with $\overrightarrow{PQ}$, as in Figure 11.20. Then it is natural to define work as the component of $\mathbf{F}$ along $\overrightarrow{PQ}$ times the distance; that is,

$$W = (\text{Comp}_{\overrightarrow{PQ}} \mathbf{F})|\overrightarrow{PQ}|$$

Substituting from Equation 11.8, we get

$$W = \mathbf{F} \cdot \overrightarrow{PQ} \tag{11.9}$$

Section 11.2 The Dot Product   843

EXAMPLE 11.10 The force $\mathbf{F} = 3\mathbf{i} + 5\mathbf{j}$ moves an object along the line segment from $(-1, 2)$ to $(3, 5)$. If the magnitude of $\mathbf{F}$ is in newtons and distance is measured in meters, find the work done by $\mathbf{F}$. (See Figure 11.21.)

Solution   Let $P = (-1, 2)$ and $Q = (3, 5)$. Then $\overrightarrow{PQ} = 4\mathbf{i} + 3\mathbf{j}$, and

$$W = \mathbf{F} \cdot \overrightarrow{PQ} = (3\mathbf{i} + 5\mathbf{j}) \cdot (4\mathbf{i} + 3\mathbf{j})$$
$$= 12 + 15 = 27 \text{ joules} \qquad \blacksquare$$

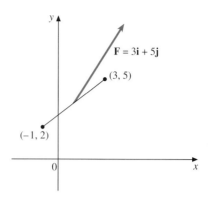

FIGURE 11.21

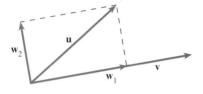

FIGURE 11.22

Vector Projections

When we write a vector $\mathbf{u} = \langle a, b \rangle$ in the form $\mathbf{u} = a\mathbf{i} + b\mathbf{j}$, we are in effect expressing $\mathbf{u}$ as the sum of two mutually perpendicular vectors, one acting horizontally and the other vertically. Sometimes it is desirable to express $\mathbf{u}$ as the sum of two mutually perpendicular vectors, one in a prescribed nonhorizontal direction and the other perpendicular to this direction. We show how to find these vectors geometrically in Figure 11.22. There we show the given vector $\mathbf{u}$ and a direction as determined by the vector $\mathbf{v}$. We construct a rectangle with sides $\mathbf{w}_1$ and $\mathbf{w}_2$ having $\mathbf{u}$ as a diagonal and one side along $\mathbf{v}$. Then $\mathbf{u} = \mathbf{w}_1 + \mathbf{w}_2$, as required. We call the vector $\mathbf{w}_1$ the **projection of u on v** and designate it by $\text{Proj}_\mathbf{v} \mathbf{u}$. The vector $\mathbf{w}_2$ is called the **projection of u orthogonal to v** and is designated by $\text{Proj}_\mathbf{v}^\perp \mathbf{u}$. So we always have, for any vector $\mathbf{u}$ and $\mathbf{v} \neq 0$,

$$\mathbf{u} = \text{Proj}_\mathbf{v} \mathbf{u} + \text{Proj}_\mathbf{v}^\perp \mathbf{u} \qquad (11.10)$$

To find algebraic representatives of these projections, observe that $\mathbf{w}_1$ can be obtained by multiplying the component of $\mathbf{u}$ along $\mathbf{v}$ by a unit vector in the direction of $\mathbf{v}$. Thus,

$$\text{Proj}_\mathbf{v} \mathbf{u} = (\text{Comp}_\mathbf{v} \mathbf{u}) \frac{\mathbf{v}}{|\mathbf{v}|} \qquad (11.11)$$

If we replace $\text{Comp}_\mathbf{v} \mathbf{u}$ by its value from Equation 11.8, we obtain

$$\text{Proj}_\mathbf{v} \mathbf{u} = \left(\frac{\mathbf{u} \cdot \mathbf{v}}{|\mathbf{v}|} \right) \frac{\mathbf{v}}{|\mathbf{v}|} = \left(\frac{\mathbf{u} \cdot \mathbf{v}}{|\mathbf{v}|^2} \right) \mathbf{v} \qquad (11.12)$$

and from Equation 11.9, we get

$$\text{Proj}_\mathbf{v}^\perp \mathbf{u} = \mathbf{u} - \text{Proj}_\mathbf{v} \mathbf{u} = \mathbf{u} - \left(\frac{\mathbf{u} \cdot \mathbf{v}}{|\mathbf{v}|^2}\right)\mathbf{v} \tag{11.13}$$

EXAMPLE 11.11 Let $\mathbf{u} = \langle 3, 4 \rangle$ and $\mathbf{v} = \langle -4, 8 \rangle$. Express $\mathbf{u}$ as the sum of two vectors, one parallel to $\mathbf{v}$ and the other perpendicular to $\mathbf{v}$. Show the results geometrically.

Solution Note that the desired vectors are $\text{Proj}_\mathbf{v}\mathbf{u}$ and $\text{Proj}_\mathbf{v}^\perp \mathbf{u}$. By Equation 11.12,

$$\text{Proj}_\mathbf{v}\mathbf{u} = \left(\frac{\mathbf{u} \cdot \mathbf{v}}{|\mathbf{v}|^2}\right)\mathbf{v} = \frac{\langle 3, 4 \rangle \cdot \langle -4, 8 \rangle}{|\langle -4, 8 \rangle|^2}\langle -4, 8 \rangle$$

$$= \frac{-12 + 32}{80}\langle -4, 8 \rangle = \frac{1}{4}\langle -4, 8 \rangle$$

$$= \langle -1, 2 \rangle$$

By Equation 11.13,

$$\text{Proj}_\mathbf{v}^\perp \mathbf{u} = \mathbf{u} - \text{Proj}_\mathbf{v}\mathbf{u} = \langle 3, 4 \rangle - \langle -1, 2 \rangle = \langle 4, 2 \rangle$$

We illustrate the results in Figure 11.23.

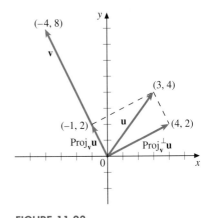

FIGURE 11.23

REMARK
- $\text{Proj}_\mathbf{v}\mathbf{u}$ and $\text{Proj}_\mathbf{v}^\perp \mathbf{u}$ are sometimes called **vector components** of $\mathbf{u}$ in the direction of $\mathbf{v}$ and orthogonal to $\mathbf{v}$, respectively, and when these are found, the vector $\mathbf{u}$ is said to have been *resolved into vector components* in these directions.

Exercise Set 11.2

In Exercises 1–4, find the dot product of $\mathbf{u}$ and $\mathbf{v}$.

1. $\mathbf{u} = \langle 4, 7 \rangle$, $\mathbf{v} = \langle -5, 2 \rangle$
2. $\mathbf{u} = \langle -3, -6 \rangle$, $\mathbf{v} = \langle 5, -2 \rangle$
3. $\mathbf{u} = 2\mathbf{i} - 3\mathbf{j}$, $\mathbf{v} = \mathbf{i} + 2\mathbf{j}$
4. $\mathbf{u} = -\mathbf{i} + 4\mathbf{j}$, $\mathbf{v} = 3\mathbf{i} + \mathbf{j}$

In Exercises 5–8, show that $\mathbf{u}$ and $\mathbf{v}$ are orthogonal.

5. $\mathbf{u} = \langle 4, -8 \rangle$, $\mathbf{v} = \langle -4, -2 \rangle$
6. $\mathbf{u} = \langle -10, 6 \rangle$, $\mathbf{v} = \langle 12, 20 \rangle$
7. $\mathbf{u} = 2\mathbf{i} - 3\mathbf{j}$, $\mathbf{v} = 9\mathbf{i} + 6\mathbf{j}$
8. $\mathbf{u} = -3\mathbf{i} + 4\mathbf{j}$, $\mathbf{v} = -12\mathbf{i} - 9\mathbf{j}$

In Exercises 9–15, find the cosine of the angle θ between $\mathbf{u}$ and $\mathbf{v}$.

9. $\mathbf{u} = \langle 4, -4 \rangle$, $\mathbf{v} = \langle 1, 7 \rangle$
10. $\mathbf{u} = \langle 1, -2 \rangle$, $\mathbf{v} = \langle -1, 1 \rangle$
11. $\mathbf{u} = -\mathbf{i} + 3\mathbf{j}$, $\mathbf{v} = -2\mathbf{i} - \mathbf{j}$
12. $\mathbf{u} = 4\mathbf{i} + 6\mathbf{j}$, $\mathbf{v} = 4\mathbf{i} - 2\mathbf{j}$
13. $\mathbf{u} = \langle 6, 8 \rangle$, $\mathbf{v} = \langle -3, 4 \rangle$
14. $\mathbf{u} = \langle 4, 8 \rangle$, $\mathbf{v} = \langle -1, 3 \rangle$; also find θ.
15. $\mathbf{u} = \mathbf{i} + \sqrt{3}\mathbf{j}$, $\mathbf{v} = 2\mathbf{i}$; also find θ.
16. Find all values of x so that $\mathbf{u} = \langle 3x, 1 - x \rangle$ and $\mathbf{v} = \langle x, -4 \rangle$ will be orthogonal.
17. Find all values of x so that the angle between $\mathbf{u} = \langle 4, -3 \rangle$ and $\mathbf{v} = \langle x, 1 \rangle$ will be $\frac{\pi}{4}$.

In Exercises 18–21, find $\text{Comp}_\mathbf{v}\mathbf{u}$.

18. $\mathbf{u} = \langle 7, -4 \rangle$, $\mathbf{v} = \langle -3, 4 \rangle$

19. $\mathbf{u} = \langle -2, -3 \rangle$, $\mathbf{v} = \langle 1, 1 \rangle$

20. $\mathbf{u} = \mathbf{i} - 2\mathbf{j}$, $\mathbf{v} = 2\mathbf{i} - \mathbf{j}$

21. $\mathbf{u} = 3\mathbf{i} - 4\mathbf{j}$, $\mathbf{v} = \mathbf{i} + 7\mathbf{j}$

22. Show that if $\mathbf{u} = \langle a, b \rangle$ and $\mathbf{v} = k\mathbf{i}$ for $k > 0$, then $\text{Comp}_\mathbf{v}\mathbf{u} = a$. Also show that if $\mathbf{v} = k\mathbf{j}$ for $k > 0$, then $\text{Comp}_\mathbf{v}\mathbf{u} = b$.

In Exercises 23–26, find the work done by the force $\mathbf{F}$ *acting on a particle along a line segment from the first point to the second. Assume* $|\mathbf{F}|$ *is in newtons and distance is in meters.*

23. $\mathbf{F} = 2\mathbf{i} + 3\mathbf{j}$; (1, 2) to (6, 8)

24. $\mathbf{F} = -\mathbf{i} + 4\mathbf{j}$; (−2, 3) to (3, 5)

25. $\mathbf{F} = 10\mathbf{i} + 20\mathbf{j}$; (2, 3) to (1, 5)

26. $\mathbf{F} = 5\mathbf{i} - 7\mathbf{j}$; (−4, −1) to (6, −6)

In Exercises 27–30, find $\text{Proj}_\mathbf{v}\mathbf{u}$ *and* $\text{Proj}_\mathbf{v}^\perp\mathbf{u}$.

27. $\mathbf{u} = \langle 3, -2 \rangle$, $\mathbf{v} = \langle 2, 4 \rangle$

28. $\mathbf{u} = \langle -2, -1 \rangle$, $\mathbf{v} = \langle -3, 4 \rangle$

29. $\mathbf{u} = 6\mathbf{i} + 2\mathbf{j}$, $\mathbf{v} = 3\mathbf{i} - 4\mathbf{j}$

30. $\mathbf{u} = -2\mathbf{i} + 3\mathbf{j}$, $\mathbf{v} = 7\mathbf{i} + \mathbf{j}$

31. A block that weighs 1000 lb and is on an inclined plane that makes a 30° angle with the horizontal is being held in place by a person pulling on a rope attached to the block and passing over a pulley, as shown in the figure. Assuming no friction, what is the magnitude of the force $\mathbf{F}$ that must be exerted?

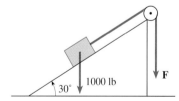

32. Resolve the vector $\mathbf{w} = 6\mathbf{i} - 4\mathbf{j}$ into vector components parallel and perpendicular, respectively, to the line that joins (−1, −2) and (2, 2).

33. Use vector methods to show that the points (2, 1), (6, 9), and (−2, 3) are vertices of a right triangle. What is the area of the triangle?

34. Use vector methods to show that the points (3, −1), (5, 4), (−5, 8), and (−7, 3) are vertices of a rectangle. What is the area of the rectangle?

35. Use vector methods to show that the points (−5, 2), (−3, −2), (6, 1), and (4, 5) are vertices of a parallelogram. Find the interior angles of the parallelogram.

Prove the identities in Exercises 36 and 37.

36. (a) $(\mathbf{u} + \mathbf{v}) \cdot (\mathbf{u} - \mathbf{v}) = |\mathbf{u}|^2 - |\mathbf{v}|^2$
    (b) $(\mathbf{u} + \mathbf{v}) \cdot (\mathbf{u} + \mathbf{v}) = |\mathbf{u}|^2 + 2\mathbf{u} \cdot \mathbf{v} + |\mathbf{v}|^2$

37. (a) $|\mathbf{u} + \mathbf{v}|^2 + |\mathbf{u} - \mathbf{v}|^2 = 2(|\mathbf{u}|^2 + |\mathbf{v}|^2)$
    (b) $|\mathbf{u} + \mathbf{v}|^2 - |\mathbf{u} - \mathbf{v}|^2 = 4\mathbf{u} \cdot \mathbf{v}$

38. Prove that $\mathbf{u}$ and $\mathbf{v}$ are orthogonal if and only if
$$|\mathbf{u} + \mathbf{v}| = |\mathbf{u} - \mathbf{v}|$$

39. Give an algebraic proof that $\text{Proj}_\mathbf{v}^\perp\mathbf{u}$ is orthogonal to $\text{Proj}_\mathbf{v}\mathbf{u}$.

40. Prove that the vector $\mathbf{n} = a\mathbf{i} + b\mathbf{j}$ is perpendicular to the line $ax + by + c = 0$. (*Hint:* Consider two points on the line.)

41. Let $P_0(x_0, y_0)$ be any point on the line $ax + by + c = 0$ and let $P_1(x_1, y_1)$ be any point not on the line. Show that the distance d from the line to the point P_1 is $d = |\text{Comp}_\mathbf{n} \overrightarrow{P_0 P_1}|$, where $\mathbf{n} = a\mathbf{i} + b\mathbf{j}$. From this result, verify the Distance Formula
$$d = \frac{|ax_1 + by_1 + c|}{\sqrt{a^2 + b^2}}$$
(*Hint:* Use the result of Exercise 40.)

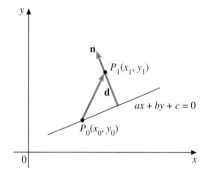

42. Use vector methods to prove that any triangle inscribed in a semicircle, with one side coinciding with the diameter, is a right triangle. (*Hint:* In the figure, find $\overrightarrow{AB}$ and $\overrightarrow{BC}$ in terms of $\mathbf{u}$ and $\mathbf{v}$, and use $|\mathbf{u}| = |\mathbf{v}|$ together with the result of Exercise 36(a).)

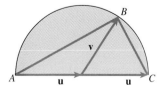

11.3 VECTORS IN SPACE

Three-Dimensional Coordinate Systems

In this section we extend the vector concept to three-dimensional space. It is necessary first to introduce a rectangular coordinate system. We begin with a horizontal plane that has a two-dimensional rectangular coordinate system with x- and y-axes in their usual orientation, the positive y-axis being $90°$ counterclockwise from the positive x-axis. Through the origin we introduce a vertical axis, called the z-axis, directed positively upward, with its origin coinciding with that of the x- and y-axes, as illustrated in Figure 11.24. We now have three mutually perpendicular axes oriented according to what is called the *right-hand rule*: if you point the index finger of your right hand in the positive x direction and the middle finger in the positive y direction, as in Figure 11.25, then your thumb will point in the positive z direction.

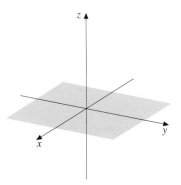

FIGURE 11.24

Each pair of axes determines a plane. We call these the **xy-plane**, the **xz-plane**, and the **yz-plane**. Frequently we will refer to the xy-plane as the **horizontal plane**. These three planes are the **coordinate planes**. Now let P denote any point in space. Through P pass planes parallel to each of the coordinate planes. If these cut the x-axis, y-axis, and z-axis at x_0, y_0, and z_0, respectively, then these three numbers are called the *coordinates* of P, and we write them as the ordered triple (x_0, y_0, z_0). We illustrate a typical such point in Figure 11.26. If we begin with the ordered triple (x_0, y_0, z_0), we locate P by proceeding x_0 units from the origin along the x-axis, then y_0 units parallel to the y-axis, and then z_0 units parallel to the z-axis, in each case using directed distances. In this way we establish a one-to-one correspondence between all points in three-dimensional space and all ordered triples of real numbers. We often will not distinguish between a point and its coordinates, saying, for example, "the point $(2, 3, -4)$" rather than "the point whose coordinates are $(2, 3, -4)$."

FIGURE 11.25

The three coordinate planes divide space into eight regions, called **octants**. The octant in which all coordinates are positive is called the first octant. There is no need to number the others. In plotting points it is useful to show lines, as we have done in plotting $P(3, 4, 6)$ in Figure 11.27. These help to make it appear that P is not in the plane of the paper. In this case we have shown the positive axes only, since the point is in the first octant.

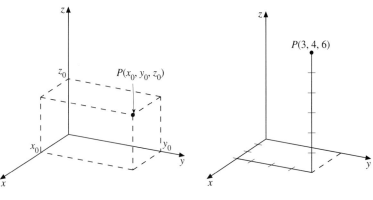

FIGURE 11.26   **FIGURE 11.27**

The Distance Formula

To determine a formula for the length of a vector, we need to know the distance between two points in space. Let $P_1(x_1, y_1, z_1)$ and $P_2(x_2, y_2, z_2)$ be any two such points. Construct a rectangular box with sides parallel to the coordinate planes so that P_1 and P_2 are at opposite corners of the box, as in Figure 11.28.

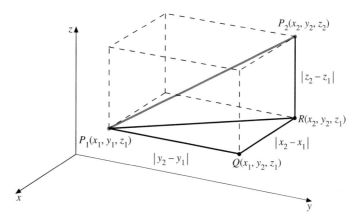

FIGURE 11.28

With vertices Q and R as shown, triangle P_1QR is a right triangle in a horizontal plane, and triangle P_1RP_2 is a right triangle in a vertical plane. Using $d(P_1, P_2)$ to mean the distance from P_1 to P_2, we have, from the first triangle, by the Pythagorean Theorem,

$$[d(P_1, R)]^2 = [d(P_1, Q)]^2 + [d(Q, R)]^2$$

and from the second,

$$[d(P_1, P_2)]^2 = [d(P_1, R)]^2 + [d(R, P_2)]^2$$

So

$$[d(P_1, P_2)]^2 = [d(P_1, Q)]^2 + [d(Q, R)]^2 + [d(R, P_2)]^2$$

But $d(P_1, Q) = |y_2 - y_1|$, $d(Q, R) = |x_2 - x_1|$, and $d(R, P_2) = |z_2 - z_1|$. Making these substitutions, we get the Distance Formula:

Distance Formula in Three Dimensions

$$d(P_1, P_2) = \sqrt{(x_2 - x_1)^2 + (y_2 - y_1)^2 + (z_2 - z_1)^2} \qquad (11.14)$$

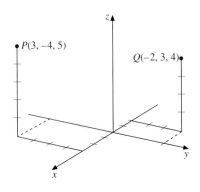

FIGURE 11.29

EXAMPLE 11.12 Plot the points $P(3, -4, 5)$ and $Q(-2, 3, 4)$, and find the distance between them.

Solution We show the points in Figure 11.29. By Equation 11.14,

$$d(P, Q) = \sqrt{(3+2)^2 + (-4-3)^2 + (5-4)^2}$$
$$= \sqrt{25 + 49 + 1}$$
$$= \sqrt{75}$$
$$= 5\sqrt{3}$$

■

Vectors in Three Dimensions

With this background we are ready to extend vectors to three dimensions. The notion of a geometric vector as a directed line segment is exactly as it was for two dimensions, with operations on vectors done in exactly the same way. Suppose a geometric vector has its initial point at $P_1(x_1, y_1, z_1)$ and its terminal point at $P_2(x_2, y_2, z_2)$. Then, analogous to the two-dimensional case, we identify the ordered triple $\langle x_2 - x_1, y_2 - y_1, z_2 - z_1 \rangle$ with the vector $\overrightarrow{P_1 P_2}$. Conversely, if we are given an ordered triple $\langle a, b, c \rangle$, we identify with this triple any geometric vector that has x displacement a, y displacement b, and z displacement c. The simplest such vector is the one with initial point at the origin and terminal point at (a, b, c), called the **position vector** of (a, b, c). In keeping with the geometric definitions of equivalence, addition, and multiplication by a scalar, we have the following:

1. $\langle a_1, a_2, a_3 \rangle = \langle b_1, b_2, b_3 \rangle$ if and only if $a_1 = b_1$, $a_2 = b_2$, and $a_3 = b_3$
2. $\langle a_1, a_2, a_3 \rangle + \langle b_1, b_2, b_3 \rangle = \langle a_1 + b_1, a_2 + b_2, a_3 + b_3 \rangle$
3. $k \langle a_1, a_2, a_3 \rangle = \langle k a_1, k a_2, k a_3 \rangle$ for any scalar k

With these definitions the set of all such ordered triples is called a **vector space of dimension 3**, and each element of this space is called a **three-dimensional vector**. If $\mathbf{u} = \langle u_1, u_2, u_3 \rangle$, then u_1, u_2, and u_3 are called the **components** of $\mathbf{u}$. The negative of $\mathbf{u}$ is $-\mathbf{u} = \langle -u_1, -u_2, -u_3 \rangle$, and the zero vector is $\langle 0, 0, 0 \rangle$. Subtraction is defined by $\mathbf{u} - \mathbf{v} = \mathbf{u} + (-\mathbf{v})$.

The **magnitude** of a vector $\mathbf{u} = \langle u_1, u_2, u_3 \rangle$ is defined by

$$|\mathbf{u}| = \sqrt{u_1^2 + u_2^2 + u_3^2} \tag{11.15}$$

Magnitude is, by Equation 11.14, the length of a geometric representative of $\mathbf{u}$. The **dot product** of two vectors $\mathbf{u} = \langle u_1, u_2, u_3 \rangle$ and $\mathbf{v} = \langle v_1, v_2, v_3 \rangle$ is defined by

$$\mathbf{u} \cdot \mathbf{v} = u_1 v_1 + u_2 v_2 + u_3 v_3 \tag{11.16}$$

The angle θ between two nonzero vectors $\mathbf{u}$ and $\mathbf{v}$ is defined as for two-dimensional vectors, and $\cos \theta$ is again given by

$$\cos \theta = \frac{\mathbf{u} \cdot \mathbf{v}}{|\mathbf{u}| \, |\mathbf{v}|} \tag{11.17}$$

The proof is the same as before. With the agreement again that **0** is orthogonal to every vector, we have that $\mathbf{u}$ *and* $\mathbf{v}$ *are orthogonal if and only if* $\mathbf{u} \cdot \mathbf{v} = 0$.

For any nonzero vector $\mathbf{u}$, $\mathbf{u}/|\mathbf{u}|$ is a *unit* vector, since its magnitude is 1. The unit vectors $\mathbf{i} = \langle 1, 0, 0 \rangle$, $\mathbf{j} = \langle 0, 1, 0 \rangle$, and $\mathbf{k} = \langle 0, 0, 1 \rangle$ form a *basis* for three-dimensional vector space, since

$$\langle a, b, c \rangle = a\mathbf{i} + b\mathbf{j} + c\mathbf{k} \tag{11.18}$$

means that every three-dimensional vector is expressible as a linear combination of $\mathbf{i}$, $\mathbf{j}$, and $\mathbf{k}$. Note that $\mathbf{i} \cdot \mathbf{j} = \mathbf{i} \cdot \mathbf{k} = \mathbf{j} \cdot \mathbf{k} = 0$, so that $\mathbf{i}$, $\mathbf{j}$, and $\mathbf{k}$ are mutually orthogonal. Geometrically, when placed with initial points at the origin, they are unit vectors directed along the positive x-axis, y-axis, and z-axis, respectively, as shown in Figure 11.30.

All other definitions and theorems in Sections 11.1 and 11.2 have natural extensions to three-dimensional vectors, and we will not repeat them. Proofs of the theorems are in many cases identical with proofs for the two-dimensional case, and at most require obvious modifications. Some of the results are illustrated in the examples that follow.

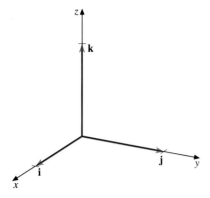

FIGURE 11.30

EXAMPLE 11.13 Let $\mathbf{u} = \langle 1, -2, 2 \rangle$ and $\mathbf{v} = \langle -3, -4, 5 \rangle$. Find each of the following.

(a) $|3\mathbf{u} - 2\mathbf{v}|$

(b) The angle between $\mathbf{u}$ and $\mathbf{v}$

Solution

(a) $3\mathbf{u} - 2\mathbf{v} = 3\langle 1, -2, 2 \rangle - 2\langle -3, -4, 5 \rangle$
$= \langle 3, -6, 6 \rangle - \langle -6, -8, 10 \rangle$
$= \langle 3, -6, 6 \rangle + \langle 6, 8, -10 \rangle$
$= \langle 9, 2, -4 \rangle$

So

$$|3\mathbf{u} - 2\mathbf{v}| = \sqrt{81 + 4 + 16} = \sqrt{101}$$

(b) $\cos \theta = \dfrac{\mathbf{u} \cdot \mathbf{v}}{|\mathbf{u}| |\mathbf{v}|} = \dfrac{\langle 1, -2, 2 \rangle \cdot \langle -3, -4, 5 \rangle}{\sqrt{1 + 4 + 4} \sqrt{9 + 16 + 25}}$

$= \dfrac{-3 + 8 + 10}{3\sqrt{50}} = \dfrac{15}{3(5\sqrt{2})} = \dfrac{1}{\sqrt{2}}$

So $\theta = \pi/4$. ∎

EXAMPLE 11.14 Find the work done by the force $\mathbf{F} = 4\mathbf{i} + 5\mathbf{j} - 8\mathbf{k}$ in moving a particle from $P(-1, 2, 4)$ to $Q(3, 6, -8)$. Assume $|\mathbf{F}|$ is in newtons and distance is in meters.

Solution

$$\overrightarrow{PQ} = \langle 4, 4, -12 \rangle = 4\mathbf{i} + 4\mathbf{j} - 12\mathbf{k}$$

so by Equation 11.9,

$$W = \mathbf{F} \cdot (\overrightarrow{PQ}) = (4\mathbf{i} + 5\mathbf{j} - 8\mathbf{k}) \cdot (4\mathbf{i} + 4\mathbf{j} - 12\mathbf{k})$$
$$= 16 + 20 + 96 = 132 \text{ joules}$$

∎

EXAMPLE 11.15 Find $\text{Comp}_v \mathbf{u}$ and $\text{Proj}_v \mathbf{u}$ if $\mathbf{u} = 4\mathbf{i} - 6\mathbf{j} + \mathbf{k}$ and $\mathbf{v} = -3\mathbf{i} - 2\mathbf{j} + 5\mathbf{k}$.

Solution From Equation 11.8,

$$\text{Comp}_v \mathbf{u} = \frac{\mathbf{u} \cdot \mathbf{v}}{|\mathbf{v}|} = \frac{4(-3) + (-6)(-2) + (1)(5)}{\sqrt{9 + 4 + 25}} = \frac{5}{\sqrt{38}}$$

By Equation 11.11,

$$\text{Proj}_v \mathbf{u} = (\text{Comp}_v \mathbf{u}) \frac{\mathbf{v}}{|\mathbf{v}|} = \frac{5}{\sqrt{38}} \frac{-3\mathbf{i} - 2\mathbf{j} + 5\mathbf{k}}{\sqrt{38}}$$

$$= -\frac{15}{38}\mathbf{i} - \frac{5}{19}\mathbf{j} + \frac{25}{38}\mathbf{k} \qquad \blacksquare$$

Direction Angles and Direction Cosines

The angles a nonzero vector $\mathbf{u}$ makes with $\mathbf{i}$, $\mathbf{j}$, and $\mathbf{k}$ are called **direction angles** of $\mathbf{u}$ and are designated by α, β, and γ, respectively. We illustrate these angles in Figure 11.31. The cosines of these angles are called **direction cosines** of $\mathbf{u}$. If $\mathbf{u} = \langle u_1, u_2, u_3 \rangle$, we have

$$\cos\alpha = \frac{\mathbf{u} \cdot \mathbf{i}}{|\mathbf{u}||\mathbf{i}|} = \frac{u_1}{|\mathbf{u}|} \quad \cos\beta = \frac{\mathbf{u} \cdot \mathbf{j}}{|\mathbf{u}||\mathbf{j}|} = \frac{u_2}{|\mathbf{u}|} \quad \cos\gamma = \frac{\mathbf{u} \cdot \mathbf{k}}{|\mathbf{u}||\mathbf{k}|} = \frac{u_3}{|\mathbf{u}|} \quad (11.19)$$

If we square and add, we get

$$\cos^2\alpha + \cos^2\beta + \cos^2\gamma = \frac{u_1^2}{|\mathbf{u}|^2} + \frac{u_2^2}{|\mathbf{u}|^2} + \frac{u_3^2}{|\mathbf{u}|^2}$$

or

$$\cos^2\alpha + \cos^2\beta + \cos^2\gamma = 1 \qquad (11.20)$$

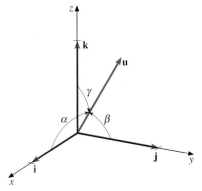

FIGURE 11.31

If $\mathbf{u}$ is a unit vector, then by Equation 11.19 its components are precisely its direction cosines:

$$\mathbf{u} = \langle \cos\theta, \cos\beta, \cos\gamma \rangle \qquad \text{if } |\mathbf{u}| = 1$$

EXAMPLE 11.16 Find the direction cosines of the vector with initial point $P(7, -2, 4)$ and terminal point $Q(5, 3, 0)$.

Solution Let $\mathbf{u} = \overrightarrow{PQ} = \langle -2, 5, -4 \rangle$. Then $|\mathbf{u}| = \sqrt{4 + 25 + 16} = \sqrt{45} = 3\sqrt{5}$. So by Equation 11.19,

$$\cos\alpha = \frac{-2}{3\sqrt{5}} \qquad \cos\beta = \frac{5}{3\sqrt{5}} \qquad \cos\gamma = \frac{-4}{3\sqrt{5}} \qquad \blacksquare$$

EXAMPLE 11.17 A unit vector $\mathbf{u}$ makes an angle of $60°$ with the positive x-axis and with the positive y-axis. What angle does it make with the positive z-axis? What are the components of $\mathbf{u}$?

Solution By the given information, $\alpha = \beta = \frac{\pi}{3}$, and by Equation 11.20,

$$\cos^2\gamma = 1 - \cos^2\alpha - \cos^2\beta = 1 - \frac{1}{4} - \frac{1}{4} = \frac{1}{2}$$

So $\cos\gamma = \pm 1/\sqrt{2}$. Thus, $\gamma = \frac{\pi}{4}$ or $\frac{3\pi}{4}$. There are therefore two possibilities for **u**:

$$\mathbf{u} = \left\langle \frac{1}{2}, \frac{1}{2}, \frac{1}{\sqrt{2}} \right\rangle \quad \text{or} \quad \mathbf{u} = \left\langle \frac{1}{2}, \frac{1}{2}, -\frac{1}{\sqrt{2}} \right\rangle \qquad \blacksquare$$

Exercise Set 11.3

1. Plot each of the following points.
   (a) $(3, 2, 4)$
   (b) $(4, -2, 1)$
   (c) $(-3, 2, 4)$
   (d) $(0, -5, -2)$
   (e) $(-4, -3, -6)$

2. Find the distance between P and Q.
   (a) $P(2, 0, -1)$, $Q(3, 5, 7)$
   (b) $P(-3, 5, 2)$, $Q(-1, -1, 4)$
   (c) $P(8, 2, 0)$, $Q(7, 6, -3)$
   (d) $P(4, -2, -3)$, $Q(-1, -3, 2)$

3. Identify each of the following three-dimensional point sets.
   (a) The set of all points for which $z = 0$
   (b) The set of all points for which $x = 0$
   (c) The set of all points for which $y = 0$
   (d) All points of the form $(x, 0, 0)$
   (e) All points for which $xyz = 0$

4. Let $P(x, y, z)$ be an arbitrary point in space and Q be the fixed point (h, k, l). Write an equation expressing the fact that $d(P, Q) = a$, where a is a positive constant. Clear the equation of radicals. How would you describe the set of all points P that satisfy this equation?

5. Express the vector $\overrightarrow{PQ}$ in terms of its components.
   (a) $P(7, 3, -1)$, $Q(5, -1, 2)$
   (b) $P(2, -3, -4)$, $Q(-1, -2, 0)$
   (c) $P(0, -2, 7)$, $Q(3, 0, 5)$
   (d) $P(-4, 6, 10)$, $Q(2, 4, 8)$

6. Find the magnitude of $\overrightarrow{PQ}$ for each part of Exercise 5.

7. Let $\mathbf{u} = \langle 3, 1, -2 \rangle$, $\mathbf{v} = \langle -1, 0, 4 \rangle$, and $\mathbf{w} = \langle 4, 1, 5 \rangle$. Find the following.
   (a) $\mathbf{u} \cdot \mathbf{w} - |\mathbf{v}|^2$
   (b) $\mathbf{v} \cdot (\mathbf{u} - \mathbf{w})$
   (c) $|\mathbf{u} - 2\mathbf{v}|$
   (d) $3\mathbf{u} + 2\mathbf{v} - \mathbf{w}$

8. Let $\mathbf{u} = 3\mathbf{i} - 2\mathbf{j} - \mathbf{k}$, $\mathbf{v} = 5\mathbf{i} - 4\mathbf{k}$, and $\mathbf{w} = -4\mathbf{i} + 6\mathbf{j} + 2\mathbf{k}$. Find the following.
   (a) $\mathbf{u} \cdot (\mathbf{v} - \mathbf{w})$
   (b) $|3\mathbf{u} + 2\mathbf{w}|$
   (c) $|\mathbf{u}| |\mathbf{w}| - |\mathbf{u} \cdot \mathbf{w}|$
   (d) $(\mathbf{v} + \mathbf{w}) \cdot (\mathbf{v} - \mathbf{w})$

9. Find a unit vector in the direction of **u**.
   (a) $\mathbf{u} = \langle 2, -1, 2 \rangle$
   (b) $\mathbf{u} = \langle 4, 3, -5 \rangle$
   (c) $\mathbf{u} = \mathbf{i} - \mathbf{j} + \mathbf{k}$
   (d) $\mathbf{u} = 2\mathbf{i} - 4\mathbf{j} + 5\mathbf{k}$

10. Find the cosine of the angle between **u** and **v**.
    (a) $\mathbf{u} = \langle 3, -2, 6 \rangle$, $\mathbf{v} = \langle 1, 1, 1 \rangle$
    (b) $\mathbf{u} = 4\mathbf{i} + 2\mathbf{j} - 2\mathbf{k}$, $\mathbf{v} = -7\mathbf{i} + 4\mathbf{j} + 5\mathbf{k}$

11. (a) Show that $\mathbf{u} = 2\mathbf{i} - 3\mathbf{j} + \mathbf{k}$ and $\mathbf{v} = 4\mathbf{i} + 2\mathbf{j} - 2\mathbf{k}$ are orthogonal.
    (b) Find x so that $\mathbf{u} = \langle x, -1, 2 \rangle$ and $\mathbf{v} = \langle 6, 4, x \rangle$ will be orthogonal.

12. Use vector methods to show that the points $A(-1, 2, -3)$, $B(1, -1, 2)$, and $C(0, 5, 6)$ are vertices of a right triangle.

13. Find the direction cosines of **u**.
    (a) $\mathbf{u} = 2\mathbf{i} - \mathbf{j} + 2\mathbf{k}$
    (b) $\mathbf{u} = \langle 3, -5, 4 \rangle$

14. Find the vector **u** with magnitude 3 whose z component is positive, if $\cos\alpha = \frac{1}{3}$ and $\cos\beta = -\sqrt{2}/3$.

15. Find the unit vector for which $\beta = \pi/3$ and $\gamma = \pi/4$ and whose x component is negative.

16. If a vector makes equal acute angles with **i**, **j**, and **k**, what is this angle?

17. Show that if the direction angles α, β, and γ of a vector are all acute and $\alpha \geq \pi/3$ and $\beta \geq \pi/3$, then $\gamma \leq \pi/4$.

*In Exercises 18–21, find the component of **u** along **v**.*

18. $\mathbf{u} = \langle -3, 1, 4 \rangle$, $\mathbf{v} = \langle 2, -1, -2 \rangle$

19. $\mathbf{u} = \langle 5, 4, -4 \rangle$, $\mathbf{v} = \langle 3, 0, -4 \rangle$

20. $u = 7i - 2k$, $v = 3i - 5j + 4k$

21. $u = i + 2j - k$, $v = i - j + k$

In Exercises 22–24, find the work done by the force F in moving a particle along the line segment from P to Q.

22. $F = 10i + 12j - 8k$; $P(2, -1, 4)$, $Q(3, 5, 2)$; $|F|$ in newtons, distance in meters

23. $F = 20i - 12j + 6k$; $P(3, 4, 6)$, $Q(8, -1, 10)$; $|F|$ in dynes, distance in centimeters

24. $F = 6i + 2j + 8k$; $P(-1, 3, 5)$, $Q(4, -1, 9)$; $|F|$ in pounds, distance in feet

In Exercises 25–28, find $\text{Proj}_v u$ and $\text{Proj}_v^\perp u$.

25. $u = \langle 5, -1, 3 \rangle$, $v = \langle 2, 6, -4 \rangle$

26. $u = \langle 2, -3, 0 \rangle$, $v = \langle -5, 1, -2 \rangle$

27. $u = 2i - 3j - 5k$, $v = i + 2j - 3k$

28. $u = 4j - 5k$, $v = 3i - 5j + 4k$

In Exercises 29–33, prove that the indicated theorem continues to hold true for three-dimensional vectors.

29. Theorem 11.1

30. Theorem 11.2

31. Theorem 11.3

32. Theorem 11.4

33. Theorem 11.5

34. Find a nonzero vector $x = \langle x_1, x_2, x_3 \rangle$ that is perpendicular to each of the vectors $u = \langle 2, 1, -3 \rangle$ and $v = \langle -1, 1, 2 \rangle$. (*Hint:* Obtain two equations with three unknowns. Choose one of the unknowns arbitrarily.)

35. Find a unit vector orthogonal to each of the vectors $u = i - j + 2k$ and $v = 3i + 2j - 2k$. (See the hint in Exercise 34.)

36. Find scalars a and b such that $w = au + bv$, where $u = \langle 3, -2, 4 \rangle$, $v = \langle 1, 1, -2 \rangle$, and $w = \langle 6, 1, -2 \rangle$. Interpret the result geometrically.

37. Let $u_1 = \langle 1, -1, 0 \rangle$, $u_2 = \langle 0, 1, -1 \rangle$, and $u_3 = \langle 1, 1, 1 \rangle$. Find scalars a, b, and c such that $v = au_1 + bu_2 + cu_3$, where $v = \langle 3, 2, -1 \rangle$.

38. With u_1, u_2, and u_3 as in Exercise 37, show that *every* three-dimensional vector v can be expressed as a linear combination of u_1, u_2, and u_3. (*Note:* This shows that u_1, u_2, and u_3 constitute a basis for three-dimensional vector space.)

39. Prove that the sum of any two of the three direction angles α, β, and γ must be greater than or equal to $\pi/2$. (*Hint:* Suppose, for example, that $\alpha + \beta < \pi/2$, so that $\alpha < \pi/2 - \beta$. Show that $\cos^2 \alpha + \cos^2 \beta > 1$.)

Use a CAS in Exercises 40–51. In Exercises 40–43, find the direction cosines and the direction angles, in degrees and radians, of the given vectors. Give the direction angles correct to three decimal places.

40. $\langle 2, 3, 5 \rangle$

41. $\langle 2.1, 3.2, -2 \rangle$

42. $-4i + 3j - 5k$

43. $i - 8j + 10k$

44. Verify that the points $(4, 6, 8)$, $(2, 3, 9)$, and $(7, 2, 2)$ form the vertices of a right triangle.

45. Use the dot product to find two unit vectors that are perpendicular to the vectors $\langle 1, 2, 3 \rangle$ and $\langle 2, 4, 3 \rangle$.

46. Let $u = 3i + 4j - 2k$ and $v = i + 2j + ak$.
    (a) Find a so that u and v are perpendicular.
    (b) Is it possible to find a so that u and v are parallel? Explain.
    (c) Approximate a so that the angle between u and v is $\pi/6$.

47. Let u and v be vectors in three space. Use a CAS to verify the identities:
    (a) $(u + v) \cdot (u - v) = |u|^2 - |v|^2$
    (b) $(u + v) \cdot (u + v) = |u|^2 + 2u \cdot v + |v|^2$

48. Let u and v be vectors in three space. Use a CAS to verify the identities:
    (a) $|u + v|^2 + |u - v|^2 = 2(|u|^2 + |v|^2)$
    (b) $|u + v|^2 - |u - v|^2 = 4u \cdot v$

49. Let u and v be any two distinct nonzero vectors in three space. Use a CAS to show that the vector $w = u - (u \cdot v)v/|v|^2$ is orthogonal to v.

50. Two tugboats are pulling a barge through a channel.
    (a) One tugboat is pulling with a force of magnitude 350 N at an angle of 25° NE with the horizontal and the second is pulling with a force of magnitude 500 N at an angle of 35° SW with the horizontal. If the barge is being moved 2 km, estimate the work done by each tugboat.
    (b) One tugboat is pulling with a force of magnitude 250 N at an angle of 30° NE with the horizontal and the second is pulling with a force of magnitude w N at an angle of 25° SW with the horizontal. If the barge is moving horizontally, find the magnitude w of the force at which the second tugboat is pulling the barge.

51. Define two vectors u and v as $u = \langle 2, -1, 5 \rangle$ and $v = \langle -4, 2, 7 \rangle$. Find
    (a) $u + v$, $u - v$, and $-2u$
    (b) the length of v
    (c) a unit vector in the direction of v
    (d) $u \cdot v$
    (e) the angle between u and v
    (f) the component of u in the direction of v
    (g) the projection of u on v

11.4 THE CROSS PRODUCT

For three-dimensional vectors **u** and **v** there is a second type of product, called the **cross product**, written **u** × **v**, that results in another vector, rather than a scalar as with the dot product. For this reason the cross product is sometimes called the **vector product**.

Definition 11.7
The Cross Product

The **cross product** of $\mathbf{u} = \langle u_1, u_2, u_3 \rangle$ and $\mathbf{v} = \langle v_1, v_2, v_3 \rangle$ is the vector

$$\mathbf{u} \times \mathbf{v} = \langle u_2 v_3 - u_3 v_2, u_3 v_1 - u_1 v_3, u_1 v_2 - u_2 v_1 \rangle \tag{11.21}$$

Determinant Notation for the Cross Product

Definition 11.7 can be remembered more easily using determinant notation, which we review briefly. A second-order determinant is defined by

$$\begin{vmatrix} a_1 & a_2 \\ b_1 & b_2 \end{vmatrix} = a_1 b_2 - a_2 b_1$$

For example,

$$\begin{vmatrix} 3 & 2 \\ -1 & 4 \end{vmatrix} = 3(4) - (2)(-1) = 12 + 2 = 14$$

A third-order determinant can be evaluated as follows:

$$\begin{vmatrix} a_1 & a_2 & a_3 \\ b_1 & b_2 & b_3 \\ c_1 & c_2 & c_3 \end{vmatrix} = a_1 \begin{vmatrix} b_2 & b_3 \\ c_2 & c_3 \end{vmatrix} - a_2 \begin{vmatrix} b_1 & b_3 \\ c_1 & c_3 \end{vmatrix} + a_3 \begin{vmatrix} b_1 & b_2 \\ c_1 & c_2 \end{vmatrix}$$

Each second-order determinant is then evaluated as above. The formula we have given is sometimes referred to as *expansion by the first row*. It is possible to expand by any row or column, but for our purposes the first row is the most convenient. To illustrate, consider the following:

$$\begin{vmatrix} 2 & -1 & -3 \\ 4 & 2 & 1 \\ 0 & 5 & -4 \end{vmatrix} = 2 \begin{vmatrix} 2 & 1 \\ 5 & -4 \end{vmatrix} - (-1) \begin{vmatrix} 4 & 1 \\ 0 & -4 \end{vmatrix} + (-3) \begin{vmatrix} 4 & 2 \\ 0 & 5 \end{vmatrix}$$

$$= 2(-8 - 5) + (-16) - 3(20) = -26 - 16 - 60 = -102$$

Now observe that Equation 11.21 can be written in the form

$$\mathbf{u} \times \mathbf{v} = \left\langle \begin{vmatrix} u_2 & u_3 \\ v_2 & v_3 \end{vmatrix}, -\begin{vmatrix} u_1 & u_3 \\ v_1 & v_3 \end{vmatrix}, \begin{vmatrix} u_1 & u_2 \\ v_1 & v_2 \end{vmatrix} \right\rangle$$

as you can verify by evaluating the second-order determinants and comparing the result with Equation 11.21. Equivalently, we can write $\mathbf{u} \times \mathbf{v}$ as

$$\mathbf{u} \times \mathbf{v} = \begin{vmatrix} u_2 & u_3 \\ v_2 & v_3 \end{vmatrix} \mathbf{i} - \begin{vmatrix} u_1 & u_3 \\ v_1 & v_3 \end{vmatrix} \mathbf{j} + \begin{vmatrix} u_1 & u_2 \\ v_1 & v_2 \end{vmatrix} \mathbf{k} \tag{11.22}$$

A convenient way of remembering this formula is to write the third-order determinant

$$\begin{vmatrix} \mathbf{i} & \mathbf{j} & \mathbf{k} \\ u_1 & u_2 & u_3 \\ v_1 & v_2 & v_3 \end{vmatrix}$$

Since the first row consists of vectors instead of numbers, this is not a proper determinant. Nevertheless, if we formally expand it by the first row, we get (writing the scalars times the vectors instead of the reverse)

$$\begin{vmatrix} u_2 & u_3 \\ v_2 & v_3 \end{vmatrix} \mathbf{i} - \begin{vmatrix} u_1 & u_3 \\ v_1 & v_3 \end{vmatrix} \mathbf{j} + \begin{vmatrix} u_1 & u_2 \\ v_1 & v_2 \end{vmatrix} \mathbf{k}$$

and the result is seen, by comparison with Equation 11.22, to be the $\mathbf{i}$, $\mathbf{j}$, $\mathbf{k}$ notation for $\mathbf{u} \times \mathbf{v}$. Thus, with this understanding of what is meant by the determinant with vector entries in the first row, we have

$$\mathbf{u} \times \mathbf{v} = \begin{vmatrix} \mathbf{i} & \mathbf{j} & \mathbf{k} \\ u_1 & u_2 & u_3 \\ v_1 & v_2 & v_2 \end{vmatrix} \quad (11.23)$$

EXAMPLE 11.18 Find $\mathbf{u} \times \mathbf{v}$, where $\mathbf{u} = \langle 3, -1, 4 \rangle$ and $\mathbf{v} = \langle -2, 2, 5 \rangle$.

Solution By Equation 11.23,

$$\mathbf{u} \times \mathbf{v} = \begin{vmatrix} \mathbf{i} & \mathbf{j} & \mathbf{k} \\ 3 & -1 & 4 \\ -2 & 2 & 5 \end{vmatrix} = \begin{vmatrix} -1 & 4 \\ 2 & 5 \end{vmatrix} \mathbf{i} - \begin{vmatrix} 3 & 4 \\ -2 & 5 \end{vmatrix} \mathbf{j} + \begin{vmatrix} 3 & -1 \\ -2 & 2 \end{vmatrix} \mathbf{k}$$

$$= (-5 - 8)\mathbf{i} - (15 + 8)\mathbf{j} + (6 - 2)\mathbf{k}$$

$$= -13\mathbf{i} - 23\mathbf{j} + 4\mathbf{k}$$

Equivalently, $\mathbf{u} \times \mathbf{v} = \langle -13, -23, 4 \rangle$. ∎

Geometric Interpretation of the Cross Product

One of the most important properties of the cross product is given by the following theorem.

THEOREM 11.6 The vector $\mathbf{u} \times \mathbf{v}$ is orthogonal to both $\mathbf{u}$ and $\mathbf{v}$.

Proof We will show that $\mathbf{u} \cdot (\mathbf{u} \times \mathbf{v}) = 0$ and leave it as an exercise to show that $\mathbf{v} \cdot (\mathbf{u} \times \mathbf{v}) = 0$. By Equation 11.21,

$$\mathbf{u} \cdot (\mathbf{u} \times \mathbf{v}) = \langle u_1, u_2, u_3 \rangle \cdot \langle (u_2 v_3 - u_3 v_2), (u_3 v_1 - u_1 v_3), (u_1 v_2 - u_2 v_1) \rangle$$

$$= u_1(u_2 v_3 - u_3 v_2) + u_2(u_3 v_1 - u_1 v_3) + u_3(u_1 v_2 - u_2 v_1)$$

$$= u_1 u_2 v_3 - u_1 u_3 v_2 + u_2 u_3 v_1 - u_2 u_1 v_3 + u_3 u_1 v_2 - u_3 u_2 v_1$$

$$= 0$$

So $\mathbf{u}$ and $\mathbf{u} \times \mathbf{v}$ are orthogonal. ∎

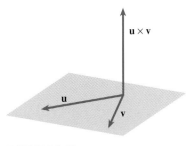

FIGURE 11.32

If **u** and **v** are nonzero and are not parallel, then we know from the preceding theorem that $\mathbf{u} \times \mathbf{v}$ is orthogonal to both **u** and **v**. Suppose geometric representatives of **u**, **v**, and $\mathbf{u} \times \mathbf{v}$ are drawn with the same initial point. Then $\mathbf{u} \times \mathbf{v}$ is perpendicular to the plane containing **u** and **v**, as shown in Figure 11.32. The direction of $\mathbf{u} \times \mathbf{v}$ is determined according to the following right-hand rule: if you curl the fingers of your right hand in the direction that would rotate **u** into **v** (through an angle of less than π), then your extended thumb will point in the direction of $\mathbf{u} \times \mathbf{v}$.

The magnitude of $\mathbf{u} \times \mathbf{v}$ is related to the magnitudes of **u** and **v** as given in the following theorem.

THEOREM 11.7

If θ is the angle between the nonzero vectors **u** and **v**, then

$$|\mathbf{u} \times \mathbf{v}| = |\mathbf{u}| |\mathbf{v}| \sin \theta \tag{11.24}$$

Proof By Equation 11.21 we have

$$|\mathbf{u} \times \mathbf{v}|^2 = (u_2 v_3 - u_3 v_2)^2 + (u_3 v_1 - u_1 v_3)^2 + (u_1 v_2 - u_2 v_1)^2$$

We leave it as an exercise for you to show that if the right-hand side is expanded and terms are appropriately grouped, it can be written in the form

$$(u_1^2 + u_2^2 + u_3^2)(v_1^2 + v_2^2 + v_3^2) - (u_1 v_1 + u_2 v_2 + u_3 v_3)^2$$

Thus,

$$|\mathbf{u} \times \mathbf{v}|^2 = |\mathbf{u}|^2 |\mathbf{v}|^2 - (\mathbf{u} \cdot \mathbf{v})^2$$

Since $\mathbf{u} \cdot \mathbf{v} = |\mathbf{u}| |\mathbf{v}| \cos \theta$, we have

$$|\mathbf{u} \times \mathbf{v}|^2 = |\mathbf{u}|^2 |\mathbf{v}|^2 - |\mathbf{u}|^2 |\mathbf{v}|^2 \cos^2 \theta$$
$$= |\mathbf{u}|^2 |\mathbf{v}|^2 (1 - \cos^2 \theta)$$
$$= |\mathbf{u}|^2 |\mathbf{v}|^2 \sin^2 \theta$$

Taking square roots, we get the desired result. ∎

COROLLARY 11.7

Two three-dimensional vectors **u** and **v** are parallel if and only if $\mathbf{u} \times \mathbf{v} = \mathbf{0}$.

Proof If either **u** or **v** is **0**, the result is trivial. If they are both nonzero, they are parallel if and only if $\theta = 0$ or $\theta = \pi$ or, equivalently, $\sin \theta = 0$. Thus, by Equation 11.24, they are parallel if and only if $|\mathbf{u} \times \mathbf{v}| = 0$, and hence, if and only if $\mathbf{u} \times \mathbf{v} = \mathbf{0}$. ∎

Area of a Parallelogram

Equation 11.24 has an interesting geometric interpretation. Let **u** and **v** be nonzero, with angle θ between them. Choose geometric representatives of **u** and **v** that have the same initial point. Complete the parallelogram with **u**

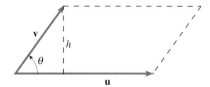

FIGURE 11.33

and **v** as adjacent sides, as in Figure 11.33. From that figure we see that the height h of the parallelogram from the base **u** is $h = |\mathbf{v}| \sin \theta$. Thus, its area is $|\mathbf{u}|h = |\mathbf{u}| |\mathbf{v}| \sin \theta$. So by Theorem 11.7,

$$|\mathbf{u} \times \mathbf{v}| = \text{the area of the parallelogram with adjacent sides } \mathbf{u} \text{ and } \mathbf{v}$$

EXAMPLE 11.19 Find the area of the triangle with vertices $A(2, 1, 4)$, $B(3, -1, 7)$, and $C(-1, 2, 5)$.

Solution Let $\mathbf{u} = \overrightarrow{AB}$ and $\mathbf{v} = \overrightarrow{AC}$. Then the area of the triangle is one-half the area of the parallelogram determined by **u** and **v**, or

$$\text{Area} = \frac{1}{2}|\mathbf{u} \times \mathbf{v}|$$

We first find $\mathbf{u} \times \mathbf{v}$:

$$\mathbf{u} = \overrightarrow{AB} = \langle 1, -2, 3 \rangle$$

$$\mathbf{v} = \overrightarrow{AC} = \langle -3, 1, 1 \rangle$$

$$\mathbf{u} \times \mathbf{v} = \begin{vmatrix} \mathbf{i} & \mathbf{j} & \mathbf{k} \\ 1 & -2 & 3 \\ -3 & 1 & 1 \end{vmatrix} = \begin{vmatrix} -2 & 3 \\ 1 & 1 \end{vmatrix} \mathbf{i} - \begin{vmatrix} 1 & 3 \\ -3 & 1 \end{vmatrix} \mathbf{j} + \begin{vmatrix} 1 & -2 \\ -3 & 1 \end{vmatrix} \mathbf{k}$$

$$= -5\mathbf{i} - 10\mathbf{j} - 5\mathbf{k}$$

Thus, the area of the triangle is

$$\text{Area} = \frac{1}{2}|\mathbf{u} \times \mathbf{v}| = \frac{1}{2}\sqrt{25 + 100 + 25} = \frac{1}{2}\sqrt{150} = \frac{5}{2}\sqrt{6} \quad \blacksquare$$

Algebraic Properties of the Cross Product

The next theorem provides some other properties of the cross product. The proof of each part can be shown by direct application of Definition 11.7. In some cases the proof can be facilitated, however, using Equation 11.23 and the following two properties of determinants:

1. If two rows in a determinant are identical, the value of the determinant is 0.
2. If two rows in a determinant are interchanged, the result is the negative of the original determinant.

In the exercises you will be asked to verify these properties for third-order determinants, and you will also be asked to verify the theorem.

THEOREM 11.8

For three-dimensional vectors **u**, **v**, and **w**,

1. $\mathbf{u} \times \mathbf{v} = -(\mathbf{v} \times \mathbf{u})$ Anticommutative property
2. $k(\mathbf{u} \times \mathbf{v}) = (k\mathbf{u}) \times \mathbf{v} = \mathbf{u} \times (k\mathbf{v})$ for any scalar k
3. $\mathbf{u} \times \mathbf{0} = \mathbf{0}$
4. $\mathbf{u} \times \mathbf{u} = \mathbf{0}$
5. $\mathbf{u} \times (\mathbf{v} + \mathbf{w}) = \mathbf{u} \times \mathbf{v} + \mathbf{u} \times \mathbf{w}$ Left distributive property
6. $(\mathbf{u} + \mathbf{v}) \times \mathbf{w} = \mathbf{u} \times \mathbf{w} + \mathbf{v} \times \mathbf{w}$ Right distributive property
7. $\mathbf{u} \times (\mathbf{v} \times \mathbf{w}) = (\mathbf{u} \cdot \mathbf{w})\mathbf{v} - (\mathbf{u} \cdot \mathbf{v})\mathbf{w}$
8. $\mathbf{u} \cdot (\mathbf{v} \times \mathbf{w}) = (\mathbf{u} \times \mathbf{v}) \cdot \mathbf{w}$

It is useful to learn the various cross products involving pairs of the basis vectors **i**, **j**, and **k**. Direct application of Definition 11.7 gives

$$\mathbf{i} \times \mathbf{j} = \mathbf{k} \qquad \mathbf{j} \times \mathbf{k} = \mathbf{i} \qquad \mathbf{k} \times \mathbf{i} = \mathbf{j}$$

By Property 1 in Theorem 11.8, if the factors on the left are reversed, the sign on the right becomes negative. That is,

$$\mathbf{j} \times \mathbf{i} = -\mathbf{k} \qquad \mathbf{k} \times \mathbf{j} = -\mathbf{i} \qquad \mathbf{i} \times \mathbf{k} = -\mathbf{j}$$

FIGURE 11.34

One way to remember these relationships is by the diagram in Figure 11.34. Going clockwise, crossing a vector with the following one produces the next one. Going counterclockwise produces the negative of the next one.

The cross product is neither commutative nor associative in general. Noncommutativity follows from Property 1 of Theorem 11.8, and nonassociativity can be seen, for example, by the following calculations:

$$(\mathbf{i} \times \mathbf{j}) \times \mathbf{j} = \mathbf{k} \times \mathbf{j} = -\mathbf{i} \qquad \text{but} \qquad \mathbf{i} \times (\mathbf{j} \times \mathbf{j}) = \mathbf{i} \times \mathbf{0} = \mathbf{0}$$

Triple Scalar Product

The product $\mathbf{u} \cdot (\mathbf{v} \times \mathbf{w})$ is called the **triple scalar product** of **u**, **v**, and **w**. Applying Definition 11.7 to $\mathbf{v} \times \mathbf{w}$, we get

$$\mathbf{u} \cdot (\mathbf{v} \times \mathbf{w}) = u_1(v_2 w_3 - v_3 w_2) + u_2(v_3 w_1 - v_1 w_3) + u_3(v_1 w_2 - v_2 w_1)$$

$$= u_1 \begin{vmatrix} v_2 & v_3 \\ w_2 & w_3 \end{vmatrix} - u_2 \begin{vmatrix} v_1 & v_3 \\ w_1 & w_3 \end{vmatrix} + u_3 \begin{vmatrix} v_1 & v_2 \\ w_1 & w_2 \end{vmatrix}$$

The right-hand side is the result of expanding by the first row the determinant whose rows, in order, are the components of **u**, **v**, and **w**, respectively. Thus,

$$\mathbf{u} \cdot (\mathbf{v} \times \mathbf{w}) = \begin{vmatrix} u_1 & u_2 & u_3 \\ v_1 & v_2 & v_3 \\ w_1 & w_2 & w_3 \end{vmatrix} \qquad (11.25)$$

Volume of a Parallelepiped

The triple scalar product has an interesting geometric interpretation. Let **u**, **v**, and **w** be nonzero vectors that do not lie in the same plane. Take geometric representatives of **u**, **v**, and **w** that have the same initial point and construct a parallelepiped, with these vectors as edges, as in Figure 11.35. Using as a base the parallogram determined by **v** and **w**, the altitude h is the absolute value of the component of **u** perpendicular to this base; that is,

$$h = |\text{Comp}_{\mathbf{v} \times \mathbf{w}} \mathbf{u}|$$

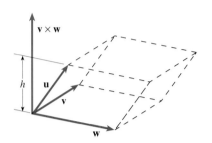

FIGURE 11.35

As we have seen, the area of the base is $|\mathbf{v} \times \mathbf{w}|$, and the volume of the parallelepiped is the area of the base times the altitude. So we have

$$\text{vol} = h|\mathbf{v} \times \mathbf{w}| = |\text{Comp}_{\mathbf{v} \times \mathbf{w}} \mathbf{u}| \, |\mathbf{v} \times \mathbf{w}|$$
$$= \frac{|\mathbf{u} \cdot (\mathbf{v} \times \mathbf{w})|}{|\mathbf{v} \times \mathbf{w}|} |\mathbf{v} \times \mathbf{w}|$$
$$= |\mathbf{u} \cdot (\mathbf{v} \times \mathbf{w})|$$

We have shown, then, that

The volume of the parallelepiped with adjacent edges **u**, **v**, and **w** is

$$\text{vol} = |\mathbf{u} \cdot (\mathbf{v} \times \mathbf{w})|$$

EXAMPLE 11.20 Given the points $A(3, -1, 1)$, $B(2, 3, -2)$, $C(0, 1, 3)$, and $D(-1, 2, 4)$, find the volume of the parallelepiped determined by the vectors $\overrightarrow{AB}$, $\overrightarrow{AC}$, and $\overrightarrow{AD}$.

Solution Let

$$\mathbf{u} = \overrightarrow{AB} = \langle -1, 4, -3 \rangle$$
$$\mathbf{v} = \overrightarrow{AC} = \langle -3, 2, 2 \rangle$$
$$\mathbf{w} = \overrightarrow{AD} = \langle -4, 3, 3 \rangle$$

By Equation 11.25,

$$\mathbf{u} \cdot (\mathbf{v} \times \mathbf{w}) = \begin{vmatrix} -1 & 4 & -3 \\ -3 & 2 & 2 \\ -4 & 3 & 3 \end{vmatrix} = -1 \begin{vmatrix} 2 & 2 \\ 3 & 3 \end{vmatrix} - 4 \begin{vmatrix} -3 & 2 \\ -4 & 3 \end{vmatrix} - 3 \begin{vmatrix} -3 & 2 \\ -4 & 3 \end{vmatrix}$$
$$= 0 - 4(-1) - 3(-1) = 7$$

The volume is $|\mathbf{u} \cdot (\mathbf{v} \times \mathbf{w})| = |7| = 7$. ∎

Coplanar Vectors

In arriving at the volume of the parallelepiped determined by **u**, **v**, and **w** as $|\mathbf{u} \cdot (\mathbf{v} \times \mathbf{w})|$, we assumed the vectors were not coplanar. However, an analysis of the computations we made will show that for any nonzero vectors **u**, **v**, and **w**,

$$|\mathbf{u} \cdot (\mathbf{v} \times \mathbf{w})| = |\text{Comp}_{\mathbf{v} \times \mathbf{w}} \mathbf{u}| \, |\mathbf{v} \times \mathbf{w}|$$

and if **u** is in the same plane with **v** and **w**, $\text{Comp}_{\mathbf{v} \times \mathbf{w}} \mathbf{u} = 0$, since $\mathbf{v} \times \mathbf{w}$ is orthogonal to **u**. Conversely, if $\text{Comp}_{\mathbf{v} \times \mathbf{w}} \mathbf{u} = 0$, then **u** is orthogonal to $\mathbf{v} \times \mathbf{w}$ and hence in the plane of **v** and **w**. We therefore conclude that if **u**, **v**, and **w** have the same initial point,

> **u**, **v**, and **w** are coplanar if and only if $\mathbf{u} \cdot (\mathbf{v} \times \mathbf{w}) = 0$.

VECTOR OPERATIONS USING COMPUTER ALGEBRA SYSTEMS

The examples considered in this section use Maple and Mathematica* to examine most of the vector operations that have been discussed in the first four sections of this chapter.

CAS 39

Define the vectors **u** and **v** as $\mathbf{u} = \langle 3, 2, 4 \rangle$ and $\mathbf{v} = \langle -1, 4, 2 \rangle$. Using Maple and Mathematica, calculate each of the following vector and/or scalar quantities.

(a) $\mathbf{u} + \mathbf{v}$, $\mathbf{u} - \mathbf{v}$ and $-3\mathbf{u}$.
(b) the length of **v**
(c) A unit vector in the direction of **v**
(d) the dot product of **u** and **v**
(e) the angle between **u** and **v**
(f) the component of **u** in the direction of **v**
(g) the projection of **u** on **v**
(h) the cross product of **u** and **v**

---

*From this chapter on, we no longer show the DERIVE commands for solving these examples. They are lengthy, and space considerations do not allow their inclusion.

Maple:[†]

In Maple, you will need to load the libraries plots and linalg using

with(plots):with(linalg):
u:=[3,2,4];
v:=[−1,4,2];
(a)
add(u,v);

Output: [2 6 6]

add(u,−v);

Output: [4 −2 2]

scalarmul(u,−3);

Output: [−9 −6 −12]

(b)
mag_v:=sqrt(v[1]^2+v[2]^2+v[3]^2);

Output: $mag\_v := \sqrt{21}$

(c)
unit_v:=scalarmul(v,1/mag_v);

Output: $unit\_v := \left[-\frac{1}{21}\sqrt{21} \quad \frac{4}{21}\sqrt{21} \quad \frac{2}{21}\sqrt{21}\right]$

(d)
dot_uv:=dotprod(u,v);

Output: $dot\_uv := 13$

(e)
mag_u:=sqrt(u[1]^2+u[2]^2+u[3]^2);

Output: $mag\_u := \sqrt{29}$

angle_uv:=arccos(dot_uv/(mag_u*mag_v));

Output: $angle\_uv := \arccos\left(\frac{13}{609}\sqrt{29}\sqrt{21}\right)$

evalf(");

Output: 1.015980710

(f)
comp_uv:=mag_u*cos(");

Output: $comp\_uv := \frac{13}{21}\sqrt{21}$

(g)
Proj_uv:=scalarmul(v,dot_uv/(mag_v)^2);

Output: $Proj\_uv := \left[\frac{-13}{21} \quad \frac{52}{21} \quad \frac{26}{21}\right]$

(h)
cross_uv:=crossprod(u,v);

Output: $cross\_uv := [-12 \;\; -10 \;\; 14]$

Mathematica:

(a)
u={3,2,4}
v={−1,4,2}
u+v
u−v
−3u

(b)
mv=Sqrt[(v[[1]])^2+(v[[2]])^2+(v[[3]])^2]

(c)
unitv = (1/mv)v

(d)
dotuv = u.v

(e)
mu = Sqrt[(u[[1]])^2+(u[[2]])^2+(u[[3]])^2]
angleuv = ArcCos[dotuv/mu*mv]
N[%]

(f)
compuv = mu*Cos[angleuv]

(g)
Projuv = dotuv/mv^2*v

(h)
cuv = CrossProduct[u,v]

[†]To define vectors in Maple, enclose the coordinates in square brackets [ ], separated by commas; and in Mathematica, enclose the coordinates in curly braces { }, also separated by commas.

Maple:

In Maple, we continue and give the commands necessary to plot **u**, **v**, and the cross product **u** × **v**. Recall the vector **u** × **v** is perpendicular to both vectors **u** and **v**.

u_1:=convert(u,list);

Output: $u\_1 := [3, 2, 4]$

p1:=polygonplot3d([[0,0,0],u_1]);
v_1:=convert(v,list);

Output: $v\_1 := [-1, 4, 2]$

p2:=polygonplot3d([[0,0,0],v_1]);
cross_uv1:=convert(cross_uv,list);

Output: $cross\_uv1 := [-12, -10, 14]$

p3:=polygonplot3d([[0,0,0],cross_uv1]);
display3d({p1,p2,p3},axes=boxed,orientation=[-54,49]);

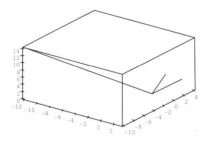

FIGURE 11.4.1

CAS 40

(a) Given the three points in space $A(-1, 2, 3)$, $B(2, 3, -1)$, and $C(3, -2, 4)$, find the area of the triangle that has these three points as vertices.

(b) Given the points $A(3, 2, -1)$, $B(2, 3, -2)$, $C(0, -2, 4)$, and $D(-1, 2, 3)$, find the volume of the parallelepiped determined by the vectors $\overrightarrow{AB}$, $\overrightarrow{AC}$, and $\overrightarrow{AD}$.

To compute the specified area and volume we will use the equations

$$\text{Area} = |\mathbf{u} \times \mathbf{v}|$$

$$\text{Volume} = |\mathbf{u} \cdot (\mathbf{v} \times \mathbf{w})|$$

where **u**, **v**, and **w** define the sides of the objects.

Maple:

(a)
A:=[−1,2,3];
B:=[2,3,−1];
C:=[3,−2,4];
A_B:=add(A,−B);
A_C:=add(A,−C);
Cross:=crossprod(A_B,A_C);
Area:=1/2*sqrt(Cross[1]^2+Cross[2]^2+Cross[3]^2);

Output: $Area := \frac{1}{2}\sqrt{842}$

(b)
A:=[3,2,−1];
B:=[2,3,−2];
C:=[0,−2,4];
D:=[−1,2,3];
u:=add(A,−B);
v:=add(A,−C);
w:=add(A,−D);
vol:=abs(dotprod(u,crossprod(v,w)));

Output: $vol := 24$

Mathematica:

(a)
A = {−1,2,3}
B = {2,3,−1}
C1 = {3,−2,4}
AB = A−B
AC = A−C1
CP = CrossProduct[AB,AC]
Area = 1/2*Sqrt[(CP[[1]])^2+(CP[[2]])^2+(CP[[3]])^2]

(b)
A = {3,2,−1}
B = {2,3,−2}
C1 = {0,−2,4}
D1 = {−1,2,3}
u = A−B
v = A−C1
w = A−D1
vol = Abs[u.CrossProduct[v,w]]

Exercise Set 11.4

In Exercises 1–8, find $\mathbf{u} \times \mathbf{v}$.

1. $\mathbf{u} = \langle 3, 1, -2 \rangle$, $\mathbf{v} = \langle -1, 1, 1 \rangle$

2. $\mathbf{u} = \langle 2, 0, -1 \rangle$, $\mathbf{v} = \langle 0, 2, 1 \rangle$

3. $\mathbf{u} = 4\mathbf{i} - 2\mathbf{j} + \mathbf{k}$, $\mathbf{v} = \mathbf{i} + \mathbf{j} - 2\mathbf{k}$

4. $\mathbf{u} = 3\mathbf{i} - 2\mathbf{j}$, $\mathbf{v} = 2\mathbf{i} + 3\mathbf{k}$

5. $\mathbf{u} = \langle 5, -3, -2 \rangle$, $\mathbf{v} = \langle -2, -3, 1 \rangle$

6. $\mathbf{u} = \langle 2, -1, 2 \rangle$, $\mathbf{v} = \langle -3, 4, -1 \rangle$

7. $\mathbf{u} = \mathbf{i} - 3\mathbf{j} + 4\mathbf{k}$, $\mathbf{v} = 2\mathbf{i} - \mathbf{j} - 5\mathbf{k}$

8. $\mathbf{u} = 6\mathbf{i} - 5\mathbf{j} + 4\mathbf{k}$, $\mathbf{v} = 4\mathbf{i} - 3\mathbf{j} - \mathbf{k}$

In Exercises 9–12, find a vector orthogonal to each of the given vectors.

9. $\mathbf{u} = \langle 0, 1, -3 \rangle$, $\mathbf{v} = \langle 2, 4, -1 \rangle$

10. $\mathbf{u} = \langle 3, 2, -3 \rangle$, $\mathbf{v} = \langle 2, 1, -4 \rangle$

11. $\mathbf{u} = 3\mathbf{i} - 2\mathbf{j} - 5\mathbf{k}$, $\mathbf{v} = \mathbf{i} + 4\mathbf{j} + 3\mathbf{k}$

12. $\mathbf{u} = 2\mathbf{i} - 3\mathbf{j} - \mathbf{k}$, $\mathbf{v} = 3\mathbf{i} - 2\mathbf{j} - \mathbf{k}$

In Exercises 13–19, $\mathbf{u} = 2\mathbf{i} - \mathbf{j} + \mathbf{k}$, $\mathbf{v} = \mathbf{i} + 2\mathbf{j} - 3\mathbf{k}$, *and* $\mathbf{w} = 3\mathbf{i} + 2\mathbf{j} - \mathbf{k}$. *Compute the value of the given expressions in Exercises 13–18.*

13. $\mathbf{u} \cdot (\mathbf{v} \times \mathbf{w})$

14. $(\mathbf{u} \times \mathbf{v}) \cdot \mathbf{w}$

15. $\mathbf{u} \times (\mathbf{v} \times \mathbf{w})$

16. $(\mathbf{u} \times \mathbf{v}) \times \mathbf{w}$

17. $(\mathbf{u} \times \mathbf{v}) \cdot (\mathbf{u} \times \mathbf{w})$

18. $(\mathbf{u} \times \mathbf{v}) \times (\mathbf{u} \times \mathbf{w})$

19. Show that $\mathbf{v}$ and $\mathbf{u} \times \mathbf{v}$ are orthogonal.

In Exercises 20–23, find the area of the parallelogram that has $\overrightarrow{AB}$ and $\overrightarrow{AC}$ as adjacent sides.

20. $A(3, 1, 0)$, $B(2, 2, -1)$, $C(4, 0, 2)$

21. $A(-1, 1, 3)$, $B(1, 3, 2)$, $C(-2, 2, -1)$

22. $A(4, -2, -7)$, $B(3, 1, -5)$, $C(-1, 2, 0)$

23. $A(0, 2, -1)$, $B(4, 0, 2)$, $C(3, -1, -4)$

24. Find the area of the triangle with vertices $A(4, -2, 3)$, $B(6, 1, -1)$, and $C(5, 2, 3)$.

25. Find the area of the triangle with vertices $A(1, 0, -2)$, $B(-3, 2, 1)$, and $C(4, -2, -3)$.

26. Find a vector perpendicular to the plane that contains the points $A(3, 4, 5)$, $B(-1, 2, 4)$, and $C(2, 3, 1)$.

27. Find a unit vector perpendicular to the plane that contains the points $P(0, -1, 3)$, $Q(1, 3, 2)$, and $R(2, -1, 4)$.

In Exercises 28 and 29 find the volume of the parallelepiped that has $\overrightarrow{AB}$, $\overrightarrow{AC}$, and $\overrightarrow{AD}$ as edges.

28. $A(3, 2, -5)$, $B(1, 4, -2)$, $C(-2, 3, 0)$, $D(4, 3, -8)$

29. $A(-2, 0, 4)$, $B(1, 1, 2)$, $C(0, 3, -1)$, $D(-3, -2, 4)$

30. Show that the vectors $\mathbf{u} = 2\mathbf{i} - 3\mathbf{j} + 4\mathbf{k}$, $\mathbf{v} = \mathbf{i} + 2\mathbf{j} - \mathbf{k}$, and $\mathbf{w} = 7\mathbf{i} + 5\mathbf{k}$ are coplanar.

31. Show that the points $A(1, -1, 2)$, $B(3, -4, 1)$, $C(0, 1, 2)$, and $D(1, 0, 1)$ all lie in the same plane.

32. Supply the missing steps in the proof of Theorem 11.7.

33. Prove that for a third-order determinant if two rows are identical, the value of the determinant is 0.

34. Prove that for a third-order determinant if two rows are interchanged, the resulting determinant is the negative of the original.

35. Prove Properties 1, 2, and 3 in Theorem 11.8.

36. Prove Properties 4, 5, and 6 in Theorem 11.8.

37. Prove Properties 7 and 8 in Theorem 11.8.

In Exercises 38–42, prove the given identities based on the properties given in Theorem 11.8, where $\mathbf{u}$, $\mathbf{v}$, $\mathbf{w}$, and $\mathbf{z}$ are three-dimensional vectors.

38. $\mathbf{u} \cdot (\mathbf{u} \times \mathbf{v}) = 0$

39. $(\mathbf{u} + \mathbf{v}) \times (\mathbf{u} - \mathbf{v}) = 2(\mathbf{v} \times \mathbf{u})$

40. $(\mathbf{u} \times \mathbf{v}) \times \mathbf{w} = (\mathbf{u} \cdot \mathbf{w})\mathbf{v} - (\mathbf{v} \cdot \mathbf{w})\mathbf{u}$

41. $\mathbf{u} \times (\mathbf{v} + \mathbf{w}) + \mathbf{v} \times (\mathbf{w} + \mathbf{u}) + \mathbf{w} \times (\mathbf{u} + \mathbf{v}) = \mathbf{0}$

42. $(\mathbf{u} \times \mathbf{v}) \cdot (\mathbf{w} \times \mathbf{z}) = \begin{vmatrix} \mathbf{u} \cdot \mathbf{w} & \mathbf{u} \cdot \mathbf{z} \\ \mathbf{v} \cdot \mathbf{w} & \mathbf{v} \cdot \mathbf{z} \end{vmatrix}$ (*Hint:* First apply Property 8 in Theorem 11.8 to the left-hand side, then Property 7, and use properties of the dot product.)

43. (a) Let $P_1(x_1, y_1, z_1)$, $P_2(x_2, y_2, z_2)$, and $P_3(x_3, y_3, z_3)$ be any three noncollinear points in space. Show that the area of the triangle that has these points as vertices is
$$A = \frac{1}{2} \left| \overrightarrow{P_1 P_2} \times \overrightarrow{P_1 P_3} \right|$$

(b) By treating points in two dimensions as points in three dimensions with the z-coordinate equal to 0, use the result of part (a) to show that the area of a triangle in a two-dimensional coordinate system that has $P_1(x_1, y_1)$, $P_2(x_2, y_2)$, and $P_3(x_3, y_3)$ as vertices can be put in the form
$$A = \pm \frac{1}{2} \begin{vmatrix} x_1 & y_1 & 1 \\ x_2 & y_2 & 1 \\ x_3 & y_3 & 1 \end{vmatrix}$$
where the sign is chosen so that the result is nonnegative.

44. Use a CAS to verify each of the following identities for vectors $\mathbf{u}$, $\mathbf{v}$, and $\mathbf{w}$ in three space.
(a) $|\mathbf{u} \times \mathbf{v}|^2 = |\mathbf{u}|^2 |\mathbf{v}|^2 - (\mathbf{u} \cdot \mathbf{v})^2$
(b) $\mathbf{u} \times (\mathbf{v} \times \mathbf{w}) = (\mathbf{u} \cdot \mathbf{w})\mathbf{v} - (\mathbf{u} \cdot \mathbf{v})\mathbf{w}$

45. Define the vectors $\mathbf{u}$ and $\mathbf{v}$ as $\mathbf{u} = \langle 2, 1, 4 \rangle$ and $\mathbf{v} = \langle -1, -3, 5 \rangle$. Find the cross product of $\mathbf{u}$ and $\mathbf{v}$ and show the three vectors in space.

46. Given the three points $(1, -2, 3)$, $(2, 1, 4)$, and $(1, -2, 1)$, find the area of the triangle that has these three points as vertices.

47. Given the four points $A(2, 4, -2)$, $B(1, -3, 6)$, $C(-1, 2, 4)$, and $D(1, 3, -2)$, find the volume of the parallelepiped determined by the vectors $\overrightarrow{AB}$, $\overrightarrow{AC}$, and $\overrightarrow{AD}$.

48. Find all vectors $\mathbf{v}$ for which $\langle 1, 1, 2 \rangle \times \mathbf{v} = \mathbf{0}$.

11.5 LINES IN SPACE

Vector Equation

A line in space can be described by a point on the line and a direction for the line. The direction is specified by means of a vector, called a **direction vector**, that is parallel to the line. This way of characterizing a line is similar to the two-dimensional case where a point and a slope are given. Suppose l is a line that passes through $P_0(x_0, y_0, z_0)$ and has direction vector $\mathbf{v} = \langle a, b, c \rangle$. Position $\mathbf{v}$ so that its initial point is at P_0, as in Figure 11.36. A point $P(x, y, z)$ will be on l if and only if

$$\overrightarrow{P_0 P} = t\mathbf{v} \qquad (11.26)$$

for some scalar t. As we allow t to range over all real numbers, P traces out the entire line.

Equation 11.26 can be put in another form using position vectors. Recall that if $P(x, y, z)$ is a point in space, then its position vector is the vector $\overrightarrow{OP}$ that has initial point at the origin and terminal point at P. The components of the position vector for P are precisely the coordinates of P, namely, $\langle x, y, z \rangle$. Now let $\mathbf{r}_0$ and $\mathbf{r}$ be the position vectors of the points P_0 and P, respectively, on the line l. Then, as illustrated in Figure 11.37, $\overrightarrow{P_0 P} = \mathbf{r} - \mathbf{r}_0$. So Equation 11.26 can be written in the form

$$\mathbf{r} = \mathbf{r}_0 + t\mathbf{v}, \qquad -\infty < t < \infty \qquad (11.27)$$

and we call this equation a **vector equation for the line** l.

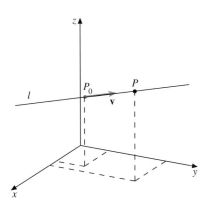

FIGURE 11.36

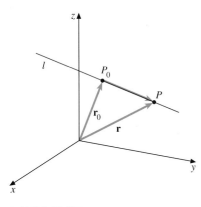

FIGURE 11.37

EXAMPLE 11.21

(a) Find a vector equation of the line l passing through the point $(1, 5, -4)$ and parallel to the vector $\langle 3, -2, 1 \rangle$.

(b) Given the equation

$$\mathbf{r} = (2 - t)\mathbf{i} + (4 + 3t)\mathbf{j} + (-2 + 5t)\mathbf{k}, \qquad -\infty < t < \infty$$

describe its graph.

Solution

(a) Let $\mathbf{r}_0$ be the position vector $\langle 1, 5, -4 \rangle$ of the given point. The vector $\langle 3, -2, 1 \rangle$ is a direction vector for l. So by Equation 11.27, the position vector $\mathbf{r}$ of any point on l is

$$\mathbf{r} = \langle 1, 5, -4 \rangle + t \langle 3, -2, 1 \rangle = \langle 1 + 3t, 5 - 2t, -4 + t \rangle$$

We show the line l along with the direction vector $\langle 3, -2, 1 \rangle$ in Figure 11.38.

(b) We can rewrite the equation as

$$\mathbf{r} = (2\mathbf{i} + 4\mathbf{j} - 2\mathbf{k}) + t(-\mathbf{i} + 3\mathbf{j} + 5\mathbf{k})$$

By comparison with Equation 11.27, we see that this equation is a vector equation for the line passing through $(2, 4, -2)$ and parallel to the vector $-\mathbf{i} + 3\mathbf{j} + 5\mathbf{k}$. (See Figure 11.39.) ∎

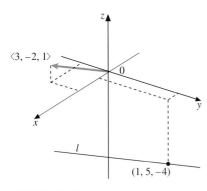

FIGURE 11.38

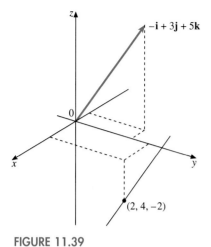

FIGURE 11.39

REMARK
■ Notice that we call Equation 11.27 *a* vector equation for l rather than *the* vector equation, because neither $\mathbf{r}_0$ nor $\mathbf{v}$ is unique. We are free to use *any* point on l as P_0 and *any* nonzero vector parallel to l as its direction vector $\mathbf{v}$.

EXAMPLE 11.22 Find a vector equation of the line l passing through the points $P(2, -1, 4)$ and $Q(3, 2, -1)$.

Solution We can use either point P or Q as the fixed point with position vector $\mathbf{r}_0$ of Equation 11.27. Letting this point be P, we have $\mathbf{r}_0 = \langle 2, -1, 4 \rangle$. The direction vector $\mathbf{v}$ can be taken as $\overrightarrow{PQ} = \langle 1, 3, -5 \rangle$. Thus, a vector equation for l is

$$\mathbf{r} = \langle 2, -1, 4 \rangle + t \langle 1, 3, -5 \rangle = \langle 2 + t, -1 + 3t, 4 - 5t \rangle$$

Notice that when $t = 0$, $\mathbf{r}$ is the position vector of P and when $t = 1$, $\mathbf{r}$ is the position vector of Q. ■

Parametric Equations of a Line

If we write Equation 11.27 in component form, we get

$$\langle x, y, z \rangle = \langle x_0, y_0, z_0 \rangle + t \langle a, b, c \rangle$$
$$= \langle x_0 + at, y_0 + bt, z_0 + ct \rangle$$

and equating components yields

$$\begin{cases} x = x_0 + at \\ y = y_0 + bt \\ z = z_0 + ct \end{cases} \quad -\infty < t < \infty \quad (11.28)$$

Equations 11.28 are **parametric equations for the line** l. If we know a vector equation for l, we can write parametric equations, and conversely.

EXAMPLE 11.23 Find parametric equations of the line in Example 11.22.

Solution In Example 11.22, we found that

$$\mathbf{r} = \langle 2 + t, -1 + 3t, 4 - 5t \rangle$$

So parametric equations are

$$\begin{cases} x = 2 + t \\ y = -1 + 3t \\ z = 4 - 5t \end{cases} \quad -\infty < t < \infty \quad ■$$

EXAMPLE 11.24 Give two points and a direction vector for the line l that has parametric equations $x = 3 - 5t$, $y = -4 + 7t$, and $z = 10 - 8t$.

Solution One point on the line is $(3, -4, 10)$ corresponding to $t = 0$. We get another point using a different value of t, say $t = 1$, giving $(-2, 3, 2)$. A direction vector is $\langle -5, 7, -8 \rangle$. ■

Direction Numbers

The numbers a, b, and c in Equations 11.28 that are the components of a direction vector $\mathbf{v}$ for l are also called **direction numbers** for l. Since $k\mathbf{v} = \langle ka, kb, kc \rangle$ is also a direction vector for l for any nonzero scalar k, it follows that ka, kb, and kc also are direction numbers for l. Knowing a set of direction numbers for l is equivalent to knowing a direction vector for l.

EXAMPLE 11.25 Find a set of direction numbers for the line
$$\mathbf{r} = \langle 3 - 6t, 2 + 4t, -5 - 8t \rangle$$

Solution One set of direction numbers is $-6, 4, -8$. Another set is $3, -2, 4$, obtained by multiplying the first set by $-\frac{1}{2}$. ∎

The Angle Between Two Lines

By the *angle between two lines* l_1 and l_2, we mean the angle between a direction vector for l_1 and a direction vector for l_2, or its supplement, whichever does not exceed $\pi/2$. If $\mathbf{u}$ is any direction vector for l_1 and $\mathbf{v}$ is any direction vector for l_2, then by our definition and by Equation 11.5, the angle θ between l_1 and l_2 satisfies

$$\cos\theta = \frac{|\mathbf{u} \cdot \mathbf{v}|}{|\mathbf{u}||\mathbf{v}|}, \qquad 0 \leq \theta \leq \frac{\pi}{2}$$

The absolute value of $\mathbf{u} \cdot \mathbf{v}$ is needed to ensure that $\cos\theta \geq 0$. The lines are *parallel* if $\mathbf{u}$ and $\mathbf{v}$ are parallel, and they are *orthogonal* if $\mathbf{u}$ and $\mathbf{v}$ are orthogonal. Lines that do not intersect and are not parallel are called **skew lines**.

EXAMPLE 11.26 Find the angle between the lines l_1 and l_2, defined by $\mathbf{r} = \langle 1 - 2t, 3 + t, -2 + 3t \rangle$ and $\mathbf{r} = \langle -2 + t, 4, 3 - t \rangle$, respectively.

Solution A direction vector for l_1 is $\mathbf{u} = \langle -2, 1, 3 \rangle$, and a direction vector for l_2 is $\mathbf{v} = \langle 1, 0, -1 \rangle$. So for the angle θ between l_1 and l_2,

$$\cos\theta = \frac{|\mathbf{u} \cdot \mathbf{v}|}{|\mathbf{u}||\mathbf{v}|} = \frac{|\langle -2, 1, 3 \rangle \cdot \langle 1, 0, -1 \rangle|}{\sqrt{4+1+9}\sqrt{1+1}} = \frac{5}{2\sqrt{7}}$$

$$\theta = \cos^{-1}\frac{5}{2\sqrt{7}} \approx 0.3335 \text{ radian} \approx 19.11°$$
∎

EXAMPLE 11.27 Let l_1 be the line with vector equation $\mathbf{r} = \langle 3 + 3t, 5 - t, -1 + 2t \rangle$ and l_2 be the line with vector equation $\mathbf{r} = \langle -1 + 4t, 7 + 2t, 3 - 5t \rangle$.

(a) Show that l_1 and l_2 are orthogonal.
(b) Find a vector equation of a line passing through $(4, -2, 0)$ that is parallel to l_1.

Solution

(a) Direction vectors for l_1 and l_2 are $\mathbf{u} = \langle 3, -1, 2 \rangle$ and $\mathbf{v} = \langle 4, 2, -5 \rangle$, respectively. Since $\mathbf{u} \cdot \mathbf{v} = 12 - 2 - 10 = 0$, it follows that $\mathbf{u}$ and $\mathbf{v}$, and hence l_1 and l_2, are orthogonal.

(b) We may use the same direction vector $\mathbf{u} = \langle 3, -1, 2 \rangle$ as for l_1. Only the fixed point need be changed. So the equation is

$$\mathbf{r} = \langle 4, -2, 0 \rangle + t \langle 3, -1, 2 \rangle = \langle 4 + 3t, -2 - t, 2t \rangle$$

∎

EXAMPLE 11.28 Find a vector equation of the line passing through $(7, -1, 2)$ that intersects the line $\mathbf{r} = \langle 2 - t, 4 + 3t, 5 - 2t \rangle$ orthogonally.

Solution Let l_1 denote the given line and Q the given point. We need to determine the point on l_1, call it P_1, so that the line l_2 passing through P_1 and Q intersects l_1 in a right angle. (See Figure 11.40.) Equivalently, we need to find P_1, so that the dot product of the direction vector $\overrightarrow{P_1 Q}$ for the line l_2 and a direction vector for l_1 is zero.

Since P_1 lies on l_1, its coordinates can be written as $(2 - t_1, 4 + 3t_1, 5 - 2t_1)$ for some value of t_1, yet to be determined. Then

$$\overrightarrow{P_1 Q} = \langle 5 + t_1, -5 - 3t_1, -3 + 2t_1 \rangle$$

A direction vector for l_1 is

$$\mathbf{v} = \langle -1, 3, -2 \rangle$$

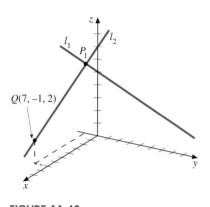

FIGURE 11.40

For orthogonality we must have

$$\mathbf{v} \cdot \overrightarrow{P_1 Q} = 0$$

or,

$$-5 - t_1 - 15 - 9t_1 + 6 - 4t_1 = 0$$
$$-14 t_1 = 14$$
$$t_1 = -1$$

So the lines intersect at $(3, 1, 7)$, and the direction vector $\overrightarrow{P_1 Q}$ for l_2 is $\langle 4, -2, -5 \rangle$. An equation for l_2 is therefore

$$\mathbf{r} = \langle 3, 1, 7 \rangle + t \langle 4, -2, -5 \rangle$$

or

$$\mathbf{r} = \langle 3 + 4t, 1 - 2t, 7 - 5t \rangle$$

∎

EXAMPLE 11.29 Let l_1 and l_2 be defined parametrically by

$$l_1 : \begin{cases} x = 3 - 2t \\ y = -1 + t \\ z = 2 + 3t \end{cases} \qquad l_2 : \begin{cases} x = 1 - s \\ y = 2 + 3s \\ z = 7 + 4s \end{cases}$$

Determine whether l_1 and l_2 intersect, are parallel, or are skew. If they intersect, find the point of intersection.

Solution Notice that different letters are used to designate the parameters. This distinction is important, since otherwise we would disguise the fact that, if the lines intersect, the point of intersection might occur for different values of the parameters for l_1 or l_2.

Direction vectors for l_1 and l_2 are $\mathbf{u} = \langle -2, 1, 3 \rangle$ and $\mathbf{v} = \langle -1, 3, 4 \rangle$, and since $\mathbf{u} \neq k \mathbf{v}$, the lines are not parallel. To see whether they intersect, our approach will be to determine values of s and t that give the same values of x and y and then test to see whether the z values also are the same.

Setting the x and y values equal to each other gives

$$\begin{cases} 3 - 2t = 1 - s \\ -1 + t = 2 + 3s \end{cases} \quad \text{or} \quad \begin{cases} -2t + s = -2 \\ t - 3s = 3 \end{cases}$$

The simultaneous solution is easily found to be $t = \frac{3}{5}$, $s = -\frac{4}{5}$. The corresponding values of x and y are $x = \frac{9}{5}$, $y = -\frac{2}{5}$. The critical test is for z. For l_1, we have

$$z = 2 + 3t = 2 + 3\left(\frac{3}{5}\right) = \frac{19}{5}$$

and for l_2,

$$z = 7 + 4s = 7 + 4\left(-\frac{4}{5}\right) = \frac{19}{5}$$

The fact that we get the same value of z tells us that the lines do intersect. The point of intersection is $\left(\frac{9}{5}, -\frac{2}{5}, \frac{19}{5}\right)$.

If the z values had been different, we would have concluded that the lines were skew. ∎

Exercise Set 11.5

In Exercises 1–4, a point P_0 and a vector $\mathbf{v}$ are given. Find (a) a vector equation and (b) parametric equations for the line through P_0 that has direction vector $\mathbf{v}$.

1. $P_0(2, 5, -1)$, $\mathbf{v} = \langle -3, 1, 2 \rangle$

2. $P_0(-1, -2, 4)$, $\mathbf{v} = \langle 2, 5, -7 \rangle$

3. $P_0(5, 8, -6)$, $\mathbf{v} = 2\mathbf{i} - 3\mathbf{j} + 4\mathbf{k}$

4. $P_0(3, -9, 4)$, $\mathbf{v} = 3\mathbf{i} + 2\mathbf{j} - 5\mathbf{k}$

In Exercises 5–8, find a vector equation for the line through P and Q.

5. $P(4, -1, 8)$, $Q(3, 2, 5)$

6. $P(-1, 5, -6)$, $Q(2, 3, -1)$

7. $P(7, -2, -4)$, $Q(3, 1, -2)$

8. $P(4, 6, 9)$, $Q(1, -1, 5)$

In Exercises 9–12, find the angle between l_1 and l_2.

9. l_1: $\mathbf{r} = \langle 1 - 2t, 3 + t, 4 - 5t \rangle$
   l_2: $\mathbf{r} = \langle 2 - t, 1 - 2t, 3 + 2t \rangle$

10. l_1: $\mathbf{r} = (3 + 4t)\mathbf{i} + (2 - t)\mathbf{j} + (2 + 3t)\mathbf{k}$
    l_2: $\mathbf{r} = (1 - 3t)\mathbf{i} + (4 + t)\mathbf{j} + (7 - 2t)\mathbf{k}$

11. l_1: $x = 5 + 3t$, $y = 7 + 4t$, $z = 11 - 2t$
    l_2: $x = 4 - t$, $y = 5 + 2t$, $z = -1 + 3t$

12. l_1 has direction numbers $-1, 2, 3$, and l_2 has direction numbers $3, 5, -2$.

In Exercises 13 and 14, show that l_1 and l_2 are orthogonal.

13. l_1: $\mathbf{r} = \langle 8 + 3t, -6 - 2t, 7 + t \rangle$
    l_2: $\mathbf{r} = \langle 11 + 7t, 9 + 8t, 3 - 5t \rangle$

14. l_1: $x = 13 - 4t$, $y = -7 - 3t$, $z = 4 + 3t$
    l_2: $x = 6 + 6t$, $y = 8 - 5t$, $z = 12 + 3t$

15. Find parametric equations of the line through $(3, -1, 2)$ that is parallel to the line $\mathbf{r} = \langle 2 - 3t, 7 + t, 8 + 5t \rangle$.

16. Find a vector equation of the line through $P_1(4, -1, 3)$ that is parallel to the line through $P_2(-1, 0, 4)$ and $P_3(1, 3, 2)$.

In Exercises 17 and 18, find a vector equation of the line that passes through P_0 and has a direction vector that is orthogonal to both lines whose equations are given.

17. $P_0 = (5, 2, -3)$; $\mathbf{r} = \langle 2+t, 3-2t, 4-5t \rangle$,
    $\mathbf{r} = \langle 1-t, 2t, 3+4t \rangle$

18. $P_0 = (0, -1, 2)$; $\mathbf{r} = (3+2t)\mathbf{i} + (4-3t)\mathbf{j} + (-2-t)\mathbf{k}$,
    $\mathbf{r} = (2-4t)\mathbf{i} + (-1+t)\mathbf{j} + 2\mathbf{k}$

19. Show that l_1 and l_2 intersect, and find parametric equations of a line orthogonal to both l_1 and l_2 at their point of intersection.

$$l_1 : \begin{cases} x = 2 - 3t \\ y = 1 + t \\ z = 5 - 4t \end{cases} \quad l_2 : \begin{cases} x = 5 + 3s \\ y = -2 - 2s \\ z = 3 + s \end{cases}$$

In Exercises 20–24, determine whether l_1 and l_2 are parallel, intersecting, or skew. If they intersect, find their point of intersection.

20. l_1: $\mathbf{r} = \langle 11-t, 7+2t, 8-3t \rangle$
    l_2: $\mathbf{r} = \langle 4+3s, 2-6s, 5+9s \rangle$

21. l_1: $x = 4-t$, $y = 2t$, $z = 3+4t$
    l_2: $x = 2+3s$, $y = 1-s$, $z = 4+s$

22. l_1: $\mathbf{r} = (3-4t)\mathbf{i} + (2+3t)\mathbf{j} + (1-t)\mathbf{k}$
    l_2: $\mathbf{r} = (2+2s)\mathbf{i} + (5-3s)\mathbf{j} + s\mathbf{k}$

23. l_1: $\mathbf{r} = \langle 3-4t, 2+t, 2t \rangle$
    l_2: $\mathbf{r} = \langle 3+2s, 1-s, 8+3s \rangle$

24. l_1: $\mathbf{r} = \langle 3-4t, 7+2t, -8-3t \rangle$
    l_2: $\mathbf{r} = \left\langle 11+2s, 3-s, \dfrac{-4+3s}{2} \right\rangle$

25. If a, b, and c are all nonzero, show that the line $\mathbf{r} = \langle x_0 + at, y_0 + bt, z_0 + ct \rangle$ can be described by the equations

$$\frac{x - x_0}{a} = \frac{y - y_0}{b} = \frac{z - z_0}{c} \quad (11.29)$$

These are called **symmetric equations** for the line. (*Hint:* Write parametric equations and eliminate the parameter.)

26. Referring to Exercise 25, suppose a line l has the symmetric equations

$$\frac{x-2}{3} = \frac{y+1}{2} = \frac{z-3}{-5}$$

(a) Give two points on l.
(b) Find a unit direction vector for l.
(c) Give parametric equations for l.
(d) Give a vector equation for l.

27. A line l has direction numbers $-2, 4, 3$ and it contains the point $(3, -1, 4)$.
    (a) Find symmetric equations for l. (See Exercise 25.)
    (b) Find parametric equations for l.
    (c) Determine a vector equation for l.
    (d) Find the points where l pierces each of the coordinate planes.

28. Find a vector equation of the line passing through $(4, 0, -5)$ that intersects $\mathbf{r} = \langle 3-t, 2t, 4+3t \rangle$ orthogonally.

29. Let l_1 be the line $\mathbf{r} = (3+t)\mathbf{i} + (4-2t)\mathbf{j} + (5+2t)\mathbf{k}$ and let l_2 be a line passing through $(-1, 2, 8)$ that intersects l_1 so that the angle between l_1 and l_2 is $\frac{\pi}{4}$. Find a vector equation for l_2. (There are two solutions.)

30. Let l_1 and l_2 be the lines with vector equations $\mathbf{r} = \langle 3-2t, 4+3t, 1-t \rangle$ and $\mathbf{r} = \langle 5+4t, 2-6t, 3+2t \rangle$, respectively.
    (a) Show there is a line l_3 passing through $(2, 5, 0)$ that intersects l_1 and l_2 orthogonally, and find a vector equation for l_3.
    (b) Find a vector equation of a line l_4 passing through the point $(5, -1, 7)$ that is perpendicular to the plane containing l_1 and l_2.

31. Let l be a line with direction vector $\mathbf{v}$. Show that the distance d from l to a point P not on l is

$$d = \frac{|\overrightarrow{PQ} \times \mathbf{v}|}{|\mathbf{v}|}$$

where Q is any point on l.

32. Find a vector equation of the line passing through the point $(1, -3, 5)$ and parallel to the vector $\langle 2, -3, 1 \rangle$. Draw the curve in space.

33. Find the angle between the lines defined by $\mathbf{r} = \langle 2+t, 1-t, -1+2t \rangle$ and $\mathbf{r} = \langle 2-t, 1+t, -2-3t \rangle$. Sketch the two lines in space and find their point of intersection.

34. Find the distance from the point $(1, \sqrt{2})$ to the line $2x - 3y = 2$.

35. Use the formula $\mathbf{r} = (1-t)(\overrightarrow{OP}) + t(\overrightarrow{OQ})$, $0 \leq t \leq 1$ to find the parametric equations of the line segment from $P(1, -2, 4)$ to $Q(-2, 5, 6)$. Plot the line segment in space.

11.6 PLANES

Equations of Planes

Given a point $P_0(x_0, y_0, z_0)$ and a nonzero vector $\mathbf{n} = \langle a, b, c \rangle$, the set of all points P such that $\overrightarrow{P_0P}$ and $\mathbf{n}$ are orthogonal is a plane (see Figure 11.41). The vector $\mathbf{n}$ is called a **normal vector** (or simply a **normal**) to the plane. The condition for orthogonality can be written as

$$\mathbf{n} \cdot \overrightarrow{P_0P} = 0 \tag{11.30}$$

As with equations of lines, if we let $\mathbf{r}_0$ and $\mathbf{r}$ be the position vectors of P_0 and P, respectively, then $\overrightarrow{P_0P} = \mathbf{r} - \mathbf{r}_0$, so Equation 11.30 becomes

$$\mathbf{n} \cdot (\mathbf{r} - \mathbf{r}_0) = 0 \tag{11.31}$$

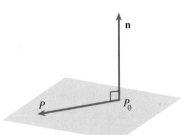

FIGURE 11.41

We call Equation 11.31 a **vector equation of the plane**. In terms of components, $\mathbf{r} - \mathbf{r}_0 = \langle x - x_0, y - y_0, z - z_0 \rangle$, so Equation 11.31 becomes

$$a(x - x_0) + b(y - y_0) + c(z - z_0) = 0 \tag{11.32}$$

Equation 11.32 is known as a **standard form** of the equation of the plane with normal $\langle a, b, c \rangle$ and containing the point (x_0, y_0, z_0). Neither the vector equation nor the standard form is unique, since we may use any point P_0 on the plane and any normal vector $\mathbf{n}$. Note, however, that all normal vectors to a given plane are parallel.

EXAMPLE 11.30 Find the equation of the plane passing through the point $(3, -1, 4)$ and having normal vector $\langle 2, 5, -3 \rangle$.

Solution From Equation 11.32, we obtain

$$2(x - 3) + 5(y + 1) - 3(z - 4) = 0$$

Simplifying, we get

$$2x + 5y - 3z + 11 = 0 \qquad \blacksquare$$

If we carry out the indicated multiplications in Equation 11.32 and simplify, as we did in Example 11.30, we get an equation of the form

$$ax + by + cz + d = 0 \qquad (11.33)$$

where $d = -(ax_0 + by_0 + cz_0)$. We can also reverse this procedure. Suppose we are given an equation in the form of Equation 11.33. We find a point (x_0, y_0, z_0) that satisfies the equation (for example, by choosing x_0 and y_0 arbitrarily and solving for z_0). Then, $ax_0 + by_0 + cz_0 + d = 0$, so that $d = -ax_0 - by_0 - cz_0$. Substituting this value of d into Equation 11.33, we get

$$ax + by + cz + (-ax_0 - by_0 - cz_0) = 0$$

or

$$a(x - x_0) + b(y - y_0) + c(z - z_0) = 0$$

This equation is the standard form (11.32), so we know it represents a plane. An equation of the form of Equation 11.33 is called **linear** (meaning first degree) in x, y, and z. So what we have is that *every linear equation in x, y, and z represents a plane in space*. Furthermore, the coefficients of x, y, and z, in order, are components of a normal vector to the plane. We also refer to Equation 11.33 as a **general form** of the equation of a plane.

Finding Normal Vectors

A normal vector is not always given directly but can be found from the given information. Frequently, finding a normal involves finding a cross product, as the next two examples show.

EXAMPLE 11.31 Find an equation for the plane that contains the points $P(1, 0, -3)$, $Q(2, -5, -6)$, and $R(6, 3, -4)$.

Solution As we show in Figure 11.42, the vectors $\overrightarrow{PQ}$ and $\overrightarrow{PR}$ lie in the plane, so a normal can be found by taking their cross product:

$$\overrightarrow{PQ} = \langle 1, -5, -3 \rangle \quad \text{and} \quad \overrightarrow{PR} = \langle 5, 3, -1 \rangle$$

$$\overrightarrow{PQ} \times \overrightarrow{PR} = \begin{vmatrix} \mathbf{i} & \mathbf{j} & \mathbf{k} \\ 1 & -5 & -3 \\ 5 & 3 & -1 \end{vmatrix} = \begin{vmatrix} -5 & -3 \\ 3 & -1 \end{vmatrix} \mathbf{i} - \begin{vmatrix} 1 & -3 \\ 5 & -1 \end{vmatrix} \mathbf{j} + \begin{vmatrix} 1 & -5 \\ 5 & 3 \end{vmatrix} \mathbf{k}$$

$$= 14\mathbf{i} - 14\mathbf{j} + 28\mathbf{k}$$

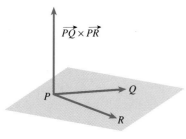

FIGURE 11.42

Since any nonzero vector perpendicular to the plane is a suitable normal, we choose $\mathbf{n} = \frac{1}{14}(14\mathbf{i} - 14\mathbf{j} + 28\mathbf{k}) = \langle 1, -1, 2 \rangle$. We can use any one of P, Q, or R as the point (x_0, y_0, z_0). Choosing P, we get for the equation in standard form

$$(x - 1) - (y - 0) + 2(z + 3) = 0$$

or in general form

$$x - y + 2z + 5 = 0$$

EXAMPLE 11.32 Find an equation of the plane that contains the line $\mathbf{r} = \langle 2 - t, 3 + 4t, -1 - 2t \rangle$ and the point $(5, -2, 7)$.

Solution To use Equation 11.32 we need a normal vector to the plane and a point in the plane. To find a normal, we can take the cross product of two vectors in the plane. One vector in the plane is a direction vector of the given line. Since the equation of the line is $\mathbf{r} = \langle 2 - t, 3 + 4t, -1 - 2t \rangle$, a direction vector for it is

$$\mathbf{v} = \langle -1, 4, -2 \rangle$$

To find another vector in the plane, we can select any point P_0 on the given line and then use the vector $\overrightarrow{P_0 Q}$, where Q is the given point $(5, -2, 7)$. (See Figure 11.43.) A convenient point P_0 on the line is found by taking $t = 0$, giving $P_0 = (2, 3, -1)$. Thus,

$$\overrightarrow{P_0 Q} = \langle 3, -5, 8 \rangle$$

So a normal $\mathbf{n}$ is

$$\mathbf{v} \times \overrightarrow{P_0 Q} = \langle -1, 4, -2 \rangle \times \langle 3, -5, 8 \rangle = \langle 22, 2, -7 \rangle$$

(You should supply the missing steps.) We can now write the equation of the plane as

$$22(x - 2) + 2(y - 3) - 7(z + 1) = 0$$

or, equivalently,

$$22x + 2y - 7z - 57 = 0 \qquad \blacksquare$$

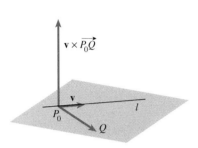

FIGURE 11.43

Parallel and Perpendicular Planes; The Angle Between Two Planes

Two planes are parallel if their respective normals are parallel, and *two planes are perpendicular* if their respective normals are orthogonal. A line and a plane are parallel if a direction vector for the line and a normal to the plane are orthogonal. By the angle between two planes with normals $\mathbf{n}_1$ and $\mathbf{n}_2$, we mean the angle between $\mathbf{n}_1$ and $\mathbf{n}_2$ or its supplement, whichever does not exceed $\pi/2$. If we call this angle θ, then

$$\cos \theta = \frac{|\mathbf{n}_1 \cdot \mathbf{n}_2|}{|\mathbf{n}_1| |\mathbf{n}_2|}, \qquad 0 \leq \theta \leq \frac{\pi}{2} \qquad (11.34)$$

In Figure 11.44 we show planes that are parallel, planes that are perpendicular, and planes intersecting at an angle θ.

EXAMPLE 11.33 Find the angle between the planes

$$2x - 3y + 4z - 7 = 0 \quad \text{and} \quad x + y - 2z + 9 = 0$$

Solution A normal vector to a plane of the form $ax + by + cz + d = 0$ is $\langle a, b, c \rangle$. So normals to the given planes are $\langle 2, -3, 4 \rangle$ and $\langle 1, 1, -2 \rangle$,

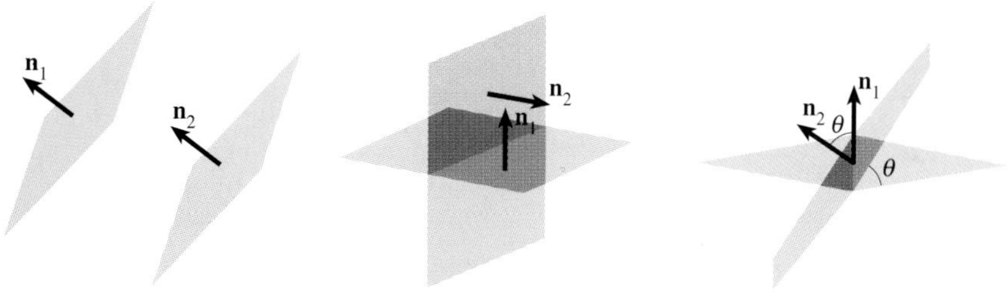

(a) Parallel planes
$\mathbf{n}_1$ and $\mathbf{n}_2$ are parallel

(b) Perpendicular planes
$\mathbf{n}_1$ and $\mathbf{n}_2$ are orthogonal

(c) The angle between the planes is θ if the acute angle between their normals is θ

FIGURE 11.44

respectively. So by Equation 11.34,

$$\cos\theta = \frac{|\langle 2, -3, 4\rangle \cdot \langle 1, 1, -2\rangle|}{|\langle 2, -3, 4\rangle| |\langle 1, 1, -2\rangle|}$$

$$= \frac{|2 - 3 - 8|}{\sqrt{29}\sqrt{6}} = \frac{9}{\sqrt{174}}$$

Using a calculator, we find that $\theta \approx 46.98°$. ■

EXAMPLE 11.34 Find an equation of the plane that contains the line

$$\mathbf{r} = (2 + t)\mathbf{i} + (-3 + 4t)\mathbf{j} + (1 - t)\mathbf{k}$$

and that is perpendicular to the plane $3x - 4y + 5z + 7 = 0$.

Solution Since the plane we want contains the line

$$\mathbf{r} = (2 + t)\mathbf{i} + (-3 + 4t)\mathbf{j} + (1 - t)\mathbf{k}$$
$$= \langle 2, -3, 1\rangle + t\langle 1, 4, -1\rangle$$

its normal must be orthogonal to the direction vector of the line,

$$\langle 1, 4, -1\rangle = \mathbf{i} + 4\mathbf{j} - \mathbf{k}$$

The plane we want is also perpendicular to the plane $3x - 4y + 5z + 7 = 0$, with normal vector $\langle 3, -4, 5\rangle = 3\mathbf{i} - 4\mathbf{j} + 5\mathbf{k}$. So the normal to the desired plane is orthogonal to this normal vector.

We therefore can obtain a normal to the plane we want by taking the cross product

$$(\mathbf{i} + 4\mathbf{j} - \mathbf{k}) \times (3\mathbf{i} - 4\mathbf{j} + 5\mathbf{k}) = 16\mathbf{i} - 8\mathbf{j} + 8\mathbf{k}$$

We use for $\mathbf{n}$ the simpler normal $2\mathbf{i} - \mathbf{j} + \mathbf{k}$. A point in the plane can be taken as any point on the given line. The one corresponding to $t = 0$ is $(2, -3, 1)$. Thus, the desired equation is

$$2(x - 2) - (y + 3) + (z - 1) = 0$$

or, in general form,

$$2x - y + z - 8 = 0$$

■

Sketching Planes

We will find it helpful later to be able to show planes graphically. Since planes are unbounded, it is impossible to graph an entire plane, but we can give an indication of its graph by showing its **traces** on the coordinate planes. These are the lines of intersection of the given plane with the coordinate planes. Finding *intercepts* on the coordinate axes is helpful in getting the traces.

For example, the plane

$$2x + 3y + 4z - 12 = 0$$

has x-intercept 6 (set $y = 0$ and $z = 0$), y-intercept 4 (set $x = 0$ and $z = 0$), and z-intercept 3 (set $x = 0$ and $y = 0$). Connecting these points gives the traces, as shown in Figure 11.45. Two-dimensional equations of the traces are found by setting one of the variables equal to 0. For example, if we set $z = 0$ in the equation of this plane, we get

$$2x + 3y = 12$$

which is the equation of the trace in the xy-plane.

Not all planes have intercepts on each axis. In the next example we illustrate some of these planes.

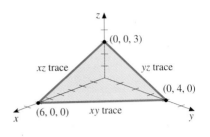

FIGURE 11.45

EXAMPLE 11.35 Describe the planes that have the following equations, and sketch their graphs.

(a) $x = 3$

(b) $z = 4$

(c) $2x + 3y = 6$

Solution Recall that a normal vector to the plane $ax + by + cz + d = 0$ is $\langle a, b, c \rangle$.

(a) The equation $x = 3$ is of the form $ax + by + cz + d = 0$ with $a = 1$, $b = 0$, $c = 0$, and $d = -3$, so a normal vector is $\langle 1, 0, 0 \rangle = \mathbf{i}$. The plane is perpendicular to the x-axis, and hence parallel to the yz-plane. The x-intercept is 3. A portion of its graph is shown in Figure 11.46(a). The plane consists of all points (x, y, z) where x is always equal to 3, and y and z can have all possible values.

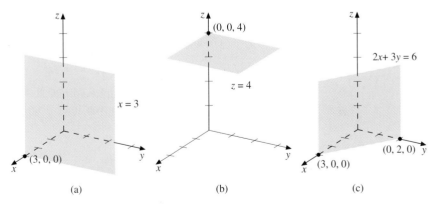

FIGURE 11.46

(b) A normal vector is $\langle 0, 0, 1 \rangle = \mathbf{k}$. So this plane is perpendicular to the z-axis, crossing it at $z = 4$. It is parallel to the xy-plane and 4 units above it. A portion of the graph is shown in Figure 11.46(b).

(c) The x-intercept is 3, and the y-intercept is 2. The xy trace has the same equation as that of the plane, since letting $z = 0$ does not alter the equation. A normal vector is $\langle 2, 3, 0 \rangle$, which is a vector in a horizontal plane, so the plane itself is vertical—that is, parallel to the z-axis. A portion of its graph is shown in Figure 11.46(c). ∎

Line of Intersection of Two Planes

Two nonparallel planes intersect in a line (see Figure 11.47), and we can find an equation of the line from the equations of the planes by the procedure shown in the next example.

EXAMPLE 11.36 Find a vector equation of the line of intersection of the planes $3x - 4y + 2z = 7$ and $x + 2y - 3z = 4$.

Solution To find the equation of the line of intersection, we need a direction vector for the line and a point on it.

Note first that the planes do intersect, since their normal vectors $\mathbf{n}_1 = \langle 3, -4, 2 \rangle$ and $\mathbf{n}_2 = \langle 1, 2, -3 \rangle$ are not parallel (neither is a multiple of the other). So the planes themselves are not parallel. The line of intersection lies in both planes, hence is orthogonal to both of their normals, $\mathbf{n}_1$ and $\mathbf{n}_2$. A direction vector $\mathbf{v}$ of the line is therefore given by the cross product of these normals:

$$\mathbf{v} = \mathbf{n}_1 \times \mathbf{n}_2 = \langle 3, -4, 2 \rangle \times \langle 1, 2, -3 \rangle = \langle 8, 11, 10 \rangle$$

We can find a point on the line by solving simultaneously the equations of the planes. Since there are three variables and two equations, we can select one variable arbitrarily and solve for the other two. Letting $z = 0$, we have

$$\begin{cases} 3x - 4y = 7 \\ x + 2y = 4 \end{cases}$$

The solution is found to be $x = 3$, $y = \frac{1}{2}$. So a point on the line is $\left(3, \frac{1}{2}, 0\right)$. A vector equation of the line is therefore

$$\mathbf{r} = \left\langle 3, \frac{1}{2}, 0 \right\rangle + t \langle 8, 11, 10 \rangle$$

or

$$\mathbf{r} = \left\langle 3 + 8t, \frac{1}{2} + 11t, 10t \right\rangle$$ ∎

Distance Between a Plane and a Point

We conclude this section by deriving a formula for the distance between a plane and a point not on the plane. Let $ax + by + cz + d = 0$ be the equation of the plane, and let the point be $P_1(x_1, y_1, z_1)$. Let $P_0(x_0, y_0, z_0)$ be any point on the plane. Then, as Figure 11.48 shows, the distance D between the plane and the point P_1 is the length of the projection $\overrightarrow{P_0P_1}$ along the normal $\mathbf{n} = \langle a, b, c \rangle$. The length of this projection is $|\text{Comp}_\mathbf{n} \overrightarrow{P_0P_1}|$. So by Equation 11.8 we have

$$D = |\text{Comp}_\mathbf{n} \overrightarrow{P_0P_1}| = \frac{|\mathbf{n} \cdot \overrightarrow{P_0P_1}|}{|\mathbf{n}|} = \frac{|\langle a, b, c \rangle \cdot \langle x_1 - x_0, y_1 - y_0, z_1 - z_0 \rangle|}{\sqrt{a^2 + b^2 + c^2}}$$

$$= \frac{|a(x_1 - x_0) + b(y_1 - y_0) + c(z_1 - z_0)|}{\sqrt{a^2 + b^2 + c^2}}$$

$$= \frac{|ax_1 + by_1 + cz_1 - (ax_0 + by_0 + cz_0)|}{\sqrt{a^2 + b^2 + c^2}}$$

Since P_0 is on the plane, its coordinates satisfy the equation of the plane. So $ax_0 + by_0 + cz_0 = -d$. Thus, we obtain the following formula.

The distance D between the plane $ax + by + cz + d = 0$ and the point (x_1, y_1, z_1) is

$$D = \frac{|ax_1 + by_1 + cz_1 + d|}{\sqrt{a^2 + b^2 + c^2}} \qquad (11.35)$$

REMARK
■ Compare this result with the two-dimensional case of the distance between a line and a point in Exercise 41 of Exercise Set 11.2.

EXAMPLE 11.37 Find the distance between the plane $3x - 4y + 5z - 8 = 0$ and the point $(2, 1, -1)$.

Solution By Equation 11.35,

$$D = \frac{|3(2) - 4(1) + 5(-1) - 8|}{\sqrt{9 + 16 + 25}} = \frac{11}{5\sqrt{2}} \qquad ■$$

INVESTIGATING LINES AND PLANES IN SPACE USING COMPUTER ALGEBRA SYSTEMS

CAS 41

(a) Find a vector equation for the line passing through the point $(-1, 2, 5)$ and parallel to the vector $\langle 2, -2, 1 \rangle$.

(b) Find the angle between the lines defined by $\mathbf{r} = \langle 1 + t, 2 + t, -1 + 2t \rangle$ and $\mathbf{r} = \langle 1 - t, 2 + t, -1 + 3t \rangle$.

(a) From Equation 11.27 the vector equation for the line is given by
$\langle -1, -2, 5 \rangle + t\langle 2, -2, 1 \rangle$.

Maple:

A:=[-1,2,5];
B:=[2,-2,1];
L:=add(A,scalarmul(B,t));
L:=convert(L,list);
spacecurve(L,t=-10..10,scaling=unconstrained,
   orientation=[-7,66],axes=framed);

Mathematica:

A = {-1,2,5}
B = {2,-2,1}
L = A+t*B
ParametricPlot3D[{-1+2*t,2-2*t,5+t},{t,-10,10}]

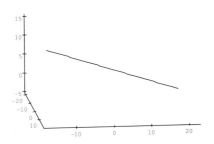

FIGURE 11.6.1

(b) To find the angle θ between the two lines we use the formula $\cos\theta = \dfrac{|\mathbf{u} \cdot \mathbf{v}|}{|\mathbf{u}||\mathbf{v}|}$ where $\mathbf{u}$ and $\mathbf{v}$ are direction vectors for the two lines respectively.

Maple:

L1:=add([1,2,-1],scalarmul([1,1,2],t));
L1:=convert(L1,list);
L2:=add([1,2,-1],scalarmul([-1,1,3],t));
L2:=convert(L2,list);
spacecurve({L1,L2},t=-10..10,scaling=unconstrained,
   orientation=[-35,65],axes=framed);

Mathematica:

L1 = {1,2,-1}+t*{1,1,2}
L2 = {1,2,-1}+t*{-1,1,3}
P1 = ParametricPlot3D[L1,{t,-10,10},
   DisplayFunction->Identity]
P2 = ParametricPlot3D[L2,{t,-10,10},
   DisplayFunction->Identity]
Show[{P1,P2},DisplayFunction->$DisplayFunction]
Solve[L1 == L2 , t]
x[t_] = L1
x[0]
u = {1,1,2}
v = {-1,1,3}
mu = Sqrt[(u[[1]])^2+(u[[2]])^2+(u[[3]])^2]
mv = Sqrt[(v[[1]])^2+(v[[2]])^2+(v[[3]])^2]
angle = ArcCos[Abs[u.v/(mu*mv)]]
anglerad = N[%]
angledeg = N[anglerad*180/Pi]

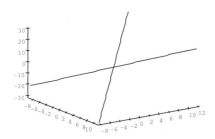

FIGURE 11.6.2

Now find the point of intersection of the two lines.

solve({L1[1]=L2[1],L1[2]=L2[2],L1[3]=L2[3]},t);

Output: $\{t = 0\}$

subs(t=0,L1);

Output: [1, 2, −1]

Direction vectors for the two lines are, respectively,

u:=[1,1,2];
v:=[−1,1,3];
mag_u:=sqrt(u[1]^2+u[2]^2+u[3]^2);

Output: $mag\_u := \sqrt{6}$

mag_v:=sqrt(v[1]^2+v[2]^2+v[3]^2);

Output: $mag\_v := \sqrt{11}$

angle:=arccos(abs(dotprod(u,v)/(mag_u*mag_u)));

Output: $angle := \arccos\left(\frac{1}{11}\sqrt{6}\sqrt{11}\right)$

anglerad:=evalf(angle);

Output: $anglerad := .7398807745$

angledeg:=evalf(anglerad*180/Pi);

Output: $angledeg := 42.39204571$

CAS 42

Make a plot displaying the planes $2x - y + 3z = 2$ and $3x - 2y + z = 6$ and find the vector equation for the line of intersection.

Maple:

plot3d({(2−2*x+y)/3,6−3*x+2*y},x=−10..10,y=−10..10,
  axes=boxed,orientation=[15,62]);

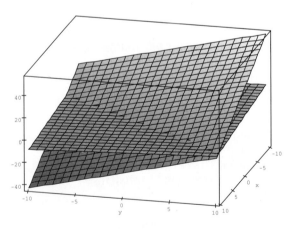

FIGURE 11.6.3

Mathematica:

P1 = Plot3D[(2−2*x+y)/3,{x,−10,10},{y,−10,10},
  DisplayFunction−>Identity]
P2 = Plot3D[6−3*x+2*y,{x,−10,10},{y,−10,10},
  DisplayFunction−>Identity]
Show [{P1,P2},DisplayFunction−>$DisplayFunction]

To find the equation of the line of intersection of the two planes find two points on the line. That is two solutions of the system consisting of the two planes. First set $z = 0$ and then $z = 1$.

Solve[{2*x−y==2 , 3*x−2*y==6},{x,y}]
Solve[{2*x−y==−1 , 3*x−2*y==5},{x,y}]

Then the equation of the line is, in vector form,
$\{-2 - 7t, -6 - 13t, t\}$

We can let Maple find the equation of the line of intersection.

plane(p1,[2*x−y+3*z=2]),plane(p2,[3*x−2*y+z=6]):
inter(p1,p2,l);
l[equation];

Output: $[-2 + 5\_t, -6 + 7\_t, -\_t]$

CAS 43

Planes that are perpendicular to one of the axes can only be plotted using parametric form. Plot each of the following planes on the same coordinate axes:

$$x = 2, \ y = -2 \text{ and } z = 4$$

Maple:

plot3d({[2,t,s],[s,−2,t],[s,t,4]},t=−10..10,s=−10..10,axes=boxed);

Mathematica:

P1 = ParametricPlot3D[{2,s,t},{s,−10,10},{t,−10,10},
DisplayFunction−>Identity]

P2 = ParametricPlot3D[{s,−2,t},{s,−10,10},{t,−10,10},
DisplayFunction−>Identity]

P3 = ParametricPlot3D[{s,t,4},{s,−10,10},{t,−10,10},
DisplayFunction−>Identity]
Show[{P1,P2,P3},DisplayFunction−>$DisplayFunction]

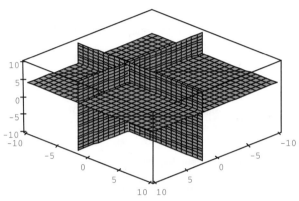

FIGURE 11.6.4

Exercise Set 11.6

In Exercises 1–4, find an equation of the plane through the given point perpendicular to the given vector in (a) vector form, (b) standard form, and (c) general form.

1. $(4, 2, 6)$, $\langle 3, 2, -1 \rangle$

2. $(1, 0, -3)$, $\langle -1, 2, 4 \rangle$

3. $(5, -3, -4)$, $2\mathbf{i} + 3\mathbf{j} - 4\mathbf{k}$

4. $(6, 1, 2)$, $\mathbf{i} - 2\mathbf{j} + 2\mathbf{k}$

In Exercises 5–12, find a general form of the equation of the plane that satisfies the given conditions.

5. Perpendicular to the line $\mathbf{r} = \langle 2 - t, 3 + 2t, 1 + 4t \rangle$ at the point $(2, 3, 1)$

6. Perpendicular to the line $\mathbf{r} = (4 + 5t)\mathbf{i} + 3t\mathbf{j} + (1 - 2t)\mathbf{k}$ at the point $(4, 0, 1)$

7. Containing the point $(1, 4, -5)$ and parallel to the plane $3x - 2y + 4z = 7$

8. Containing the point $(-2, 3, 4)$ and parallel to the plane $2x - 3z = 4$

9. Containing the point $(3, -2, 4)$ and perpendicular to the z-axis

10. Containing the point $(-1, 8, 11)$ and perpendicular to the y-axis

11. Parallel to the plane $3x - 4y + 5z + 8 = 0$ and passing through the origin

12. Perpendicular to the line $\mathbf{r} = \langle 2 + t, 3 - 2t, 4 + 5t \rangle$ and passing through the origin

In Exercises 13–16, find a general equation of the plane that contains the three given points.

13. $(2, 4, -5)$, $(1, -3, 4)$, $(3, -1, 2)$

14. $(0, 3, -1)$, $(2, 4, 2)$, $(-1, 2, -3)$

15. $(1, 0, -1)$, $(2, 3, 1)$, $(4, -3, 2)$

16. $(5, 4, -3)$, $(2, -1, -2)$, $(4, 2, 3)$

In Exercises 17 and 18, find the angle between the two planes.

17. $2x - 3y - 4z = 8$
$3x + 2y - z = 4$

18. $3x + 4y - 2z = 3$
$2x - y - 3z = 5$

In Exercises 19 and 20, show that the two planes are perpendicular.

19. $3x - 4y + 2z = 5$
$2x + 3y + 3z = 7$

20. $5x + 3y - 4z = 8$
$2x - 6y - 2z = 15$

In Exercises 21–24, find a general equation of the plane that satisfies the given conditions.

21. Containing the line $\mathbf{r} = \langle 3 - 2t, 2 + t, 4 + 3t \rangle$ and the point $(-1, 2, 4)$

22. Containing the points $(4, -2, 1)$ and $(3, 1, 2)$ and perpendicular to the plane $3x + 2y - 4z = 5$

23. Perpendicular to each of the planes $3x + 5y - 4z = 4$ and $2x - 3y - z = 2$ and containing the point $(3, -3, 1)$

24. Containing the line $\mathbf{r} = 2\mathbf{i} + (3 - t)\mathbf{j} + (4 + 2t)\mathbf{k}$ and parallel to the line $\mathbf{r} = (3 + 2t)\mathbf{i} + (1 - t)\mathbf{j} + (-2 + 3t)\mathbf{k}$

In Exercises 25–34, use intercepts and traces on the coordinate planes to sketch the given plane.

25. $3x + 2y + z - 6 = 0$

26. $9x + 2y + 6z = 18$

27. $2x - y + z = 4$

28. $x + y - z + 4 = 0$

29. $3x + 4y = 12$

30. $2x + z = 8$

31. $y = 2z$

32. $x + 1 = 0$

33. $y = 4$

34. $x + y - 2z = 0$

In Exercises 35 and 36, find parametric equations of the line of intersection of the two planes.

35. $3x - y - 2z = 4$
$5x + y + z = -2$

36. $x + 4y + 3z = 3$
$2x - 7y + z = 11$

In Exercises 37–38, find the distance between the plane and the point.

37. $3x - 4y + 10z = 5$; $(1, -1, 2)$

38. $2x - y - 2z + 3 = 0$; $(6, -1, -4)$

39. Show that planes $x - 2y + 2z = 3$ and $3x - 6y + 6z + 5 = 0$ are parallel, and find the distance between them. (*Hint:* Find the distance from one of the planes to a point on the other.)

40. Show that l_1 and l_2 are parallel, and find an equation of the plane that contains them.
l_1: $\mathbf{r} = \langle 2 - 3t, 4 + 2t, -3 + t \rangle$
l_2: $\mathbf{r} = \langle 6t, 3 - 4t, 5 - 2t \rangle$

41. Show that l_1 and l_2 intersect, and find an equation of the plane that contains them.
l_1: $\mathbf{r} = \langle 1 + 2t, -1 + 3t, 2 - t \rangle$
l_2: $\mathbf{r} = \langle 5 + 3t, 6 + 5t, -1 - 2t \rangle$

42. Find an equation of the plane perpendicular to the line of intersection of the planes $3x - y - 4z = 2$ and $x + 2y - z = 3$ at the point where this line pierces the xy-plane.

43. Find the point where the line $\mathbf{r} = (2 + t)\mathbf{i} + (3 - 2t)\mathbf{j} + (1 - t)\mathbf{k}$ pierces the plane $3x - 2y + 4z + 5 = 0$.

44. Find the minimum distance between the two skew lines
l_1: $x = 3 + t$, $y = 2 - t$, $z = 4 + 3t$
l_2: $x = -1 - 2t$, $y = 5 + 4t$, $z = -3t$
(*Hint:* First find a plane that contains one of the lines and is parallel to the other.)

45. Make a plot displaying the two planes $x - 2y + 3z = 3$ and $3x - y - z = 4$ and find the equation of the line of intersection.

46. Find the equation of the plane that passes through the three points $A(-1, 2, 3)$, $B(2, 4, 5)$, and $C(1, 2, -3)$. Show the plane.

47. Consider the plane $2x + 3y + z = 3$ and the line L given parametrically by $x = t - 1$, $y = -t + 2$, and $z = t + 3$. By plotting several views of the line and the plane, show convincingly that they are parallel. Why are they parallel?

48. Do the plane $x + 2y + z = 1$ and the line given parametrically by $x = -3t - 1$, $y = 2t - 2$, and $z = t - 1$ intersect? Plot the line and plane first to answer the question; then give an argument for your answer.

49. Find the distance between the point $(-1, 2, 5)$ and the plane $-x + 2y + z = 7$.

50. Find the distance between the parallel planes $x + y - z = 5$ and $x + y - z = 10$. Draw the two planes.

51. A surveyor wishes to measure the height of a tall landmark. Using a rectangular coordinate system, the surveyor selects two points $P(81, 1, 0)$ and $Q(54, 131, 0)$ at ground level and sights from the points to the point T at the top of the landmark. Given that the direction vector for PT is $\langle -48/61, 11/61, 36/61 \rangle$ and the direction vector of QT is $\langle -23/49, -36/49, 24/49 \rangle$, determine the coordinates of the point T and hence deduce the height of the landmark.*

*From Chi-Keung Cheung and John Harer, *A Guidebook to Multivariable Calculus with Maple V* (John Wiley & Sons).

Chapter 11 Review Exercises

Exercises 1 and 2 refer to the vectors $\mathbf{u} = 3\mathbf{i} - 4\mathbf{j}$, $\mathbf{v} = \mathbf{i} + 2\mathbf{j}$, and $\mathbf{w} = 5\mathbf{i} - 7\mathbf{j}$. Find the specified quantities.

1. (a) $(2\mathbf{u} - 3\mathbf{v}) \cdot (\mathbf{v} + 2\mathbf{w})$
  (b) A vector in the direction of $\mathbf{v}$ with length $|3\mathbf{u} - 2\mathbf{w}|^2$

2. (a) $\text{Proj}_\mathbf{v} \mathbf{u}$ and $\text{Proj}_\mathbf{v}^{\perp} \mathbf{u}$
  (b) Scalars a and b such that $\mathbf{w} = a\mathbf{u} + b\mathbf{v}$

Exercises 3–6 refer to the vectors $\mathbf{u} = \langle 1, 1, -4 \rangle$, $\mathbf{v} = \langle -2, 0, 3 \rangle$, and $\mathbf{w} = \langle -2, 1, 2 \rangle$. Find the specified quantities.

3. (a) $|\mathbf{u}|^2 - (\mathbf{v} \cdot \mathbf{w})^2$
  (b) The angle between $\mathbf{u}$ and $\mathbf{w}$

4. (a) $(\mathbf{u} \times \mathbf{v}) \cdot \mathbf{w}$
  (b) $\mathbf{u} \times (\mathbf{v} \times \mathbf{w})$

5. (a) $|2\mathbf{u} - 3\mathbf{v} + \mathbf{w}|$
  (b) $\text{Proj}_\mathbf{w} \mathbf{v}$

6. (a) The component of $\mathbf{v}$ in the direction of $\mathbf{w} \times \mathbf{u}$
  (b) The volume of the parallelepiped with adjacent sides $\mathbf{u}$, $\mathbf{v}$, and $\mathbf{w}$

7. Using vector methods, prove that the line segment joining the midpoints of two sides of a triangle is parallel to the third side and one-half its length.

8. Forces $\mathbf{F}_1$, $\mathbf{F}_2$, and $\mathbf{F}_3$ of magnitudes 25.3 N, 14.8 N, and 19.6 N, respectively, are acting on an object as shown in the figure. Find the magnitude and direction of the resultant force.

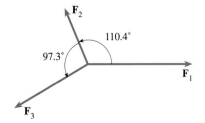

9. A pilot wants to fly from town A to town B, 400 mi due south of town A. A 60-mph wind is blowing from 210°. To make the trip in 2 hr, at what heading and at what average speed should the pilot fly?

10. Prove that for arbitrary three-dimensional vectors $\mathbf{u}$, $\mathbf{v}$, and $\mathbf{w}$, $(\mathbf{u} \times \mathbf{v}) \times \mathbf{w} + (\mathbf{v} \times \mathbf{w}) \times \mathbf{u} + (\mathbf{w} \times \mathbf{u}) \times \mathbf{v} = \mathbf{0}$. (This is known as *Jacobi's identity*.)

In Exercises 11–13, find both vector and parametric equations of the line that satisfies the given conditions.

11. Through $(3, 1, -2)$, parallel to $\mathbf{r} = \langle 2 - t, 4 + 2t, 3 + 5t \rangle$

12. Formed by the intersection of the planes $2x - 3y - z = 4$ and $x + y + 2z + 3 = 0$

13. Through $(4, 2, -1)$, perpendicular to the plane $5x - 3y + 4z = 7$

14. Find the point where the line $\mathbf{r} = \langle 1 - 3t, 2t, 3 + t \rangle$ pierces the plane $3x - 5y + 4z = 5$. Also find the angle between the line and the plane (defined as the complement of the angle between the line and a normal to the plane).

15. Determine whether the lines $\mathbf{r}_1 = \langle 3 - t, 4 + 2t, -1 + t \rangle$ and $\mathbf{r}_2 = \langle 2s, 1 - s, 3 + 4s \rangle$ intersect. If so, find their point of intersection.

16. Show that the lines $\mathbf{r}_1 = \langle 2 - t, 4 + 2t, -3 + 4t \rangle$ and $\mathbf{r}_2 = \langle 1 + t, 5 - 3t, 3 - 2t \rangle$ intersect, and find an equation of the plane that contains them.

In Exercises 17 and 18, find an equation of the plane that satisfies the given conditions.

17. Containing the points $(2, 0, -1)$, $(3, 2, 0)$, and $(-4, -2, 3)$

18. Perpendicular to the line of intersection of the planes $3x - y + z + 3 = 0$ and $x + 2y - z = 9$ at the point where this line pierces the xy-plane

19. (a) Find the angles the vector $\mathbf{u} = -2\mathbf{i} + 4\mathbf{j} + 5\mathbf{k}$ makes with the coordinate axes.
    (b) For a certain vector $\mathbf{v}$ of length 12, $\cos \beta = \frac{7}{9}$ and $\cos \gamma = \frac{4}{9}$. Find $\mathbf{v}$, given that it has a negative x component.

20. A force $\mathbf{F}$ of magnitude 30 N acts in a direction perpendicular to the plane $3x - 4y + 5z = 10$, and its z component is positive. Find the work done by $\mathbf{F}$ in moving an object from $A(4, 1, -2)$ to $B(2, -5, -1)$, if distance is in meters.

21. (a) Find the distance from the line $3x - 4y = 7$ to the point $(5, -2)$.
    (b) Find the distance from the plane $x - 2y + 2z = 7$ to the point $(2, -1, 4)$.

22. Find x, y, and z so that the vectors $x\mathbf{i} + \mathbf{j} - 3\mathbf{k}$, $3\mathbf{i} - y\mathbf{j} - 7\mathbf{k}$, and $8\mathbf{i} + 5\mathbf{j} + z\mathbf{k}$ will be mutually orthogonal.

23. Prove that if $\mathbf{u}$, $\mathbf{v}$, and $\mathbf{w}$ are mutually orthogonal nonzero vectors, then they are *linearly independent*; that is, the equation
$$a\mathbf{u} + b\mathbf{v} + c\mathbf{w} = \mathbf{0}$$
is satisfied only if $a = b = c = 0$. (*Hint:* In turn, find the dot product of $\mathbf{u}$, $\mathbf{v}$, and $\mathbf{w}$ with both sides.)

24. Unit vectors that are mutually orthogonal are said to form an **orthonormal** set. Prove that if $\mathbf{u}$, $\mathbf{v}$, and $\mathbf{w}$ are any three noncoplanar three-dimensional vectors, then the vectors $\mathbf{e}_1$, $\mathbf{e}_2$, and $\mathbf{e}_3$, defined by
$$\mathbf{e}_1 = \frac{\mathbf{u}}{|\mathbf{u}|} \qquad \mathbf{e}_2 = \frac{\mathbf{v} - (\mathbf{v} \cdot \mathbf{e}_1)\mathbf{e}_1}{|\mathbf{v} - (\mathbf{v} \cdot \mathbf{e}_1)\mathbf{e}_1|}$$
$$\mathbf{e}_3 = \frac{\mathbf{w} - (\mathbf{w} \cdot \mathbf{e}_1)\mathbf{e}_1 - (\mathbf{w} \cdot \mathbf{e}_2)\mathbf{e}_2}{|\mathbf{w} - (\mathbf{w} \cdot \mathbf{e}_1)\mathbf{e}_1 - (\mathbf{w} \cdot \mathbf{e}_2)\mathbf{e}_2|}$$
form an orthonormal set.

Chapter 11 Concept Quiz

1. Define each of the following:
    (a) The magnitude of a vector
    (b) The dot product of two vectors
    (c) The cross product of two vectors
    (d) A unit vector
    (e) The unit vectors $\mathbf{i}$, $\mathbf{j}$, and $\mathbf{k}$
    (f) The vectors $\mathbf{u}$ and $\mathbf{v}$ are orthogonal
    (g) $\text{Comp}_\mathbf{v} \mathbf{u}$
    (h) $\text{Proj}_\mathbf{v} \mathbf{u}$

2. Let $P_0 = (x_0, y_0, z_0)$ and $\mathbf{v} = \langle a, b, c \rangle$. Write the following:
    (a) A vector equation of the line through P_0 and parallel to $\mathbf{v}$
    (b) Parametric equations of the line in part (a)
    (c) An equation of the plane containing P_0 and perpendicular to $\mathbf{v}$

3. Let $\mathbf{u}$, $\mathbf{v}$, and $\mathbf{w}$ be three two-dimensional or three-dimensional vectors. State five properties, each dealing with addition or scalar multiplication involving one or more of these vectors.

4. Draw two vectors $\mathbf{u}$ and $\mathbf{v}$ having the same initial point. Show each of the following vectors:
    (a) $\mathbf{u} + \mathbf{v}$
    (b) $\mathbf{u} - \mathbf{v}$
    (c) $\frac{1}{2}\mathbf{u}$
    (d) $2\mathbf{v}$
    (e) $2\mathbf{v} - \frac{1}{2}\mathbf{u}$

5. State each of the following:
    (a) A necessary and sufficient condition for two vectors to be orthogonal
    (b) The triangle inequality for vectors
    (c) The Cauchy-Schwarz inequality
    (d) A formula for the volume of the parallelepiped having the vectors $\mathbf{u}$, $\mathbf{v}$, and $\mathbf{w}$ as adjacent edges
    (e) A formula for the work done by a constant force $\mathbf{F}$ in moving a particle in a line from P to Q

6. Explain in each of the following cases how you would find an equation of a plane satisfying the given conditions:
    (a) Contains three noncollinear points
    (b) Passes through a given point and is parallel to a given plane
    (c) Contains a line and a point not on the line
    (d) Is the perpendicular bisector of the line segment joining two points

7. State which of the following are valid operations on three-dimensional vectors and which are not.
    (a) $(\mathbf{u} \times \mathbf{v}) \times \mathbf{w}$
    (b) $(\mathbf{u} \times \mathbf{v}) \cdot \mathbf{w}$
    (c) $(\mathbf{u} \cdot \mathbf{v}) \times \mathbf{w}$
    (d) $(\mathbf{u} \cdot \mathbf{v}) \cdot \mathbf{w}$
    (e) $(\mathbf{u} \cdot \mathbf{v})\mathbf{w}$
    (f) $(\mathbf{u} \times \mathbf{v})\mathbf{w}$

APPLYING CALCULUS

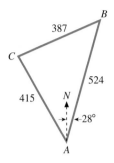

1. The pilot of a small airplane flies in the triangular path shown in the figure from A to B to C and back to A. The distances shown are in kilometers. If there had been no wind, the pilot would have flown at a heading of 28° (measured clockwise from north) to go on the straight path from A to B. Throughout the trip the pilot maintained an average airspeed of 332 kph. There was a constant wind of average velocity 113 kph blowing from 223°.
   (a) Use trigonometry to find the angles in the triangle at A, B, and C.
   (b) Find the pilot's heading on each leg of the trip.
   (c) What was the groundspeed (speed relative to the ground) on each leg?
   (d) How long did the trip take?

2. In crystallography a linear combination of vectors $\mathbf{v}_1$, $\mathbf{v}_2$, and $\mathbf{v}_3$ of the form
$$m_1\mathbf{v}_1 + m_2\mathbf{v}_2 + m_3\mathbf{v}_3$$
where m_1, m_2, and m_3 are positive integers, is called a **lattice** for a crystal. The **reciprocal lattice** is
$$m_1\mathbf{w}_1 + m_2\mathbf{w}_2 + m_3\mathbf{w}_3$$
where m_1, m_2, and m_3 are again positive integers, and where
$$\mathbf{w}_1 = \frac{\mathbf{v}_2 \times \mathbf{v}_3}{\mathbf{v}_1 \cdot (\mathbf{v}_2 \times \mathbf{v}_3)}, \quad \mathbf{w}_2 = \frac{\mathbf{v}_3 \times \mathbf{v}_1}{\mathbf{v}_1 \cdot (\mathbf{v}_2 \times \mathbf{v}_3)}, \quad \text{and} \quad \mathbf{w}_3 = \frac{\mathbf{v}_1 \times \mathbf{v}_2}{\mathbf{v}_1 \cdot (\mathbf{v}_2 \times \mathbf{v}_3)}$$
Show that
(a) $\mathbf{v}_i \cdot \mathbf{w}_j = \delta_{ij}$, where δ_{ij} is the *Kronecker delta* whose value is 0 if $i \neq j$ and is 1 if $i = j$.
(b) $\mathbf{w}_1 \cdot (\mathbf{w}_2 \times \mathbf{w}_3) = \dfrac{1}{\mathbf{v}_1 \cdot (\mathbf{v}_2 \times \mathbf{v}_3)}$

3. If a force $\mathbf{F}$ acts on a rigid body at a point with position vector $\mathbf{r}$, then the vector
$$\boldsymbol{\tau} = \mathbf{r} \times \mathbf{F}$$
is called the **torque** produced by $\mathbf{F}$.
(a) A rod as shown in the figure is free to pivot about one end, and a force of magnitude 6.0 N is applied at the other end in the direction shown. Find the magnitude of the torque about the pivot point.
(b) In the accompanying figure, a continuously varying force is applied, with magnitude at the distance x from the left end given by
$$F(x) = F_0\left[1 - \left(\frac{x}{L}\right)^2\right]$$
Find the magnitude of the torque about $x = L$.

4. In chemistry the *bond angle* is the angle between the vectors from an atom to two atoms to which it is bonded. A *tetrahedral bond angle* is the angle from an atom at the centroid of a tetrahedron to two atoms at vertices of the tetrahedron. (See the figure.) This arrangement of atoms occurs in the methane molecule and in diamond crystals, for example. In methane, CH_4, the carbon atom is at the centroid and is bonded to four hydrogen atoms at the vertices. Find the bond angle AMD in the figure. The points A, B, C, and D have coordinates $(a, 0, 0)$, $(0, a, 0)$, (a, a, a), and $(0, 0, a)$. The centroid M has coordinates $\left(\frac{a}{2}, \frac{a}{2}, \frac{a}{2}\right)$.

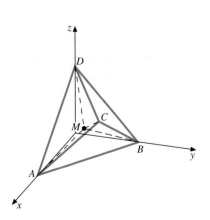

5.
(a) Let l_1 and l_2 be two skew lines with vector equations $\mathbf{r} = \mathbf{a}_1 + t\mathbf{v}_1$, and $\mathbf{r} = \mathbf{a}_2 + t\mathbf{v}_2$, respectively. Derive the following formula for the minimum distance d between the lines:

$$d = \frac{|(\mathbf{a}_2 - \mathbf{a}_1) \cdot (\mathbf{v}_1 \times \mathbf{v}_2)|}{|\mathbf{v}_1 \times \mathbf{v}_2|}$$

(b) Suppose two airplanes are flying on the paths l_1 and l_2 of part (a), where $\mathbf{a}_1 = 5\mathbf{i} + 2\mathbf{j} + 8\mathbf{k}$, $\mathbf{v}_1 = 3\mathbf{i} - 2\mathbf{j} + 2\mathbf{k}$, $\mathbf{a}_2 = -\mathbf{i} + 2\mathbf{j} + 5\mathbf{k}$, and $\mathbf{v}_2 = -2\mathbf{i} + 3\mathbf{j} + \mathbf{k}$ (with coordinates in kilometers). What is the closest the airplanes could come to each other?

6. A light ray emanates from the point (2, 3, 1) and follows a path in the direction of the vector $\mathbf{i} - \mathbf{j} - 2\mathbf{k}$.
(a) Find the point at which the light ray strikes the plane $2x + 3y + z = 8$ and the angle it makes with this plane.

(b) What is the minimum distance between the light ray and the origin?

(c) If the light ray strikes a vertical reflector that is perpendicular to the x-axis, what is the path of the reflected ray? (Note that the angle of incidence equals the angle of reflection.)

(d) Suppose that after hitting the vertical reflector in part (c), the reflected ray strikes a horizontal plane. What is the path of the reflected ray?

This laser beam sent to the Moon reflects off corner reflectors left there by astronauts so that its time of travel to the Moon and back can be measured.

7. An incoming light ray hitting the inside of a cube reflects off three faces of the cube in such a way that it winds up heading in the opposite direction from which it came. The technique is used to reflect laser beams from earth off corner-cube reflectors placed on the moon by Apollo astronauts. By timing the light-travel time from earth to moon and back, we can now measure the distance to the moon to within a few centimeters. Further, we find out accurately where the telescope is on the earth; the technique is used to measure continental drift.

Using unit-vector notation, prove that the internal reflection off three adjacent faces of a cube indeed returns the light ray in the opposite direction.

8. Points P and Q are directly opposite each other on the shores of a straight river that flows at a uniform speed of v miles per hour. A boat whose speed is u miles per hour must cross the river from P to Q.
(a) In what direction should the boat head from P and what is the boat's actual speed; that is, what is the component of the velocity in the direction of the line from P to Q?

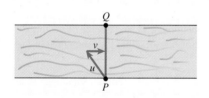

(b) Is the trip across the river from P to Q always possible? If not, give conditions under which the trip is possible.

(c) Suppose that it is only necessary for the boat to leave from P and to reach the other shore, it does not matter where. If the boat leaves P on a heading that makes an angle α with the line connecting P and Q, will the boat reach the opposite shore, and if so, at what point will it land?

A PERSONAL VIEW

Allan Cormack

How did mathematics play a role in the work that led to your Nobel Prize?

It was vital. The whole problem is a mathematical one.

Please tell us about the tomography project.

The problem is to infer the absorption coefficient of tissue from X rays from absorption measurements taken outside the object being X-rayed. In a single absorption measurement, what you are getting is a measurement of the line integral of the absorption coefficient along the line of the X-ray beam, or, if you like, the average X-ray absorption coefficient along that line. Now, if you do that for a number of lines in different directions, can you then infer the variation of the absorption coefficient from point to point? The answer is yes, and there are many different algorithms for doing that.

How did you come to that problem?

Well, it was just by an accident. The hospital physicist—and there was only one in those days—at the University hospital for the University of Cape Town Medical School quit. And incidentally, this was the Groot Schuur Hospital, where Christiaan Barnard subsequently did his heart transplants. This hospital physicist quit and went to Canada, and I was the only person in South Africa at the time who knew anything about handling of radioactive isotopes. There was no medical physics section of the hospital at the time (I'm talking about 1956) and so I was put in the radiology department. And there I couldn't avoid seeing the way they did radiotherapy-treatment planning. I was appalled by what I saw, even though it was a place where you got as good treatment as you would anywhere in the world. So my thought was that with a map of the point-by-point absorption coefficient in the body, one would be able to design radiotherapy treatments much more accurately. That was the motivation. And it occurred to me that such a map would be useful of itself, but I didn't realize just how useful it would be. Without taking anything away from physicians, the layman doesn't realize how almost impossible it is to see a tumor in the soft tissues of the head, where the tumor is obscured by all the images of the bones in the head that you get in an ordinary X ray. And so this is one of the reasons why it was the radiologists in the soft tissues who were the ones who were so delighted when the first commercial CT scanners came out, because they were seeing things with a clarity that they had never come close to before. But there are other people, like a distinguished physician at Mass General Hospital in Boston, who said "oh yes, this might be very well for the head, but it will never work in the chest." And, of course, he was dead wrong.

How did your training make you the person who was chosen to work at the hospital?

Well, I was the only person in Cape Town, certainly, and probably in all of South Africa who had been trained in handling radioactive isotopes, measuring absorption coefficients, and so on. I learned those skills when I was a graduate student in nuclear physics at Cambridge. I had done three years of engineering in Cape Town, and then I quit and changed to physics and went to Cambridge in England for a Ph.D.

Did you have any special mathematical training that led you to this problem?

No, and unfortunately, I took the mathematics sequence intended for engineers as an undergraduate and that leaves out a lot of the more interesting and fundamental parts of mathematics in the interest of getting to usable results quickly. For example, mathematicians in those days would spend a long time on the discussion of convergence of series and they probably do still. The engineering syllabus kind of brushed over those things, and I've regretted all my life not having had a more fundamental training in mathematics. So there's a plug for mathematics.

How did the mathematics get introduced to the CT problem?

Well, if you simply ask what you measure when you pass a beam of X rays through an inhomogeneous material, it turns out that the pertinent quantity is the line integral of the absorption coefficient along the line of the beam. It is Beer's Law, which has been known for a long time. In the early days of nuclear physics, for example, you got some information about a gamma-ray energy by measuring its absorption—not an accurate way by modern standards at all. So I knew about measuring gamma-ray absorption coefficients in a homogeneous medium, so the next question is what happens if the medium is inhomogeneous, as in the head.

Was this a three-dimensional problem?

Ideally, yes, three dimensions, but you can simplify by cutting into a series of two-dimensional slices. And tomography gets its name from that fact. A *tomogram* is a picture of a slice and it comes from the Greek *tomon*, which is the same root that occurs in *atom*, which means "that which cannot be cut [or sliced]."

You said CT, but wasn't it called CAT at one time?

People were trying to do tomography by different means as early as 1920. What they did was CAT, which originally meant "computed axial tomography," from an old-fashioned meaning, and then it became just CT, computer tomography. Now CAT stands for "computer-assisted tomography."

What relation did the CT work have to the development of MRI, medical-research imaging?

In the original development of MRI, people used CT algorithms to interpret their data, because they were also averaging over an NMR [nuclear magnetic resonance] signal over lines, or the three-dimensional version of this. The problem is, "Can you determine a function in an area knowing the line integral everywhere"—a two-dimensional problem. The three-dimensional problem is, "If you know the averages over planes, can you determine the three-dimensional function?" It turns out that the three-dimensional version is, in fact, simpler than the two-dimensional version. This is true for all odd-dimensional spaces as opposed to even-dimensional spaces. In the case of the even-dimensional spaces, such as two dimensions, you have to know the line integrals through all lines intersecting the plane in order to find the value of the function at a point, whereas in the three-dimensional case, you need to know only the averages over the planes that intersect a neighborhood of the point. All this was known in 1917 by Johann Radon, an Austrian mathematician. Radon had himself been anticipated in the three-dimensional case by Lorentz, in about 1905, who hadn't bothered to publish his result. Now Lorentz was interested in it, I think, because of the propagation of waves in crystals. One of the nice features of the so-called Radon transform is that if you take the wave equation and average over planes parallel to a given direction, you reduce the three-dimensional problem to a one-dimensional problem in that direction. So the Radon transformation is a very useful technique in the study of partial differential equations. And if I had known that, I would have been saved a great deal of work.

In view of the developments in mathematics and science over the last decades, is calculus still a good way to introduce students to mathematics?

I think it is, but there is a strong movement that says the contrary, why bother with classical analysis at all when you can do it all on computers. I think you would lose a great deal, but there are people who maintain this. Hounsfield, who won the Prize jointly with me, simply saw the problem as a successive approximation problem. His algorithm consisted of backprojecting an average density along every line and progressively correcting, so you get a better fit to the data. That is a perfectly good way to do it. It works.

Allan Cormack is University Professor at Tufts University. He was born and raised in South Africa, went to Cambridge, England, for graduate study, and then returned to South Africa. He was on a sabbatical at Harvard in 1956–57 when he was offered a position at Tufts, where he became professor of physics. He shared in the 1979 Nobel Prize in Physiology or Medicine with Godfrey Hounsfield for the development of "computer-assisted tomography."

CHAPTER 12

VECTOR-VALUED FUNCTIONS

When we are describing the motion of an object such as a space vehicle or a planet moving in its orbit, the position of the object, its velocity, and its acceleration are usually given as vectors. These vectors are dependent on time, and as we will see, they provide a useful means of describing motion. The position of the object at any time t can be given by its position vector $\mathbf{r}(t)$ in reference to some coordinate system. Since values of $\mathbf{r}(t)$ are vectors, $\mathbf{r}$ is called a **vector-valued function,** or more briefly, a **vector function**. As we shall see, derivatives of $\mathbf{r}(t)$ give us information about the velocity and acceleration of the moving object. Similarly, to find the distance covered by the object in a given time interval, we will see that integration is needed.

We begin by discussing the nature of vector-valued functions in general, not just in relation to moving objects. While our emphasis will be on three-dimensional vector functions, most of our theory (except where cross products are involved) applies also to the two-dimensional case. Vector functions in two dimensions provide an alternative way of looking at the parametric representation of curves that we studied in Chapter 9.

Next, we discuss the calculus of vector functions and show how the length of a curve can be given as an integral involving a vector function. We introduce the concepts of the **unit tangent vector**, **unit normal vector**, and **curvature**, which are useful in describing a curve.

We then return to the important topic of motion along a curve and show how to apply the ideas we have introduced. We conclude the chapter by considering Kepler's Laws of Planetary Motion.

12.1 VECTOR-VALUED FUNCTIONS

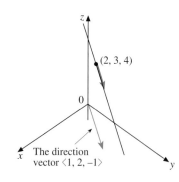

FIGURE 12.1
The graph of $\mathbf{r}(t) = \langle 2 + t, 3 + 2t, 4 - t \rangle$ is the line through $(2, 3, 4)$, parallel to the direction vector $\langle 1, 2, -1 \rangle$

We have already encountered one special type of vector function in studying straight lines in space. For example, we know that for each real t,

$$\mathbf{r}(t) = \langle 2 + t, 3 + 2t, 4 - t \rangle$$

is the position vector of a point on the line that passes through the point $(2, 3, 4)$ and is parallel to the direction vector $\langle 1, 2, -1 \rangle$. As t varies over all real numbers, the tip of the vector $\mathbf{r}(t)$ traces out the entire line. We show the line in Figure 12.1.

As another example, consider the vector function $\mathbf{r}$ for which

$$\mathbf{r}(t) = \langle 4\cos t, 4\sin t, t \rangle$$

As t varies, the tip of the position vector $\mathbf{r}(t)$ traces out a curve in space. In this case, the graph is not a straight line. We will analyze its graph shortly.

More generally, we have the following definition.

Definition 12.1
Vector-Valued Function

A **vector-valued function**, or, more briefly, a **vector function**, is a function whose domain is a set of real numbers and whose range is a set of vectors.

If $\mathbf{r}$ is a vector function with values given by

$$\mathbf{r}(t) = \langle f(t), g(t) \rangle$$

then we call $\mathbf{r}$ a two-dimensional vector function with *component functions* f and g. Similarly, if

$$\mathbf{r}(t) = \langle f(t), g(t), h(t) \rangle$$

then $\mathbf{r}$ is a three-dimensional vector function having component functions f, g, and h. We will concentrate on the three-dimensional case, since we can think of a two-dimensional vector function as having three component functions, with the third component being the zero function.

REMARK
■ The letter $\mathbf{r}$ is frequently used to name a vector-valued function, because it suggests *radius vector,* which is another name for position vector. When we need to discuss more than one vector function, we will usually use letters such as $\mathbf{u}$, $\mathbf{v}$, $\mathbf{w}$.

The domain of a vector function, unless otherwise specified, will be understood to mean the common domain of its component functions.

To distinguish vector functions from ordinary real-valued functions, we call the latter **scalar functions.** So the component functions of a vector function are scalar functions. In $\mathbf{r}(t) = \langle 4\cos t, 4\sin t, t \rangle$, the component functions are the scalar functions $f(t) = 4\cos t$, $g(t) = 4\sin t$, and $h(t) = t$.

Addition, subtraction, and multiplication by a scalar (or by a scalar function) are all carried out component by component, just as with constant vectors. The dot product of two vector functions is a scalar function, whereas the cross product is a vector function.

EXAMPLE 12.1   Find the domain of the vector function **r** defined by

$$\mathbf{r}(t) = \langle \ln(1-t), \sqrt{t}, t^2 \rangle$$

Solution   Write $f(t) = \ln(1-t)$, $g(t) = \sqrt{t}$, and $h(t) = t^2$. The domain of f is the set of t values for which $1 - t > 0$, or equivalently, $t < 1$. The domain of g is the set of all t values such that $t \geq 0$. The domain of h is all of $\mathbb{R}$. The domain of **r** is therefore the set of t values for which $0 \leq t < 1$. We can indicate this set more briefly as the half-open interval $[0, 1)$. ∎

Definition 12.2
Graph of a Vector Function

The **graph** of a vector function **r** is the set of points taken on by the tip of the position vector $\mathbf{r}(t)$ as t varies over the domain of **r**.

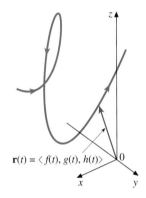

FIGURE 12.2
Graph of the vector function **r**

If

$$\mathbf{r}(t) = \langle f(t), g(t), h(t) \rangle$$

and f, g, and h are *continuous* on an interval I, then the graph of **r** is called a **space curve**. The curve consists of all points (x, y, z) of the form $(f(t), g(t), h(t))$ for t in I. In Figure 12.2, we show such a space curve. We assign a *direction* to the curve according to the manner in which it is traced out by the tip of $\mathbf{r}(t)$ as t increases through the domain values. We sometimes indicate the direction by arrows, as in Figure 12.2.

Since the tip of the vector $\mathbf{r}(t) = \langle f(t), g(t), h(t) \rangle$ lies on the graph of **r** for each value of t, we can give the coordinates (x, y, z) of such a point by the equations

$$\begin{cases} x = f(t) \\ y = g(t) \quad t \in I \\ z = h(t) \end{cases}$$

We call these three equations **parametric equations** of the curve C that is the graph of **r**. If we know the vector equation for C, we know the parametric equations, and conversely.

EXAMPLE 12.2   **The Circular Helix**   Sketch the graph of the vector function

$$\mathbf{r}(t) = \langle 4\cos t, 4\sin t, t \rangle$$

for $t \geq 0$.

Solution   The parametric equations of the graph are $x = 4\cos t$, $y = 4\sin t$, and $z = t$. Since $x^2 + y^2 = 16\cos^2 t + 16\sin^2 t = 16$, we see that the projection of the point (x, y, z) on the xy-plane lies on the circle $x^2 + y^2 = 16$ of radius 4. As t increases, the z-coordinate increases. The result is the climbing circular curve shown in Figure 12.3. Its direction is indicated by the arrows. The initial point, corresponding to $t = 0$, is $(4, 0, 0)$. ∎

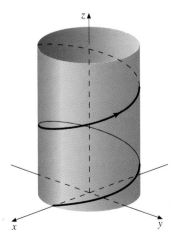

FIGURE 12.3
The circular helix $\mathbf{r}(t) = \langle 4\cos t, 4\sin t, t \rangle$ for $t \geq 0$.

The curve in Figure 12.3 is called a **circular helix**. The general equation for a helix is of the form $\mathbf{r}(t) = \langle a\cos\omega t, b\sin\omega t, ct\rangle$. The helix is circular if $a = b$ and elliptical if $a \neq b$.

EXAMPLE 12.3 Discuss and sketch the graph of

$$\mathbf{r}(t) = (2t)\mathbf{i} + (3t^2)\mathbf{j} + (t^3)\mathbf{k}$$

for $t \geq 0$.

Solution The parametric equations of the curve are $x = 2t$, $y = 3t^2$, and $z = t^3$. If we eliminate the parameter t between x and y, we get $y = (3x^2)/4$. Thus, the xy-projection of a point on the curve lies on this parabola. When $t = 0$, the point is at the origin. As t increases, the point moves upward, always in such a way that its xy-projection is on the parabola $y = (3x^2)/4$. We therefore have the graph shown in Figure 12.4. ∎

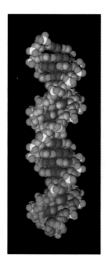

The DNA molecule is a double helix with a radius of 1 nanometer, 1×10^{-9} m.

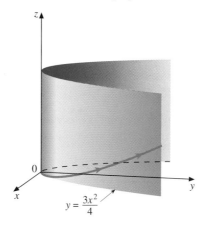

FIGURE 12.4
The twisted cubic $\mathbf{r}(t) = \langle 2t, 3t^2, t^3\rangle, t \geq 0$

The curve in Figure 12.4 is called a **twisted cubic**. The general equation is of the form $\mathbf{r}(t) = \langle at, bt^2, ct^3\rangle$.

If $\mathbf{r}(t) = \langle f(t), g(t)\rangle$ is a two-dimensional vector, with f and g continuous on an interval I, then the graph of $\mathbf{r}$ is a **plane curve**. Its parametric equations are

$$\begin{cases} x = f(t) \\ y = g(t) \end{cases} \quad t \in I$$

We studied such curves in Chapter 9. So the vector equation $\mathbf{r}(t) = \langle f(t), g(t)\rangle$ is simply an alternative way of describing curves that we represented by parametric equations in Chapter 9.

EXAMPLE 12.4 Identify and draw the graph of the curve C defined by the vector equation

$$\mathbf{r}(t) = \langle 3\cos t, 2\sin t\rangle$$

for $0 \leq t \leq 2\pi$.

Solution The parametric equations of C are $x = 3\cos t$ and $y = 2\sin t$. Thus,

$$\frac{x}{3} = \cos t \quad \text{and} \quad \frac{y}{2} = \sin t$$

Section 12.1 Vector-Valued Functions 891

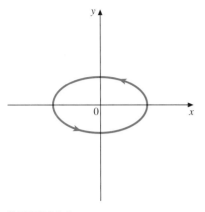

By squaring and adding, we get the rectangular equation

$$\frac{x^2}{9} + \frac{y^2}{4} = 1$$

The graph is the ellipse shown in Figure 12.5. Note that when $t = 0$, the tip of the vector $\mathbf{r}(t)$ is at the point $(3, 0)$, and as t increases, the curve is traced out in a counterclockwise direction, as shown. When $t = 2\pi$, the tip of $\mathbf{r}(t)$ is again at $(3, 0)$, so the ellipse has been traced out once in the interval $0 \leq t \leq 2\pi$. ∎

FIGURE 12.5
The graph of $r(t) = \langle 3\cos t, 2\sin t \rangle$ for $0 \leq t \leq 2\pi$

GRAPHING VECTOR-VALUED FUNCTIONS USING COMPUTER ALGEBRA SYSTEMS

Computer algebra systems are very useful for viewing curves in three space that are otherwise either very difficult or impossible to sketch and analyze. In this section, we define and sketch several curves in three space using Maple and Mathematica.

CAS 44

Make a sketch of the helix given in Example 12.2.
The helix is given by the vector-valued function $\mathbf{r}(t) = \langle 4\cos t, 4\sin t, t \rangle$ for $t \geq 0$. We will also display the cylinder on which the helix travels.

Maple:

with(plots):

r:=t->[4*cos(t),4*sin(t),t];
p1 :=plot3d([4*cos(t),4*sin(t),x],t=0..2*Pi,
  x=0..15,color=grey,grid=[20,5]);
p2 :=spacecurve(r(t),t=0..4*Pi,
  color=black,numpoints=200);
display3d(p1,p2,style=wireframe);

Mathematica:

x[t_] = {4*Cos[t],4*Sin[t],t}
ParametricPlot3D[{4*Cos[t],4*Sin[t],t},{t,0,2*Pi}]

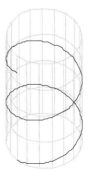

FIGURE 12.1.1

The general helix can be written in the form $\mathbf{r}(t) = \langle a \cos \omega t, a \sin \omega t, bt \rangle$. If you have access to a CAS, explore the effect the constants a, b, and ω have on the graph of the helix. For example, try the Maple command

spacecurve(r(t),t=0..4*Pi,color=black,numpoints=200);

CAS 45

(a) Sketch the curve defined by $\mathbf{r}(t) = \langle e^{-t^2} \cos t, e^{-t^2} \sin t, t \rangle$.

(b) Sketch the curve defined by

$$\mathbf{r}(t) = \left\langle \left(3 + \cos\left(\frac{3t}{2}\right)\right) \cos t, \left(3 + \cos\left(\frac{3t}{2}\right)\right) \sin t, \sin\left(\frac{3t}{2}\right) \right\rangle$$

Maple:

(a)
r:=t->[exp(-t^2)*cos(t),exp(-t^2)*sin(t),t];
spacecurve(r(t),t=-3..3,axes=boxed);

See Figure 12.1.2.

(b)
r:=t->[(3+cos(3*t/2))*cos(t),(3+cos(3*t/2))*sin(t),sin(3*t/2)];
spacecurve(r(t),t=0..4*Pi,color=black,numpoints=200, axes=boxed);

See Figure 12.1.3.

Mathematica:

(a)
ParametricPlot3D[{Exp[-t^2]*Cos[t],Exp[-t^2]*Sin[t],t}, {t,-3,3}]

(b)
ParametricPlot3D[{(3+Cos[3*t/2])*Cos[t], (3+Cos[3*t/2])*Sin[t],Sin[3*t/2]},{t,0,4*Pi}]

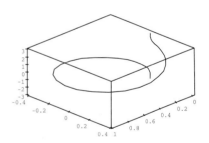

FIGURE 12.1.2

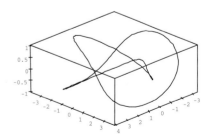

FIGURE 12.1.3

Exercise Set 12.1

In Exercises 1–6, give the domain of each of the vector functions.

1. $\mathbf{r}(t) = \left\langle \dfrac{t}{t-1}, \sqrt{1-t} \right\rangle$

2. $\mathbf{r}(t) = (\ln t)\mathbf{i} + \sqrt{1-t}\,\mathbf{j}$

3. $\mathbf{r}(t) = \left\langle 2t, \dfrac{1}{t}, \sqrt{t-1} \right\rangle$

4. $\mathbf{r}(t) = \langle \ln(t+2), e^{-t}, \ln(1-t) \rangle$

5. $\mathbf{r}(t) = (\tan t)\mathbf{i} + (\cot t)\mathbf{j} + \left(\sin \sqrt{t}\right)\mathbf{k}$

6. $\mathbf{r}(t) = \left(\sqrt{1-t^2}\right)\mathbf{i} + (\ln t)\mathbf{j} + e^{\sin t}\mathbf{k}$

In Exercises 7–18, sketch the graph of $\mathbf{r}$. *Indicate the direction for increasing t values with arrows.*

7. $\mathbf{r}(t) = \langle t^2, t^3 \rangle$

8. $\mathbf{r}(t) = (t-2)\mathbf{i} + t^2\mathbf{j}$

9. $\mathbf{r}(t) = \left(\dfrac{1}{t}\right)\mathbf{i} + t\mathbf{j}, \quad t > 0$

10. $\mathbf{r}(t) = \langle \sin^2 t, \cos t - 1 \rangle, \quad 0 \le t \le 2\pi$

11. $\mathbf{r}(t) = \langle 1+t, 2t, 3t+2 \rangle$

12. $\mathbf{r}(t) = \langle 1-t^2, t, 2 \rangle$

13. $\mathbf{r}(t) = (\sin t)\mathbf{i} + (\cos t)\mathbf{j} + t\mathbf{k}$

14. $\mathbf{r}(t) = \langle \cos t, t, \sin t \rangle$

15. $\mathbf{r}(t) = \langle 3\cos t, 2\sin t, t \rangle$

16. $\mathbf{r}(t) = t\mathbf{i} + t^2\mathbf{j} + t^3\mathbf{k}$

17. $\mathbf{r}(t) = \langle t, t, \sin t \rangle$

18. $\mathbf{r}(t) = \langle e^t, e^{-t}, t \rangle$

19. Plot a sample of curves from the family of curves defined by $\mathbf{r}(t) = \langle \cos(nt), \sin(nt), \cos(pt) \rangle$, where n and p are positive integers.

20. Repeat Exercise 19 with $\mathbf{r}(t) = \langle \cos(mt), \sin(nt), \sin(pt) \rangle$ for different values of m, n, and p.

21. Plot a sample of curves from the family of curves given by $\mathbf{r}(t) = \langle (a + \cos(bt/2))\cos(t), (c + \sin(dt/2))\sin(t), \cos(et/2) \rangle$.

12.2 THE CALCULUS OF VECTOR FUNCTIONS

The calculus concepts of limit, continuity, differentiation, and integration are all defined for vector functions in terms of the component functions. In particular, we have the following.

Definition 12.3
The Limit of a Vector Function

Let $\mathbf{r}(t) = \langle f(t), g(t), h(t) \rangle$. The **limit** of $\mathbf{r}(t)$ as t approaches a exists if and only if the limits of the component functions,

$$\lim_{t \to a} f(t) = l_1, \quad \lim_{t \to a} g(t) = l_2, \quad \text{and} \quad \lim_{t \to a} h(t) = l_3$$

all exist. In this case,

$$\lim_{t \to a} \mathbf{r}(t) = \mathbf{L} \quad \text{where } \mathbf{L} = \langle l_1, l_2, l_3 \rangle$$

In particular, if $\lim_{t \to a} \mathbf{r}(t) = \mathbf{r}(a)$, then $\mathbf{r}$ is **continuous** at $t = a$, and conversely.

REMARK

■ To say that $\lim_{t \to a} \mathbf{r}(t) = \mathbf{L}$ means that the vectors $\mathbf{r}(t)$ come arbitrarily close to the constant vector $\mathbf{L}$ in *both magnitude and direction* for values of t sufficiently close to a (but not equal to a). From the definition, we can conclude that *$\mathbf{r}$ is continuous at $t = a$ if and only if its component functions are continuous there.*

EXAMPLE 12.5

(a) Let $\mathbf{u}(t) = \left\langle t^2, \cos t, \dfrac{\sin t}{t} \right\rangle$. Find $\lim_{t \to 0} \mathbf{u}(t)$.

(b) Let $\mathbf{v}(t) = \langle e^t, \ln t, \sinh t \rangle$. Show that $\mathbf{v}$ is continuous at all points of its domain.

Solution

(a) By Definition 12.3,

$$\lim_{t \to 0} \mathbf{u}(t) = \left\langle \lim_{t \to 0} t^2, \lim_{t \to 0} \cos t, \lim_{t \to 0} \frac{\sin t}{t} \right\rangle$$
$$= \langle 0, 1, 1 \rangle$$

(b) The only limitation on the domain of $\mathbf{v}$ is that t be positive, so that $\ln t$ is defined. If $t = a$, where a is any positive number, then each of the component functions is continuous there. So we have

$$\lim_{t \to a} \mathbf{v}(t) = \left\langle \lim_{t \to a} e^t, \lim_{t \to a} \ln t, \lim_{t \to a} \sinh t \right\rangle$$
$$= \langle e^a, \ln a, \sinh a \rangle = \mathbf{v}(a)$$

Thus, $\mathbf{v}$ is continuous at $t = a$. ■

The Derivative of a Vector Function

Definition 12.4
The Derivative of a Vector Function

The **derivative $\mathbf{r}'$** of a vector function $\mathbf{r}$ is defined by

$$\mathbf{r}'(t) = \lim_{h \to 0} \frac{\mathbf{r}(t + h) - \mathbf{r}(t)}{h}$$

provided this limit exists.

Combining Definition 12.4 with the limit of a vector-valued function given in Definition 12.3, we get an easy method of computing the derivative of a vector-valued function. We simply differentiate each of the component functions. We set off this important result for emphasis.

The Derivative of a Vector Function

If $\mathbf{r}(t) = \langle f(t), g(t), h(t) \rangle$, and if $f'(t)$, $g'(t)$, and $h'(t)$ exist, then
$$\mathbf{r}'(t) = \langle f'(t), g'(t), h'(t) \rangle$$

EXAMPLE 12.6 Find $\mathbf{r}'(\frac{\pi}{3})$ if
$$\mathbf{r}(t) = (\sin 2t)\mathbf{i} + (2\cos t)\mathbf{j} + (\tan t)\mathbf{k}$$

Solution Each of the component functions is differentiable except where $\tan t$ is undefined (at odd multiples of $\frac{\pi}{2}$). At any value of t where they are differentiable, we have
$$\mathbf{r}'(t) = (2\cos 2t)\mathbf{i} + (-2\sin t)\mathbf{j} + (\sec^2 t)\mathbf{k}$$

For $t = \frac{\pi}{3}$, we have
$$\mathbf{r}'\left(\frac{\pi}{3}\right) = \left(2\cos\frac{2\pi}{3}\right)\mathbf{i} + \left(-2\sin\frac{\pi}{3}\right)\mathbf{j} + \left(\sec^2\frac{\pi}{3}\right)\mathbf{k}$$
$$= 2\left(-\frac{1}{2}\right)\mathbf{i} + (-2)\left(\frac{\sqrt{3}}{2}\right)\mathbf{j} + (2)^2\mathbf{k}$$
$$= -\mathbf{i} - \sqrt{3}\mathbf{j} + 4\mathbf{k} \qquad \blacksquare$$

Geometric Interpretation of the Derivative: Tangent Vectors and Tangent Lines

It should come as no surprise that the derivative of a vector function $\mathbf{r}$ has some relationship to the tangent line to the graph of $\mathbf{r}$. We show this relationship in Figure 12.6. The position vectors of the points P and Q are $\mathbf{r}(t)$ and $\mathbf{r}(t+h)$. The vector $\overrightarrow{PQ} = \mathbf{r}(t+h) - \mathbf{r}(t)$ is a direction vector for the secant line through P and Q (we are supposing that $h \neq 0$). The vector
$$\frac{\mathbf{r}(t+h) - \mathbf{r}(t)}{h} = \frac{1}{h}[\mathbf{r}(t+h) - \mathbf{r}(t)]$$

is just a scalar multiple of $\overrightarrow{PQ}$, so it is also a direction vector for this secant line (in the same direction if $h > 0$). It appears that as $h \to 0$, this direction vector for the secant line approaches a vector lying on the tangent line at P. It is natural, then, to define the tangent line at P as the line through P with direction vector equal to the limit
$$\lim_{h \to 0} \frac{\mathbf{r}(t+h) - \mathbf{r}(t)}{h}$$

But by Definition 12.4 this limit is $\mathbf{r}'(t)$. We must restrict $\mathbf{r}'(t)$ to be different from the zero vector, since otherwise there would be no well-defined direction for the tangent line. We therefore have the following definition.

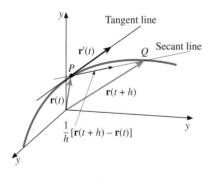

FIGURE 12.6

Definition 12.5
Tangent Vector and Tangent Line to a Space Curve

Let C be the curve with vector equation $\mathbf{r}(t) = \langle f(t), g(t), h(t) \rangle$ for $t \in I$, and suppose that $\mathbf{r}'(t)$ exists and is not the zero vector. Let $P = (f(t), g(t), h(t))$. Then the vector $\mathbf{r}'(t)$ is called a **tangent vector** to C at P, and the **tangent line** to C at P is the line through P with direction vector $\mathbf{r}'(t)$.

EXAMPLE 12.7 Find a tangent vector and a vector equation of the tangent line to the circular helix $\mathbf{r}(t) = \langle 4\cos t, 4\sin t, t \rangle$ at $t = \frac{\pi}{3}$.

Solution Since $\mathbf{r}'(t) = \langle -4\sin t, 4\cos t, 1 \rangle$, a tangent vector at $t = \frac{\pi}{3}$ is

$$\mathbf{r}'\left(\frac{\pi}{3}\right) = \left\langle -4\sin\frac{\pi}{3}, 4\cos\frac{\pi}{3}, 1 \right\rangle = \left\langle -2\sqrt{3}, 2, 1 \right\rangle$$

The point on the curve at which $t = \frac{\pi}{3}$ is the tip of the position vector $\mathbf{r}\left(\frac{\pi}{3}\right)$, namely,

$$\left(4\cos\frac{\pi}{3}, 4\sin\frac{\pi}{3}, \frac{\pi}{3}\right) = \left(2, 2\sqrt{3}, \frac{\pi}{3}\right)$$

The tangent line through this point has the direction vector $\mathbf{r}'\left(\frac{\pi}{3}\right)$, so its equation can be written as

$$\mathbf{r}(t) = \left\langle 2, 2\sqrt{3}, \frac{\pi}{3} \right\rangle + t\left\langle -2\sqrt{3}, 2, 1 \right\rangle$$
$$= \left\langle 2 - 2\sqrt{3}t, 2\sqrt{3} + 2t, \frac{\pi}{3} + t \right\rangle$$

We show the graph in Figure 12.7, along with the tangent vector and tangent line at the given point. ∎

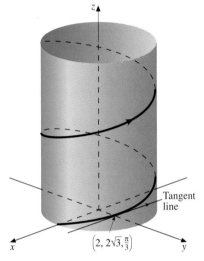

FIGURE 12.7

Properties of Derivatives of Vector-Valued Functions

The following properties of differentiation follow from the analogous results for derivatives of the component functions.

Properties of the Derivative

If $\mathbf{u}$ and $\mathbf{v}$ are differentiable vector functions and f is a differentiable scalar function, then the following properties hold true.

1. $\frac{d}{dt}[\mathbf{u}(t) \pm \mathbf{v}(t)] = \mathbf{u}'(t) \pm \mathbf{v}'(t)$
2. $\frac{d}{dt}[c\mathbf{u}(t)] = c\mathbf{u}'(t)$ for any constant c
3. $\frac{d}{dt}[f(t)\mathbf{u}(t)] = f'(t)\mathbf{u}(t) + f(t)\mathbf{u}'(t)$
4. $\frac{d}{dt}[\mathbf{u}(t) \cdot \mathbf{v}(t)] = \mathbf{u}'(t) \cdot \mathbf{v}(t) + \mathbf{u}(t) \cdot \mathbf{v}'(t)$
5. $\frac{d}{dt}[\mathbf{u}(t) \times \mathbf{v}(t)] = \mathbf{u}'(t) \times \mathbf{v}(t) + \mathbf{u}(t) \times \mathbf{v}'(t)$
6. $\frac{d}{dt}[\mathbf{u}(f(t))] = \mathbf{u}'(f(t))f'(t)$ Chain Rule

---

 In Property 5 it is essential that the order of the factors not be reversed, since the cross product is noncommutative.

EXAMPLE 12.8 Let $\mathbf{u}(t) = \langle t, 2 - t^3, 2t \rangle$ and $\mathbf{v}(t) = \langle 1, t^2, t^3 \rangle$. Find each of the following.
(a) $(\mathbf{u} \cdot \mathbf{v})'(1)$ (b) $(\mathbf{u} \times \mathbf{v})'(1)$

Solution

(a) By Property 4 of the derivative,
$$(\mathbf{u} \cdot \mathbf{v})'(t) = \mathbf{u}'(t) \cdot \mathbf{v}(t) + \mathbf{u}(t) \cdot \mathbf{v}'(t)$$

So for $t = 1$,
$$(\mathbf{u} \cdot \mathbf{v})'(1) = \mathbf{u}'(1) \cdot \mathbf{v}(1) + \mathbf{u}(1) \cdot \mathbf{v}'(1)$$

To evaluate the right-hand side, we first compute each of the vectors involved.

$$\begin{aligned}
\mathbf{u}(t) &= \langle t, 2 - t^3, 2t \rangle & \mathbf{u}(1) &= \langle 1, 1, 2 \rangle \\
\mathbf{u}'(t) &= \langle 1, -3t^2, 2 \rangle & \mathbf{u}'(1) &= \langle 1, -3, 2 \rangle \\
\mathbf{v}(t) &= \langle 1, t^2, t^3 \rangle & \mathbf{v}(1) &= \langle 1, 1, 1 \rangle \\
\mathbf{v}'(t) &= \langle 0, 2t, 3t^2 \rangle & \mathbf{v}'(1) &= \langle 0, 2, 3 \rangle
\end{aligned}$$

Thus,
$$\begin{aligned}
(\mathbf{u} \cdot \mathbf{v})'(1) &= \langle 1, -3, 2 \rangle \cdot \langle 1, 1, 1 \rangle + \langle 1, 1, 2 \rangle \cdot \langle 0, 2, 3 \rangle \\
&= (1 - 3 + 2) + (0 + 2 + 6) = 8
\end{aligned}$$

(b) We make use of Derivative Property 5 and the values computed in part (a) to get
$$\begin{aligned}
(\mathbf{u} \times \mathbf{v})'(1) &= \mathbf{u}'(1) \times \mathbf{v}(1) + \mathbf{u}(1) \times \mathbf{v}'(1) \\
&= \langle 1, -3, 2 \rangle \times \langle 1, 1, 1 \rangle + \langle 1, 1, 2 \rangle \times \langle 0, 2, 3 \rangle \\
&= \langle -5, 1, 4 \rangle + \langle -1, -3, 2 \rangle = \langle -6, -2, 6 \rangle
\end{aligned}$$ ∎

EXAMPLE 12.9 Prove that if $\mathbf{r}(t)$ is a vector function such that for all t in the domain, $\mathbf{r}'(t)$ exists and $|\mathbf{r}(t)|$ is constant, then $\mathbf{r}(t)$ and $\mathbf{r}'(t)$ are orthogonal vectors for every t in the domain.

Solution To show that the vectors $\mathbf{r}(t)$ and $\mathbf{r}'(t)$ are perpendicular, we will show that their dot product is zero. Denote the constant value of $|\mathbf{r}(t)|$ by c. That is, $|\mathbf{r}(t)| = c$. Recall that for any vector $\mathbf{v}$, we have $|\mathbf{v}|^2 = \mathbf{v} \cdot \mathbf{v}$. Thus,
$$\mathbf{r}(t) \cdot \mathbf{r}(t) = c^2$$

Now we take the derivative of each side, applying Property 4.
$$\mathbf{r}'(t) \cdot \mathbf{r}(t) + \mathbf{r}(t) \cdot \mathbf{r}'(t) = 0$$

But dot products are commutative. So we have
$$2\mathbf{r}(t) \cdot \mathbf{r}'(t) = 0$$

Thus, $\mathbf{r}(t) \cdot \mathbf{r}'(t) = 0$, which shows that $\mathbf{r}(t)$ and $\mathbf{r}'(t)$ are always orthogonal. ∎

Integrals of Vector Functions

Antiderivatives and definite integrals of vector functions can also be defined in terms of the component functions.

Definition 12.6
Integrals of Vector Functions

Let $\mathbf{r}(t) = \langle f(t), g(t), h(t) \rangle$. Then

(a) $\int \mathbf{r}(t)dt = \left\langle \int f(t)dt, \int g(t)dt, \int h(t)dt \right\rangle$
provided each of the indefinite integrals on the right exists.

(b) $\int_a^b \mathbf{r}(t)dt = \left\langle \int_a^b f(t)dt, \int_a^b g(t)dt, \int_a^b h(t)dt \right\rangle$
provided each of the definite integrals on the right exists.

Note that an antiderivative (indefinite integral) of a vector function is again a vector function, whereas a definite integral of a vector function is a constant vector.

If $\mathbf{R}'(t) = \mathbf{r}(t)$, we may write the result of part (a) of Definition 12.6 in the form

$$\int \mathbf{r}(t)dt = \mathbf{R}(t) + \mathbf{C}$$

where $\mathbf{C}$ is any constant vector.

The First and Second Fundamental Theorems of Calculus hold true for vector functions, as can be seen by a consideration of components. We state the two results in the following theorem.

THEOREM 12.1

(a) If $\mathbf{r}$ is continuous on the closed interval $[a, b]$, then for any t in $[a, b]$,

$$\frac{d}{dt} \int_a^t \mathbf{r}(u)du = \mathbf{r}(t) \qquad \text{First Fundamental Theorem}$$

(b) If $\mathbf{r}$ is continuous on $[a, b]$ and $\mathbf{R}$ is any antiderivative of $\mathbf{r}$, then

$$\int_a^b \mathbf{r}(t)dt = \mathbf{R}(b) - \mathbf{R}(a) \qquad \text{Second Fundamental Theorem}$$

EXAMPLE 12.10 Let $\mathbf{r}(t) = \langle 2t, 3, t^2 \rangle$. Find each of the following.

(a) $\int \mathbf{r}(t)dt$ (b) $\int_1^4 \mathbf{r}(t)dt$

Solution

(a) By Definition 12.6(a),

$$\int \mathbf{r}(t)\,dt = \left\langle \int 2t\,dt, \int 3\,dt, \int t^2\,dt \right\rangle$$

$$= \left\langle t^2 + C_1, 3t + C_2, \frac{t^3}{3} + C_3 \right\rangle$$

$$= \left\langle t^2, 3t, \frac{t^3}{3} \right\rangle + \langle C_1, C_2, C_3 \rangle$$

We can write the answer as $\mathbf{R}(t) + \mathbf{C}$, where $\mathbf{R}(t) = \langle t^2, 3t, t^3/3 \rangle$ and $\mathbf{C} = \langle C_1, C_2, C_3 \rangle$.

(b) Using Definition 12.6(b) and Theorem 12.1(b), with the antiderivative $\mathbf{R}$ found in part (a), we have

$$\int_1^4 \mathbf{r}(t)\,dt = \mathbf{R}(4) - \mathbf{R}(1) = \left\langle t^2, 3t, \frac{t^3}{3} \right\rangle \bigg]_1^4$$

$$= \left\langle 16, 12, \frac{64}{3} \right\rangle - \left\langle 1, 3, \frac{1}{3} \right\rangle$$

$$= \langle 15, 9, 21 \rangle \qquad \blacksquare$$

We list below the most important properties of the definite integral of a vector function. In most cases the proofs are a direct consequence of the analogous properties of the component functions. An exception is Property 6, which we will prove. You will be asked to prove the others in the exercises.

Properties of the Integral

If $\mathbf{u}$ and $\mathbf{v}$ are integrable vector functions on $[a, b]$, the following properties hold true.

1. $\int_a^b [\mathbf{u}(t) \pm \mathbf{v}(t)]\,dt = \int_a^b \mathbf{u}(t)\,dt \pm \int_a^b \mathbf{v}(t)\,dt$
2. $c \int_a^b \mathbf{u}(t)\,dt = \int_a^b c\mathbf{u}(t)\,dt$ for any scalar c
3. $\int_a^b \mathbf{u}(t)\,dt = \int_a^c \mathbf{u}(t)\,dt + \int_c^b \mathbf{u}(t)\,dt$ for $a < c < b$
4. $\mathbf{K} \cdot \int_a^b \mathbf{u}(t)\,dt = \int_a^b \mathbf{K} \cdot \mathbf{u}(t)\,dt$ for any constant vector $\mathbf{K}$
5. $\mathbf{K} \times \int_a^b \mathbf{u}(t)\,dt = \int_a^b \mathbf{K} \times \mathbf{u}(t)\,dt$ for any constant vector $\mathbf{K}$
6. $\left| \int_a^b \mathbf{u}(t)\,dt \right| \leq \int_a^b |\mathbf{u}(t)|\,dt$ if $|\mathbf{u}(t)|$ is integrable on $[a, b]$

Proof of Property 6   Denote the integral $\int_a^b \mathbf{u}(t)\,dt$ by $\mathbf{K}$. Note that $\mathbf{K}$ is a constant vector. If $\mathbf{K} = \mathbf{0}$, then $|\mathbf{K}| = 0$, so the inequality of Property 6 is satisfied in a trivial way. Assume, then, that $\mathbf{K} \neq \mathbf{0}$. Then we have

$$|\mathbf{K}|^2 = \mathbf{K} \cdot \mathbf{K} = \mathbf{K} \cdot \int_a^b \mathbf{u}(t)\,dt = \int_a^b \mathbf{K} \cdot \mathbf{u}(t)\,dt \qquad \text{Property 4}$$

Since any real number cannot exceed its own absolute value, we have $\mathbf{K} \cdot \mathbf{u}(t) \leq |\mathbf{K} \cdot \mathbf{u}(t)|$. Thus,

$$\int_a^b \mathbf{K} \cdot \mathbf{u}(t)\,dt \leq \int_a^b |\mathbf{K} \cdot \mathbf{u}(t)|\,dt \leq \int_a^b |\mathbf{K}||\mathbf{u}(t)|\,dt$$

since $|\mathbf{K} \cdot \mathbf{u}(t)| \leq |\mathbf{K}||\mathbf{u}(t)|$ by the Cauchy-Schwarz Inequality (see Equation 11.4). By Property 2 we can factor out the constant $|\mathbf{K}|$ in the last integral. Combining our results, we have

$$|\mathbf{K}|^2 \leq |\mathbf{K}| \int_a^b |\mathbf{u}(t)|\,dt$$

Since $|\mathbf{K}| > 0$, we can divide both sides by $|\mathbf{K}|$ to get

$$|\mathbf{K}| \leq \int_a^b |\mathbf{u}(t)|\,dt$$

That is, $\left|\int_a^b \mathbf{u}(t)\,dt\right| \leq \int_a^b |\mathbf{u}(t)|\,dt$, which is what we wanted to prove. ∎

Exercise Set 12.2

In Exercises 1–4, evaluate the limits.

1. $\lim\limits_{t \to 0}\langle e^{-t}, 2e^{3t}, 3e^t \rangle$

2. $\lim\limits_{t \to 1}\left\langle \dfrac{t-1}{t^2-1}, \dfrac{1-t}{1+t} \right\rangle$

3. $\lim\limits_{t \to 0}\left(\dfrac{\sin t}{t}\mathbf{i} + \dfrac{t^2}{1-\cos t}\mathbf{j} \right)$

4. $\lim\limits_{t \to 0^+}\left[(t \ln t)\mathbf{i} + t^2\left(1 - \dfrac{1}{t}\right)\mathbf{j} + 3t\mathbf{k} \right]$

In Exercises 5–8, determine the set of t values for which $\mathbf{r}$ is continuous.

5. $\mathbf{r}(t) = \left\langle t, \dfrac{1}{t-1}, \sqrt{1-t} \right\rangle$

6. $\mathbf{r}(t) = \langle t^{3/2}, e^{-t}, \ln t \rangle$

7. $\mathbf{r}(t) = \sqrt{\dfrac{1-t}{1+t}}\,\mathbf{i} + \dfrac{1}{(t-1)^2}\,\mathbf{j}$

8. $\mathbf{r}(t) = (\ln \cosh t)\mathbf{i} + (\ln \sinh t)\mathbf{j}$

In Exercises 9–11, evaluate the given expressions for $\mathbf{u}(t) = \langle 2t, t, 3 \rangle$, $\mathbf{v}(t) = \langle 1-t, 2, t \rangle$, and $f(t) = 1-t$.

9. (a) $\mathbf{u}(t) - \mathbf{v}(t)$
   (b) $2\mathbf{u}(t) + 3\mathbf{v}(t)$

10. (a) $f(t)\mathbf{u}(t)$
    (b) $\mathbf{u}(t) \cdot \mathbf{v}(t)$

11. (a) $\mathbf{u}(t) \times \mathbf{v}(t)$
    (b) $\mathbf{v}(f(t))$

In Exercises 12–14, repeat Exercises 9–11 for $\mathbf{u}(t) = (\cos t)\mathbf{i} + (\sin t)\mathbf{j} + (\sin t)\mathbf{k}$, $\mathbf{v}(t) = (\sin t)\mathbf{i} + (\cos t)\mathbf{j} + \mathbf{k}$, and $f(t) = 2t$.

In Exercises 15–22, find $\mathbf{r}'(t)$.

15. $\mathbf{r}(t) = \left\langle t^2, e^{2t}, \dfrac{1}{t} \right\rangle$

16. $\mathbf{r}(t) = \langle \ln t, \sin t, \cos 2t \rangle$

17. $\mathbf{r}(t) = \langle t \sin t, t \ln t \rangle$

18. $\mathbf{r}(t) = \langle \sin^{-1} t, \tan^{-1} t, \ln(1+t) \rangle$

19. $\mathbf{r}(t) = (1-e^{-t})\mathbf{i} + te^{2t}\mathbf{j}$

20. $\mathbf{r}(t) = (\sinh t)\mathbf{i} + (\cosh t)\mathbf{j} + (\tanh t)\mathbf{k}$

21. $\mathbf{r}(t) = \dfrac{t}{\sqrt{1-t^2}}\mathbf{i} + (\sin^{-1} t)\mathbf{j} + \sqrt{1-t^2}\,\mathbf{k}$

22. $\mathbf{r}(t) = \dfrac{1-t}{1+t}\mathbf{i} + \dfrac{t}{1-t^2}\mathbf{j}$

In Exercises 23–28, find parametric equations of the tangent line at t_0.

23. $\mathbf{r}(t) = \langle 2t, 3t^2, t^3 \rangle;\ t_0 = 1$

24. $\mathbf{r}(t) = \langle 2\sin t, 3\cos t, 4t\rangle$; $t_0 = \dfrac{\pi}{2}$

25. $\mathbf{r}(t) = \left(\dfrac{1}{t}\right)\mathbf{i} + \sqrt{t}\mathbf{j} + (4t^2)\mathbf{k}$; $t_0 = \dfrac{1}{4}$

26. $\mathbf{r}(t) = e^t\mathbf{i} + e^{-t}\mathbf{j} + e^{2t}\mathbf{k}$; $t_0 = \ln 2$

27. $\mathbf{r}(t) = \left\langle \dfrac{1}{1-t}, t\sqrt{t+2}, \dfrac{1}{t-1}\right\rangle$; $t_0 = 2$

28. $\mathbf{r}(t) = \left\langle t\ln 2, 2\ln t, \dfrac{\ln t}{t}\right\rangle$; $t_0 = 1$

In Exercises 29–32, draw the graph of $\mathbf{r}$. Find the tangent vector $\mathbf{r}'(t)$ at the specified point, and show it on the graph.

29. $\mathbf{r}(t) = \langle 4\cos t, 2\sin t\rangle$; $t = \dfrac{\pi}{3}$

30. $\mathbf{r}(t) = \langle 1-t, t^2\rangle$; $t = 2$

31. $\mathbf{r}(t) = t^2\mathbf{i} + t^3\mathbf{j}$; $t = -1$

32. $\mathbf{r}(t) = (\sec t)\mathbf{i} + (\tan t)\mathbf{j}$; $t = \dfrac{3\pi}{4}$

In Exercise 33 and 34, find a tangent vector to the graph of $\mathbf{r}$ at the specified point.

33. $\mathbf{r}(t) = \langle t^2, 1-t^3, 3t\rangle$; $(1, 0, 3)$

34. $\mathbf{r}(t) = \left\langle 2 + \ln t, 1 - t\ln t, \dfrac{2\ln t}{t}\right\rangle$; $(2, 1, 0)$

In Exercises 35–42, evaluate the integrals.

35. $\displaystyle\int \langle \sin t, 1 - \cos t, t\rangle\, dt$

36. $\displaystyle\int \left\langle te^{-t^2}, \dfrac{\ln t}{t}, \dfrac{1}{t\ln t}\right\rangle dt$

37. $\displaystyle\int_0^1 \left\langle \dfrac{1}{1+t^2}, \sqrt{1-t}, t\sqrt{1-t^2}\right\rangle dt$

38. $\displaystyle\int_0^{\ln 2} \langle e^{-t}, e^{2t}, 6e^{-3t}\rangle\, dt$

39. $\displaystyle\int_0^{\pi/2} \left[(\sin^2 t)\mathbf{i} + (\sin t\cos^2 t)\mathbf{j}\right] dt$

40. $\displaystyle\int_1^{\sqrt{3}} \left(\dfrac{1}{1+t^2}\mathbf{i} + \dfrac{1}{\sqrt{4-t^2}}\mathbf{j}\right) dt$

41. $\displaystyle\int_{2\pi/3}^{\pi} \left[(\cos^3 t)\mathbf{i} + (\tan^2 t)\mathbf{j} + \mathbf{k}\right] dt$

42. $\displaystyle\int_0^1 \left[\dfrac{t-1}{t+1}\mathbf{i} + \dfrac{1}{\sqrt{t}(\sqrt{t}+1)}\mathbf{j} + \dfrac{t}{(t+1)^2}\mathbf{k}\right] dt$

43. Evaluate the integral

$$\int_1^{2\sqrt{2}} |\mathbf{r}(t)|\, dt$$

where $\mathbf{r}(t) = \langle t^2\sin t, t^2\cos t, t^2\rangle$.

44. Prove that if $\mathbf{r}(t) = \mathbf{C}$, where $\mathbf{C} = \langle C_1, C_2, C_3\rangle$ is a constant vector, then $\mathbf{r}'(t) = \mathbf{0}$ for all t.

45. Verify the inequality $\left|\int_a^b \mathbf{r}(t)\, dt\right| \leq \int_a^b |\mathbf{r}(t)|\, dt$ for $\mathbf{r}(t) = \langle \sin t, \cos t, 1\rangle$.

In Exercises 46–56, prove the specified property.

46. Derivative Property 1

47. Derivative Property 2

48. Derivative Property 3

49. Derivative Property 4

50. Derivative Property 5

51. Derivative Property 6

52. Integral Property 1

53. Integral Property 2

54. Integral Property 3

55. Integral Property 4

56. Integral Property 5

57. Prove that if $\mathbf{u}'(t) = \mathbf{v}'(t)$ for all t on an interval I, then $\mathbf{u}(t) = \mathbf{v}(t) + \mathbf{C}$ on I.

58. Suppose the position of a particle moving in space is given by

$$\mathbf{r}(t) = \left\langle e^{(t-1)^2} \cos(t-1), e^{(t-1)^2} \sin(t-1), t \right\rangle$$

(a) Plot the curve in space for $0 \leq t \leq 2$.
(b) Calculate the velocity and speed of the particle at any time t.
(c) Calculate the acceleration at time t.

59. Suppose the position of a particle moving in space is given by $\mathbf{r}(t) = \langle t, t^2, t^3 \rangle$. Find the tangent line to the curve at the point $(2, 4, 8)$ and plot the curve and the tangent line.

60. Find the tangent line to the space curve $\mathbf{r}(t) = \langle \sin 2t, 2\sin t, \cos t \rangle$ at the point where $t = 0$ and the point where $t = \pi/6$. Sketch the curve and the two tangent lines and find the point of intersection.

61. A stable camera is mounted on the nose of a spy airplane, so it can point directly ahead of the airplane. The path of the airplane is described by the curve $\mathbf{r}(t) = \langle 3\cos t, 4\sin t, 6+\sin 3t \rangle$. Ground level is the plane $z = 0$. At time $t = \pi/3$, determine the coordinates of the ground-level point seen by the camera.*

---

*From Chi-Keung Cheung and John Harer, *A Guidebook to Multivariable Calculus with Maple V.*

12.3 ARC LENGTH

The distance traveled around one leaf of a freeway intersection can be found by parameterization of the appropriate function.

The **length** of a space curve C on an interval $I = [a, b]$ is defined analogously to the length of a plane curve given in Chapter 9. When C has a finite length, it is said to be **rectifiable**. For a plane curve C, with parameterization $x = f(t)$ and $y = g(t)$, we showed in Chapter 9 (Equation 9.6) that if f' and g' are continuous on $[a, b]$, then C is rectifiable and its length L is given by

$$L = \int_a^b \sqrt{[f'(t)]^2 + [g'(t)]^2}\, dt = \int_a^b \sqrt{\left(\frac{dx}{dt}\right)^2 + \left(\frac{dy}{dt}\right)^2}\, dt$$

Letting $\mathbf{r}(t) = \langle f(t), g(t) \rangle$, we see that the integrand is the length $|\mathbf{r}'(t)|$ of the tangent vector $\mathbf{r}'(t)$, so we can write L in the succinct form

$$L = \int_a^b |\mathbf{r}'(t)|\, dt$$

The following theorem generalizes this result.

THEOREM 12.2

Let C be the graph of the continuous vector function $\mathbf{r}$ on $[a, b]$, and suppose $\mathbf{r}'$ also is continuous on $[a, b]$. Then C is rectifiable, and if C is traversed exactly once as t increases from a to b, its length L is given by

$$L = \int_a^b |\mathbf{r}'(t)|\, dt \qquad (12.1)$$

REMARK

■ Just as in Definition 9.2, we say that the graph of $\mathbf{r}$ is **smooth** if $\mathbf{r}'$ is continuous and $\mathbf{r}'(t) \neq 0$. So by Theorem 12.2, we see that every smooth curve is rectifiable.

When C is a plane curve, Equation 12.1 is merely a restatement of Equation 9.6, but it holds true for space curves as well. If $\mathbf{r}(t) = \langle f(t), g(t), h(t) \rangle$, then Equation 12.1 can be written in the following form.

The Length of a Space Curve

$$L = \int_a^b \sqrt{[f'(t)]^2 + [g'(t)]^2 + [h'(t)]^2}\, dt$$

$$= \int_a^b \sqrt{\left(\frac{dx}{dt}\right)^2 + \left(\frac{dy}{dt}\right)^2 + \left(\frac{dz}{dt}\right)^2}\, dt$$

EXAMPLE 12.11 Find the length of the arc of the circular helix $\mathbf{r}(t) = \langle \cos t, \sin t, t \rangle$ for t varying from $t = 0$ to $t = 2\pi$.

Solution

$$\mathbf{r}'(t) = \langle -\sin t, \cos t, 1 \rangle$$

So by Equation 12.1,

$$L = \int_0^{2\pi} |\mathbf{r}'(t)|\, dt = \int_0^{2\pi} \sqrt{\sin^2 t + \cos^2 t + 1}\, dt$$

$$= \int_0^{2\pi} \sqrt{1+1}\, dt = 2\sqrt{2}\,\pi \quad \blacksquare$$

EXAMPLE 12.12 Find the length of the curve

$$\mathbf{r}(t) = 3t^2 \mathbf{i} + (1 - 4t^2)\mathbf{j} + 2t^3 \mathbf{k}$$

from the point given by $t = 0$ to the point given by $t = 4$.

Solution We have $\mathbf{r}'(t) = 6t\mathbf{i} - 8t\mathbf{j} + 6t^2 \mathbf{k}$, so

$$|\mathbf{r}'(t)| = \sqrt{36t^2 + 64t^2 + 36t^4} = \sqrt{100t^2 + 36t^4} = 2t\sqrt{25 + 9t^2}$$

since $t \geq 0$. Thus, by Equation 12.1,

$$L = \int_0^4 2t\sqrt{25 + 9t^2}\, dt = \frac{1}{9} \cdot \frac{2}{3}\left[(25 + 9t^2)^{3/2}\right]_0^4 \quad \text{Let } u = 25 + 9t^2.$$

$$= \frac{2}{27}\left[(169)^{3/2} - (25)^{3/2}\right] = \frac{4144}{27} \quad \blacksquare$$

The Arc Length Function

If C has the vector equation $\mathbf{r}(t) = \langle f(t), g(t), h(t) \rangle$ for t in $[a, b]$, and if $\mathbf{r}'$ is continuous on $[a, b]$, we know that C is rectifiable. For an arbitrary t in $[a, b]$ we designate the length of C from $\mathbf{r}(a)$ to $\mathbf{r}(t)$ by $s(t)$; that is,

$$s(t) = \int_a^t |\mathbf{r}'(u)|\, du, \qquad a \leq t \leq b \tag{12.2}$$

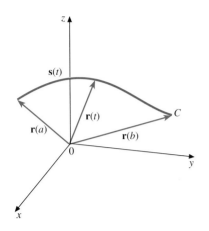

FIGURE 12.8

(where u is used as the variable of integration to avoid confusion with the upper limit t). We show the arc length $s(t)$ from $\mathbf{r}(a)$ to $\mathbf{r}(t)$ in Figure 12.8. In this way we are defining a scalar function s, called the **arc length function** for C. Note that $s(a) = 0$ and $s(b) = L$, the total length of C on $[a, b]$. Since $|\mathbf{r}'|$ is continuous on $[a, b]$, it follows by the First Fundamental Theorem of Calculus (Theorem 5.2) that

$$s'(t) = |\mathbf{r}'(t)| \tag{12.3}$$

Equivalently,

$$\frac{ds}{dt} = \sqrt{\left(\frac{dx}{dt}\right)^2 + \left(\frac{dy}{dt}\right)^2 + \left(\frac{dz}{dt}\right)^2} \tag{12.4}$$

or, in terms of differentials,

$$ds = \sqrt{\left(\frac{dx}{dt}\right)^2 + \left(\frac{dy}{dt}\right)^2 + \left(\frac{dz}{dt}\right)^2}\, dt \tag{12.5}$$

EXAMPLE 12.13 Find $s(t)$ for the circular helix

$$\mathbf{r}(t) = \langle \cos t, \sin t, t \rangle, \qquad t \geq 0$$

Solution First, we calculate $|\mathbf{r}'(t)|$.

$$|\mathbf{r}'(t)| = \sqrt{\sin^2 t + \cos^2 t + 1} = \sqrt{2}$$

By Equation 12.2,

$$s(t) = \int_0^t |\mathbf{r}'(u)|\, du = \int_0^t \sqrt{2}\, du = \sqrt{2}\, t \qquad \blacksquare$$

Curves Parameterized by Arc Length

It is sometimes useful to make a change of parameter in the equation of a curve C. Under appropriate conditions, the length of the curve as found from the new parameterization will be the same as that found from the original equation (see Exercise 27 in Exercise Set 12.3). For example, if C is the graph of $\mathbf{r}(t) = \langle t^2, 1 - t^2, t^3 \rangle$ from $t = 1$ to $t = 2$, and we let $t = e^u$, so that $u = \ln t$, then the equation of C with parameter u is $\mathbf{R}(u) = \langle e^{2u}, 1 - e^{2u}, e^{3u} \rangle$ from $u = 0$ to $u = \ln 2$. In Exercise 21 of Exercise Set 12.3, you will be asked to verify that the same length L is given by the two integrals

$$\int_1^2 |\mathbf{r}'(t)|\, dt \quad \text{and} \quad \int_0^{\ln 2} |\mathbf{R}'(u)|\, du$$

It is particularly useful to consider a curve parameterized by arc length s, measured along the curve. If C has length L, then its vector equation with s as a parameter has the form

$$\mathbf{r}(s) = \langle f(s), g(s), h(s) \rangle, \qquad 0 \le s \le L$$

Equation 12.2 then gives

$$s = \int_0^s |\mathbf{r}'(u)|\, du$$

where we are assuming $\mathbf{r}'$ is continuous for s on $[0, L]$. If we differentiate both sides of this equation with respect to s, we get

$$1 = |\mathbf{r}'(s)|$$

by the First Fundamental Theorem of Calculus. Since $\mathbf{r}'(s)$ is a tangent vector to the curve, we have shown the following:

When a smooth curve C is parameterized by arc length, the tangent vector $\mathbf{r}'(s)$ is a unit vector at every point on C.

We will make use of this result in Section 12.4.

EXAMPLE 12.14 Obtain the parameterization by arc length of the circular helix in Example 12.13. Show that the tangent vector obtained by differentiation with respect to s is a unit vector.

Solution In Example 12.13 we found that

$$s = \sqrt{2}\, t$$

Thus, $t = s/\sqrt{2}$. On substituting this value of t in the equation $\mathbf{r}(t) = \langle \cos t, \sin t, t \rangle$, we get

$$\mathbf{r}\left(\frac{s}{\sqrt{2}}\right) = \left\langle \cos\left(\frac{s}{\sqrt{2}}\right), \sin\left(\frac{s}{\sqrt{2}}\right), \frac{s}{\sqrt{2}} \right\rangle$$

as the parameterization with respect to arc length. Let $\mathbf{R}(s) = \mathbf{r}(s/\sqrt{2})$. Then

$$\mathbf{R}'(s) = \left\langle -\frac{1}{\sqrt{2}} \sin\left(\frac{s}{\sqrt{2}}\right), \frac{1}{\sqrt{2}} \cos\left(\frac{s}{\sqrt{2}}\right), \frac{1}{\sqrt{2}} \right\rangle$$

and

$$|\mathbf{R}'(s)| = \sqrt{\frac{1}{2}\sin^2\left(\frac{s}{\sqrt{2}}\right) + \frac{1}{2}\cos^2\left(\frac{s}{\sqrt{2}}\right) + \frac{1}{2}} = \sqrt{\frac{1}{2} + \frac{1}{2}} = 1 \qquad \blacksquare$$

Exercise Set 12.3

In Exercises 1–6, find the length of the curve on the specified interval.

1. $\mathbf{r}(t) = \langle 2-t, 3+2t, 5-3t \rangle$, $1 \leq t \leq 3$
2. $\mathbf{r}(t) = \langle 3t, 2t^{3/2}, 4 \rangle$, $0 \leq t \leq 8$
3. $\mathbf{r}(t) = (3t^2 + 1)\mathbf{i} + 3t^2\mathbf{j} + 2t^3\mathbf{k}$, $0 \leq t \leq 2$
4. $\mathbf{r}(t) = 2\sin^2 t\, \mathbf{i} + \cos^3 t\, \mathbf{j} + \sin^3 t\, \mathbf{k}$, $0 \leq t \leq \dfrac{\pi}{2}$
5. $x = 3t$, $y = 2\cos 3t$, $z = 2\sin 3t$, $0 \leq t \leq \dfrac{\pi}{3}$
6. $x = 2e^t$, $y = e^{-t}$, $z = 2t$, $-1 \leq t \leq 1$

In Exercises 7–12, find $s(t)$.

7. $\mathbf{r}(t) = \langle 3-2t, 4+6t, 5t \rangle$, $-1 \leq t \leq 10$
8. $\mathbf{r}(t) = \langle 2\sin t, 4t, 2\cos t \rangle$, $t \geq 0$
9. $x = 3t \sin t$, $y = 3t \cos t$, $z = (2t)^{3/2}$, $t \geq 0$
10. $x = 5e^{-t}$, $y = (3e^{-t} + 1)$, $z = -4e^{-t}$, $t \geq 0$
11. $\mathbf{r}(t) = 7\mathbf{i} + t^3\mathbf{j} - t^2\mathbf{k}$, $t \geq 0$
12. $\mathbf{r}(t) = \ln t^2 \mathbf{i} + \dfrac{1}{t}\mathbf{j} + 2t\mathbf{k}$, $t \geq 1$

In Exercises 13–16, find an equivalent vector equation for the curve under the specified change in parameter. What is the parameter interval for u? State whether the direction is unchanged or reversed.

13. $\mathbf{r}(t) = \left\langle t, 2t, \dfrac{1}{t} \right\rangle$, $t > 0$; $u = t^2$
14. $\mathbf{r}(t) = \langle e^{-2t}, 1+e^t, 2e^{2t} \rangle$, $-1 \leq t \leq 1$; $u = e^t$
15. $\mathbf{r}(t) = \dfrac{t}{t+1}\mathbf{i} + \dfrac{1}{t+1}\mathbf{j} + \dfrac{t+2}{t+1}\mathbf{k}$, $0 \leq t \leq 1$;

$u = \dfrac{1}{t+1}$

16. $\mathbf{r}(t) = \sqrt{4-t^2}\,\mathbf{i} + \dfrac{1}{\sqrt{4-t^2}}\mathbf{j} + \dfrac{t}{\sqrt{4-t^2}}\mathbf{k}$,

$0 \leq t \leq 1$; $u = \sqrt{4-t^2}$

In Exercises 17–20, re-parameterize the given curve with respect to arc length.

17. $\mathbf{r}(t) = \langle a\cos t, a\sin t \rangle$, $0 \leq t \leq 2\pi$, where a is a positive constant
18. $\mathbf{r}(t) = \langle 2t-1, 3t, 1-t \rangle$, $t \geq 0$
19. $\mathbf{r}(t) = (4\cos t)\mathbf{i} + 3t\mathbf{j} + (4\sin t)\mathbf{k}$, $t \geq 0$
20. $\mathbf{r}(t) = t^2\mathbf{i} + (1-t^2)\mathbf{j} + \left(\dfrac{t^2+3}{2}\right)\mathbf{k}$, $t \geq 0$

21. Let C be the graph of $\mathbf{r}(t) = \langle t^2, 1-t^2, t^3 \rangle$ for $1 \leq t \leq 2$. Find the length of C. Make the change of parameter $u = \ln t$, and let $\mathbf{R}(u)$ be the resulting function. What is the domain of $\mathbf{R}$ so that its graph is also C? Find the length of C using $\mathbf{R}(u)$, and verify that the answer is the same as that found using $\mathbf{r}(t)$.

22. Let $\mathbf{r}(t) = (2t-1)\mathbf{i} + (2e^t + 3)\mathbf{j} + e^{-t}\mathbf{k}$ for t in the interval $0 \leq t \leq \ln 2$. Follow the instructions for Exercise 21 with the change of parameter $u = e^{-t}$.

23. Find the length of the curve $\mathbf{r}(t) = \langle t^2, t^3, -3t^2 \rangle$ from $(1, -1, -3)$ to $(1, 1, -3)$.

24. Let C be the curve defined parametrically by $x = t$, $y = t^2/2$, and $z = 2t$.
    (a) Find a unit vector tangent to C at the point $(2, 2, 4)$.
    (b) Find the length of C from $(-2, 2, -4)$ to $(2, 2, 4)$.
    (c) Sketch C over the range from $t = 0$ to $t = 4$, and show the unit tangent vector found in part (a).

25. Let C be the curve with position vector $\mathbf{r}(t) = \langle 3t^2/2, 4 + 2t^2, 3 \rangle$ on $[1, \infty)$. Make the change of parameter $u = s(t)$, where s is the arc length function, and let $\mathbf{R}(u)$ be the resulting position vector for C. Verify that for all $u \geq 0$, $|\mathbf{R}'(u)| = 1$.

26. Let C be the graph of the differentiable vector function $\mathbf{r}$ on the interval I. Let α be a differentiable scalar function on an interval J that has range I, suppose $\alpha'(u) \neq 0$ for all u in J, and let $\mathbf{R} = \mathbf{r} \circ \alpha$. If $t_0 = \alpha(u_0)$, prove that $\mathbf{r}'(t_0) = k\mathbf{R}'(u_0)$ for some scalar k. What is the geometric significance of this result?

27. Under the hypotheses of Exercise 26, prove that the length of the graph of $\mathbf{r}$ over I equals the length of the graph of $\mathbf{R}$ over J. (This result shows that the length of a curve is invariant under a parameter change of the type described.)

28. Let C be the graph of the continuously differentiable function $\mathbf{r}(t)$ for t in $[a, b]$, and let the length of C be L. Change the parameter to arc length $s = s(t)$ for s in $[0, L]$, and denote the result by $\mathbf{R}(s)$. Show that for every s in $[0, L]$, $|\mathbf{R}'(s)| = 1$. (This is an alternative proof to the one at the end of this section that when a smooth curve is parameterized by arc length, the tangent vector is always a unit vector.) (*Hint:* Let $t = t(s)$ be the inverse of $s = s(t)$, and use the Chain Rule, along with the fact that $dt/ds = 1/(ds/dt)$.)

29. Estimate the arc length of the space curve $\mathbf{r}(t) = \langle 2t, t^2, 3t^3 \rangle$, for t in $[-1, 1]$. Use Simpson's Rule with $n = 10$.

30. Estimate the arc length of the space curve $\mathbf{r}(t) = \langle e^t \cos t, e^t \sin t, t \rangle$, for t in $[-1, 1]$. Use Simpson's Rule with $n = 10$.

12.4 UNIT TANGENT AND NORMAL VECTORS; CURVATURE

Unit Tangent and Normal Vectors

We saw in Section 12.3 that when a smooth curve C is parameterized by arc length, the tangent vector $\mathbf{r}'(s)$ always has length 1. We call this vector the **unit tangent vector** for C and designate it by $\mathbf{T}$.

The Unit Tangent Vector

$$\mathbf{T} = \frac{d\mathbf{r}}{ds} \tag{12.6}$$

Since $|\mathbf{T}| = 1$, we have $\mathbf{T} \cdot \mathbf{T} = |\mathbf{T}|^2 = 1$, and if $d\mathbf{T}/ds$ exists, we can differentiate both sides of $\mathbf{T} \cdot \mathbf{T} = 1$ and use Property 4 for derivatives to get

$$\frac{d\mathbf{T}}{ds} \cdot \mathbf{T} + \mathbf{T} \cdot \frac{d\mathbf{T}}{ds} = 0$$

$$2\mathbf{T} \cdot \frac{d\mathbf{T}}{ds} = 0$$

Thus, $\mathbf{T} \cdot d\mathbf{T}/ds = 0$, so $\mathbf{T}$ and $d\mathbf{T}/ds$ are orthogonal. If $d\mathbf{T}/ds$ is nonzero and is not a unit vector, we make it into a unit vector by dividing it by its length. We designate the result by $\mathbf{N}$ and call it the **principal unit normal vector** (or **unit normal** for short). The unit tangent vector $\mathbf{T}$ and unit normal $\mathbf{N}$ to a representative curve $\mathbf{C}$ are shown in Figure 12.9.

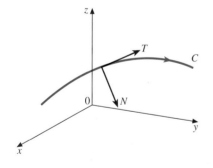

FIGURE 12.9

The Unit Normal Vector

$$\mathbf{N} = \frac{\dfrac{d\mathbf{T}}{ds}}{\left|\dfrac{d\mathbf{T}}{ds}\right|} \tag{12.7}$$

Curvature

Since **T** has constant length 1, the derivative $d\mathbf{T}/ds$ reflects the change in *direction* of **T** only. Its magnitude $|d\mathbf{T}/ds|$ thus provides a measure of how rapidly the unit tangent vector **T** is turning as arc length is traversed. We call this magnitude the **curvature** of C and designate it by the Greek letter kappa, κ.

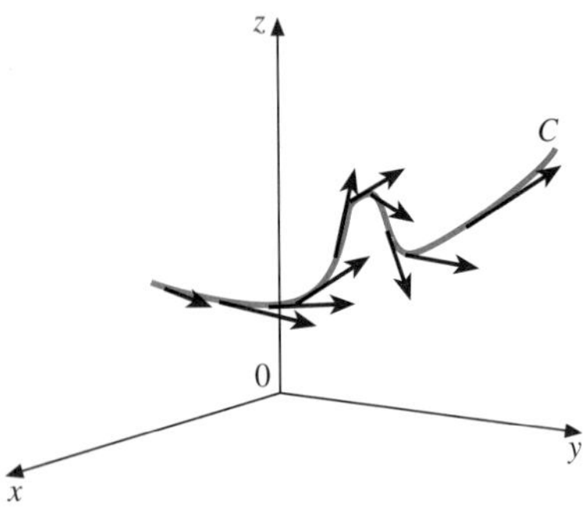

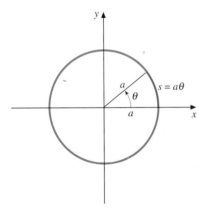

FIGURE 12.10
Unit tangent vector changes direction slowly when the curve is relatively flat, and it changes direction rapidly when the curve bends sharply.

Curvature
$$\kappa = \left|\frac{d\mathbf{T}}{ds}\right| \qquad (12.8)$$

In Figure 12.10 we show how the unit tangent vector changes direction slowly when the curve is fairly straight (curvature is small) but changes direction rapidly when the curve bends more sharply (curvature is large).

We can now write Equation 12.7 in the form

$$\mathbf{N} = \frac{1}{\kappa}\frac{d\mathbf{T}}{ds}$$

so that

$$\frac{d\mathbf{T}}{ds} = \kappa \mathbf{N} \qquad (12.9)$$

FIGURE 12.11

EXAMPLE 12.15 Find **T**, **N**, and κ at an arbitrary point on a circle of radius a in the xy-plane.

Solution For convenience we take the center of the circle at the origin. From Figure 12.11 we see that parametric equations for the circle are $x = a\cos\theta$ and $y = a\sin\theta$, where $0 \leq \theta \leq 2\pi$. Equivalently, the circle, parameterized in terms of θ, has the vector equation $\mathbf{r}(\theta) = (a\cos\theta)\mathbf{i} + (a\sin\theta)\mathbf{j}$. Since the arc length s is given by $s = a\theta$, we have $\theta = s/a$. So the parameterization of the circle with respect to arc length s is given by

$$\mathbf{r}\left(\frac{s}{a}\right) = a\cos\left(\frac{s}{a}\right)\mathbf{i} + a\sin\left(\frac{s}{a}\right)\mathbf{j}, \qquad 0 \leq s \leq 2\pi a$$

Let $\mathbf{R}(s) = \mathbf{r}(s/a)$. Then, from Equation 12.6,

$$\mathbf{T} = \frac{d\mathbf{R}}{ds} = -\sin\left(\frac{s}{a}\right)\mathbf{i} + \cos\left(\frac{s}{a}\right)\mathbf{j}$$

To find **N** and κ we calculate $d\mathbf{T}/ds$:

$$\frac{d\mathbf{T}}{ds} = -\frac{1}{a}\cos\left(\frac{s}{a}\right)\mathbf{i} - \frac{1}{a}\sin\left(\frac{s}{a}\right)\mathbf{j}$$

$$\kappa = \left|\frac{d\mathbf{T}}{ds}\right| = \sqrt{\frac{1}{a^2}\cos^2\left(\frac{s}{a}\right) + \frac{1}{a^2}\sin^2\left(\frac{s}{a}\right)} = \frac{1}{a}$$

$$\mathbf{N} = -\left[\cos\left(\frac{s}{a}\right)\mathbf{i} + \sin\left(\frac{s}{a}\right)\mathbf{j}\right] = -\frac{1}{a}\mathbf{R}(s)$$

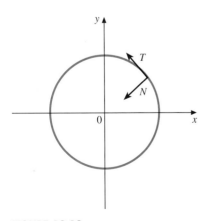

FIGURE 12.12

Observe that the direction of **N** is opposite that of $\mathbf{R}(s)$, so **N** points toward the center of the circle (Figure 12.12). ■

REMARK
■ The idea that the curvature of a circle is the reciprocal of the radius makes sense on intuitive grounds. When the radius is small, the unit tangent vector changes direction rapidly as we progress around the circle and so the curvature is large, whereas for a circle with a large radius, this change is slow and the curvature is small.

T, N, and κ in Terms of Other Parameters

As we saw in Example 12.15, it is easy to parameterize the circle in terms of arc length. Unfortunately, for most curves doing so is difficult. So we need computational formulas for **T**, **N**, and κ in terms of other parameters. If t is any other parameter and $\mathbf{r}(t)$ is the position vector for C, we know that in theory at least, we can introduce as the parameter $s = s(t)$ given by Equation 12.2, provided $\mathbf{r}'$ is continuous on the t interval $[a, b]$ and $\mathbf{r}'(t) \neq 0$ there (that is, C is smooth). It follows that $s(t) = |\mathbf{r}'(t)| > 0$, so that s is an increasing function of t and hence has an inverse, say $t(s)$, that is also increasing. Under these conditions,

$$\frac{dt}{ds} = \frac{1}{\frac{ds}{dt}}$$

So, using the Chain Rule, we have

$$\mathbf{T} = \frac{d\mathbf{r}}{ds} = \frac{d\mathbf{r}}{dt}\frac{dt}{ds} = \frac{\frac{d\mathbf{r}}{dt}}{\frac{ds}{dt}}$$

and since $ds/dt = |\mathbf{r}'(t)|$, we can write

$$\mathbf{T} = \frac{\mathbf{r}'(t)}{|\mathbf{r}'(t)|} \qquad (12.10)$$

To find **N** also as a function of t, we assume $\mathbf{r}''(t)$ exists and get, again by the Chain Rule,

$$\mathbf{N} = \frac{\frac{d\mathbf{T}}{ds}}{\left|\frac{d\mathbf{T}}{ds}\right|} = \frac{\frac{d\mathbf{T}}{dt}\frac{dt}{ds}}{\left|\frac{d\mathbf{T}}{dt}\frac{dt}{ds}\right|} = \frac{\frac{d\mathbf{T}}{dt}\frac{dt}{ds}}{\left|\frac{d\mathbf{T}}{dt}\right|\frac{dt}{ds}} = \frac{\frac{d\mathbf{T}}{dt}}{\left|\frac{d\mathbf{T}}{dt}\right|}$$

since $dt/ds > 0$. Thus,

$$\mathbf{N} = \frac{\mathbf{T}'(t)}{|\mathbf{T}'(t)|} \qquad (12.11)$$

Equations 12.10 and 12.11 are valid both for plane curves and for space curves. For space curves, the cross product $\mathbf{T} \times \mathbf{N}$ is a vector orthogonal to both $\mathbf{T}$ and $\mathbf{N}$. It is called the **binormal** for C, and we designate it by $\mathbf{B}$:

$$\mathbf{B} = \mathbf{T} \times \mathbf{N} \qquad (12.12)$$

The binormal is also a unit vector, as we see by

$$|\mathbf{B}| = |\mathbf{T} \times \mathbf{N}| = |\mathbf{T}||\mathbf{N}| \sin \frac{\pi}{2} = 1 \qquad \text{Since } \mathbf{T} \text{ and } \mathbf{N} \text{ are unit vectors}$$

The triple $\mathbf{T}, \mathbf{N}, \mathbf{B}$ thus is a set of mutually orthogonal unit vectors, much like the triple $\mathbf{i}, \mathbf{j}, \mathbf{k}$. But whereas the latter triple is fixed in direction, $\mathbf{T}, \mathbf{N}, \mathbf{B}$ can vary at different points on the curve. This triple is often called the *moving trihedral* for C (see Figure 12.13).

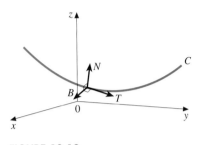

FIGURE 12.13

To find a more useful computational formula for κ than Equation 12.8, we derive some preliminary results. From Equation 12.10, we can write $\mathbf{r}'(t)$ in the form

$$\frac{d\mathbf{r}}{dt} = \left|\frac{d\mathbf{r}}{dt}\right| \mathbf{T} = \frac{ds}{dt} \mathbf{T}$$

so that

$$\frac{d^2\mathbf{r}}{dt^2} = \frac{d^2s}{dt^2} \mathbf{T} + \frac{ds}{dt} \frac{d\mathbf{T}}{dt}$$

By the Chain Rule and Equation 12.9,

$$\frac{d\mathbf{T}}{dt} = \frac{d\mathbf{T}}{ds} \frac{ds}{dt} = \kappa \frac{ds}{dt} \mathbf{N}$$

Thus,

$$\frac{d^2\mathbf{r}}{dt^2} = \frac{d^2s}{dt^2} \mathbf{T} + \kappa \left(\frac{ds}{dt}\right)^2 \mathbf{N} \qquad (12.13)$$

We assume now that C is a space curve so that all vectors are three-dimensional. The calculations are valid as well, however, for curves in the xy-plane if we treat the vectors as being three-dimensional with third component 0. We take the cross product of $\mathbf{r}'(t)$ and $\mathbf{r}''(t)$, getting

$$\frac{d\mathbf{r}}{dt} \times \frac{d^2\mathbf{r}}{dt^2} = \frac{ds}{dt} \mathbf{T} \times \left[\frac{d^2s}{dt^2} \mathbf{T} + \kappa \left(\frac{ds}{dt}\right)^2 \mathbf{N}\right]$$

$$= \left(\frac{ds}{dt} \frac{d^2s}{dt^2}\right) (\mathbf{T} \times \mathbf{T}) + \kappa \left(\frac{ds}{dt}\right)^3 (\mathbf{T} \times \mathbf{N})$$

$$= \kappa \left(\frac{ds}{dt}\right)^3 \mathbf{B}$$

since $\mathbf{T} \times \mathbf{T} = \mathbf{0}$. Since $|\mathbf{B}| = 1$ and both κ and ds/dt are nonnegative,

$$\left|\frac{d\mathbf{r}}{dt} \times \frac{d^2\mathbf{r}}{dt^2}\right| = \kappa \left(\frac{ds}{dt}\right)^3$$

Finally, we solve for κ, writing $\mathbf{r}'$ and $\mathbf{r}''$ for $d\mathbf{r}/dt$ and $d^2\mathbf{r}/dt^2$, respectively, and replacing ds/dt by $|\mathbf{r}'|$:

Curvature in Terms of $\mathbf{r}(t)$

$$\kappa = \frac{|\mathbf{r}' \times \mathbf{r}''|}{|\mathbf{r}'|^3} \qquad (12.14)$$

EXAMPLE 12.16 Find $\mathbf{T}$, $\mathbf{N}$, and κ at an arbitrary point on the circular helix $\mathbf{r}(t) = \langle 2\cos 3t, 2\sin 3t, 8t \rangle$.

Solution We will need $\mathbf{r}'$, $\mathbf{r}''$, and $|\mathbf{r}'|$:

$$\mathbf{r}'(t) = \langle -6\sin 3t, 6\cos 3t, 8 \rangle$$

$$\mathbf{r}''(t) = \langle -18\cos 3t, -18\sin 3t, 0 \rangle$$

$$|\mathbf{r}'(t)| = \sqrt{36\sin^2 3t + 36\cos^2 3t + 64} = \sqrt{36 + 64} = 10$$

By Equation 12.10,

$$\mathbf{T} = \frac{\mathbf{r}'(t)}{|\mathbf{r}'(t)|} = \frac{\langle -6\sin 3t, 6\cos 3t, 8 \rangle}{10} = \left\langle -\frac{3}{5}\sin 3t, \frac{3}{5}\cos 3t, \frac{4}{5} \right\rangle$$

To find $\mathbf{N}$, we first calculate $\mathbf{T}'(t)$:

$$\mathbf{T}'(t) = \left\langle -\frac{9}{5}\cos 3t, -\frac{9}{5}\sin 3t, 0 \right\rangle$$

Then, by Equation 12.11,

$$\mathbf{N} = \frac{\mathbf{T}'(t)}{|\mathbf{T}'(t)|} = \frac{\left\langle -\frac{9}{5}\cos 3t, -\frac{9}{5}\sin 3t, 0 \right\rangle}{\sqrt{\frac{81}{25}\cos^2 3t + \frac{81}{25}\sin^2 3t}} = \frac{\left\langle -\frac{9}{5}\cos 3t, -\frac{9}{5}\sin 3t, 0 \right\rangle}{\frac{9}{5}}$$

$$= \langle -\cos 3t, -\sin 3t, 0 \rangle$$

Finally, we calculate κ, using Equation 12.14:

$$\mathbf{r}' \times \mathbf{r}'' = \begin{vmatrix} \mathbf{i} & \mathbf{j} & \mathbf{k} \\ -6\sin 3t & 6\cos 3t & 8 \\ -18\cos 3t & -18\sin 3t & 0 \end{vmatrix}$$

$$= 144\sin 3t\,\mathbf{i} - 144\cos 3t\,\mathbf{j} + (108\sin^2 3t + 108\cos^2 3t)\mathbf{k}$$

$$= 36(4\sin 3t\,\mathbf{i} - 4\cos 3t\,\mathbf{j} + 3\mathbf{k})$$

$$|\mathbf{r}' \times \mathbf{r}''| = 36\sqrt{16\sin^2 3t + 16\cos^2 3t + 9} = 36\sqrt{25} = 180$$

Thus,

$$\kappa = \frac{|\mathbf{r}' \times \mathbf{r}''|}{|\mathbf{r}'|^3} = \frac{180}{1000} = \frac{9}{50}$$

Notice that κ is constant in this case. ∎

For reference, we summarize here Equations 12.10, 12.11, and 12.14, for **T**, **N**, and κ.

$$\mathbf{T} = \frac{\mathbf{r}'(t)}{|\mathbf{r}'(t)|} \qquad \text{Unit tangent vector}$$

$$\mathbf{N} = \frac{\mathbf{T}'(t)}{|\mathbf{T}'(t)|} \qquad \text{Unit normal vector}$$

$$\kappa = \frac{|\mathbf{r}' \times \mathbf{r}''|}{|\mathbf{r}'|^3} \qquad \text{Curvature}$$

Curvature in Two Dimensions; Radius of Curvature

For a curve C in the xy-plane, we can find the unit tangent and normal vectors by using Equations 12.10 and 12.11, respectively. In this case, however, there is no binormal vector. Equation 12.14 is valid, as remarked earlier, if we write two-dimensional vectors as three-dimensional vectors with third component 0. When we write the vectors in this way, we arrive at a formula for κ that is applicable to plane curves only, as follows. Suppose C has position vector $\mathbf{r}(t) = \langle f(t), g(t) \rangle$. Then we treat C as a space curve lying in the xy-plane, so that $\mathbf{r}(t) = \langle f(t), g(t), 0 \rangle$. For brevity, we can write $x = f(t)$ and $y = g(t)$, so $\mathbf{r}' = \langle x', y', 0 \rangle$ and $\mathbf{r}'' = \langle x'', y'', 0 \rangle$. Then $\mathbf{r}' \times \mathbf{r}'' = \langle 0, 0, x'y'' - x''y' \rangle$. (You should verify this result.) Finally, $|\mathbf{r}' \times \mathbf{r}''| = |x'y'' - x''y'|$, and we have the following equation.

Curvature for the Plane Curve $\mathbf{r}(t) = \langle x(t), y(t) \rangle$

$$\kappa = \frac{|x'y'' - x''y'|}{[(x')^2 + (y')^2]^{3/2}} \qquad (12.15)$$

In case the equation of C is in the form $y = f(x)$, we can use the parameterization $x = t$ and $y = f(t)$, and since $x' = 1$ and $x'' = 0$, Equation 12.15 becomes

Curvature for the Plane Curve $y = f(x)$

$$\kappa = \frac{|y''|}{[1 + (y')^2]^{3/2}} \qquad (12.16)$$

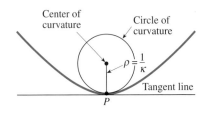

FIGURE 12.14

At a point P on a curve C where $\kappa \neq 0$, we define $\rho = 1/\kappa$ as the **radius of curvature** of C at P. If C is a plane curve, the circle of radius ρ that is tangent to C at P on its concave side (the direction of **N**) is called the **circle of curvature**, or **osculating circle**. Its center is called the **center of curvature**. (See Figure 12.14.) If C itself is a circle, Example 12.15 shows that its curvature is the reciprocal of its radius, $\kappa = 1/r$. Hence, $r = 1/\kappa$. That is, the radius of the circle is the radius of curvature. So C is its own circle of curvature. This result shows that for any plane curve C at a point P where $\kappa \neq 0$, the circle of curvature has the same curvature as C at P. In the sense that C and its circle of curvature have the same tangent line and the same curvature at P, the circle of curvature is the circle that best "fits" C in a neighborhood of P.

EXAMPLE 12.17 Find the curvature and center of curvature at the point $P(2, 1)$ on the curve $\mathbf{r}(t) = \langle 2t^2, 2 - t^3 \rangle$, where $t > 0$.

Solution The parametric equations corresponding to $\mathbf{r}(t)$ are $x = 2t^2$ and $y = 2 - t^3$. So we have

$$x' = 4t \qquad y' = -3t^2$$
$$x'' = 4 \qquad y'' = -6t$$

The point P is given by $t = 1$. At this point, $x' = 4$, $x'' = 4$, $y' = -3$, and $y'' = -6$. So by Equation 12.15,

$$\kappa = \frac{|x'y'' - x''y'|}{[(x')^2 + (y')^2]^{3/2}} = \frac{|4(-6) - (4)(-3)|}{(16 + 9)^{3/2}} = \frac{12}{125}$$

The radius of curvature ρ is therefore $\frac{125}{12}$. The center of curvature is ρ units in the direction of the normal from P. So we need $\mathbf{N}$. Since $t > 0$, we have

$$\mathbf{T} = \frac{\mathbf{r}'(t)}{|\mathbf{r}'(t)|} = \frac{\langle 4t, -3t^2 \rangle}{\sqrt{16t^2 + 9t^4}}$$

$$= \frac{t\langle 4, -3t^2 \rangle}{t\sqrt{16 + 9t^2}} = \left\langle \frac{4}{\sqrt{16 + 9t^2}}, \frac{-3t}{\sqrt{16 + 9t^2}} \right\rangle$$

Thus, we find (omitting some details)

$$\mathbf{T}'(t) = \left\langle \frac{-36}{(16 + 9t^2)^{3/2}}, \frac{-48}{(16 + 9t^2)^{3/2}} \right\rangle$$

At $t = 1$,

$$\mathbf{T}'(1) = \left\langle \frac{-36}{125}, \frac{-48}{125} \right\rangle = \frac{-12}{125} \langle 3, 4 \rangle$$

and

$$|\mathbf{T}'(1)| = \frac{12}{25}$$

Thus, by Equation 12.14, the unit normal at P is

$$\mathbf{N}(1) = \frac{\mathbf{T}'(1)}{|\mathbf{T}'(1)|} = \left\langle -\frac{3}{5}, -\frac{4}{5} \right\rangle$$

Now let $Q(h, k)$ denote the center of curvature. Then, as seen in Figure 12.15, $\overrightarrow{PQ} = \rho \mathbf{N}$, or

$$\langle h - 2, k - 1 \rangle = \frac{125}{12} \left\langle -\frac{3}{5}, -\frac{4}{5} \right\rangle = \left\langle -\frac{25}{4}, -\frac{25}{3} \right\rangle$$

So $h - 2 = -\frac{25}{4}$ and $k - 1 = -\frac{25}{3}$, from which we get the center,

$$(h, k) = \left(\frac{-17}{4}, \frac{-22}{3} \right)$$

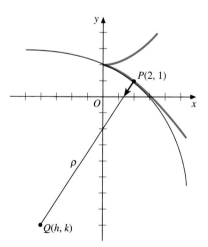

FIGURE 12.15

THE CALCULUS OF VECTOR-VALUED FUNCTIONS USING COMPUTER ALGEBRA SYSTEMS

All the calculus operations for vector-valued functions can be easily carried out by a CAS, including symbolic operations such as differentiation and integration. Analyzing the standard features of space curves, such as curvature, is generally very complicated, and, as we will see, they can be more easily examined using a computer algebra system such as Maple or Mathematica.

CAS 46

Suppose the position of a particle moving in space at time t is given by $\mathbf{r}(t) = \langle e^t \cos t, e^t \sin t, t \rangle$.

(a) Plot the path of the particle.

(b) Calculate the velocity, speed, and acceleration of the particle at time t.

(c) Calculate the curvature, κ, of the path and make a sketch of the curvature.

(d) Calculate the unit tangent vector and unit normal vector of the position.

(e) Calculate the tangential and normal components of the acceleration.

Maple:

(a) with(plots):with(linalg):
r:=t–>[exp(t)*cos(t),exp(t)*sin(t),t];
spacecurve(r(t),t=0..4*Pi,numpoints=200),axes=boxed;

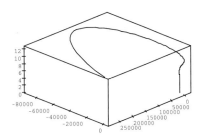

FIGURE 12.4.1

(b) v:=diff(r(t),t);

Output: $v := \left[e^t \cos(t) - e^t \sin(t), \; e^t \sin(t) + e^t \cos(t), 1 \right]$

a:=diff(v,t);

Output: $a = \left[-2e^t \sin(t), 2e^t \cos(t), 0 \right]$

Now write a function to compute the length of a vector in three space. The length of a vector $\mathbf{w}$ can be written as $|\mathbf{w}| = \sqrt{\mathbf{w} \cdot \mathbf{w}}$.

Mathematica:

<<Calculus`VectorAnalysis`
ParametricPlot3D[{Exp[t]*Cos[t],Exp[t]*Sin[t],t},
 {t,0,4*Pi}]
r[t_] = {Exp[t]*Cos[t],Exp[t]*Sin[t],t}
v = Simplify[r'[t]]
START
a = Simplify[D[v,t]]
m[w_List] = Simplify[Sqrt[w.w],Trig–>True]
speed = Simplify[m[v]]
k = Simplify[m[CrossProduct[v,a]]/(m[v])^3]
Plot[k,{t,–10,10},PlotRange–>{0,1}]
T = Simplify[v/m[v]]
DT = Simplify[D[T,t]]
N = Simplify[DT/m[DT]]
Ta = Simplify[v.a/m[a]]
Na = Simplify[m[CrossProduct[v,a]]/m[v]]

Maple:

```
magnitude:=w–>sqrt(dotprod(w,w));
speed := simplify (magnitude(v));
```

Output: $speed := \sqrt{2e^{2t} + 1}$

(c) For the curvature we use Equation 12.14
```
k:=simplify(magnitude(crossprod(v,a))/(magnitude(v))^3);
```

Output:

$$k := 2\frac{\sqrt{e^{2t} + e^{4t}}}{\left(2e^{2t} + 1\right)^{3/2}}$$

```
plot(k,t=-10..10,y=0..1,numpoints=200);
```

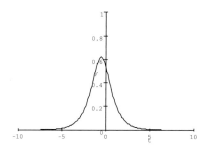

FIGURE 12.4.2

From the plot, we can easily see where the curvature is a maximum.

(d) `T:=simplify(v/magnitude(v));`

Output:

$$T = \frac{\left[e^t \cos(t) - e^t \sin(t), e^t \sin(t) + e^t \cos(t), 1\right]}{\sqrt{2e^{2t} + 1}}$$

Check that this is a unit vector:
```
simplify(magnitude(T));
```
Output: 1
```
DT:=simplify(diff(T,t));
N:=Simplify(DT/magnitude(DT));
```
(e)
```
T_a:=simplify(dotprod(v,a)/magnitude(a));
```

Output:

$$T\_a := e^t$$

```
N_a:=simplify(magnitude(crossprod(v,a))/magnitude(v));
```

Output:

$$N\_a := 2\frac{\sqrt{e^{2t} + e^{4t}}}{\sqrt{2e^{2t} + 1}}$$

Exercise Set 12.4

In Exercises 1–10, find $\mathbf{T}$ and $\mathbf{N}$ at the prescribed point.

1. $\mathbf{r}(t) = \langle t, t^2 \rangle$; $t = 2$
2. $\mathbf{r}(t) = \langle 1 - t^2, 2 + 3t \rangle$; $t = 1$
3. $\mathbf{r}(t) = 2\cos t\,\mathbf{i} + 2\sin t\,\mathbf{j}$; $t = \frac{\pi}{4}$
4. $\mathbf{r}(t) = 3\cos t\,\mathbf{i} + 4\sin t\,\mathbf{j}$; $t = \frac{\pi}{2}$
5. $\mathbf{r}(t) = \langle 2\cos t, 2\sin t, 4t \rangle$; $t = \frac{3\pi}{4}$
6. $\mathbf{r}(t) = \langle t, t^2, 3 - t^2 \rangle$; at $(1, 1, 2)$
7. $\mathbf{r}(t) = 2t^2\mathbf{i} + t^3\mathbf{j} + 3\mathbf{k}$; at $(2, 1, 3)$
8. $\mathbf{r}(t) = 2t^3\mathbf{i} + 3t^2\mathbf{j} + 3t\mathbf{k}$; $t = -1$
9. $\mathbf{r}(t) = \langle e^t, 2e^{-t}, 2t \rangle$; at $(1, 2, 0)$
10. $\mathbf{r}(t) = \langle 2t, \frac{1}{t}, 2\ln t \rangle$; at $(2, 1, 0)$

In Exercises 11–25, find κ at the prescribed point.

11. $\mathbf{r}(t) = \langle t^2, t^3 \rangle$; $t = 1$
12. $\mathbf{r}(t) = \langle 1 + t, 1 - t^2 \rangle$; $t = 0$
13. $\mathbf{r}(t) = (t \ln t)\mathbf{i} + \frac{1}{t}\mathbf{j}$; $t = 1$
14. $\mathbf{r}(t) = \sqrt{t^2 - 3}\,\mathbf{i} + \frac{t}{\sqrt{t^2 - 3}}\mathbf{j}$; $t = 2$
15. $x = \sin 2t$, $y = 4\cos 2t$; $t = \frac{\pi}{3}$
16. $x = t - \cos t$, $y = 1 - \sin t$; $t = \frac{\pi}{6}$
17. $y = 2x - x^2$; at $(0, 0)$
18. $y = x^3 - 2x^2 + 3$; at $(1, 2)$
19. $y = \frac{2}{x}$; at $(2, 1)$
20. $y = \sec x$; at $x = \frac{3\pi}{4}$
21. $\mathbf{r}(t) = \langle t^2, t^3, 1 - 2t \rangle$; $t = -1$
22. $\mathbf{r}(t) = \langle 4\cos t, 4\sin t, 3t \rangle$; $t = \frac{\pi}{3}$
23. $\mathbf{r}(t) = \langle 2e^t, e^{-t}, 2t \rangle$; at $(2, 1, 0)$
24. $\mathbf{r}(t) = t^2\mathbf{i} + t^3\mathbf{j} - 3t^2\mathbf{k}$; $t = \frac{1}{2}$
25. $\mathbf{r}(t) = (e^t \sin t)\mathbf{i} + (e^t \cos t)\mathbf{j} + e^t\mathbf{k}$; $t = 0$

In Exercises 26–30, find the center of the circle of curvature at P. Sketch the graph in a neighborhood of P, showing the circle of curvature.

26. $\mathbf{r}(t) = \langle t, 1 + t^2 \rangle$; $P(0, 1)$
27. $\mathbf{r}(t) = \langle 4\cos t, 3\sin t \rangle$; $P(4, 0)$
28. $\mathbf{r}(t) = t\mathbf{i} + \sqrt{2t}\,\mathbf{j}$; $P(\frac{1}{2}, 1)$
29. $\mathbf{r}(t) = t\mathbf{i} + e^t\mathbf{j}$; $P(0, 1)$
30. $\mathbf{r}(t) = \langle t - \sin t, 1 - \cos t \rangle$; $t = \frac{\pi}{3}$

31. Let C be a plane curve that is the graph of $\mathbf{r}(s)$, where s is the arc length parameter, and suppose $\mathbf{r}''(s)$ exists on $[0, L]$, where L is the length of C. Let $\theta = \theta(s)$ be the angle between the vectors $\mathbf{i}$ and $\mathbf{T}$ at an arbitrary point on C. Show the following:
    (a) $\mathbf{T} = \langle \cos\theta, \sin\theta \rangle$
    (b) $\kappa = \left| \dfrac{d\theta}{ds} \right|$
    (c) $\mathbf{N} = \langle -\sin\theta, \cos\theta \rangle$ if $d\theta/ds > 0$
    $\mathbf{N} = \langle \sin\theta, -\cos\theta \rangle$ if $d\theta/ds < 0$
    Use the result of part (c) to show that $\mathbf{N}$ is always directed toward the concave side of C.

32. Suppose $\mathbf{N} = \langle n_1, n_2 \rangle$ is the unit normal vector at a point $P(x, y)$ on the plane curve C. Prove that the center of curvature at P is the point $(x + \rho n_1, y + \rho n_2)$, where ρ is the radius of curvature.

33. Find $\mathbf{T}$, $\mathbf{N}$, $\mathbf{B}$, and κ at an arbitrary point on the circular helix $\mathbf{r}(t) = \langle a\cos bt, a\sin bt, ct \rangle$.

34. Find $\mathbf{T}$, $\mathbf{N}$, $\mathbf{B}$, and κ at an arbitrary point on the curve $\mathbf{r}(t) = \langle t, t^2, 2t^3/3 \rangle$.

35. Use Equation 12.13 to show that $(\mathbf{r}' \times \mathbf{r}'') \times \mathbf{r}' = |\mathbf{r}'|^2 (ds/dt)^2 \kappa \mathbf{N}$, and hence obtain the formula
$$\mathbf{N} = \frac{(\mathbf{r}' \times \mathbf{r}'') \times \mathbf{r}'}{\kappa\,|\mathbf{r}'|^4}$$

In Exercises 36 and 37, assume C is a smooth curve parameterized by arc length, and $d\mathbf{T}/ds$, $d\mathbf{N}/ds$, and $d\mathbf{B}/ds$ all exist.

36. By differentiating both sides of $\mathbf{B} \cdot \mathbf{B} = 1$, show that $\mathbf{B}$ and $d\mathbf{B}/ds$ are orthogonal. Explain why it follows that

$$\frac{d\mathbf{B}}{ds} = \alpha \mathbf{T} + \beta \mathbf{N}$$

for some scalars α and β. By differentiating both sides of $\mathbf{B} \cdot \mathbf{T} = 0$, show that $\alpha = 0$.

37. (Continuation of Exercise 36) Write $\tau = -\beta$, so that $d\mathbf{B}/ds = -\tau \mathbf{N}$. (The scalar τ is called the **torsion**.) Show that $\mathbf{N} = \mathbf{B} \times \mathbf{T}$. Prove that

$$\frac{d\mathbf{N}}{ds} = \tau \mathbf{B} - \kappa \mathbf{T}$$

(*Note:* The formulas

$$\frac{d\mathbf{T}}{ds} = \kappa \mathbf{N} \qquad \frac{d\mathbf{N}}{ds} = \tau \mathbf{B} - \kappa \mathbf{T} \qquad \frac{d\mathbf{B}}{ds} = -\tau \mathbf{N}$$

are called the **Frenet Formulas**.)

38. Find the torsion τ as a function of t for the curve of Exercise 34. Verify the Frenet Formulas for this curve. (See Exercise 37.)

39. Suppose the position of a particle moving in space at time t is given by $\mathbf{r}(t) = \langle t \cos t, t \sin t, t \rangle$.
  (a) Plot the space curve.
  (b) Calculate the velocity, speed, and acceleration of the particle at time t.
  (c) Calculate the curvature, κ, of the path and make a sketch of the curvature. From the sketch, estimate the point at which the curvature is a maximum.
  (d) Calculate the unit tangent vector and unit normal vector of the position.
  (e) Calculate the tangential and normal components of the acceleration.

40. Calculate the unit tangent and unit normal vectors to the curve $\mathbf{r}(t) = \cos(t)\mathbf{i} + \sin(t)\mathbf{j}$ at $t = \pi/4$. Plot the curve, the unit tangent, and the unit normal. In which direction does the unit normal point?

41. Calculate the unit tangent, unit normal, and curvature at $t = 2$, to the curve $\mathbf{r}(t) = (t^3/2 - t)\mathbf{i} - t^2 \mathbf{j}$. Plot the curve, the unit tangent, and the unit normal vectors for $t = 2$.

12.5 MOTION ALONG A CURVE

Runners rounding a curve

In the preceding section, we considered curves as sets of terminal points of position vectors $\mathbf{r}(t)$ for continuous vector functions $\mathbf{r}$. No concept of movement was involved. We were looking at curves from the *static* point of view. Now we want to look at them from the *dynamic* point of view; that is, we want to consider a curve as the path traced out by a particle moving in space (or in a plane).

We let t denote time, measured in whatever units we choose, from some convenient time origin, and we let $\mathbf{r}(t) = \langle f(t), g(t), h(t) \rangle$ be the position vector of the moving particle at time t. Since a moving particle always goes in a continuous path, it follows that $\mathbf{r}$ is continuous, so that the path of the particle is a curve. Suppose the time interval of interest is I. We assume further that $\mathbf{r}'$ and $\mathbf{r}''$ both exist for all t in I. We know that

$$\mathbf{r}'(t) = \lim_{h \to 0} \frac{\mathbf{r}(t+h) - \mathbf{r}(t)}{h}$$

The vector $\mathbf{r}(t+h) - \mathbf{r}(t)$ is the displacement vector from the particle's position at time t to its position at time $t + h$. When we divide by the elapsed time h, we obtain the *average velocity*. Its limit as $h \to 0$ is the *velocity* $\mathbf{v}(t)$ at time t. The direction of the velocity vector indicates the (instantaneous) direction of movement of the particle, and its magnitude is the speed of the particle. A similar analysis applies to the derivative,

$$\mathbf{v}'(t) = \lim_{h \to 0} \frac{\mathbf{v}(t+h) - \mathbf{v}(t)}{h}$$

which we call the *acceleration* at time t.

In summary, we have the following definition.

> **Definition 12.7**
> **Velocity, Acceleration, and Speed**
>
> Let $\mathbf{r}(t)$ be the position vector of a moving particle at time t, where t varies over the interval I, and suppose $\mathbf{r}'$ and $\mathbf{r}''$ both exist in I. Then the **velocity** $\mathbf{v}(t)$ and **acceleration** $\mathbf{a}(t)$ are defined as
>
> $$\mathbf{v}(t) = \mathbf{r}'(t)$$
>
> $$\mathbf{a}(t) = \mathbf{v}'(t) = \mathbf{r}''(t)$$
>
> The **speed** of the particle is defined as the magnitude of the velocity vector:
>
> $$\text{Speed} = |\mathbf{v}(t)| = |\mathbf{r}'(t)|$$

REMARK
■ Observe that both velocity and acceleration are vectors, whereas speed is a scalar. Velocity is the instantaneous rate of change of position, and acceleration is the instantaneous rate of change of velocity. These descriptions agree with our intuitive understanding of the concepts of velocity and acceleration. We know from Equation 12.3 that the arc length $s(t)$ satisfies $s'(t) = |\mathbf{r}'(t)|$; that is, speed $= ds/dt$. Thus, speed can be interpreted as the rate at which the arc length s is changing with time or, equivalently, the rate at which the distance along the path covered by the particle is changing with time. We observe further that the velocity vector is directed along the tangent line and points in the direction of motion.

EXAMPLE 12.18 A particle moves on the curve $\mathbf{r}(t) = \langle t, \frac{1}{t}, 2\sqrt{t} \rangle$ for $t > 0$. Find its velocity, acceleration, and speed when $t = 1$. Show the results graphically.

Solution

$$\mathbf{v}(t) = \mathbf{r}'(t) = \left\langle 1, -\frac{1}{t^2}, \frac{1}{\sqrt{t}} \right\rangle$$

$$\mathbf{a}(t) = \mathbf{v}'(t) = \left\langle 0, \frac{2}{t^3}, -\frac{1}{2t^{3/2}} \right\rangle$$

So when $t = 1$, $\mathbf{v}(1) = \langle 1, -1, 1 \rangle$ and $\mathbf{a}(1) = \langle 0, 2, -\frac{1}{2} \rangle$. The speed at this instant is therefore $|\mathbf{v}(1)| = \sqrt{3}$. From the parametric equations $x = t$, $y = \frac{1}{t}$, and $z = 2\sqrt{t}$, we see that for $t > 0$, all three coordinates are positive. Also, since $xy = 1$, the curve lies above the first-quadrant branch of this equilateral hyperbola. By plotting points corresponding to $t = \frac{1}{9}, \frac{1}{4}, 1, 4,$ and 9, we get a reasonably accurate sketch, as shown in Figure 12.16. ∎

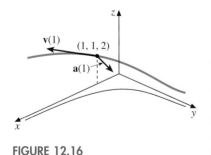

FIGURE 12.16

EXAMPLE 12.19 Find the force acting on a particle that moves in a circular path of radius r with constant speed v_0.

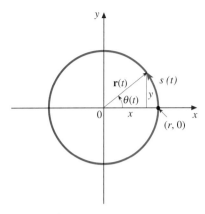

FIGURE 12.17

Solution We introduce a coordinate system so that the center of the circle is at the origin and when $t = 0$ the particle is at $(r, 0)$. We assume the motion is in the positive (counterclockwise) direction. Let $\theta(t)$ be the polar angle to the position vector $\mathbf{r}(t)$ at time t, as shown in Figure 12.17, and let $s(t)$ be the arc length covered by the particle, measured from $(r, 0)$. Then, $x = r \cos \theta(t)$ and $y = r \sin \theta(t)$ are the parametric equations of the circle. In vector form, the equation is

$$\mathbf{r}(t) = (r \cos \theta(t))\mathbf{i} + (r \sin \theta(t))\mathbf{j}$$

To find an explicit expression for $\theta(t)$, we use the relationship $s = r\theta$ to get $ds/dt = r\, d\theta/dt$. But ds/dt is the speed of the particle, which by hypothesis is equal to the constant v_0. So $d\theta/dt = v_0/r$, which is also a constant. We designate this constant by ω (the Greek letter omega) and call it the **angular speed** because it is the rate of change of θ with respect to time. Thus,

$$\frac{d\theta}{dt} = \omega$$

and integrating both sides with respect to t, we obtain

$$\int \left(\frac{d\theta}{dt}\right) dt = \int \omega\, dt$$

$$\theta(t) = \omega t + C$$

Since $\theta(0) = 0$, it follows that $C = 0$. So $\theta(t) = \omega t$. The equations of motion now become

$$\mathbf{r}(t) = (r \cos \omega t)\mathbf{i} + (r \sin \omega t)\mathbf{j}$$
$$\mathbf{v}(t) = (-r\omega \sin \omega t)\mathbf{i} + (r\omega \cos \omega t)\mathbf{j}$$
$$\mathbf{a}(t) = (-r\omega^2 \cos \omega t)\mathbf{i} + (-r\omega^2 \sin \omega t)\mathbf{j}$$
$$= -\omega^2 [(r \cos \omega t)\mathbf{i} + (r \sin \omega t)\mathbf{j}]$$
$$= -\omega^2 \mathbf{r}(t)$$

We can calculate the speed as the magnitude of the velocity $\mathbf{v}(t)$:

$$|\mathbf{v}(t)| = \sqrt{r^2\omega^2 \sin^2 \omega t + r^2\omega^2 \cos^2 \omega t} = r\omega$$

This result confirms that the constant speed v_0 is $r\omega$: $v_0 = r\omega$.

Since $|\mathbf{r}(t)| = \sqrt{r^2 \cos^2 \omega^2 t + r^2 \sin^2 \omega t} = r$, we see that $|\mathbf{a}(t)| = |-\omega^2| |\mathbf{r}(t)| = r\omega^2$. So each of the vectors $\mathbf{r}(t)$, $\mathbf{v}(t)$, and $\mathbf{a}(t)$ has constant magnitude.

According to Newton's Second Law of Motion, the force $\mathbf{F}$ acting on the body satisfies $\mathbf{F} = m\mathbf{a}$, where m is the mass of the body. Thus,

$$\mathbf{F} = -m\omega^2 \mathbf{r}(t)$$

This force, which is of constant magnitude $mr\omega^2$, is directed toward the center of the circle since it is opposite in sign to the position vector $\mathbf{r}(t)$. It is called **centripetal force**.   ∎

EXAMPLE 12.20

(a) Find a formula for the speed necessary to maintain a satellite of mass m in a fixed orbit h km above the earth's surface.

(b) Find a formula for the time required for the satellite of part (a) to complete one revolution around the earth.

Communication satellite in geostationary orbit

(c) Taking the radius of the earth as approximately 6,370 km, find the speed of a satellite in orbit 1500 km above the earth's surface. Find the number of hours required to complete one revolution.

Solution We make the simplifying assumptions that the satellite's orbit is circular, that the acceleration caused by gravity at the earth's surface is constant at $g \approx 9.81$ m/sec^2, and that the gravitational attraction of other bodies is negligible.

(a) Let R denote the radius of the earth. Then the radius of the orbit of the satellite is $r = R + h$. By what we found in Example 12.19, the centripetal force necessary to keep the satellite in orbit has magnitude $mr\omega^2$, and since $v = r\omega$, we get

$$|\mathbf{F}| = mr\left(\frac{v}{r}\right)^2 = \frac{mv^2}{r}$$

This force is produced by the earth's gravitational pull, which according to Newton's Law of Universal Gravitation is given by

$$|\mathbf{F}| = \frac{GMm}{r^2}$$

where M is the mass of the earth and G is a constant, called the universal gravitational constant. Equating the two expressions for $|\mathbf{F}|$ gives

$$v^2 = \frac{GM}{r}$$

A more useful form is obtained by observing that when the satellite is on the earth's surface, so that $r = R$, the attractive force is just the weight, mg, of the satellite. So by Newton's Gravitational Law with $r = R$,

$$mg = \frac{GMm}{R^2}$$

and thus $GM = R^2 g$. Making this substitution in the formula for v^2 gives $v^2 = R^2 g/r = R^2 g/(R + h)$. So

$$v = R\sqrt{\frac{g}{R+h}}$$

(b) The angular speed ω is the number of radians through which the position vector turns per unit of time. Since one revolution is equivalent to 2π radians, the time T required for the satellite to complete one revolution is

$$T = \frac{2\pi}{\omega} = \frac{2\pi r}{v} = \frac{2\pi(R+h)}{v} \qquad v = r\omega, \text{ so } \omega = v/r$$

From the result of part (a), we can substitute for v to obtain T in the alternative form

$$T = \frac{2\pi(R+h)^{3/2}}{R\sqrt{g}}$$

(c) We will use distances in kilometers and time in hours. So $R = 6370$ and $h = 1500$. To convert $g \approx 9.81$ m/sec^2 to km/hr^2, we divide by 1000 and multiply by $(3600)^2$, getting $g \approx 127,140$ km/hr^2. From part (a), the velocity is

$$v = R\sqrt{\frac{g}{R+h}} \approx 6370\sqrt{\frac{127,140}{7870}} \approx 25,600 \text{ km/hr}$$

and from part (b), the time T for one revolution is

$$T = \frac{2\pi(R+h)}{v} \approx \frac{2\pi(7870)}{25,600} \approx 1.93 \text{ hr} \qquad \blacksquare$$

Projectile Motion

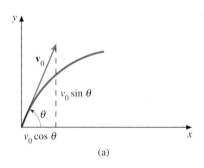

As a further application of motion on a curve, consider a projectile fired from the ground at an angle θ from the horizontal, with an initial velocity $\mathbf{v}_0$. We introduce x- and y-axes with the origin coinciding with the point from which the projectile is fired, as in Figure 12.18(a). We make the assumptions that the curvature of the earth is negligible in the interval in question, that air resistance and wind can be neglected, and that the acceleration caused by gravity, g, is constant. (It should be noted that the mathematical model we chose is an idealization, since none of these assumptions is, strictly speaking, valid, but for moderate distances the errors introduced by making them are small.) For simplicity we also assume that the projectile is fired from ground level. The questions of interest are the following:

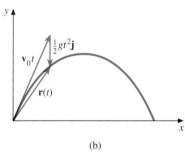

1. What is the path of the projectile?
2. How high does it rise?
3. What is its range—that is, how far does it go in a horizontal direction?
4. What is its velocity at impact?

FIGURE 12.18

We consider each of these questions in order.

1. If $\mathbf{r}(t)$ is the position vector of the projectile (considered as a point mass) at time t, then $\mathbf{r}(0) = \mathbf{0}$ and $\mathbf{v}(0) = \mathbf{v}_0$, since the initial position is at ground level and the initial velocity is $\mathbf{v}_0$. By our assumptions, the only force that acts on the projectile is the pull of gravity toward the earth. The magnitude of this force is the weight of the projectile—namely, mg, where m is its mass. Since the force is downward, we have $\mathbf{F} = -mg\mathbf{j}$. From Newton's Second Law of Motion we know that $\mathbf{F} = m\mathbf{a}$, where $\mathbf{a}$ is the acceleration of the projectile. Thus, $m\mathbf{a} = -mg\mathbf{j}$, or

$$\mathbf{a}(t) = -g\mathbf{j}$$

To find the velocity $\mathbf{v}(t)$ at time t, we integrate $\mathbf{a}(t)$, since acceleration and velocity are related by the differential equation $\mathbf{a}(t) = \mathbf{v}'(t)$:

$$\mathbf{v}(t) = \int \mathbf{a}(t)dt = -gt\mathbf{j} + \mathbf{C}_1$$

where $\mathbf{C}_1$ is a constant vector. Substituting $t = 0$, we get $\mathbf{v}(0) = \mathbf{C}_1$, and since $\mathbf{v}(0) = \mathbf{v}_0$,

$$\mathbf{v}(t) = -gt\mathbf{j} + \mathbf{v}_0$$

Now we integrate again to find $\mathbf{r}(t)$:

$$\mathbf{r}(t) = \int \mathbf{v}(t)dt = -\frac{1}{2}gt^2\mathbf{j} + \mathbf{v}_0 t + \mathbf{C}_2$$

Again letting $t = 0$, we find that $\mathbf{C}_2 = \mathbf{r}(0) = \mathbf{0}$. Thus,

$$\mathbf{r}(t) = -\frac{1}{2}gt^2\mathbf{j} + \mathbf{v}_0 t$$

If we denote the initial speed $|\mathbf{v}_0|$ by v_0, then we see from Figure 12.18 that $\mathbf{v}_0 = (v_0 \cos\theta)\mathbf{i} + (v_0 \sin\theta)\mathbf{j}$. So for $\mathbf{r}(t)$ we have

$$\mathbf{r}(t) = -\frac{1}{2}gt^2\mathbf{j} + [(v_0\cos\theta)\mathbf{i} + (v_0\sin\theta)\mathbf{j}]t$$

which can be rewritten as

$$\mathbf{r}(t) = (v_0 t \cos\theta)\mathbf{i} + \left(v_0 t \sin\theta - \frac{1}{2}gt^2\right)\mathbf{j} \tag{12.17}$$

Similarly, $\mathbf{v}(t)$ becomes

$$\mathbf{v}(t) = (v_0 \cos\theta)\mathbf{i} + (v_0 \sin\theta - gt)\mathbf{j} \tag{12.18}$$

Equation 12.17 is a vector equation of the path of the projectile. To analyze it further, we use the parametric equations for the path

$$\begin{cases} x = v_0 t \cos\theta \\ y = v_0 t \sin\theta - \frac{1}{2}gt^2 \end{cases} \tag{12.19}$$

By eliminating the parameter t, we obtain the rectangular equation (verify)

$$y = x\tan\theta - \frac{gx^2}{2v_0^2\cos^2\theta}, \qquad 0 \le \theta < \frac{\pi}{2}$$

whose graph is the parabola shown in Figure 12.18(b).

2. Since $\mathbf{v}(t)$ is tangent to the path, the maximum height occurs when $\mathbf{v}(t)$ is horizontal, that is, when the y component of $\mathbf{v}(t)$ is 0. From Equation 12.18, this situation occurs when $v_0\sin\theta - gt = 0$, or

$$t = \frac{v_0 \sin\theta}{g}$$

To find the maximum height, we substitute this value of t into the second of Equations 12.19:

$$y_{\max} = v_0\left(\frac{v_0\sin\theta}{g}\right)\sin\theta - \frac{1}{2}g\left(\frac{v_0\sin\theta}{g}\right)^2 = \frac{1}{2}\frac{v_0^2\sin^2\theta}{g}$$

3. The projectile's maximum horizontal distance occurs when it strikes the ground, that is, when $y = 0$. So to find the range, we set $y = 0$ in the second of Equations 12.19 and solve for t:

$$t\left(v_0\sin\theta - \frac{1}{2}gt\right) = 0$$

$$t = 0 \quad \text{or} \quad t = \frac{2v_0\sin\theta}{g}$$

Clearly, the second value is the one we want, since $t = 0$ is when the projectile was fired. The range is found by substituting this value of t in the first of Equations 12.19:

$$\text{Range} = x_{\max} = v_0\left(\frac{2v_0\sin\theta}{g}\right)\cos\theta = \frac{v_0^2\sin 2\theta}{g}$$

4. The velocity at impact with the ground occurs when $t = (2v_0\sin\theta)/g$. Putting this value of t in Equation 12.18 and writing $\mathbf{v}_I$ for impact velocity, we get

$$\mathbf{v}_I = (v_0\cos\theta)\mathbf{i} + (v_0\sin\theta - 2v_0\sin\theta)\mathbf{j}$$
$$= (v_0\cos\theta)\mathbf{i} - (v_0\sin\theta)\mathbf{j}$$

Also, if $v_I = |\mathbf{v}_I|$, then

$$v_I = \sqrt{(v_0 \cos\theta)^2 + (-v_0 \sin\theta)^2} = v_0$$

So the speed at impact is the same as the initial speed.

Tangential and Normal Components of Acceleration

It is sometimes useful to express the acceleration of a moving particle as the sum of two orthogonal vectors, one parallel to the unit tangent vector $\mathbf{T}$ and the other parallel to the unit normal vector $\mathbf{N}$. The lengths of these two vectors are called the tangential and normal components, respectively, of the acceleration. In this section we derive formulas for these components.

Our starting point is Equation 12.13 that we obtained in Section 12.4:

$$\frac{d^2\mathbf{r}}{dt^2} = \frac{d^2s}{dt^2}\mathbf{T} + \kappa\left(\frac{ds}{dt}\right)^2 \mathbf{N}$$

But $d^2\mathbf{r}/dt^2 = \mathbf{a}$, the acceleration of the particle moving along the curve with position vector $\mathbf{r}(t)$. So we have

$$\mathbf{a} = \frac{d^2s}{dt^2}\mathbf{T} + \kappa\left(\frac{ds}{dt}\right)^2 \mathbf{N} \tag{12.20}$$

The vectors $\mathbf{T}$ and $\mathbf{N}$ at a point on the path determine a plane, and since the right-hand side of Equation 12.20 is a linear combination of $\mathbf{T}$ and $\mathbf{N}$, it is a vector in that plane. That is, the acceleration vector at any point lies in the same plane as $\mathbf{T}$ and $\mathbf{N}$ at that point.

If we take the dot product of each side of Equation 12.20, first with $\mathbf{T}$ and then with $\mathbf{N}$, we get

$$\mathbf{a} \cdot \mathbf{T} = \frac{d^2s}{dt^2}\mathbf{T} \cdot \mathbf{T} + \kappa\left(\frac{ds}{dt}\right)^2 \mathbf{N} \cdot \mathbf{T} = \frac{d^2s}{dt^2} \tag{12.21}$$

$$\mathbf{a} \cdot \mathbf{N} = \frac{d^2s}{dt^2}\mathbf{T} \cdot \mathbf{N} + \kappa\left(\frac{ds}{dt}\right)^2 \mathbf{N} \cdot \mathbf{N} = \kappa\left(\frac{ds}{dt}\right)^2 \tag{12.22}$$

where we have used the facts that $\mathbf{T}$ and $\mathbf{N}$ are orthogonal unit vectors so that $\mathbf{T} \cdot \mathbf{T} = |\mathbf{T}|^2 = 1$, $\mathbf{N} \cdot \mathbf{N} = |\mathbf{N}|^2 = 1$, and $\mathbf{T} \cdot \mathbf{N} = 0$. Now from Section 11.3 we know that the components of $\mathbf{a}$ along $\mathbf{T}$ and $\mathbf{N}$, respectively, are

$$\text{Comp}_\mathbf{T}\mathbf{a} = \frac{\mathbf{a} \cdot \mathbf{T}}{|\mathbf{T}|}$$

and

$$\text{Comp}_\mathbf{N}\mathbf{a} = \frac{\mathbf{a} \cdot \mathbf{N}}{|\mathbf{N}|}$$

But $|\mathbf{T}| = 1$ and $|\mathbf{N}| = 1$, since $\mathbf{T}$ and $\mathbf{N}$ are unit vectors. Thus, $\text{Comp}_\mathbf{T}\mathbf{a} = \mathbf{a} \cdot \mathbf{T}$ and $\text{Comp}_\mathbf{N}\mathbf{a} = \mathbf{a} \cdot \mathbf{N}$.

So from Equations 12.21 and 12.22, the components of $\mathbf{a}$ along $\mathbf{T}$ and $\mathbf{N}$, respectively, are

$$\text{Comp}_\mathbf{T}\mathbf{a} = \frac{d^2s}{dt^2}$$

$$\text{Comp}_\mathbf{N}\mathbf{a} = \kappa\left(\frac{ds}{dt}\right)^2$$

We designate these components by the symbols a_T and a_N, respectively, and call them the **tangential** and **normal components of acceleration**.

Tangential and Normal Components of Acceleration

$$a_T = \frac{d^2s}{dt^2} \qquad a_N = \kappa \left(\frac{ds}{dt}\right)^2 \qquad (12.23)$$

Equation 12.20 can now be written

$$\mathbf{a} = a_T \mathbf{T} + a_N \mathbf{N} \qquad (12.24)$$

By Equation 12.11, we can write $a_T \mathbf{T} = \text{Proj}_T \mathbf{a}$ and $a_N \mathbf{N} = \text{Proj}_N \mathbf{a}$. We show a typical acceleration vector and its tangential and normal projections in Figure 12.19.

From Equation 12.24, we also have

$$\mathbf{a} \cdot \mathbf{a} = (a_T \mathbf{T} + a_N \mathbf{N}) \cdot (a_T \mathbf{T} + a_N \mathbf{N})$$
$$= a_T^2 \mathbf{T} \cdot \mathbf{T} + 2a_T a_N \mathbf{T} \cdot \mathbf{N} + a_N^2 \mathbf{N} \cdot \mathbf{N}$$
$$= a_T^2 + a_N^2$$

since $\mathbf{T} \cdot \mathbf{T} = 1$, $\mathbf{T} \cdot \mathbf{N} = 0$, and $\mathbf{N} \cdot \mathbf{N} = 1$. That is,

$$|\mathbf{a}|^2 = a_T^2 + a_N^2 \qquad (12.25)$$

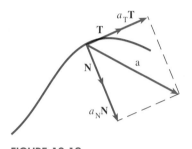

FIGURE 12.19

(This result can also be seen from Figure 12.19, using the Pythagorean Theorem.) In practice it is usually easier to find a_T than a_N from Equations 12.23. Then Equation 12.25 can be used to get a_N, as shown in the next example.

EXAMPLE 12.21 Find the tangential and normal components of acceleration of a particle whose position vector is $\mathbf{r}(t) = 3t\mathbf{i} + 2t^3\mathbf{j} + 3t^2\mathbf{k}$. Also find the curvature κ of the path at an arbitrary time t.

Solution We have

$$\mathbf{v} = \mathbf{r}'(t) = 3\mathbf{i} + 6t^2\mathbf{j} + 6t\mathbf{k} = 3(\mathbf{i} + 2t^2\mathbf{j} + 2t\mathbf{k})$$
$$\mathbf{a} = \mathbf{r}''(t) = 12t\mathbf{j} + 6\mathbf{k} = 6(2t\mathbf{j} + \mathbf{k})$$
$$\frac{ds}{dt} = |\mathbf{v}| = 3\sqrt{1 + 4t^4 + 4t^2} = 3\sqrt{(2t^2+1)^2} = 3(2t^2 + 1)$$
$$\frac{d^2s}{dt^2} = 12t$$
$$|\mathbf{a}| = 6\sqrt{4t^2 + 1}$$

So by the first of Equations 12.23, $a_T = 12t$. Then, from Equation 12.25,

$$a_N = \sqrt{|\mathbf{a}|^2 - a_T^2} = \sqrt{36(4t^2+1) - 144t^2} = 6$$

We can now obtain κ using the formula for a_N in Equations 12.23. Since $a_N = \kappa (ds/dt)^2$, we have

$$\kappa = \frac{a_N}{\left(\dfrac{ds}{dt}\right)^2} = \frac{6}{9(2t^2+1)^2} = \frac{2}{3(2t^2+1)^2}$$

(You may wish to compare this "backdoor" method of finding κ with the direct method.) ■

The tangential component of acceleration is $a_T = d^2s/dt^2$. So a_T is the rate of change of speed, which agrees with our intuitive idea of acceleration. To help understand the normal component, $a_N = \kappa(ds/dt)^2$, consider an automobile going around a curve. Since force is mass times acceleration, the normal component of the force necessary to hold the car on the road (the centripetal force) is ma_N, where m is the mass of the car. This normal component is the magnitude of the force of friction between the tires and the road. If the curve is sharp, so that the curvature κ is large, this frictional force has a large magnitude. Similarly, if the speed ds/dt is large, the magnitude of the force is large. In fact, it increases as the square of the speed. Of course, if the curve is sharp *and* the speed is great, it is unlikely the car will stay on the road.

We conclude this section by giving alternative forms for a_T and a_N that are sometimes easier to use than those given in Equations 12.23. Note first that since by Equation 12.10,

$$\mathbf{T} = \frac{\mathbf{r}'(t)}{|\mathbf{r}'(t)|}$$

and $\mathbf{v}(t) = \mathbf{r}'(t)$, we have

$$\mathbf{T} = \frac{\mathbf{v}(t)}{|\mathbf{v}(t)|} \quad (12.26)$$

or, $\mathbf{v}(t) = |\mathbf{v}(t)|\mathbf{T}$. For simplicity, we write this equation as $\mathbf{v} = |\mathbf{v}|\mathbf{T}$.

Thus, from Equation 12.24,

$$\mathbf{v} \cdot \mathbf{a} = |\mathbf{v}|\,\mathbf{T} \cdot (a_T\mathbf{T} + a_N\mathbf{N}) = |\mathbf{v}|\,a_T$$

and

$$\mathbf{v} \times \mathbf{a} = |\mathbf{v}|\,\mathbf{T} \times (a_T\mathbf{T} + a_N\mathbf{N}) = |\mathbf{v}|\,a_N\mathbf{B}$$

where we have used $\mathbf{T} \cdot \mathbf{T} = 1$, $\mathbf{T} \cdot \mathbf{N} = 0$, $\mathbf{T} \times \mathbf{T} = \mathbf{0}$, and $\mathbf{T} \times \mathbf{N} = \mathbf{B}$. Thus, since $|\mathbf{B}| = 1$ and $a_N \geq 0$,

---

Alternative Formulas for Tangential and Normal Components of Acceleration

$$a_T = \frac{\mathbf{v} \cdot \mathbf{a}}{|\mathbf{v}|}$$

$$a_N = \frac{|\mathbf{v} \times \mathbf{a}|}{|\mathbf{v}|} \quad (12.27)$$

EXAMPLE 12.22 Use Equations 12.27 to find a_T and a_N at time $t = 1$ for a particle with position vector

$$\mathbf{r}(t) = \left\langle 4\sqrt{t},\, 1 - 2t^2,\, \frac{8(t-1)}{\sqrt{t+3}} \right\rangle$$

Solution First we calculate $\mathbf{v}(t)$ and $\mathbf{a}(t)$:

$$\mathbf{v}(t) = \left\langle \frac{2}{\sqrt{t}}, -4t, \frac{4(t+7)}{(t+3)^{3/2}} \right\rangle$$

$$\mathbf{a}(t) = \left\langle -\frac{1}{t^{3/2}}, -4, \frac{-2(t+15)}{(t+3)^{5/2}} \right\rangle$$

So

$$\mathbf{v}(1) = \langle 2, -4, -4 \rangle, \quad |\mathbf{v}(1)| = \sqrt{4+16+16} = 6,$$

and

$$\mathbf{a}(1) = \langle -1, -4, -1 \rangle$$

Then, by Equations 12.27,

$$a_T = \frac{|\mathbf{v} \cdot \mathbf{a}|}{|\mathbf{v}|} = \frac{\langle 2, -4, 4 \rangle \cdot \langle -1, -4, -1 \rangle}{6} = \frac{-2+16-4}{6} = \frac{5}{3}$$

$$a_N = \frac{|\mathbf{v} \times \mathbf{a}|}{|\mathbf{v}|} = \frac{|\langle 2, -4, 4 \rangle \times \langle -1, -4, -1 \rangle|}{6} = \frac{|\langle 20, -2, -12 \rangle|}{6}$$

$$= \frac{\sqrt{137}}{3}$$ ∎

Exercise Set 12.5

In Exercises 1–10, assume a particle moves so that its position vector at time t is $\mathbf{r}(t)$. Find its velocity, acceleration, and speed at t_0. Draw the graph of the curve followed by the particle, showing $\mathbf{v}(t_0)$ and $\mathbf{a}(t_0)$, drawn from the tip of $\mathbf{r}(t_0)$.

1. $\mathbf{r}(t) = \langle 2\cos t, 3\sin t \rangle; t_0 = \frac{\pi}{4}$

2. $\mathbf{r}(t) = t^2\mathbf{i} + t^3\mathbf{j}; t_0 = -1$

3. $\mathbf{r}(t) = 2e^t\mathbf{i} + 3e^{-t}\mathbf{j}; t_0 = 0$

4. $\mathbf{r}(t) = \langle \cosh t, \sinh t \rangle; t_0 = 0$

5. $\mathbf{r}(t) = \langle 2t-1, t^2+3 \rangle; t_0 = 2$

6. $\mathbf{r}(t) = 2\cos t\mathbf{i} + 2\sin t\mathbf{j} + 3t\mathbf{k}; t_0 = \frac{\pi}{3}$

7. $\mathbf{r}(t) = 2t\mathbf{i} + t^2\mathbf{j} + t^3\mathbf{k}; t_0 = 1$

8. $\mathbf{r}(t) = \langle t+1, 3t, t^2 \rangle; t_0 = 1$

9. $\mathbf{r}(t) = \langle e^t, 2t, e^{-t} \rangle; t_0 = 0$

10. $\mathbf{r}(t) = \cos^2 t\mathbf{i} + 2\sin t\mathbf{j} + 2t\mathbf{k}; t_0 = \frac{\pi}{4}$

In Exercises 11–15, find $\mathbf{v}(t)$, $\mathbf{a}(t)$, and the speed at an arbitrary t in the given domain.

11. $\mathbf{r}(t) = (5-2t)^{3/2}\mathbf{i} + \frac{1}{2}(t^2+4t)\mathbf{j}; t < \frac{5}{2}$

12. $\mathbf{r}(t) = (\ln t^2)\mathbf{i} + \frac{1}{t}\mathbf{j} + 2t\mathbf{k}; t > 0$

13. $\mathbf{r}(t) = \left\langle t\cos t \sin t, \frac{(2t)^{3/2}}{3} \right\rangle; t > 0$

14. $\mathbf{r}(t) = \langle \cos^3 t, \sin^3 t, \cos 2t \rangle; 0 \leq t \leq \frac{\pi}{2}$

15. $\mathbf{r}(t) = \langle e^t \cos t, e^t \sin t, e^t \rangle; -\infty < t < \infty$

In Exercises 16–20, use the results of Examples 12.19 and 12.20.

16. A 2-kg mass attached to one end of a rope 3 m long is being whirled around horizontally in a circular path by a child holding the other end of the rope. If the speed of the mass is 4 m/s, find the force exerted by the child on the rope. Through how many revolutions per minute is the mass turning?

17. Using Exercise 16, find the effect on the force exerted by the child if (a) the speed is doubled and (b) the rope is half as long.

18. A satellite moves in a circular orbit 400 km above the earth. What is its speed? How long does it take to complete one orbit?

19. A satellite is in a circular orbit h km above the earth. If its speed is 28,000 km/hr, find h.

20. If a satellite circles the earth once every 90 min, find its height above the earth and its velocity.

21. A projectile is fired from the earth's surface with an initial speed of 600 m/s at an angle of 30° with the horizontal. Find the maximum height and range of the projectile.

22. At what angle θ should the projectile of Exercise 21 be fired for the range to be 30 km?

23. If a projectile is fired from the ground at an angle of 42° and attains a maximum height of 1500 m, what is its initial speed?

24. In Exercise 23 if, instead of the known height, we are given that the range is 36 km, what is the initial speed?

25. Prove that if a particle moves along a curve with constant speed, then its velocity and acceleration vectors are always orthogonal. (*Hint:* Use the fact that $|\mathbf{v}|^2 = \mathbf{v} \cdot \mathbf{v} = C$, and differentiate.)

26. Prove that if the position vector of a particle moving in space is of the form

$$\mathbf{r}(t) = (\cos \omega t)\mathbf{A} + (\sin \omega t)\mathbf{B}$$

where $\mathbf{A}$ and $\mathbf{B}$ are arbitrary constant vectors, then $\mathbf{a}(t) = -\omega^2 \mathbf{r}(t)$.

27. Prove that the position vector of a moving particle is of the form $\mathbf{r}(t) = t\mathbf{A} + \mathbf{B}$, where $\mathbf{A}$ and $\mathbf{B}$ are constant vectors if and only if $\mathbf{a}(t) = \mathbf{0}$ for all t. Describe the motion in this case.

28. A communications satellite is located above the equator, and its speed and altitude are such that it remains stationary relative to the earth. What are its speed and altitude?

29. A projectile is fired at an angle of 25° with the horizontal from the top of a hill 1000 m above the plain below. If the initial speed of the projectile is 500 m/s, find the range and the speed at impact.

30. In order to feed cattle in winter, a rancher drops bales of hay from a light airplane. If a bale is dropped from a height of 200 m while the airplane is flying horizontally at 50 m/s, how far is it from the point on the ground below the airplane when the bale is dropped to the point where it hits the ground? What is its speed at impact? (Neglect air resistance and assume the ground is flat.)

31. For a certain particle moving in space, it is known that $\mathbf{a}(t) = -t\mathbf{k}$ and that $\mathbf{v}(0) = 2\mathbf{i} - 3\mathbf{j} + \mathbf{k}$ and $\mathbf{r}(0) = 4\mathbf{i} + 2\mathbf{j}$. Find $\mathbf{v}(t)$ and $\mathbf{r}(t)$.

32. Redo Exercise 31 if $\mathbf{a}(t) = e^{-t}\mathbf{i} + 2e^t\mathbf{j} + te^t\mathbf{k}$, $\mathbf{v}(0) = 2\mathbf{i} + 6\mathbf{j} - \mathbf{k}$, and $\mathbf{r}(0) = \mathbf{i} + 2\mathbf{j} - 2\mathbf{k}$.

33. In Example 12.19, imbed the problem in a three-dimensional coordinate system, with the circle lying in the xy-plane, and write $\mathbf{r}(t)$ as a three-dimensional vector with third component 0. Let $\boldsymbol{\omega} = \omega \mathbf{k}$. Prove that $\mathbf{v}(t) = \boldsymbol{\omega} \times \mathbf{r}(t)$. Show that the same result holds true if the circle is in any plane parallel to the xy-plane, with the center on the z-axis.

In Exercises 34–39, find the tangential and normal components of acceleration at the indicated time. Draw the graph of $\mathbf{r}$, showing the acceleration, together with its tangential and normal projections at the point in question.

34. $\mathbf{r}(t) = 2t\mathbf{i} - t^2\mathbf{j}$; $t = 2$

35. $\mathbf{r}(t) = \langle t^2, t^3 \rangle$; $t = 1$

36. $\mathbf{r}(t) = \langle 2\cos t, 4\sin t \rangle$; $t = \dfrac{\pi}{3}$

37. $\mathbf{r}(t) = \cosh t\,\mathbf{i} + \sinh t\,\mathbf{j}$; $t = \ln 2$

38. $\mathbf{r}(t) = (t-1)\mathbf{i} + 4\sqrt{t}\,\mathbf{j}$; $t = 1$

39. $\mathbf{r}(t) = \langle e^{2t} - 1, e^t \rangle$; $t = 0$

In Exercises 40–45, use the results of the specified problem to find the curvature at the indicated point in Exercises 34–39, respectively.

In Exercises 46–51, find a_T and a_N at an arbitrary value of t in the domain.

46. $\mathbf{r}(t) = \langle a\cos \omega t, a\sin \omega t, bt \rangle$; $t \geq 0$, $a > 0$

47. $\mathbf{r}(t) = \left\langle 2\ln t, \dfrac{t-1}{t}, 2t \right\rangle$; $t > 0$

48. $\mathbf{r}(t) = t\mathbf{i} + t^2\mathbf{j} + \tfrac{2}{3}t^3\mathbf{k}$; $-\infty < t < \infty$

49. $\mathbf{r}(t) = (3t\sin t)\mathbf{i} + (3t\cos t)\mathbf{j} + (2t)^{3/2}\mathbf{k}$; $t > 0$

50. $\mathbf{r}(t) = \langle 2t, 2e^t, e^{-t} \rangle$; $-\infty < t < \infty$

51. $\mathbf{r}(t) = \langle \sin t - t\cos t, \cos t + t\sin t, t^2 \rangle$; $t \geq 0$

52. Show that if $a_N \neq 0$,

$$\mathbf{N} = \dfrac{\mathbf{a} - a_T \mathbf{T}}{a_N}$$

In Exercises 53–57, find $\mathbf{T}$, $\mathbf{N}$, and κ for the curve in Exercises 47–51, using a_T and a_N as previously found, together with the result of Exercise 52.

58. Let $v = |\mathbf{v}|$. Show that

$$\mathbf{a} = \dfrac{dv}{dt}\mathbf{T} + \dfrac{v^2}{\rho}\mathbf{N}$$

where ρ is the radius of curvature.

59. Use the formulas for a_N from Equations 12.23 and 12.27 to obtain the following formula for the curvature κ:

$$\kappa = \frac{|\mathbf{v} \times \mathbf{a}|}{|\mathbf{v}|^3}$$

60. Prove that if the normal component of acceleration of a particle is constantly 0, the particle moves in a straight line.

61. Prove that if the force on a particle is always centripetal (directed along the normal to the path), its speed is constant.

Exercises 62–66 form a sequential unit.

62. Let a particle move in the xy-plane in which a polar coordinate system is superimposed, with the polar axis coinciding with the positive x-axis. Let $\mathbf{r} = \mathbf{r}(t)$ be its position vector at time t, and let $r = |\mathbf{r}|$. If $\theta = \theta(t)$ is the polar angle to the vector $\mathbf{r}$ at time t, define $\mathbf{u}_r = \langle \cos\theta, \sin\theta \rangle$ and $\mathbf{u}_\theta = \langle -\sin\theta, \cos\theta \rangle$. Show each of the following.
(a) $\mathbf{u}_r$ and $\mathbf{u}_\theta$ are orthogonal unit vectors, and $\mathbf{u}_\theta$ is rotated $90°$ counterclockwise from $\mathbf{u}_r$.
(b) $\mathbf{r} = r\mathbf{u}_r$
(c) $\dfrac{d\mathbf{u}_r}{dt} = \mathbf{u}_\theta \dfrac{d\theta}{dt}$ and $\dfrac{d\mathbf{u}_\theta}{dt} = -\mathbf{u}_r \dfrac{d\theta}{dt}$

(*Note:* $\mathbf{u}_r$ and $\mathbf{u}_\theta$ are called the **radial** and **transverse** unit vectors, respectively.)

63. (a) Show that $\mathbf{v} = \dfrac{dr}{dt}\mathbf{u}_r + r\dfrac{d\theta}{dt}\mathbf{u}_\theta$.

(b) Show that speed $= \sqrt{\left(\dfrac{dr}{dt}\right)^2 + r^2\left(\dfrac{d\theta}{dt}\right)^2}$.

64. Show that

$$\mathbf{a} = \left[\frac{d^2r}{dt^2} - r\left(\frac{d\theta}{dt}\right)^2\right]\mathbf{u}_r + \left[r\frac{d^2\theta}{dt^2} + 2\frac{dr}{dt}\frac{d\theta}{dt}\right]\mathbf{u}_\theta$$

65. Show the following:
(a) $\text{Comp}_{\mathbf{u}_r}\mathbf{a} = \dfrac{d^2r}{dt^2} - r\left(\dfrac{d\theta}{dt}\right)^2$
(b) $\text{Comp}_{\mathbf{u}_\theta}\mathbf{a} = r\dfrac{d^2\theta}{dt^2} + 2\dfrac{dr}{dt}\dfrac{d\theta}{dt}$

66. Let $a_r = \text{Comp}_{\mathbf{u}_r}\mathbf{a}$ and $a_\theta = \text{Comp}_{\mathbf{u}_\theta}\mathbf{a}$. Show that $|\mathbf{a}|^2 = a_r^2 + a_\theta^2$.

67. A particle moves in the horizontal plane so that its polar coordinates at time t are $r = 1 + \cos t^2$ and $\theta = t^2$. Use the results of Exercises 63 and 64 to resolve $\mathbf{v}$ and $\mathbf{a}$ into radial and transverse vector components.

68. A particle moves in the xy-plane with position vector $\mathbf{r}(t) = (t^2 \cos t)\mathbf{i} + (t^2 \sin t)\mathbf{j}$ at time t. Find the polar coordinates of a point on the path at time t. Find the radial component a_r of acceleration and the transverse component a_θ, at time t. (See Exercise 65.)

 69. A projectile is fired with an initial speed of 400 m/s and an angle of elevation of $30°$. Find the range of the projectile, the maximum height the projectile attains, and the speed at impact. Plot the path of the projectile.

 70. Repeat Exercise 69 with an initial speed of 600 m/s.

 71. Repeat Exercise 69 with the initial speed 500 m/s and the angle of elevation $50°$.

12.6 KEPLER'S LAWS

In the early seventeenth century, the German mathematician and astronomer Johannes Kepler (1571–1630) postulated the following three laws governing the orbits of planets around the sun.

Kepler's Laws of Planetary Motion

1. The orbit of each planet is an ellipse with the sun at one focus.
2. The radius vector from the sun to a planet sweeps out area at a constant rate.
3. The square of the time for a planet to complete one revolution around its elliptical orbit is proportional to the cube of the length of the semimajor axis of the ellipse.

Section 12.6 Kepler's Laws   929

Johannes Kepler, ca. 1627

Artist's rendition of the orbits of the planets

Kepler deduced these laws based on the astronomical observations of his mentor Tycho Brahe. His deductions required years of analyzing massive amounts of data with laborious calculations. Kepler's discoveries rank as one of the outstanding achievements in the history of science. Based as they were on empirical evidence, however, they lacked a sound theoretical basis until Newton, some fifty years later, deduced Kepler's Laws using his newly invented calculus. Newton published his findings in his book *Principia Mathematica* (1687), generally considered to be the most important scientific book ever written. Newton based his proofs on the following two principles:

1. *Newton's Second Law of Motion.* For a body of constant mass m moving under the action of a force $\mathbf{F}$,

$$\mathbf{F} = m\mathbf{a} \qquad (12.28)$$

where $\mathbf{a}$ is the acceleration of the body.

2. *Newton's Law of Gravitation.* The force $\mathbf{F}$ of attraction between two bodies of masses M and m, respectively, is proportional to the product of the masses and inversely proportional to the square of the distance, r, between them:

$$\mathbf{F} = -\frac{GMm}{r^2}\mathbf{u}_r \qquad (12.29)$$

where $\mathbf{u}_r$ is a unit vector directed from one mass toward the other.

The constant G in Equation 12.29 is called the *universal gravitational constant* and has the approximate value

$$G = 6.672 \times 10^{-11} \frac{\text{N} \cdot \text{m}^2}{\text{kg}^2}$$

To demonstrate the power of the vector calculus we have studied, we will prove the first of Kepler's Laws. In the exercises we outline proofs of the other two laws. You should follow the proof below using pencil and paper, verifying all steps. Note carefully the use of properties of the dot and cross products.

We take the origin as the center of mass of the sun and let M be its mass. Let $\mathbf{r}(t)$ be the position vector at time t of the center of mass of a planet that has mass m. Let $r = |\mathbf{r}|$, and let $\mathbf{u}_r$ be the unit vector $\mathbf{r}/r$, so that $\mathbf{r} = r\mathbf{u}_r$. We assume that the forces of attraction between this planet and all other bodies are negligible compared with the gravitational attraction between it and the sun.

We first show that the planet moves in one plane. Equating the right-hand sides of Equations 12.28 and 12.29, we get the acceleration in the form

$$\mathbf{a} = -\frac{GM}{r^2}\mathbf{u}_r \qquad (12.30)$$

Then

$$\mathbf{r} \times \mathbf{a} = r\mathbf{u}_r \times \left(-\frac{GM}{r^2}\mathbf{u}_r\right) = -\frac{GM}{r}(\mathbf{u}_r \times \mathbf{u}_r) = \mathbf{0}$$

and since $\mathbf{v} \times \mathbf{v} = \mathbf{0}$, we also have

$$\frac{d}{dt}(\mathbf{r} \times \mathbf{v}) = \mathbf{r} \times \mathbf{a} + \mathbf{v} \times \mathbf{v} = \mathbf{0}$$

Thus,

$$\mathbf{r} \times \mathbf{v} = \mathbf{c} \qquad (12.31)$$

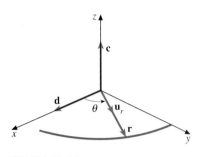

FIGURE 12.20

where **c** is a constant vector. Since **r** is orthogonal to $\mathbf{r} \times \mathbf{v}$ and hence to **c**, this result says that at all times, the vector $\mathbf{r}(t)$ is perpendicular to the fixed vector **c**. So the path of the planet lies in a plane. There is no loss of generality in assuming that this plane is the xy-plane and that **c** is directed along the positive z-axis. We illustrate the path of the planet in Figure 12.20.

We will show that the planet moves in an elliptical orbit with the sun as a focus (Kepler's First Law) by showing that an equation for the graph of $\mathbf{r}(t)$ is the polar form of the equation of an ellipse with a focus at the pole. Let $\theta = \theta(t)$ be the polar angle from the positive x-axis to $\mathbf{r}(t)$. Then, $\mathbf{u}_r = \langle \cos\theta, \sin\theta, 0 \rangle$, and if we define $\mathbf{u}_\theta = \langle -\sin\theta, \cos\theta, 0 \rangle$, $\mathbf{u}_\theta$ is a unit vector such that $\mathbf{u}_r \cdot \mathbf{u}_\theta = 0$ and $\mathbf{u}_r \times \mathbf{u}_\theta = \langle 0, 0, 1 \rangle = \mathbf{k}$. As in Exercises 62 and 63 of Exercise Set 12.5, we obtain, from differentiating both sides of $\mathbf{r} = r\mathbf{u}_r$,

$$\mathbf{v} = \frac{dr}{dt}\mathbf{u}_r + r\frac{d\theta}{dt}\mathbf{u}_\theta$$

Thus, using Equation 12.31,

$$\mathbf{c} = \mathbf{r} \times \mathbf{v} = (r\mathbf{u}_r) \times \left(\frac{dr}{dt}\mathbf{u}_r + r\frac{d\theta}{dt}\mathbf{u}_\theta\right)$$

$$= r^2\frac{d\theta}{dt}(\mathbf{u}_r \times \mathbf{u}_\theta) = r^2\frac{d\theta}{dt}\mathbf{k} \quad (12.32)$$

From Equations 12.30 and 12.32, we get

$$\frac{d}{dt}(\mathbf{v} \times \mathbf{c}) = \mathbf{a} \times \mathbf{c} = \left(-\frac{GM}{r^2}\mathbf{u}_r\right) \times \left[r^2\frac{d\theta}{dt}\mathbf{k}\right]$$

$$= -GM\frac{d\theta}{dt}[\mathbf{u}_r \times \mathbf{k}]$$

$$= GM\frac{d\theta}{dt}\mathbf{u}_\theta$$

$$= \frac{d}{dt}(GM\mathbf{u}_r)$$

The equality of these two derivatives implies that

$$\mathbf{v} \times \mathbf{c} = GM\mathbf{u}_r + \mathbf{d} \quad (12.33)$$

The constant vector **d** is a linear combination of $\mathbf{u}_r$ and $\mathbf{v} \times \mathbf{c}$, each of which lies in the xy-plane, so **d** also lies in the xy-plane. We may assume, again without loss of generality, that **d** is directed along the positive x-axis. The relationship among the vectors **c**, **d**, $\mathbf{u}_r$, and **r** is shown in Figure 12.20.

Let $c = |\mathbf{c}|$ and $d = |\mathbf{d}|$. Then from Equations 12.31 and 12.33,

$$c^2 = \mathbf{c} \cdot \mathbf{c} = (\mathbf{r} \times \mathbf{v}) \cdot \mathbf{c}$$

$$= \mathbf{r} \cdot (\mathbf{v} \times \mathbf{c})$$

$$= (r\mathbf{u}_r) \cdot (GM\mathbf{u}_r + \mathbf{d})$$

$$= rGM + r\mathbf{u}_r \cdot \mathbf{d}$$

$$= r(GM + d\cos\theta)$$

Solving for r gives

$$r = \frac{c^2}{GM + d\cos\theta}$$

The comet of 1680 as drawn in Newton's *Principia*.

If we let $e = d/GM$ and $p = c^2/GMe$, this equation becomes, on dividing numerator and denominator by GM,

$$r = \frac{ep}{1 + e\cos\theta} \quad (12.34)$$

We know from Section 9.8 that Equation 12.34 is the polar equation of an ellipse if $0 < e < 1$, a hyperbola if $e > 1$, and a parabola if $e = 1$, each having a focus at the pole. Since it is known that planets travel in closed orbits, it follows that the equation is that of an ellipse. Kepler's First Law is therefore proved.

Exercise Set 12.6

1. Write the polar equation (12.34) for $0 < e < 1$ in the standard rectangular form
$$\frac{(x-h)^2}{a^2} + \frac{(y-k)^2}{b^2} = 1$$

2. From the result of Exercise 1, show the following:
   (a) $b = a\sqrt{1 - e^2}$
   (b) The center of the ellipse is at $\left(\frac{-e^2 p}{1 - e^2}, 0\right)$.
   (c) The distance from the center to each focus is $\frac{e^2 p}{1 - e^2}$.
   (d) One focus is at $(0, 0)$ and the other is at $\left(\frac{-2e^2 p}{1 - e^2}, 0\right)$.

In Exercises 3 and 4, each equation was used in the proof of Kepler's First Law. Verify each one.

3. (a) $\mathbf{u}_r \cdot \mathbf{u}_\theta = 0$
   (b) $\mathbf{u}_r \times \mathbf{u}_\theta = \mathbf{k}$
   (c) $\mathbf{v} = \frac{dr}{dt}\mathbf{u}_r + r\frac{d\theta}{dt}\mathbf{u}_\theta$

4. (a) $(r\mathbf{u}_r) \times \left(\frac{dr}{dt}\mathbf{u}_r + r\frac{d\theta}{dt}\mathbf{u}_\theta\right) = r^2\frac{d\theta}{dt}(\mathbf{u}_r \times \mathbf{u}_\theta)$
   (b) $-GM\frac{d\theta}{dt}(\mathbf{u}_r \times \mathbf{k}) = GM\frac{d\theta}{dt}\mathbf{u}_\theta$
   (c) $GM\frac{d\theta}{dt}\mathbf{u}_\theta = \frac{d}{dt}(GM\mathbf{u}_r)$

5. Referring to the proof of Kepler's First Law, explain fully the justification for the following assertions:
   (a) $\mathbf{v} \times \mathbf{c}$ lies in the xy-plane.
   (b) $\mathbf{d}$ lies in the xy-plane.

In Exercises 6 and 7, fill in the details of the outlines given of proofs of Kepler's Second and Third Laws.

6. (a) Using the formula $A = \frac{1}{2}\int_\alpha^\beta r^2\, d\theta$ for area in polar coordinates, show that the area between any fixed angle $\theta_0 = \theta(t_0)$ and the angle $\theta = \theta(t)$, bounded by the ellipse of Equation 12.34, is
$$A(t) = \frac{1}{2}\int_{t_0}^{t} r^2 \frac{d\theta}{d\tau}\, d\tau$$
   (b) By using part (a) together with Equation 12.32, show that
$$\frac{dA}{dt} = \frac{c}{2}$$
   (c) Conclude that area is swept out by $\mathbf{r}(t)$ at a constant rate, proving Kepler's Second Law.

7. Let T be the time required for a planet to complete one revolution around the sun (the *period*).
   (a) Using Exercise 6, show that the total area enclosed by the ellipse is
$$A = \frac{1}{2}cT$$
   (b) Recall from Example 8.18 that the area also is given by $A = \pi ab$, where a and b are the lengths of the semimajor and semiminor axes of the ellipse. Combining this with part (a) and Exercise 2, part (a), show that
$$T = \frac{2\pi a^2}{c}\sqrt{1 - e^2}$$
   (c) From the result of Exercise 1, show that $1 - e^2 = \frac{ep}{a}$, and hence obtain
$$T^2 = \frac{4\pi^2 ep}{c^2}a^3$$
   thus proving Kepler's Third Law. By replacing p and e with the values assigned to them, rewrite the result in the form
$$T^2 = \frac{4\pi^2}{GM}a^3 \quad (12.35)$$

8. A reasonable approximation to the period of a planet is obtained by replacing the semimajor axis a in Equation 12.35 by the mean distance of the planet from the Sun.
  (a) The mean distance of Mars from the Sun is approximately $1\frac{1}{2}$ times that of the Earth. Find the approximate time (in "Earth days") it takes Mars to complete one revolution around the Sun.
  (b) It takes Mercury approximately 88 Earth days to orbit the Sun. If the mean distance of the Earth from the Sun is approximately 93 million miles, find the approximate mean distance of Mercury from the Sun.

9. Let r_0 denote the minimum value of the distance r of a planet from the sun and let v_0 be its speed when $r = r_0$.
  (a) Show that $r_0 = ep/(1+e)$ when $\theta = 0$.
  (b) Use Equation 12.33 to show that
  $$v_0 = \frac{(GM + d)}{c}$$
  (*Hint:* Use $|\mathbf{v} \times \mathbf{c}| = |\mathbf{v}||\mathbf{c}|$. Why is this equation true?)
  (c) Show that
  $$v_0 = \frac{c}{r_0}$$

10. (a) Use the result of Exercise 7, part (b), to express v_0 in terms of a, e, and T.
  (b) The earth takes approximately 365.26 days to complete its orbit around the sun. Its semimajor axis a is approximately 1.4959×10^8 km, and the eccentricity of the orbit is approximately 0.016732. Find r_0 and v_0 for the earth. (Express v_0 in kilometers per hour.)

Chapter 12 Review Exercises

1. Let $\mathbf{u}(t) = \langle t^2, t^3, 1-t \rangle$, $\mathbf{v}(t) = \langle t \ln t, -2, t+3 \rangle$, and $\alpha(t) = e^t$. Find the following:
  (a) $(\mathbf{u} \cdot \mathbf{v})(t)$
  (b) $\mathbf{u}(\alpha(t))$
  (c) $(\mathbf{u} - \mathbf{v})(t)$
  (d) $(\mathbf{v} \circ \alpha)(0)$
  (e) $(\mathbf{u}' \times \mathbf{v}')(1)$

2. Find $\mathbf{r}'(t)$ if:
  (a) $\mathbf{r}(t) = \langle t \ln \sqrt{t}, t^2 e^{-t} \rangle$, $t > 0$
  (b) $\mathbf{r}(t) = \left(\frac{t}{\sqrt{t^2 - 1}} \right) \mathbf{i} + \left(\sin^{-1} \frac{1}{t} \right) \mathbf{j} + \left(t\sqrt{t^2 - 1} \right) \mathbf{k}$

3. Evaluate the integrals:
  (a) $\int \left\langle \frac{1}{\sqrt{1-t^2}}, \frac{t}{\sqrt{1-t^2}}, \frac{1}{1-t^2} \right\rangle dt$
  (b) $\int_0^{\pi/3} \left[(\cos^2 t)\mathbf{i} - (\sin^3 t)\mathbf{j} + (\tan^2 t)\mathbf{k} \right] dt$

4. Let $\mathbf{r}(t) = t^2 \mathbf{i} + (3t - 2)\mathbf{j} + (1 - t^2)\mathbf{k}$ and $\alpha(t) = \cos t$. Find:
  (a) $\int (\alpha \mathbf{r})(t) dt$
  (b) $\int (\mathbf{r} \circ \alpha)(t) dt$

5. Let $\mathbf{r}(t) = (\ln t^2)\mathbf{i} - \frac{2}{\sqrt{t}}\mathbf{j} + 4\sqrt{t}\,\mathbf{k}$. Find:
  (a) $\int_1^4 \mathbf{r}'(t) dt$
  (b) $\int_1^4 |\mathbf{r}'(t)| dt$

6. Verify Property 6 for integrals for $\mathbf{r}(t) = \langle 20t, 9t^2, 12t^2 \rangle$ on $[0, 1]$.

7. Let C be the graph of $\mathbf{r}(t) = (2t^3/3)\mathbf{i} + (1 - 2t^2)\mathbf{j} + 4t\mathbf{k}$ for $0 \le t \le 3$. Find the length of C.

8. Sketch the graph of $\mathbf{r}(t) = \langle \cos t, t, \sin t \rangle$ for $0 \le t \le 4\pi$. Find parametric equations of the tangent line to the graph at $t = 4\pi/3$.

9. Let C be the graph of $\langle \ln(\cosh t), 2\tan^{-1} e^t, \sqrt{3}\,t \rangle$ on $[-1, 2]$.
  (a) Find the length of C.
  (b) Make the change of variables $u = e^{-t}$, and let $\mathbf{R}(u)$ be the new position vector for C. Find $\mathbf{R}(u)$ and the u interval. Is the orientation of C preserved or reversed?
  (c) Find the length of C using $\mathbf{R}(u)$ and show it is the same as that found in part (a).

10. Let C be the graph of $\mathbf{r}(t) = \langle 2e^t \sin t, 2e^t \cos t, e^t \rangle$ on $[0, \infty)$. Introduce arc length s as a parameter and let $\mathbf{R}(s)$ be the resulting position vector for C. Show that for all $s > 0$, $|\mathbf{R}'(s)| = 1$.

11. Use the arc length parameterization of C in Exercise 10 to find $\mathbf{T}$, $\mathbf{N}$, and κ at an arbitrary $s \ge 0$. Evaluate each of these at the point $(0, 2, 1)$.

12. A particle moves in the xy-plane so that its position vector at time t is $\mathbf{r}(t) = \langle 3 - 2\sqrt{t}, t + 1 \rangle$. Find its velocity, acceleration, and speed when $t = 4$. Identify the curve.

13. A particle moves so that its position vector at time t is $\mathbf{r}(t) = \langle e^{-t} \sin 2t, e^{-t} \cos 2t, 2e^{-t} \rangle$. Find $\mathbf{v}(t)$, $\mathbf{a}(t)$, and ds/dt at an arbitrary t.

14. A projectile is fired from ground level with an initial velocity $\mathbf{v}_0 = 0.4\mathbf{i} + 0.3\mathbf{j}$, with the magnitude in kilometers per second. Find the range of the projectile ($x_{\max}$) and the maximum height it attains ($y_{\max}$).

15. (a) A satellite is in orbit 240 km above the earth's surface. What is its speed?
    (b) A satellite completes one orbit around the earth every 2 hr. Find its altitude and speed.

16. Let $\mathbf{r}(t) = t^3\mathbf{i} - 4t^2\mathbf{j}$ for $t \in [1, 4]$. At the point $(8, -16)$, find $\mathbf{T}$, $\mathbf{N}$, κ, and the center of curvature.

17. (a) Find the curvature of $y = \ln|\csc x|$ for any $x \neq n\pi$.
    (b) Find the curvature of $x^3 + 3xy - y^3 = 3$ at the point $(2, -1)$.

18. Let C be defined by
$$\mathbf{r}(t) = \langle -1 + 5\sin t, 3\cos t, 1 - 4\cos t \rangle$$
for t in $[0, 2\pi]$. Find $\mathbf{T}$, $\mathbf{N}$, and $\mathbf{B}$ at an arbitrary point on C.

19. A particle moves so that its position vector at time $t \geq 0$ is $\mathbf{r}(t) = \langle \ln(t + 1), 2t, t^2 + 2t \rangle$. Find $\mathbf{a}$, $a_\mathbf{T}$, and $a_\mathbf{N}$.

20. A particle moves in the horizontal plane so that its polar coordinates at time t are $r = 1 + 2\sin e^t$ and $\theta = e^t$. Describe its path for $t \geq 0$. Find the radial and transverse components of acceleration. (See Exercise 65 in Exercise Set 12.5.)

21. If a planet has the elliptical orbit $r = ep/(1 + e\cos\theta)$, show that its speed v_m at the point on its orbit farthest from the sun is
$$v_m = \frac{2\pi a}{T}\sqrt{\frac{1-e}{1+e}}$$
where a is the length of the semimajor axis and e is the eccentricity of the ellipse and T is the period. Use the data given for the earth in Exercise 10, part (b), of Exercise Set 12.6 to find v_m for the earth.

Chapter 12 Concept Quiz

1. Define each of the following:
   (a) the position vector of a point
   (b) the derivative of the vector function $\mathbf{F}$ defined by $\mathbf{F}(t) = \langle f(t), g(t), h(t) \rangle$
   (c) a space curve
   (d) a simple closed curve C that is the graph of a vector function $\mathbf{r}(t)$ on $[a, b]$
   (e) a smooth curve

2. Let C be the curve that is the graph of the twice-differentiable vector function $\mathbf{r}(t)$ for $a \leq t \leq b$. Give formulas for each of the following:
   (a) the length of C
   (b) the unit tangent vectors $\mathbf{T}$ and $\mathbf{N}$ at an arbitrary point on C
   (c) the curvature of C at an arbitrary point

3. If the position vector of a moving point is $\mathbf{r}(t)$ at time t, give formulas for each of the following:
   (a) its velocity
   (b) its speed
   (c) its acceleration
   (d) the tangential and normal components of its acceleration

4. (a) If a curve C that is the graph of a vector function $\mathbf{r}$ is parameterized by arc length s, what can you say about $\mathbf{r}'(s)$?
   (b) If the plane curve C has the equation $\mathbf{r}(t) = \langle x(t), y(t) \rangle$, express its curvature in terms of x', x'', y', and y''.
   (c) If C is the graph of $y = f(x)$, express its curvature in terms of y' and y''.
   (d) State a sufficient condition for the graph of a vector function $\mathbf{r}(t)$ on $[a, b]$ to be rectifiable (have finite length).

5. State Kepler's three laws of planetary motion.

APPLYING CALCULUS

1. In an oil refinery, one of the cylindrical storage tanks is 40 m in diameter and 15 m high. A staircase in the form of a helix goes around the outside of the tank and reaches the top after one revolution. Find the equation of the helix. What is the approximate length of handrail required along the outer edge of the staircase if the staircase is 60 cm wide?

2. A curve in a railroad track follows the parabola $25y = x^2$ from (0, 0) to (5, 1) and then follows the circle having the same curvature as the parabola at (5, 1). The x- and y-coordinates are measured in kilometers. The circular part of the track is 4 kilometers long, and then the track becomes straight in the direction tangent to the circle. Find:
   (a) The center and radius of the circular arc;
   (b) The direction of the straight track, where the positive y-axis points north.

3.
   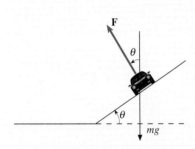

   (a) Use the result of Example 12.19 to show that the magnitude of the centripetal force **F** required to keep an object of mass m on a circular path of radius r, with constant speed, can be written in the form
   $$|\mathbf{F}| = \frac{m|\mathbf{v}|^2}{r}$$
   where **v** is the velocity of the object.

   (b) Use part (a) to show that the maximum safe speed v a car can travel around a curve of radius r without skidding, if the highway is banked at an angle θ (see the figure), is given by
   $$v = \sqrt{rg \tan \theta}$$
   (*Hint*: Show that $|\mathbf{F}| \cos \theta = mg$ and $|\mathbf{F}| \sin \theta = mv^2/r$. Then find $\tan \theta$).

   (c) If a curve of radius 120 m is banked at an angle of 15°, what is the maximum safe speed for traversing the curve?

4. A golfer hits a golf ball so that the initial angle of elevation of its path is θ and its initial velocity is v_0. The fairway is sloping upward at an angle α, as shown in the figure.

   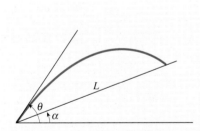

   (a) If $\alpha = 10°$, $\theta = 45°$, and $v_0 = 120$ ft/s, find the distance L the ball will go along the fairway.
   (b) For constant values of α and v_0, show that the angle θ that maximizes the distance L satisfies
   $$\tan 2\theta = -\cot \alpha$$

5. A particle in space moves under the influence of a force that is always directed toward the origin. Let **r**, **v**, and **a** be the position, velocity, and acceleration vectors of the particle at time t.
   (a) Show that $\mathbf{a}(t) = f(t)\mathbf{r}(t)$ for some scalar function f.
   (b) Show that $(\mathbf{r} \times \mathbf{v})' = \mathbf{0}$ from which it follows that $\mathbf{r} \times \mathbf{v} = \mathbf{C}$ for some constant vector **C**. Conclude that the path of the particle lies in a plane. What is the plane if $\mathbf{C} \neq \mathbf{0}$? What is the path of the particle if $\mathbf{C} = \mathbf{0}$?
   (c) Suppose that $\mathbf{a}(t) = -k^2 \mathbf{r}(t)$ where k is a constant. From part (b), the path of the particle lies in a plane. Assume that the plane is the xy-plane and that when $t = 0$,

the particle is on the *x*-axis with velocity vector parallel to the *y*-axis. Show that the path of the particle is an ellipse centered at the origin.

6. A new swimming pool was recently installed in the basement of a large urban high school, with the bottom resting on a floor below. Through a viewing window, it was observed that water was leaking out of the side of the pool (see the figure). It can be shown that the speed v at which water goes out of the hole is given by the formula

$$v = \sqrt{2g(h-y)}$$

(a) If $h = 2$ m and $y = 1.3$ m, find x.

(b) Assume h is constant (water is added at the same rate it is leaking out). If the hole is at the height y for which x is a maximum, find both x and y.

7. A disk of radius 1 is rotating in the counterclockwise direction at a constant angular speed ω. A particle starts at the center of the disk and moves toward the edge along a fixed radius so that its position at time t, $t > 0$, is given by $\mathbf{r}(t) = t\mathbf{R}(t)$ where

$$\mathbf{R}(t) = (\cos \omega t)\mathbf{i} + (\sin \omega t)\mathbf{j}$$

(a) Show that the velocity $\mathbf{v}$ of the particle is

$$\mathbf{v} = (\cos \omega t)\mathbf{i} + (\sin \omega t)\mathbf{j} + t\mathbf{v}_d$$

where $\mathbf{v}_d = \mathbf{R}'(t)$ is the velocity of a point on the edge of the disk.

(b) Show that the acceleration $\mathbf{a}$ of the particle is

$$\mathbf{a} = 2\mathbf{v}_d + t\mathbf{a}_d$$

where $\mathbf{a}_d = \mathbf{R}''(t)$ is the acceleration of a point on the edge of the disk. The extra term $2\mathbf{v}_d$ is called the *Coriolis acceleration*; it is the result of the interaction of the rotation of the disk and the motion of the particle. One can obtain a physical demonstration of this acceleration by walking toward the edge of a moving merry-go-round.

(c) Determine the Coriolis acceleration of a particle that moves on a rotating disk according to

$$\mathbf{r}(t) = e^{-t}(\cos \omega t)\mathbf{i} + e^{-t}(\sin \omega t)\mathbf{j}$$

A PERSONAL VIEW

Joseph Newhouse

At what point in your studies did you realize that calculus would be useful to you?

When I took my first course in microeconomic theory, I was a junior in college and I already had a year of calculus at that point. I discovered that it was much easier to write an equation than it was to write out long strings of words. There are some classic books in economics from the 1930s that are exceedingly difficult to understand because they try to write equations in words when it is much simpler just to write out the equations in mathematical notation.

For example, in economics, probably the most famous condition for the theory of the firm is *marginal revenue = marginal cost*, which means that the additional revenue you get from the last unit you sell equals the cost of producing it. Students taking their first economics course will understand this at about the level I just described, because it cannot be assumed that people who are studying economics know the calculus. However, what this equation really says is that profit is the total revenue minus the total cost. If I just differentiate both with respect to quantity, which are both functions of the quantity of goods the firm produces, I get the first derivative of total revenue minus the first derivative of total cost. There might be other things in those functions, in which case I have to take the cost function. Those derivatives are the marginal revenue and the marginal cost. So this condition that every first-year economics student learns is just a simple first-order condition. Although that condition is pretty easy to understand verbally, as one progresses, it gets harder and harder to verbalize and easier and easier to use the math. In fact, the things that economists get interested in, such as what happens when something changes, what happens at the optimum, frequently can only be done with the mathematics. It would be almost impossible to verbalize one's way through to a solution.

That is really why anybody pursuing economics on a more advanced level must learn calculus. Some people think that economists only use calculus because they want to obfuscate the evidence, but in fact it is impossible to derive answers for anything but the most simple problems without it (in continuous cases, where calculus applies; there is also a whole economics of discrete things in which calculus is not used).

You are an expert in health-care policy and economics. How is calculus useful in your field?

Broadly speaking it is useful in two ways. First of all, much of microeconomic theory is built on the calculus because it has an assumption of maximization or minimization. For example, for profit firms try to maximize profits or consumers try to maximize their utility or their well being, all of these subject to some constraints. So, traditional theories of consumer behavior and firm behavior are really problems in constrained maximization and how firms and consumers act as the constraints they face change. For example, how household incomes change behavior will be analyzed by perturbing some kind of first-order condition and solving for the result. That is one way in which all of economics uses the calculus. What I do is no exception to that.

The second way in which I use the calculus is in statistical analysis of data to estimate certain population values (parameters). A useful technique is maximum-likelihood estimation, which involves finding the parameters that maximize the probability of observing the sample of data that one has.

Let's talk about maximum-likelihood estimation.

A simple idea is that I might want to estimate a person's demand for medical care as a function of the price of medical care the person faced and the person's income. In the simplest form, I might want to estimate use or demand as a linear function of price and income. So I would write an equation that says Use = constant + another constant times price plus another constant times income plus random error (written as $\beta_0 + \beta_1$ Price $+ \beta_2$ Income + error). The estimation problem would be to estimate the three constants. There are various ways to do this, but if the random error were normally distributed, the least-squares estimator, which is the most popular estimator, is also the maximum-likelihood estimator. Maximum likelihood means that for each person I have, each observation, I write down the probability-density function for the error term. (See p. 463.) The likelihood of the error term will be a function of the three constants. I choose the three constants, in effect, so as to maximize that likelihood function. I multiply the probability of each observation to get a giant product, which I call the likelihood that I observe a particular sample. That likelihood is a function of the three constants, and it is convenient to take the logarithm of that product so that the result is a sum rather than a product. Then I will simply take the partial derivatives of the log likelihood function, which is the sum of the logs, with respect to each of the constants, and set that equal to zero and solve those equations for the constants.

It turns out that, under fairly general conditions, this method of going about estimating the constants gives me a best—in a sense that can be made precise—estimator. "Best" in this context means that it is best among a class of estimates that are asymptotically normally distributed. These will have the minimum variance. If I repeat this calculation in different samples, I will get a normal distribution of my estimate.

Those are the two main domains in which I personally use the calculus. There are certainly other applications of the calculus in economics, particularly in analysis of population dynamics or dynamic systems where control theory is used.

When we're dealing with multiple variables, do we need to use partial differential equations?

Yes. Because economists tend to be interested in partial effects—what happens if something changes and all else is constant—that is analogous to a partial derivative that holds everything else constant. Occasionally one is interested in a total derivative but most often in economics one is interested in a partial derivative.

You are a professor in the Medical School, the Kennedy School of Government, the School of Public Health, and the Faculty of Arts and Sciences. Do you see a difference in the mathematical training of students in those schools?

Students pursuing a Ph.D. in economics probably won't be admitted without calculus, including multivariable calculus. They may take a kind of refresher course to brush up, but right away in the first year the curriculum makes so many demands on the calculus that they simply wouldn't get in. So they tend to be well prepared. Masters-level students are a much more heterogeneous group, but in both the faculty of government and the faculty of public health, students can progress faster and go further who know the calculus in both economics and in statistics courses, which are used heavily by both faculties. Indeed, a masters degree in public policy at the school of government requires one year of economics and one year of statistics. Advanced courses are built on the calculus, and those courses go further with students knowing more at the end of the year, simply because you can go faster with the calculus than without it.

Joseph Newhouse is the John D. MacArthur Professor of Health Policy and Management at Harvard University and Director of the Division of Health Policy Research and Education. He is a member of the faculties of the John F. Kennedy School of Government, the Harvard Medical School, the Harvard School of Public Health, and the Faculty of Arts and Sciences. He is the editor of the *Journal of Health Economics* and an associate editor of the *Journal of Economic Perspectives*. He serves on the Council of the Institute of Medicine and is the President of the Association for Health Services Research.

CHAPTER 13

FUNCTIONS OF MORE THAN ONE VARIABLE

In this chapter, we begin the study of real-valued functions of two or more independent variables. You already are familiar with some functions of this type. For example, the formula

$$V = \pi r^2 h$$

expresses the volume of a right circular cylinder (such as a can) as a function of the base radius r and the height h. So we can say that V is a function of the two independent variables r and h. Or consider the formula

$$A = P(1 + r)^t$$

where A is the amount of money that has accumulated after t years from an initial investment of P dollars at an interest rate r (expressed as a decimal), compounded annually. In this case A is a function of the three independent variables P, r, and t.

As another example, consider the Earth's surface temperature T as a function of position. If an xy-coordinate system is set up on a map, then at any fixed time, the temperature is a function of the two variables x and y. In this case, finding a formula for T in terms of x and y would be difficult, but by obtaining satellite data, we can show the temperature graphically. By connecting points where the temperature is the same, we get what are called **isotherms** ("equal temperature" curves). Figure 13.1 is a photograph that shows such isotherms for the Earth. This isothermic image is an example of what is called a **contour map** of a function. Contour maps provide one means of describing a function of two variables graphically.

Our emphasis in this chapter will be on functions of two or three variables, especially the former, since they can be pictured geometrically. We should mention, however, that functions involving large numbers of variables are often used in applications. For example, cosmologists are now considering the universe to be an eleven-dimensional space, and in economics as many as 100 independent variables are sometimes needed. Much of our theory can be extended to these higher dimensions.

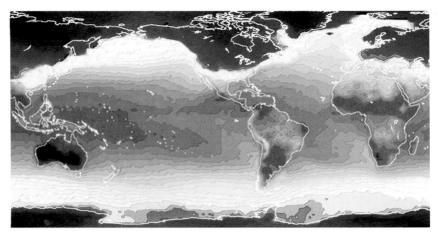

FIGURE 13.1
Isotherms (equal-temperature curves) for the Earth

In the first section we give definitions of functions of two variables and of three variables, and then we discuss how to find their domains.

We discuss how to represent functions graphically, and we show a variety of graphs generated by a computer algebra system. These graphs are three-dimensional *surfaces*, which are hard to draw without the aid of a CAS, except in certain special cases. These special cases are included in a broad class called *quadric surfaces*, which we examine in Section 13.2.

We then define the notions of limit and continuity for functions of two or three variables. We conclude the chapter by considering what are called *partial derivatives*, and we indicate some of their applications.

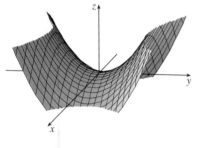

An example of a quadric surface

13.1 FUNCTIONS OF TWO AND THREE VARIABLES

In Chapter 1, we defined a function from a set A to a set B as a rule that assigns to each element of A one, and only one, element of B. By specifying the set A (the domain) to be a set of ordered pairs of real numbers and the set B (the codomain) to be a subset of the set $\mathbb{R}$ of real numbers, we have the following definition of a real-valued function of two real variables.

Definition 13.1
A Real-Valued Function of Two Real Variables

Let A denote a set of ordered pairs of real numbers. A function f that assigns a unique real number z to each ordered pair (x, y) in A is called a **real-valued function of two real variables**, and we write

$$z = f(x, y)$$

The set A is the domain of f, and the subset B of $\mathbb{R}$ consisting of all numbers $f(x, y)$ for (x, y) in A is the range of f.

REMARK
■ For brevity, we will often refer to such a function simply as a function of two variables.

In Figure 13.2 we indicate how such a function f maps a point (x, y) in the domain A to a point z on the real line.

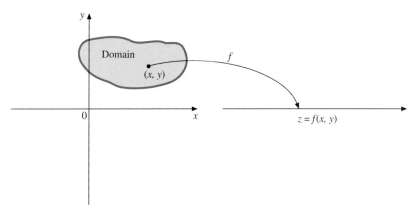

FIGURE 13.2
The function f maps the point (x, y) in the domain to a real number z in the range.

The set of all ordered pairs of real numbers (geometrically, the xy-plane) is frequently denoted by $\mathbb{R}^2$. So we can say that a function of two variables is from $\mathbb{R}^2$ to $\mathbb{R}$ and indicate this relationship in symbols by

$$f : \mathbb{R}^2 \to \mathbb{R}$$

If the domain A consists of ordered triples (x, y, z), then we call f a **real-valued function of three real variables** (or more simply, a function of three variables) and we write

$$f : \mathbb{R}^3 \to \mathbb{R}$$

The extension to a higher number of variables should be clear. To give one simple example of a function of many variables, the function f defined by

$$f(x_1, x_2, \ldots, x_n) = \frac{x_1 + x_2 + \cdots + x_n}{n}$$

is called the *mean* of the variables $x_1, x_2, \ldots, x_n$ and is commonly denoted by $\bar{x}$.

We might note that the vector-valued functions we studied in Chapter 12 assigned two-dimensional or three-dimensional vectors to each real number t in the domain. Since the coordinates of the tip of a position vector uniquely determine the vector, vector functions can be thought of as functions from $\mathbb{R}$ to $\mathbb{R}^2$ or $\mathbb{R}^3$. In this sense, then, the functions we are studying now reverse the roles of the domain and range of vector-valued functions.

The domain of a function of two variables is a subset of $\mathbb{R}^2$—that is, it is a subset of the xy-plane. In the next two examples we illustrate how to find domains and how to show them graphically.

EXAMPLE 13.1 Let $f(x, y) = \sqrt{x - y}$. Find the domain of f and show it graphically. Show that the point $(3, -1)$ lies in this domain, and find the value of $f(3, -1)$.

Solution In order that $\sqrt{x - y}$ be a real number, we must have $x - y \geq 0$, or equivalently, $y \leq x$. The set of points satisfying this inequality are those that

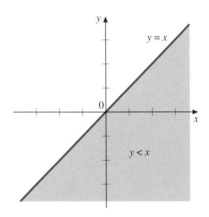

FIGURE 13.3
The domain of $f(x, y) = \sqrt{x - y}$

lie either *on* the line $y = x$ or *below* this line. This set is the shaded region in Figure 13.3.

Since $-1 < 3$, the point $(3, -1)$ satisfies the inequality $y < x$ and so lies in the domain of f. Furthermore,

$$f(3, -1) = \sqrt{3 - (-1)} = \sqrt{4} = 2$$

∎

EXAMPLE 13.2 Find the domain of each of the following functions and show it graphically.

(a) $f(x, y) = \dfrac{1}{\sqrt{1 - x^2 - y^2}}$ (b) $f(x, y) = \ln(y - x^2)$

Solution

(a) For $\sqrt{1 - x^2 - y^2}$ to be real, x and y must satisfy $1 - x^2 - y^2 \geq 0$. All values of x and y for which $1 - x^2 - y^2 = 0$ must be ruled out, since they would make the denominator zero. Thus, the domain of f is the set of all points (x, y) for which $1 - x^2 - y^2 > 0$, or equivalently, $x^2 + y^2 < 1$. A point (x, y) satisfies this inequality if its distance from the origin is less than 1. So the domain consists of all points (x, y) *inside* the unit circle $x^2 + y^2 = 1$. We show this set in Figure 13.4. We indicate the bounding unit circle with a broken line to signify that it is not a part of the domain.

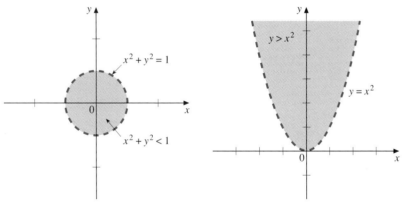

FIGURE 13.4
The domain of $f(x, y) = \dfrac{1}{\sqrt{1 - x^2 - y^2}}$

FIGURE 13.5
The domain of $f(x, y) = \ln(y - x^2)$

(b) Since the natural logarithm is defined only for positive values, x and y must satisfy $y - x^2 > 0$, or equivalently, $y > x^2$. In order to show the set of points satisfying this inequality, we first draw the parabola $y = x^2$, again with a broken line since it is not included, and then shade the region *above* this parabola. We show the resulting region in Figure 13.5. ∎

Graphs of Functions

In Figure 13.2 we indicated by an arrow from the xy-plane to the real line how a function of two variables associates a point (x, y) with a real number

Section 13.1 Functions of Two and Three Variables 943

$z = f(x, y)$. A more useful way to picture a function is by its graph, defined as follows:

Definition 13.2
The Graph of a Function of Two Variables

The **graph of a function f of two variables** is the set of all points (x, y, z) in $\mathbb{R}^3$ for which $z = f(x, y)$.

In general, the graph of a function of two variables is called a **surface**. We show such a surface in Figure 13.6.

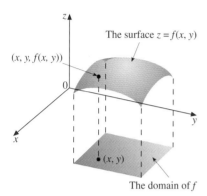

FIGURE 13.6

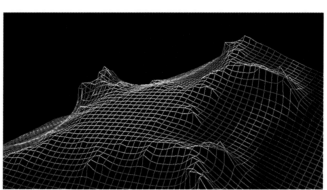

By plotting many points, a computer can generate a reasonably accurate description of a surface.

EXAMPLE 13.3 Let
$$f(x, y) = 6 - x - 2y$$
Show the portion of the graph of f that lies in the first octant.

Solution Let $z = f(x, y)$. Then by Definition 13.2, the graph of f is the set of all points (x, y, z) satisfying $z = 6 - x - 2y$. From Chapter 11, we recognize this equation as being the equation of a plane. Its intercepts on the x-axis, y-axis, and z-axis are $x = 6$, $y = 3$, and $z = 6$, respectively. (We find the intercepts by setting two of the variables equal to 0 and solving for the other.) The traces of the plane on the coordinate planes are the lines connecting the intercepts. We show the graph in Figure 13.7. ∎

EXAMPLE 13.4 Identify and draw the graph of the function f defined by
$$f(x, y) = \sqrt{4 - x^2 - y^2}$$

Solution The domain of f is the set of all points (x, y) satisfying $4 - x^2 - y^2 \geq 0$, or equivalently,
$$x^2 + y^2 \leq 4$$
We recognize this set as the circle $x^2 + y^2 = 4$, of radius 2, centered at the origin, together with all points inside the circle. To find the graph of f, let $z = f(x, y)$. Then $z = \sqrt{4 - x^2 - y^2}$. Note that $z \geq 0$ for all (x, y) in the domain. By squaring the equation and rearranging, we get
$$x^2 + y^2 + z^2 = 4$$

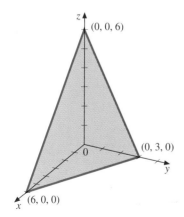

FIGURE 13.7
First-octant portion of the graph of $f(x, y) = 6 - x - 2y$

The left-hand side of the equation is the square of the distance of the point (x, y, z) from the origin. Hence, the equation is satisfied by points (x, y, z) if, and only if, the distance from the origin is 2. In other words, the graph is given by points on the *sphere* of radius 2 centered at the origin. For our function, $z \geq 0$. Hence, its graph is the upper half of the sphere, as we show in Figure 13.8. ∎

Most surfaces are difficult to draw by hand. In the next section we will discuss certain special surfaces that *can* be drawn by hand, and we will show how to obtain graphs of functions based on these special surfaces.

In the next example we show some graphs generated by a computer algebra system.

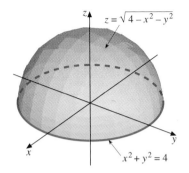

FIGURE 13.8
Graph of $f(x, y) = \sqrt{4 - x^2 - y^2}$ is a hemisphere of radius 3.

EXAMPLE 13.5 Use a CAS to obtain the graphs of each of the following functions.

(a) $f(x, y) = x^2 + \dfrac{y^2}{2}$ 　　(b) $f(x, y) = \dfrac{1}{\sqrt{1 - x^2 - y^2}}$

(c) $f(x, y) = \ln(y - x^2)$ 　　(d) $f(x, y) = \dfrac{\sin xy}{x^2 + y^2}$

Solution We show the graphs in Figure 13.9. The surface in part (a) is called an *elliptical paraboloid* and is one of the special types we will consider in the next section. The functions in parts (b) and (c) are those for which we found the domains in Example 13.2. The graphs in parts (b), (c), and (d) would be difficult to obtain without a CAS. ∎

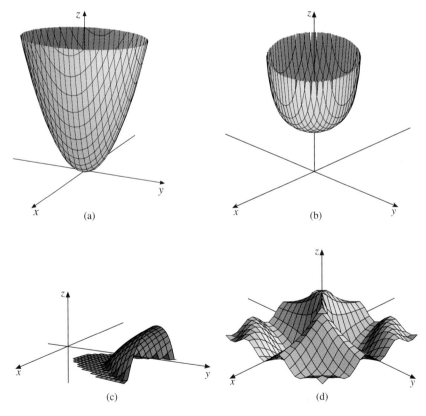

FIGURE 13.9

Level Curves and Contour Maps

Another way to gain insight into the nature of a function of two variables is by means of its *level curves,* defined as follows:

Definition 13.3
Level Curves and Contour Maps

A **level curve** for a function f of two variables is the graph of the equation $f(x, y) = c$ in the xy-plane, where c is any constant value in the range of f. A collection of level curves for different values of c is called a **contour map** for f.

An isotherm on a weather map is a level curve for the temperature function. Along each such curve the temperature is constant. The collection of isotherms on the weather map is a contour map of the temperature function.

Another example of a contour map is a *topographical map* such as the one shown in Figure 13.10. The level curves are the curves of constant elevation. This topographical map is of Mount Shasta and the surrounding region, in California. To help understand the meaning of a level curve, consider the curve marked 13,800. Imagine a horizontal plane 13,800 feet above sea level, slicing through the mountain. The curve of intersection of the plane with the surface of the mountain is called a *trace* of the mountain surface on the plane. If you walked around the mountain on this trace, you would always be at the 13,800-foot level. The corresponding level curve on the topographical map is the *projection* of this trace on the xy-plane.

FIGURE 13.10

We illustrate these ideas further in Figure 13.11. There we show the planes $z = 5$, $z = 7$, $z = 9$, and $z = 11$, cutting the surface $z = f(x, y)$, along with the traces on these planes. The contour map in the xy-plane is the collection of projections of these traces. Note that with a constant difference in values of c (in this case a difference of 2), the closer together the level curves are, the steeper is the graph of the function.

Now suppose we were given only the contour map in Figure 13.11 and wanted to try to determine the corresponding surface. We could raise each curve $f(x, y) = c$ up to the height $z = c$ (or if c were negative, lower it to $z = c$). Then we could sketch the surface with these traces as guides. An experienced reader of topographical maps, for example, can visualize the general shape of the landscape in the manner described.

EXAMPLE 13.6 Let $f(x, y) = x^2 + (y^2/2)$. Draw a contour map showing the level curves $f(x, y) = c$ for $c = 2, 4, 6,$ and 8.

Solution We showed the CAS-generated graph of f in Figure 13.9(a). The equations of the level curves are

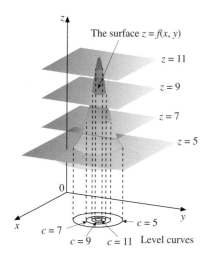

FIGURE 13.11

$$x^2 + \frac{y^2}{2} = 2, \quad \text{or} \quad \frac{x^2}{2} + \frac{y^2}{4} = 1$$

$$x^2 + \frac{y^2}{2} = 4, \quad \text{or} \quad \frac{x^2}{4} + \frac{y^2}{8} = 1$$

$$x^2 + \frac{y^2}{2} = 6, \quad \text{or} \quad \frac{x^2}{6} + \frac{y^2}{12} = 1$$

$$x^2 + \frac{y^2}{2} = 8, \quad \text{or} \quad \frac{x^2}{8} + \frac{y^2}{16} = 1$$

Each curve is an ellipse with major axis along the y-axis. In Figure 13.12(a) we repeat the graph of f (a paraboloid), and in Figure 13.12(b) we show the level curves. ∎

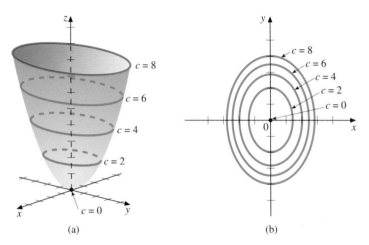

(a)   (b)

FIGURE 13.12

EXAMPLE 13.7 Draw the level curves $f(x, y) = c$ for $c = \pm 1, \pm 3$, and ± 5, for the function $f(x, y) = y^2 - x^2$. From the contour map obtained, sketch the graph of f.

Solution For $c > 0$, the curves $y^2 - x^2 = c$ are hyperbolas with transverse axis along the y-axis. As c increases, the vertices move away from the origin. For $c < 0$, the level curves $y^2 - x^2 = c$ are hyperbolas with transverse axis along the x-axis. For example, with $c = -1$, we have

$$y^2 - x^2 = -1$$

or equivalently,

$$x^2 - y^2 = 1$$

We show the contour map in Figure 13.13(a).

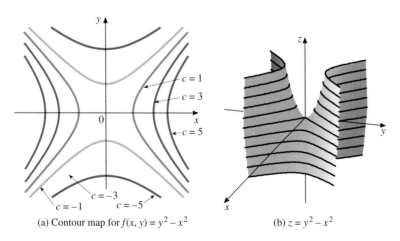

(a) Contour map for $f(x, y) = y^2 - x^2$ (b) $z = y^2 - x^2$

FIGURE 13.13

If we imagine the horizontal planes $z = c$ cutting the surface $z = f(x, y)$ at each of the given values of c, we can form a mental image of the surface. Above the xy-plane, the traces are hyperbolas with transverse axes parallel to the y-axis, with vertices moving away from the z-axis as c increases. Below the xy-plane the traces are hyperbolas with transverse axes parallel to the x-axis, with vertices again moving away from the z-axis for larger absolute values of c. The resulting surface is shown in Figure 13.13(b). It is called a **hyperbolic paraboloid** and is one of the special surfaces we will consider in the next section. ∎

Functions of Three Variables

When we go from functions of two variables to functions of three variables, the notion of a graph becomes more complicated. Although we can say that the graph of $w = f(x, y, z)$ is the set of "points" (x, y, z, w) in four-dimensional space for which w satisfies the given equation, it is impossible to draw such a graph. However, the extension of level curves to three dimensions is possible. In three dimensions, instead of level curves, we have the **level surfaces** $f(x, y, z) = c$ for constants c in the range of f. As an example, consider the function defined by

$$f(x, y, z) = \sqrt{16 - x^2 - y^2 - z^2}$$

The level surfaces $f(x, y, z) = c$ for $0 \leq c \leq 4$ are the spheres

$$x^2 + y^2 + z^2 = 16 - c^2$$

The function assumes its smallest value, 0, on the sphere of radius 4, and as c increases, the spheres on which $f(x, y, z) = c$ shrink in size, finally contracting to the single point $(0, 0, 0)$ when $c = 4$, which is the maximum value the function assumes.

SKETCHING SURFACES USING COMPUTER ALGEBRA SYSTEMS

Being able to make a mental image of a surface can make all the difference in understanding the essential nature of a problem. As we have seen, sketching surfaces, even in the simplest cases, can be very difficult. Computer algebra systems have the capability of quickly rendering accurate sketches of most surfaces we will encounter and many that we could not possibly envision without the aid of technology. Plotting two-dimensional projections of three-dimensional surfaces is one of the most interesting capabilities of computer algebra systems.

In this section, we will demonstrate how Maple and Mathematica sketch surfaces. The output displayed will be from the Maple system.

CAS 47

Sketch the surface $z = f(x, y) = e^{-x^2 - y^2}$.

Maple:

plot3d(exp(–x^2–y^2),x=–2..2,y=–2..2,style=patch,
  axes=boxed,orientation=[45,57]);

Mathematica:

Plot3D[Exp[–x^2–y^2],{x–2,2},{y,–2,2}]

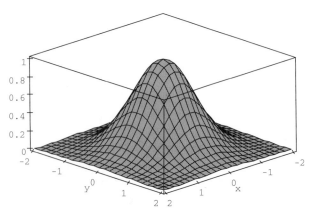

FIGURE 13.1.1

CAS 48

Sketch the surface $z = f(x, y) = \dfrac{-4x}{x^2 + y^2 + 2}$.

Maple:

plot3d(–4*x/(x^2+y^2+2),x=–10..10,y=–10..10,
  grid=[50,50],style=patch,axes=normal,
  tickmarks=[0,0,0],orientation=[14,54]);

Mathematica:

Plot3D[–4*x/(x^2+y^2+2),{x,–10,10},{y,–10,10}]

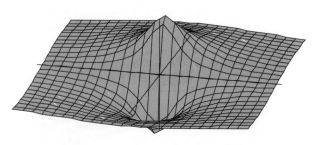

FIGURE 13.1.2

Maple:

Some insight into the nature of a surface can be obtained from a contour map that is a collection of level curves of the form $f(x, y) = c$. In Maple enter:

contourplot(–4*x/(x^2+y^2+2),x=–10..10,y=–10..10,
  numpoints=2000,axes=normal,tickmarks=[0,0,0],
  orientation=[90,0],scaling=constrained);

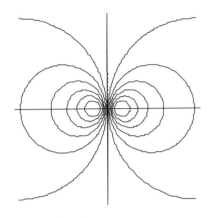

FIGURE 13.1.3

CAS 49

Sketch the surface $z = f(x, y) = \sin x + \sin y$.

Maple:

plot3d(sin(x)+sin(y),x=–4*Pi..4*Pi,y=–4*Pi..4*Pi,
  numpoints=1500,style=patch,axes=none,
  scaling=constrained);

Mathematica:

Plot3D[Sin[x]+Sin[y],{x,–4*Pi,4*Pi},
  {y,–4*Pi,4*Pi},PlotPoints–>40]

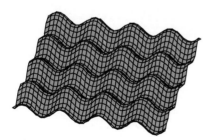

FIGURE 13.1.4

CAS 50

Sketch the sphere of radius 1 and center (1, 2, 1) given by
$$x^2 - 2x + y^2 - 4y + z^2 - 2z = -5$$

The surface this time is given implicitly by the equation involving the three variables x, y, and z. It is often easiest to represent such a surface in parametric form when using a CAS to generate the plot. However, many CAS include the capability of generating implicit plots.

Maple:

implicitplot3d(x^2–2*x+y^2–4*y+z^2–2*z=–5,
　x=0..2,y=1..3,z=0..3,view=[–1..3,0..4,0..3],
　scaling=constrained);

Mathematica:

ParametricPlot3D[{1+(Cos[t])*Cos[u],2+(Sin[t])*Cos[u],
　1+Sin[u]},{t,0,2*Pi},{u,–Pi/2,Pi/2}]

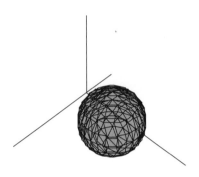

FIGURE 13.1.5

CAS 51

In Chapter 6 we considered finding volumes of solids of revolution. Computer algebra systems can be used to generate the solids of revolution using 3D plot capabilities.

(a) Generate the solid of revolution when the region bounded by $y = x + 4$ and $y = x^2/2$ is rotated about the y-axis.

(b) Generate the solid of revolution when the region bounded by $y = x^2 + 1$ and $\sqrt{x}$ is rotated about the x-axis.

Maple:

(a)

f:=x–>(x+4)–(x^2/2);
First plot the region in the plane:
plot({x+4,x^2/2},x=–5..5,y=0..8,scaling=constrained);

Now rotate the region about the y-axis:

plot3d([x*cos(t),x*sin(t),f(x)],x=–2..4,t=0..2*Pi,
　axes=normal,tickmarks=[0,0,0],labels=['x','y','z'],
　scaling=constrained);

Now change the orientation of the surface to view the inside:*

plot3d([x*cos(t),x*sin(t),f(x)],x=–2..4,t=0..2*Pi,orientation=
　[62,131],axes=normal,tickmarks=[0,0,0],
　scaling=constrained);

Mathematica:

(a)

f[x_] = (x+4)–(x^2/2)

Plot[{x+4,x^2/2},{x,–5,5},PlotRange–>{0,8},
　AspectRatio–>Automatic]

ParametricPlot3D[{x*Cos[t],x*Sin[t],f[x]},{x,–2,4},
　{t,0,2*Pi}]

ParametricPlot3D[{x*Cos[t],x*Sin[t],f[x]},{x,–2,4},
　{t,0,2*Pi},ViewPoint–>{–2,2,–2}]

---

*In Maple, the **orientation** command changes the view of the plot, and specifying the scaling to be constrained makes the aspect ratios of the axes the same.

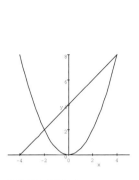

FIGURE 13.1.6a

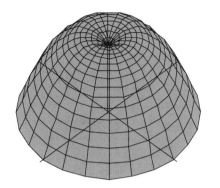

FIGURE 13.1.6b

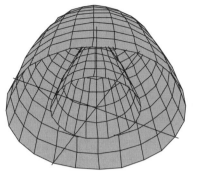

FIGURE 13.1.6c

Maple:

(b)

f:=x–>x^2+1–sqrt(x);
plot({x^2+1,sqrt(x),[1,x,x=0..2]},x=0..1,y=0..2,
  xtickmarks=2,ytickmarks=2,scaling=constrained);
plot3d([x,f(x)*cos(t),f(x)*sin(t)],x=0..1,t=0..2*Pi,
  axes=normal,tickmarks=[0,0,0],orientation=[69,81],
  scaling=constrained);

Mathematica:

(b)

f[x_] = x^2+1–Sqrt[x]

Plot[{x^2+1,Sqrt[x]},{x,0,1},PlotRange–>{0,2},
  AspectRatio–>Automatic]

ParametricPlot3D[{x,f[x]*Cos[t],f[x]*Sin[t]},{x,0,1},
  {t,0,2*Pi}]

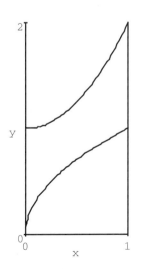

FIGURE 13.1.7a

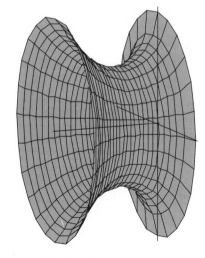

FIGURE 13.1.7b

Exercise Set 13.1

1. If $f(x, y) = \dfrac{2xy - y^2}{x^2 + 3xy}$, find
   (a) $f(-1, 1)$ (b) $f(1, 0)$
   (c) $f(2, 3)$ (d) $f(a, a);\ a \neq 0$

2. If $g(x, y) = 2 \tan^{-1} \dfrac{y}{x} - \sin^{-1} \dfrac{x}{\sqrt{x^2 + y^2}}$, find
   (a) $g(-1, 0)$ (b) $g(1, -1)$
   (c) $g(-1, \sqrt{3})$ (d) $g(-3, \sqrt{3})$

3. If $f(x, y, z) = \sqrt{12 - x^2 - y^2 - z^2}$, find
   (a) $f(1, -3, 1)$ (b) $f(-1, 1, -1)$
   (c) $f(2, 0, -2)$ (d) $f(-3, 1, \sqrt{2})$

4. If $f(x, y) = \ln(2x + y)$, find
   (a) $f(-1, 3)$ (b) $f(0, e^2)$
   (c) $f(x + h, y)$ (d) $f(x, y + k)$

5. If $g(x, y) = \dfrac{x^2 - y}{x - y^2}$, find
   (a) $g(-1, 1)$ (b) $g\left(a, \dfrac{1}{a}\right);\ a \neq 0$
   (c) $g(x + \Delta x, y)$ (d) $g(x, y + \Delta y)$

6. If $f(x, y) = x^2 - 2xy + 3y^2$, find each of the following and simplify:
   (a) $\dfrac{f(x + h, y) - f(x, y)}{h};\ h \neq 0$
   (b) $\dfrac{f(x, y + k) - f(x, y)}{k};\ k \neq 0$

7. If $f(x, y) = \dfrac{x + y}{x - y}$, find each of the following and simplify:
   (a) $\dfrac{f(-2 + h, 1) - f(-2, 1)}{h};\ h \neq 0$
   (b) $\dfrac{f(-2, 1 + k) - f(-2, 1)}{k};\ k \neq 0$

8. Let $f(x, y) = e^{x-y} \sin(x + y)$, $g(t) = 3t$, and $h(t) = t$. Find $f(g(t), h(t))$.

9. Let $f(x, y) = \dfrac{2xy}{x^2 - y^2} + \dfrac{1}{x^2 + y^2}$, $g(t) = \cos t$, and $h(t) = \sin t$. Find $f(g(t), h(t))$.

10. Let $f(x, y) = xy - \dfrac{x}{y}$, $g(u, v) = uv$, and $h(u, v) = \dfrac{u}{v}$. Find $f(g(u, v), h(u, v))$.

In Exercises 11–21, find the domain of f and show it graphically.

11. (a) $f(x, y) = \dfrac{1}{x^2 - y}$ (b) $f(x, y) = \dfrac{1}{x^2 - y^2}$

12. $f(x, y) = \sqrt{x + y}$

13. $f(x, y) = \sqrt{xy}$

14. $f(x, y) = \ln(2x - y)$

15. $f(x, y) = \dfrac{xy}{x^2 + y^2}$

16. $f(x, y) = \sqrt{x + y} - \sqrt{x - y}$

17. $f(x, y) = \ln(x^2 - y^2)$

18. $f(x, y) = \dfrac{x + y}{\sqrt{x^2 + y^2 - 1}}$

19. $f(x, y) = \ln \sinh(x^2 - 2y)$

20. $f(x, y) = \ln\left(\dfrac{2x + y}{2x - y}\right)$

21. $f(x, y) = \ln \sqrt{xy}$

In Exercises 22–25, give the domain of f and describe it geometrically.

22. $f(x, y, z) = \dfrac{\sin(xyz)}{\sqrt{x^2 + y^2 + z^2}}$

23. $f(x, y, z) = \dfrac{2x - 3y + z}{(x - 1)(y + 2)(z - 3)}$

24. $f(x, y, z) = \sqrt{xyz}$

25. $f(x, y, z) = e^{-(x^2 + y^2)} \ln(x + y - z)$

26. Let $f(x, y) = \sqrt{4x + 3y}$. Show that if $h \neq 0$,
$$\dfrac{f(1 + h, 4) - f(1, 4)}{h} = \dfrac{2}{2 + \sqrt{h + 4}}$$

27. Let $f(x, y) = x^2 - 3xy + 2y^2$. Show that
$$f(x + \Delta x, y + \Delta y) = f(x, y) + (2x - 3y)\Delta x$$
$$+ (4y - 3x)\Delta y + \varepsilon_1 \Delta x + \varepsilon_2 \Delta y$$
where $\varepsilon_1 = g(\Delta x, \Delta y)$ and $\varepsilon_2 = h(\Delta x, \Delta y)$. Give the explicit forms of $g(x, y)$ and $h(x, y)$.

In Exercises 28 and 29, find the domain of f and show it graphically.

28. $f(x, y) = \sqrt{\dfrac{x - 3y}{x - y}}$

29. $f(x, y) = \ln(1 - |x| - |y|)$

30. (a) Express the volume of a right circular cone as a function of its base radius r and its altitude h.
(b) Express the surface area of an open-top box as a function of its length l, width w, and depth d.

31. A water tank is to be constructed in the form of a right circular cylinder of radius r and altitude h, with the top in the form of a hemisphere. The hemispherical top costs twice as much per unit area as the lateral surface and bottom. Express the total cost as a function of r, h, and the price p per square unit for the lateral surface and bottom.

32. A company manufactures two types of washing machines: deluxe model A and standard model B. When the price of each A model is p dollars and the price of each B model is q dollars, x model-A machines and y model-B machines can be sold each week. These price functions (called *demand functions*) are found by experience to be approximated by the equations

$$p(x, y) = 600 - 0.4x - 0.2y$$
$$q(x, y) = 400 - 0.3x - 0.5y$$

(a) Find the weekly revenue $R(x, y)$ from producing x model-A machines and y model-B machines.
(b) If the weekly cost of producing x model-A machines and y model-B machines is $C(x, y) = 120x + 90y + 600$, find the profit function $P(x, y)$ for this weekly production.

In Exercises 33–36, use a CAS to obtain the graph of f.

33. $f(x, y) = x^2 + y^2$
34. $f(x, y) = \sqrt{4 - x^2 - y^2}$
35. $f(x, y) = \sqrt{x^2 + y^2}$
36. $f(x, y) = x^2 - y^2$

In Exercises 37–47, draw a contour map with at least six level curves. Use both positive and negative values of c where appropriate.

37. $f(x, y) = 3x - 5y$
38. $f(x, y) = \dfrac{x}{y}$
39. $f(x, y) = x^2 - 2y$
40. $f(x, y) = ye^{-x}$
41. $f(x, y) = x^2 y$
42. $f(x, y) = y - \cos x$
43. $f(x, y) = (x^2 - 1)/y$
44. $f(x, y) = \sqrt{4 - x^2 - y^2}$
45. $f(x, y) = x^2 - y^2$
46. $f(x, y) = y - \ln x$
47. $f(x, y) = y - \sin \pi x$

48. Below are six contour maps labeled (a)–(f) and on the next page are six surfaces labeled (1)–(6). Match up each contour map with the correct surface.

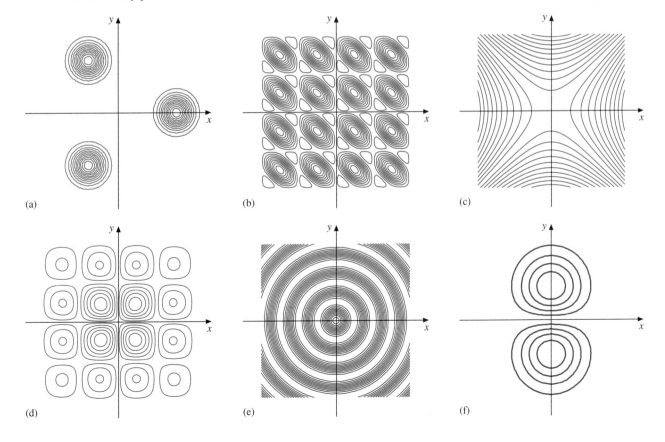

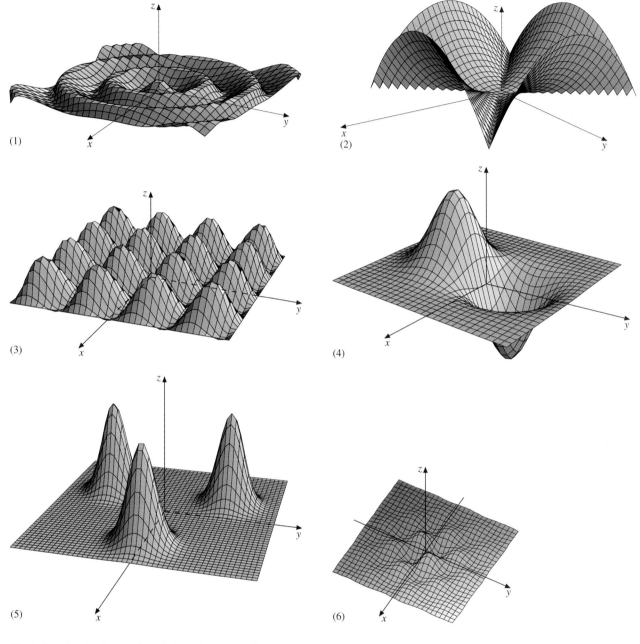

(1) (2) (3) (4) (5) (6)

49. A flat plate in the xy-plane is heated from a point source at the origin. The temperature T in degrees Celsius at a point on the plate varies inversely as the square of its distance from the origin. Describe the isotherms. Suppose $T = 50$ at the point $(3, 4)$. Find all points at which $T = 30$.

50. The *ideal gas law* states that the temperature T of a gas, the volume V it occupies, and its pressure P are related by the equation $PV = kT$, where k is a constant. Draw several isothermal curves in the PV-plane and interpret the results. Express P as a function of T and V, and draw several isobars (lines of equal pressure) in the TV-plane. Interpret the results.

51. The speed of sound in an ideal gas is given by

$$v = \sqrt{\frac{kp}{d}}$$

where p is the pressure of the gas, d is its density, and k is a positive constant. Draw some level curves for v in the pd-plane. Solve the equation for p as a function of v and d, and draw some isobars in the vd-plane. (See Exercise 50.)

52. The accompanying figure shows a cross section of a circular cylinder lying on a horizontal plane. The cylinder and the plane are held at two different electric potentials. The electric potential in the shaded region is given by

$$V(x, y) = \frac{ky}{x^2 + y^2}$$

Draw several equipotentials (lines of equal potential) for this function.

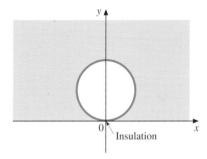

53. In hydrodynamics the level curves for the *stream function* ψ for two-dimensional fluid flow are called *streamlines*. A streamline is the path along which a given particle of the fluid moves. In the accompanying figure a fluid (such as water) flows from the negative x-axis toward the positive x-axis, with $y > 0$. There is a semicircular obstruction, as shown, centered at the origin. (You can think of the hump as a half-buried pipe at the bottom of a stream.) The stream function in this case is

$$\psi(x, y) = y - \frac{y}{x^2 + y^2}$$

Draw several streamlines. (*Hint:* Write the equations of the level curves in polar coordinates.)

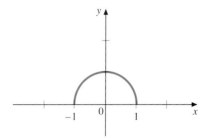

54. Plot the surface for the function $f(x, y) = \dfrac{x^2}{3} - \dfrac{y^2}{4}$ along with the horizontal planes $z = 0.5$, $z = 1$, $z = 2$, and $z = 4$. Then generate a contour plot for the surface.

55. For each of the following surfaces, make a plot of the surface and make contour plots. In plotting the surfaces, be sure to determine a scale and orientation that shows the important information.

(a) $f(x, y) = \dfrac{x^2 y}{x^4 + y^2}$ (b) $f(x, y) = x^2 y e^{-x^2 - y^2}$

(c) $f(x, y) = \dfrac{-x}{x^2 + y^2 + 2}$ (d) $f(x, y) = \dfrac{100}{\sqrt{x^2 - y^2}}$

(e) $f(x, y) = \sin(x) + \cos(y)$

56. Plot level surfaces for each of the following functions of three variables.

(a) $f(x, y, z) = \dfrac{x^2 + y^2}{z}$

(b) $f(x, y, z) = z - x - \sin\sqrt{x^2 + y^2}$

13.2 SKETCHING SURFACES

In this section, we will illustrate some commonly occurring surfaces and give some guidelines for sketching them. Each of the surfaces is the *graph of an equation* in three variables, by which we mean the set of all points (x, y, z) whose coordinates satisfy the equation. For example, we have seen already (in Chapter 11) that the graph of every equation of the form $ax + by + cz + d = 0$ is a plane.

Not all the surfaces we consider are graphs of functions. In fact, we can identify which ones are functions by the **Vertical Line Test**: *for a surface to be the graph of a function, each vertical line must intersect the surface in at most one point.* This test follows from the fact that a function assigns one and only one value to each point in its domain.

One of the chief aids in sketching surfaces are **traces**. A trace of a surface on a plane is the curve of intersection of the surface with the plane. Of primary interest are traces on the coordinate planes and on planes parallel to the coordinate planes. To obtain the trace on the xy-plane, we set $z = 0$ in the equation of the surface. Similarly, setting $x = 0$ gives the yz-trace, and setting $y = 0$ gives the xz-trace. Traces on planes parallel to the coordinate planes are obtained by setting one of the variables equal to a constant. For example, if we set $z = c$, we get the trace on the plane $z = c$, parallel to the xy-plane.

Planes

First, we show planes parallel to the coordinate planes. For definiteness, we use $x = 3$, $y = 2$, and $z = 4$, as shown in Figure 13.14. For the plane $x = 3$ in Figure 13.14(a), we first mark the point three units out on the positive x-axis. Then from this point we draw a vertical line (the xz-trace), and a horizontal line (the xy-trace). In this case, there is no yz-trace because the plane is parallel to the yz-plane. Now we form a rectangle with portions of the two traces we have drawn as adjacent sides. The length and width of the rectangle are arbitrary. The region enclosed by the rectangle, of course, is only a part of the plane, but it is impossible to show the entire plane.

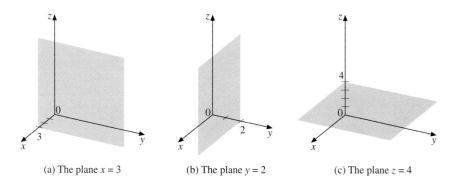

(a) The plane $x = 3$ (b) The plane $y = 2$ (c) The plane $z = 4$

FIGURE 13.14

The planes $y = 2$ and $z = 4$ are obtained in a similar way, and are shown in Figures 13.14(b) and (c), respectively.

For planes not parallel to any of the coordinate planes, we typically show a triangular portion of the plane bounded by the traces. The equations of the traces are found by setting just one variable at a time equal to 0. For example, consider the plane $2x + 3y + 4z = 12$. The equation of the xy-trace is the line $2x + 3y = 12$, that of the yz-trace is $3y + 4z = 12$, and that of the xz-trace is $2x + 4z = 12$, or equivalently, $x + 2y = 6$. A quick way to graph the traces is to connect the *intercepts* of the plane on the coordinate axes. The intercepts of the plane $2x + 3y + 4z = 12$ are found by setting two variables at a time equal to zero, giving the x-intercept 6, y-intercept 4, and z-intercept 3. The traces are the lines joining the intercepts. We show the portion of this plane bounded by the traces in Figure 13.15.

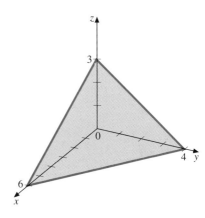

FIGURE 13.15
The plane $2x + 3y + 4z = 12$

EXAMPLE 13.8 Sketch the following planes.
(a) $z = x + y$ (b) $4x + 3y = 12$

Solution

(a) When all three variables are set equal to zero, the equation is satisfied. Therefore, the plane passes through the origin. The xz-trace is the line $z = x$, and the yz-trace is the line $z = y$. Since the xy-trace is the single point at the origin, we find a trace on a plane parallel to the xy-plane to aid in sketching a part of the plane. To do so, we set $z = k$, for an arbitrary $k > 0$, giving $x + y = k$. Together, the traces we have found outline the triangular portion of the plane shown in Figure 13.16.

(b) The xy-trace is the line $4x + 3y = 12$, since setting $z = 0$ does not alter the equation. If the point (x, y) is on this line, then (x, y, z) is on the plane for any value of z. That is, the three-dimensional graph of $4x + 3y = 12$ is the plane formed by extending the line $4x + 3y = 12$ vertically. So the plane is parallel to the z-axis. The xz-trace is the vertical line $x = 4$, and the yz-trace is the vertical line $y = 4$. We again show a trace k units up on the plane $z = k$, with $k > 0$. The trace is simply a replica of the trace in the xy-plane in this case. We show the portion of the plane formed by these traces in Figure 13.17. ■

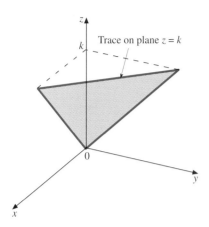

FIGURE 13.16
The plane $x + y - z = 0$

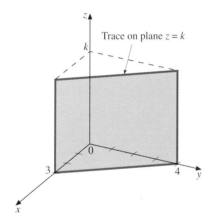

FIGURE 13.17
The plane $4x + 3y = 12$

REMARK

■ When you are sketching surfaces by hand, it is probably easiest to show the yz-plane in the plane of the paper, so that the yz-trace can be shown full-size, as with any other two-dimensional drawing. The positive x-axis should be shown at an angle of about 135° from the positive y-axis, and a unit of distance on it should be about two-thirds of a unit in the y and z directions. Following these conventions gives the illusion that the positive x-axis projects outward from the plane of the paper. For surfaces that are obtained using a CAS, the axes are oriented somewhat differently in order to show the surfaces from the most advantageous perspective.

Cylinders

A surface that is generated by a line moving along a plane curve C so that the line always remains parallel to some fixed line l not in or parallel to the plane of C is called a **cylindrical surface**, or more briefly, a **cylinder**. In Figure 13.18 we show such a surface, where the curve C is in the xy-plane and the line l is parallel to the z-axis. The use of the term *cylinder* here is more general than the usual notion of a cylinder in which the curve C is a circle and l is perpendicular to the plane of the circle (the correct name in this case is *right circular cylinder*).

We will restrict our consideration of cylinders to those for which the curve C is in one of the coordinate planes and l is perpendicular to that plane. Thus, l may be taken as one of the coordinate axes. In all such cases, the equation of the cylinder will simply be the equation of the curve C. As an example,

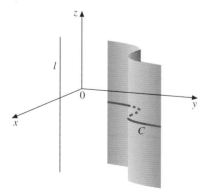

FIGURE 13.18
A cylinder generated by a line parallel to l moving along the curve C.

consider the equation $y = x^2$. Its graph in the xy-plane is a parabola, but as a three-dimensional surface, it is the **parabolic cylinder** shown in Figure 13.19, generated by a line parallel to the z-axis. To understand why, consider any point on the parabola in the xy-plane, say, $(2, 4, 0)$. When viewed as an equation in three variables, all points of the form $(2, 4, z)$ satisfy the equation, regardless of the value of z. Thus, the entire vertical line through the point $(2, 4, 0)$ lies on the surface. A similar result holds for all other points on the parabola. More generally, we have the following result.

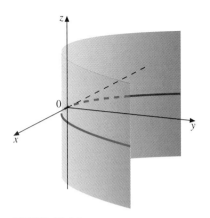

FIGURE 13.19
The parabolic cylinder $y = x^2$

Equations of Cylinders Parallel to a Coordinate Axis

If one of the variables x, y, or z is missing from an equation, then the graph of the equation is a cylinder generated by a line parallel to the axis of the missing variable.

To sketch a cylinder of this type, first draw the two-dimensional curve in the plane of the variables present in the equation. Then extend the curve parallel to the axis of the missing variable, using traces where useful. We illustrate this technique in the next three examples.

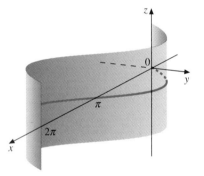

EXAMPLE 13.9 Identify and sketch the graph of the equation $y = \sin x$ for $0 \leq x \leq \pi$.

Solution Since z is missing from the equation, the graph is a cylinder with generating line parallel to the z-axis. To sketch it, we draw one cycle of the sine curve in the xy-plane and then extend it vertically. To give a better picture, we then show traces on two horizontal planes, one above and one below the xy-plane. These traces duplicate the original curve. We show the result in Figure 13.20. ∎

FIGURE 13.20
The cylinder $y = \sin x$

EXAMPLE 13.10 Identify and sketch the graph of the equation $9y^2 + 4z^2 = 36$.

Solution In the yz-plane, the curve is an ellipse with standard form

$$\frac{y^2}{4} + \frac{z^2}{9} = 1$$

Since x is missing, the graph is a cylinder, called an **elliptical cylinder**, with generating line parallel to the x-axis. To sketch it, we first sketch the ellipse in the yz-plane. The y-intercepts are ± 2, and the z-intercepts are ± 3. Then we extend the ellipse in the x direction, showing a trace in a plane $x = k$ (where k is an arbitrary positive number). Again, the trace is a replica of the original curve. We show the sketch in Figure 13.21. ∎

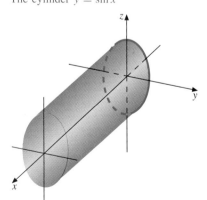

FIGURE 13.21
The elliptic cylinder
$9y^2 + 4z^2 = 36$

EXAMPLE 13.11 Identify and sketch the graph of the equation $z^2 - x^2 = 4$.

Solution In two dimensions, the graph is a hyperbola in the xz-plane. In three dimensions it is the cylinder sketched in Figure 13.22, generated by a line parallel to the y-axis. We use a trace in a plane perpendicular to the y-axis to aid in visualizing the surface. This cylinder is called a **hyperbolic cylinder**. ∎

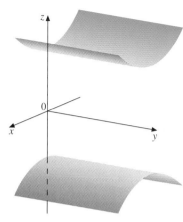

FIGURE 13.22
The hyperbolic cylinder $z^2 - x^2 = 4$

REMARK

■ Suppose you are given an equation such as $y = x^2$, involving only two variables. How do you know if the graph is a parabola or a parabolic cylinder? The answer is that without further information, there is no way to tell. The context is essential. The situation is even more uncertain with an equation such as $x = 4$. In one dimension the graph is a point, in two dimensions it is a vertical line, and in three dimensions it is a plane parallel to the yz-plane. *Throughout this chapter and the two that follow, all equations should be viewed as having graphs in three-dimensional space, unless otherwise specified.*

Spheres

A sphere is the set of all points in $\mathbb{R}^3$ equidistant from a fixed point (the center). If the center is (h, k, l) and the common distance is a (that is, the radius is a), then a point (x, y, z) lies on the sphere if and only if

$$\sqrt{(x-h)^2 + (y-k)^2 + (z-l)^2} = a$$

or equivalently,

$$(x-h)^2 + (y-k)^2 + (z-l)^2 = a^2 \tag{13.1}$$

If the center is at the origin, the equation simplifies to

$$x^2 + y^2 + z^2 = a^2$$

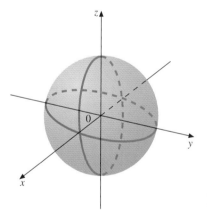

FIGURE 13.23
The sphere $x^2 + y^2 + z^2 = a^2$

We show the graph of a sphere with center at the origin in Figure 13.23. Each of the traces with the coordinate planes is a circle of radius a. By sketching these traces, using broken lines for the parts hidden from view, we get the figure. It is best to draw the yz-trace first, since it provides the outline of the sphere.

The next example shows how an equation of the form

$$x^2 + y^2 + z^2 + ax + by + cz + d = 0 \tag{13.2}$$

can, under certain conditions on the coefficients, be put in the form of Equation 13.1.

EXAMPLE 13.12 Discuss the graph of the equation

$$x^2 + y^2 + z^2 - 4x - 10y - 6z + k = 0$$

where
(a) $k = 34$ (b) $k = 38$ (c) $k = 42$

Solution

(a) We complete the squares on x, y, and z by adding appropriate constants to both sides:

$$(x^2 - 4x + 4) + (y^2 - 10y + 25) + (z^2 - 6z + 9) = -34 + 4 + 25 + 9$$

or

$$(x - 2)^2 + (y - 5)^2 + (z - 3)^2 = 4$$

This equation is in the form of Equation 13.1, so it represents a sphere with center (2, 5, 3) and radius 2. We show a sketch of its graph in Figure 13.24.

(b) The left-hand side is the same as in part (a), but the right-hand side is 0:

$$(x - 2)^2 + (y - 5)^2 + (z - 3)^2 = 0$$

The only point satisfying this equation is (2, 5, 3). This is an example of a *degenerate sphere*.

(c) Again, the only change is on the right-hand side. The result is

$$(x - 2)^2 + (y - 5)^2 + (z - 3)^2 = -4$$

Since the left-hand side cannot be negative, there is no graph. ∎

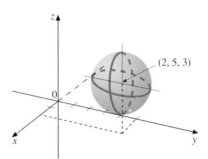

FIGURE 13.24
The sphere
$(x - 2)^2 + (y - 5)^2 + (z - 3)^2 = 4$

Quadric Surfaces

Equation 13.2 is a special case of the general second-degree equation in three variables,

$$Ax^2 + By^2 + Cz^2 + Dxy + Exz + Fyz + Gx + Hy + Iz + J = 0 \quad (13.3)$$

When the coefficients in Equation 13.3 are appropriately restricted, the graph is a sphere. Parabolic, elliptic, and hyperbolic cylinders also occur as special cases of Equation 13.3. For example, if $A = 1$, $H = -1$, and all other coefficients are 0, the equation reduces to $x^2 - y = 0$, or $y = x^2$, whose graph is the parabolic cylinder in Figure 13.19. There are six other possibilities (excluding degenerate cases) for graphs of Equation 13.3. (We are assuming that at least one of the coefficients A through F is nonzero, since otherwise the equation would not be of the second degree.) We will consider these other six types below. They, together with parabolic, elliptic, and hyperbolic cylinders, are known as **quadric surfaces**. Quadric surfaces can be thought of as three-dimensional analogues of the conic sections.

To describe the six quadric surfaces mentioned above, we write each equation in *standard form*. By rearranging, if necessary, you can verify that each equation is a special case of Equation 13.3. For simplicity, we take the coefficients D, E, and F to be 0. Later, we will show a computer-generated graph in which one of these coefficients is nonzero. It can be proved that when one or more of D, E, and F is nonzero, the graph, if it exists and is nondegenerate, is a rotation of one of the types we illustrate.

The equations we give are for surfaces conveniently placed with respect to the origin and the coordinate axes. When interchanges are made among the variables x, y, and z, the resulting surface is of the same type as illustrated but the orientation is changed. If x, y, and z are replaced by $x - h$, $y - k$, and $z - l$, respectively, the given surface is translated h units in the x direction, k units in the y direction, and l units in the z direction.

Ellipsoid

$$\frac{x^2}{a^2} + \frac{y^2}{b^2} + \frac{z^2}{c^2} = 1 \tag{13.4}$$

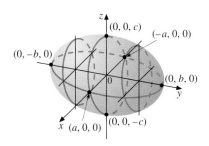

FIGURE 13.25
The ellipsoid $\frac{x^2}{a^2} + \frac{y^2}{b^2} + \frac{z^2}{c^2} = 1$

xy-trace: ellipse $\quad \frac{x^2}{a^2} + \frac{y^2}{b^2} = 1$

xz-trace: ellipse $\quad \frac{x^2}{a^2} + \frac{z^2}{c^2} = 1$

yz-trace: ellipse $\quad \frac{y^2}{b^2} + \frac{z^2}{c^2} = 1$

Traces on planes perpendicular to each coordinate axis between intercepts are ellipses. A special case of the ellipsoid is the sphere, in which $a = b = c$. The graph of Equation 13.4 is shown in Figure 13.25.

The key to recognizing an ellipsoid is that its standard equation involves the sum of the squares of all three variables.

Elliptic Paraboloid

$$\frac{x^2}{a^2} + \frac{y^2}{b^2} = cz \tag{13.5}$$

We illustrate the case $c > 0$.

xy-trace: the origin

xz-trace: parabola $\quad \frac{x^2}{a^2} = cz$

yz-trace: parabola $\quad \frac{y^2}{b^2} = cz$

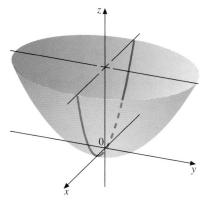

FIGURE 13.26
The paraboloid $\frac{x^2}{a^2} + \frac{y^2}{b^2} = cz \quad (c > 0)$

Traces on planes perpendicular to the positive z-axis are ellipses. For the given standard form, if $c > 0$, the paraboloid opens upward as in Figure 13.26, and if $c < 0$, it opens downward.

The key to recognizing an elliptic paraboloid is that its equation can be written so that one side involves the sum of the squares of two of the variables and the other side involves the first power of the third variable. The axis corresponds to the first-degree variable.

Elliptic Hyperboloid of One Sheet

$$\frac{x^2}{a^2} + \frac{y^2}{b^2} - \frac{z^2}{c^2} = 1 \tag{13.6}$$

xy-trace: ellipse $\quad \frac{x^2}{a^2} + \frac{y^2}{b^2} = 1$

xz-trace: hyperbola $\quad \frac{x^2}{a^2} - \frac{z^2}{c^2} = 1$

yz-trace: hyperbola $\quad \frac{y^2}{b^2} - \frac{z^2}{c^2} = 1$

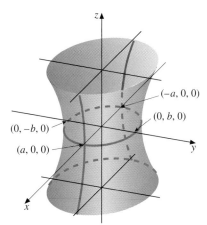

FIGURE 13.27
The hyperboloid of one sheet
$\frac{x^2}{a^2} + \frac{y^2}{b^2} - \frac{z^2}{c^2} = 1$

Traces on planes perpendicular to the z-axis are ellipses. We show the graph in Figure 13.27.

Elliptic Hyperboloid of Two Sheets

$$\frac{z^2}{c^2} - \frac{x^2}{a^2} - \frac{y^2}{b^2} = 1 \tag{13.7}$$

xy-trace: none

xz-trace: hyperbola $\quad \dfrac{z^2}{c^2} - \dfrac{x^2}{a^2} = 1$

yz-trace: hyperbola $\quad \dfrac{z^2}{c^2} - \dfrac{y^2}{b^2} = 1$

For $|k| > c$, traces on planes $z = k$ are ellipses. We show the graph in Figure 13.28.

The key to recognizing the equation of a hyperboloid of two sheets is that the standard form involves the squares of all three variables, one with a positive sign and two with negative signs. The axis is that of the variable with a positive sign.

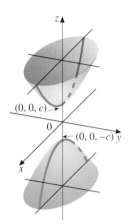

FIGURE 13.28
The hyperboloid of two sheets
$\dfrac{z^2}{c^2} - \dfrac{x^2}{a^2} - \dfrac{y^2}{b^2} = 1$

Elliptic Cone

$$\frac{x^2}{a^2} + \frac{y^2}{b^2} = \frac{z^2}{c^2} \tag{13.8}$$

xy-trace: origin

xz-trace: two lines $\quad \dfrac{x}{a} = \pm \dfrac{z}{c}$

yz-trace: two lines $\quad \dfrac{y}{b} = \pm \dfrac{z}{c}$

Traces on planes perpendicular to the z-axis are ellipses. We show the graph in Figure 13.29.

For the cone the equation can be written so that one side involves the sum of the squares of two of the variables and the other side involves the square of the third variable, with the axis that of the latter variable.

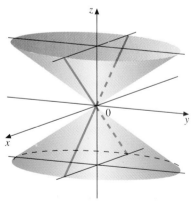

FIGURE 13.29
The cone $\dfrac{x^2}{a^2} + \dfrac{y^2}{b^2} = \dfrac{z^2}{c^2}$

REMARK
■ The upper and lower parts of the cone are called *nappes*. In customary usage when one refers to a cone (or to a right circular cone) only one nappe of the total cone is intended.

Section 13.2 Sketching Surfaces 963

Hyperbolic Paraboloid

$$\frac{y^2}{b^2} - \frac{x^2}{a^2} = cz \qquad (13.9)$$

xy-trace: two lines $\dfrac{y}{b} = \pm \dfrac{x}{a}$

xz-trace: parabola $-\dfrac{x^2}{a^2} = cz$, opens downward

yz-trace: parabola $\dfrac{y^2}{b^2} = cz$, opens upward

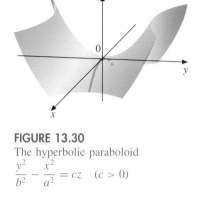

FIGURE 13.30
The hyperbolic paraboloid
$\dfrac{y^2}{b^2} - \dfrac{x^2}{a^2} = cz \quad (c > 0)$

Traces on planes perpendicular to the z-axis are hyperbolas, and traces on planes perpendicular to the x-axis or y-axis are parabolas. We show a hyperbolic paraboloid with $c > 0$ in Figure 13.30.

The hyperbolic paraboloid has the appearance of a saddle near the origin. For this reason, the origin is called a **saddle point** of the surface.

The key to recognizing the hyperbolic paraboloid is that the first power of one of the variables in the standard equation equals the *difference* of the squares of the other two variables. In contrast, in the standard equation of the elliptic paraboloid (Figure 13.26), the first power of one of the variables equals the *sum* of the squares of the other two variables.

REMARK

■ When traces on planes perpendicular to an axis of a surface are circular, the surface is called a *surface of revolution*, since it could be formed by revolving a plane curve about the axis. An ellipsoid of revolution can be recognized from its standard equation when any two of the numbers a, b, or c are equal (if all three are equal, the ellipsoid is a sphere). For the paraboloid, the hyperboloids, and the cone in the orientations shown in Figures 13.26 through 13.29, a surface of revolution results when $a = b$. (Obvious modifications are required when the axis is not the z-axis.) A cone of revolution is also called a *circular cone*.

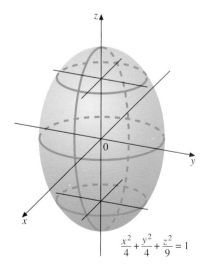

FIGURE 13.31
Ellipsoid of revolution

EXAMPLE 13.13   Identify and sketch the graph of

$$9x^2 + 9y^2 + 4z^2 = 36$$

Solution   We divide both sides by 36 to obtain the standard form

$$\frac{x^2}{4} + \frac{y^2}{4} + \frac{z^2}{9} = 1$$

xy-trace: $\dfrac{x^2}{4} + \dfrac{y^2}{4} = 1$ or $x^2 + y^2 = 4$, circle

yz-trace: $\dfrac{y^2}{4} + \dfrac{z^2}{9} = 1$, ellipse

xz-trace: $\dfrac{x^2}{4} + \dfrac{z^2}{9} = 1$, ellipse

The graph is the ellipsoid of revolution in Figure 13.31. Since the z-intercepts are ± 3 and the x- and y-intercepts are ± 2, the major axis of the ellipsoid is the z-axis. ■

EXAMPLE 13.14 Identify and sketch the surface whose equation is $4x^2 + z^2 = 4y$.

Solution First, we divide by 4 to obtain the standard form

$$\frac{x^2}{1} + \frac{z^2}{4} = y$$

By comparison with Equation 13.5, we recognize the surface as an elliptic paraboloid with axis on the y-axis. To sketch it, we find the traces on the coordinate planes.

xy-trace: $x^2 = y$, a parabola

yz-trace: $z^2 = 4y$, a parabola

xz-trace: $\frac{x^2}{1} + \frac{z^2}{4} = 0$, the origin

After drawing the xy-trace and yz-trace, it appears that a trace on a plane perpendicular to the y-axis to the right of $y = 0$ would be helpful. If we let $y = k$, for an arbitrary positive k, we get

$$\frac{x^2}{1} + \frac{z^2}{4} = k$$

which we recognize as an ellipse. We can now sketch the graph as in Figure 13.32.

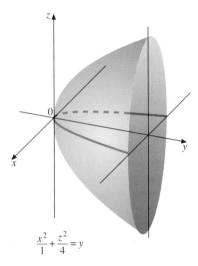

$\frac{x^2}{1} + \frac{z^2}{4} = y$

FIGURE 13.32
Elliptic paraboloid

EXAMPLE 13.15 Identify and sketch the graph of the equation

$$x^2 - 2y^2 - 3z^2 - 6 = 0$$

Solution When we write the equation as

$$\frac{x^2}{6} - \frac{y^2}{3} - \frac{z^2}{2} = 1$$

we recognize it as representing a hyperboloid of two sheets with axis along the x-axis (compare with Equation 13.7).

xy-trace: $\frac{x^2}{6} - \frac{y^2}{3} = 1$, hyperbola

yz-trace: none

xz-trace: $\frac{x^2}{6} - \frac{z^2}{2} = 1$, hyperbola

For $k > \sqrt{6}$, the trace on each of the planes $x = \pm k$ is the ellipse

$$\frac{y^2}{3} + \frac{z^2}{2} = \frac{k^2}{6} - 1$$

With these traces we obtain the sketch in Figure 13.33.

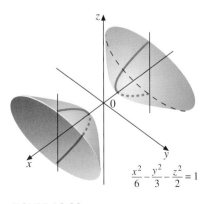

$\frac{x^2}{6} - \frac{y^2}{3} - \frac{z^2}{2} = 1$

FIGURE 13.33
Elliptic hyperboloid of two sheets

In the next two examples, we show how we can sometimes use knowledge of quadric surfaces to obtain graphs of functions of two variables.

EXAMPLE 13.16 Identify and sketch the graph of the function

$$f(x, y) = \sqrt{x^2 + y^2}$$

Solution Let $z = f(x, y)$. If we square both sides, we get

$$z^2 = x^2 + y^2$$

which, by comparison with Equation 13.8, is a cone with axis on the z-axis. It is circular, since the coefficients of x^2 and y^2 are equal. The equation $z = f(x, y)$ represents the upper nappe of the cone only, since z is nonnegative. The traces are

xy-trace: $x^2 + y^2 = 0$, the origin
yz-trace: $z = \pm y$, two lines
xz-trace: $z = \pm x$, two lines

Traces on planes $z = \pm k$ are the circles $x^2 + y^2 = k^2$.
We can now sketch the graph (Figure 13.34). ∎

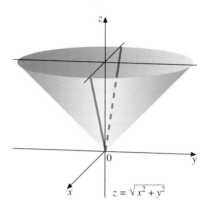

FIGURE 13.34
Upper nappe of circular cone

EXAMPLE 13.17 Identify and sketch the graph of the function

$$f(x, y) = \sqrt{1 + x^2 + y^2}$$

Give the domain and range of f.

Solution Let $z = \sqrt{1 + x^2 + y^2}$. If we square both sides and rearrange, we get $z^2 - x^2 - y^2 = 1$. The graph of this latter equation is a hyperboloid of revolution of two sheets, with axis along the z-axis. Since $z > 0$, the graph of the original function is the upper sheet only, shown in Figure 13.35. The domain of f is all of $\mathbb{R}^2$, and the range is the set of z values for which $z \geq 1$. ∎

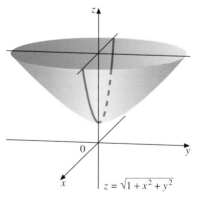

FIGURE 13.35
Upper sheet of hyperboloid of revolution of two sheets.

For reference, we summarize in Chart 13.1 the six quadric surfaces given by Equations 13.4 through 13.9. The figures shown were generated by a CAS and show many traces to help in visualizing the shapes.

CHART 13.1

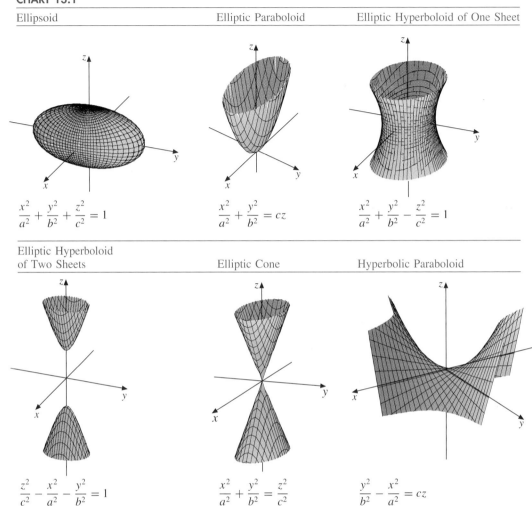

Exercise Set 13.2

In Exercises 1–10, draw the graph of the given equation in three dimensions.

1. $y^2 = 2x$
2. $x^2 + y^2 = 4$
3. $y = 4 - x^2$
4. $z = \dfrac{x^2}{4}$
5. $z = \sqrt{4 - y^2}$
6. $4x^2 + 9y^2 = 36$
7. $x^2 - y^2 = 1$
8. $xy = 2$
9. $z = e^{-x}$
10. $z = \ln y$

In Exercises 11–15, determine whether the graph is a sphere or a degenerate sphere, or if there is no graph. If the graph is a sphere, give its center and radius and draw the graph.

11. $x^2 + y^2 + z^2 - 2x - 6y - 8z + 10 = 0$
12. $x^2 + y^2 + z^2 - 8y - 4z + 11 = 0$
13. $x^2 + y^2 + z^2 + 10x - 2y + 6z + 35 = 0$
14. $x^2 + y^2 + z^2 - 10x + 4y + 8z + 47 = 0$
15. $x^2 + y^2 + z^2 - 6x + 8y - 8z + 33 = 0$

In Exercises 16–24, identify the quadric surface and draw its graph, showing traces where useful.

16. $36x^2 + 9y^2 + 16z^2 = 144$
17. $36x^2 - 9y^2 + 16z^2 = 144$
18. $36x^2 - 9y^2 + 16z^2 + 144 = 0$
19. $36x^2 - 9y^2 + 16z^2 = 0$
20. $36z = 9x^2 + 4y^2$
21. $16x - 4y^2 - 9z^2 = 0$
22. $9x^2 - 4y^2 + 36z = 0$
23. $y = \sqrt{4x^2 + z^2}$
24. $3x^2 + 2z^2 - 6y = 0$

In Exercises 25–43, identify and draw the graph of f. State the domain and range.

25. $f(x, y) = 4 - x - y$
26. $f(x, y) = 4 - x$
27. $f(x, y) = x + 2y$
28. $f(x, y) = y^2$
29. $f(x, y) = \sqrt{x}$
30. $f(x, y) = \sqrt{4 - x^2}$
31. $f(x, y) = 9 - y^2$
32. $f(x, y) = \sqrt{1 + x^2}$
33. $f(x, y) = 2x^2 + y^2$
34. $f(x, y) = 4 - x^2 - y^2$
35. $f(x, y) = \sqrt{12 - 4x^2 + 3y^2}$
36. $f(x, y) = \sqrt{36 - 4x^2 - 9y^2}$
37. $f(x, y) = 1 + x^2 + y^2$
38. $f(x, y) = \sqrt{x^2 + y^2}$
39. $f(x, y) = \sqrt{x^2 + y^2 - 4}$
40. $f(x, y) = \sqrt{4 - x^2 + y^2}$
41. $f(x, y) = \sqrt{4y - x^2}$
42. $f(x, y) = \sqrt{4 + 4x^2 + y^2}$
43. $f(x, y) = y^2 - x^2$

In Exercises 44–49, make an appropriate translation to identify and draw the graph.

44. $25x^2 + 9y^2 + 15x^2 - 100x - 54y - 60z + 16 = 0$
45. $4x^2 + y^2 - 2z^2 - 8x - 4y - 8z + 8 = 0$
46. $9x^2 - 4y^2 + 9z^2 - 54x - 16y - 18z + 38 = 0$
47. $4x^2 + y^2 - 24x - 4y - 4z + 20 = 0$
48. $4x^2 + 3y^2 - z^2 - 32x - 12y + 2z + 75 = 0$
49. $2x^2 - 3y^2 - 8x - 12y + 12z - 52 = 0$

In Exercises 50–55, show the volume in the first octant bounded by the given surfaces.

50. $x^2 + z^2 = 4$, $y = x$, $y = 0$, $z = 0$
51. $z = x^2 + y^2$, $x + y + z = 4$
52. $x^2 + z^2 = 4$, $y^2 + z^2 = 4$, $x = 0$, $y = 0$, $z = 0$
53. $z = 4 - y^2$, $x^2 = 2y$, $x = 0$, $z = 0$
54. $z = 4 - x^2 - y^2$, $z^2 = x^2 + y^2$
55. $4x^2 + 2y^2 + 3z^2 = 48$, $y = 2x$, $y = 4$, $x = 0$, $z = 0$

13.3 LIMITS AND CONTINUITY

The concept of limit is as important in the context of functions of two, three, or even more variables as it is in the case of functions of a single variable. In particular, once we define the limit of a function of two variables, we will be able to discuss the slope of the tangent line to a curve on the surface, as well as rates of change in the function in different directions. The definition of the limit in this context is a natural extension of the limit concept for a function of one variable.

The Limit of a Function

In order to motivate the definition, let us begin with an example. Let

$$f(x, y) = \sqrt{1 - x^2 - y^2}$$

Setting $z = \sqrt{1 - x^2 - y^2}$ and squaring gives $z^2 = 1 - x^2 - y^2$, or equivalently, $x^2 + y^2 + z^2 = 1$. We recognize this equation as that of a sphere of radius 1, centered at the origin. For our original function, $z \geq 0$, so the graph is the upper hemisphere, shown in Figure 13.36. Suppose we now want to know what value $f(x, y)$ approaches as (x, y) approaches $(0, 0)$. That is, we want to know the limit

$$\lim_{(x,y) \to (0,0)} f(x, y)$$

We have not yet defined what such a limit means, but we can proceed intuitively.

Suppose we consider points in the domain of f that are close to $(0, 0)$. That is, we consider points (x, y) inside some small circle centered at the origin, as indicated in Figure 13.37(a). Then, as we see from Figure 13.37(b), the corresponding z values are close to 1. That is, $f(x, y)$ is close to 1.

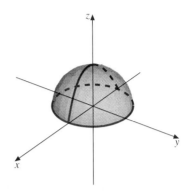

FIGURE 13.36
The graph of
$f(x, y) = \sqrt{1 - x^2 - y^2}$ is a hemisphere.

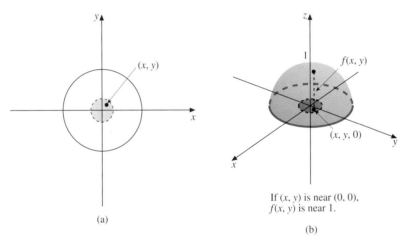

If (x, y) is near $(0, 0)$, $f(x, y)$ is near 1.

(a)   (b)

FIGURE 13.37

(Along $y = 0$)

(x, y)	$f(x, y)$
$(\pm 0.5, 0)$	0.86603
$(\pm 0.1, 0)$	0.99499
$(\pm 0.01, 0)$	0.99995

(Along $x = 0$)

(x, y)	$f(x, y)$
$(0, \pm 0.5)$	0.86603
$(0, \pm 0.1)$	0.99499
$(0, \pm 0.01)$	0.99995

(Along $y = x$)

(x, y)	$f(x, y)$
$(\pm 0.5, \pm 0.5)$	0.70711
$(\pm 0.1, \pm 0.1)$	0.98995
$(\pm 0.01, \pm 0.01)$	0.99990

(Along $y = -x$)

(x, y)	$f(x, y)$
$(\pm 0.5, \mp 0.5)$	0.70711
$(\pm 0.1, \mp 0.1)$	0.98995
$(\pm 0.01, \mp 0.01)$	0.9990

One major difference between the limit idea for a function of two (or more) variables and that for a function of one variable has to do with the manner of approach. In the case of

$$\lim_{x \to a} f(x) = L$$

there are only two possible ways for x to approach a—from the right and from the left. For the limit to exist, both one-sided limits have to exist and be the same value. In the case of the limit of a function of two variables, there are infinitely many ways (x, y) can approach the point in question. In the margin we show tables in which (x, y) approaches $(0, 0)$ along the paths $y = 0$ (the x-axis), $x = 0$ (the y-axis), the line $y = x$, and the line $y = -x$, respectively. In each case we show the corresponding values of $f(x, y) = \sqrt{1 - x^2 - y^2}$ (rounded to five places). Although these tables strongly suggest that the function is approaching 1 as (x, y) approaches $(0, 0)$, they do not provide a proof of this fact. We could never *prove* this limit is 1 by considering different paths of approach, because we could never consider all such paths.

The important thing is that for *all* points (x, y) sufficiently close to the origin, other than the origin itself, the function values must be as close to 1 as we please.

REMARK

■ Although we cannot prove that the limit of $f(x, y)$ exists as (x, y) approaches a point (x_0, y_0), by considering different paths of approach, we can prove the limit *does not* exist if we get different limiting values along two different paths. We will say more later about this method of showing the failure of the limit to exist.

Definition 13.4
Informal Limit Definition

Let f be a function of two variables whose domain includes all points inside some circle centered at (x_0, y_0), except possibly (x_0, y_0) itself. Then the **limit of $f(x, y)$ as (x, y) approaches (x_0, y_0) is L**, written

$$\lim_{(x,y) \to (x_0, y_0)} f(x, y) = L$$

means that $f(x, y)$ will be as close as we please to L for all points (x, y) sufficiently close to (x_0, y_0), but not equal to (x_0, y_0).

REMARK

■ The corresponding definition for the limit of a function of three (or more) variables should be clear. If f is a function of three variables defined inside some sphere centered at (x_0, y_0, z_0) except possibly at that point itself, then in Definition 13.4, we simply replace (x, y) by (x, y, z) and (x_0, y_0) by (x_0, y_0, z_0).

EXAMPLE 13.18   Find

$$\lim_{(x,y) \to (1,-2)} (8 - 2x^2 - y^2)$$

Solution   It seems intuitively evident that for x close to 1 and y close to -2, $f(x, y) = 8 - 2x^2 - y^2$ is close to $8 - 2(1)^2 - (-2)^2 = 2$. That is, by taking (x, y) close enough to $(1, -2)$, we can force $f(x, y)$ to be as close to 2 as we choose. Thus, by Definition 13.4, we conclude that

$$\lim_{(x,y) \to (1,-2)} (8 - 2x^2 - y^2) = 2$$

We show the graph of f in Figure 13.38. The surface is an elliptic paraboloid with vertex $(0, 0, 8)$. The trace on the xy-plane is the ellipse $2x^2 + y^2 = 8$, and the traces on the xz- and yz-planes are the parabolas $z = 8 - 2x^2$ and $z = 8 - y^2$, respectively. ■

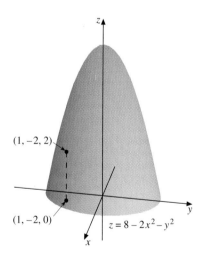

FIGURE 13.38

EXAMPLE 13.19   Find the limit

$$\lim_{(x,y) \to (0,0)} \frac{x^4 - y^4}{x^2 + y^2}$$

or show that it fails to exist.

Solution   At $(0, 0)$ the function

$$f(x, y) = \frac{x^4 - y^4}{x^2 + y^2}$$

is not defined. But it is defined for all other points. If we restrict (x, y) to be different from $(0, 0)$, we can factor the numerator and divide out the common factor to get

$$f(x, y) = \frac{(x^2 - y^2)(x^2 + y^2)}{x^2 + y^2} = x^2 - y^2 \quad \text{if } (x, y) \neq (0, 0)$$

Since we are not concerned with the point $(0, 0)$ in finding the limit, we have

$$\lim_{(x,y) \to (0,0)} f(x, y) = \lim_{(x,y) \to (0,0)} (x^2 - y^2) = 0$$

We show the graph of f in Figure 13.39. It is identical to the paraboloid $z = x^2 + y^2$ except that the point $(0, 0, 0)$ is missing. ∎

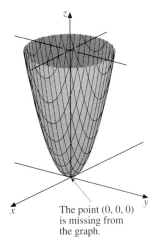

The point $(0, 0, 0)$ is missing from the graph.

FIGURE 13.39
$f(x, y) = \dfrac{x^4 - y^4}{x^2 + y^2} = x^2 - y^2$
if $(x, y) \neq (0, 0)$

Showing the Limit Fails to Exist

We remarked earlier that if we can find two different paths in the domain of f along which (x, y) approaches (x_0, y_0), for which $f(x, y)$ approaches two different limits, then we can conclude that

$$\lim_{(x,y) \to (x_0, y_0)} f(x, y)$$

does not exist. This conclusion follows from the fact that there are some points (x, y) arbitrarily close to (x_0, y_0) for which $f(x, y)$ will be close to one value and other points for which $f(x, y)$ will be close to a different value. So it is not the case that for *all* points sufficiently close (but not equal) to (x_0, y_0), the corresponding function values are as close as we please to a single limiting value. In the next two examples, we illustrate this method of showing the limit fails to exist.

EXAMPLE 13.20   Show that

$$\lim_{(x,y) \to (0,0)} \frac{x^2 - y^2}{x^2 + y^2}$$

does not exist.

Solution   Again, we restrict (x, y) to be different from $(0, 0)$. If (x, y) lies on the x-axis, with $x \neq 0$, then

$$f(x, y) = f(x, 0) = \frac{x^2}{x^2} = 1$$

whereas if (x, y) is on the y-axis, with $y \neq 0$, then

$$f(x, y) = f(0, y) = \frac{-y^2}{y^2} = -1$$

We can now see that however close to the origin (x, y) is, if the point approaches $(0, 0)$ on the x-axis, $f(x, y) = 1$, and if it approaches $(0, 0)$ on the y-axis, $f(x, y) = -1$. We conclude that there is no *single* value L (neither 1 nor -1) that $f(x, y)$ approaches as (x, y) approaches $(0, 0)$, without restriction as to the manner of approach. Thus, $\lim_{(x,y) \to (0,0)} f(x, y)$ does not exist.

We show the graph of f (generated by a CAS) in Figure 13.40. ∎

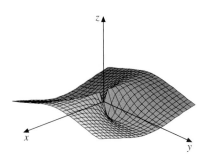

FIGURE 13.40

EXAMPLE 13.21 Show that $\lim_{(x,y)\to(0,0)} f(x, y)$ fails to exist, where

(a) $f(x, y) = \dfrac{2xy}{x^2 + y^2}$ (b) $f(x, y) = \dfrac{x^2 y}{x^4 + y^2}$

Solution

(a) If we let (x, y) approach $(0, 0)$ along the x-axis, then $y = 0$, so we see that $f(x, y) = 0$. Similarly, if (x, y) approaches $(0, 0)$ along the y-axis, $x = 0$, so $f(x, y)$ is again equal to 0. So we get the same limit for $f(x, y)$ as (x, y) approaches $(0, 0)$ along either axis. Suppose, however, that (x, y) approaches $(0, 0)$ along the line $y = x$. Then $(x, y) = (x, x)$, with $x \neq 0$. So

$$f(x, y) = \frac{2x^2}{x^2 + x^2} = \frac{2x^2}{2x^2} = 1$$

Now we see that $\lim_{(x,y)\to(0,0)} f(x, y)$ does not exist, since $f(x, y)$ approaches 0 along one path (either the x-axis or y-axis) and $f(x, y)$ approaches 1 along another path (the line $y = x$).

We show the CAS-generated graph of f in Figure 13.41.

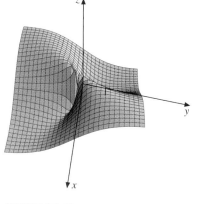

FIGURE 13.41

(b) As in part (a), $f(x, y)$ approaches 0 along either axis. It also approaches 0 along the line $y = x$. In fact, (x, y) approaches $(0, 0)$ along any line $y = mx$ through the origin with $m \neq 0$. To see why, when $y = mx$, then $(x, y) = (x, mx)$, so that with $x \neq 0$,

$$f(x, y) = \frac{x^2(mx)}{x^4 + (mx)^2} = \frac{mx^3}{x^4 + m^2 x^2} = \frac{x^2(mx)}{x^2(x^2 + m^2)} \to \frac{0}{m^2} = 0$$

Now consider the situation where (x, y) approaches the origin along the parabola $y = x^2$. Then $(x, y) = (x, x^2)$. So with $x \neq 0$,

$$f(x, y) = \frac{x^2(x^2)}{x^4 + (x^2)^2} = \frac{x^4}{x^4 + x^4} = \frac{x^4}{2x^4} = \frac{1}{2}$$

We can now conclude that $\lim_{(x,y)\to(0,0)} f(x, y)$ does not exist, since as (x, y) approaches the origin along the x-axis (or the y-axis or any other line through the origin), $f(x, y)$ approaches 0, whereas along the parabola $y = x^2$, $f(x, y)$ approaches $\frac{1}{2}$.

We show the CAS-generated surface in Figure 13.42. ∎

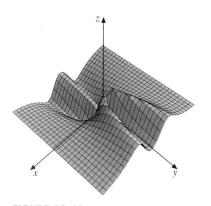

FIGURE 13.42

For emphasis, we restate the technique we have illustrated for showing a limit fails to exist.

If $f(x, y)$ approaches different limits as (x, y) approaches (x_0, y_0) along two different paths, then $\lim_{(x,y)\to(x_0,y_0)} f(x, y)$ does not exist.

If you get the *same* limit for $f(x, y)$ as (x, y) approaches (x_0, y_0) along two or more different paths, you cannot conclude that the limit exists. In Example 13.21(b), we showed that $f(x, y)$ approached 0 along any straight-line path through the origin, suggesting perhaps that $\lim_{(x,y)\to(0,0)} f(x, y) = 0$. Yet we saw that $f(x, y)$ had the value $\frac{1}{2}$ when (x, y) approached $(0, 0)$ along the parabolic path $y = x^2$. Consequently, we concluded that the limit did not exist. Getting different values on different paths shows that the limit *fails* to exist. But we must use some other technique to show a limit *does* exist.

Limit Properties

We can often find limits by making use of the following properties, analogous to those for functions of one variable, given in Section 2.2.

Properties of Limits

If $\lim_{(x,y) \to (x_0,y_0)} f(x, y)$ and $\lim_{(x,y) \to (x_0,y_0)} g(x, y)$ both exist, then the following properties hold true.

1. $\lim_{(x,y) \to (x_0,y_0)} cf(x, y) = c \left[\lim_{(x,y) \to (x_0,y_0)} f(x, y) \right]$ for any constant c

2. $\lim_{(x,y) \to (x_0,y_0)} [f(x, y) \pm g(x, y)]$
   $= \lim_{(x,y) \to (x_0,y_0)} f(x, y) \pm \lim_{(x,y) \to (x_0,y_0)} g(x, y)$

3. $\lim_{(x,y) \to (x_0,y_0)} [f(x, y) \cdot g(x, y)]$
   $= \left[\lim_{(x,y) \to (x_0,y_0)} f(x, y) \right] \cdot \left[\lim_{(x,y) \to (x_0,y_0)} g(x, y) \right]$

4. $\lim_{(x,y) \to (x_0,y_0)} \dfrac{f(x, y)}{g(x, y)} = \dfrac{\lim_{(x,y) \to (x_0,y_0)} f(x, y)}{\lim_{(x,y) \to (x_0,y_0)} g(x, y)}$,

   provided $\lim_{(x,y) \to (x_0,y_0)} g(x, y) \neq 0$

From Properties 1, 2, and 3, it follows that for $P(x, y)$ a polynomial in x and y,

$$\lim_{(x,y) \to (x_0,y_0)} P(x, y) = P(x_0, y_0)$$

We make use of this result in the next example.

EXAMPLE 13.22 Evaluate the limit

$$\lim_{(x,y) \to (2,-3)} \frac{x^2 y - 3y^3}{x^3 + 2xy^2}$$

Solution We use Property 4, taking limits on numerator and denominator separately and dividing:

$$\lim_{(x,y) \to (2,-3)} \frac{x^2 y - 3y^3}{x^3 + 2xy^2} = \frac{\lim_{(x,y) \to (2,-3)} (x^2 y - 3y^3)}{\lim_{(x,y) \to (2,-3)} (x^3 + 2xy^2)} = \frac{(2)^2(-3) - 3(-3)^3}{2^3 + 2(2)(-3)^2}$$

$$= \frac{-12 + 81}{8 + 36} = \frac{69}{44} \quad \blacksquare$$

The Formal Limit Definition

For some purposes, such as for proving the limit properties and for proving other theorems concerning limits as well as for making the concept precise and unambiguous, we need a more precise statement of the definition of the limit of a function than that given in Definition 13.4. The following definition is analogous to Definition 2.5 for the one-variable case.

Definition 13.5
Formal Limit Definition

Let f be a function of two variables defined at all points inside some circle centered at (x_0, y_0), except possibly at (x_0, y_0) itself. Then we say that **the limit of $f(x, y)$ as (x, y) approaches (x_0, y_0) is L** and write

$$\lim_{(x,y) \to (x_0, y_0)} f(x, y) = L$$

provided that corresponding to any positive number ε there exists a positive number δ such that

$$\text{if } 0 < \sqrt{(x - x_0)^2 + (y - y_0)^2} < \delta, \text{ then } |f(x, y) - L| < \varepsilon \qquad (13.10)$$

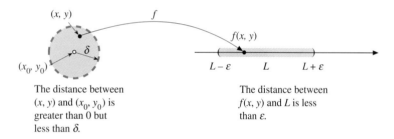

FIGURE 13.43

In Figure 13.43 we show a diagram to help you understand this definition. The inequality $0 < \sqrt{(x - x_0)^2 + (y - y_0)^2} < \delta$ says that the distance between the point (x, y) and the fixed point (x_0, y_0) is greater than 0 and less than δ. So (x, y) is not equal to (x_0, y_0) but can be any other point inside the circle of radius δ centered at (x_0, y_0). Its image $f(x, y)$ will then lie within ε units of L. Since ε can be made as small as we want, the definition says that $f(x, y)$ comes arbitrarily close to L when (x, y) is close enough to (x_0, y_0) (within δ units of it) but is different from (x_0, y_0). The number δ usually depends on the value of ε.

REMARK
■ It *may* happen that the inequality $|f(x, y) - L| < \varepsilon$ is true even when $(x, y) = (x_0, y_0)$, but the definition does not *require* the inequality to hold true in this case. In fact, $f(x_0, y_0)$ need not be defined, as was the case in Example 13.19.

To extend Definition 13.5 to the three-variable case, we require that $f(x, y, z)$ be defined at all points inside a sphere centered at (x_0, y_0, z_0), except possibly at (x_0, y_0, z_0) itself. Then

$$\lim_{(x,y,z) \to (x_0,y_0,z_0)} f(x, y, z) = L$$

means that given any $\varepsilon > 0$, there is a $\delta > 0$ such that

if $0 < \sqrt{(x - x_0)^2 + (y - y_0)^2 + (z - z_0)^2} < \delta$,

then $|f(x, y, z) - L| < \varepsilon$ \hfill (13.11)

Continuity

The definition of continuity for a function of two variables is the same as for a function of one variable. We require not only that the limit of $f(x, y)$ exist as $(x, y) \to (x_0, y_0)$ but also that it be equal to $f(x_0, y_0)$.

Definition 13.6
Continuity of a Function of Two Variables

Let f be a function of two variables that is defined at all points inside a circle centered at (x_0, y_0). Then f is **continuous** at (x_0, y_0) provided that

$$\lim_{(x,y) \to (x_0,y_0)} f(x, y) = f(x_0, y_0) \quad (13.12)$$

REMARK

■ Continuity of a function of three variables is defined in an analogous way, with (x_0, y_0) replaced by (x_0, y_0, z_0) and (x, y) replaced by (x, y, z). We assume that f is defined inside some sphere centered at (x_0, y_0, z_0).

Equation 13.12 implies three things:

1. $\lim_{(x,y) \to (x_0,y_0)} f(x, y)$ exists.
2. $f(x_0, y_0)$ exists.
3. The values in 1 and 2 are the same.

If any one of these conditions fails, the function is discontinuous at (x_0, y_0).

If f is continuous at all points of some set S, we say that f **is continuous on** S. When all points inside some circle centered at (x_0, y_0) lie in a set, the point (x_0, y_0) is called an **interior point** of the set. So in Definition 13.6 the point (x_0, y_0) is an interior point of the domain of f. A point belonging to a set that is not an interior point is called a **boundary point** of the set. More generally, a boundary point of a set (which may or may not be in the set) is a point such that every circle centered on it contains at least one point in the set and one point not in the set. The set of all boundary points of a set is called its **boundary**. We can alter Definition 13.6 to permit (x_0, y_0) to be a boundary point of the domain D of f (see Figure 13.44), provided (x_0, y_0) is in D, by adding the requirement that $(x, y) \in D$ to the definition. That is, we require only that points *of the domain* that are closer to (x_0, y_0) than δ units have function values closer to L than ε units.

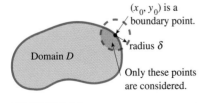

FIGURE 13.44

EXAMPLE 13.23 Show that the function

$$f(x, y) = 2x^2 + y^2$$

is continuous at all points of $\mathbb{R}^2$.

Solution Since f is a polynomial function, for any point (x_0, y_0) in $\mathbb{R}^2$, we have

$$\lim_{(x,y) \to (x_0, y_0)} (2x^2 + y^2) = 2x_0^2 + y_0^2$$

Furthermore, $f(x_0, y_0) = 2x_0^2 + y_0^2$. So by Definition 13.6, f is continuous at (x_0, y_0). ∎

REMARK
■ We have seen that if P is any polynomial function of two variables,

$$\lim_{(x,y) \to (x_0, y_0)} P(x, y) = P(x_0, y_0)$$

It follows by Definition 13.6 that P is continuous at (x_0, y_0). Thus, every polynomial function of two variables is continuous on all of $\mathbb{R}^2$. Using Limit Property 4, we also then see that *every rational function of two variables is continuous at all points of $\mathbb{R}^2$ in its domain*, that is, wherever the denominator is nonzero.

EXAMPLE 13.24 Let

$$f(x, y) = \begin{cases} \dfrac{2xy}{x^2 + y^2} & \text{if } (x, y) \neq (0, 0) \\ 0 & \text{if } (x, y) = (0, 0) \end{cases}$$

Determine all points at which f is continuous.

Solution At all points other than the origin, f is a rational function with nonzero denominator. So by the remark above, f is continuous everywhere except at the origin.

To check continuity at the origin, we check to see if $\lim_{(x,y) \to (0,0)} f(x, y) = f(0, 0)$. We are given that $f(0, 0) = 0$. But in Example 13.21(a) we showed that $f(x, y)$ does not approach a unique value as $(x, y) \to (0, 0)$. (We got different limits along two different paths.) Thus,

$$\lim_{(x,y) \to (0,0)} f(x, y)$$

does not exist, so f is discontinuous at $(0, 0)$. ∎

Based on the properties of limits, we can prove that the *sum*, *difference*, *product*, and *quotient* of two continuous functions are also continuous, provided that in the case of the quotient the denominator is nonzero. Furthermore, *a continuous function of a continuous function is continuous*. For example, the composite function $\sin(x^2 + y^2)$ is continuous at all points of $\mathbb{R}^2$, since $u = x^2 + y^2$ is continuous on all of $\mathbb{R}^2$, and $z = \sin u$ is continuous on all of $\mathbb{R}$.

Exercise Set 13.3

In Exercises 1–12, find the limit.

1. $\lim_{(x,y)\to(2,-1)} (3x - 2y + 5)$
2. $\lim_{(x,y)\to(-1,-2)} (2x^2y - 3xy^3)$
3. $\lim_{(x,y)\to(-2,3)} \dfrac{x - 2y}{x + y}$
4. $\lim_{(x,y)\to(1,3)} \dfrac{x^2 - y^2}{x^2 + y^2}$
5. $\lim_{(x,y,z)\to(1,-1,2)} (xz - yz + xy)$
6. $\lim_{(x,y,z)\to(-4,5,3)} \dfrac{x^2 - 2yz + z^2}{x^2 + y^2 + z^2}$
7. $\lim_{(x,y)\to(0,0)} e^{x^2+y^2}$
8. $\lim_{(x,y)\to(\pi/2,1)} y\cos(xy)$
9. $\lim_{(x,y,z)\to(1,0,0)} \ln(x^2 + y^2 + z^2)$
10. $\lim_{(x,y)\to(0,0)} \dfrac{\sin(x^2 + y^2)}{x^2 + y^2}$ (Hint: Let $u = x^2 + y^2$.)
11. $\lim_{(x,y)\to(0,0)} f(x, y)$, where

$$f(x, y) = \begin{cases} \sqrt{x^2 + y^2} & \text{if } (x, y) \neq (0, 0) \\ 1 & \text{if } (x, y) = (0, 0) \end{cases}$$

12. $\lim_{(x,y)\to(1,-1)} f(x, y)$, where

$$f(x, y) = \begin{cases} \dfrac{x^2 - 2xy}{2x + 3y} & \text{if } (x, y) \neq (1, -1) \\ 3 & \text{if } (x, y) = (1, -1) \end{cases}$$

In Exercises 13–18, show that $\lim_{(x,y)\to(0,0)} f(x, y)$ does not exist.

13. $f(x, y) = \dfrac{2x^2 - 3y^2}{x^2 + y^2}$
14. $f(x, y) = \dfrac{3xy}{2x^2 + 5y^2}$
15. $f(x, y) = \dfrac{x + y}{x^2 + y^2}$
16. $f(x, y) = \dfrac{x^2 - xy + y^2}{x^2 + y^2}$
17. $f(x, y) = \dfrac{xy^2}{x^2 + y^4}$
18. $f(x, y) = \dfrac{x^3 y}{x^6 + y^2}$

In Exercises 19–30, determine the largest subset of $\mathbb{R}^2$ on which f is continuous.

19. $f(x, y) = 2x^3 - 3xy + 7y^2$
20. $f(x, y) = \dfrac{x^2 - y^2}{x^2 + y^2}$
21. $f(x, y) = \dfrac{x^2 - y^2}{x - y}$
22. $f(x, y) = \dfrac{x^2 - 4y^2}{x - 2y}$
23. $f(x, y) = \dfrac{x - y}{x + y}$
24. $f(x, y) = \dfrac{2x - 3y}{x^2 + y^2 - 1}$
25. $f(x, y) = \dfrac{x^2 + y^2}{x^2 - y^2}$
26. $f(x, y) = \dfrac{2x^2 - 3xy + y^2}{x^2 - y + 1}$
27. $f(x, y) = e^{2xy}$
28. $f(x, y) = e^{-x/y}$
29. $f(x, y) = \ln\sqrt{x^2 + y^2}$
30. $f(x, y) = \tan^{-1}\dfrac{y}{x}$

In Exercises 31–35, determine if f is continuous at $(0, 0)$.

31. $f(x, y) = \begin{cases} \dfrac{x^4 - y^4}{x^2 + y^2} & \text{if } (x, y) \neq (0, 0) \\ 0 & \text{if } (x, y) = (0, 0) \end{cases}$

32. $f(x, y) = \begin{cases} \dfrac{x - y}{x^2 + y^2} & \text{if } (x, y) \neq (0, 0) \\ 0 & \text{if } (x, y) = (0, 0) \end{cases}$

33. $f(x, y) = \begin{cases} \dfrac{\sin\sqrt{x^2 + y^2}}{\sqrt{x^2 + y^2}} & \text{if } (x, y) \neq (0, 0) \\ 1 & \text{if } (x, y) = (0, 0) \end{cases}$

34. $f(x, y) = \begin{cases} \dfrac{x^2 + 2xy + y^2}{x^2 + y^2} & \text{if } (x, y) \neq (0, 0) \\ 1 & \text{if } (x, y) = (0, 0) \end{cases}$

35. $f(x, y) = \begin{cases} \dfrac{x^3 - y^3}{x^2 + xy + y^2} & \text{if } (x, y) \neq (0, 0) \\ 0 & \text{if } (x, y) = (0, 0) \end{cases}$

36. Let f and g be functions of two variables, each of which is continuous at (x_0, y_0). Use Definition 13.6 and the properties of limits to prove that each of the following functions is also continuous at (x_0, y_0).
    (a) cf, where c is a constant
    (b) $f \pm g$
    (c) fg
    (d) $\dfrac{f}{g}$, provided $g(x_0, y_0) \neq 0$

37. Use Definition 13.5 to prove that if
$$\lim_{(x,y)\to(x_0,y_0)} f(x, y) = L \text{ and } \lim_{(x,y)\to(x_0,y_0)} g(x, y) = M,$$
then $\lim_{(x,y)\to(x_0,y_0)} [f(x, y) + g(x, y)] = L + M$.

38. Let f be a function of two variables with domain D. A point (a, b) is said to be an **isolated point** of D if (a, b) is in D and there is a circle with center (a, b) with no other point of D inside the circle.
   (a) Show that an isolated point of D is always a boundary point of D.
   (b) Show that any function is continuous at an isolated point of its domain.
   (*Hint:* Choose δ so small that the circle of radius δ centered at (a, b) has no point of the domain D except (a, b) inside it. Then explain why the inequality $|f(x, y) - f(a, b)| < \varepsilon$ is satisfied regardless of what the value of the positive number ε is, provided (x, y) is in D and $\sqrt{(x-a)^2 + (y-b)^2} < \delta$.)

 39. Let $z = f(x, y) = \dfrac{x^2 + y}{3x}$.
   (a) Plot the surface.
   (b) Plot each of the following curves: $z = f(x, x)$, $z = f(x, 0)$, $z = f(x, 1)$, and $z = f(x, x^2)$. Does the limit as (x, y) approaches $(0, 0)$ exist?

 40. For each of the following functions, conjecture whether the function has a limit as (x, y) approaches $(0, 0)$.
   (a) $f(x, y) = \dfrac{x^2 + 3xy + y^2}{x^2 + y^2}$
   (b) $f(x, y) = \dfrac{x^3 + y^2}{x^3 + y^3}$
   (c) $f(x, y) = |x|^y$
   (d) $f(x, y) = \dfrac{\sin(x^2 + y^2)}{x^2 + y^2}$

13.4 PARTIAL DERIVATIVES

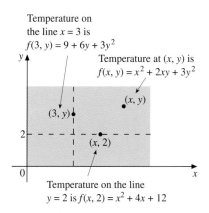

FIGURE 13.45

If f is a function of two variables x and y and we hold the variable y fixed, then f becomes a function of the single variable x. Similarly, if we hold x fixed but allow y to vary, then f is a function of the single variable y. For example, suppose that

$$f(x, y) = x^2 + 2xy + 3y^2$$

represents the temperature at a point (x, y) on a flat metal plate (see Figure 13.45). If we hold y fixed, say, $y = 2$, we get

$$f(x, 2) = x^2 + 4x + 12$$

This function of x gives the temperature at points along the horizontal line $y = 2$. If, instead, we hold x fixed, say, $x = 3$, we get

$$f(3, y) = 9 + 6y + 3y^2$$

which gives the temperature along the vertical line $x = 3$.

Since $f(x, 2) = x^2 + 4x + 12$ is a function of the single variable x, we can find its derivative, $2x + 4$, and interpret it as giving the rate of change of temperature in the x direction along the line $y = 2$. Similarly, the derivative of $f(3, y)$ is $6 + 6y$, which gives the rate of change of temperature in the y direction along the line $x = 3$.

More generally, for any function $f(x, y)$, when we hold y fixed, the resulting function of x may be differentiable. If so, we call the derivative the **partial derivative of f with respect to x** and denote it by $f_x(x, y)$. Similarly, if we hold x fixed and the resulting function of y is differentiable, we call its derivative the **partial derivative of f with respect to y** and denote it by $f_y(x, y)$.

To illustrate these partial derivatives, let us consider again the function

$$f(x, y) = x^2 + 2xy + 3y^2$$

Rather than assign a specific value to y, we treat it as a constant and differentiate with respect to x, to get

$$f_x(x, y) = 2x + 2y$$

Similarly,
$$f_y(x, y) = 2x + 6y$$

In the first case, since y is a constant, the derivative of $2xy$ with respect to x is $2y$, and the derivative of $3y^2$ is 0 (since $3y^2$ is a constant when differentiating with respect to x). Also, when we differentiate with respect to y, the term x^2 is constant, so its derivative is 0.

When you see f_x you will know to do two things: hold y fixed, and differentiate with respect to x. Similar remarks apply to the symbol f_y.

Before stating the formal definition of partial derivatives, we remind you of the definition of the derivative of a function of one variable. If f is a function of the single variable x, then its derivative at x is

$$f'(x) = \lim_{h \to 0} \frac{f(x+h) - f(x)}{h} \tag{13.13}$$

provided the limit on the right exists.

The partial derivatives $f_x(x, y)$ and $f_y(x, y)$ are defined exactly as in Equation 13.13.

Definition 13.7
The Partial Derivatives f_x and f_y

Let f be a function of two variables x and y, with domain D. The functions f_x and f_y defined by

$$f_x(x, y) = \lim_{h \to 0} \frac{f(x+h, y) - f(x, y)}{h} \tag{13.14}$$

and

$$f_y(x, y) = \lim_{k \to 0} \frac{f(x, y+k) - f(x, y)}{k} \tag{13.15}$$

at all points D where these limits exist are called, respectively, the **partial derivative of f with respect to x** and **the partial derivative of f with respect to y**.

Other Notations

Just as with ordinary derivatives, different notations for partial derivatives are commonly used. Analogous to the Leibniz notation df/dx for the derivative of a function of one variable, the symbols $\partial f/\partial x$ and $\partial f/\partial y$ are alternative notations for f_x and f_y, respectively, when f is a function of two variables x and y. The symbol ∂ (sometimes referred to as "curly d") replaces the d in ordinary differentiation to signify that more than one variable is involved. The symbols $\partial/\partial x$ and $\partial/\partial y$ can be regarded as *partial derivative operators*, which instruct you to take the partial derivative of whatever follows. For example,

$$\frac{\partial}{\partial x}(2x^2 y + x^3) = 4xy + 3x^2 \quad \text{and} \quad \frac{\partial}{\partial y}(2x^2 y + x^3) = 2x^2$$

When a dependent variable is introduced, say, $z = f(x, y)$, the partial derivatives with respect to x and y, respectively, can be written as z_x or $\partial z/\partial x$ and z_y or $\partial z/\partial y$.

EXAMPLE 13.25 Let $f(x, y) = \tan^{-1} y/x$. Find $f_x(x, y)$, $f_y(x, y)$, $f_x(4, -3)$, and $f_y(4, -3)$.

Solution Holding y fixed, we get

Recall that $\dfrac{d}{dx} \tan^{-1} u = \dfrac{1}{1+u^2} \dfrac{du}{dx}$.

$$f_x(x, y) = \frac{1}{1 + \left(\dfrac{y}{x}\right)^2} \left(-\frac{y}{x^2}\right) = \frac{-y}{x^2 + y^2}$$

Holding x fixed, we get

$$f_y(x, y) = \frac{1}{1 + \left(\dfrac{y}{x}\right)^2} \left(\frac{1}{x}\right) = \frac{x}{x^2 + y^2}$$

Substituting $x = 4$ and $y = -3$, we have

$$f_x(4, -3) = \frac{3}{25} \quad \text{and} \quad f_y(4, -3) = \frac{4}{25}$$ ∎

EXAMPLE 13.26 Find $\dfrac{\partial}{\partial x}(e^{-xy} \cos y)$ and $\dfrac{\partial}{\partial y}(e^{-xy} \cos y)$.

Solution

$$\frac{\partial}{\partial x}(e^{-xy} \cos y) = -ye^{-xy} \cos y$$

$$\frac{\partial}{\partial y}(e^{-xy} \cos y) = -xe^{-xy} \cos y + e^{-xy}(-\sin y) \quad \text{Product Rule}$$

$$= -e^{-xy}(x \cos y + \sin y)$$ ∎

EXAMPLE 13.27 If $z = xy/(x + y)$, show that

$$x \frac{\partial z}{\partial x} + y \frac{\partial z}{\partial y} = z$$

Solution

$$\frac{\partial z}{\partial x} = \frac{y(x+y) - xy}{(x+y)^2} = \frac{y^2}{(x+y)^2} \quad \text{Quotient rule}$$

$$\frac{\partial z}{\partial y} = \frac{x(x+y) - xy}{(x+y)^2} = \frac{x^2}{(x+y)^2}$$

$$x \frac{\partial z}{\partial x} + y \frac{\partial z}{\partial y} = \frac{xy^2}{(x+y)^2} + \frac{x^2 y}{(x+y)^2} = \frac{xy(y+x)}{(x+y)^2} = \frac{xy}{x+y} = z$$ ∎

Geometric Interpretation

If f is a function of two variables x and y, we have already seen that the partial derivatives f_x and f_y are really derivatives of functions of one variable. For example, we can write $f_x(x, y) = g'(x)$, where $g(x) = f(x, y)$ with y held fixed. So the geometric interpretation of a partial derivative is the same as in

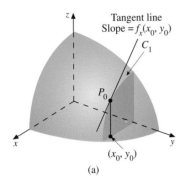

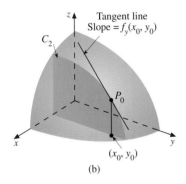

FIGURE 13.46

the single variable case, namely, the slope of the tangent line to a plane curve. To illustrate, let $P_0 = (x_0, y_0, z_0)$ be a point of the surface $z = f(x, y)$. Let C_1 denote the plane curve passing through P_0 that is the trace of the surface on the plane $y = y_0$. Also, let C_2 be the plane curve passing through P_0 that is the trace of the surface on the plane $x = x_0$. (See Figure 13.46.) Then $f_x(x_0, y_0)$ is the slope of the tangent line to C_1 at P_0, and $f_y(x_0, y_0)$ is the slope of the tangent line to C_2 at P_0, as we illustrate in Figure 13.46.

EXAMPLE 13.28 Find the slope of the tangent line to the curve of intersection of the surface $z = \sqrt{10 - 2x^2 - y^2}$ and the plane (a) $y = 1$ and (b) $x = 2$ at the point $(2, 1, 1)$. Show the results graphically.

Solution By the preceding discussion, the slope of the tangent line in part (a) is $\partial z/\partial x$ and in part (b) is $\partial z/\partial y$, each evaluated at $(2, 1)$:

$$\frac{\partial z}{\partial x} = \frac{-2x}{\sqrt{10 - 2x^2 - y^2}} \quad \text{and} \quad \frac{\partial z}{\partial y} = \frac{-y}{\sqrt{10 - 2x^2 - y^2}}$$

So for the curve of part (a),

$$\text{slope} = \left.\frac{\partial z}{\partial x}\right|_{(2,1)} = \frac{-4}{1} = -4$$

and for the curve of part (b),

$$\text{slope} = \left.\frac{\partial z}{\partial y}\right|_{(2,1)} = \frac{-1}{1} = -1$$

We illustrate the results in Figure 13.47. Notice that the given surface is the upper half of the ellipsoid c

$$\frac{x^2}{5} + \frac{y^2}{10} + \frac{z^2}{10} = 1$$

We show the first-octant portion only. ∎

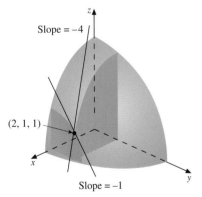

FIGURE 13.47

Three or More Variables

If f is a function of three or more variables, we define partial derivatives in a similar way to their definitions for the two-variable case. All variables except the one in question are held fixed, and the derivative is taken with respect to the one that is not held fixed (which is the single independent variable). We illustrate how to find the first-order partials for a function of three variables in the next two examples.

EXAMPLE 13.29 Let

$$f(x, y, z) = z^2 \sin(xy)$$

Find f_x, f_y, and f_z.

Solution

$$\begin{aligned} f_x(x, y, z) &= yz^2 \cos(xy) & &\text{y and z are held fixed.} \\ f_y(x, y, z) &= xz^2 \cos(xy) & &\text{x and z are held fixed.} \\ f_z(x, y, z) &= 2z \sin(xy) & &\text{x and y are held fixed.} \end{aligned}$$

∎

EXAMPLE 13.30 Under certain conditions, the temperature at a point (x, y) on a flat metal plate at time $t > 0$ is given by

$$T(x, y, t) = \frac{k}{t^{3/2}} e^{-(x^2+y^2)/ct}$$

where k and c are constants. Find $\partial T/\partial x$, $\partial T/\partial y$, and $\partial T/\partial t$.

Solution Holding y and t fixed, we get

$$\frac{\partial T}{\partial x} = \frac{k}{t^{3/2}} e^{-(x^2+y^2)/ct} \left[-\frac{2x}{ct} \right] = -\frac{2kx}{ct^{5/2}} e^{-(x^2+y^2)/ct}$$

Since the given function is symmetric in x and y, we can immediately write

$$\frac{\partial T}{\partial y} = -\frac{2ky}{ct^{5/2}} e^{-(x^2+y^2)/ct}$$

If we hold x and y fixed, we have a function of t alone, whose derivative can be found by the Product Rule:

$$\frac{\partial T}{\partial t} = -\frac{3k}{2t^{5/2}} e^{-(x^2+y^2)/ct} + \frac{k}{t^{3/2}} e^{-(x^2+y^2)/ct} \left[\frac{x^2+y^2}{ct^2} \right]$$

$$= e^{-(x^2+y^2)/ct} \left[-\frac{3k}{2t^{5/2}} + \frac{k(x^2+y^2)}{ct^{7/2}} \right]$$

$$= \frac{k}{2ct^{7/2}} e^{-(x^2+y^2)/ct} \left[2(x^2+y^2) - 3ct \right] \qquad \blacksquare$$

Higher-Order Partials

Since f_x and f_y are themselves functions of the two variables x and y, it is possible that they too have partial derivatives, called *second-order partial derivatives of f* (or just "second partials" for short). These are denoted as follows:

1. f_{xx}, second partial of f with respect to x
2. f_{xy}, second mixed partial of f, first with respect to x and then with respect to y
3. f_{yy}, second partial of f with respect to y
4. f_{yx}, second mixed partial of f, first with respect to y and then with respect to x

Note carefully the order of differentiation in the mixed partials. For example, f_{xy} means $(f_x)_y$, so we first differentiate with respect to x and then differentiate the result with respect to y.

Using the Leibniz notation, we write

$$\frac{\partial^2 f}{\partial x^2} \qquad \text{Same as } f_{xx}$$

$$\frac{\partial^2 f}{\partial y \, \partial x} \qquad \text{Same as } f_{xy}$$

$$\frac{\partial^2 f}{\partial y^2} \qquad \text{Same as } f_{yy}$$

$$\frac{\partial^2 f}{\partial x \, \partial y} \qquad \text{Same as } f_{yx}$$

These notations are suggested by applying the partial differential operators twice—for example, as in

$$\frac{\partial}{\partial x}\left(\frac{\partial f}{\partial x}\right) = \frac{\partial^2 f}{\partial x^2}$$

and

$$\frac{\partial}{\partial y}\left(\frac{\partial f}{\partial x}\right) = \frac{\partial^2 f}{\partial y\,\partial x}$$

Again, observe carefully the order of differentiation. Compare the following, for example:

$$\frac{\partial^2 f}{\partial y\,\partial x} = f_{xy}$$
$$\phantom{\frac{\partial^2 f}{\partial y\,\partial x}}\ \ \uparrow\ \uparrow\ \ \ \ \uparrow\ \nwarrow$$
$$\phantom{\frac{\partial^2 f}{\partial y\,\partial x}}\ \ \text{2nd 1st}\ \ \text{1st 2nd}$$

We could continue to higher-order partials, using notations such as

$$f_{xxy}\quad \text{or}\quad \frac{\partial^3 f}{\partial y\,\partial x^2}$$

You should verify that there are eight such third-order partials, and in general 2^n partials of nth order. In applications, partials of orders higher than 2 are seldom used.

EXAMPLE 13.31 Let $f(x, y) = 2x^3 y^2 - 3xy^4$. Find $f_{xx}(x, y)$, $f_{xy}(x, y)$, $f_{yy}(x, y)$, and $f_{yx}(x, y)$.

Solution

$f_x(x, y) = 6x^2 y^2 - 3y^4$ \qquad $f_y(x, y) = 4x^3 y - 12xy^3$
$f_{xx}(x, y) = 12xy^2$ \qquad\qquad $f_{yy}(x, y) = 4x^3 - 36xy^2$
$f_{xy}(x, y) = 12x^2 y - 12y^3$ \qquad $f_{yx}(x, y) = 12x^2 y - 12y^3$ ∎

Note that in this example $f_{xy} = f_{yx}$. The equality of these mixed partials is no accident. Although it is not always true, it is true for "well-behaved" functions. (See Exercise 55 in Exercise Set 13.4 for a function where $f_{xy} \neq f_{yx}$.) The following theorem, whose proof is given in Appendix 5, gives sufficient conditions for the equality of f_{xy} and f_{yx}.

THEOREM 13.1

Let f be a function of x and y. If f_{xy} and f_{yx} both exist at all points inside some circle centered at (x_0, y_0) and they are continuous at (x_0, y_0), then

$$f_{xy}(x_0, y_0) = f_{yx}(x_0, y_0)$$

One advantage of Theorem 13.1 is that differentiating in one order can sometimes be much easier than in the other. For example, consider

$$f(x, y) = x^3 y + \frac{e^y}{y^2 \sin y}$$

Since $f_x(x, y) = 3x^2y$, we have
$$f_{xy}(x, y) = 3x^2$$

If we were to first calculate $f_y(x, y)$ and then $f_{yx}(x, y)$, we would get the same result, but it would be much harder. (Try it.)

Partial Differential Equations

An equation such as
$$k\frac{\partial^2 u}{\partial x^2} = \frac{\partial u}{\partial t} \qquad (k > 0) \qquad (13.16)$$

is an example of a **partial differential equation** because it involves partial derivatives. Equation 13.16 is called the one-dimensional **heat equation** and occurs in the theory of heat flow in a rod or thin wire. The function $u(x, t)$ is the temperature in the rod at time t at a distance x from one end. The constant k is called the *thermal diffusivity* of the rod or wire. The same equation also arises in the study of the flow of electricity in a transmission line. For this reason it is sometimes called a *telegraph equation*.

Two other important partial differential equations of physics are the **wave equation**,
$$\frac{\partial^2 y}{\partial t^2} = a^2 \frac{\partial^2 y}{\partial x^2} \qquad (13.17)$$

and **Laplace's Equation**,
$$\frac{\partial^2 u}{\partial x^2} + \frac{\partial^2 u}{\partial y^2} + \frac{\partial^2 u}{\partial z^2} = 0 \qquad (13.18)$$

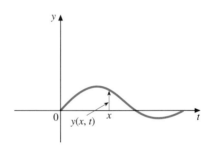

FIGURE 13.48
$y(x, t)$ is the vertical displacement at x at time t.

The wave equation relates the vertical displacements $y(x, t)$ in a vibrating string, where t represents time and x the horizontal distance from one end of the string (see Figure 13.48). Laplace's Equation holds true for the steady-state (time-independent) temperature $u(x, y, z)$ at points interior to a solid in which no heat is generated. Any function $u(x, y, z)$ that satisfies Laplace's Equation is said to be **harmonic**.

In the next three examples, we verify that certain functions are solutions to Equations 13.16, 13.17, and 13.18, respectively.

EXAMPLE 13.32 Show that the function
$$u(x, t) = e^{-kt} \cos x$$

satisfies the heat equation (Equation 13.16).

Solution First, we compute the partials with respect to x.
$$\frac{\partial u}{\partial x} = -e^{-kt} \sin x$$
$$\frac{\partial^2 u}{\partial x^2} = -e^{-kt} \cos x$$

Now we obtain the partial of u with respect to t.
$$\frac{\partial u}{\partial t} = -ke^{-kt} \cos x$$

Comparing $\partial^2 u/\partial x^2$ and $\partial u/\partial t$, we see that

$$k\frac{\partial^2 u}{\partial x^2} = \frac{\partial u}{\partial t}$$

So the given function does satisfy the heat equation. ∎

EXAMPLE 13.33 Show that if f is any twice-differentiable function of one variable, then

$$y(x,t) = \frac{1}{2}\left[f(x+at) + f(x-at)\right]$$

satisfies the wave equation (Equation 13.17).

Solution Let $u = x + at$ and $v = x - at$. If we hold t fixed, then we can apply the Chain Rule to $f(u)$, with $u = x + at$, and $f(v)$, with $v = x - at$, to get

$$\frac{\partial y}{\partial x} = \frac{1}{2}\left[f'(u)\cdot 1 + f'(v)\cdot 1\right] = \frac{1}{2}\left[f'(u) + f'(v)\right]$$

Similarly,

$$\frac{\partial^2 y}{\partial x^2} = \frac{1}{2}\left[f''(u) + f''(v)\right]$$

Now we hold x fixed and apply the Chain Rule to get

$$\frac{\partial y}{\partial t} = \frac{1}{2}\left[f'(u)\cdot a + f'(v)(-a)\right]$$

Applying the Chain Rule another time gives

$$\frac{\partial^2 y}{\partial t^2} = \frac{1}{2}\left[f''(u)a^2 + f''(v)(-a)^2\right]$$
$$= \frac{a^2}{2}\left[f''(u) + f''(v)\right]$$

Comparing this result with $\partial^2 y/\partial x^2$, we see that

$$\frac{\partial^2 y}{\partial t^2} = a^2\frac{\partial^2 y}{\partial x^2}$$ ∎

REMARK
■ The solution $y(x,t)$ to the wave equation given in Example 13.33 is known as *d'Alembert's Solution*, after the French mathematician, Jean Le Rond d'Alembert (1717–1783).

EXAMPLE 13.34 The function

$$\psi(x,y,z) = \frac{-GMm}{\sqrt{x^2 + y^2 + z^2}}$$

is called the gravitational potential function for the force exerted by a body of mass M on a body of mass m, where the vector **r** from the first mass to the second is $\mathbf{r} = x\mathbf{i} + y\mathbf{j} + z\mathbf{k}$. The constant G is the universal gravitational constant. Show that ψ is a harmonic function.

Solution Recall that a harmonic function is a function that satisfies Laplace's Equation (Equation 13.18). We first calculate $\partial \psi / \partial x$ and $\partial^2 \psi / \partial x^2$.

$$\frac{\partial \psi}{\partial x} = -GMm\left(-\frac{1}{2}\right)(x^2 + y^2 + z^2)^{-3/2}(2x) = \frac{GMmx}{(x^2 + y^2 + z^2)^{3/2}}$$

$$\frac{\partial^2 \psi}{\partial x^2} = GMm \frac{(x^2 + y^2 + z^2)^{3/2} - x \cdot \frac{3}{2}(x^2 + y^2 + z^2)^{1/2} \cdot 2x}{(x^2 + y^2 + z^2)^3} \quad \text{Quotient Rule}$$

$$= GMm \frac{(x^2 + y^2 + z^2) - 3x^2}{(x^2 + y^2 + z^2)^{5/2}} \quad \text{We divided by } (x^2 + y^2 + z^2)^{1/2}.$$

$$= GMm \frac{-2x^2 + y^2 + z^2}{(x^2 + y^2 + z^2)^{5/2}}$$

Since $\psi(x, y, z)$ is symmetric in x, y, and z, we can immediately conclude that

$$\frac{\partial^2 \psi}{\partial y^2} = GMm \frac{-2y^2 + x^2 + z^2}{(x^2 + y^2 + z^2)^{5/2}}$$

and

$$\frac{\partial^2 \psi}{\partial z^2} = GMm \frac{-2z^2 + x^2 + y^2}{(x^2 + y^2 + z^2)^{5/2}}$$

Thus,

$$\frac{\partial^2 \psi}{\partial x^2} + \frac{\partial^2 \psi}{\partial y^2} + \frac{\partial^2 \psi}{\partial z^2}$$

$$= GMm \frac{(-2x^2 + y^2 + z^2) + (-2y^2 + x^2 + z^2) + (-2z^2 + x^2 + y^2)}{(x^2 + y^2 + z^2)^{5/2}}$$

All the terms in the numerator cancel out, so we have

$$\frac{\partial^2 \psi}{\partial x^2} + \frac{\partial^2 \psi}{\partial y^2} + \frac{\partial^2 \psi}{\partial z^2} = 0$$

which shows that ψ is harmonic. ∎

Applications in Economics

The cost C of producing a given commodity typically is a function of unit cost of raw materials, say x, and the average hourly wage rate, say y, of the work force. (It may depend on other variables as well, but we will consider this simplified model.) Then the partial derivatives $\partial C/\partial x$ and $\partial C/\partial y$ are called **the marginal cost of raw materials** and **marginal cost of labor**, respectively. For given values of x and y, $\partial C/\partial x$ is approximately the added cost if the unit price of raw materials were increased by \$1. Similarly, $\partial C/\partial y$ is approximately the added cost if the hourly wage rate were increased by \$1. We illustrate these marginal functions in the next example.

EXAMPLE 13.35 Suppose that the cost function (in dollars) for producing some commodity is given by

$$C(x, y) = 180 + x^2 + 2y^2 - xy$$

where the fixed cost is $180, x is the cost per pound of raw materials, and y is the hourly wage rate. The current cost of materials is $8 per pound, and the current hourly wage rate is $12 per hour. Find each of the following:

(a) the approximate change in C if the cost of raw materials goes up to $9 per pound;

(b) the approximate change in C if the hourly wage rate goes up to $13 per hour.

Solution   We first calculate the partial derivatives C_x and C_y:

$$C_x(x, y) = 2x - y \quad \text{and} \quad C_y(x, y) = 4y - x$$

Thus, with $x = 8$ and $y = 12$, we have

$$C_x(8, 12) = 16 - 12 = 4 = \text{Marginal Cost of Raw Materials}$$

$$C_y(8, 12) = 48 - 8 = 40 = \text{Marginal Cost of Labor}$$

An increase of $1 in the cost per pound of raw materials would result in approximately a $4 increase in the cost of production, whereas an increase of $1 in the hourly wage rate would result in approximately a $40 increase in the cost of production. Clearly, in this case C is much more sensitive to a change in labor costs than to a change in the cost of raw materials. ∎

REMARK

■ Marginal revenue functions, marginal profit functions, marginal production functions, and the like are defined similarly to marginal cost functions.

The function

$$Q(K, L) = AK^\alpha L^{1-\alpha} \qquad (0 < \alpha < 1) \qquad (13.19)$$

is known as the **Cobb-Douglas Production Function**. Here Q represents the number of units of some commodity that are produced, A is a positive constant, K is the amount of capital investment, and L is the size of the labor force. This function is used widely in economics. The variables can also have different meanings. For example, $Q(K, L)$ might represent the demand function for a commodity, K the disposable income, and L the amount spent on advertising.

EXAMPLE 13.36   For the Cobb-Douglas Production Function given by Equation 13.19, evaluate the ratio Q_K/Q_L when $K = 250$ (in thousands of dollars), $L = 500$, and $\alpha = \frac{2}{3}$. Interpret the result.

Solution   With $\alpha = \frac{2}{3}$, Equation 13.19 becomes

$$Q(K, L) = AK^{2/3}L^{1/3}$$

So

$$Q_K(K, L) = \frac{2}{3}AK^{-1/3}L^{1/3}$$

and

$$Q_L(K, L) = \frac{1}{3}AK^{2/3}L^{-2/3}$$

The ratio of Q_K to Q_L is

$$\frac{Q_K(K,L)}{Q_L(K,L)} = \frac{\frac{2}{3}AK^{-1/3}L^{1/3}}{\frac{1}{3}AK^{2/3}L^{-2/3}} = \frac{2L}{K}$$

Thus, when $K = 250$ and $L = 500$,

$$\frac{Q_K(250, 500)}{Q_L(250, 500)} = \frac{2(500)}{250} = 4$$

The partial derivative Q_K is the *marginal productivity of capital*, and Q_L is the *marginal productivity of labor*. At $K = 250$ and $L = 500$, we have found that $Q_K = 4Q_L$. Thus, an increase of 1 (thousand) in capital, holding the labor force fixed, will result in about four times as much additional production as increasing the labor force by 1, keeping the capital fixed. ∎

LIMITS AND PARTIAL DERIVATIVES OF FUNCTIONS OF SEVERAL VARIABLES USING COMPUTER ALGEBRA SYSTEMS

Determining whether limits exist for functions of several variables is typically much more difficult than for functions of one variable. The reason lies in the fact that there are infinitely many paths of approach in the multivariable case. When showing that a limit does not exist, one typically tries to select different paths that yield different limits. This algebraic process is perfectly suited for computer algebra systems. In this section we investigate limits and partial derivatives for functions of several variables.

CAS 52

Investigate the limits of each of the following:

(a) $\lim\limits_{(x,y)\to(0,0)} \dfrac{x^2 - y^2}{x^2 + y^2}$ (b) $\lim\limits_{(x,y)\to(0,0)} \dfrac{x^2 y}{x^4 + y^2}$

Maple:

(a)

First, define f as a function of the two variables x and y:

f:=(x,y)–>(x^2–y^2)/(x^2+y^2);

Now compute the limit:

limit(f(x,y), {x=0,y=0});

Output: *undefined*

Mathematica:

Our version of Mathematica does not have a limit command for functions of several variables.

(a)
f[x_,y_] = (x^2–y^2)/(x^2+y^2)
Plot 3D[f[x,y],{x,–1,1},{y,–1,1}]
t=m*x
g=f[x,t]
Limit[g,{x–>0}]

Maple:

A plot indicates that as the origin is approached along different lines, the value of the function approaches different values.

plot3d(f(x,y),x=−1..1,y=−1..1,orientation=[45, 45],axes=boxed);

Mathematica:

(b)
f[x_,y_] = (x^2*y)/(x^4+y^2)
t = m*x
g = f[x,t]
Limit[g,{x->0}]
t = x^2
g = f[x,t]
Limit[g,{x->0}]

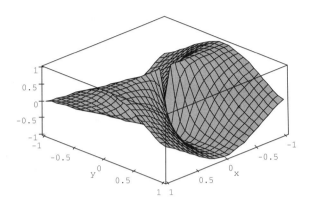

FIGURE 13.4.1

To verify this, we let (x, y) approach the origin along straight lines of varying slopes, all of which pass through the origin.

t:=m*x;
g:=f(x,t);

Output: $g := \dfrac{x^2 - m^2 x^2}{x^2 + m^2 x^2}$

limit(g, {x=0});

Output: $-\dfrac{-1 + m^2}{1 + m^2}$

(b)
f:=(x,y)−>x^2*y/(x^4+y^2);
limit(f(x,y),{x=0,y=0});

Output: $\text{limit}\left(\dfrac{x^2 y}{x^4 + y^2}, \{x = 0, y = 0\}\right)$

A limit is not computed in this case, so we proceed as in part (a).

t:=m*x;
g:=f(x,t);

Output: $g := \dfrac{x^3 m}{x^4 + m^2 x^2}$

limit(g,{x=0});

Output: 0

So if (x, y) approaches along lines through the origin, $f(x, y)$ does approach 0. But this does not say the limit exists. This time we need to consider other paths.

t:=x^2;
g:=f(x,t);

Maple:

Output: $g := \dfrac{1}{2}$

limit(g,{x=0});

Output: $\dfrac{1}{2}$

In this case, as (x, y) approaches $(0, 0)$ along the parabolic path $y = x^2$, the limit is 1/2, which does not agree with the limit along straight line paths through the origin, and again the limit does not exist.

CAS 53

Find the slope of the curve of intersection of the upper nappe of the cone $\dfrac{z^2}{4} = \dfrac{x^2}{4} + \dfrac{y^2}{9}$ and the plane $x = -2$ at the point $\left(-2, 1, \dfrac{\sqrt{40}}{3}\right)$.

Maple:

We first make a plot of the cone and the intersecting plane.

with(plots):
cone :=solve((x/2)^2+(y/3)^2=(z/2)^2,z);

Output: $cone := -\dfrac{1}{3}\sqrt{9x^2 + 4y^2}, \dfrac{1}{3}\sqrt{9x^2 + 4y^2}$

The upper nappe is given by the second expression with the positive square root.

pl:=plot3d({cone[2]},x=−6..6,y=−8..8,view=−5..5,
    scaling=constrained,axes=boxed,tickmarks=[3,3,3]);
p2:=plot3d({[−2,s,t]},s=−10..10,t=−10..10);
display3d({pl,p2},orientation=[20,77]);

Mathematica:

cone = Solve[(x/2)^2+(y/3)^2==(z/2)^2,z]
P1 = Plot3D[Sqrt[9*x^2+4*y^2]/3,{x,−6,6},{y,−8,8},
    PlotRange−>{−5,5},DisplayFunction−>Identity]
P2 = ParametricPlot3D[{−2,s,t},{t,−10,10},{s,−10,10},
    DisplayFunction−>Identity]
Show[{P1,P2},DisplayFunction−>$DisplayFunction]
f[x_,y_] = Sqrt[9*x^2+4*y^2]/3
fp[x_,y_] = D[f[x,y],y]
fp[−2,1]

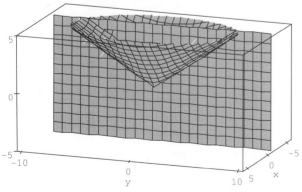

FIGURE 13.4.2

Maple:

Now define a function of two variables f as the upper nappe of the cone.

f:=(x,y)->cone[2];

Find the partial of f with respect to y and substitute the x- and y-coordinates of the point $\left(-2, 1, \dfrac{\sqrt{40}}{3}\right)$ to find the slope of the curve.

m:=diff(f(x,y),y);

Output: $m := \dfrac{4}{3} \dfrac{y}{\sqrt{9x^2 + 4y^2}}$

subs(x=-2,y=1,m);

Output: $\dfrac{1}{30}\sqrt{40}$

Exercise Set 13.4

In Exercises 1–10, find $f_x(x, y)$ and $f_y(x, y)$.

1. $f(x, y) = x^2 + y^2$
2. $f(x, y) = \sqrt{x^2 + y^2}$
3. $f(x, y) = \sin xy$
4. $f(x, y) = e^x \cos y$
5. $f(x, y) = \dfrac{x + y}{x - y}$
6. $f(x, y) = \ln \dfrac{x}{y}$
7. $f(x, y) = x \ln y - y \ln x$
8. $f(x, y) = \tan^{-1} \dfrac{x}{y}$
9. $f(x, y) = \dfrac{xy}{\sqrt{x^2 - y^2}}$
10. $f(x, y) = e^{x-y} \sin(x - y)$

In Exercises 11–16, find $\partial z/\partial x$ and $\partial z/\partial y$.

11. $z = (1 - x^2 - y^2)^{-1/2}$
12. $z = \sin x \cosh y$
13. $z = \ln \cos(x - y)$
14. $z = \ln \left(\dfrac{x^2 - 2xy}{3xy - y^2}\right)$
15. $z = \sin^{-1} \sqrt{1 - x^2 y^2}$, $x > 0$, $y > 0$, $xy < 1$
16. $z = \sqrt{\dfrac{x - 2y}{x + 2y}}$

17. Find $f_x(3, -2)$ and $f_y(3, -2)$ if $f(x, y) = x^2 y - 2y^2$.

18. If $g(x, y) = 2xy/(x - y)$, find $g_x(-1, 1)$ and $g_y(-1, 1)$.

19. If $f(r, \theta) = e^r \cos \theta$, find $f_r\left(0, \dfrac{\pi}{3}\right)$ and $f_\theta\left(0, \dfrac{\pi}{3}\right)$.

20. If $w = \dfrac{u}{v} - \dfrac{v}{u}$, find $\partial w/\partial u$ and $\partial w/\partial v$ when $u = -2$ and $v = 2$.

21. If $w = 1/(2s - t^2)^2$, find $\partial w/\partial s$ and $\partial w/\partial t$ when $s = 3$ and $t = -2$.

22. Let $w = (u + v)/(u - v)$. Show that
$$v\dfrac{\partial w}{\partial u} + u\dfrac{\partial w}{\partial v} = 2w$$

23. Let $z = \tan^{-1} \dfrac{y}{x}$. Show that $x(\partial z/\partial y) - y(\partial z/\partial x) = 1$.

24. Find the slope of the curve of intersection of the cone $z = \sqrt{4x^2 + 3y^2}$ and
    (a) the plane $y = 4$
    (b) the plane $x = -2$ at the point $(-2, 4, 8)$.

25. Let $f(x, y) = 2x^2 - 3xy^2 + 3y^3$, and let C_1 and C_2 be the curves of intersection of the graph of f and the planes $y = 2$ and $x = 3$, respectively. Show that both C_1 and C_2 have horizontal tangent lines at the point $(3, 2, 6)$.

In Exercises 26–31, find all second-order partial derivatives of f.

26. $f(x, y) = \sqrt{x - 2y}$

27. $f(x, y) = \ln(x^2 + 3y^2)$

28. $f(x, y) = \sin xy$

29. $f(x, y) = e^{x^2 y}$

30. $f(x, y) = \dfrac{x - y}{x + y}$

31. $f(x, y) = e^{-x} \sin y + e^{-y} \cos x$

32. Let $f(x, y, z) = \dfrac{x - y}{y - z}$. Find f_x, f_y, and f_z.

In Exercises 33 and 34, verify that the given function satisfies the heat equation $ku_{xx} = u_t$.

33. $u(x, t) = 2(\cos 3x)e^{-9kt}$

34. $u(x, t) = (A \cos nx + B \sin nx)e^{-n^2 kt}$, where A, B, and n are constants.

In Exercises 35 and 36, verify that the given function satisfies the wave equation $y_{tt} = a^2 y_{xx}$.

35. $y(x, t) = \sin x \cos at$

36. $y(x, t) = e^x(A \cosh at + B \sinh at)$, where A and B are constants

*A function $u(x, y)$ is called **harmonic** if it satisfies Laplace's Equation in two variables, $u_{xx} + u_{yy} = 0$. In Exercises 37–40, show that u is harmonic at all points where u_{xx} and u_{yy} are defined.*

37. $u(x, y) = \cos x \cosh y$

38. $u(x, y) = \ln \sqrt{x^2 + y^2}$

39. $u(x, y) = \tan^{-1} \dfrac{y}{x}$

40. $u(x, y) = \dfrac{x}{x^2 + y^2}$

*If $u(x, y)$ is harmonic and $v(x, y)$ is a harmonic function such that $u_x = v_y$ and $u_y = -v_x$, then v is said to be a **harmonic conjugate** of u. In Exercises 41–44, show that v is a harmonic conjugate of u.*

41. $v(x, y) = -\sin x \sinh y$; $u(x, y)$ in Exercise 37

42. $v(x, y) = \tan^{-1} \dfrac{y}{x}$; $u(x, y)$ in Exercise 38

43. $v(x, y) = -\ln \sqrt{x^2 + y^2}$; $u(x, y)$ in Exercise 39

44. $v(x, y) = -\dfrac{y}{x^2 + y^2}$; $u(x, y)$ in Exercise 40

45. Let $u(x, y) = x^2 - y^2 + 2x + 1$ and $v(x, y) = 2xy + 2y$.
(a) Show that u and v are harmonic.
(b) Show that v is a harmonic conjugate of u. (See the instructions for Exercises 41–44.)
(c) Show that the level curves of u and the level curves of v intersect at right angles, except for $u(x, y) = 0$ and $v(x, y) = 0$.
(d) Draw several level curves for u and v on the same graph.

46. Verify that the function
$$f(x, y, z) = x^4 + y^4 + z^4 - 3(x^2 y^2 + x^2 z^2 + y^2 z^2)$$
satisfies Laplace's Equation in three variables.

47. Let $f(x, y, z) = xy \ln z + xz \ln y + yz \ln x$. Show the following:
(a) $f_{xxy} = f_{xyx} = f_{yxx}$
(b) $f_{xyy} = f_{yxy} = f_{yyx}$
(c) $f_{xzz} = f_{zxz} = f_{zzx}$
(d) The equality of all third-order partials of f with respect to x, y, and z, taken in any order

48. Suppose the cost function $C(x, y)$ for producing a certain commodity is given by
$$C(x, y) = 72 + x^2 + 2y^2 - xy + 3x + 4y$$
where x is the unit price of raw materials and y is the unit cost of labor. Find
(a) the marginal cost of raw materials;
(b) the marginal cost of labor.

49. Suppose the profit function $P(x, y, z)$ from the sale of a certain product is
$$P(x, y, z) = 30x + 50y + 80z - x^2 - 2yz - 3z^2$$
where x is the unit cost of raw materials, y is the unit cost of labor, and z is the unit cost of shipping and handling. If the current values of x, y, and z are $x = \$5$ per pound, $y = \$9$ per hour, and $z = \$10$ per item, find the approximate change in profit if
(a) x is increased by $\$1$;
(b) y is increased by $\$1$;
(c) z is increased by $\$1$.

50. Suppose the production function $Q(K, L)$ where K represents available capital and L the size of the labor force, is given by
$$Q(K, L) = 325 K^{0.72} L^{0.28}$$
Find the marginal productivity of capital and of labor.

51. In the Cobb-Douglas Production Function $Q(K, L) = A K^\alpha L^{1-\alpha}$, show that
$$K^2 Q_{KK} = L^2 Q_{LL}$$

52. Let $f(x, y) = (x^2 + y^2)^{1/3}$. Show that f is continuous at $(0, 0)$ but $f_x(0, 0)$ and $f_y(0, 0)$ do not exist.

53. Let
$$f(x, y) = \begin{cases} \dfrac{x^3 - y^3}{x^2 + y^2} & \text{if } (x, y) \neq (0, 0) \\ 0 & \text{if } (x, y) = (0, 0) \end{cases}$$

Use Definition 13.7 to show that $f_x(0, 0)$ and $f_y(0, 0)$ both exist. What are their values?

54. A Discontinuous Function Whose Partial Derivatives Exist

Let
$$f(x, y) = \begin{cases} \dfrac{xy}{x^2 + y^2} & \text{if } (x, y) \neq (0, 0) \\ 0 & \text{if } (x, y) = (0, 0) \end{cases}$$

Prove that $f_x(0, 0)$ and $f_y(0, 0)$ both exist but that f is discontinuous at $(0, 0)$.

55. A Function for Which $f_{xy} \neq f_{yx}$

Let
$$f(x, y) = \begin{cases} \dfrac{xy(x^2 - y^2)}{x^2 + y^2} & (x, y) \neq (0, 0) \\ 0 & (x, y) = (0, 0) \end{cases}$$

Show that $f_{xy}(0, 0) = -1$ but $f_{yx}(0, 0) = 1$.

56. Find the slope of the tangent line at $(1, 2, 1/e^3)$ to the curve of intersection of the surface $f(x, y) = e^{-x^2 - y^2}$ and the planes
(a) $x = 1$ and (b) $y = 2$
In each case, make a plot of the surface and the intersecting plane.

57. Find the slope of the tangent line at $(0, 1, 1)$ to the curve of intersection of the surface $f(x, y) = x^2 + 2xy + y^2$ and the planes
(a) $x = 0$ and (b) $y = 1$
In each case, make a plot of the surface and the intersecting plane.

58. Suppose $u(x, y) = f(x)g(y)$. Show that u satisfies the partial differential equation $uu_{xy} - u_x u_y = 0$.

Chapter 13 Review Exercises

1. Give the domain of f, and show it graphically.
  (a) $f(x, y) = \ln(x^2 + y^2 - 1)$
  (b) $f(x, y) = xy/\sqrt{x^2 - y^2}$

In Exercises 2–5, identify the surface and draw its graph.

2. (a) $4x^2 + y^2 + 4z = 8$
  (b) $y = \sqrt{x^2 + y^2}$

3. (a) $y^2 - 2x - 4y + 4 = 0$
  (b) $x^2 + y^2 + z^2 - 6x - 4y - 2z + 10 = 0$

4. (a) $x^2 + 4z^2 = 4$
  (b) $9(x^2 + z^2) = 4(y^2 + 9)$

5. (a) $x^2 - 4y^2 + 4z = 0$
  (b) $z = 1 + \tan^{-1} y$

6. Show the region in the first octant bounded by the given surfaces.
  (a) $4x^2 + 3y^2 + 6z^2 = 48$, $x = 3$, $y = 2$, $z = 0$
  (b) $4x^2 + y^2 = 4$, $z = 4 - y^2$, and the coordinate planes

7. Give the domain and range of f, identify its graph, and draw it.
  (a) $f(x, y) = \sqrt{4 + x^2 + 4y^2}$
  (b) $f(x, y) = 4 - x^2 - y^2$

8. Draw several level curves in part (a) and level surfaces in part (b) for both positive and negative values of the constant c.
  (a) $f(x, y) = y^2 - x^2$
  (b) $f(x, y, z) = \dfrac{x^2 + y^2}{z}$

9. A coordinate system is set up on a round metal plate with the origin at its center. A point source of heat is at the origin, and the temperature $T(x, y)$ in degrees Celsius at any point on the plate is given by
$$T(x, y) = 100 e^{-(x^2 + y^2)/2}$$
Draw the isotherms $T(x, y) = C$ for $C = 80, 60, 40$, and 20. How rapidly is the temperature changing in the direction of the positive y-axis at the point $(1, 3)$? Is it increasing or decreasing?

In Exercises 10 and 11, find f_x and f_y.

10. (a) $f(x, y) = e^{\sin xy}$
  (b) $f(x, y) = \ln \dfrac{x^2 + 2y^2}{\sqrt{3x + 4y}}$

11. (a) $f(x, y) = \dfrac{y}{x} \cosh^2 \dfrac{x}{y}$

  (b) $f(x, y) = \sin^{-1}\left(\dfrac{\sqrt{y^2 - 2x}}{y}\right)$, $y > 0$

12. Let $w = (2u - v)/(u + 2v)$. Show that
$$u\dfrac{\partial w}{\partial u} + v\dfrac{\partial w}{\partial v} = 0$$

13. Show that $u(x, y) = \sin x \cosh y$ and $v(x, y) = \cos x \sinh y$ are harmonic conjugates. That is, show that each function satisfies Laplace's Equation in two variables and that $u_x = v_y$ and $u_y = -v_x$.

14. Let $f(x, y) = xy \ln \dfrac{x}{y}$. What is the domain of f? Find all second-order partials and show that $f_{xy} = f_{yx}$.

15. Let $w = \sqrt{z^2 - x^2 - y^2}$. Find each of the following:

  (a) $\dfrac{\partial^2 w}{\partial z^2}$
  (b) $\dfrac{\partial^2 w}{\partial x\, \partial z}$
  (c) $\dfrac{\partial^2 w}{\partial y\, \partial z}$
  (d) $\dfrac{\partial^2 w}{\partial y\, \partial x}$

16. Show that
$$\lim_{(x,y)\to(0,0)} \dfrac{\sin xy}{\sqrt{x^2 + y^2}} = 0$$

  (Hint: $|\sin xy| \le |xy| \le x^2 + y^2$.)

17. Let $f(x, y) = (x^3 - 3x^2 y)/(x^2 + 2y^2)$ if $(x, y) \ne (0, 0)$. What value should be assigned to $f(0, 0)$ to make f continuous at the origin? Prove your result.

18. For $(x, y) \ne (0, 0)$ let
$$f(x, y) = (x^2 + y^2)\left(\sin\dfrac{1}{x^2 + y^2} + \cos\dfrac{1}{x^2 + y^2}\right)$$
and let $f(0, 0) = 0$. Use Definition 13.7 to show that $f_x(0, 0)$ and $f_y(0, 0)$ both exist and equal 0.

19. Prove that $\lim_{(x,y)\to(0,0)} f(x, y)$ does not exist for the following functions.

  (a) $f(x, y) = \dfrac{x^3 - 2y^3}{3x^3 + 4y^3}$

  (b) $f(x, y) = \dfrac{3xy^2}{2x^2 + 5y^4}$

20. When three resistors, with resistances of R_1 ohm, R_2 ohm, and R_3 ohm, respectively, are connected in parallel, their combined resistance R satisfies the equation
$$\dfrac{1}{R} = \dfrac{1}{R_1} + \dfrac{1}{R_2} + \dfrac{1}{R_3}$$

Find $\partial R/\partial R_1$. What can you conclude about $\partial R/\partial R_2$ and $\partial R/\partial R_3$?

21. The Ideal Gas Law relates the pressure P, the volume V, and the temperature T of a gas by the equation $PV = nRT$, where n and R are constant. Show that
$$\dfrac{\partial P}{\partial V}\dfrac{\partial V}{\partial T}\dfrac{\partial T}{\partial P} = -1$$

22. Show that each of the following functions satisfies the heat equation $ku_{xx} = u_t$.

  (a) $u(x, t) = e^{-n^2 \pi^2 kt} \sin n\pi x$
  (b) $u(x, t) = \sin x(\cosh kt - \sinh kt)$

23. Show that each of the following functions satisfies the wave equation $y_{tt} = a^2 y_{xx}$.

  (a) $y(x, t) = \ln(x^2 - a^2 t^2)$
  (b) $y(x, t) = Ae^{x+at} + Be^{x-at}$

24. Show that each of the following functions satisfies Laplace's Equation in three variables.

  (a) $u(x, y, z) = \sin 3x \sin 4y \sinh 5z$
  (b) $u(x, y, z) = \cos 5x \cos 12y \cosh 13z$

25. For the Cobb-Douglas Production Function $Q(K, L) = 30K^{0.6}L^{0.4}$, find and interpret $Q_K(240, 150)$ and $Q_L(240, 150)$.

26. Suppose the revenue $R(x, y)$ of a company as a function of advertising expenditures in printed form and on TV is given by
$$R(x, y) = 20x^{3/2} + 48y^{1/2} + 2xy$$
where x is the amount spent on TV and y is the amount spent on print advertising (both in hundreds of thousands of dollars). If the current expenditures are $x = 25$ and $y = 16$, use marginal analysis to approximate the additional revenue that would result from

  (a) changing x to 26, leaving y alone;
  (b) changing y to 17, leaving x alone.

27. Let
$$f(x, y) = \begin{cases} \dfrac{2x^3 - y^3}{x^2 + 2y^2} & \text{if } (x, y) \ne (0, 0) \\ 0 & \text{if } (x, y) = (0, 0) \end{cases}$$

  (a) Use Definition 13.7 to find $f_x(0, 0)$.
  (b) Find $f_x(x, y)$ for $(x, y) \ne (0, 0)$ using the Quotient Rule.
  (c) Show that f_x is not continuous at the origin.
  (d) Carry out steps (a), (b), and (c) for f_y.

28. Let

$$f(x, y) = \begin{cases} \dfrac{x^3 y - xy^3}{x^2 + y^2} & \text{if } (x, y) \neq (0, 0) \\ 0 & \text{if } (x, y) = (0, 0) \end{cases}$$

(a) Prove that f is continuous at $(0, 0)$.
(b) Find $f_x(0, y)$ and $f_y(x, 0)$ using Definition 13.7.
(c) From part (b), find $f_{xy}(0, 0)$ and $f_{yx}(0, 0)$.
(d) What conclusion about the continuity of f_{xy} and f_{yx} can you draw from part (c)?

29. A set is said to be **open** if it does not contain any of its boundary points. Prove that a nonempty set S in $\mathbb{R}^2$ or $\mathbb{R}^3$ is open if and only if, for every point P in S, all points inside some circle (in $\mathbb{R}^2$) or sphere (in $\mathbb{R}^3$) centered at P lie in S.

30. If S is a set, its **complement** is the set of all points not in S. A set is said to be **closed** if it contains all of its boundary points. Prove that a set S in $\mathbb{R}^2$ or $\mathbb{R}^3$ is closed if and only if its complement is open.

Chapter 13 Concept Quiz

1. Define each of the following:
   (a) The graph of a function f of two variables
   (b) A level curve of f
   (c) The partial derivatives $f_x(x, y)$ and $f_y(x, y)$
   (d) $\lim_{(x, y) \to (x_0, y_0)} f(x, y) = L$
   (e) Continuity of a function f of two variables at (x_0, y_0)

2. Identify each of the following surfaces, where a, b, c, and d are positive constants.
   (a) $x = ax^2 + by^2$
   (b) $z = ax^2$
   (c) $x^2 + y^2 + z^2 = a^2$
   (d) $ax^2 - by^2 + cz^2 = d$
   (e) $ax^2 + by^2 + cz^2 = d$
   (f) $ax^2 - by^2 - cz^2 = d$
   (g) $ax^2 - by^2 - cz^2 = 0$
   (h) $z = ax^2 - by^2$

3. (a) State a sufficient condition for $f_{xy}(x_0, y_0)$ to equal $f_{yx}(x_0, y_0)$.
   (b) Explain a way of showing that $\lim_{(x, y) \to (x_0, y_0)} f(x, y)$ does not exist.

4. (a) Explain the geometric significance of the partial derivative $f_x(x_0, y_0)$ and $f_y(x_0, y_0)$.
   (b) Explain the geometric nature of a contour map of a function of three variables.

5. Determine which of the following statements are true and which are false.
   (a) If the limit of $f(x, y)$ is L as (x, y) approaches (x_0, y_0) along every straight line through (x_0, y_0), then
   $$\lim_{(x, y) \to (x_0, y_0)} f(x, y) = L$$
   (b) If $f_x(x_0, y_0)$ and $f_y(x_0, y_0)$ both exist, then f is continuous at (x_0, y_0).
   (c) If f and g are both continuous at (x_0, y_0), then so is f/g.
   (d) A three-dimensional surface is the graph of a function $z = f(x, y)$ if and only if every line parallel to the z-axis intersects the surface in at most one point.
   (e) The trace of the surface $z = f(x, y)$ in the plane $y = y_0$ has slope $f_x(x_0, y_0)$.

APPLYING CALCULUS

1. To estimate the value of a tree for lumber, the probable number of board feet in various sections of the tree are calculated. (A board 1 ft long with a cross-section of 12 in.² is one board-foot.) The following formula, called the *Doyle Log Rule*, is used for this calculation:

$$N(d, L) = \left(\frac{d-4}{4}\right)^2 L$$

where d is the diameter of the log in inches, L is its length in feet, and $N(d, L)$ is the number of board feet of lumber that can be obtained from the log.

(a) Use a CAS to obtain the graph of $N(d, L)$ for $5 \le d \le 50$ and $1 \le L \le 40$.
(b) The lower part of a tree, before the first branch, is the most valuable, since it will have fewer knots and its diameter is greatest. Suppose that the first 17 ft of a tree has minimum diameter 24 in. and is valued at $1 per board foot, the next 17 ft has minimum diameter 18 in. and is valued at 30¢ per board foot, and the next 25 ft has minimum diameter 12 in. and is valued at 20¢ per board foot. Determine how many board feet there are in each section. What is the estimated value of the tree?

2. The larvae of certain insects have approximately the shape of a circular cylinder capped at each end by a hemisphere (see the figure). Let the length of the cylinder be h and its diameter d.
(a) Express the volume V and the surface area S as functions of h and d.
(b) The rate R of absorption of a chemical substance into the larvae is directly proportional to S and inversely proportional to V. Show that $\dfrac{\partial R}{\partial h} < 0$ and $\dfrac{\partial R}{\partial d} < 0$.

3. The temperature T in degrees Celsius at the point (x, y) on a flat plate in the form of the disk $\{(x, y): x^2 + y^2 \le 100\}$ is given by

$$T(x, y) = 100e^{-\sqrt{x^2+y^2}}$$

Suppose a bug is on the plate at the point $(4, -3)$.
(a) If the bug moves upward on the line $x = 4$, does the temperature increase or decrease? What is the instantaneous rate of change?
(b) If, instead, the bug moves to the left along the line $y = -3$, answer the questions in part (a).
(c) If the bug wants to travel in a path where the temperature doesn't change, what path should it take?
(d) If the bug heads on a straight path toward the origin, find the instantaneous rate of change of temperature in that direction. (*Hint:* Find a unit vector pointing toward the origin and use it to find a vector equation of the line. Then write T in terms of the parameter.)

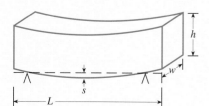

4. A rectangular beam of height h, width w, and length L is supported at each end and is subjected to a uniform load p. The deflection of the beam at its midpoint is called the sag, s, and is given by

$$s = C \frac{pL^3}{wh^3}$$

where C is a constant depending on the type of material and the units of measurement.
(a) Find $\partial s/\partial L$, $\partial s/\partial w$, and $\partial s/\partial h$. How would you interpret the results?

(b) Verify each of the following relationships:

$$L\frac{\partial s}{\partial L} + h\frac{\partial s}{\partial h} = 0, \qquad L\frac{\partial s}{\partial L} + 3w\frac{\partial s}{\partial w} = 0, \qquad h\frac{\partial s}{\partial h} = 3w\frac{\partial s}{\partial w}$$

(c) Suppose the length and the load are fixed. If one of the dimensions w or h can be increased by one unit, which one would you choose in order to reduce the sag the most? Justify your answer.

5. In the wave equation (Equation 13.17), giving the vertical displacement in a stretched string at time t at a point x units from the origin, it can be shown that the constant a^2 is F_0/μ, where F_0 is the tension in the string and μ is its mass per unit length. Show that

$$y(x, t) = y_0 \cos(kx - \omega t)$$

satisfies the equation provided ω and k are related to μ and F_0 by the equation

$$\frac{\omega}{k} = \sqrt{\frac{F_0}{\mu}}$$

(*Note:* The constants w and k are called *angular frequency* and *wave number*, respectively. The ratio ω/k is the wave speed.)

6. The equation

$$-\frac{h^2}{8\pi^2 m}\left(\frac{\partial^2 \psi}{\partial x^2} + \frac{\partial^2 \psi}{\partial y^2} + \frac{\partial^2 \psi}{\partial z^2}\right) + V\psi = E\psi$$

is called the three-dimensional, time-independent Schrödinger equation and is the basis for quantum mechanics. It was discovered by Erwin Schrödinger in 1926. The equation describes the motion of a particle of mass m moving in space with potential energy V and kinetic energy E. The function ψ has the property that ψ^2 is the probability density function for the position of the particle. The constant h is known as *Planck's constant* and is equal to 6.63×10^{-34} J · s. If the particle is confined to the cube $\{(x, y, z): 0 \leq x \leq L, 0 \leq y \leq L, 0 \leq z \leq L\}$ and the potential energy is zero, show that a solution to Schrödinger's equation is

$$\psi(x, y, z) = A \sin\left(\frac{n_x \pi x}{L}\right) \sin\left(\frac{n_y \pi y}{L}\right) \sin\left(\frac{n_z \pi z}{L}\right)$$

where the *quantum numbers* $n_x, n_y,$ and n_z satisfy

$$E = \frac{h^2}{8mL^2}\left(n_x^2 + n_y^2 + n_z^2\right)$$

7. In an economy in competitive equilibrium, the real wage = the marginal product of labor and the real interest rate = the marginal product of capital. If Y is the aggregate production function, K is the capital factor of production, and L is the labor factor of production, then $Y = Y(K, L)$. (Aggregate means that all kinds of goods are combined as a single "good.")
  (a) Since the real wage is the marginal product of labor, write the formula for the real wage W in terms of the production function Y.
  (b) For the Cobb-Douglas Production Function

$$Y = K^{1-\alpha}L^\alpha, \quad 0 < \alpha < 1,$$

find W in terms of α, K, and L.
  (c) Show that the labor share of the national product, WL/Y, is constant. (*Note:* This constant labor share is now about 3/4.)
  (d) Show that the same calculation for the capital share of the national market gives a value that, when added to the constant in part (c), gives 1.

8. The electric field in a region can be expressed as the plane wave

$$\mathbf{E}(x, t) = E_0 \sin(kx - \omega t)\mathbf{j}$$

and the magnetic field as the plane wave

$$\mathbf{B}(x,t) = B_0 \sin(kx - \omega t)\mathbf{k}$$

where E_0 and B_0 are constants.

How the magnitude E of the electric field changes with position is related to the way the magnitude B of the magnetic field changes with time by **Faraday's Law**, named after the nineteenth-century British physicist Michael Faraday:

$$\frac{\partial E}{\partial x} = -\frac{\partial B}{\partial t}$$

The electric and magnetic fields are also related by **Ampère's Law**, named after the nineteenth-century French physicist André Marie Ampère:

$$\frac{\partial B}{\partial x} = -\mu_0 \varepsilon_0 \frac{\partial E}{\partial t}$$

where ε_0 is an electric constant, the *permittivity constant*, and μ_0 is a magnetic constant, the *permeability*.

(a) The speed of a wave is ω/k. Express the speed of electromagnetic waves—waves of simultaneously changing electric and magnetic fields—in terms of ε_0 and μ_0.

(b) Given that $\varepsilon_0 = 8.85 \times 10^{-12}$ F/m and $\mu_0 = 4\pi \times 10^{-7}$ H/m, where F/m is farads/meter and H/m is henries/meter, solve for the wave speed in part (a) numerically. You may use the fact that $1 \text{ F} \cdot \text{H} = 1 \text{ s}^2$.

(c) Relate the result of part (c) to the observed value of the speed of light, $c = 3.00 \times 10^8$ m/s. James Clerk Maxwell, in the 1860s, used this result to show that light is an electromagnetic wave.

(d) The angular velocity is $2\pi f$, where f is the frequency, and the wave constant is $k = 2\pi/\lambda$, where λ is the wave length. How are f, λ, and c related?

A PERSONAL VIEW

David Lieberman

Is calculus still the best way to start studying mathematics in college?

I think calculus is fundamental to understanding all of the theoretical and computational mathematical developments that go on today. A thorough knowledge of calculus is required to understand algorithms that computers are implementing and to know how one might use those algorithms to adapt to particular problems one faces. Also, to check whether computer results are reasonable, the techniques of calculus are valuable for approximating the answers that one expects. Just on the side, I think that more and more of the calculations—even symbolic calculations—are going to be done by computer, and the computer will replace the purely calculational aspect of calculus; but I think to understand how to use the computer, it is still going to require a deep appreciation and understanding of the methodology of the calculus.

Just what is it that you do?

I work on the mathematical foundations of cryptology and related subjects, such as processing speech by computer. I also do statistical analyses, trying to find statistical structure in data.

What kind of mathematics is involved with cryptology and speech processing?

Many modern cryptographic schemes attempt to use intractable mathematical problems as the basis for their security. One well-known scheme builds its security on the intractability of the problem of factoring large integers into prime factors. (Currently, to find the primes dividing a 500-digit number would require centuries running on the world's fastest computers.) Consequently, many areas of pure mathematics, particularly number theory and algebraic geometry, are playing an increasing role in the design and analysis of communications systems.

Mathematics also continues to play a central role in statistical analysis of data sets. A standard problem is that one has a set of data and needs to identify which of a large number of possible hypotheses best explain the properties of the data.

A simple case of this kind of problem is that in which the data are a collection of numbers, normally distributed, and the different hypotheses are distinguished by the expected values of the mean and standard deviation of the observed data set. Such problems are analyzed at length in any elementary statistics book. This analysis already requires interesting techniques from calculus.

A more complex problem concerns multivariate data, i.e., the case in which one observes n-tuples of numbers. You ask how the ith number varies when the jth number changes. And if you just make those measurements—that is, the correlation between the ith number and the jth number—you get a matrix. Those statistics—that is, the mean vector of the data and the covariance matrix—are all the parameters one needs to find the best fit—gaussian or multivariate gaussian—to the data. Then, when other data come in, you can ask how well do they fit. You can calculate the probability that the data look like the old data by measuring them and using the covariance matrix to define a distance measure. That kind of simple analysis you will find in most statistics books. It already involves multiple variables, fitting gaussian distributions, and so on—so it involves a lot of multivariable calculus. But if one tries to build more complicated models for the data that take into account higher-order effects on the data, then one needs to use multivariable calculus to derive all the corresponding theorems—properties of the new sets of models. In particular, it is frequently much more difficult to estimate the parameters of the model directly from the data. The problem becomes an implicit problem in which one has to find the parameters that give the highest probability of an observed data set—the maximum-likelihood fitting of the parameters of the data. That typically involves a great deal of multivariable calculus, particularly maximization by iterative techniques. One studies how quickly such techniques converge, how often they converge, and so on. Another area has to do with formulation of hypotheses about data. That is, if you observe trends in data, you may build a model that takes account of those trends. And then you want to know whether you have taken account of all the information—whether the data exhibit roughness not predicted by the model you had so far. And you need a measure of how much additional information might be present in the data. That method is known as "maximal-entropy modeling." Fitting maximal-entropy models to data and analyzing residual information involves difficult analytic calculations. Calculus provides the theoretical framework but the ultimate calculations have to be done using a computer. Frequently you need to develop the computer algorithm yourself based on theoretical calculations made using multivariable calculus.

Another area of mathematics that is very hot—that is, very active—today is closely related to hard questions of combinatorial optimization. Finding, say, the shortest path through graphs or the best order in which to put objects in order to maximize some quantity. What one has is a large number of discrete variables, rather than continuously varying quantities. They take on a large but finite number of values. You must try all the values to find which values give the best answer. But the general ideas that one has for trying to solve such problems are based on continuous models that look like the discrete problem. And then you try to use the techniques of optimization, searching for improvement along gradient-like directions in order to find the answers—good approximate answers—to the intractable problems.

What uses does cryptography have today?

More and more messages and contracts are sent over computer; contract information and financial information are being sent digitally over telephone lines. To protect that information from eavesdroppers or tampering, banks and financial institutions are eager to encode and encrypt the information to prevent unauthorized listeners from hearing the information or being able to falsely identify themselves as people with access to that information.

Cryptography could be used to permit one to sign documents digitally so that you can prove who you are and prove that you signed the documents over a computer line without actually being there. Thus, one could provide authorization for financial transactions by computer, while protecting oneself from forgery and fraud. Fundamentally, cryptography is used as a lock and a key for protecting information.

How did you become interested in cryptography?

Originally, I applied for a summer job when I was an undergraduate without knowing what the work was going to be, but only knowing that they were looking for mathematicians. When I got there, the subject was extremely mathematical, and used a great deal of multivariate calculus and linear algebra, and I found the subject extremely intriguing. I served as a consultant for many years while a professor at a university, helping develop new tools for the government to use in its cryptologic program.

Is this a growing field with opportunities for new graduates?

Very much so. As more and more information is transmitted electronically, the need to protect that information and to permit long-distance contractual transactions will replace all kinds of written communications and paper checks and contracts. I believe it will all be done in the future electronically with greater assuredness of authenticity and greater safeguards than can be provided by traditional means. This type of remote contractual transaction is increasing at a great rate today. It is becoming ever more important, as the world is being wired up.

Do you have hope for speech recognition improving? Will I be able to stop typing at my computer and start talking to it?

A lot of the work we did here, early on, concerned the development of statistical models of speech in the hopes of allowing the computer to recognize and automatically transcribe human speech. Using mathematical models we have created here, a lot of theoretical and developmental work has been carried out by IBM, Bell Laboratories, BBN, and several universities. There now exist several speech-recognition systems that work with a 20,000 word vocabulary—and without the necessity for the speaker to pause in between words. But there still remains a lot to do to improve the performance of those systems, both in terms of vocabulary and in terms of making them independent of the speaker, to eliminate the necessity for each person to train the computer to recognize the way he pronounces words. Another area where multivariate calculus is used is in the design of codes to correct and detect errors in transmitted or stored data—the subject is called "the theory of error-correcting codes." And mathematics plays two roles there—major roles. One is the theory of how you actually find the best reconstruction of an original message given a corrupted message, and secondly, how do you design the set of messages so that they are not likely to be corrupted into confusion with one another. And those questions involve again combinatorial optimization, searching over large finite collections of things for ways of organizing them, with searches guided by analogous continuous questions.

Do you still find that mathematics is fun?

Very much. I enjoy thinking about mathematical problems. I get a tremendous satisfaction in seeing how mathematics will solve what appears to be a very difficult problem by simply organizing it and exposing its essential elements. I think that we continue to be surprised by the connections between the many fields of mathematics—geometry, analysis (including calculus), algebra. Most remarkable is the use of calculus techniques to obtain profound information about the theory of numbers. For example, the recent work of Wiles on Fermat's Last Theorem, which is apparently a purely number-theoretic question, might be thought to require discrete rather than continuous methods. But the method of solution grows out of work done 200 years ago on evaluating integrals for calculating the arc length of ellipses. These integrals led to the study of elliptic curves in algebraic geometry. This work was done long ago by Abel and Gauss and was then developed by the analysts and number theorists of the last two centuries.

David Lieberman is a Research Mathematician on the staff of the Center for Communications Research in Princeton, NJ. He majored in mathematics at Harvard and obtained his Ph.D. in 1966 from M.I.T., writing his thesis in the field of algebraic geometry. After two years at the Institute for Advanced Study at Princeton, he taught mathematics at Brandeis University for 10 years. He served as deputy director and director of the Institute for Defense Analyses at Princeton.

CHAPTER 14

MULTIVARIABLE DIFFERENTIAL CALCULUS

We begin this chapter by defining what it means for a function of two variables to be *differentiable*. Most of the remainder of the chapter is devoted to analyzing some of the important consequences of differentiability. In particular, we will define the *differential* of a differentiable function, and we will see that near a given point the differential can be used to approximate the change in the function produced by a small change in x and a small change in y. We will also consider, for compositions of differentiable functions, various forms of chain rules for calculating derivatives, or partial derivatives, depending on the nature of the functions involved.

As we have seen in Section 13.4, the partial derivatives f_x and f_y give the rate of change of the function f at a given point in the x direction and y direction, respectively. These two partials are special cases of a more general derivative, called the *directional derivative*, that measures the rate of change of f in any prescribed direction.

We know that a function of one variable has a tangent line at each point on its graph at which the function is differentiable. Similarly, we will see that the graph of a function of two variables has a *tangent plane* at each point where the function is differentiable.

We conclude the chapter by analyzing how to find maximum and minimum values of a function of two variables. Here, again, we will see parallels with the one-variable case.

14.1 DIFFERENTIABILITY

In order to define the notion of differentiability for a function of two variables, let us first restate its meaning for a function of one variable. Let f be such a

function, and suppose its derivative $f'(x_0)$ at x_0 exists. Then, by the definition of the derivative,
$$f'(x_0) = \lim_{\Delta x \to 0} \frac{f(x_0 + \Delta x) - f(x)}{\Delta x}$$

For brevity, we denote the numerator of the difference quotient on the right by Δf:
$$\Delta f = f(x_0 + \Delta x) - f(x_0)$$

We call Δf the *increment* in the function that corresponds to the increment Δx in x. So we can write
$$f'(x_0) = \lim_{\Delta x \to 0} \frac{\Delta f}{\Delta x}$$

We now denote the difference between $\Delta f / \Delta x$ and its limiting value $f'(x_0)$ by ε:
$$\varepsilon = \frac{\Delta f}{\Delta x} - f'(x_0)$$

Then $\varepsilon \to 0$ as $\Delta x \to 0$. Solving for Δf gives
$$\Delta f = f'(x_0)\Delta x + \varepsilon \Delta x \tag{14.1}$$

where $\varepsilon \to 0$ as $\Delta x \to 0$. Equation 14.1, along with the stated condition on ε, can be taken as an alternative definition of differentiability for the function f at x_0, equivalent to the condition that $f'(x_0)$ exists.

Recall that we defined the *differential* of a differentiable function f of one variable by
$$df = f'(x)dx$$

where dx is an independent variable. (We can also denote the differential of $y = f(x)$ by dy.) If we let $dx = \Delta x$ and take $x = x_0$, then we see that the first term on the right-hand side of Equation 14.1 is df, so we can write that equation as
$$\Delta f = df + \varepsilon \Delta x$$

Since $\varepsilon \to 0$ as $\Delta x \to 0$, the differential df approximates the increment Δf for small values of Δx.

In Figure 14.1, we illustrate the relationships among the quantities involved in Equation 14.1. Observe that whereas Δf is the change in the y value of the curve $y = f(x)$ caused by changing x from x_0 to $x_0 + \Delta x$, df is the change in the y value of the *tangent line* at $(x_0, f(x_0))$ caused by this change in x. In the equation $\Delta f = df + \varepsilon \Delta x$, when we substitute the values of Δf and df and omit the "error term" $\varepsilon \Delta x$, we can solve for $f(x_0 + \Delta x)$ to obtain the *linear approximation* formula,
$$f(x_0 + \Delta x) \approx f(x_0) + f'(x_0)\Delta x$$

We can also write this result in the form
$$f(x) \approx f(x_0) + f'(x_0)(x - x_0)$$

where $x = x_0 + \Delta x$. The equation $y = f(x_0) + f'(x_0)(x - x_0)$ is the equation of the tangent line at $(x_0, f(x_0))$, which accounts for the name *linear* approximation.

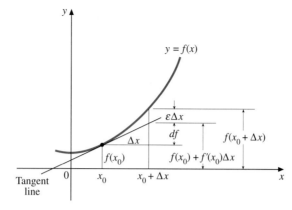

FIGURE 14.1

Now we extend these ideas to two dimensions. We begin by stating a theorem establishing an increment formula analogous to Equation 14.1. We give a proof in Appendix 5.

THEOREM 14.1

Fundamental Increment Formula

If the partial derivatives f_x and f_y of a function of two variables both exist inside some circle centered at (x_0, y_0) and are continuous at (x_0, y_0), then for all sufficiently small values of Δx and Δy, the increment $\Delta f = f(x_0 + \Delta x, y_0 + \Delta y) - f(x_0, y_0)$ can be written in the form

$$\Delta f = f_x(x_0, y_0)\Delta x + f_y(x_0, y_0)\Delta y + \varepsilon_1 \Delta x + \varepsilon_2 \Delta y \quad (14.2)$$

where $\varepsilon_1 \to 0$ and $\varepsilon_2 \to 0$ as $(\Delta x, \Delta y) \to (0, 0)$.

By analogy with the one-variable case, we take Equation 14.2 as the definition of differentiability.

Definition 14.1
Differentiability of a Function of Two Variables

A function f of two variables is **differentiable** at (x_0, y_0) provided the increment $\Delta f = f(x_0 + \Delta x, y_0 + \Delta y) - f(x_0, y_0)$ can be written in the form

$$\Delta f = f_x(x_0, y_0)\Delta x + f_y(x_0, y_0)\Delta y + \varepsilon_1 \Delta x + \varepsilon_2 \Delta y$$

where $\varepsilon_1 \to 0$ and $\varepsilon_2 \to 0$ as $(\Delta x, \Delta y) \to (0, 0)$.

One immediate consequence of differentiability of f at a point is that f is continuous there. In order to see why, write the condition for continuity (Definition 13.6),

$$\lim_{(x,y) \to (x_0, y_0)} f(x, y) = f(x_0, y_0)$$

in the equivalent form

$$\lim_{(x,y) \to (x_0, y_0)} [f(x, y) - f(x_0, y_0)] = 0$$

If we let $\Delta x = x - x_0$ and $\Delta y = y - y_0$, so that $f(x, y) = f(x_0 + \Delta x, y_0 + \Delta y)$, we see that the condition for continuity becomes

$$\lim_{(\Delta x, \Delta y) \to (0,0)} \Delta f = 0$$

which is true if f is differentiable, since each term on the right-hand side of Equation 14.2 approaches 0 as Δx and Δy approach 0.

We state this result again for emphasis.

Differentiability Implies Continuity

If a function of two variables is differentiable at a point, then it is continuous there.

REMARKS
- The converse of this result is not true, since there are continuous functions that are not differentiable (see, for example, Exercise 37 in Exercise Set 14.1).
- Although differentiability at a point requires that f_x and f_y exist there, mere existence of these partial derivatives does not imply differentiability. In fact, the function in Exercise 54 of Exercise Set 13.4 is discontinuous, hence not differentiable, at $(0, 0)$, even though $f_x(0, 0)$ and $f_y(0, 0)$ both exist.

It can be difficult to test for differentiability directly from Definition 14.1. Theorem 14.1, however, provides sufficient conditions that often enable us to conclude that a function is differentiable without having to apply the definition. We state the result as Theorem 14.2.

THEOREM 14.2 If f is a function of two variables such that f_x and f_y exist at all points inside some circle centered at (x_0, y_0) and are continuous at (x_0, y_0), then f is differentiable at (x_0, y_0).

The key word here is *continuous*. It is not enough that f_x and f_y merely exist at (x_0, y_0) and at all nearby points, but if these partials are also continuous at (x_0, y_0), differentiability is assured.

By Theorem 14.2, we can conclude immediately that all *polynomial functions* in two variables are differentiable everywhere in the xy-plane, since the partial derivatives are also polynomials and hence are continuous everywhere. Similarly, all *rational functions* of two variables are differentiable except where the denominator is zero. Furthermore, if $f(x, y)$ is differentiable and g is a differentiable function of one variable whose domain includes the range of f, then the composite function $g(f(x, y))$ is a differentiable function of two variables. For example, e^{xy}, $\sin(2x - 3y)$, and $\ln(x^2 + y^2)$ are all differentiable (in the case of $\ln(x^2 + y^2)$ we must exclude $(0, 0)$).

Geometric Interpretation

When Δx and Δy are sufficiently small, we can conclude by Equation 14.2 that if f is differentiable at (x_0, y_0), then

$$f(x_0 + \Delta x, y_0 + \Delta y) \approx f(x_0, y_0) + f_x(x_0, y_0)\Delta x + f_y(x_0, y_0)\Delta y \quad (14.3)$$

as can be seen by writing $\Delta f = f(x_0 + \Delta x, y_0 + \Delta y) - f(x_0, y_0)$ and deleting the terms $\varepsilon_1 \Delta x$ and $\varepsilon_2 \Delta y$. In order to understand the geometric significance of this approximation, let us again write $x = x_0 + \Delta x$ and $y = y_0 + \Delta y$. Then Equation 14.3 becomes

$$f(x, y) \approx f(x_0, y_0) + f_x(x_0, y_0)(x - x_0) + f_y(x_0, y_0)(y - y_0)$$

The function on the right is linear (first-degree) in x and y, so its graph is a plane. If we set z equal to this function, and $z_0 = f(x_0, y_0)$, we can write the equation of this plane as

$$z - z_0 = f_x(x_0, y_0)(x - x_0) + f_y(x_0, y_0)(y - y_0) \quad (14.4)$$

Note that this plane passes through the point $P_0 = (x_0, y_0, z_0)$ and so coincides with the surface $z = f(x, y)$ there. In Section 14.4, we will define what is meant by a *tangent plane* to a surface and will show that Equation 14.4 is in fact the tangent plane to the surface $z = f(x, y)$ at the point (x_0, y_0, z_0). Now we can interpret the linear approximation given by Equation 14.3, as shown in Figure 14.2. Observe the analogy with Figure 14.1, in which the tangent plane takes the place of the tangent line. Since the right-hand side of Equation 14.3 approximates the value of f at a point (x, y) near (x_0, y_0) by the value of z on the tangent plane (the graph of a linear function), Equation 14.3 is called a *linear approximation* formula.

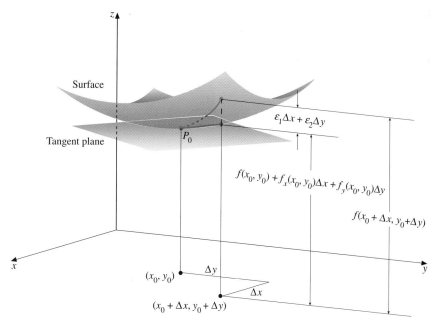

FIGURE 14.2

The Differential

The differential df of a function of one variable was defined in such a way that the increment formula (Equation 14.1) became

$$\Delta f = df + \varepsilon \Delta x$$

so that for Δx small, $df \approx \Delta f$. We define the differential of a function of two variables analogously, so that Equation 14.2 can be written as

$$\Delta f = df + \varepsilon_1 \Delta x + \varepsilon_2 \Delta y$$

Then it will follow that $df \approx \Delta f$ for Δx and Δy small.

Definition 14.2
The Differential of a Function of Two Variables

If f is differentiable at (x, y), then its **differential** df is given by

$$df = f_x(x, y)dx + f_y(x, y)dy \tag{14.5}$$

where dx and dy are independent variables.

REMARK
■ If we set $z = f(x, y)$ then we can use dz instead of df for the differential. We can also use the Leibniz notation for the partial derivatives, writing

$$dz = \frac{\partial z}{\partial x}dx + \frac{\partial z}{\partial y}dy$$

If we evaluate the differential at (x_0, y_0) with $dx = \Delta x$ and $dy = \Delta y$, then we see that df is the same as the right-hand side of Equation 14.2 except for the terms $\varepsilon_1 \Delta x$ and $\varepsilon_2 \Delta y$. Thus, as we anticipated,

$$\Delta f = df + \varepsilon_1 \Delta x + \varepsilon_2 \Delta y$$

and $df \approx \Delta f$ if Δx and Δy are small. The error in this approximation is $\varepsilon_1 \Delta x + \varepsilon_2 \Delta y$. We show the relationship between Δf and df in Figure 14.3.

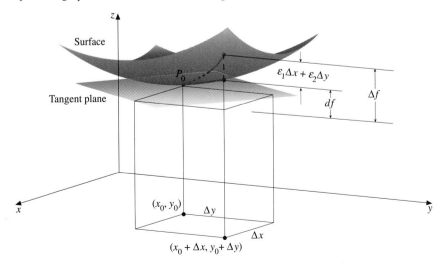

FIGURE 14.3

Applications of the Differential

As one application of the differential, suppose x and y are measured quantities, subject to possible errors of dx and dy, respectively. Suppose also that $z = f(x, y)$. Then the error in calculating z is approximately dz. The **relative error** is the true error divided by the calculated value. Thus, the relative error is approximated by dz/z. The relative error is often more important than the error itself. For example, an error of 0.1 cm in a calculated value of 2 cm gives a relative error of $\frac{0.1}{2} = 0.05$, whereas an error of 0.1 cm in a measured value of 2 m = 200 cm is $\frac{0.1}{200} = 0.0005$. Clearly, in the second case the relative error is much less serious than in the first case, even though the absolute error of 0.1 cm is the same in both cases. We summarize this discussion of errors below.

Suppose $z = f(x, y)$ is differentiable and that x and y are in error by amounts dx and dy, respectively. Then the approximate absolute error, relative error, and percentage error in z are as follows:

Approximate Absolute, Relative, and Percentage Errors

$$\text{Absolute error in } z \approx dz.$$

$$\text{Relative error in } z \approx \frac{dz}{z}.$$

$$\text{Percentage error in } z \approx 100\frac{dz}{z}.$$

Estimating error tolerances is very important in manufacturing barrels to keep hazardous waste contained.

EXAMPLE 14.1 A can is in the shape of a right circular cylinder. The radius and height are measured as 0.40 m and 1.20 m, respectively. If the error in the radius is at most 0.005 m and the error in the height is at most 0.002 m, find the approximate maximum values of (a) the absolute error, (b) the relative error, and (c) the percentage error in the calculated value of the volume. Also find the actual absolute error in the volume, assuming the maximum errors in the radius and the height.

Solution

(a) The volume of the can (see Figure 14.4) is given by

$$V = \pi r^2 h$$

We will use the differential dV to approximate the actual error ΔV. By Equation 14.5,

$$dV = \frac{\partial V}{\partial r}dr + \frac{\partial V}{\partial h}dh$$

$$= (2\pi r h)dr + (\pi r^2)dh$$

We take $dr = 0.005$ and $dh = 0.002$. Then, with the measured values of $r = 0.40$ and $h = 1.20$, we obtain

$$dV = 2\pi(0.40)(1.20)(0.005) + \pi(0.40)^2(0.002)$$

$$= 0.00512\pi \approx 0.01608$$

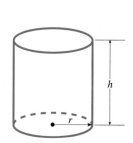

FIGURE 14.4
Volume =
(Area of Base) × (Altitude)
$= \pi r^2 h$

Thus, with the given measurements and errors in those measurements, the approximate absolute error in measuring the volume is 0.0161 m^3.

(b) With $r = 0.40$ and $h = 1.20$, the volume is

$$V = \pi(0.40)^2(1.20) = 0.192\pi \approx 0.603 \text{ m}^3$$

The approximate relative error is therefore

$$\frac{dV}{V} = \frac{0.00512\pi}{0.192\pi} \approx 0.0267$$

(c) The approximate percentage error is

$$100\frac{dV}{V} = 2.67\%$$

The actual absolute error is ΔV, found by calculating V using $r = 0.405$ and $h = 1.202$ and subtracting the value of V using $r = 0.40$ and $h = 1.20$.

$$\Delta V = \pi(0.405)^2(1.202) - \pi(0.40)^2(1.20)$$
$$= 0.005158\pi \approx 0.0162044$$

In part (a) we found $dV \approx 0.01608$. The approximation of ΔV by dV does not appear to be as close as we might hope for. However, when we calculate the actual percentage error,

$$100\frac{\Delta V}{V} = 100\frac{0.005158\pi}{0.192\pi} \approx 2.69\%$$

we see that the difference between this percentage error and the approximation in part (c) is only about 0.02%. ∎

EXAMPLE 14.2 Let $f(x, y) = \sqrt{9 - x^2 - y^2}$. Use the differential of f to approximate $f(0.92, 2.12)$. Show the result graphically.

Solution We choose (x_0, y_0) as $(1, 2)$, since this is the point nearest $(0.92, 2.12)$ at which we easily can obtain the exact value of f. We have

$$f(x_0, y_0) = f(1, 2) = \sqrt{9 - (1)^2 - (2)^2} = 2$$

Also, we take $dx\,(=\Delta x)$ as -0.08 and $dy\,(=\Delta y)$ as 0.12, in order that $x_0 + \Delta x = 0.92$ and $y_0 + \Delta y = 2.12$.

The differential of f is

$$df = \frac{\partial f}{\partial x}dx + \frac{\partial f}{\partial y}dy = \frac{-x}{\sqrt{9 - x^2 - y^2}}dx + \frac{-y}{\sqrt{9 - x^2 - y^2}}dy$$

Evaluating df at $(1, 2)$ with $dx = -0.08$ and $dy = 0.12$, we obtain

$$df = -\left(\frac{1}{2}\right)(-0.08) - \left(\frac{2}{2}\right)(0.12) = -0.08$$

The linear approximation given by Equation 14.3 can be written as

$$f(x_0 + \Delta x, y_0 + \Delta y) \approx f(x_0, y_0) + df$$

Thus,

$$f(0.92, 2.12) \approx 2 - 0.08 = 1.92$$

By calculator, the value (correct to four places) is 1.9129.

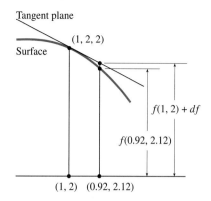

FIGURE 14.5

The graph of f is the upper half of the sphere $x^2+y^2+z^2 = 9$. In Figure 14.5 we show a portion of the trace of this surface with the vertical plane containing the points $(1, 2)$ and $(0.92, 2.12)$ in the xy-plane, and the line of intersection of this plane and the tangent plane to the surface drawn at $(1, 2, 2)$. Since we approximated the z value on the hemisphere by the z value to the tangent plane, it is clear that our estimate is somewhat too high. ∎

If $z = f(x, y)$, then we have seen that a change of dx in x and dy in y produces an approximate change of dz in z (assuming f is differentiable). We can use the notions of relative error and percentage error in this situation, where we replace *error* by *change*. That is,

$$\frac{dz}{z} = \text{approximate relative change in } z$$

$$100\frac{dz}{z} = \text{approximate percentage change in } z$$

We use these ideas in the next example, which is taken from economics.

EXAMPLE 14.3 The Cobb-Douglas Production Function, Equation 13.19, can be written in the form

$$Q(K, L) = AK^\alpha L^\beta$$

where $\beta = 1 - \alpha$ and $0 < \alpha < 1$. Here $Q(K, L)$ represents the number of units of some commodity that a company produces, A is a positive constant, K is the amount of capital investment, and L is the size of the labor force.

(a) By treating Q as a continuous function for $K > 0$ and $L > 0$ (it actually only takes on integer values, as does L), show that the relative change in Q caused by changes dK in K and dL in L is approximated by

$$\frac{dQ}{Q} = \alpha\left(\frac{dK}{K}\right) + \beta\left(\frac{dL}{L}\right)$$

(b) If $\alpha = 2/3$, find the approximate percentage change in Q caused by a 3% increase in K and a 2% decrease in L.

Solution

(a) First, we calculate dQ, using Equation 14.5.

$$dQ = \frac{\partial Q}{\partial K}dK + \frac{\partial Q}{\partial L}dL = \left(A\alpha K^{\alpha-1}L^\beta\right)dK + \left(A\beta K^\alpha L^{\beta-1}\right)dL$$

Thus,

$$\frac{dQ}{Q} = \frac{\left(A\alpha K^{\alpha-1}L^\beta\right)dK + \left(A\beta K^\alpha L^{\beta-1}\right)dL}{AK^\alpha L^\beta}$$

$$= \frac{A\alpha K^{\alpha-1}L^\beta}{AK^\alpha L^\beta}dK + \frac{A\beta K^\alpha L^{\beta-1}}{AK^\alpha L^\beta}dL = \frac{\alpha}{K}dK + \frac{\beta}{L}dL$$

$$= \alpha\left(\frac{dK}{K}\right) + \beta\left(\frac{dL}{L}\right)$$

(b) If $\alpha = 2/3$, then $\beta = 1/3$. A percentage increase of 3% in K means that

$$100\frac{dK}{K} = 3$$

so
$$\frac{dK}{K} = 0.03$$
Similarly, a percentage decrease of 2% in L means that
$$\frac{dL}{L} = -0.02$$
Substituting these values in the formula for dQ/Q from part (a) gives
$$\frac{dQ}{Q} = \frac{2}{3}(0.03) + \frac{1}{3}(-0.02) \approx 0.013$$
Thus, there is an approximate increase in production of 1.3%. ∎

Functions of Three or More Variables

The notions of differentiability and the differential extend in natural ways to functions of more than two variables. For three variables, for example, a function $f(x, y, z)$ is differentiable at the point (x, y, z) provided that the change Δf in the function caused by a change Δx in x, Δy in y, and Δz in z can be written in the form
$$\Delta f = f_x(x, y, z)\Delta x + f_y(x, y, z)\Delta y + f_z(x, y, z)\Delta z + \varepsilon_1 \Delta x + \varepsilon_2 \Delta y + \varepsilon_3 \Delta z$$
where $\varepsilon_1, \varepsilon_2$, and ε_3 all approach 0 as $(\Delta x, \Delta y, \Delta z) \to (0, 0, 0)$. If f is differentiable at a point (x, y, z), we define the differential of f as
$$df = f_x(x, y, z)dx + f_y(x, y, z)dy + f_z(x, y, z)dz$$
Just as in the two-variable case, we can set $dx = \Delta x$, $dy = \Delta y$, and $dz = \Delta z$, so that Δf becomes
$$\Delta f = df + \varepsilon_1 \Delta x + \varepsilon_2 \Delta y + \varepsilon_3 \Delta z$$
Thus, $\Delta f \approx df$ when Δx, Δy, and Δz are small in absolute value.

Exercise Set 14.1

In Exercises 1–6, use Theorem 14.2 to show that the given function is differentiable on the specified domain.

1. $f(x, y) = x^3 + 3x^2 y - 2y^4$, on $\mathbb{R}^2$

2. $f(x, y) = \dfrac{x + y}{x - y}$, $x \neq y$

3. $f(x, y) = e^x \sin y$, on $\mathbb{R}^2$

4. $f(x, y) = \sin^{-1} \dfrac{y}{x}$, $x^2 > y^2$

5. $f(x, y) = \ln \dfrac{xy}{x^2 + y^2}$, $xy > 0$

6. $f(x, y) = x \ln(\cos y)$, $-\pi/2 < y < \pi/2$

In Exercises 7–16, find df.

7. $f(x, y) = 2x^2 - 3xy + y^3$

8. $f(x, y) = x^4 y^2 - 2xy^3$

9. $f(x, y) = e^{xy}$

10. $f(x, y) = \ln \sqrt{x^2 + y^2}$

11. $f(x, y) = \dfrac{x - y}{x + y}$

12. $f(x, y) = x \sin y + y \cos x$

13. $f(x, y) = \tan^{-1} \dfrac{x}{y}$

14. $f(x, y) = \ln\left(\dfrac{2x^2 y}{x + y}\right)$

15. $f(x, y, z) = \sqrt{x^2 + y^2 + z^2}$

16. $f(x, y, z) = \ln(x^2 y^3 z^4)$

In Exercises 17–22, use df to approximate the change Δf in f from P_0 to P_1.

17. $f(x, y) = 2x^3 - 3x^2 y + y^2$; $P_0 = (2, -1)$, $P_1 = (2.02, -0.99)$

18. $f(x, y) = \sqrt{\dfrac{2x - y}{x + 3y}}$; $P_0 = (4, -1)$, $P_1 = (3.97, -0.95)$

19. $f(x, y) = \ln(x - y)^2$; $P_0 = (3, -2)$, $P_1 = (3.04, -1.99)$

20. $f(x, y) = \tan^{-1} \dfrac{y}{x}$; $P_0 = (-3, 4)$, $P_1 = (-3.02, 3.98)$

21. $f(x, y) = \ln\sqrt{x^2 + y^2}$; $P_0 = (-4, 3)$, $P_1 = (-3.98, 2.99)$

22. $f(x, y, z) = (x^2 + y^2 + z^2)^{-1}$; $P_0 = (1, -1, 2)$, $P_1 = (0.97, -1.01, 2.05)$

23. The dimensions of a room are measured to be as follows: length 21 ft, width 12 ft, and height 8 ft. If each measurement is accurate only to the nearest tenth of a foot, find the approximate maximum error in volume calculated from the measured values, making use of differentials. What is the approximate percentage error?

24. In Example 14.1, find the approximate maximum error in total surface area, making use of differentials. What is the approximate relative error?

25. Suppose the temperature at a point in a thin, rectangular metal sheet is given by

$$T(x, y) = \dfrac{x^2 y^2}{x^2 + y^2}$$

where the origin is at the lower left corner. Using differentials, find the approximate difference in temperature at the points $(1.15, 2.05)$ and $(1, 2)$. Compare your answer with the actual change. What is the approximate percentage change?

26. A company manufactures and sells two models of automatic ice cream makers, the standard model, A, and the deluxe model, B. Through an analysis of sales over a period of time, it is found that the weekly profit from producing and selling x Model-A and y Model-B machines is approximately

$$P(x, y) = 200x + 300y - 0.2x^2 - 0.3y^2 - 0.4xy - 50$$

where P is in dollars. On average, 50 Model-A and 20 Model-B machines are sold each week. Use differentials to approximate the effect on profit of selling one fewer Model-A and two more Model-B machines per week than the average. What is the approximate percentage change in profit?

27. The cost of producing one unit of a certain manufactured item is given by

$$C(x, y) = 2x^2 + 3xy + y^2 + 15x + 6y + 50$$

where x is the hourly wage rate for the workers and y is the cost per pound of raw materials. Currently, the hourly wage is \$9.50, and raw materials cost \$6.00 per pound. Use differentials to estimate the increase in cost if the labor cost goes up by \$0.50 per hour and the cost of raw materials increases by \$0.40 per pound. By approximately what percentage does the cost go up?

28. At a certain factory, the weekly production is given by the Cobb-Douglas Production Function as

$$Q(K, L) = 120 K^{0.6} L^{0.4}$$

units, where K represents the capital investment (in thousands of dollars) and L represents the size of the labor force. The current values of K and L are 450 and 260, respectively. Estimate, by differentials, the change in production that will result if K is increased by 2 and L is increased by 3. What is the approximate relative change in Q?

29. For the Cobb-Douglas Production Function in Exercise 28, estimate the percentage change in Q if K increases by 2% and L decreases by 1%.

In Exercises 30–33, find an expression for
$$\Delta f = f(x + \Delta x, y + \Delta y) - f(x, y)$$
and show that it can be written in the form of Equation 14.2. Identify ε_1 and ε_2 and show they both approach 0 as Δx and Δy approach 0.

30. $f(x, y) = 2x + 3y$

31. $f(x, y) = xy$

32. $f(x, y) = 2x^2 y$

33. $f(x, y) = (x - 3y)^2$

34. Prove that $f(x, y) = \sqrt{x^2 + y^2}$ is not differentiable at the origin. (*Hint:* Show that f_x does not exist at the origin.)

35. Let
$$f(x, y) = \begin{cases} \dfrac{x^2 y^2}{x^2 + y^2} & \text{if } (x, y) \neq (0, 0) \\ 0 & \text{if } (x, y) = (0, 0) \end{cases}$$

Prove that f is differentiable on all of $\mathbb{R}^2$. (*Hint:* Apply Theorem 14.2, paying particular attention to the origin.)

36. Let
$$f(x, y) = \begin{cases} \dfrac{xy}{x^2 + y^2} & \text{if } (x, y) \neq (0, 0) \\ 0 & \text{if } (x, y) = (0, 0) \end{cases}$$

(a) Show that f is discontinuous at $(0, 0)$.
(b) Show that $f_x(0, 0)$ and $f_y(0, 0)$ both exist.
(c) What can you conclude from part (a) about differentiability of f at the origin? Explain.
(d) What can you conclude from part (c) about continuity of f_x and f_y at the origin? Explain.

37. Let $f(x, y) = \sqrt{|xy|}$. Prove the following:
(a) f is continuous at $(0, 0)$
(b) $f_x(0, 0) = f_y(0, 0) = 0$
(c) f is not differentiable at $(0, 0)$
(*Hint:* Show that at $(0, 0)$ Equation 14.2 becomes $\sqrt{|\Delta x \Delta y|} = \varepsilon_1 \Delta x + \varepsilon_2 \Delta y$, and by taking $\Delta x = \Delta y$, conclude that ε_1 and ε_2 do not approach 0 as $(\Delta x, \Delta y) \to (0, 0)$.)

14.2 CHAIN RULES

For a function f of one variable x, if x in turn is a function of t, say $x = g(t)$, then the Chain Rule for differentiation (Equation 3.13) can be written as

$$(f \circ g)'(t) = f'(g(t))g'(t) \tag{14.6}$$

This formula holds true at all points t in the domain of g at which g' exists and for which $f'(g(t))$ exists. When we introduce the dependent variable $y = f(x)$, Equation 14.6 becomes, in Leibniz notation,

$$\frac{dy}{dt} = \frac{dy}{dx} \frac{dx}{dt} \tag{14.7}$$

Equation 14.7 is the familiar form of the Chain Rule. Care should be taken to distinguish between dy/dt and dy/dx. On the left, it is understood that x is first replaced by $g(t)$ and then the derivative is taken, so that

$$\frac{dy}{dt} \quad \text{means} \quad \frac{d}{dt} f(g(t))$$

On the right, dy/dx means that we first calculate $f'(x)$ and then replace x by $g(t)$, so that

$$\frac{dy}{dx} \quad \text{means} \quad f'(x)\bigg|_{x=g(t)} \quad \text{or} \quad f'(g(t))$$

It is convenient in this context to call x the *intermediate variable* and t the *final (independent) variable*.

The form of the extension of Equation 14.6 or 14.7 to two or more variables depends on the number of intermediate variables and final variables. We begin with the simplest case, in which $z = f(x, y)$, $x = g(t)$, and $y = h(t)$. Here x and y are intermediate variables and t is the single, final, independent variable. If we first substitute for x and y, we get $z = f(g(t), h(t))$, so that z is finally a function of the single independent variable t. We want to find a formula for dz/dt. The secret lies in the definition of differentiability. We assume that $g'(t)$ and $h'(t)$ both exist and that f is differentiable at $(x, y) = (g(t), h(t))$. Let

$$\Delta x = g(t + \Delta t) - g(t)$$
$$\Delta y = h(t + \Delta t) - h(t)$$

Then, because g and h are continuous at t (why?), Δx and Δy both approach 0 as $\Delta t \to 0$. Now, by Definition 14.1, we can write

$$\Delta z = f_x(x, y)\Delta x + f_y(x, y)\Delta y + \varepsilon_1 \Delta x + \varepsilon_2 \Delta y \qquad (14.8)$$

where $\varepsilon_1 \to 0$ and $\varepsilon_2 \to 0$ as $(\Delta x, \Delta y) \to (0, 0)$ and hence also as $\Delta t \to 0$. Here Δz means

$$\Delta z = \Delta f = f(x + \Delta x, y + \Delta y) - f(x, y)$$
$$= f(g(t + \Delta t), h(t + \Delta t)) - f(g(t), h(t))$$

Dividing Equation 14.8 by Δt on both sides gives

$$\frac{\Delta z}{\Delta t} = f_x(x, y)\frac{\Delta y}{\Delta t} + f_y(x, y)\frac{\Delta y}{\Delta t} + \varepsilon_1 \frac{\Delta x}{\Delta t} + \varepsilon_2 \frac{\Delta y}{\Delta t}$$

Now we let $\Delta t \to 0$ to obtain the result

$$\frac{d}{dt}f(x, y) = f_x(x, y)g'(t) + f_y(x, y)h'(t)$$

or, in Leibniz notation, since $z = f(x, y)$,

$$\frac{dz}{dt} = \frac{\partial f}{\partial x}\frac{dx}{dt} + \frac{\partial f}{\partial y}\frac{dy}{dt} \qquad (14.9)$$

It is understood that $\partial f/\partial x$ and $\partial f/\partial y$ are to be evaluated at the point $(g(t), h(t))$. We can also use $\partial z/\partial x$ in place of $\partial f/\partial x$ and $\partial z/\partial y$ in place of $\partial f/\partial y$ to obtain the equivalent equation

$$\frac{dz}{dt} = \frac{\partial z}{\partial x}\frac{dx}{dt} + \frac{\partial z}{\partial y}\frac{dy}{dt}$$

We summarize our result as follows.

If $z = f(x, y)$ is a differentiable function of x and y, where $x = g(t)$ and $y = h(t)$ are differentiable functions of the single variable t, then z is a differentiable function of t, and

$$\frac{\partial z}{\partial t} = \frac{\partial z}{\partial x}\frac{dx}{dt} + \frac{\partial z}{\partial y}\frac{dy}{dt} \qquad (14.10)$$

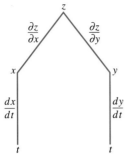

FIGURE 14.6

As an aid in obtaining Equation 14.10, refer to Figure 14.6. This figure is an example of a **tree diagram**. Starting from the dependent variable z, we move on the branch leading to the intermediate variable x, writing $\partial z/\partial x$. Then we move from x on the (only) branch from it leading to t, writing dx/dt. Note that we write dx/dt rather than $\partial x/\partial t$, since x is a function of just one variable, t. When we move from any one variable, if there are two (or more) branches, we use *partial* derivatives along each branch, whereas if there is only one branch, we use the *total* derivative (that is, the ordinary derivative of a function of one variable).

Now we multiply together the derivatives $\partial z/\partial x$ and dx/dt that we have obtained:

$$\frac{\partial z}{\partial x}\frac{dx}{dt}$$

Then we repeat the process, branching from z to y and from y to t, obtaining the product

$$\frac{\partial z}{\partial y}\frac{dy}{dt}$$

Finally, we add the results we have obtained, getting the right-hand side of Equation 14.10.

We will use tree diagrams for other combinations of intermediate and final variables, but for simplicity we will in the future not show the derivatives on the diagram.

EXAMPLE 14.4   Let

$$z = \frac{x - 2y}{2x + 3y} \quad \text{and} \quad \begin{cases} x = 2t - 3 \\ y = t^2 + 1 \end{cases}$$

Use the Chain Rule to find dz/dt when $t = -1$.

Solution   By Equation 14.10,

$$\frac{dz}{dt} = \frac{\partial z}{\partial x}\frac{dx}{dt} + \frac{\partial z}{\partial y}\frac{dy}{dt}$$

$$= \frac{7y}{(2x + 3y)^2}(2) + \frac{-7}{(2x + 3y)^2}(2t) \quad \text{Verify.}$$

When $t = -1$, we find that $x = -5$ and $y = 2$, so

$$\left.\frac{dz}{dt}\right|_{t=-1} = \frac{14}{(-4)^2}(2) + \frac{(35)}{(-4)^2}(-2) = -\frac{21}{8} \quad \blacksquare$$

The next case we consider is that in which $z = f(x, y)$, and each of the variables x and y is a function of two other variables, say u and v (there are two intermediate variables, x and y, and two final variables u and v). The derivation goes in much the same way as for the previous case, but we obtain two equations, one for $\partial z/\partial u$ and one for $\partial z/\partial v$. We summarize the results as follows.

If $z = f(x, y)$ is a differentiable function of x and y, where $x = g(u, v)$ and $y = h(u, v)$, and the partial derivatives $\partial x/\partial u$, $\partial x/\partial v$, $\partial y/\partial u$, and $\partial y/\partial v$ all exist, then

$$\frac{\partial z}{\partial u} = \frac{\partial z}{\partial x}\frac{\partial x}{\partial u} + \frac{\partial z}{\partial y}\frac{\partial y}{\partial u}$$

$$\frac{\partial z}{\partial v} = \frac{\partial z}{\partial x}\frac{\partial x}{\partial v} + \frac{\partial z}{\partial y}\frac{\partial y}{\partial v}$$

(14.11)

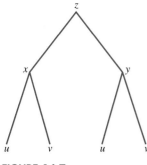

FIGURE 14.7

The tree diagram in Figure 14.7 illustrates a way to arrive at each of Equations 14.11. Since there are two branches from each of the variables z, x, and y, all derivatives are partial derivatives. To get the formula for $\partial z/\partial u$, we go along each branch leading from z to u. Remember that as we go from one variable to another, we take the partial derivative of the first with respect to the second. So, going from z to x to u, we get

$$\frac{\partial z}{\partial x}\frac{\partial x}{\partial u}$$

Similarly, going from z to y to u, we get

$$\frac{\partial z}{\partial y}\frac{\partial y}{\partial u}$$

Then we add these results to get $\partial z/\partial u$, as given by the first of Equations 14.11. When we repeat the process, going along all branches leading to v, we get $\partial z/\partial v$, as in the second of Equations 14.11.

EXAMPLE 14.5 If $z = x^2 - 2xy + 3y^2$, $x = uv$, and $y = u^2 - v^2$, find $\partial z/\partial u$ and $\partial z/\partial v$.

Solution By Equations 14.11, or from the tree diagram in Figure 14.7,

$$\frac{\partial z}{\partial u} = \frac{\partial z}{\partial x}\frac{\partial x}{\partial u} + \frac{\partial z}{\partial y}\frac{\partial y}{\partial u} = (2x - 2y)(v) + (-2x + 6y)(2u)$$
$$= 2(uv - u^2 + v^2)v + 4(-uv + 3u^2 - 3v^2)u$$
$$= 12u^3 - 6u^2v - 10uv^2 + 2v^3$$
$$\frac{\partial z}{\partial v} = \frac{\partial z}{\partial x}\frac{\partial x}{\partial v} + \frac{\partial z}{\partial y}\frac{\partial y}{\partial v} = (2x - 2y)(u) + (-2x + 6y)(-2v)$$
$$= 2(uv - u^2 + v^2)u - 4(-uv + 3u^3 - 3v^2)v$$
$$= -2u^3 - 10u^2v + 6uv^2 + 12v^3$$
∎

REMARK

■ Note that in the preceding example we first obtained $\partial z/\partial u$ and $\partial z/\partial v$ as a mixed expression involving both intermediate and final variables. Then we replaced the intermediate variables x and y by their values in terms of the final variables u and v. In Example 14.4, we avoided the last step, since we were evaluating the derivative at a specific point, and we substituted the values of dx/dt and dy/dt at that point.

EXAMPLE 14.6 If $z = f(u/v, v/u)$, show that

$$u\frac{\partial z}{\partial u} + v\frac{\partial z}{\partial v} = 0$$

Solution We can consider the variable z as the composition of $z = f(x, y)$ and $x = u/v$, $y = v/u$. Equations 14.11 then give

$$\frac{\partial z}{\partial u} = \frac{\partial z}{\partial x}\frac{1}{v} + \frac{\partial z}{\partial y}\left(-\frac{v}{u^2}\right)$$

$$\frac{\partial z}{\partial v} = \frac{\partial z}{\partial x}\left(-\frac{u}{v^2}\right) + \frac{\partial z}{\partial y}\left(\frac{1}{u}\right)$$

Even though we do not know the values of $\partial z/\partial x$ and $\partial z/\partial y$, we see that multiplying the first equation by u and the second by v and adding gives

$$u\frac{\partial z}{\partial u} + v\frac{\partial z}{\partial v} = 0 \qquad \blacksquare$$

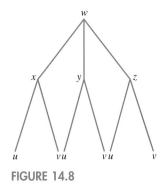

FIGURE 14.8

It should be clear now that the number of intermediate and final variables can be any number greater than or equal to one. For example, if $w = f(x, y, z)$ and, in turn, $x = x(u, v)$, $y = y(u, v)$, and $z = z(u, v)$, then the tree diagram in Figure 14.8 can be used to write the appropriate chain rule. In this case, each of the equations for $\partial w/\partial u$ and $\partial w/\partial v$ would contain three terms. For example,

$$\frac{\partial w}{\partial u} = \frac{\partial w}{\partial x}\frac{\partial x}{\partial u} + \frac{\partial w}{\partial y}\frac{\partial y}{\partial u} + \frac{\partial w}{\partial z}\frac{\partial z}{\partial u}$$

Rather than list formulas for various other combinations of intermediate and final variables, we state the following Generalized Chain Rule, where f is a function of n variables, each of which is a function of m variables. We assume differentiability of all functions involved and that the composite function is defined in the domain under consideration.

Generalized Chain Rule

Let $z = f(x_1, x_2, \ldots, x_n)$, and for each i from 1 to n, let $x_i = g_i(u_1, u_2, \ldots, u_m)$. Then

$$\frac{\partial z}{\partial u_j} = \frac{\partial z}{\partial x_1}\frac{\partial x_1}{\partial u_j} + \frac{\partial z}{\partial x_2}\frac{\partial x_2}{\partial u_j} + \cdots + \frac{\partial z}{\partial x_m}\frac{\partial x_m}{\partial u_j} \qquad (14.12)$$

for $j = 1, 2, \ldots, m$.

Section 14.2 Chain Rules 1017

Here there are n intermediate variables $x_1, x_2, \ldots, x_n$ and m final variables $u_1, u_2, \ldots, u_m$. So there are m equations, each with n terms on the right-hand side.

There is nothing special about the particular letters used to designate the dependent variable or the intermediate and final variables. Consider the following example.

EXAMPLE 14.7   Let $w = r^2 + s^2 - 2t^2$ and $r = e^{-y}\cos z$, $s = e^{-y}\sin z$, and $t = e^{-y}$. Find $\partial w/\partial y$ and $\partial w/\partial z$.

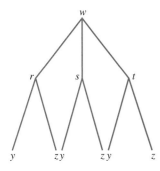

FIGURE 14.9

Solution   We can use Equation 14.12, with appropriate changes in variable names, or we can use the tree diagram in Figure 14.9.

$$\frac{\partial w}{\partial y} = \frac{\partial w}{\partial r}\frac{\partial r}{\partial y} + \frac{\partial w}{\partial s}\frac{\partial s}{\partial y} + \frac{\partial w}{\partial t}\frac{\partial t}{\partial y}$$

$$= (2r)(-e^{-y}\cos z) + (2s)(-e^{-y}\sin z) + (-4t)(-e^{-y})$$

$$= (2e^{-y}\cos z)(-e^{-y}\cos z) + (2e^{-y}\sin z)(-e^{-y}\sin z) + (-4e^{-y})(-e^{-y})$$

$$= 2e^{-2y}(-\cos^2 z - \sin^2 z + 2)$$

$$= 2e^{-2y}$$

$$\frac{\partial w}{\partial z} = \frac{\partial w}{\partial r}\frac{\partial r}{\partial z} + \frac{\partial w}{\partial s}\frac{\partial s}{\partial z} + \frac{\partial w}{\partial t}\frac{\partial t}{\partial z}$$

$$= (2r)(-e^{-y}\sin z) + (2s)(e^{-y}\cos z) + (-4t)(0)$$

$$= (2e^{-y}\cos z)(-e^{-y}\sin z) + (2e^{-y}\sin z)(-e^{-y}\cos z)$$

$$= 0$$

■

Second Derivatives by the Chain Rule

The next example shows how the Chain Rule given by Equation 14.10 can be used to compute the second derivative d^2z/dt^2 when $z = f(x, y)$ and $x = g(t)$, $y = h(t)$. In the exercises you will be asked to derive similar formulas for second-order partial derivatives where there are two final variables.

EXAMPLE 14.8   Let $z = f(x, y)$ and $x = g(t)$, $y = h(t)$. Assuming suitable differentiability conditions, derive a formula for d^2z/dt^2.

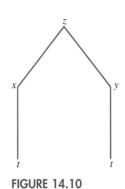

FIGURE 14.10

Solution   For the first derivative, we have from Equation 14.10 or from the tree diagram in Figure 14.10,

$$\frac{dz}{dt} = \frac{\partial z}{\partial x}\frac{dx}{dt} + \frac{\partial z}{\partial y}\frac{dy}{dt}$$

Using the Product Rule, we obtain

$$\frac{d^2z}{dt^2} = \frac{d}{dt}\left(\frac{\partial z}{\partial x}\right)\frac{dx}{dt} + \frac{\partial z}{\partial x}\frac{d^2x}{dt^2} + \frac{d}{dt}\left(\frac{\partial z}{\partial y}\right)\frac{dy}{dt} + \frac{\partial z}{\partial y}\frac{d^2y}{dt^2} \quad (14.13)$$

Now $\partial z/\partial x$ and $\partial z/\partial y$ are initially functions of the intermediate variables x and y. So to find their derivatives with respect to the final variable t, we use the Chain Rule given by Equation 14.10 again, with z replaced by $\partial z/\partial x$ in the one case and by $\partial z/\partial y$ in the other (see also Figure 14.11).

$$\frac{d}{dt}\left(\frac{\partial z}{\partial x}\right) = \frac{\partial^2 z}{\partial x^2}\frac{dx}{dt} + \frac{\partial^2 z}{\partial y \partial x}\frac{dy}{dt}$$

$$\frac{d}{dt}\left(\frac{\partial z}{\partial y}\right) = \frac{\partial^2 z}{\partial x \partial y}\frac{dx}{dt} + \frac{\partial^2 z}{\partial y^2}\frac{dy}{dt}$$

If we assume equality of the second-order mixed partials, then on substitution of these expressions into Equation 14.13, we obtain

$$\frac{d^2 z}{dt^2} = \frac{\partial z}{\partial x}\frac{d^2 x}{dt^2} + \frac{\partial z}{\partial y}\frac{d^2 y}{dt^2} + \frac{\partial^2 z}{\partial x^2}\left(\frac{dx}{dt}\right)^2 + 2\frac{\partial^2 z}{\partial y \partial x}\left(\frac{dx}{dt}\frac{dy}{dt}\right) + \frac{\partial^2 z}{\partial y^2}\left(\frac{dy}{dt}\right)^2$$

■

EXAMPLE 14.9 Let $z = f(x, y)$ and $x = t^2$, $y = \ln t$. Find $d^2 z/dt^2$.

Solution We could use the formula found in Example 14.8, but it is probably more instructive to go through all the steps again.

Using the Chain Rule, we find

$$\frac{dz}{dt} = (f_x)(2t) + (f_y)\left(\frac{1}{t}\right)$$

To obtain the second derivative, we differentiate each term on the right-hand side as a product of two functions of t. For example, to differentiate the product $(f_x)(2t)$ with respect to t, we use the Product Rule to get

$$\frac{d}{dt}\left[(f_x)(2t)\right] = \left(\frac{df_x}{dt}\right)(2t) + (f_x)(2)$$

But to find df_x/dt, we must use the Chain Rule:

$$\frac{df_x}{dt} = f_{xx}\frac{dx}{dt} + f_{xy}\frac{dy}{dt} = (f_{xx})(2t) + (f_{xy})\left(\frac{1}{t}\right)$$

Thus,

$$\frac{d}{dt}\left[(f_x)(2t)\right] = (f_{xx})(4t^2) + 2f_{xy} + 2f_x$$

Similarly, for the product $(f_y)\left(\frac{1}{t}\right)$, we find that

$$\frac{d}{dt}\left[(f_y)\left(\frac{1}{t}\right)\right] = (f_{yx})(2t)\cdot\frac{1}{t} + (f_{yy})\left(\frac{1}{t}\right)^2 - \frac{1}{t^2}f_y$$

Combining these results gives

$$\frac{d^2 z}{dt^2} = 4t^2 f_{xx} + 4f_{xy} + \frac{1}{t^2}f_{yy} + 2f_x - \frac{1}{t^2}f_y$$

You can verify that this result agrees with the formula found in Example 14.8. ■

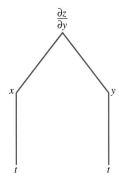

FIGURE 14.11

In the next example, we illustrate how to find second-order partials when there are two final variables. In the exercises (Exercise 37), you will be asked to develop general formulas for this case.

EXAMPLE 14.10 Let $z = f(x, y)$ and let x and y be expressed in polar coordinates. Find $\partial^2 z/\partial r^2$ and $\partial^2 z/\partial \theta^2$.

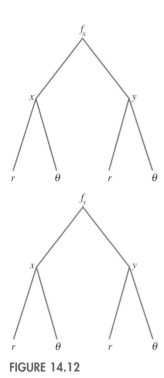

FIGURE 14.12

Solution In terms of polar coordinates, $x = r\cos\theta$ and $y = r\sin\theta$. Applying the Chain Rule, we get

$$\frac{\partial z}{\partial r} = f_x \frac{\partial x}{\partial r} + f_y \frac{\partial y}{\partial r} = f_x \cos\theta + f_y \sin\theta$$

and

$$\frac{\partial z}{\partial \theta} = f_x \frac{\partial x}{\partial \theta} + f_y \frac{\partial y}{\partial \theta} = f_x(-r\sin\theta) + f_y(r\cos\theta)$$

To find the second-order partials, we will need to apply the Chain Rule again to f_x and f_y. The tree diagrams in Figure 14.12 are helpful in this regard. Since $\cos\theta$ and $\sin\theta$ are independent of r, we find that

$$\frac{\partial^2 z}{\partial r^2} = (f_{xx}\cos\theta + f_{xy}\sin\theta)\cos\theta + (f_{yx}\cos\theta + f_{yy}\sin\theta)\sin\theta$$

$$= f_{xx}\cos^2\theta + 2f_{xy}\sin\theta\cos\theta + f_{yy}\sin^2\theta \quad \text{Assume } f_{xy} = f_{yx}.$$

For $\partial^2 z / \partial \theta^2$, we must use the Product Rule as well as the Chain Rule:

$$\frac{\partial^2 z}{\partial \theta^2} = \left[f_{xx}(-r\sin\theta) + f_{xy}(r\cos\theta)\right](-r\sin\theta) - rf_x \cos\theta$$

$$+ \left[f_{yx}(-r\sin\theta) + f_{yy}(r\cos\theta)\right](r\cos\theta) + rf_y(-\sin\theta)$$

$$= r^2 \sin^2\theta f_{xx} - 2r^2 \sin\theta \cos\theta f_{xy} + r^2 \cos^2\theta f_{yy}$$

$$- rf_x \cos\theta - rf_y \sin\theta \qquad\blacksquare$$

Derivative Formulas for Implicit Functions

The Chain Rule can be used to derive a very convenient formula for the derivative of a function defined implicitly. As in Section 2.8, suppose that $F(x, y) = 0$ defines y as a differentiable function of x, say $y = f(x)$, on some domain D. We want to find a formula for dy/dx. We can think of this problem as a chain rule situation with intermediate variables x and y and final variable x. To distinguish the two roles played by x, however, we will use the letter t (temporarily) as the final variable. So we have

$$F(x, y) = 0 \quad \text{and} \quad \begin{cases} x = t \\ y = f(t) \end{cases}$$

Since $F(t, f(t)) = 0$ for all t in D, it follows that

$$\frac{d}{dt}F(t, f(t)) = 0$$

there also. By Equation 14.10 or by Figure 14.13,

$$\frac{d}{dt}F(t, f(t)) = \frac{\partial F}{\partial x}\frac{dx}{dt} + \frac{\partial F}{\partial y}\frac{dy}{dt}$$

$$= \frac{\partial F}{\partial x}(1) + \frac{\partial F}{\partial y}\frac{dy}{dt} = 0$$

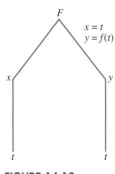

FIGURE 14.13

Thus, if $\partial F/\partial y \neq 0$, we can solve for dy/dt to get

$$\frac{dy}{dt} = -\frac{\dfrac{\partial F}{\partial x}}{\dfrac{\partial F}{\partial y}}$$

Now we can replace t by x and write the answer in the following alternative form:

$$\frac{dy}{dx} = -\frac{F_x}{F_y} \quad \text{if } F_y \neq 0 \tag{14.14}$$

EXAMPLE 14.11 Find dy/dx if

$$x^3 - 2xy + y^3 - 4 = 0$$

Solution Let $F(x, y) = x^3 - 2xy + y^3 - 4$. Then the given equation becomes $F(x, y) = 0$. Assuming this equation defines y as a differentiable function of x on some domain D, we can use Equation 14.14 to get the result

$$\frac{dy}{dx} = -\frac{F_x}{F_y} = -\frac{3x^2 - 2y}{-2x + 3y^2} = \frac{2y - 3x^2}{3y^2 - 2x}$$

at all points of D for which $3y^2 - 2x \neq 0$. This same problem was solved in Example 3.26, part (a). You might be interested in comparing the new method with the one used in Chapter 3. ∎

A similar method can be applied when an equation $F(x, y, z) = 0$ defines z implicitly as a differentiable function of x and y, say $z = f(x, y)$, on some domain D of $\mathbb{R}^2$. We can obtain formulas for $\partial z/\partial x$ and $\partial z/\partial y$, as follows.

Treat x, y, and z as intermediate variables and u and v as final variables, where

$$\begin{cases} x = u \\ y = v \\ z = f(u, v) \end{cases}$$

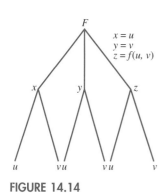

FIGURE 14.14

Then for (u, v) in D, $F(u, v, f(u, v)) = 0$. So $\partial F/\partial u = 0$ and $\partial F/\partial v = 0$. By Equations 14.11, or by the tree diagram in Figure 14.14, we have

$$\frac{\partial F}{\partial u} = \frac{\partial F}{\partial x}\frac{\partial x}{\partial u} + \frac{\partial F}{\partial y}\frac{\partial y}{\partial u} + \frac{\partial F}{\partial z}\frac{\partial z}{\partial u} = \frac{\partial F}{\partial x}(1) + \frac{\partial F}{\partial y}(0) + \frac{\partial F}{\partial z}\frac{\partial z}{\partial u}$$

and

$$\frac{\partial F}{\partial v} = \frac{\partial F}{\partial x}\frac{\partial x}{\partial v} + \frac{\partial F}{\partial y}\frac{\partial y}{\partial v} + \frac{\partial F}{\partial z}\frac{\partial z}{\partial v} = \frac{\partial F}{\partial x}(0) + \frac{\partial F}{\partial y}(1) + \frac{\partial F}{\partial z}\frac{\partial z}{\partial v}$$

Since $\partial F/\partial u$ and $\partial F/\partial v$ both equal 0, we can solve for $\partial z/\partial u$ and $\partial z/\partial v$ provided $\partial F/\partial z \neq 0$, getting

$$\frac{\partial z}{\partial u} = -\frac{\frac{\partial F}{\partial x}}{\frac{\partial F}{\partial z}} \quad \text{and} \quad \frac{\partial z}{\partial v} = -\frac{\frac{\partial F}{\partial y}}{\frac{\partial F}{\partial z}}$$

Finally, since $u = x$ and $v = y$,

$$\frac{\partial z}{\partial x} = -\frac{F_x}{F_z} \quad \text{and} \quad \frac{\partial z}{\partial y} = -\frac{F_y}{F_z} \quad \text{if } F_z \neq 0 \qquad (14.15)$$

Note the similarity between these equations and Equation 14.14.

EXAMPLE 14.12 If z is defined implicitly as a differentiable function of x and y by the equation

$$z^3 - 2xz + y^2 - x^3 - 13 = 0$$

find $\partial z/\partial x$ and $\partial z/\partial y$ at the point $(-1, 3, 1)$.

Solution Here $F(x, y, z)$ is the left-hand side of the given equation. So we have, by Equations 14.15,

$$\frac{\partial z}{\partial x} = -\frac{F_x}{F_z} = -\frac{-2z - 3x^2}{3z^2 - 2x}$$

$$\frac{\partial z}{\partial y} = -\frac{F_y}{F_z} = -\frac{2y}{3z^2 - 2x}$$

On substituting $(-1, 3, 1)$ and simplifying, we get

$$\left.\frac{\partial z}{\partial x}\right|_{(-1,3,1)} = 1 \quad \text{and} \quad \left.\frac{\partial z}{\partial y}\right|_{(-1,3,1)} = -\frac{6}{5} \qquad \blacksquare$$

Exercise Set 14.2

In Exercises 1–16, make use of a chain rule. In Exercises 1–4, find dz/dt at the specified value of t.

1. $z = x^2 + y^2$, $x = t^2 + 1$, $y = 2 - t^2$; at $t = 2$

2. $z = 2x^2 - 3xy$, $x = e^t$, $y = te^t$; at $t = 0$

3. $z = xe^y - ye^x$, $x = \ln t$, $y = \ln \frac{1}{t}$; at $t = 1$

4. $z = x \cos xy$, $x = 2t^2$, $y = \frac{1}{t}$; at $t = \frac{\pi}{4}$

In Exercises 5–8, find dz/dt. Express answers in terms of t.

5. $z = \ln \frac{y}{x}$, $x = \sin t$, $y = \cos t$

6. $z = \tan^{-1} \frac{y}{x}$, $x = \sin 2t$, $y = \cos 2t$

7. $z = r^2(1 - \cos \theta)$, $r = \sqrt{t}$, $\theta = t^2$

8. $z = uv^2w^3$, $u = e^t$, $v = e^{-t}$, $w = te^t$

In Exercises 9–12, find $\partial z/\partial u$ and $\partial z/\partial v$ at the specified point.

9. $z = x^2 - 2xy$, $x = \dfrac{u}{v}$, $y = uv$; at $(u, v) = (2, -1)$

10. $z = \ln\sqrt{x^2 + y^2}$, $x = u + v$, $y = 2u - 3v$; at $(u, v) = (3, 1)$

11. $z = e^{xy}$, $x = \dfrac{u}{v}$, $y = 2v$; at $(u, v) = \left(1, \dfrac{1}{2}\right)$

12. $z = \sin(x + y)$, $x = u^2 - v^2$, $y = 2uv$; at $(u, v) = (\sqrt{\pi}, \sqrt{\pi}/2)$

In Exercises 13–16, find $\partial z/\partial u$ and $\partial z/\partial v$ at an arbitrary point (u, v).

13. $z = \ln\dfrac{x+y}{x-y}$; $x = \cos^2 uv$, $y = \sin^2 uv$

14. $z = x^2 + 2xy - y^2$; $x = \cosh u + \sinh v$, $y = \cosh u - \sinh v$

15. $z = r^2(3\sin^2\theta - 2\cos^2\theta)$; $r = \sqrt{u^2 + v^2}$, $\theta = \tan^{-1}(v/u)(u \neq 0)$

16. $z = \sqrt{rst}$; $r = u^2v$, $s = u/v$, $t = v^2/u$ $(uv > 0)$

In Exercises 17–22, assume the given equation defines y as a differentiable function of x on some domain, and use Equation 14.14 to find dy/dx there.

17. $x^3 - 2xy^2 + y^4 = 0$

18. $2x^2y^2 - 3xy^3 + 4y - 5 = 0$

19. $x \sin y - y \sin x = 1$

20. $x \ln y + y \ln x = xy$

21. $e^{-xy}(x^2 - y^2) = 4$

22. $xy \tan xy + 1 = 0$

In Exercises 23–28, assume the given equation defines z as a differentiable function of x and y on some domain, and use Equations 14.15 to find $\partial z/\partial x$ and $\partial z/\partial y$ there.

23. $x^2yz - 2y^2 - 3z^2 = 0$

24. $x^2 + y^2 + z^2 - 2xy + 3xz - 4yz = 1$

25. $\ln\left[(x+y)/\sqrt{z}\right] - 2xyz = 4$

26. $\tan^{-1}\dfrac{x}{y} - \tan^{-1}\dfrac{y}{z} = 2$

27. $\sin xz + \ln \cos yz = 0$

28. $z - xy \ln z = 1$

29. If $w = (u + v)/u$ and $u = \sqrt{x^2 + y^2 + z^2}$, $v = 2xyz$, find $\partial w/\partial x$, $\partial w/\partial y$, and $\partial w/\partial z$ as functions of x, y, and z.

30. Let $f(x, y, z) = xyz$ and $F(s, t) = f(s+t, s-t, st)$. Using an appropriate chain rule, find $F_s(2, -1)$ and $F_t(2, -1)$.

31. Let $z = f(x, y, z)$, and let the xy coordinates be transformed to polar coordinates by the equations $x = r\cos\theta$, $y = r\sin\theta$. Show that
$$\left(\dfrac{\partial z}{\partial r}\right)^2 + \dfrac{1}{r^2}\left(\dfrac{\partial z}{\partial \theta}\right)^2 = \left(\dfrac{\partial z}{\partial x}\right)^2 + \left(\dfrac{\partial z}{\partial y}\right)^2$$

32. The relationship between the pressure P, volume V, and temperature T of a certain gas is given by $PV = 12T$. If the pressure is decreasing at the constant rate of 3 psi/min and the temperature is constantly increasing at $4°$K per minute, find the rate at which the volume is changing when $P = 10$ psi and $T = 298°$K.

33. An oil slick in the Gulf of Mexico from a ruptured oil tanker is approximately triangular in shape. When the height of the triangle is 2 km, the base is 3 km, and at that instant the height and base are increasing at the rates of 200 m/hr and 320 m/hr, respectively. Find the rate at which the area is increasing at that instant.

34. Let $z = f(x - y, y - x)$. Show that
$$\dfrac{\partial z}{\partial x} + \dfrac{\partial z}{\partial y} = 0$$
(Hint: Let $u = x - y$ and $v = y - x$.)

35. Let $z = f(x/y)$. Show that
$$x\dfrac{\partial z}{\partial x} + y\dfrac{\partial z}{\partial y} = 0$$

36. Prove that $y = f(x + ct) + g(x - ct)$ is a solution of the wave equation

$$\frac{\partial^2 y}{\partial t^2} = c^2 \frac{\partial^2 y}{\partial x^2}$$

for any twice-differentiable functions f and g.

37. Let $z = f(x, y)$, $x = g(u, v)$, and $y = h(u, v)$. Derive formulas for $\partial^2 z/\partial u^2$ and $\partial^2 z/\partial v^2$.

38. Assume that the equation $F(x, y) = 0$ defines y as a twice-differentiable function, $y = f(x)$, on a domain D. Derive the following formula.

$$f''(x) = -\frac{F_{xx} F_y^2 - 2F_x F_y F_{xy} + F_{yy} F_x^2}{F_y^3} \quad (F_y \neq 0)$$

39. For the function in Example 14.10, find $\partial^2 z/(\partial \theta \, \partial r)$.

40. If $z = f(x, y)$, the expression $\partial^2 z/\partial x^2 + \partial^2 z/\partial y^2$ is called the *Laplacian* of z. Show that on changing the xy coordinates to polar coordinates by means of the transformation $x = r \cos \theta$, $y = r \sin \theta$, the Laplacian of z becomes

$$\frac{\partial^2 z}{\partial r^2} + \frac{1}{r} \frac{\partial z}{\partial r} + \frac{1}{r^2} \frac{\partial^2 z}{\partial \theta^2}$$

41. In a grain elevator, grain deposited through a chute at the rate of 60 ft³/min assumes the form of a cone as it accumulates on the floor. After 5 min the base radius of the cone is 10 ft, and at that instant it is increasing at the rate of 0.5 ft/min. How fast is the height of the cone changing at that instant?

42. Part of a certain hydraulic lift has a triangular shape that varies in size as the lift is actuated. At the instant when two adjacent sides of the triangle are 3.1 m and 1.7 m long and the angle between them is 30°, these sides and this angle are increasing at the respective rates of 0.5 m/s, 0.8 m/s, and 2°/s. Find the rate at which the area of the triangle is changing at that instant.

CAS *Use a CAS for Exercises 43 through 46.*

43. Express a formula for the Chain Rule for a differentiable function $w = f(x, y, z)$, where x, y, and z are differentiable functions of t.

44. Use the formula obtained in Exercise 43 to find dw/dt in each of the following:

(a) $w = f(x, y, z) = xy + \sin\left(\dfrac{x^2}{z}\right) + x^3 z$, $x(t) = t$, $y(t) = t^2$, $z(t) = t + 1$

(b) $w = f(x, y, z) = x \sin(y + z)$, $x(t) = 3t$, $y(t) = t^3$, $z(t) = e^t$

45. If $z = f(x, y)$ and x and y are both functions of two variables u and v, establish formulas for $\partial f/\partial u$ and $\partial f/\partial v$. Use the formula to compute $\partial f/\partial u$ and $\partial f/\partial v$ when $f(x, y) = e^{xy}$ and $x = u - 3uv + \sin v$, $y = u + e^{u^2 v}$.

46. Let $z = f(x, y)$ and show that $x \partial z/\partial x - y \partial z/\partial y = 0$.

14.3 DIRECTIONAL DERIVATIVES

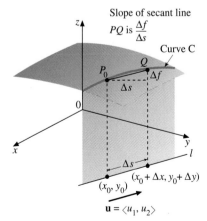

FIGURE 14.15

In Section 13.4 we defined the partial derivatives f_x and f_y for a function of two variables $z = f(x, y)$ as

$$f_x(x_0, y_0) = \lim_{\Delta x \to 0} \frac{f(x_0 + \Delta x, y_0) - f(x_0, y_0)}{\Delta x}$$

and

$$f_y(x_0, y_0) = \lim_{\Delta y \to 0} \frac{f(x_0, y_0 + \Delta y) - f(x_0, y_0)}{\Delta y}$$

The first partials describe the rate of change of z in the x direction and y direction, respectively—that is, in the direction of the unit vectors $\mathbf{i} = \langle 1, 0 \rangle$ and $\mathbf{j} = \langle 0, 1 \rangle$. We now want to investigate the rate of change of z at a point (x_0, y_0) in the direction of an arbitrary unit vector. We will see that this rate of change can be expressed in a way that involves both partials f_x and f_y.

Suppose, then, that we want to determine the rate of change of $z = f(x, y)$ at (x_0, y_0) in the direction of the unit vector $\mathbf{u} = \langle u_1, u_2 \rangle$. The vertical plane that passes through the point $P_0 = (x_0, y_0, z_0)$ on the surface $z = f(x, y)$ in the direction $\mathbf{u}$ intersects the surface in some curve C, as we show in Figure 14.15.

The slope of the tangent line to the curve C at P_0 is the rate of change of z in the direction **u**.

We can write a vector equation of the line l through (x_0, y_0) in the direction **u** as

$$\mathbf{r}(s) = \langle x_0 + su_1, y_0 + su_2 \rangle$$

where the parameter s is the directed distance along l, starting from (x_0, y_0), with positive direction given by **u**.

A point on the line l that is Δs units from (x_0, y_0) has the position vector

$$r(\Delta s) = \langle x_0 + (\Delta s)u_1, y_0 + (\Delta s)u_2 \rangle$$

If we let $\Delta x = (\Delta s)u_1$, and $\Delta y = (\Delta s)u_2$, then the coordinates of this point are $x_0 + \Delta x$ and $y_0 + \Delta y$. Let Q be the point on the surface directly above $(x_0 + \Delta x, y_0 + \Delta y)$, as in Figure 14.15. As a point (x, y) moves from (x_0, y_0) to $(x_0 + \Delta x, y_0 + \Delta y)$ in the xy-plane, $f(x, y)$ changes by the amount

$$\begin{aligned}\Delta f &= f(x_0 + \Delta x, y_0 + \Delta y) - f(x_0, y_0) \\ &= f(x_0 + u_1 \Delta s, y_0 + u_2 \Delta s) - f(x_0, y_0)\end{aligned}$$

The quotient

$$\frac{\Delta f}{\Delta s}$$

is the slope of the secant line through P and Q. (See Figure 14.15.) If this quotient approaches a limit as Δs approaches 0, we obtain the rate of change of z in the direction **u** (the slope of the tangent line to the curve C at P_0). We call this limit the *directional derivative* of f at (x_0, y_0) in the direction **u**, and we denote it either by $D_\mathbf{u} f(x_0, y_0)$ or by df/ds. In summary, we have the following definition.

Definition 14.3
The Directional Derivative

The **directional derivative** of $f(x, y)$ at (x_0, y_0) in the direction of the unit vector $\mathbf{u} = \langle u_1, u_2 \rangle$ is

$$D_\mathbf{u} f(x_0, y_0) = \frac{df}{ds} = \lim_{\Delta s \to 0} \frac{\Delta f}{\Delta s} \quad (14.16)$$

provided this limit exists, where

$$\Delta f = f(x_0 + u_1 \Delta s, y_0 + u_2 \Delta s) - f(x_0, y_0)$$

If we consider the special case of the directional derivative in which $\mathbf{u} = \langle 1, 0 \rangle$, then $\Delta x = \Delta s$ and $\Delta y = 0$. So, by Equation 14.16,

$$D_\mathbf{u} f(x_0, y_0) = \lim_{\Delta x \to 0} \frac{f(x_0 + \Delta x, y_0) - f(x_0, y_0)}{\Delta x} = f_x(x_0, y_0)$$

That is, the directional derivative in the direction of the x-axis is the same as the partial derivative with respect to x. Similarly, we can show that when $\mathbf{u} = \langle 0, 1 \rangle$, $D_\mathbf{u} f(x_0, y_0) = f_y(x_0, y_0)$. Thus, the partial derivatives of f with respect to x and y are special cases of the directional derivative.

Calculating the Directional Derivative

When f is *differentiable* at (x_0, y_0), we can show that $D_{\mathbf{u}} f(x_0, y_0)$ does exist, and we can obtain a useful computational formula for the directional derivative. By Definition 14.1, we can write $\Delta f / \Delta s$ in the form

$$\frac{\Delta f}{\Delta s} = f_x(x_0, y_0) \frac{\Delta x}{\Delta s} + f_y(x_0, y_0) \frac{\Delta y}{\Delta s} + \varepsilon_1 \frac{\Delta x}{\Delta s} + \varepsilon_2 \frac{\Delta y}{\Delta s}$$

But $\Delta x = u_1 \Delta s$ and $\Delta y = u_2 \Delta s$, so $\Delta x / \Delta s = u_1$ and $\Delta y / \Delta s = u_2$. Furthermore, as $\Delta s \to 0$, both Δx and Δy approach 0, and hence $\varepsilon_1 \to 0$ and $\varepsilon_2 \to 0$. Thus,

$$D_{\mathbf{u}} f(x_0, y_0) = \lim_{\Delta s \to 0} \frac{\Delta f}{\Delta s} = f_x(x_0, y_0) u_1 + f_y(x_0, y_0) u_2$$

We restate this result for emphasis.

The Directional Derivative of a Differentiable Function

If f is differentiable at (x_0, y_0), then for any unit vector $\mathbf{u} = \langle u_1, u_2 \rangle$, $D_{\mathbf{u}} f(x_0, y_0)$ exists and is given by the formula

$$D_{\mathbf{u}} f(x_0, y_0) = f_x(x_0, y_0) u_1 + f_y(x_0, y_0) u_2 \tag{14.17}$$

REMARK

■ For Equation 14.17 to be valid, it is essential that $\mathbf{u}$ be a *unit* vector. If the direction is given by some non-unit-vector $\mathbf{v}$, then we replace $\mathbf{v}$ with the unit vector $\mathbf{u} = \mathbf{v}/|\mathbf{v}|$.

EXAMPLE 14.13 Find the directional derivative of the function $f(x, y) = x^2 - 2xy^3$ at $(-2, 1)$ in the direction of the vector $\mathbf{v} = \langle 3, 4 \rangle$.

Solution First, we obtain a unit direction vector:

$$\mathbf{u} = \frac{\mathbf{v}}{|\mathbf{v}|} = \left\langle \frac{3}{5}, \frac{4}{5} \right\rangle$$

Now we use Equation 14.17:

$$D_{\mathbf{u}} f(x, y) = f_x(x, y) \cdot \left(\frac{3}{5}\right) + f_y(x, y) \cdot \left(\frac{4}{5}\right)$$

$$= (2x - 2y^3) \left(\frac{3}{5}\right) + (-6xy^2) \left(\frac{4}{5}\right)$$

Setting $(x, y) = (-2, 1)$, we get

$$D_{\mathbf{u}} f(-2, 1) = (-6) \left(\frac{3}{5}\right) + (12) \left(\frac{4}{5}\right) = 6 \qquad ■$$

EXAMPLE 14.14 A coordinate system is established on a flat metal plate, and it is determined that at a point (x, y) on the plate, other than the origin, the temperature $T(x, y)$ in degrees Celsius is given by $T(x, y) = 100(x^2 + y^2)^{-1/2}$. Find the instantaneous rate of change of temperature at the point $P(2, 6)$ in the direction from P toward $Q(4, 2)$.

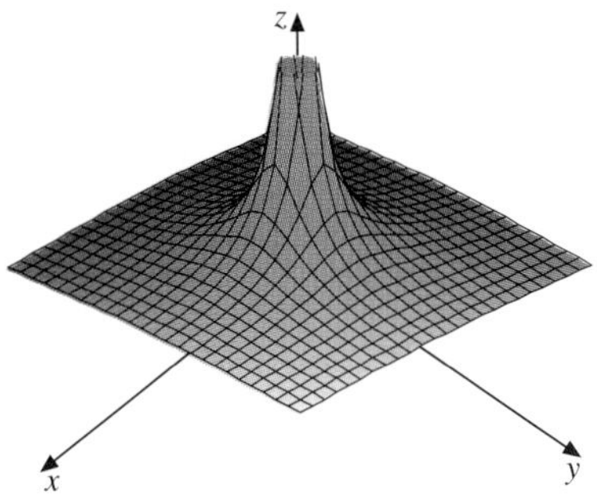

Solution We show the graph of the temperature function $z = T(x, y)$ in Figure 14.16. The instantaneous rate of change of T in the direction of $\overrightarrow{PQ}$ is the directional derivative at P in the direction of the unit vector

$$\mathbf{u} = \frac{\overrightarrow{PQ}}{|\overrightarrow{PQ}|} = \frac{\langle 2, -4 \rangle}{\sqrt{20}} = \left\langle \frac{1}{\sqrt{5}}, \frac{-2}{\sqrt{5}} \right\rangle$$

So we have, by Equation 14.17,

$$D_{\mathbf{u}} T(x, y) = \left[\frac{-100x}{(x^2 + y^2)^{3/2}} \right] \left(\frac{1}{\sqrt{5}} \right) + \left[\frac{-100y}{(x^2 + y^2)^{3/2}} \right] \left(\frac{-2}{\sqrt{5}} \right)$$

and after simplification, we obtain

$$D_{\mathbf{u}} T(2, 6) = \left(\frac{-200}{80\sqrt{10}} \right) \left(\frac{1}{\sqrt{5}} \right) + \left(\frac{-600}{80\sqrt{10}} \right) \left(\frac{-2}{\sqrt{5}} \right) = \frac{5\sqrt{2}}{4} \approx 1.768$$

Thus, as a point moves from P one unit on a line toward Q, one would expect an increase in temperature of approximately $1.768°$. (Remember, though, that this rate of change is instantaneous, so after moving away from P the rate changes.) ∎

FIGURE 14.16

The Gradient of a Function

The right-hand side of Equation 14.17 has the appearance of the dot product of two vectors. In fact,

$$D_{\mathbf{u}} f(x, y) = f_x(x, y) u_1 + f_y(x, y) u_2$$
$$= \langle f_x(x, y), f_y(x, y) \rangle \cdot \langle u_1, u_2 \rangle$$
$$= \langle f_x(x, y), f_y(x, y) \rangle \cdot \mathbf{u}$$

The vector $\langle f_x(x, y), f_y(x, y) \rangle$ occurs in a number of contexts and is given the special name, the *gradient* of f.

Definition 14.4
The Gradient

The **gradient** of f at a point $P(x, y)$ is denoted by $\nabla f(P)$ or $\nabla f(x, y)$ and is defined as the vector

$$\nabla f(x, y) = \langle f_x(x, y), f_y(x, y) \rangle \qquad (14.18)$$

provided these partials exist.

REMARK

■ The symbol ∇ is read "del." Another notation for the gradient of f is **Grad**. It is an operator, in the same sense that d/dx is an operator. So ∇f and **Grad** f mean the same thing. We will use ∇f exclusively in this book.

Using Definition 14.4, we can rewrite Equation 14.17 in the following form:

$$D_{\mathbf{u}} f(x, y) = \nabla f(x, y) \cdot \mathbf{u} \qquad (14.19)$$

EXAMPLE 14.15 Let $f(x, y) = \ln(x/y)$. Find ∇f and use it to find the directional derivative of f at $(1, 2)$ in the direction of $\mathbf{u} = \langle 1/2, -\sqrt{3}/2 \rangle$.

Solution To simplify the computation of the partials f_x and f_y, we write $\ln \frac{x}{y}$ in the equivalent form $\ln x - \ln y$, so that $f(x, y) = \ln x - \ln y$. Then we see that

$$\nabla f = \langle f_x, f_y \rangle = \left\langle \frac{1}{x}, -\frac{1}{y} \right\rangle$$

At the point $(1, 2)$, we therefore have

$$\nabla f(1, 2) = \left\langle 1, -\frac{1}{2} \right\rangle$$

By Equation 14.19,

$$D_{\mathbf{u}} f(1, 2) = \left\langle 1, -\frac{1}{2} \right\rangle \cdot \left\langle \frac{1}{2}, -\frac{\sqrt{3}}{2} \right\rangle = \frac{1}{2} + \frac{\sqrt{3}}{4} = \frac{2 + \sqrt{3}}{4} \qquad \blacksquare$$

Extreme Values of the Directional Derivative

Sometimes we want to know the direction in which a function changes most rapidly and what this maximum rate of change is. The answer is provided in the following theorem.

THEOREM 14.3 Let f have continuous partial derivatives f_x and f_y in a region containing a point (x_0, y_0) at which $\nabla f \neq 0$. Then the maximum value of $D_{\mathbf{u}} f(x_0, y_0)$ is $|\nabla f(x_0, y_0)|$, which occurs when $\mathbf{u}$ is in the direction of $\nabla f(x_0, y_0)$. The minimum value of $D_{\mathbf{u}} f(x_0, y_0)$ is $-|\nabla f(x_0, y_0)|$ and occurs when $\mathbf{u}$ is in the direction of $-\nabla f(x_0, y_0)$.

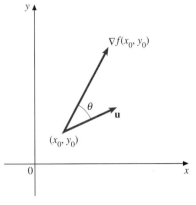

FIGURE 14.17

Proof In Figure 14.17 we show the gradient vector $\nabla f(x_0, y_0)$ and a unit vector **u**, both drawn from the point (x_0, y_0). If θ denotes the angle between these two vectors, then

$$D_{\mathbf{u}} f(x_0, y_0) = \nabla f(x_0, y_0) \cdot \mathbf{u} \quad \text{By Equation 14.19}$$
$$= |\nabla f(x_0, y_0)||\mathbf{u}| \cos \theta \quad \text{By Equation 11.4}$$
$$= |\nabla f(x_0, y_0)| \cos \theta \quad \text{Since } |\mathbf{u}| = 1 \quad (14.20)$$

Now the maximum value $\cos \theta$ can assume is 1, which occurs when $\theta = 0$. Thus, to obtain the maximum value of $D_{\mathbf{u}} f(x_0, y_0)$, we take the angle between **u** and $\nabla f(x_0, y_0)$ to be 0 (the cosine is 1 at 0 and is never greater than 1); that is, we take **u** in the same direction as $\nabla f(x_0, y_0)$, namely,

$$\mathbf{u} = \frac{\nabla f(x_0, y_0)}{|\nabla f(x_0, y_0)|}$$

Furthermore, when we use this value of **u**, we see from Equation 14.20 that

$$\max D_{\mathbf{u}} f(x_0, y_0) = |\nabla f(x_0, y_0)|$$

since $\cos 0 = 1$.

Similarly, the minimum value of $\cos \theta$ is -1, which occurs when $\theta = \pi$. In this case **u** is the opposite of $\nabla f(x_0, y_0)$. That is,

$$\mathbf{u} = -\frac{\nabla f(x_0, y_0)}{|\nabla f(x_0, y_0)|}$$

and from Equation 14.20 we see that

$$\min D_{\mathbf{u}} f(x_0, y_0) = -|\nabla f(x_0, y_0)| \quad \blacksquare$$

EXAMPLE 14.16 For the function $f(x, y) = \ln(x/y)$ of Example 14.15, find the maximum value of the directional derivative at the point $(1, 2)$, and give the direction **u** that produces this maximum value.

Solution In Example 14.15, we found that

$$\nabla f(1, 2) = \left\langle 1, -\frac{1}{2} \right\rangle$$

Thus, by Theorem 14.3,

$$\max D_{\mathbf{u}} f(1, 2) = \left| \left\langle 1, -\frac{1}{2} \right\rangle \right| = \sqrt{1 + \frac{1}{4}} = \frac{\sqrt{5}}{2}$$

This maximum value occurs in the direction of the vector

$$\mathbf{u} = \frac{\nabla f(1, 2)}{|\nabla f(1, 2)|} = \frac{2}{\sqrt{5}} \left\langle 1, -\frac{1}{2} \right\rangle = \left\langle \frac{2}{\sqrt{5}}, \frac{-1}{\sqrt{5}} \right\rangle \quad \blacksquare$$

EXAMPLE 14.17 Refer to Example 14.14. In what direction from the point $P(2, 6)$ does the temperature $T(x, y) = 100(x^2 + y^2)^{-1/2}$ change most rapidly, and what is this maximum rate of change?

Solution In Example 14.14, the temperature function $T(x, y)$ is given as

$$T(x, y) = 100(x^2 + y^2)^{-1/2}$$

The gradient vector $\nabla T = \langle T_x(x, y), T_y(x, y)\rangle$ is given by

$$\nabla T = 100 \left\langle \frac{-x}{(x^2 + y^2)^{3/2}}, \frac{-y}{(x^2 + y^2)^{3/2}} \right\rangle$$

Evaluating ∇T at (2, 6), we have

$$\nabla T(2, 6) = 100 \left\langle \frac{-2}{(40)^{3/2}}, \frac{-6}{(40)^{3/2}} \right\rangle$$

and

$$|\nabla T(2, 6)| = 100 \sqrt{\frac{4 + 36}{(40)^3}} = \frac{100}{40} = \frac{5}{2}$$

We therefore have, by Theorem 14.3,

$$\max D_{\mathbf{u}} T(2, 6) = \frac{5}{2}$$

and the maximum occurs in the direction of the vector

$$\mathbf{u} = \frac{\nabla T(2, 6)}{|\nabla T(2, 6)|} = \frac{100 \left\langle -\frac{2}{(40)^{3/2}}, -\frac{6}{(40)^{3/2}} \right\rangle}{5/2} = 40 \left\langle -\frac{2}{(40)^{3/2}}, -\frac{6}{(40)^{3/2}} \right\rangle$$

$$= \left\langle -\frac{2}{(40)^{1/2}}, -\frac{6}{(40)^{1/2}} \right\rangle = \left\langle -\frac{1}{\sqrt{10}}, -\frac{3}{\sqrt{10}} \right\rangle \qquad \blacksquare$$

Relation of ∇f to Level Curves

The relationship between the gradient vector $\nabla f(x_0, y_0)$ and the level curve to the surface $z = f(x, y)$ that passes through (x_0, y_0) is helpful in understanding the maximum value of the directional derivative $D_{\mathbf{u}} f(x_0, y_0)$. We suppose f is differentiable at (x_0, y_0) and that $\nabla f(x_0, y_0) \neq 0$. If $c = f(x_0, y_0)$, then the level curve $f(x, y) = c$ passes through (x_0, y_0). Suppose this level curve is parameterized by the differentiable functions $x = x(t)$, $y = y(t)$. Let $x_0 = x(t_0)$ and $y_0 = y(t_0)$. Then, since $f(x(t), y(t)) = c$, we have, by the Chain Rule (Equation 14.10),

$$f_x(x_0, y_0)x'(t_0) + f_y(x_0, y_0)y'(t_0) = 0$$

or, equivalently,

$$\nabla f(x_0, y_0) \cdot \langle x'(t_0), y'(t_0)\rangle = 0$$

The vector $\langle x'(t_0), y'(t_0)\rangle$ is a tangent vector to the level curve $f(x, y) = c$ at the point (x_0, y_0). We conclude, therefore, that the gradient vector $\nabla f(x_0, y_0)$ is perpendicular to this tangent vector at (x_0, y_0). We illustrate this result in Figure 14.18. We summarize our results as follows.

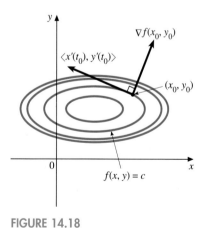

FIGURE 14.18

Orthogonality of the Gradient Vector and Level Curves

$\nabla f(x_0, y_0)$ is orthogonal to the level curve $f(x, y) = c$ passing through (x_0, y_0).

Combining this result with Theorem 14.3, we conclude that from any point in its domain a function $f(x, y)$ increases (or decreases) most rapidly in a direction perpendicular to the level curve through that point. For example, on a weather map, to move in the direction of the most rapid change in temperature, we would move in a direction perpendicular to the isotherms.

EXAMPLE 14.18   Let $f(x, y) = x^2/4 + y^2$. Sketch the level curve through the point (3, 2), and show the gradient vector at that point.

Solution   Since $f(3, 2) = 9/4 + 4 = 25/4$, the level curve has the equation $x^2/4 + y^2 = 25/4$, or in standard form,

$$\frac{x^2}{25} + \frac{y^2}{25/4} = 1$$

The graph is an ellipse with semimajor axis 5 and semiminor axis 5/2.

We also have

$$\nabla f = \left\langle \frac{x}{2}, 2y \right\rangle$$

so $\nabla f(3, 2) = \langle 3/2, 4 \rangle$. We show the level curve and the gradient orthogonal to it in Figure 14.19. ∎

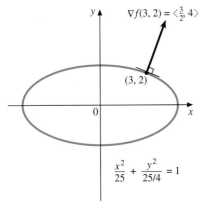

FIGURE 14.19

Extension to Three Variables

All of the concepts in this section can be extended in natural ways to functions of three (or more) variables. We summarize the results as follows.

Directional Derivative for a Function of Three Variables

If f is a differentiable function of three variables at the point $P_0 = (x_0, y_0, z_0)$ and **u** is a unit vector, then the directional derivative of f at P_0 in the direction **u** is given by

$$D_{\mathbf{u}} f(P_0) = \nabla f(P_0) \cdot \mathbf{u}$$

where $\nabla f = \langle f_x, f_y, f_z \rangle$. The maximum value of this directional derivative, which is $|\nabla f(P_0)|$, occurs when **u** is in the direction of $\nabla f(P_0)$. Furthermore, the direction of maximum change in f is orthogonal to the level surface $f(x, y, z) = c$ passing through P_0. The minimum value, which is $-|\nabla f(P_0)|$, occurs when **u** is directed opposite to $\nabla f(P_0)$.

Exercise Set 14.3

In Exercises 1–14, find the directional derivative of f at the given point in the direction of the given vector.

1. $f(x, y) = 2x^2y - 3xy^2$ at (3, 1); $\mathbf{u} = \langle \frac{3}{5}, \frac{4}{5} \rangle$

2. $f(x, y) = x^3 - 2xy + y^3$ at (1, 0); $\mathbf{u} = \langle 1/\sqrt{10}, 3/\sqrt{10} \rangle$

3. $f(x, y) = \ln[xy/(x^2 + y^2)]$ at (1, 2); $\mathbf{v} = \langle -2, 2 \rangle$

4. $f(x, y) = xe^y - ye^x$ at (0, 0); $\mathbf{v} = \langle 8, -1 \rangle$

5. $f(x, y) = x(\cos y - \sin y)$ at $\left(-1, \frac{\pi}{2}\right)$; $\mathbf{v} = 4\mathbf{i} - 3\mathbf{j}$

6. $f(x, y) = \tan^{-1} \frac{y}{x}$ at $(2, 1)$; $\mathbf{v} = \mathbf{i} + \mathbf{j}$

7. $f(x, y) = x \cosh y + y \cosh x$ at $(0, 0)$; $\mathbf{v} = -2\mathbf{i} + 4\mathbf{j}$

8. $f(x, y) = y^2 \ln \sqrt{x}$ at $(1, -2)$; $\mathbf{v} = 3\mathbf{i} + 2\mathbf{j}$

9. $f(x, y) = \sqrt{9 - x^2 - y^2}$ at $(2, -1)$; $\mathbf{v} = \langle 1, -1 \rangle$

10. $f(x, y) = e^{(x-y)/(x+y)}$ at $(1, 1)$; $\mathbf{v} = \langle 5, 12 \rangle$

11. $f(x, y, z) = x^2 - 2y^2 + 3z^2$ at $(2, 0, -1)$;
    $\mathbf{u} = \langle \frac{1}{3}, \frac{2}{3}, -\frac{2}{3} \rangle$

12. $f(x, y, z) = \frac{x+y}{x+z}$ at $(3, -2, -1)$;
    $\mathbf{u} = \langle \frac{1}{\sqrt{6}}, -\frac{1}{\sqrt{6}}, \frac{2}{\sqrt{6}} \rangle$

13. $f(x, y, z) = \ln[x^2/(yz^3)]$ at $(1, 1, 3)$; $\mathbf{v} = 3\mathbf{i} - 4\mathbf{j} + 5\mathbf{k}$

14. $f(x, y, z) = x \cosh(y + z)$ at $(3, 2, -1)$; $\mathbf{v} = \mathbf{i} + \mathbf{j} - \mathbf{k}$

15. Find the directional derivative of $f(x, y) = 3x^2 - 2xy^2$ at $(3, -2)$ in the direction from $(3, -2)$ toward $(5, 4)$.

16. Find the directional derivative of $f(x, y) = \ln \sqrt{x^2 - 2y^2}$ at $(2, 1)$ in the direction from $(2, 1)$ toward $(5, -3)$.

In Exercises 17–20, find the unit vector $\mathbf{u}$ for which $D_\mathbf{u} f(P_0)$ is a maximum, and give this maximum value.

17. $f(x, y) = \sqrt{\frac{x-y}{x+y}}$; $P_0 = (5, 4)$

18. $f(x, y) = \ln \cos(x + 2y)$; $P_0 = \left(\frac{\pi}{4}, \frac{\pi}{4}\right)$

19. $f(x, y) = y^2 + e^{(\sin x)/y}$; $P_0 = (0, -1)$

20. $f(x, y, z) = \ln \frac{x + 2y}{z^3}$; $P_0 = (5, -2, 3)$

21. In what direction from the point $(1, -1)$ is the instantaneous rate of change of $f(x, y) = 2x^2 + 2xy - 3y^2$ equal to 2? (There are two solutions.) In what direction from $(1, -1)$ does this function increase most rapidly? What is this most rapid rate of change?

22. In what direction from the point $(4, 1)$ is the function $f(x, y) = x/(y + 1)$ stationary? From the same point, in what direction is the instantaneous rate of change of this function equal to 1? (There are two solutions.) Can the rate of change from the point $(4, 1)$ in any direction ever equal 2? Explain.

23. Two adjacent edges of a flat, rectangular, metal plate coincide, respectively, with the positive x- and y-axes. For points other than the origin, the temperature $T(x, y)$ at an arbitrary point (x, y) is inversely proportional to the distance from P to the origin. At the point $P(8, 6)$, the temperature is $10°C$. How rapidly is the temperature changing at P in the direction from P toward $Q(6, 10)$? In what direction from P does the temperature decrease most rapidly, and what is this rate of decrease?

24. Two adjacent edges of a large, square, metal plate are kept at temperatures $T = 0$ and $T = 100$, respectively, and the flat surfaces are well insulated. By taking the positive x- and y-axes along the edges held at $T = 0$ and $T = 100$, respectively, it can be shown that the temperature $T(x, y)$ at an arbitrary point in the plate is approximated by

$$T(x, y) = \frac{200}{\pi} \tan^{-1} \frac{y}{x}$$

How rapidly is the temperature changing at the point $(2, 4)$ in the direction of the vector $\mathbf{v} = 3\mathbf{i} - 4\mathbf{j}$? In what direction from $(2, 4)$ is the temperature increasing most rapidly, and what is this most rapid change?

25. A cross section of two long, coaxial, conducting cylindrical surfaces consists of the circles $x^2 + y^2 = 1$ and $x^2 + y^2 = 4$. If the smaller cylinder is held at electrostatic potential $V = 0$ and the larger at $V = 1$, then it can be shown that in the annular ring between the two, the potential $V(x, y)$ is given by

$$V(x, y) = \frac{\ln(x^2 + y^2)}{\ln 4}$$

Find the rate of change in potential at $P(\frac{3}{2}, \frac{1}{2})$ in the direction from P toward $Q(\frac{3}{4}, 1)$. In what direction from P does V increase most rapidly, and what is this rate of change?

26. In Exercise 24 find the equation of the isotherm $T(x, y) = c$ for c between 0 and 100 and identify the graph. Show that $\nabla T(x_0, y_0)$ is orthogonal to the isotherm through (x_0, y_0).

27. In Exercise 25 find the equation of the equipotential $V(x, y) = c$ for c between 0 and 1 and identify the graph. Show that $\nabla V(x_0, y_0)$ is orthogonal to the equipotential curve through (x_0, y_0).

28. Find a function $f(x, y)$ for which $\nabla f = \langle xe^x, e^{-y} \rangle$. Is this function unique? Explain.

29. Find the function f for which $\nabla f = (x \sin x)\mathbf{i} + (\cos y)\mathbf{j}$ and $f(\frac{\pi}{2}, 0) = 3$.

30. Show that if α, β, and γ are direction angles of the unit vector $\mathbf{u}$, and if f is differentiable at $P = (x, y, z)$, then
$$D_\mathbf{u} f(P) = f_x(P) \cos \alpha + f_y(P) \cos \beta + f_z(P) \cos \gamma$$

In Exercises 31–36, $u = f(x, y)$ and $v = g(x, y)$ are differentiable functions, and c and α are arbitrary real numbers. Prove each statement.

31. $\nabla(cu) = c\nabla u$

32. $\nabla(u + v) = \nabla u + \nabla v$

33. $\nabla(uv) = u\nabla v + v\nabla u$

34. $\nabla\left(\dfrac{u}{v}\right) = \dfrac{v\nabla u - u\nabla v}{v^2}$ if $v \neq 0$

35. $\nabla u^\alpha = \alpha u^{\alpha - 1} \nabla u$

36. If $w = h(u, v)$, and h is differentiable, then
$$\nabla w = \frac{\partial w}{\partial u} \nabla u + \frac{\partial w}{\partial v} \nabla v$$

CAS Use a CAS for Exercises 37 and 38.

37. Let $f(x, y) = 2x - y^2$. Plot the surface and a contour map for the surface. Find the gradient of f and use it to compute the gradient at the point $(2, -1)$ and the tangent line to the level curve $f(x, y) = 3$. Plot the tangent line and gradient vectors on the contour map.

38. Let $f(x, y) = x^2 + y^2$. Plot the surface and a contour map for the surface. Find the gradient of f and use it to compute the gradient at the point $(1, -1)$ and the tangent line to the level curve $f(x, y) = 2$. Plot the tangent line and gradient vectors on the contour map.

14.4 TANGENT PLANES AND NORMAL LINES

Denote by S the surface that is the graph of $F(x, y, z) = 0$. For example, S might be the ellipsoid whose equation is $x^2 + 2y^2 + 4z^2 = 16$. In this case, $F(x, y, z) = x^2 + 2y^2 + 4z^2 - 16$. Assume that F is differentiable at the point (x_0, y_0, z_0) on the surface S, with $\nabla F(x_0, y_0, z_0) \neq \mathbf{0}$. Let C be a curve on the surface S passing through (x_0, y_0, z_0) and defined by the vector function $\mathbf{r}(t) = \langle f(t), g(t), h(t) \rangle$. If $\mathbf{r}(t_0) = \langle x_0, y_0, z_0 \rangle$, then from Section 12.2, we know that $\mathbf{r}'(t_0)$ is a tangent vector to C at (x_0, y_0, z_0). Because C lies on the surface S, all points $(f(t), g(t), h(t))$ on C satisfy the equation of the surface; that is, $F(f(t), g(t), h(t)) = 0$, and so $dF/dt = 0$ wherever this derivative exists. It does exist at t_0 and is found by the Chain Rule:

$$\frac{dF}{dt}(f(t_0), g(t_0), h(t_0)) = F_x(x_0, y_0, z_0) f'(t_0) + F_y(x_0, y_0, z_0) g'(t_0)$$
$$+ F_z(x_0, y_0, z_0) h'(t_0)$$
$$= 0$$

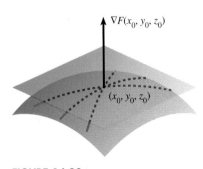

FIGURE 14.20

or, in vector form,

$$\nabla F(x_0, y_0, z_0) \cdot \mathbf{r}'(t_0) = 0$$

Since $\mathbf{r}'(t_0)$ is tangent to C, it follows that $\nabla F(x_0, y_0, z_0)$ is orthogonal to the tangent line to C at (x_0, y_0, z_0).

The argument just given applies to *every* curve C on S that passes through (x_0, y_0, z_0) and has a tangent line there. Thus, the plane through (x_0, y_0, z_0) that is perpendicular to $\nabla F(x_0, y_0, z_0)$ must contain all the tangent lines to such curves (see Figure 14.20). It is natural to call this plane the **tangent plane** to S at (x_0, y_0, z_0).

Definition 14.5
Tangent Plane and Normal Line

Let S be the surface that is the graph of $F(x, y, z) = 0$. Let (x_0, y_0, z_0) be a point on S at which F is differentiable, with $\nabla F(x_0, y_0, z_0) \neq \mathbf{0}$. Then the plane through (x_0, y_0, z_0) that has normal vector $\nabla F(x_0, y_0, z_0)$ is called the **tangent plane** to S at (x_0, y_0, z_0). The line through (x_0, y_0, z_0) with direction vector $\nabla F(x_0, y_0, z_0)$ is called the **normal line** to S at (x_0, y_0, z_0).

Since $\nabla F(x_0, y_0, z_0) = \langle F_x(x_0, y_0, z_0), F_y(x_0, y_0, z_0), F_z(x_0, y_0, z_0) \rangle$ is a normal vector to the tangent plane, we know from Section 11.6 that the equation of the tangent plane can be written in the following form:

Tangent Plane to the Surface $F(x, y, z) = 0$ at (x_0, y_0, z_0)

$$F_x(x_0, y_0, z_0)(x - x_0) + F_y(x_0, y_0, z_0)(y - y_0) + F_z(x_0, y_0, z_0)(z - z_0) = 0 \tag{14.21}$$

Also, the normal line has the following parametric equations:

The Normal Line to the Surface $F(x, y, z) = 0$ at (x_0, y_0, z_0)

$$\begin{cases} x = x_0 + F_x(x_0, y_0, z_0)\,t \\ y = y_0 + F_y(x_0, y_0, z_0)\,t \\ z = z_0 + F_z(x_0, y_0, z_0)\,t \end{cases} \tag{14.22}$$

EXAMPLE 14.19 Find equations of the tangent plane and normal line to the ellipsoid $x^2 + 2y^2 + 4z^2 = 16$ at the point $(2, -2, 1)$.

Solution We let $F(x, y, z) = x^2 + 2y^2 + 4z^2 - 16$, so that $\nabla F = \langle 2x, 4y, 8z \rangle$. Since F_x, F_y, and F_z are continuous everywhere, F is differentiable. Also, $\nabla F \neq \mathbf{0}$ for all points on the surface, since $\nabla F = \mathbf{0}$ only at the origin, which does not lie on the ellipsoid. Thus, a tangent plane exists everywhere. At $(2, -2, 1)$ the vector $\nabla F(2, -2, 1) = \langle 4, -8, 8 \rangle$ is normal to the tangent plane. So the equation of the tangent plane is

$$4(x - 2) - 8(y + 2) + 8(z - 1) = 0$$

which, on simplification, becomes

$$x - 2y + 2z = 8$$

Parametric equations of the normal line at $(2, -1, 1)$ are $x = 2 + 4t$, $y = -2 - 8t$, and $z = 1 + 8t$.

We might note that since $\nabla F(2, -2, 1) = \langle 4, -8, 8 \rangle$ is normal to the surface, so is $\frac{1}{4} \nabla F(2, -2, 1) = \langle 1, -2, 2 \rangle$, and this simpler vector could have been used to get the equations of both the tangent plane and the normal line. ■

Tangent Plane for $z = f(x, y)$

An equation in the form $z = f(x, y)$ can be written in the form $F(x, y, z) = 0$, where $F(x, y, z) = f(x, y) - z$. For example, $z = x^2 + y^2$ would be written as $x^2 + y^2 - z = 0$. So Equation 14.21 can be used to find the equation of the tangent plane. It is useful to obtain the general result for surfaces with equations in this form. We assume f is a differentiable function of two variables at (x_0, y_0). It follows that $F(x, y, z) = f(x, y) - z$ is a differentiable function of three variables at (x_0, y_0, z_0) (see Exercise 22 in Exercise Set 14.4). Furthermore, $\nabla F = \langle f_x, f_y, -1 \rangle$ is never $\mathbf{0}$. So the tangent plane at (x_0, y_0, z_0) exists and has $\langle f_x(x_0, y_0), f_y(x_0, y_0), -1 \rangle$ as a normal vector.

By Equation 14.21, the equation of the tangent plane is

$$f_x(x_0, y_0)(x - x_0) + f_y(x_0, y_0)(y - y_0) - (z - z_0) = 0$$

which can be written in the following equivalent form.

The Tangent Plane to the Surface $z = f(x, y)$ at (x_0, y_0, z_0)

$$z - z_0 = f_x(x_0, y_0)(x - x_0) + f_y(x_0, y_0)(y - y_0) \qquad (14.23)$$

You may use this result to get the tangent plane when $z = f(x, y)$, or you may use Equation 14.21 with $F(x, y, z) = f(x, y) - z$. The answers will be equivalent.

In employing Equation 14.23, it is important to remember the hypothesis that f is differentiable at (x_0, y_0), since otherwise there is no tangent plane. For example, it can be shown (see Exercise 37 in Exercise Set 14.1) that the function $f(x, y) = -\sqrt{|xy|}$ is continuous at the origin and that $f_x(0, 0) = 0$ and $f_y(0, 0) = 0$, yet f is not differentiable at $(0, 0)$. The nonexistence of a tangent plane to the surface $z = f(x, y)$ in this case can be seen clearly from the computer-generated graph in Figure 14.21.

REMARK
■ Equation 14.23 for the equation of the tangent plane to the surface $z = f(x, y)$ confirms our observation in Section 14.1 that the right-hand side of the linear approximation formula

$$f(x, y) \approx f(x_0, y_0) + f_x(x_0, y_0)(x - x_0) + f_y(x_0, y_0)(y - y_0)$$

is the vertical distance from the point (x, y) to the tangent plane to the surface at (x_0, y_0, z_0). (See Figure 14.2.)

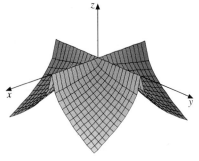
FIGURE 14.21

THE CHAIN RULE AND TANGENT PLANES USING COMPUTER ALGEBRA SYSTEMS

CAS 54

Use Maple and Mathematica to express the Chain Rule for a differentiable function $z = f(x, y)$ of x and y, where x and y are differentiable functions of t (see Equation 14.10). Then find dz/dt, where $z = f(x, y) = x^3 y + x \sin y$ and where $x = x(t) = t^2 + 1$ and $y = y(t) = 2t - 3$.

Maple:

z:=f(x,y);

Output: $z := f(x, y)$;

x:=X(t);y:=Y(t);

Here we use capital X and Y to avoid redefining the symbols x and y.

diff(z,t);

Output:

$D_{[1]}(f)(X(t), Y(t)) \left(\dfrac{\partial}{\partial t} X(t) \right) + D_{[2]}(f)(X(t), Y(t)) \left(\dfrac{\partial}{\partial t} Y(t) \right)$*

Compare this with Equation 14.10.

f:=(x,y)->x^3*y+x*sin(y);

Output: $f := (x, y) \to x^3 y + x \sin(y)$

z:=f(x,y);

Output: $z := x^3 y + x \sin(y)$

x:=t^2+1;y:=2*t-1;
diff(z,t);

Output:
$6(t^2 + 1)^2 (2t - 1)t + 2(t^2 + 1)^3 + 2t \sin(2t - 1)$
$+ 2(t^2 + 1) \cos(2t - 1)$

Mathematica:

z=f[x,y]
x=X[t]
y=Y[t]
D[z,t]
Clear[x,y,z,t]
z=x^3*y+x*Sin[y]
x=t^2+1
y=2*t-1
D[z,t]

CAS 55

Find the tangent plane to $z = f(x, y) = 5e^{-x^2 - y^2}$ at the point $(0, 1/4, 5e^{-1/16})$.

---

*In Maple, $D_{[1]}$ and $D_{[2]}$ represent the first partial derivatives with respect to x and y, respectively.

Maple:

f:=(x,y)->5*exp(-x^2-y^2);

To use Equation 14.21, we first compute the partial derivatives.

fx:=diff(f(x,y),x);

Output: $fx := -10xe^{-x^2-y^2}$

fy:=diff(f(x,y),y);

Output: $fy := -10ye^{-x^2-y^2}$

a:=0;b:=0.25;
plane:=f(a,b)+subs(x=a,y=b,fx)*(x-a)+subs(x=a,y=b,fy)*(y-b);

Output: $plane := 4.697065314 - 2.50e^{-.0625}(y - .25)$

Now we plot the surface and the tangent plane.

p1:=plot3d(f(x,y),x=-5..5,y=-5..5,scaling=unconstrained,
    numpoints=2000);
p2:=plot3d(plane,x=-1..1,y=-0.5..0.5,scaling=unconstrained,
    numpoints=2000);
display3d({p1,p2},style=patch,axes=boxed,orientation=[40,77]);

Mathematica:

Clear[x,y,z,t]
f[x_, y_]=5*Exp[-x^2-y^2]
fx=D[f[x,y],x]
fy=D[f[x,y],y]
a=0
b=0.25
plane=f[a,b]+ReplaceAll[fx,{x->a,y->b}]*(x-a)
    +ReplaceAll[fy,{x->a},y->b}]*(y-b)
P1=Plot3D[f[x,y],{x,-5,5},{y,-5,5},
    DisplayFunction->Identity]
P2=Plot3D[plane,{x,-1,1},{y,-0.5,0.5},
    DisplayFunction->Identity]
Show[P1,P2,DisplayFunction->$DisplayFunction]

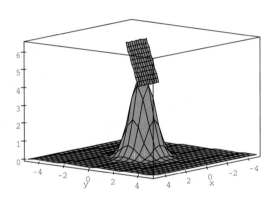

FIGURE 14.4.1

Exercise Set 14.4

In Exercises 1–14, find equations of the tangent plane and normal line to the given surface at the specified point.

1. $2x^2 + 3y^2 - z^2 = 5$; $(3, -2, 5)$

2. $z^2 = 3x^2 + 4y^2$; $(-2, 1, -4)$

3. $4x^2 + 3y^2 + 2z^2 = 12$; $(1, 0, -2)$

4. $xy + 2yz + 3xz = 16$; $(4, -2, 3)$

5. $(x + y)^2 + (y + z)^2 + (x + z)^2 = 10$; $(1, -1, 2)$

6. $xe^{2y-z} - 3 = 0$; $(3, 1, 2)$

7. $y = \ln\left(\dfrac{x + 2y}{y + 2z}\right) - 1$; $(3, -1, 1)$

8. $\sin\left(\dfrac{x}{y}\right) + \cos\left(\dfrac{y}{z}\right) = 0$; $\left(\pi, 1, \dfrac{2}{\pi}\right)$

9. $z = x^2 - 2y^2$; $(5, 4, -7)$

10. $z = \ln\sqrt{x+y}$; $(3, -2, 0)$

11. $z = \tan^{-1}\dfrac{y}{x}$; $\left(1, -1, -\dfrac{\pi}{4}\right)$

12. $z = \dfrac{x+y}{x-y}$; $(4, 3, 7)$

13. $z = e^{2x}\sin 3y$; $\left(0, \dfrac{\pi}{2}, -1\right)$

14. $z = \sqrt{\dfrac{x-2y}{x+2y}}$; $(5, -2, 3)$

15. Find the point on the hyperbolic paraboloid $z = 2x^2 - 3y^2$ at which the tangent plane is parallel to the plane $4x + 9y - 2z = 11$.

16. Find the point on the elliptic paraboloid $z = 3x^2 + 4y^2$ at which the tangent plane is perpendicular to the line through the points $(1, -2, 4)$ and $(-2, 0, 3)$.

In Exercises 17 and 18, assume F and G are differentiable functions at $P_0 = (x_0, y_0, z_0)$ and have nonzero gradients there.

17. The surfaces defined by $F(x, y, z) = 0$ and $G(x, y, z) = 0$ are said to be *tangent* at P_0 if they have the same tangent plane there.
    (a) Prove that the surfaces are tangent at P_0 if and only if $\nabla F(P_0) = k\nabla G(P_0)$ for some nonzero scalar k.
    (b) Find all points P_0 at which the surfaces $x^2 + 2y^2 - 2z^2 = 20$ and $xy - yz + 2xz = 5$ are tangent to each other.

18. The surfaces defined by $F(x, y, z) = 0$ and $G(x, y, z) = 0$ are said to be *orthogonal* at P_0 if their normal lines at P_0 are perpendicular to each other.
    (a) Prove that the surfaces are orthogonal at P_0 if and only if $\nabla F(P_0) \cdot \nabla G(P_0) = 0$.
    (b) Find two points P_0 at which the surfaces defined by $2x^2 - 3y^2 + 4z^2 = 10$ and $2x^2 + y^2 - 4z^2 + 2y - z = 21$ are orthogonal to each other.

19. Find all points on the surface $z = 2x^3 - 6x^2y + 9y^2 + 2y^3$ at which the tangent plane is horizontal.

20. The angle between a line l and a surface S is defined as the complement of the angle between l and the normal line to S at the point where l pierces S. Find the angle between the line $\mathbf{r}(t) = \langle t, 2t, 2-t \rangle$ and the elliptic cone $2z^2 = 4x^2 + y^2$ at each point of intersection.

21. The angle between a curve C and a surface S is defined as the angle between the tangent line to C and the surface at each point of intersection. (See Exercise 20.) Find the angle between the curve $\mathbf{r}(t) = \langle 1-t, 2+t, t^2 \rangle$ and the paraboloid $9z = 4x^2 + y^2$ at each of their points of intersection.

22. Prove that if f is a function of two variables that is differentiable at (x_0, y_0), then $F(x, y, z) = f(x, y) - z$ is differentiable at (x_0, y_0, z_0), where $z_0 = f(x_0, y_0)$.

CAS *Use a CAS for Exercises 23 through 25.*

23. Find the equation of the tangent plane to $f(x, y) = \dfrac{x^2}{2} + \dfrac{y^2}{3}$ at the point $\left(1, 1, \dfrac{5}{6}\right)$. Plot the surface and the tangent plane to the surface at the point.

24. Find the equation of the tangent plane to $f(x, y) = xe^{-x^2-y^2}$ at the point $(1, 1, 1)$. Also find the normal line at this point. Plot the surface, the tangent plane, and the normal line to the surface at the point.

25. The angle of inclination of a plane is defined to be the angle θ, $0 \leq \theta \leq \pi/2$, between the given plane and the xy-plane as shown in the figure. Find the angle of inclination of the tangent plane found in Exercise 24.

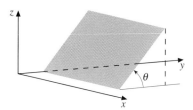

14.5 EXTREME VALUES

Just as in the case of one variable (see Chapter 4), some of the most important applications of multivariable differential calculus involve finding maximum and minimum values of functions. We will concentrate in this section on the two-variable case, but much of the theory can be extended to functions of three or more variables.

The definitions of local and absolute maxima and minima (which are referred to collectively as extreme values, or extrema) parallel those for functions of one variable.

Definition 14.6
Local and Absolute Extrema for Functions of Two Variables

Let f be a function of two variables with domain D. Then f is said to have a **local maximum** at a point (x_0, y_0) in D if there exists some circle centered at (x_0, y_0) such that $f(x_0, y_0) \geq f(x, y)$ for all points (x, y) of D that lie inside this circle. If this inequality holds true for all points (x, y) in D, then f is said to have an **absolute maximum** at (x_0, y_0). When the reverse inequality holds, f has a **local minimum** at (x_0, y_0) in the first instance and an **absolute minimum** in the second.

Relative to the surface of the Earth, Mt. Everest's peak is an absolute maximum.

The terms *relative maximum* and *relative minimum* are often used instead of *local* maximum and minimum. If f has a local maximum at (x_0, y_0), then $f(x_0, y_0)$ is called a **local maximum value** of f, and the point $(x_0, y_0, f(x_0, y_0))$ is a **local maximum point** on the graph of f. This point is the highest point on the graph in its immediate vicinity. Similar remarks apply for a local minimum. The absolute maximum value of f, if it exists, is the largest of its local maximum values, and the absolute minimum value is the smallest of its local minimum values.

Figure 14.22 shows the computer-generated graph of the function

$$f(x, y) = 4e^{-\sqrt{x^2+y^2}/4} \sin x \sin y$$

having many local maxima and minima.

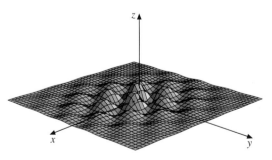

FIGURE 14.22

Sometimes it is possible to determine maximum and minimum values without using calculus. (In Chapter 1 we saw how we could find maximum or minimum values of quadratic functions of one variable without calculus.) For polynomial functions of degree 2, the technique of completing the square is especially useful in this regard, as we illustrate in the following example.

Section 14.5 Extreme Values   1039

EXAMPLE 14.20 Find the absolute extrema and where they occur for the function
$$f(x, y) = x^2 + y^2 - 2x + 4y + 10$$

Solution We complete the squares on x and y, getting
$$f(x, y) = (x^2 - 2x + 1) + (y^2 + 4y + 4) + 10 - 1 - 4$$
$$= (x-1)^2 + (y+2)^2 + 5$$

The two squared terms are positive except when $x = 1$ and $y = -2$, when each is 0. So the absolute minimum value of f occurs at $(1, -2)$, and this minimum value is 5. There is no maximum value. We show the graph in Figure 14.23. It is a paraboloid with vertex $(1, -2, 5)$. ∎

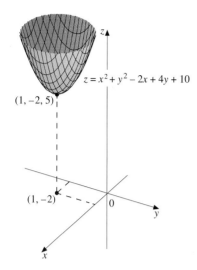

FIGURE 14.23

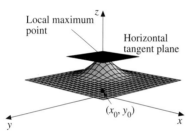

FIGURE 14.24

Since problems of this type are rather specialized, we need to develop methods for handling a wider class of problems. The first task is to find a systematic way of determining points in the domain at which maximum or minimum values *might* occur. Then we must test to see the actual nature of the function at these points.

If $z = f(x, y)$, and f is differentiable at a point (x_0, y_0) of its domain where f attains a local maximum or minimum value, then it is geometrically evident that the graph of f has a horizontal tangent plane at the point (x_0, y_0, z_0), where $z_0 = f(x_0, y_0)$. We illustrate this fact for a local maximum in Figure 14.24.

A horizontal tangent plane through the point (x_0, y_0, z_0) has the equation $z = z_0$. It follows, therefore, from Equation 14.23 that $f_x(x_0, y_0) = 0$ and $f_y(x_0, y_0) = 0$. That is, $\nabla f(x_0, y_0) = \mathbf{0}$. As the following argument shows, the condition $\nabla f(x_0, y_0) = \mathbf{0}$ holds true at a local maximum or minimum even if f is not differentiable, provided $f_x(x_0, y_0)$ and $f_y(x_0, y_0)$ both exist.

Suppose $\nabla f(x_0, y_0)$ exists—that is, $f_x(x_0, y_0)$ and $f_y(x_0, y_0)$ both exist—and suppose for definiteness that f has a local maximum at (x_0, y_0). Then, as we see in Figure 14.25, the curves $z = f(x, y_0)$ and $z = f(x_0, y)$ formed by the intersection of the surface $z = f(x, y)$ with the planes $y = y_0$ and $x = x_0$, respectively, have maximum points at (x_0, y_0, z_0), where $z_0 = f(x_0, y_0)$. Their slopes at (x_0, y_0, z_0) therefore both equal 0. But these slopes are $f_x(x_0, y_0)$ and $f_y(x_0, y_0)$, respectively. So $\nabla f(x_0, y_0) = \mathbf{0}$. A similar argument can be given when f has a local minimum at (x_0, y_0).

We see, then, that if $\nabla f(x_0, y_0)$ exists at a local maximum or minimum point, it must be $\mathbf{0}$. The only other possibility is that $\nabla f(x_0, y_0)$ does not exist. We therefore have the following theorem.

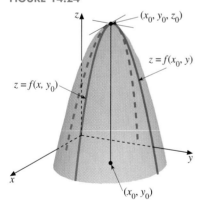

FIGURE 14.25

THEOREM 14.4

> If f is a function of two variables that has a local maximum or local minimum at (x_0, y_0), then either $\nabla f(x_0, y_0) = \mathbf{0}$ or $\nabla f(x_0, y_0)$ does not exist. That is, either $f_x(x_0, y_0)$ and $f_y(x_0, y_0)$ both equal 0, or at least one of these partials does not exist.

Points in the domain of f where either the partial derivatives f_x and f_y both equal 0 or at least one of these partials fails to exist are *candidates* for points where f has a local maximum or minimum. We give such points a name in the following definition.

Definition 14.7
Critical Point

A point (x_0, y_0) in the domain of f for which $\nabla f(x_0, y_0) = \mathbf{0}$ or $\nabla f(x_0, y_0)$ fails to exist is called a **critical point** of f.

Note that this definition is analogous to that of a critical point for a function of one variable, with ∇f replacing f'.

EXAMPLE 14.21 Find all critical points of
$$f(x, y) = x^3 - 3x^2 y + 6y^2 + 24y$$

Solution To calculate the gradient, ∇f, we calculate the first partials.
$$\frac{\partial f}{\partial x} = 3x^2 - 6xy \quad \text{and} \quad \frac{\partial f}{\partial y} = -3x^2 + 12y + 24$$

Thus, $\nabla f = \langle 3x^2 - 6xy, -3x^2 + 12y + 24 \rangle$. Since ∇f exists everywhere, the only critical points are those for which $\nabla f = \mathbf{0}$. We set $\partial f/\partial x = 0$ and $\partial f/\partial y = 0$ and solve simultaneously. Setting $\partial f/\partial x = 0$ gives
$$3x^2 - 6xy = 0 \quad \text{or} \quad 3x(x - 2y) = 0$$

Thus, $x = 0$ or $x = 2y$. Setting $\partial f/\partial y = 0$ gives
$$-3x^2 + 12y + 24 = 0$$

or, equivalently (dividing by -3),
$$x^2 - 4y - 8 = 0$$

Substituting $x = 0$ gives $-4y - 8 = 0$, so that $y = -2$. Thus, $(0, -2)$ is a critical point. Substituting $x = 2y$ gives $4y^2 - 4y - 8 = 0$, whose solutions are readily found to be $y = 2$ and $y = -1$. The corresponding x values are 4 and -2. The critical points of f are therefore $(0, -2)$, $(4, 2)$, and $(-2, -1)$. ∎

EXAMPLE 14.22 Find all critical points of
$$f(x, y) = (3x^2 + 4y^2)^{1/2}$$

and determine the nature of the function at each point.

Solution The gradient in this case is
$$\nabla f = \left\langle \frac{3x}{\sqrt{3x^2 + 4y^2}}, \frac{4y}{\sqrt{3x^2 + 4y^2}} \right\rangle$$

which is never $\mathbf{0}$. It is undefined only when $x = 0$ and $y = 0$. So $(0, 0)$ is the only critical point. We see that $f(0, 0) = 0$, and for all other points (x, y), $f(x, y) > 0$. Thus, $f(0, 0)$ is the absolute minimum value. The graph of f is the upper nappe of the elliptical cone $z^2 = 3x^2 + 4y^2$, pictured in Figure 14.26. At the minimum point the graph comes to a sharp point, and there is no tangent plane there. ∎

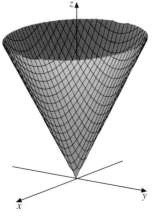

FIGURE 14.26

According to Theorem 14.4, a function can have a local maximum or minimum at a point of its domain only if that point is a critical point. Note carefully, however, it *does not* say that f *will* have a local maximum or minimum at each critical point. The next example illustrates this fact.

EXAMPLE 14.23 Show that for the function $f(x, y) = xy$ the point $(0, 0)$ is a critical point but $f(0, 0)$ is neither a local maximum nor a local minimum value of f.

Solution Since $\nabla f = \langle y, x \rangle$, we see that $\nabla f(0, 0) = \mathbf{0}$. So $(0, 0)$ is a critical point. Also, $f(0, 0) = 0$.

If x and y are both nonzero and they are like in sign (first or third quadrants), then $f(x, y) > 0$. If both x and y are nonzero and unlike in sign (second and fourth quadrants), then $f(x, y) < 0$. So $f(0, 0)$ is neither larger nor smaller than all other values of f for points in the immediate vicinity of $(0, 0)$. That is, $f(0, 0)$ is neither a local maximum nor a local minimum value. (This situation is similar to $f(x) = x^3$ in the single-variable case, in which $f'(0) = 0$, but the origin is neither a maximum nor a minimum point.)

We show the graph of f in Figure 14.27. It is a hyperbolic paraboloid (rotated 45° from the position we showed in Chapter 13). The origin is a saddle point. ∎

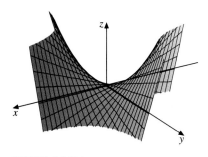
FIGURE 14.27

A Test for Local Extrema

Just as in the single variable case, finding that a function of two variables has a critical point does not guarantee that the function will have a maximum or minimum value there. We need a test that will enable us to determine the nature of the function at its critical points. The test we give is analogous to the Second Derivative Test for a function of one variable. It applies only to critical points for which the gradient exists and equals **0**. The proof can be found in most advanced calculus texts.

THEOREM 14.5

A Test for Local Extrema

Let (x_0, y_0) be a critical point of the function f for which $\nabla f(x_0, y_0) = \mathbf{0}$, and let f have continuous second partial derivatives at all points inside some circle centered at (x_0, y_0). Define

$$D(x, y) = f_{xx}(x, y) f_{yy}(x, y) - \left(f_{xy}(x, y)\right)^2$$

1. If $D(x_0, y_0) > 0$ and $f_{xx}(x_0, y_0) < 0$, then $f(x_0, y_0)$ is a local maximum.
2. If $D(x_0, y_0) > 0$ and $f_{xx}(x_0, y_0) > 0$, then $f(x_0, y_0)$ is a local minimum.
3. If $D(x_0, y_0) < 0$, then f has a saddle point at (x_0, y_0). (That is, $f(x_0, y_0)$ is neither a local maximum nor a local minimum.)
4. If $D(x_0, y_0) = 0$, then the test is inconclusive.

If the test is inconclusive or not applicable, then it may be possible to determine the nature of f by examining its values near the critical point.

EXAMPLE 14.24 Find all local maximum and minimum values of the function f defined by

$$f(x, y) = x^2 - 2xy + 4y^2 - 2x - 4y + 1$$

Solution First, we find the critical points by setting $\nabla f(x, y) = \mathbf{0}$.

$$\nabla f = \langle 2x - 2y - 2, \; -2x + 8y - 4 \rangle$$

By setting each component equal to 0 and dividing both sides of each equation by 2, we obtain the two equations

$$x - y = 1 \quad \text{and} \quad -x + 4y = 2$$

The simultaneous solution is $(2, 1)$. To test this point by Theorem 14.5, we need the second partials:

$$f_{xx} = 2 \quad f_{xy} = -2 \quad f_{yy} = 8$$

So $D(x, y) = 2(8) - (-2)^2 = 16 - 4 = 12$. Since $D(x, y)$ is constant in this case, its value at the critical point $(2, 1)$ is 12 also, which is positive. So we see that $D(2, 1) > 0$ and $f_{xx}(2, 1) = 2 > 0$. By part 2 of Theorem 14.5, we conclude that $f(2, 1)$ is a local minimum value. To find this minimum value, we substitute $x = 2$ and $y = 1$ into the formula for $f(x, y)$ and obtain $f(2, 1) = 3$. A computer-generated graph of f is shown in Figure 14.28. ∎

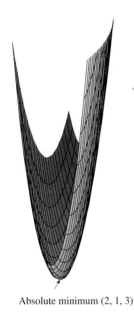

Absolute minimum (2, 1, 3)

FIGURE 14.28

EXAMPLE 14.25 Find and classify all extrema of $f(x, y) = x^3 - 3x^2y + 6y^2 + 24y$.

Solution This is the function from Example 14.21, and we found the critical points to be $(0, -2)$, $(4, 2)$, and $(-2, -1)$. To test them we need the second partials f_{xx}, f_{xy}, and f_{yy}:

$$\begin{aligned} f_x &= 3x^2 - 6xy & f_y &= -3x^2 + 12y + 24 \\ f_{xx} &= 6x - 6y & f_{xy} &= -6x & f_{yy} &= 12 \end{aligned}$$

The following table helps to keep track of things.

(x_0, y_0)	$f_{xx}(x_0, y_0)$	$f_{xy}(x_0, y_0)$	$f_{yy}(x_0, y_0)$	$D(x_0, y_0)$	Test result
$(0, -2)$	12	0	12	144	Minimum
$(4, 2)$	12	-24	12	-432	Saddle point
$(-2, -1)$	-6	12	12	-216	Saddle point

So the only local extremum is $(0, -2)$, where f has a minimum value. The minimum value is $f(0, -2) = -24$. Figure 14.29 shows a computer-generated graph of f, where the vertical scale has been compressed. We also show a contour map for f (Figure 14.30). ∎

EXAMPLE 14.26 A crate in the shape of a rectangular box is to be constructed so that its volume is 270 ft^3. The sides and top each cost \$1/ft^2 to construct, and the bottom, which must be stronger, costs \$1.50/ft^2. What are the dimensions of the crate that will yield the minimum cost? What is the minimum cost?

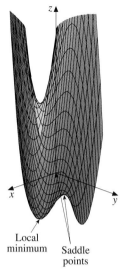

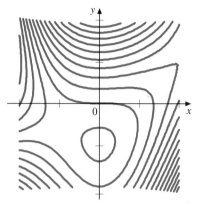

FIGURE 14.29

FIGURE 14.30
A contour map for
$f(x, y) = x^3 - 3x^2y + 6y^2 + 24y$

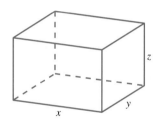

FIGURE 14.31

Solution   Denote the dimensions of the crate by x, y, and z, as shown in Figure 14.31. Then, since there are two sides of area xz and two ends of area yz, the cost function C is given by

$$C = \overbrace{2xz}^{\text{sides}} + \overbrace{2yz}^{\text{ends}} + \overbrace{1.5xy}^{\text{bottom}} + \overbrace{xy}^{\text{top}} = 2(xz + yz) + 2.5xy$$

We are also given that the volume must be 270. So the variables are related by the equation

$$xyz = 270$$

This relationship enables us to eliminate one of the variables. We solve for z and substitute in the cost function, reducing C to a function of two variables only:

$$z = \frac{270}{xy}$$

$$C(x, y) = 2\left(\frac{270}{y} + \frac{270}{x}\right) + \frac{5xy}{2} \quad \text{Write 2.5 as } \tfrac{5}{2}.$$

To apply Theorem 14.5, we need the gradient of C:

$$\nabla C = \left\langle -\frac{540}{x^2} + \frac{5y}{2}, -\frac{540}{y^2} + \frac{5x}{2} \right\rangle$$

Setting the first component of ∇C equal to 0 gives

$$-\frac{540}{x^2} + \frac{5y}{2} = 0 \quad \text{or} \quad y = \frac{216}{x^2}$$

Setting the second component equal to 0 gives

$$-\frac{540}{y^2} + \frac{5x}{2} = 0 \quad \text{or} \quad x = \frac{216}{y^2}$$

Now we substitute $x = 216/y^2$ into the equation $y = 216/x^2$ to obtain (after simplification)

$$y = \frac{y^4}{216} \quad \text{or} \quad y\left(216 - y^3\right) = 0$$

Clearly, $y \neq 0$ (since we are talking about a real crate), so $y^3 = 216$, or $y = 6$. Thus, $x = 216/36 = 6$ also. Since $(6, 6)$ is the only critical value for the cost function C, and we know from the nature of the problem that C does assume a minimum value, we can conclude that this minimum occurs at $(6, 6)$. So it is not essential in this case to apply Theorem 14.5.

With $x = 6$ and $y = 6$, we get $C(6, 6) = 2\left(\frac{270}{6} + \frac{270}{6}\right) + \frac{5(36)}{2} = 270$. Thus, since $z = 270/xy = 7.5$, the dimensions for minimum cost are 6 ft × 6 ft × 7.5 ft, and the minimum cost is \$270. ∎

In the next two examples we show how to determine absolute extrema of a function on a closed and bounded set (a closed set is one that includes its boundary). The technique works when the boundary consists of a finite number of curves on each of which the function can be expressed in terms of one variable.

EXAMPLE 14.27 Find the absolute maximum and minimum values of

$$f(x, y) = x^2 - 2xy + 3y^2 - 4x$$

on the closed trapezoidal region pictured in Figure 14.32.

Solution Name the boundary segments C_1, C_2, C_3, and C_4, as shown. First, we look for points in the interior where extrema occur. To find the critical points, we see where the gradient either is $\langle 0, 0 \rangle$ or is undefined.

$$\nabla f = \langle 2x - 2y - 4, -2x + 6y \rangle$$

In this case, ∇f is defined for all points. It is $\langle 0, 0 \rangle$ when

$$2x - 2y - 4 = 0 \quad \text{and} \quad -x + 3y = 0 \tag{14.24}$$

Thus,

$$x = y + 2 \quad \text{and} \quad x = 3y \tag{14.25}$$

Equating the two expressions for x, we get $3y = y + 2$, or $y = 1$. Thus $x = 3$. To test whether $(3, 1)$ gives a local extremum, we need the second partials $f_{xx} = 2$, $f_{xy} = -2$, and $f_{yy} = 6$. So $D(x, y) = 12 - (-2)^2 = 8 > 0$, and since $f_{xx} > 0$, we conclude from Theorem 14.5 that f has a local minimum at $(3, 1)$. The value there is found to be $f(3, 1) = -6$.

Now we consider each boundary segment. On C_1 we have $y = 0$. So, the function becomes $f(x, 0) = x^2 - 4x$, which is a function of one variable on $0 \leq x \leq 4$. By setting the derivative equal to 0 and testing, we find that $x = 2$ yields a minimum value—namely, $f(2, 0) = -4$. The endpoint values are $f(0, 0) = 0$ and $f(4, 0) = 0$.

We sketch briefly the results along C_2, C_3, and C_4. You should verify these. In each case we are working with one variable only.

C_2: $\underline{x = 4}$. $f(4, y) = 3y^2 - 8y$ on $0 \leq y \leq 5$. Minimum at $y = \frac{4}{3}$, $f\left(4, \frac{4}{3}\right) = -\frac{16}{3}$. Endpoint values: $f(4, 0) = 0$, $f(4, 5) = 35$.

C_3: $\underline{y = x + 1}$. $f(x, x+1) = 2x^2 + 3$ on $0 \leq x \leq 4$. No interior critical values. Endpoint values: $f(4, 5) = 35$, $f(0, 1) = 3$.

C_4: $\underline{x = 0}$. $f(0, y) = 3y^2$ on $0 \leq y \leq 1$. No interior critical values. Endpoint values already found.

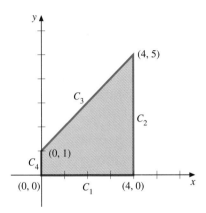

FIGURE 14.32

The extreme values are to be found among the following: $f(3, 1) = -6$, $f(2, 0) = -4$, $f(4, \frac{4}{3}) = -\frac{16}{3}$, $f(0, 0) = 0$, $f(4, 0) = 0$, $f(4, 5) = 35$, and $f(0, 1) = 3$. So f assumes the absolute maximum value of 35 at $(4, 5)$ and the absolute minimum value of -6 at $(3, 1)$. ∎

EXAMPLE 14.28 Find the absolute maximum and minimum values of

$$f(x, y) = x^2 - 4xy - 2y^2$$

on the closed disk $x^2 + y^2 \leq 5$.

Solution The critical points in the interior of the disk, if any exist, occur where $\nabla f = \langle 0, 0 \rangle$. Since $\nabla f = \langle 2x - 4y, -4x - 4y \rangle$, the coordinates of such critical points satisfy $x = 2y$ and $x = -y$. The simultaneous solution gives the point $(0, 0)$.

Rather than use the rectangular equation $x^2 + y^2 = 5$ of the boundary, it is easier to use the parametric equations

$$\begin{cases} x = \sqrt{5} \cos \theta \\ y = \sqrt{5} \sin \theta \end{cases} \quad 0 \leq \theta < 2\pi$$

On this boundary, we have

$$f(x, y) = f\left(\sqrt{5} \cos \theta, \sqrt{5} \sin \theta\right) = 5\cos^2 \theta - 20 \cos \theta \sin \theta - 10 \sin^2 \theta$$

giving f as a function of the single variable θ.

We find critical values of this function of θ by setting $df/d\theta = 0$.

$$\frac{df}{d\theta} = -10 \cos \theta \sin \theta - 20 \cos^2 \theta + 20 \sin^2 \theta - 20 \sin \theta \cos \theta$$

$$= 10(2 \sin^2 \theta - 3 \sin \theta \cos \theta - 2 \cos^2 \theta)$$

$$= 10(2 \sin \theta + \cos \theta)(\sin \theta - 2 \cos \theta)$$

Thus, $df/d\theta = 0$ when

$$2 \sin \theta + \cos \theta = 0 \quad \text{or} \quad \sin \theta - 2 \cos \theta = 0$$

From the first of these equations we get, on dividing by $\cos \theta$,

$$\tan \theta = -\frac{1}{2}$$

and from the second,

$$\tan \theta = 2$$

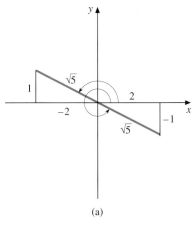

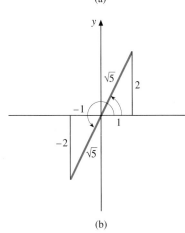

FIGURE 14.33

(x, y)	$f(x, y)$
$(0, 0)$	0
$(2, -1)$	10
$(-2, 1)$	10
$(1, 2)$	-15
$(-1, -2)$	-15

In Figure 14.33, we show the primary angles determined by these two equations. Since $x = \sqrt{5} \cos \theta$ and $y = \sqrt{5} \sin \theta$, we obtain the four critical points on the boundary: $(2, -1)$, $(-2, 1)$, $(1, 2)$, and $(-1, -2)$.

Now we calculate the value of $f(x, y)$ at the interior critical point and each critical point on the boundary. We show the results in the margin.

We conclude that the absolute maximum value of f is 10, occurring at $(2, -1)$ and $(-2, 1)$, and the absolute minimum is -15, occurring at $(1, 2)$ and $(-1, -2)$. ∎

We can summarize the procedure illustrated in Examples 14.27 and 14.28 as follows.

> **A Procedure for Finding Absolute Extrema on a Closed and Bounded Set**
>
> Let f be a continuous function of two variables on a closed and bounded region R, with the boundary consisting of finitely many curves where f can be expressed as a function of one variable.
>
> **Step 1.** Find all critical points in the interior of R and calculate $f(x, y)$ at each such point.
>
> **Step 2.** Find the local and endpoint extrema of f along each boundary curve.
>
> **Step 3.** Among all values of f found in Steps 1 and 2, select the largest and smallest values. The largest is the absolute maximum, and the smallest is the absolute minimum of f on all of R.

FINDING MAXIMUM AND MINIMUM VALUES OF FUNCTIONS OF SEVERAL VARIABLES USING COMPUTER ALGEBRA SYSTEMS

CAS 56

Find the extreme values for $f(x, y) = \dfrac{y^2}{3} - \dfrac{x^2}{4}$.

Since the function is differentiable and the domain does not contain boundary points, the function can have extreme values only where the first partials with respect to x and y are both 0.

Maple:

f:=(x,y)->y^2/3–x^2/4;

Output: $f := (x, y) \to \dfrac{1}{3}y^2 - \dfrac{1}{4}x^2$

Compute the first partials.

fx:=diff(f(x,y),x);

Output: $fx := -\dfrac{1}{2}x$

fy:=diff(f(x,y),y);

Output: $fy := \dfrac{2}{3}y$

Solve the partials simultaneously equal to 0.

solve({fx,fy},{x,y});

Mathematica:

f[x_, y_]=y^2/3–x^2/4

fx=D[f[x,y],x]

fy=D[f[x,y],y]

Solve[{fx==0, fy==0}, {x,y}]

Plot3D[f[x,y], {x,–5,5}, {y,–4,4},PlotRange–>{–1,2}, AspectRatio–>Automatic]

D[f[x,y], {x,2}]*D[f[x,y], {y,2}]–(D[f[x,y], {x,1},{y,1}])^2

Maple:

Output: $\{y = 0, x = 0\}$

So the only possible critical value is the origin (0, 0). A sketch will reveal that the origin is in fact a saddle point and not an extreme point. See Figure 14.5.1.

plot3d(f(x,y),x=–5..5,y=–4..4,scaling=constrained,axes=normal,
  tickmarks=[0,0,0],orientation=[10,80],view=–1..2);

Finally, we check algebraically that the origin is a saddle point. That is, we show that the discriminant function of f is negative at the origin (see Theorem 14.5). In fact, we see that the discriminant function is always negative in this case.

diff(f(x,y),x,x)*diff(f(x,y),y,y)–(diff(f(x,y),y,x))^2;

Output: $\dfrac{-1}{3}$

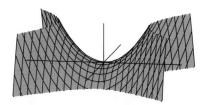

FIGURE 14.5.1

CAS 57

A thin metal plate is in the shape of the disk $x^2 + y^2 \leq 1$. The plate is heated in such a way that the temperature at the point (x, y) on the surface of the plate is given by $T(x, y) = 2x^2 + 3y^2 - 2x$. Find the temperatures of the hottest and coldest points on the plate.

Maple:

f:=(x,y)–>2*x^2+3*y^2–x;

Plot the temperature along with the region in the xy-plane corresponding to the plate.

p1:=plot3d(2*x^2+3*y^2–x,x=–1..1,y=2..2,view=–1..4,
  scaling=constrained):
p2:=plot3d({[t,sqrt(1–t^2),0],[t,–sqrt(1–t^2),0]},t=–1..1,
  scaling=constrained):
display3d({p1,p2},axes=boxed);

Mathematica:

f[x_,y_]=2*x^2+3*y^2–x

P1=Plot3D[f[x,y], {x, –2,2}, {y,–2,2},
  PlotRange–>{–1,4},DisplayFunction–>Identity]

P2=ParametricPlot3D[{t, Sqrt[1–t^2],0},{t,–1,1},
  DisplayFunction–>Identity]

P3=ParametricPlot3D[{t, –Sqrt[1–t^2],0},{t,–1,1},
  DisplayFunction–>Identity]

Show[P1,P2,P3,DisplayFunction–>$DisplayFunction]

fx=D[f[x,y],x]

fy=D[f[x,y],y]

Solve[{fx==0,fy==0}, {x,y}]

D[f[x,y],{x,2}]*D[f[x,y],{y,2}]–(D[f[x,y],{x,1},{y,1}])^2

s=–x^2+1

g=Simplify[ReplaceAll[f[x,y],{x–>x,y–>Sqrt[s]}]]

dg=D[g,x]

Solve[dg==0,x]

ReplaceAll[s,{x–>–1/2}]

Maple:

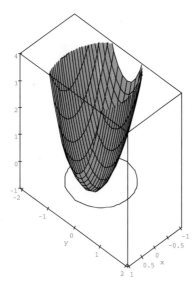

FIGURE 14.5.2

Mathematica:

h=Solve[y^2==3/4,y]

f[1/4,0]

f[3/4,Sqrt[3]/2]

f[3/4,-Sqrt[3]/2]

ContourPlot[f[x,y],{x,-2,2},{y,-2,2}]

First, locate any interior extreme values, as we did in CAS 56.

fx:=diff(f(x,y),x);

Output: $fx := 4x - 1$

fy:=diff(f(x,y),y);

Output: $fy := 6y$

solve({fx,fy},{x,y});

Output: $\left\{ y = 0, x = \dfrac{1}{4} \right\}$

Now check, using the discriminant function for f, if the point $(1/4, 0)$ is an extreme value.

diff(f(x,y),x,x)*diff(f(x,y),y,y)–(diff(f(x,y),y,x))^2;

Output: 24

Since the value of the discriminant function is greater than 0, the point is a local maximum. Now analyze the boundary points. First solve for y^2 in terms of x and then substitute into $f(x, y)$.

s:=solve(x^2+y^2=1,y^2);

Output: $s := -x^2 + 1$

g:=subs(x=x,y^2=s,f(x,y));

Output: $g := -x^2 + 3 - x$

Maple:

Notice g is now a function of only one variable, x. Apply the critical point analysis for functions of one variable.

dg:=diff(g,x);

Output: $dg := -2x - 1$

solve(dg=0,x);

Output: $\dfrac{-1}{2}$

subs(x=–1/2,s);

Output: $\dfrac{3}{4}$

h:=solve(y^2=3/4,y);

Output: $h := \dfrac{1}{2}\sqrt{3}, -\dfrac{1}{2}\sqrt{3}$

f(1/4,0);

Output: $\dfrac{-1}{8}$

f(3/4,h[1]);

Output: $\dfrac{21}{8}$

f(–3/4,h[2]);

Output: $\dfrac{33}{8}$

Thus, we see that the coldest spot on the plate is at the point $(1/4, 0)$ and has temperature of $-1/8$ degree, and the hottest spot on the plate occurs at the point $(-\frac{3}{4}, \frac{\sqrt{3}}{2})$ with a temperature of $33/8$ degrees.

Finally, we plot a contour map of the temperature surface showing curves of constant temperature. See Figure 14.5.3.

contourplot(2*x^2+3*y^2–x,x=–2..2,y=–2..2,view=–1..4,
    numpoints=1500, scaling=constrained,axes=normal,tickmarks=[0,0,0]);

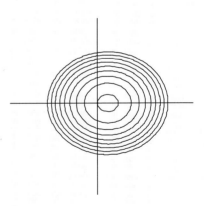

FIGURE 14.5.3

Exercise Set 14.5

In Exercises 1–4, find the extreme values of f and where they occur by completing the square.

1. $f(x, y) = x^2 + y^2 + 2x - 4y + 3$
2. $f(x, y) = 2x^2 + 3y^2 - 4x + 6y - 9$
3. $f(x, y) = x^4 + y^2 - 8y + 13$
4. $f(x, y) = x^2 + y^4 - 2y^2 + 4x + 1$

In Exercises 5–26, locate all critical points, and at each such point determine whether f has a local maximum, a local minimum, or a saddle point.

5. $f(x, y) = x^2 - 6xy + 2y^3 - 8x - 16$
6. $f(x, y) = x^2 + y^3 - 4xy - 8x + 13y + 1$
7. $f(x, y) = 2x^3 - 6xy + y^2 + 30$
8. $f(x, y) = x^3 + 3xy^2 - 3x^2 - 3y^2 + 4$
9. $f(x, y) = 2x^2 + y^4 - 4xy + 2$
10. $f(x, y) = x^4 + 2y^2 + 8xy - 7$
11. $f(x, y) = 6xy - x^2 - y^3$
12. $f(x, y) = 4xy - 2x^2 - y^3 + 3$
13. $f(x, y) = x^4 - 2x^2y + y^3 - y$
14. $f(x, y) = x^4 - y^4 - 4x^2y^2 + 20y^2$
15. $f(x, y) = xy + \dfrac{1}{x} + \dfrac{2}{y}$
16. $f(x, y) = 4 - \dfrac{2}{x} - \dfrac{1}{y} - x^2y$
17. $f(x, y) = 8x^2 - \dfrac{1}{y} + 2x - y$
18. $f(x, y) = \dfrac{8}{xy} + \dfrac{2}{x} - \dfrac{4}{y}$
19. $f(x, y) = x^3 + 2x^2y + y^3 + x$
20. $f(x, y) = 2x^3 + 3y^3 + xy^2 + 2y$
21. $f(x, y) = e^x(x^2 - y^2)$
22. $f(x, y) = e^{-y}(x^2 - 3x + 3y)$
23. $f(x, y) = \sin x \sin y$, $-\pi < x < \pi$, $-\pi < y < \pi$
24. $f(x, y) = \sin^2 x - 2\cos^2 y$, $-\dfrac{\pi}{4} < x < \dfrac{3\pi}{4}$, $-\dfrac{\pi}{4} < y < \dfrac{3\pi}{4}$
25. $f(x, y) = x^2 - 2x \cos y + 1$, $0 \le y \le 2\pi$
26. $f(x, y) = y^2 - 4y(\sin x + \cos x)$, $-\pi < x < \pi$

27. Show that $f(x, y) = 4 - x^{2/3} + 2x^{1/3}y^{1/3} - y^{2/3}$ has a critical point at $(0, 0)$ for which ∇f does not exist and that f has a local (and absolute) maximum value there.

In Exercises 28–31, find the absolute maximum and absolute minimum values of f on the closed domain bounded by the given curves.

28. $f(x, y) = x^2 + 2y^3$; the line segments joining $(0, 0)$, $(2, 0)$, and $(0, 1)$
29. $f(x, y) = x^2y - xy^2 - y$; x-axis, y-axis, $x = 1$, $y = 1$
30. $f(x, y) = 2x^3 - 3x^2y + 2y^3 - 3y$; $x + y = \pm 1$, $x - y = \pm 1$
31. $f(x, y) = x^2 - xy - x + y$; $y = 5 - x^2$, $y = 0$

32. Find the absolute maximum and minimum values of $f(x, y) = x^3 - y^3 - 3x$ on the closed unit disk $x^2 + y^2 \le 1$. At what points do these extrema occur? (Hint: Use the parameterization $x = \cos t$ and $y = \sin t$ for $0 \le t \le 2\pi$.)

33. An open-top rectangular box is to have a volume of 256 ft^3. What dimensions will require the least amount of material?

34. Find the point on the plane $3x + 2y - z = 4$ that is nearest the origin. (Hint: Minimize the *square* of the distance of the point (x, y, z) from the origin, using the fact that z satisfies the given equation.)

35. The temperature on the surface of the hemisphere $z = \sqrt{1 - x^2 - y^2}$ is given by $T(x, y, z) = 400xyz^2$. Find the hottest and coldest temperatures on the hemisphere and the points where these extrema occur.

36. A company makes two types of automatic ice cream freezers, Type A and Type B. The cost C of producing x Type-A and y Type-B machines per day is

$$C(x, y) = x^2 + xy + y^2 + 20x - 20y$$

and the revenue from selling x Type A and y Type B machines per day is $R(x, y) = 100x + 80y$. How many machines of each type should be manufactured and sold each day to maximize profit? What is the maximum profit?

37. An open-top rectangular box is to be constructed with a divider in the middle. The unit cost of the divider is half that of the bottom and sides. If the volume is to be 320 in.^{3}, find the dimensions that minimize the cost.

38. A common problem in experimental work is to find the line $y = mx + b$ that "fits" a set of data points $(x_1, y_1), (x_2, y_2), \ldots, (x_n, y_n)$ best in the sense that the sum of the squares of the vertical deviations of the data points from the line is minimum. This line is said to fit the data best in the sense of *least squares*. So the problem is to find m and b such that

$$F(m, b) = \sum_{k=1}^{n}(y_k - mx_k - b)^2$$

is a minimum. Determine the values of m and b.

39. In a chemistry experiment, the density (in gr/mL) of potassium chloride in a solution with water was measured for various solutions, with known weights of potassium chloride as a percentage of the weight of water. The results were as shown.

weight %	23.04	17.73	15.57	4.78
Density	1.170	1.163	1.115	1.078

weight %	1.80	11.21	14.36	22.75
Density	1.058	1.107	1.135	1.165

(a) Fit a straight line to the data points using the results of Exercise 38.
(b) Draw a graph showing the line and the data points.
(c) Predict the density for a solution with a weight percentage of 20.

CAS *Use a CAS for Exercises 40 and 41.*

40. Find the maximum value of the function $f(x, y) = xy^2$ on the circle $x^2 + y^2 = 1$. Parameterize the circle by $x = \cos t$ and $y = \sin t$ for $-\pi \leq t \leq \pi$. This converts f into a function of t only—say, g—which can then be maximized. Plot the surface $z = f(x, y)$ and also the function g.

41. A thin metal plate is in the shape of a rectangle in the xy-plane with vertices $(1, 1)$, $(1, -1)$, $(-1, 1)$, and $(-1, -1)$. The plate is heated in such a way that the temperature of the plate is given by $T(x, y) = x^2 + 3y^2 + 3x$. Find the temperatures of the hottest and coldest points on the plate. Plot the temperature surface and the region in the xy-plane corresponding to the plate. Finally, plot a contour map of the temperature surface showing curves of constant temperature.

14.6 CONSTRAINED EXTREMUM PROBLEMS

In extremum problems we often seek to maximize or minimize some function subject to a *constraint* on the variables, as was the case in Example 14.26. There we wanted to find the dimensions x, y, and z of a crate that minimized the cost, subject to the constraint that the volume had to be constant at 270; that is, the constraint on the variables was that $xyz = 270$. In that example we solved the constraint equation for z and substituted into the cost function, thereby reducing it to a function of two variables. In this section we consider an alternative method of solving such problems that in many cases is easier. In fact, depending on the nature of the constraint equation, it may be difficult or impossible to use the substitution method.

The method we describe is called the **Method of Lagrange Multipliers**, after the French-Italian mathematician Joseph Louis Lagrange (1736–1813). We consider first the simplest case in which a function of two variables, say $f(x, y)$, is to be maximized or minimized subject to a constraint on x and y. We assume this constraint to be expressed as an equation, say $g(x, y) = k$, whose graph is a curve C in the xy-plane. We say that f has a *constrained local maximum*

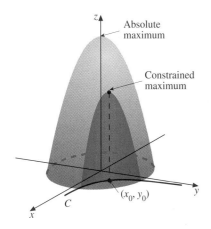

FIGURE 14.34

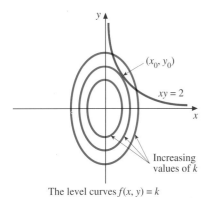

The level curves $f(x, y) = k$

FIGURE 14.35

at a point (x_0, y_0) on C provided that $f(x, y) \leq f(x_0, y_0)$ for all points near (x_0, y_0) that *lie on C* and are in the domain of f. A similar definition applies for a *constrained local minimum*.

Figure 14.34 illustrates a constrained local maximum. The function whose graph we have shown (a paraboloid) clearly has an absolute maximum when $x = 0$ and $y = 0$. But if (x, y) is constrained to lie on the curve C, then the maximum occurs at (x_0, y_0), as shown.

To be specific, suppose the paraboloid in Figure 14.34 is the graph of

$$f(x, y) = 6 - 2x^2 - y^2$$

and the curve C is the first-quadrant branch of the hyperbola $xy = 2$, so that $g(x, y) = xy$. Thus, we want to find the maximum value of the function f, where (x, y) is restricted to lie on C. In Figure 14.35 we show the graph of C, along with several level curves $f(x, y) = k$ of the function f. These level curves are the ellipses

$$2x^2 + y^2 = 6 - k \qquad (k > 0)$$

For increasing values of k the ellipses get smaller. We want the largest value of k (this value will be the maximum of the function f) for which (x, y) will be on the level curve $f(x, y) = k$ and also be on the constraint curve $g(x, y) = 2$. This value of k will be the one for which the level curve $f(x, y) = k$ and the constraint curve $g(x, y) = 2$ just touch; that is, where they are tangent to one another.

At such a point of tangency, the gradient vectors ∇f and ∇g must be parallel, since these gradient vectors are normal to the curves. Thus, if (x_0, y_0) is the point of tangency,

$$\nabla f(x_0, y_0) = \lambda \nabla g(x_0, y_0)$$

for some constant λ. The constant λ is called a **Lagrange Multiplier**.

This geometric discussion is intended to help you understand the next theorem. We will return to our example after the theorem and find the constrained maximum using the Lagrange Multiplier Method.

THEOREM 14.6

Lagrange's Theorem

Let $f(x, y)$ have a constrained local maximum or minimum at (x_0, y_0), with the constraint curve given by $g(x, y) = k$. If f and g are differentiable in some circle centered at (x_0, y_0) with $\nabla g(x_0, y_0) \neq \mathbf{0}$, then there exists a constant λ such that

$$\nabla f(x_0, y_0) = \lambda \nabla g(x_0, y_0) \tag{14.26}$$

Proof   Let the curve $g(x, y) = k$ be given by the vector equation

$$\mathbf{r}(t) = \langle x(t), y(t) \rangle$$

and let t_0 be the value of t for which

$$(x_0, y_0) = \big(x(t_0), y(t_0)\big)$$

Then, by hypothesis, $\mathbf{r}'(t)$ exists for values of t in some open interval about t_0, and $\mathbf{r}'(t_0) \neq \mathbf{0}$. Define the function F of the single variable t by

$$F(t) = f\big(x(t), y(t)\big)$$

Then F has a local maximum or minimum at t_0, so $F'(t_0) = 0$. Using the Chain Rule, we have

$$F'(t_0) = f_x(x_0, y_0)x'(t_0) + f_y(x_0, y_0)y'(t_0) = 0$$

We can write $F'(t_0)$ as the dot product

$$\nabla f(x_0, y_0) \cdot \mathbf{r}'(t_0) = 0$$

Since $\mathbf{r}'(t_0)$ is a tangent vector to the curve $g(x, y) = k$, it follows that ∇f is orthogonal to this curve at (x_0, y_0).

Since $g(x, y) = k$ can be interpreted as a level curve of $z = g(x, y)$, we know (see Section 14.3) that ∇g is orthogonal to this level curve at each point on the curve, and in particular, at the point (x_0, y_0). We have just shown that $\nabla f(x_0, y_0)$ is also orthogonal to $g(x, y) = k$. Hence, $\nabla f(x_0, y_0)$ and $g(x_0, y_0)$ are parallel vectors. So

$$\nabla f(x_0, y_0) = \lambda \nabla g(x_0, y_0)$$

for some constant λ. ∎

To make use of the theorem to find where f can assume a constrained local extreme value, we write the component equations that arise from Equation 14.26, namely,

$$f_x(x_0, y_0) = \lambda g_x(x_0, y_0)$$
$$f_y(x_0, y_0) = \lambda g_y(x_0, y_0)$$

These, together with the constraint equation

$$g(x_0, y_0) = k$$

constitute a system of three equations in the three unknowns x_0, y_0, and λ, which we solve simultaneously. Our objective is to find x_0 and y_0, and the multiplier λ is just a means to an end. So we might attempt to eliminate λ from the three equations as a first step. However, there are times when this approach is not feasible. It might be best, in fact, in some cases to solve first for λ and then find x_0 and y_0. The examples that follow illustrate some possible strategies.

EXAMPLE 14.29 Find the maximum value of the function $f(x, y) = 6 - 2x^2 - y^2$ subject to the constraint $xy = 2$.

Solution This problem is the one we discussed just prior to Theorem 14.6. We set $g(x, y) = xy$. Then the constraint equation is $g(x, y) = 2$. The gradient vectors are

$$\nabla f(x, y) = \langle -4x, -2y \rangle$$
$$\nabla g(x, y) = \langle y, x \rangle$$

By Theorem 14.6, the maximum occurs when

$$\nabla f = \lambda \nabla g$$

or

$$\langle -4x, -2y \rangle = \lambda \langle y, x \rangle$$

Equating components, we get
$$-4x = \lambda y \quad \text{and} \quad -2y = \lambda x$$

Solving the second of these equations for λ, we find that $\lambda = -2y/x$, and when this value is substituted in the first equation, we obtain
$$-4x = \left(-\frac{2y}{x}\right) y$$

or, after simplification,
$$y^2 = 2x^2$$

From the constraint equation $xy = 2$, we have $x = 2/y$. So
$$y^2 = 2\left(\frac{4}{y^2}\right)$$

or
$$y^4 = 8$$

So $y = \sqrt[4]{8} \approx 1.68$ and $x = 2/\sqrt[4]{8} \approx 1.19$. Thus, the maximum value of $f(x, y) = 6 - 2x^2 - y^2$ subject to the given constraint is
$$f\left(\frac{2}{\sqrt[4]{8}}, \sqrt[4]{8}\right) = 6 - 2\left(\frac{4}{\sqrt{8}}\right) - \sqrt{8}$$
$$= 6 - \frac{8}{\sqrt{8}} - \sqrt{8}$$
$$= 6 - 2\sqrt{8} \approx 0.343 \quad \blacksquare$$

EXAMPLE 14.30 Find the points on the ellipse $x^2 + 2y^2 = 6$ at which the function $f(x, y) = x^2 y$ assumes its largest and smallest values. What are these values?

Solution The constraint on points (x, y) is that they lie on the ellipse $x^2 + 2y^2 = 6$. Let $g(x, y) = x^2 + 2y^2$. Now $\nabla f(x, y) = \langle 2xy, x^2 \rangle$ and $\nabla g(x, y) = \langle 2x, 4y \rangle$. So from $\nabla f = \lambda \nabla g$ and $g(x, y) = 6$, we get the three equations
$$\begin{cases} 2xy = 2\lambda x \\ x^2 = 4\lambda y \\ x^2 + 2y^2 = 6 \end{cases}$$

From the first of these equations we have $2x(y - \lambda) = 0$, so either $x = 0$ or $\lambda = y$. If $x = 0$, then from the third equation $2y^2 = 6$, or $y = \pm\sqrt{3}$. When $\lambda = y$, we substitute for λ in the second equation and get $x^2 = 4y^2$, or $x = \pm 2y$. Then replacing x^2 by $4y^2$ in the third equation gives $6y^2 = 6$, or $y = \pm 1$, and so $x = \pm 2$.

Summarizing, we have found the points $(0, \pm\sqrt{3})$ and $(\pm 2, \pm 1)$ as candidates for places where f reaches extreme values. Next, we calculate $f(x, y)$ at each point:

(x_0, y_0)	$(0, \sqrt{3})$	$(0, -\sqrt{3})$	$(2, 1)$	$(-2, 1)$	$(2, -1)$	$(-2, -1)$
$f(x_0, y_0)$	0	0	4	4	-4	-4

Clearly, f has an absolute maximum of 4 at $(2, 1)$ and $(-2, 1)$ and an absolute minimum of -4 at $(2, -1)$ and $(-2, -1)$. Now consider points (x, y) on the ellipse close to $(0, \sqrt{3})$. Since $y > 0$ and $x^2 > 0$, $f(x, y) > 0$. So $f(0, \sqrt{3}) = 0$ is a local minimum value. By similar reasoning, we see that $f(0, -\sqrt{3}) = 0$ is a local maximum.

In Figure 14.36, we show the constraint curve $x^2 + 2y^2 = 6$ along with the level curves $f(x, y) = k$ for $k = 1, 4, 8, -1, -4$, and -8. Notice that as k increases through positive values, the level curves are above the y-axis and move outward from the origin. The one farthest from the origin that touches the constraint curve is for $k = 4$. That is, the largest value of $f(x, y)$ for which (x, y) lies on the constraint curve is $f(x, y) = 4$, and as we have seen, this contact occurs at $(2, 1)$ and $(-2, 1)$. At these points the level curve $f(x, y) = 4$ and the constraint curve $x^2 + 2y^2 = 6$ have a common tangent line, since their normals are parallel. A similar analysis can be given for k negative, with $f(x, y) = -4$ being the minimum value f can be with (x, y) on the constraint curve, occurring at $(-2, -1)$ and $(2, -1)$. ∎

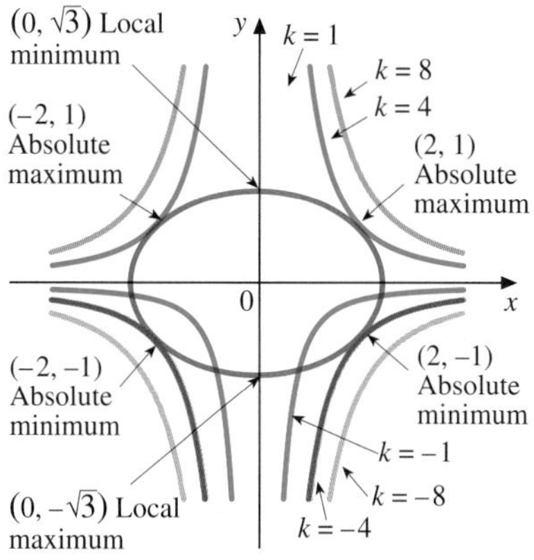

FIGURE 14.36

EXAMPLE 14.31 Find the point on the line $3x - 4y = 10$ that is nearest the origin.

Solution The distance from a point (x, y) to the origin is $\sqrt{x^2 + y^2}$, and this distance is a minimum if and only if its square is a minimum. So to simplify calculations, we use the square of the distance. We therefore want to minimize $f(x, y) = x^2 + y^2$ subject to the constraint that $3x - 4y = 10$. The constraint equation can be written as $g(x, y) = 10$, where $g(x, y) = 3x - 4y$. Proceeding as before, we have

$$\nabla f(x, y) = \lambda \nabla g(x, y)$$
$$\langle 2x, 2y \rangle = \lambda \langle 3, -4 \rangle$$

So we have three equations to solve:

$$\begin{cases} 2x = 3\lambda \\ 2y = -4\lambda \\ 3x - 4y = 10 \end{cases}$$

This time we solve for x and y in terms of λ from the first two equations and substitute into the third:

$$x = \frac{3\lambda}{2} \qquad y = -2\lambda$$

$$3\left(\frac{3\lambda}{2}\right) - 4(-2\lambda) = 10$$

Solving for λ, we get $\lambda = \frac{4}{5}$. So

$$x = \frac{3}{2}\left(\frac{4}{5}\right) = \frac{6}{5} \qquad \text{and} \qquad y = -2\left(\frac{4}{5}\right) = -\frac{8}{5}$$

Thus, the point to be tested is $(\frac{6}{5}, -\frac{8}{5})$. The geometry of the situation tells us that there is some minimum distance and no maximum distance (see Figure 14.37), and since there is only one critical point, it must be the point at which the distance is minimum. ∎

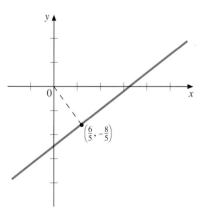

FIGURE 14.37

Let us summarize our discussion of the Method of Lagrange Multipliers for a constrained extremum of a function of two variables.

The Method of Lagrange Multipliers

Let f be a function of two variables that is to be maximized or minimized subject to the constraint $g(x, y) = k$, where f and g satisfy the conditions stated in Theorem 14.6.

Step 1. Set $\nabla f(x, y) = \lambda \nabla g(x, y)$, and write the equations obtained by equating corresponding components.

Step 2. Combine the equations from step 1 with the constraint equation $g(x, y) = k$ to obtain the system

$$\begin{cases} f_x(x, y) = \lambda g_x(x, y) \\ f_y(x, y) = \lambda g_y(x, y) \\ g(x, y) = k \end{cases}$$

of three equations with the unknowns x, y, and λ. Solve this system simultaneously.

Step 3. For each solution (x, y, λ) obtained in step 2, calculate $f(x, y)$. If f has a constrained maximum, it will be the largest of these values, and if f has a constrained minimum, it will be the smallest.

More Than Two Variables

The analogue of Theorem 14.6 for f and g functions of three (or more) variables also holds true, with the proof being virtually the same, and the procedure given above requires only slight modifications. We illustrate the procedure for the three-variable case in the next two examples.

EXAMPLE 14.32 Rework Example 14.26 using the Method of Lagrange Multipliers.

Solution The problem can be phrased as follows:
Minimize

$$C(x, y, z) = 2(xz + yz) + \frac{5}{2}xy$$

subject to the constraint $g(x, y, z) = 270$, where

$$g(x, y, z) = xyz$$

We seek solutions to the system $\nabla C = \lambda \nabla g$ and $g(x, y, z) = 270$:

$$\begin{cases} 2z + \frac{5}{2}y = \lambda yz \\ 2z + \frac{5}{2}x = \lambda xz \\ 2(x + y) = \lambda xy \\ xyz = 270 \end{cases} \quad \text{or} \quad \begin{cases} 5y + 4z = 2\lambda yz \\ 5x + 4z = 2\lambda xz \\ 2x + 2y = \lambda xy \\ xyz = 270 \end{cases}$$

This system requires a little more ingenuity to solve than those of the preceding examples. One approach is to subtract the first equation from the second:

$$5x - 5y = 2\lambda xz - 2\lambda yz$$
$$5(x - y) = 2\lambda z(x - y)$$
$$(x - y)(5 - 2\lambda z) = 0$$

So either $x = y$ or $2\lambda z = 5$. The constraint equation ensures that $z \neq 0$, so from $2\lambda z = 5$ we get $\lambda = \frac{5}{2z}$. When this value of λ is substituted into the first equation, we get

$$5y + 4z = 2\left(\frac{5}{2z}\right)yz$$
$$5y + 4z = 5y$$
$$4z = 0$$

giving $z = 0$, which is not possible. Thus, the only feasible solution is $x = y$. The third equation then gives $4x = \lambda x^2$, or $\lambda = 4/x$, since $x \neq 0$. Substituting this value of λ into the second equation of our system gives

$$5x + 4z = 2\left(\frac{4}{x}\right)xz$$
$$= 8z$$

Thus,

$$5x - 4z = 0$$
$$z = \frac{5x}{4}$$

Now that we have $y = x$ and $z = (5x)/4$, the constraint equation $xyz = 270$ gives $5x^3/4 = 270$, or $x^3 = 216$. Finally, $x = 6$, $y = 6$, and $z = \frac{15}{2}$. That a minimum value of C exists is evident from physical considerations, so $(6, 6, \frac{15}{2})$ must yield the minimum—namely,

$$C\left(6, 6, \frac{15}{2}\right) = 2(45 + 45) + \frac{5}{2}(36) = 270 \qquad \blacksquare$$

EXAMPLE 14.33 The largest box the United Parcel Service will accept is one for which the length plus the girth (distance around) is 108 in. What are the dimensions of the box of maximum volume that can be sent by UPS?

Solution Let the dimensions be x, y, and z, as shown in Figure 14.38. Then we want to maximize $V = xyz$ subject to the constraint $x + 2(y + z) = 108$. Taking $g(x, y, z) = x + 2(y + z)$, we must have for the constrained maximum $\nabla V = \lambda \nabla g$ and $g(x, y, z) = 108$. These equations give

$$\begin{cases} yz = \lambda \\ xz = 2\lambda \\ xy = 2\lambda \\ x + 2(y + z) = 108 \end{cases}$$

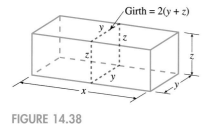

FIGURE 14.38

Eliminating λ from the first two equations yields $2yz = xz$. Since $z = 0$ is not a feasible solution, we have $x = 2y$. Again eliminating λ from the second and third equations of our system, we obtain $xz = xy$, or $z = y$ (since $x \neq 0$). Substituting $x = 2y$ and $z = y$ into the constraint equation gives

$$2y + 2(2y) = 108$$
$$6y = 108$$
$$y = 18$$

So the dimensions that give the maximum volume are $x = 36$, $y = 18$, and $z = 18$. ∎

Two or More Constraints

The Method of Lagrange Multipliers can also be applied to problems involving more than one constraint equation. We then have a multiplier for each constraint. For example, all local extrema of a function $f(x, y, z)$ subject to the constraints $g(x, y, z) = k_1$ and $h(x, y, z) = k_2$ will occur at points for which

$$\nabla f = \lambda_1 \nabla g + \lambda_2 \nabla h \qquad (14.27)$$

EXAMPLE 14.34 Find the points on the curve of intersection of the paraboloid of revolution $x^2 + y^2 + 2z = 4$ and the plane $x - y + 2z = 0$ that are closest to and farthest from the origin.

Solution We show the two surfaces and the curve of intersection in Figure 14.39. As in Example 14.31, we find the minimum and maximum values of the square of the distance from the origin. Thus, we take

$$f(x, y, z) = x^2 + y^2 + z^2$$

where the points (x, y, z) are constrained to lie on the given paraboloid and the given plane. We write the constraints in the form $g(x, y, z) = 4$ and $h(x, y, z) = 0$ by letting

$$g(x, y, z) = x^2 + y^2 + 2z$$

and

$$h(x, y, z) = x - y + 2z$$

From Equation 14.27, we have

$$\langle 2x, 2y, 2z \rangle = \lambda_1 \langle 2x, 2y, 2 \rangle + \lambda_2 \langle 1, -1, 2 \rangle$$

or, in terms of components,

$$\begin{cases} 2x = 2x\lambda_1 + \lambda_2 \\ 2y = 2y\lambda_1 - \lambda_2 \\ 2z = 2\lambda_1 + 2\lambda_2 \end{cases}$$

These three equations, together with the two constraint equations, constitute a system of five equations involving the five unknowns x, y, z, λ_1, and λ_2. By eliminating λ_2 from the first two equations, we get

$$(x + y)(1 - \lambda_1) = 0$$

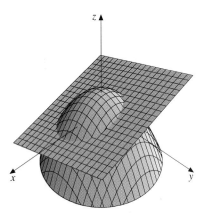

FIGURE 14.39

so that $y = -x$ or $\lambda_1 = 1$. We leave it as an exercise to show that $y = -x$ yields the points $(2, -2, -2)$ and $(-1, 1, 1)$, and that $\lambda_1 = 1$ also gives the point $(-1, 1, 1)$. Since $f(2, -2, -2) = 8$ and $f(-1, 1, 1) = 3$, we conclude that the maximum distance from the origin is $\sqrt{8} = 2\sqrt{2}$ and the minimum distance is $\sqrt{3}$. ∎

Exercise Set 14.6

In Exercises 1–8, find the local maxima and minima of f subject to the given constraint, using the Method of Lagrange Multipliers.

1. $f(x, y) = x^2 - y^2$; $x - 2y = 4$
2. $f(x, y) = 2x^2 + 3y^2$; $2x - 6y = 7$
3. $f(x, y) = x^2 + 2xy$; $y = x^2 - 2$
4. $f(x, y) = 2x - 4y$; $x^2 + y^2 = 5$
5. $f(x, y) = x^3 + 3y^2$; $xy + 4 = 0$
6. $f(x, y, z) = x^2 + 2y^2 - 2z^2$; $z = x^2 y$
7. $f(x, y, z) = x - 2y - 3z$; $xyz = 36$
8. $f(x, y, z) = 6x^2 + 3y^2 + 4z^2$; $3x^2 y + 2z^2 = 4$

In Exercises 9–16, re-do Exercises 1–8, without Lagrange Multipliers, by making a substitution to reduce the number of variables.

17. Use Lagrange Multipliers to find the distance from the line $2x + y = 3$ to the point $(1, -1)$.

18. Find the point on the plane $3x - y - 3z = 6$ that is nearest the origin, using Lagrange Multipliers.

In Exercises 19 and 20, use Lagrange Multipliers.

19. A coordinate system is set up on a flat metal plate so that the temperature $T(x, y)$ in degrees Celsius at the point (x, y) is $T(x, y) = 10x^2 y + 50$. Find the hottest and coldest spots at points on the ellipse $2x^2 + 3y^2 = 9$. What are these extreme temperatures?

20. A company determines from experience that its monthly revenue from the sale of a certain product is
$$R(x, y) = y^2 + 5xy + 20x$$

where x is the amount spent on magazine ads and y is the amount spent on television commercials, both in thousands of dollars. If the company plans to spend a total of $60,000 per month on advertising, how should it be divided to maximize R?

In Exercises 21–24, rework the specified exercises from Exercise Set 14.5 using the Method of Lagrange Multipliers.

21. Exercise 33
22. Exercise 34
23. Exercise 35
24. Exercise 37

25. Supply the details for the solution of Example 14.34.

26. Prove that a function can have a constrained local extremum at a point but not have a local extremum there. (*Hint:* Consider $f(x, y) = xy$ with an appropriate constraint.)

27. Find the dimensions of the rectangular box of greatest volume that can be inscribed in the ellipsoid $2x^2 + y^2 + 4z^2 = 12$.

28. Find the dimensions of the cone of maximum volume that can be inscribed in a sphere of radius a.

29. Find the maximum and minimum values of $f(x, y, z) = x^2 + 2y^2 - 3z^2$ subject to the two constraints $2x^2 - 3y^2 = 8$ and $y^2 - 2z = 3$.

30. Use Lagrange Multipliers to derive the formula
$$d = \frac{|Ax_0 + By_0 + Cz_0 + D|}{\sqrt{A^2 + B^2 + C^2}}$$
for the distance d from the plane $Ax + By + Cz + D = 0$ to the point (x_0, y_0, z_0).

31. Find the points on the curve of intersection of the ellipsoid $2x^2 + 3y^2 + 4z^2 = 6$ and the paraboloid $z = 4 - x^2 - 2y^2$ that are closest to the origin and farthest from the origin.

Chapter 14 Review Exercises

In Exercises 1 and 2, find df.

1. (a) $f(x, y) = \ln \dfrac{x^2}{\sqrt{1-y^2}}$

(b) $f(x, y) = \dfrac{\sin x}{\cosh y}$

2. (a) $f(x, y) = \sin^{-1} \dfrac{x}{y}, \; y > 0$

(b) $f(x, y, z) = \dfrac{4x - 2z}{z + 3y}$

3. Approximate Δf using df.

(a) $f(x, y) = \dfrac{(x-y)^2}{x^2 + y^2}$ from $(2, -4)$ to $(2.02, -3.97)$

(b) $f(x, y) = \ln\sqrt{9 - x^2 - y^2}$ from $(-2, 1)$ to $(-1.99, 0.98)$

4. Suppose the electrostatic potential at a point in $\mathbb{R}^3$ is given by

$$V(x, y, z) = \dfrac{140z}{\sqrt{x^2 + y^2 + z^2}}$$

Find the approximate change in potential from $(3, -2, 6)$ to $(2.6, -1.8, 6.5)$.

5. A company makes two types of toasters, Models A and B. It costs \$15 to produce each unit of Model A and \$21 to produce each unit of Model B. The revenue from producing and selling x Model-A units and y Model-B units is

$$R(x, y) = 42x + 56y - 0.02xy - 0.01x^2 - 0.03y^2$$

The current weekly production level is 150 Model-A and 100 Model-B units. Find the approximate increase in profit if 5 more Model-A units and 8 more Model-B units are produced each week. Approximately what is the profit at this new level?

6. Use Theorem 14.2 to show that f is differentiable except at $(0, 0)$, where

$$f(x, y) = \tan^{-1} \dfrac{x}{y}$$

7. Let

$$f(x, y) = \begin{cases} \dfrac{x^2 y}{x^4 + y^2} & \text{if } (x, y) \neq (0, 0) \\ 0 & \text{if } (x, y) = (0, 0) \end{cases}$$

Show that f_x and f_y both exist at $(0, 0)$ but that f is not differentiable there.

8. Use Definition 14.1 to show that $f(x, y) = x^2 - 2xy + 3y$ is differentiable throughout $\mathbb{R}^2$.

In Exercises 9–13, use an appropriate Chain Rule.

9. Find dz/dt at $t = 5\pi/6$ if $z = x^2 - 2xy - y^3$ and $x = \cos 2t, \; y = \sin 2t$.

10. Find $\dfrac{dz}{dt}$ at $t = 2$ if $z = \ln\sqrt{\dfrac{x+y}{x-y}}$ and $x = t + \dfrac{1}{t}$, $y = t - \dfrac{1}{t}$.

11. Find dz/dt if $z = e^{-x^2/y}$ and $x = \sinh t, \; y = 1 + \cosh t$.

12. Find $\dfrac{\partial z}{\partial u}$ and $\dfrac{\partial z}{\partial v}$ if $z = \dfrac{x^2 - y^2}{xy}$ and $x = \dfrac{u}{v}, \; y = \dfrac{1}{u}$.

13. Find $\dfrac{\partial z}{\partial t}$ at $(s, t) = \left(1, \dfrac{1}{2}\right)$ if $z = x \sin \pi y - y \cos \pi x, \; x = s^2 - t^2,$ and $y = 2st$.

14. Find the equation of the tangent line to the graph of

$$2x^4 - 3x^2 y + 4xy^2 + y^3 + 4 = 0$$

at $(-1, 1)$.

15. Find $\partial z/\partial x$ and $\partial z/\partial y$ if

$$y \ln(\cos xz) + xyz = 3$$

16. Let $z = f\left(\dfrac{x-y}{x+y}\right)$. Show that

$$x \dfrac{\partial z}{\partial x} + y \dfrac{\partial z}{\partial y} = 0$$

17. A water tank is in the form of a frustum of a cone, as shown in the figure, with bottom radius 3 ft. Water is being drained from the tank at the constant rate of

10π ft³/min. When 210π ft³ of water remain in the tank, the radius r of the upper surface of the water is 6 ft and is decreasing at the rate of 1 in./min. Find how fast the water level is falling at that instant.

In Exercises 18 and 19, find the directional derivatives of f in the direction indicated.

18. $f(x, y) = \ln(x^2/\sqrt{x - y})$ at $(1, -3)$, in the direction $6\mathbf{i} + 8\mathbf{j}$

19. $f(x, y, z) = x^2 y \cos \pi z$ at $(3, -1, 1)$, toward $(4, 1, -1)$

20. For the function
$$f(x, y, z) = \sqrt{\frac{x - 2y}{y - z}}$$
in what direction from $P_0 = (0, -2, -3)$ is $D_{\mathbf{u}} f(P_0)$ a maximum, and what is this maximum value?

21. Find $f(x, y)$ if $f(1, 1) = 1$ and
$$\nabla f = \left\langle \frac{1 - 2x^2}{x}, \frac{2y^2 - 1}{y} \right\rangle$$

In Exercises 22 and 23, find the equation of the tangent plane and normal line at the indicated point.

22. $z = x^2 - xy - 2y^2$ at $(2, -1, 4)$

23. $z = 2e^{(x-y)/z}$ at $(1, 1, 2)$

24. The accompanying figure shows a cross section of a long semicircular cylinder with a flat base. The curved surface is kept at electrostatic potential $V = 1$ and the base at $V = 0$. It can be shown that the potential $V(x, y)$ at points inside the region is
$$V(x, y) = \frac{2}{\pi} \tan^{-1}\left(\frac{2y}{1 - x^2 - y^2}\right)$$

(a) Draw several equipotential curves $V(x, y) = c$ for $0 < c < 1$.

(b) Find the rate of change of V at $(\frac{1}{2}, \frac{1}{2})$ in the direction toward $(0, 1)$.

(c) Show that $\nabla V(\frac{1}{2}, \frac{1}{2})$ is orthogonal to the level curve through $(\frac{1}{2}, \frac{1}{2})$.

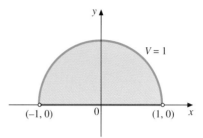

25. Find the angle between the hyperbolic paraboloid $z = x^2 - y^2$ and the line $\mathbf{r}(t) = \langle -t, 1 + 2t, 5(1 + t) \rangle$ at each of their points of intersection. (See Exercise 20 in Exercise Set 14.4.)

26. Find all maximum, minimum, and saddle points on the surface $z = x^3 - 2xy^2 - 9x + 4y^3$.

27. Find the absolute maximum and minimum values of the function
$$f(x, y) = \frac{4}{x} + \frac{1}{y} + 2xy$$
over the closed region bounded by $x = 1$, $x = 2y$, and $y = 2$, and give the points at which these extreme values occur.

28. Suppose the temperature in a three-dimensional region that contains the ellipsoid $x^2 + y^2 + 4z^2 = 12$ is given by $T(x, y, z) = xyz$. Find the hottest and coldest temperatures on the ellipsoid, and identify where they occur.

29. Use the Method of Lagrange Multipliers to find the dimensions of the right circular cylinder inscribed in a sphere of radius a that has a maximum (a) volume and (b) lateral surface area.

30. A company produces two types of electric brooms, the standard and deluxe models. Suppose the monthly profit (in thousands of dollars) from producing and selling x thousand standard models and y thousand deluxe models is given by
$$P(x, y) = 2x + 3y - 0.1x^2 - 0.2y^2 - 0.5xy - 4$$

If the combined production of the two models is to be 6000 per month, how many of each model should be produced to maximize profit?

Chapter 14 Concept Quiz

1. Define each of the following:
   (a) A differentiable function at a point (x, y)
   (b) The differential of a function f of two variables
   (c) The gradient of f
   (d) The tangent plane to the surface $F(x, y, z) = 0$ at a point P_0 where $\nabla F \neq \mathbf{0}$.
   (e) A critical point for a function of two variables

2. State the following:
   (a) A sufficient condition (other than the definition) for f to be differentiable at (x_0, y_0)
   (b) The Chain Rule for $f(x, y, z)$, where $x = x(u, v)$, $y = y(u, v)$, and $z = z(u, v)$, showing also an appropriate tree diagram
   (c) The formula for dy/dx if $F(x, y) = 0$, assuming appropriate hypotheses
   (d) A formula for the directional derivative $D_{\mathbf{u}} f(x_0, y_0)$
   (e) A test for determining the nature of a function f at a critical point (x_0, y_0), assuming appropriate hypotheses

3. (a) Give the linear approximation formula for a function f near (x_0, y_0) at which it is differentiable.
   (b) Let $z = f(x, y)$, with f differentiable at (x_0, y_0). Write the equation of the tangent plane and normal line to the graph at (x_0, y_0, z_0), where $z_0 = f(x_0, y_0)$.
   (c) Give the direction and the value of the maximum rate of change of a differentiable function f at a point (x_0, y_0).
   (d) Explain in your own words how to apply the Lagrange Multiplier Method.

4. Fill in the blanks.
   (a) If $z = f(x, y)$, f is differentiable at (x, y), and x and y are changed by small amounts dx and dy, respectively, then the approximate change in z is _____, and the percentage change in z is approximately _____.
   (b) If $w = g(s, t)$ and $s = s(x, y)$, $t = t(x, y)$, then $\dfrac{\partial w}{\partial x} = $ _____ and $\dfrac{\partial w}{\partial y} = $ _____.
   (c) The gradient of f is orthogonal to the _____ _____ of f.
   (d) The normal line to the surface $F(x, y, z) = 0$ at a point (x_0, y_0, z_0) for which $\nabla F \neq \mathbf{0}$ has direction vector _____.

5. Indicate which of the following statements are true and which are false.
   (a) If f is continuous at (x_0, y_0) and $f_x(x_0, y_0)$ and $f_y(x_0, y_0)$ both exist, then f is differentiable at (x_0, y_0).
   (b) If $\nabla f(x_0, y_0) = \mathbf{0}$, then f has either a local maximum or a local minimum at (x_0, y_0).
   (c) If $f(x, y)$ has a constrained maximum or minimum value at (x_0, y_0), with constraint $g(x, y) = k$, then the gradient vector of f is parallel to the gradient vector of g at (x_0, y_0) (assuming appropriate differentiability conditions).
   (d) If f_x and f_y are continuous at (x_0, y_0), then f is differentiable at (x_0, y_0).
   (e) The differentiable function f changes most rapidly at (x_0, y_0) in the direction of $\nabla f(x_0, y_0)$.

APPLYING CALCULUS

1. In economics, the quotient
$$\frac{f'(t)}{f(t)} = \frac{\text{marginal function}}{\text{function}}$$
is called the *relative growth rate of f*. Show that it can also be calculated as
$$\frac{d}{dt}[\ln(f(t))]$$

   (a) If the relative growth rate of the total consumption C of the population is c and the relative growth rate of the population P is p, show that the relative growth rate of per capita consumption C/P is $c - p$.

   (b) If C is the total amount of cash on deposit in banks and D is the total of all demand deposits, then the money supply M is given by $M = C + D$. Let m, c, and d denote the relative growth rates of M, C, and D, respectively. Show that
$$m = \frac{cC + dD}{C + D}$$

2. As a rocket lifts off from the earth, its mass decreases as fuel is burned. Suppose at liftoff the mass is 80,000 kg and that for the time interval in question, the mass is decreasing at the constant rate of 72 kg/s. The rocket reaches an altitude of 30 km after 45 s, and at that time it is rising at the rate of 90 km/s. How fast is the magnitude F of the force of gravity decreasing at that instant? Use Newton's Law of Gravitation
$$F = -\frac{GMm}{r^2}$$
where G is the universal gravitational constant, M is the mass of the earth, and r is the distance of the rocket from the center of the earth. (The radius of the earth is approximately 6,370 km.)

A Saturn V rocket launches Apollo 17 to the Moon.

3. A mountain has the approximate shape of the surface $2x^2 + 3y^2 - 4x + z = 1$, for $x \geq 0$ and $y \geq 0$ and units are in kilometers. Suppose a climber is at the point $P(0.6, -0.4, 2.2)$. Assume that the positive y-axis points north and the positive x-axis points east.

   (a) How high is the mountain?

   (b) What is the climber's rate of ascent when moving from P in a northeasterly direction (i.e., in the direction $(\mathbf{i} + \mathbf{j})/\sqrt{2}$)?

   (c) What direction from P should the climber take to ascend most rapidly? (Express as a unit vector.)

   (d) What direction from P should the climber take in order not to change level?

4. The work W done during a reversible adiabatic process is given by
$$W = nC_v(T_1 - T_2)$$
where n is the number of moles of gas, C_v is molar specific heat, and T_1 and T_2 are the initial and final temperatures. Show that if both T_1, and T_2 are increased by $\alpha\%$, then the work also is increased by $\alpha\%$.

5. A storage tank for propane is in the form of a circular cylinder 6 m long and 3m in diameter, with hemispherical caps at both ends (see the figure). There is a possible

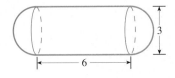

1% error in each of the measurements. Find the maximum percentage error in (a) the volume and (b) the surface area.

6. A builder is in the process of designing low-cost, one-story, tract homes. Excluding the roof, each home will have the basic shape of a rectangular box of length x, width y, and height z, and will enclose 16,000 cubic feet. The builder is concerned about heat loss through the walls, floor, and ceiling during the winter months when the temperature averages 20°F. The loss of heat through a surface is proportional to the surface area.

(a) Based on some experiments, the builder has determined that the heat loss through the ceiling will be 5 times as great as the loss through the floor and the heat loss through the walls will be 3 times as great as the heat loss through the floor. What dimensions will minimize the heat loss? Are these dimensions reasonable for the type of house being considered?

(b) The builder has decided that the height z of the walls of the house should be 10 feet. If the heat loss through the ceiling can be made arbitrarily low by adding insulation, is it possible to minimize the heat loss with $z = 10$? If so, what is the ratio of the heat loss through the ceiling to the heat loss through the floor?

7. Let L represent the number of units of labor and K the number of units of capital in the manufacture of $P(L, K)$ units of production. Suppose that labor costs a dollars per unit and capital costs b dollars per unit, and suppose that there is a total of c dollars available for production. Then $aL + bK = c$.

(a) Use Lagrange Multipliers to show that production is a maximum at the point (L_0, K_0) where

$$\frac{P_L(L_0, K_0)}{a} = \frac{P_K(L_0, K_0)}{b} = \lambda$$

Here the Lagrange Multiplier λ is called the *equimarginal productivity* of the production function P.

(b) Consider the Cobb-Douglas Production Function given by

$$P(L, K) = mL^\alpha K^\beta$$

where α and β are positive constants and $\alpha + \beta = 1$. Using the result in part (a), show that at the point of maximum production (L_0, K_0)

$$\frac{L_0}{K_0} = \frac{\alpha b}{\beta a}$$

What can you conclude from this result?

A PERSONAL VIEW

N. Scott Urquhart

What do you remember of your first calculus course?

I remember lots about my first calculus course. I worked hard and learned the difference between limits and continuity, a distinction which escaped me in an earlier brief encounter with calculus in high school. I enjoyed the course—the instructor made us work hard, but we learned a lot. It was just the starting point; over the years, I went on to three higher levels of calculus after the first sequence, which lasted four quarters.

Did taking calculus (and later mathematics) courses in any way influence your career choice?

The calculus was an absolutely required starting point on my path to becoming a consulting statistician. My skills and insights in this area were essential to subsequent opportunities even appearing. Without the calculus, the opportunities would never have appeared.

How did your interests in statistics develop?

I grew up around agriculture, but always was good at math and science. Statistics provided a way to work in areas which interested me and to use mathematics on real and important problems. I never made a conscious choice to become a statistician, it just happened. Given my interests in real problems and association with consulting statisticians who worked on relevant problems, it was natural for me to move in that direction. Early on in college, I got a job running (old!) rotary calculators doing statistical analyses, and learned to use the old electronic accounting equipment (which was punch-card based) to accumulate results for these analyses. This job paid nearly twice as well as mopping floors in the dormitories and gym so I was very willing to learn more and become more valuable to the project as a result. The computer I had access to was in Los Angeles; we mailed card decks from Fort Collins, Colorado, to Los Angeles so we could do batch jobs there. Turnaround was a week. How is that for "Remote Job Entry"?

How do biologists and statisticians typically use calculus?

Biologists and statisticians use the calculus in a variety of ways, but the most fundamental aspect is to understand what derivatives and integrals are. Occasionally I do integrations or differentiations to compute some complex areas or probability functions. For that, the concepts of calculus are absolutely critical; experience with integration across complexly shaped regions was essential. Although I rarely use the machinery of calculus at the level of this text, an understanding of that material is critical to my present job.

Can you tell us about a particular problem that you solved for which calculus was useful?

I presently spend about half of my time collaborating with researchers in EPA's Environmental Monitoring and Assessment Program. I am heavily involved in developing sampling plans for inland aquatic ecological resources such as lakes, streams and wetlands. As a part of the comparison and use of spatially distributed sampling plans, I need to evaluate individual and joint inclusion probabilities for points that might be selected by a first-stage area sample. The points could be the "label points"—a geographic coordinate associated with an individual body of water as assigned by a geographic information system. The required probabilities depend on the sampling plan, the shape of the partitions across which the area sample is taken, and, in a major way, on the distance between the points.

Specifically, consider a shape formed by combining seven hexagons of the same size. Now evaluate the area of the intersection of two such shapes as a function of the location of their centers. The resulting values can be thought of as a surface over two-dimensional space; this surface rises to the area of the shaped area as the two points converge. This surface falls off quickly as the points are moved apart, going to zero when the points are as far apart as the maximum width of the sampling shape. Near its top, its cross sections have moderately pronounced corners, but they smooth out as the points pull apart. This problem has made it obvious how important it is for a calculus student to become as competent with complex regions of integration as he or she is with complex integrands.

How is multivariable calculus useful in biological analysis?

Most interesting biological quantities are functions that depend on several independent variables. For example, fisheries biologists sometimes want to know how body weight of a particular species of fish depends on other identifiable variables. Length would be an important variable, but age, water temperature, food availability, predatory pressure, etc., might all be significant in the problem. Biologists do not usually apply multiple integrals to problems involving specific functions; however, multiple integrals of general functions are sometimes used to communicate the workings of a biological system without ever explicitly stating an integrand to integrate. The function becomes specific only for a particular point in time, or space, and only for a given set of conditions. Even then, it is frequently evaluated only in a discrete manner using numerical approximations. It's more typical to use calculus to describe a kind of general representation of an accumulative biological process.

For example, the photosynthesis of carbohydrates in a plant depends on the variables of light, nutrients, temperature, carbon dioxide concentration, and leaf surface area. Photosynthesis uses solar radiation to transform inorganic compounds into carbohydrates. If this occurs in an aquatic environment where there are phytoplankton floating at all depths to about several hundred feet, the light intensity diminishes with depth, and net photosynthetic production also declines with depth. These changes occur continuously and so general integral formulas are used to describe the total photosynthetic production. The light intensity is a function of both water depth (or height) and time, because the amount and direction of incoming radiation depends on time, and its penetration depends on depth. Under a fixed nutrient supply, the instantaneous rate of net photosynthesis depends primarily on the energy available, that is, light intensity. Therefore, if q represents the functional dependence of instantaneous net photosynthesis on intensity, $q(t) = q(I(h,t)) = p(h,t)$, so p has an equivalent representation as a function of time and depth. If p has been evaluated on a per unit area basis, then the total net production between times t_1 and t_2 occurring from the surface down to depth h_l (where l stands for light) can be written as $\iint_R p(h,t)\,dA$ where $R = \{(h,t): 0 \leq h \leq h_l, t_1 \leq t \leq t_2\}$. This double integral can be written as an iterated integral, $\iint_R p(h,t)\,dA = \int_{t_1}^{t_2}\int_0^{h_l} p(h,t)\,dh\,dt$, which indicates the involvement of time and depth. Functions of *three* or more variables present only a modest kind of notational change mathematically, but are radically different when we try to give them a visual representation simply because figures cannot be drawn in four or more dimensions. A thorough treatment of these extensions to three or more variables requires linear algebra and students who are interested in serious mathematical biology will need to take linear algebra right after calculus. Biology abounds with situations that can be described by some form of mathematical transformation, especially linear transformations as a first approximation. For example, an organism's physical condition might be described in terms of a number of variables whose values at a given point in time would be a point in a vector space. A certain influence such as stress or feeding the organism would change these variables, that is, transform one point into another. A linear transformation approximates this very well in many cases.

Do you use alternative coordinate systems to model other biological phenomena?

Yes. Polar coordinates, for instance, are useful in describing certain regions bounded by curves, more easily than a rectangular system. We use double integration in applications such as studies of populations introduced to a particular location. When animals are introduced to a particular location, for example, they frequently disperse without any particular directional orientation. One well-known study of muskrats introduced to Central Europe in 1900 used a model of the time-dependent density of the population in a polar coordinate system. This was effectively a probability density function that described a dispersal whose contours of equal density spread out from the release point in ever-expanding circles, like the ripples created by tossing a pebble into calm water. The model suggests that in the absence of some kind of force encouraging directional dispersion, the square root of the area occupied should have a linear relation to time. This same model was used by the author of the muskrat study to show that oak trees in Great Britain after the last ice age had been aided in their dispersal by animals or some other extrinsic factor.

You were coauthor of a textbook on Mathematics in Biology. Do students get the proper exposure to mathematical methods in the biology curriculum?

My past experience suggests that biology students see far too few relevant examples because few mathematicians teaching such courses have enough relevant experience with biology. It can be difficult to convince biology students that the precision of mathematics and mathematical operations is needed by biologists. The most relevant experiences of a biology student in a calculus course may have nothing to do with any of the specific functions to which they are exposed. Biology students need to learn to be precise; at the local level, because relative to the quantitative sciences, biology has lots of "fuzz"—imprecision in what is really happening or being seen. Secondly, fruitful development of computational procedures relevant to biology require an assimilation of many parts of mathematics. The value of repeated exposure to the concepts of rates of change (derivatives), accumulation processes (integration), and their relation (differential equations) cannot be emphasized enough.

Just as with other disciplines, the precision and conciseness of mathematics provides biologists with compact ways of expressing complex ideas. The tools of mathematics open up new ways of examining these ideas and their implications. Some biological problems, even apparently simple ones, require a mathematical approach for their understanding. Since more and more biologists are using mathematics to express their ideas, it's likely that understanding the basic literature of biology will require an increased understanding of basic mathematics. The trend toward more use of mathematics has occurred for a variety of reasons. One is that biologists are drawing more on ideas from chemistry and physics, which have long adopted mathematics as a tool and a language; another is the growing interface between biology and technology, particularly in the area of environmental studies.

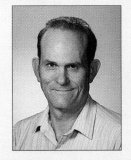

N. Scott Urquhart, professor of statistics at Oregon State University, grew up in the agricultural country of western Colorado and studied statistics, mathematics, and biology at Colorado State University, leading to a Ph.D. in 1965. He has taught, been an internal consultant, and done research affiliated with statistical aspects of agriculture at Cornell University, New Mexico State University, and Oregon State University. He has received several awards for excellence in teaching. His research interests span many areas of statistics in biology, including determining the factors affecting beef quality and associated consumer acceptance; using waste radioisotopes to convert sewage sludge to beneficial uses; developing resistance to root-rot diseases; projecting the effects of hunting regulations on deer populations; modeling plant growth in Alaska to minimize the environmental effects of energy development in fragile ecosystems; and developing a national sampling program for ecological resources, in cooperation with USEPA.

CHAPTER 15

MULTIPLE INTEGRALS

In this chapter we consider the integral calculus of functions of more than one variable. We begin by defining a *double integral*, in which a function of two variables is integrated over a region in the xy-plane. Just as the integral of a nonnegative function of one variable can be interpreted geometrically as the area under its graph, the double integral of a nonnegative function of two variables can be interpreted as the volume under its graph.

After defining the double integral and listing some of its properties, we will give an important theorem that will let us evaluate many double integrals by evaluating two single integrals in succession. (By a single integral we mean an integral of a function of one variable—the type we have studied up to now.) This theorem plays a role for double integrals similar to that of the Second Fundamental Theorem of Calculus for single integrals.

The region in the xy-plane over which a function of two variables is integrated sometimes can be better described with polar coordinates than with rectangular coordinates. For this reason, we devote a section to studying double integrals in polar coordinates.

As one application of double integrals, we revisit moments and centers of mass of two-dimensional laminas. Recall that we studied these concepts in Chapter 6 as an application of the integral. There we were restricted to homogeneous laminas (ones with constant density). With double integration, we can remove this restriction.

As another application, we develop a formula for the area of a surface. This result is more general than the formulas we developed for surface area in Chapter 6, where we considered surfaces of revolution only.

Next, we extend our results for double integrals to *triple integrals*—that is, integrals of functions of three variables over suitably restricted three-dimensional regions. There are two three-dimensional analogues of polar coordinates—cylindrical coordinates and spherical coordinates. We will study triple integrals in each of these systems.

We conclude the chapter with a section on changing variables in double and triple integrals.

15.1 DOUBLE INTEGRALS

We begin by defining the double integral of a function of two variables over a region in the xy-plane. Since the definition has much in common with the definition of the single integral $\int_a^b f(x)dx$, you may find it helpful to review this latter definition, found in Section 5.2, before proceeding.

We begin with the simplest case, in which the region of integration is rectangular. Then we extend the definition to more general regions.

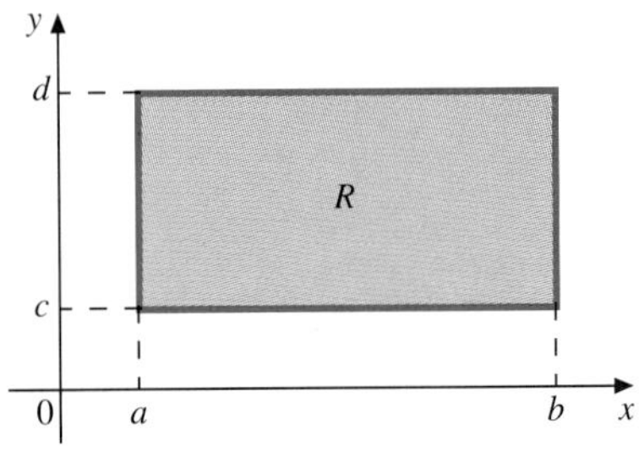

Double Integrals Over Rectangular Regions

Let f be a function of two variables that is defined on the rectangular region

$$R = \{(x, y): a \leq x \leq b, c \leq y \leq d\}$$

We have pictured such a region in Figure 15.1.

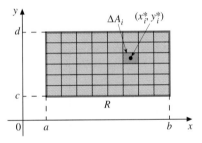

FIGURE 15.1
The rectangular region
$R = \{(x, y): a \leq x \leq b, c \leq y \leq d\}$

FIGURE 15.2
A partition of the region R

We form a rectangular grid over R as in Figure 15.2 by means of horizontal and vertical lines drawn at arbitrary points on the boundary of R. These lines divide R into finitely many smaller rectangles, which we call *subrectangles*. We call this collection of subrectangles a *partition* of R and designate the partition by P. The subrectangles need not all be of the same size. We call the length of the longest diagonal of all of them the *norm* of P and denote it by $\|P\|$.

We number the subrectangles of the partition P consecutively in any manner, starting with 1. Suppose there are n of them. In each subrectangle we select an arbitrary point. We designate the point selected in the ith subrectangle by (x_i^*, y_i^*) and the area of the ith subrectangle by ΔA_i, for $i = 1$ to $i = n$.

Now we form the sum

$$\sum_{i=1}^{n} f(x_i^*, y_i^*) \Delta A_i$$

This sum is called a *Riemann sum* for f corresponding to the partition P. If we take finer and finer partitions (that is, partitions whose norms approach 0), the corresponding Riemann sums may approach some limit L. If so, we call L the double integral of f over R and denote it by

$$\iint_R f(x, y) dA$$

In summary, we have the following definition.

Definition 15.1
The Double Integral Over a Rectangular Region R

The **double integral** of a function f of two variables over the rectangular region R is

$$\iint_R f(x, y) dA = \lim_{\|P\| \to 0} \sum_{i=1}^{n} f(x_i^*, y_i^*) \Delta A_i \qquad (15.1)$$

provided the limit on the right exists. When the limit does exist, we say that f is *integrable* on R.

REMARK

The precise meaning of the limit on the right-hand side of Equation 15.1 is as follows:

$$\lim_{\|P\|\to 0} \sum_{i=1}^{n} f(x_i^*, y_i^*) \Delta A_i = L$$

provided that, corresponding to each positive number ε, there is a positive number δ such that for all partitions P of R with $\|P\| < \delta$,

$$\left| \sum_{i=1}^{n} f(x_i^*, y_i^*) \Delta A_i - L \right| < \varepsilon$$

independently of how the point (x_i^*, y_i^*) is chosen in the ith subrectangle for $1 \leq i \leq n$.

Although we will not prove it here, it can be shown that a *continuous* function is always integrable over any rectangular region. Most of the functions we work with will be continuous. In the next section, we will see a way of evaluating double integrals that does not require us to calculate the limit in Equation 15.1 directly. In fact, evaluating most integrals by calculating this limit would be a hopeless task. However, we can approximate the integral by means of Riemann sums for particular partitions. The next example illustrates this type of approximation.

EXAMPLE 15.1 Approximate $\iint_R (x + 2y)\, dA$, where R is the rectangular region bounded by the lines $x = 0$, $x = 4$, $y = 0$, and $y = 1$. Use a partition formed by vertical lines at $x = 1, 2,$ and 3, and a horizontal line at $y = \frac{1}{2}$. Take (x_i^*, y_i^*) as the lower left-hand corner of the ith subrectangle.

Solution Figure 15.3 shows the region R and the given partition of it. We have numbered the eight subrectangles. In each case, $\Delta A_i = (1)(\frac{1}{2}) = \frac{1}{2}$. We calculate the Riemann sum using the following table.

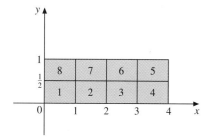

FIGURE 15.3

i	(x_i^*, y_i^*)	$f(x_i^*, y_i^*)$	$f(x_i^*, y_i^*)\Delta A_i$
1	$(0, 0)$	0	0
2	$(1, 0)$	1	$\frac{1}{2}$
3	$(2, 0)$	2	1
4	$(3, 0)$	3	$\frac{3}{2}$
5	$(3, \frac{1}{2})$	4	2
6	$(2, \frac{1}{2})$	3	$\frac{3}{2}$
7	$(1, \frac{1}{2})$	2	1
8	$(0, \frac{1}{2})$	1	$\frac{1}{2}$

$$\sum_{i=1}^{8} f(x_i^*, y_i^*) \Delta A_i = 8$$

By using the technique we will learn in the next section, we can show that the exact value of the integral is 12, so our approximation is not very good. We could improve it by using a partition with a smaller norm. ∎

Double Integrals Over General Regions

If the region R of integration is not rectangular but can be contained within some rectangle, then we partition this rectangle as before but count only those subrectangles of the partition that are completely contained within R. We illustrate a typical situation in Figure 15.4. Then we define the double integral of f over R by

$$\iint_R f(x, y) \, dA = \lim_{\|P\| \to 0} \sum_{i=1}^n f(x_i^*, y_i^*) \Delta A_i$$

provided the limit exists. So the integral is defined exactly as in Definition 15.1, with the understanding that for each partition P of the outer rectangle, we count only those subrectangles of the partition that are contained in the region R.

When f is continuous on R, the double integral will exist provided the boundary of R is not too complicated. We will not state the most general conditions on the boundary, but in the next section we will describe two commonly occurring types of regions for which the integral will exist.

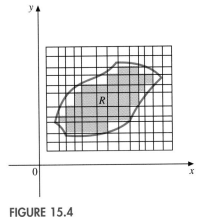

FIGURE 15.4

A Geometric Interpretation of the Double Integral

If f is a nonnegative continuous function over the region R, we can interpret the double integral of f over R as follows: Let S denote that part of the surface $z = f(x, y)$ that lies over R. Partition a rectangle containing R as before, numbering those subrectangles lying completely within R. For each i from 1 to n, build a rectangular box, called a rectangular *prism*, with the ith subrectangle as a base and having height $f(x_i^*, y_i^*)$, where (x_i^*, y_i^*) is any point in the ith subrectangle. We show a typical prism in Figure 15.5. The Riemann sum

$$\sum_{i=1}^n f(x_i^*, y_i^*) \Delta A_i$$

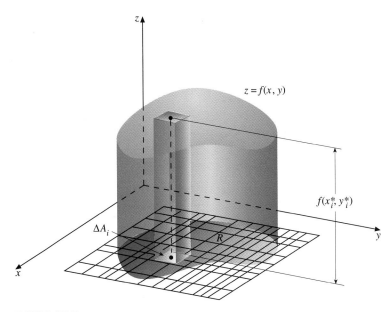

FIGURE 15.5

is the sum of the volumes of all such prisms. It is reasonable to expect that this sum approximates the volume under the surface S and above the region R. The sum may be an approximation only, since the prisms have flat tops, whereas the surface S may be curved. Also, the subrectangles inside R may not fill out all of R. Nevertheless, it seems reasonable to assume that as we take finer and finer partitions, these deviations become negligible. Since the Riemann sums approach the double integral as the norms of the partitions approach 0, we expect the double integral to give the exact volume under the surface. In fact, we *define* the volume by the double integral.

Definition 15.2
Volume Defined by a Double Integral

Let f be a nonnegative continuous function on the closed and bounded region R. If the double integral of f over R exists, then the volume V under the graph of f and above the region R is equal to this double integral:

$$V = \iint\limits_R f(x, y)\, dA \qquad (15.2)$$

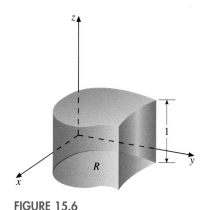

FIGURE 15.6

The special case of Equation 15.2 in which $f(x, y) = 1$ for all (x, y) in R is of particular interest. A typical volume of this type is shown in Figure 15.6. By Equation 15.2, the volume is

$$V = \iint\limits_R 1\, dA = \iint\limits_R dA$$

But we can also calculate the volume of such a slab by multiplying the area of the base by the altitude. So if we denote the area of R by A, we have $V = (A)(1) = A$. If we equate these two expressions for V, we get the following very useful formula for the area of R.

The Area of a Plane Region R

$$A = \iint\limits_R dA \qquad (15.3)$$

Although our discussion showed the plausibility of this result, we can take Equation 15.3 as the *definition* of the area of the region R, when the integral exists. In Exercise 54 of Exercise Set 15.2 you will be asked to show that when R is of a particular type, this definition is consistent with that of the area between two curves, which we studied in Chapter 6.

Properties of the Double Integral

A number of properties of the double integral can be proved based on Definition 15.1 and its extension to general regions. Some of the more important ones are stated here. We assume that each integral exists.

Properties of the Double Integral

1. $\iint_R cf(x, y)\, dA = c \iint_R f(x, y)\, dA$ for any constant c

2. $\iint_R [f(x, y) \pm g(x, y)]\, dA = \iint_R f(x, y)\, dA \pm \iint_R g(x, y)\, dA$

3. If $R = R_1 \cup R_2$, where R_1 and R_2 are nonoverlapping, then

$$\iint_R f(x, y)\, dA = \iint_{R_1} f(x, y)\, dA + \iint_{R_2} f(x, y)\, dA$$

4. If $f(x, y) \geq 0$ for all (x, y) in R, then

$$\iint_R f(x, y)\, dA \geq 0$$

5. If f is integrable over R, then so is $|f|$, and

$$\left| \iint_R f(x, y)\, dA \right| \leq \iint_R |f(x, y)|\, dA$$

Exercise Set 15.1

1. Redo Example 15.1, taking (x_i^*, y_i^*) as the midpoint of the ith subrectangle.

2. Redo Example 15.1 with a partition that divides R into 16 squares $\frac{1}{2}$ unit on each side.

In Exercises 3–6, calculate the Riemann sum for f over R where P is the partition formed by vertical lines at the specified x values and horizontal lines at the specified y values. Take (x_i^*, y_i^*) as the lower right-hand corner of the ith subrectangle.

3. $f(x, y) = 2xy$; $R = \{(x, y) : 1 \leq x \leq 3, 0 \leq y \leq 4\}$; $x = 2$; $y = 1, 2, 3$

4. $f(x, y) = x^2 - y^2$; $R = \{(x, y) : -1 \leq x \leq 2, 1 \leq y \leq 3\}$; $x = 0$; $y = 2$

5. $f(x, y) = x^2 - 2xy$; $R = \{(x, y) : 2 \leq x \leq 6, -2 \leq y \leq 4\}$; $x = 3, 4$; $y = 0, 2$

6. $f(x, y) = 2x^2 + 3y^2$; $R = \{(x, y) : -2 \leq x \leq 2, 0 \leq y \leq 3\}$; $x = 0, 1$; $y = \frac{1}{2}, \frac{3}{2}$

In Exercises 7–10, approximate the volume under the surface $z = f(x, y)$ and above the region R, using a Riemann sum for the partition formed by the specified vertical and horizontal lines. Take (x_i^*, y_i^*) as the center of the ith subrectangle.

7. $f(x, y) = 2x + 3y$; R is bounded by the triangle with vertices $(0, 0)$, $(4, 0)$, and $(0, 4)$; $x = 1, 2, 3$; $y = 1, 2, 3$

8. $f(x, y) = \sqrt{x^2 + y^2}$; $R = \{(x, y) : 0 \leq x \leq 5, -x \leq y \leq x\}$; $x = 1, 2, 3, 4$; $y = \pm 1, \pm 2, \pm 3, \pm 4$

9. $f(x, y) = 2x^2 + 4y^2$; $R = \{(x, y) : |x| + |y| \leq 2\}$; $x = 0$, $\pm\frac{1}{2}, \pm 1, \pm\frac{3}{2}$; $y = 0, \pm\frac{1}{2}, \pm 1, \pm\frac{3}{2}$

10. $f(x, y) = x^2 + 2xy + 3y^2$; $R = \{(x, y) : y-4 \leq x \leq 4-y, 0 \leq y \leq 2\}$; $x = \pm 1, \pm 2, \pm 3$; $y = \frac{1}{2}, 1, \frac{3}{2}$

11. Make use of one or more of the properties of the double integral to show that if f and g are integrable over R and $f(x, y) \leq g(x, y)$ for all (x, y) in R, then

$$\iint_R f(x, y)\, dA \leq \iint_R g(x, y)\, dA$$

12. If f is integrable over R, and m and M are numbers such that $m \leq f(x, y) \leq M$ for all (x, y) in R, prove that

$$mA \leq \iint_R f(x, y)\, dA \leq MA$$

where A is the area of R.

In Exercises 13 and 14, use a calculator, a CAS, or a spreadsheet to approximate $\iint_R f(x, y)\, dA$, where R is the rectangular region specified. Use partitions formed by m equally spaced vertical lines and n equally spaced horizontal lines. Take the point (x_i^*, y_i^*) as the center of the ith rectangle.

13. $f(x, y) = e^x \sin y$; $R = \{(x, y) : 0 \leq x \leq 2, 0 \leq y \leq 4\}$
    (a) $m = 4, n = 4$
    (b) $m = 6, n = 10$
    (c) $m = 10, n = 8$
    (d) $m = 40, n = 20$

14. $f(x, y) = \sqrt[3]{2x^2 y - 3y^2}$;
    $R = \{(x, y) : -4 \leq x \leq 6, -1 \leq y \leq 4\}$
    (a) $m = 5, n = 10$
    (b) $m = 20, n = 40$
    (c) $m = 30, n = 20$
    (d) $m = 50, n = 100$

15.2 EVALUATING DOUBLE INTEGRALS BY ITERATED INTEGRALS

Certain double integrals can be evaluated by means of *iterated integrals*, in which two single integrals are evaluated in succession. To explain what an iterated integral is and its relationship to a double integral, let us return to the simplest case, in which the region R of integration is rectangular.

One of the single integrals in an iterated integral in this case will either be of the form

$$\int_a^b f(x, y)\, dx$$

or of the form

$$\int_c^d f(x, y)\, dy$$

In the first of these integrals, the integration is with respect to x, *holding y fixed*, whereas in the second, the integration is with respect to y, *holding x fixed*. For example, consider the two integrals

$$\int_0^2 x \cos y\, dx \quad \text{and} \quad \int_0^{\pi/2} x \cos y\, dy$$

For the first integral, we have

$$\int_0^2 x \cos y\, dx = \frac{x^2}{2} \cos y \Big]_0^2 = 2 \cos y$$

Notice that $\cos y$ was treated as a constant, since we were holding y fixed.

For the second integral, we hold x fixed to get

$$\int_0^{\pi/2} x \cos y\, dy = x \sin y \Big]_0^{\pi/2} = x$$

since $\sin(\pi/2) = 1$ and $\sin 0 = 0$.

Notice the similarity between integrals of these two types and partial differentiation. In fact, the integration process we have illustrated is sometimes called *partial integration*.

As our example shows, an integral of the form

$$\int_a^b f(x, y)\, dx$$

is a function of y, since after integrating with respect to x, we substituted the upper and lower limits for x, leaving the variable y only. In our example we found that

$$\int_0^2 x \cos y \, dx = 2 \cos y$$

confirming that the result is a function of y. Suppose now that we integrate this function of y over the interval $[c, d]$. Then we have

$$\int_c^d \left[\int_a^b f(x, y)\, dx \right] dy$$

where the bracket on the inside indicates the (partial) integration with respect to x is to be performed first. For example,

$$\int_0^{\pi/2} \left[\int_0^2 x \cos y \, dx \right] dy = \int_0^{\pi/2} 2 \cos y \, dy$$
$$= 2 \sin y \Big]_0^{\pi/2} = 2$$

Similarly, the result of the integration

$$\int_c^d f(x, y)\, dy$$

is a function of x, which can then be integrated over the x interval to get

$$\int_a^b \left[\int_c^d f(x, y)\, dy \right] dx$$

For example,

$$\int_0^2 \left[\int_0^{\pi/2} x \cos y \, dy \right] dx = \int_0^2 x \, dx = x^2 \Big]_0^2 = 2$$

Notice that we got the same answer as when we integrated first with respect to x and then y. This equivalence is no accident, as we will soon see.

Each of the integrals

$$\int_c^d \left[\int_a^b f(x, y)\, dx \right] dy \quad \text{and} \quad \int_a^b \left[\int_c^d f(x, y)\, dy \right] dx$$

is called an *iterated integral*. It is customary to delete the brackets and write simply

$$\int_c^d \int_a^b f(x, y)\, dx\, dy \quad \text{and} \quad \int_a^b \int_c^d f(x, y)\, dy\, dx$$

The order in which the integration should be performed is indicated by the order of the dx and dy, from left to right.

REMARK ─────────────────────────────────
■ An iterated integral has much in common with a composite function. In a composite function $f(g(x))$, the value of the inner function is obtained first and then the result is substituted into the outer function. In an iterated integral, the inner integral is evaluated first and then the result becomes the integrand for the outer integral.
─────────────────────────────────

EXAMPLE 15.2 Evaluate the iterated integral

$$\int_0^3 \int_{-1}^1 \frac{x^2}{1+y^2} \, dy \, dx$$

Solution An antiderivative of $1/(1+y^2)$ is $\tan^{-1} y$. So on integrating first with respect to y, holding x fixed, we get

$$\int_0^3 \left[x^2 \tan^{-1} y \right]_{-1}^1 dx = \int_0^3 x^2 \left[\tan^{-1} 1 - \tan^{-1}(-1) \right] dx$$

$$= \int_0^3 x^2 \left[\frac{\pi}{4} - \left(-\frac{\pi}{4} \right) \right] dx$$

$$= \frac{\pi}{2} \int_0^3 x^2 \, dx$$

Now we integrate with respect to x to get

$$\frac{\pi}{2} \left[\frac{x^3}{3} \right]_0^3 = \frac{\pi}{2} \left(\frac{27}{3} \right) = \frac{9\pi}{2}$$

So we have the final result

$$\int_0^3 \int_{-1}^1 \frac{x^2}{1+y^2} \, dy \, dx = \frac{9\pi}{2} \quad ■$$

We are ready now for the important theorem relating a double integral over a rectangular region and two associated iterated integrals. It is named for the Italian mathematician Guido Fubini (1879–1943).

THEOREM 15.1

Fubini's Theorem

If f is continuous on the rectangular region $R = \{(x, y) : a \leq x \leq b, c \leq y \leq d\}$, then

$$\iint_R f(x, y) \, dA = \int_a^b \int_c^d f(x, y) \, dy \, dx = \int_c^d \int_a^b f(x, y) \, dx \, dy$$

REMARK ─────────────────────────────────
■ Fubini's Theorem tells us that the iterated integrals taken in either order are equal and that their common value equals the double integral over R.
─────────────────────────────────

Although we will not give a proof of Fubini's Theorem, we can show the result is plausible at least for the case in which $f(x, y)$ is nonnegative and continuous on R. We know in this case that the volume V under the surface $z = f(x, y)$ and above the region R is, according to Definition 15.2,

$$V = \iint_R f(x, y)\,dA \tag{15.4}$$

But we can also calculate the volume in another way. As indicated in Figure 15.7, we select an arbitrary x value between a and b and hold it fixed temporarily. To emphasize that it is held fixed, we designate it by x_0 (but we will soon remove the subscript). The plane $x = x_0$ intersects the given surface in a curve $z = f(x_0, y)$. Here y is the independent variable and z is the dependent variable. We are interested in that portion of this curve from $y = c$ to $y = d$. Denote the area under its graph by $A(x_0)$. Then from Chapter 5 we know that

$$A(x_0) = \int_c^d f(x_0, y)\,dy$$

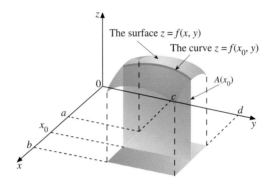

FIGURE 15.7

Observe that $A(x_0)$ is the area of a typical cross section of the solid under the surface, taken perpendicular to the x-axis. So by Definition 6.1, we get the total volume by integrating this cross-sectional area over the interval $[a, b]$. Since we now want x_0 to vary over this interval, we drop the subscript and write

$$V = \int_a^b A(x)\,dx$$

That is,

$$V = \int_a^b \left[\int_c^d f(x, y)\,dy \right] dx \tag{15.5}$$

Equating the two expressions for volume (Equations 15.4 and 15.5) gives

$$\iint_R f(x, y)\,dA = \int_a^b \int_c^d f(x, y)\,dy\,dx$$

A similar argument can be given when we pass a plane perpendicular to the y-axis at $y = y_0$ to obtain the result

$$\iint_R f(x, y)\,dA = \int_c^d \int_a^b f(x, y)\,dx\,dy$$

(See Exercise 53 in Exercise Set 15.2.)

EXAMPLE 15.3 Use Fubini's Theorem to evaluate the double integral

$$\iint_R (x + 2y)\,dA$$

where R is the rectangular region $\{(x, y) : 0 \le x \le 4, 0 \le y \le 1\}$.

Solution Note that the integral in question is the one that we estimated in Example 15.1.

By Fubini's Theorem, we can write

$$\iint_R (x + 2y)\,dA = \int_0^4 \int_0^1 (x + 2y)\,dy\,dx$$

$$= \int_0^4 \left[xy + y^2\right]_0^1 dx$$

$$= \int_0^4 (x + 1)\,dx$$

$$= \left.\frac{(x+1)^2}{2}\right|_0^4 = \frac{25}{2} - \frac{1}{2} = \frac{24}{2} = 12$$

We chose to integrate first with respect to y, then x. If we use the other order, we get

$$\iint_R (x + 2y)\,dA = \int_0^1 \int_0^4 (x + 2y)\,dx\,dy$$

$$= \int_0^1 \left[\frac{x^2}{2} + 2xy\right]_0^4 dy$$

$$= \int_0^1 (8 + 8y)\,dy$$

$$= \left.8y + 4y^2\right|_0^1 = 8 + 4 = 12$$

Of course, we expected that the results would be the same, as guaranteed by Fubini's Theorem. ∎

Extension of Fubini's Theorem to Nonrectangular Regions

In order to extend Fubini's Theorem to more general regions, we first identify two special types of regions, which we call **type I** and **type II**. A type I region is shown in Figure 15.8 and a type II region in Figure 15.9.

More precisely, we say a region R is of type I if

$$R = \{(x, y) : a \le x \le b,\ g_1(x) \le y \le g_2(x)\}$$

where g_1 and g_2 are continuous functions on the closed interval $[a, b]$. Similarly, R is a type II region if

$$R = \{(x, y) : h_1(y) \le x \le h_2(y),\ c \le y \le d\}$$

where h_1 and h_2 are continuous functions on the closed interval $[c, d]$.

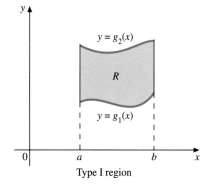

Type I region

FIGURE 15.8

When we are working with a type I region, an iterated integral will be of the form

$$\int_a^b \int_{g_1(x)}^{g_2(x)} f(x, y)\, dy\, dx$$

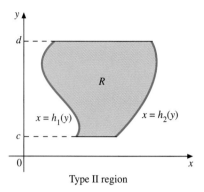

FIGURE 15.9 Type II region

REMARK

■ To help determine when a region is of type I, consider vertical lines drawn at x values between the left and right extremities of the region. Each such line must intersect the boundary of R in exactly two points—once on the lower bounding curve and once on the upper bounding curve. Similarly, for a type II region each horizontal line between the lower and upper extremities will intersect the boundary exactly twice—once on the left bounding curve and once on the right bounding curve.

When the region is type II, we will have an iterated integral of the form

$$\int_c^d \int_{h_1(y)}^{h_2(y)} f(x, y)\, dx\, dy$$

Note that these two types of iterated integrals differ from those we considered previously only in the limits of integration on the inner integral. Previously, all the limits were constants. Now we have variables as limits on the inner integral. In the next example, we illustrate how to evaluate iterated integrals of these two types.

EXAMPLE 15.4   Evaluate the following iterated integrals.

(a) $\displaystyle\int_0^2 \int_{x^2}^{4-x} (x - 2y)\, dy\, dx$ (b) $\displaystyle\int_0^1 \int_{y-1}^{\sqrt{1-y^2}} (xy + y)\, dx\, dy$

Solution

(a) We proceed just as before, integrating first with respect to y, holding x fixed.

$$\int_0^2 \int_{x^2}^{4-x} (x - 2y)\, dy\, dx = \int_0^2 \left[xy - y^2\right]_{x^2}^{4-x} dx$$

$$= \int_0^2 \left[x(4-x) - (4-x)^2 - x^3 + x^4\right] dx$$

$$= \int_0^2 \left[4x - x^2 - (4-x)^2 - x^3 + x^4\right] dx$$

$$= \left. 2x^2 - \frac{x^3}{3} + \frac{(4-x)^3}{3} - \frac{x^4}{4} + \frac{x^5}{5} \right]_0^2 = -\frac{164}{15}$$

(You should check the algebra.)

(b) $\displaystyle\int_0^1 \int_{y-1}^{\sqrt{1-y^2}} (xy+y)\,dx\,dy$

$\displaystyle= \int_0^1 \left[\frac{x^2}{2}y + xy\right]_{y-1}^{\sqrt{1-y^2}} dy$

$\displaystyle= \int_0^1 \left[\frac{(1-y^2)y}{2} + y\sqrt{1-y^2} - \frac{(y-1)^2}{2}y - y(y-1)\right] dy$

$\displaystyle= \int_0^1 \left[\frac{1}{2}(y-y^3) + y\sqrt{1-y^2} - \frac{1}{2}(y^3 - 2y^2 + y) - y^2 + y\right] dy$

$\displaystyle= \int_0^1 (-y^3 + y + y\sqrt{1-y^2})\,dy$

$\displaystyle= -\frac{y^4}{4} + \frac{y^2}{2} - \frac{1}{2}\cdot\frac{2}{3}(1-y^2)^{3/2}\Big]_0^1$

$\displaystyle= -\frac{1}{4} + \frac{1}{2} + \frac{1}{3} = \frac{7}{12}$ ∎

We can now state the following stronger form of Fubini's Theorem.

THEOREM 15.2

Fubini's Theorem (Stronger Form)

1. If f is continuous on the type I region
$$R = \{(x,y) : a \le x \le b,\ g_1(x) \le y \le g_2(x)\}$$
then
$$\iint_R f(x,y)\,dA = \int_a^b \int_{g_1(x)}^{g_2(x)} f(x,y)\,dy\,dx$$

2. If f is continuous on the type II region
$$R = \{(x,y) : h_1(y) \le x \le h_2(y),\ c \le y \le d\}$$
then
$$\iint_R f(x,y)\,dA = \int_c^d \int_{h_1(y)}^{h_2(y)} f(x,y)\,dx\,dy$$

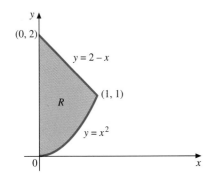

FIGURE 15.10

EXAMPLE 15.5 Evaluate the double integral
$$\iint_R (x+\sqrt{y})\,dA$$
where R is the region shown in Figure 15.10.

Solution Observe that since R is bounded below and above by the continuous functions $y = x^2$ and $y = 2-x$, respectively, for $0 \le x \le 1$, R is a type I region. It is also of type II, since it is bounded on the left by the line $x = 0$

and on the right by the continuous function consisting of the line $x = 2 - y$ when $1 \leq y \leq 2$ and the parabolic arc $x = \sqrt{y}$ when $0 \leq y \leq 1$. We choose to treat it as a type I region, however, to avoid the necessity of having to use one integral for $0 \leq y \leq 1$ and another for $1 \leq y \leq 2$, as we would have to do if we treated it as a type II region.

By part 1 of Theorem 15.2,

$$\iint_R (x + \sqrt{y})\,dA = \int_0^1 \int_{x^2}^{2-x} (x + \sqrt{y})\,dy\,dx$$

$$= \int_0^1 \left[xy + \frac{2}{3}y^{3/2}\right]_{x^2}^{2-x} dx$$

$$= \int_0^1 \left[x(2-x) + \frac{2}{3}(2-x)^{3/2} - x^3 - \frac{2}{3}x^3\right] dx$$

$$= \int_0^1 \left[2x - x^2 + \frac{2}{3}(2-x)^{3/2} - \frac{5}{3}x^3\right] dx$$

$$= \left. x^2 - \frac{x^3}{3} - \frac{2}{3} \cdot \frac{2}{5}(2-x)^{5/2} - \frac{5}{12}x^4 \right]_0^1$$

$$= 1 - \frac{1}{3} - \frac{4}{15} - \frac{5}{12} + \frac{4}{15}(2)^{5/2} = \frac{-1 + 64\sqrt{2}}{60} \approx 1.492 \quad \blacksquare$$

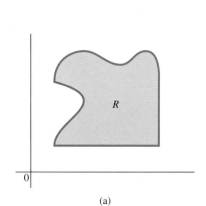

(a)

R is of type I or type II

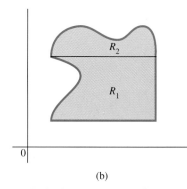

(b)

R_1 is of type II, and R_2 is of type I

FIGURE 15.11

REMARKS

■ If, as in Example 15.5, a region can be treated either as type I or type II but treating it one way requires splitting an integral into two or more parts whereas treating it the other way does not, we usually choose to treat it as the type that does not require splitting up the integral.

■ When a region is neither type I nor type II, it may be possible to divide it into two or more regions, each of which is either type I or type II. We show such a region in Figure 15.11. By dividing it as shown, we see that R_1 is of type II and R_2 is of type I. We first write

$$\iint_R f(x,y)\,dA = \iint_{R_1} f(x,y)\,dA + \iint_{R_2} f(x,y)\,dA \qquad \text{By Property 3}$$

Then we apply Theorem 15.2 to each integral on the right.

The next three examples further illustrate the use of Theorem 15.2.

EXAMPLE 15.6 Evaluate the integral $\iint_R x^2 y\,dA$, where R is bounded by $x = 0$, $x = 1$, $y = \sqrt{x}$, and $y = 2$. Draw the region.

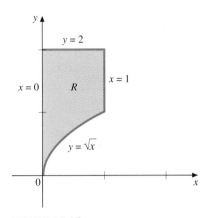

FIGURE 15.12

Solution The region is pictured in Figure 15.12. It is a type I region (also type II, but it is better to treat it as type I). So by Theorem 15.2,

$$\iint_R x^2 y \, dA = \int_0^1 \int_{\sqrt{x}}^2 x^2 y \, dy \, dx$$

$$= \int_0^1 \left[\frac{x^2 y^2}{2} \right]_{\sqrt{x}}^2 dx$$

$$= \int_0^1 \left(2x^3 - \frac{x^3}{2} \right) dx = \left. \frac{2x^3}{3} - \frac{x^4}{8} \right]_0^1 = \frac{13}{24}$$

EXAMPLE 15.7 Find the volume under the paraboloid $z = x^2 + 3y^2$ and above the region bounded by the lines $x = 0$, $y = 0$, and $x + y = 1$.

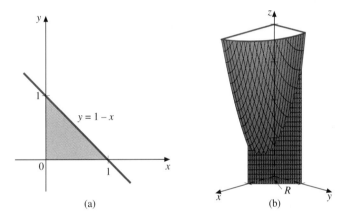

FIGURE 15.13

Solution The region R in the xy-plane is shown in Figure 15.13(a) and the three-dimensional solid is shown in Figure 15.13(b). The region R is both a type I and a type II region. Treating it as type I and using Definition 15.2 and Theorem 15.2, we have

$$V = \iint_R (x^2 + 3y^2) \, dA = \int_0^1 \int_0^{1-x} (x^2 + 3y^2) \, dy \, dx$$

$$= \int_0^1 \left[x^2 y + y^3 \right]_0^{1-x} dx$$

$$= \int_0^1 \left[x^2 - x^3 + (1-x)^3 \right] dx$$

$$= \left. \frac{x^3}{3} - \frac{x^4}{4} - \frac{(1-x)^4}{4} \right]_0^1$$

$$= \left(\frac{1}{3} - \frac{1}{4} \right) + \frac{1}{4} = \frac{1}{3}$$

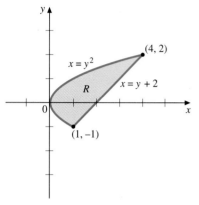

FIGURE 15.14

EXAMPLE 15.8 Using double integration, find the area of the region R bounded by the curves $y^2 = x$ and $x - y = 2$. Sketch the region.

Solution The region is shown in Figure 15.14. Solving the two equations simultaneously, we find the points of intersection $(1, -1)$ and $(4, 2)$ as shown. (Verify.) The region is of type II. So we have from Equation 15.3 and Theorem 15.2,

$$A = \iint_R dA = \int_{-1}^{2} \int_{y^2}^{y+2} dx\,dy = \int_{-1}^{2} \left[x\right]_{y^2}^{y+2} dy = \int_{-1}^{2} (y + 2 - y^2)\,dy$$

$$= \left. \frac{y^2}{2} + 2y - \frac{y^3}{3} \right]_{-1}^{2} = \left(2 + 4 - \frac{8}{3}\right) - \left(\frac{1}{2} - 2 + \frac{1}{3}\right) = \frac{9}{2}$$

Interchange of Order of Integration

Sometimes an iterated integral that is difficult (or impossible) to evaluate can be evaluated by interchanging the order of integration. Such an interchange is possible when the region described by the limits is both type I and type II. We illustrate this procedure in the next example.

EXAMPLE 15.9 Evaluate the iterated integral

$$\int_0^1 \int_x^1 \sin y^2 \, dy\, dx$$

Solution Since $\sin y^2$ has no elementary antiderivative, it seems we are stymied. But the given iterated integral, according to Theorem 15.2, equals the double integral $\iint_R \sin y^2 \, dA$, where R is the type I region $\{(x, y) : 0 \le x \le 1, x \le y \le 1\}$ pictured in Figure 15.15. But R can be viewed equally well as a type II region with left boundary $x = 0$ and right boundary $x = y$. Thus, the double integral, and hence the original iterated integral, equal

$$\int_0^1 \int_0^y \sin y^2 \, dx\, dy = \int_0^1 \left[x \sin y^2\right]_0^y dy \quad \text{Remember that } y \text{ is constant at this stage.}$$

$$= \int_0^1 y \sin y^2 \, dy$$

$$= \left. -\frac{1}{2} \cos y^2 \right]_0^1 = \frac{1}{2}(1 - \cos 1)$$

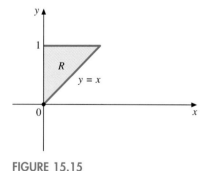

FIGURE 15.15

By treating this region R as type II, thereby changing both the order of integration and the limits on the integrals, we arrived at the integral of $y \sin y^2$. The presence of the factor y enabled us to make the (mental) substitution $u = y^2$. All that we needed for the differential $du = 2y\,dy$ was the constant factor 2, which we supplied, compensating on the outside with the factor $1/2$.

Exercise Set 15.2

In Exercises 1–10, evaluate the iterated integral. Sketch the region R determined by the limits.

1. $\int_0^1 \int_1^2 (2x - 3y)\, dy\, dx$
2. $\int_0^2 \int_{-1}^1 (3x^2 - 2xy)\, dx\, dy$
3. $\int_1^2 \int_0^x xy\, dy\, dx$
4. $\int_0^1 \int_x^{4-x} (2x - 1)\, dy\, dx$
5. $\int_0^4 \int_{-2y}^{\sqrt{y}} y(2x - 1)\, dx\, dy$
6. $\int_{-2}^2 \int_{-y}^{\sqrt{2-y}} (x^3 + 2xy)\, dx\, dy$
7. $\int_0^{\sqrt{3}} \int_{2-\sqrt{4-x^2}}^{\sqrt{4-x^2}} x\, dy\, dx$
8. $\int_0^{\sqrt{5}} \int_{y^2-2}^{\sqrt{y^2+4}} y\, dx\, dy$
9. $\int_1^2 \int_{-y}^{4-y} \frac{y - 4}{(x + 4)^2}\, dx\, dy$
10. $\int_1^4 \int_{x-2}^{\sqrt{x}} (2xy + 3)\, dy\, dx$

In Exercises 11–21, evaluate the double integral of f over R, making use of Theorem 15.1 or 15.2. Sketch the region and identify its type.

11. $f(x, y) = 4x - 3y$; $R = \{(x, y) : 0 \leq x \leq 4, 0 \leq y \leq 2\}$
12. $f(x, y) = x^2 - y^2$; $R = \{(x, y) : 1 \leq x \leq 2, -1 \leq y \leq 3\}$
13. $f(x, y) = y/(1 + x^3)$; $R = \{(x, y) : 0 \leq x \leq 1, 0 \leq y \leq 2x\}$
14. $f(x, y) = x + y$; $R = \{(x, y) : y - 1 \leq x \leq \sqrt{1 - y^2}, 0 \leq y \leq 1\}$
15. $f(x, y) = 2x - y$; $R = \{(x, y) : y \leq x \leq \sqrt{4 - y^2}, 0 \leq y \leq \sqrt{2}\}$
16. $f(x, y) = xe^y$; $R = \{(x, y) : 0 \leq x \leq 1, x^2 \leq y \leq 2 - x^2\}$
17. $f(x, y) = \sqrt{(2x^2 + 7)/y}$; $R = \{(x, y) : 1 \leq x \leq 3, x^2/4 \leq y \leq x^2\}$
18. $f(x, y) = e^{x+2y}$; R is bounded by the triangle with vertices $(0, 0)$, $(1, 1)$, and $(3, 0)$.
19. $f(x, y) = 10xy^3$; R is bounded by the parallelogram with vertices $(-1, 0)$, $(0, 1)$, $(1, 0)$, and $(2, 1)$.
20. $f(x, y) = 4xy$; R is bounded by the triangle with vertices $(-1, 1)$, $(0, 2)$, and $(1, 0)$.
21. $f(x, y) = y^2$; R is bounded by the triangle with vertices $(0, 0)$, $(2, 2)$, and $(3, -1)$.

In Exercises 22–31, make use of Equation 15.3 and Theorem 15.2 to find the area of the region bounded by the given curves. Sketch the region and identify its type.

22. $y = x^2$, $y = 2 - x^2$
23. $y = x$, $y = 3x - x^2$
24. $y^2 = 4x$, $y = 2x - 4$
25. $y = \sqrt{x}$, $x = 0$, $y = 4$
26. $y = e^x$, $y = e^{-x}$, $x = \ln 3$
27. $y = 1 - x^2$, $y = \ln x$, $y = 1$
28. $y = \cos \pi x$, $4x^2 + 4y = 1$, $-\frac{1}{2} \leq x \leq \frac{1}{2}$
29. $y = \cos^{-1}\frac{x}{2}$, $y = \tan^{-1}\frac{x}{3}$, $x = 0$
30. $y = \sqrt{x}$, $x + y = 0$, $x - y = 2$
31. Below $y = 3$ and $y = 3(x + 1)$, and above $y = x^2 - 1$

In Exercises 32–39, find the volume under the surface $z = f(x, y)$ that is above the region R bounded by the given curves.

32. $f(x, y) = 2 - x^2 - y^2$; $x = 0$, $x = 1$, $y = 0$, $y = 1$. Sketch the solid.
33. $f(x, y) = xy$; $y = \sqrt{8 - x^2}$, $y = x$, $y = 0$. Sketch the solid.
34. $f(x, y) = y\sqrt{1 + x^3}$; $y = x$, $x = 2$, $y = 0$
35. $f(x, y) = x + y$; $x = \sqrt{2 - y^2}$, $x = 0$, $x = y$. Sketch the solid.
36. $f(x, y) = x^2 e^{xy}$; $xy = 1$, $x = 2$, $y = 2$
37. $f(x, y) = 1 + x^2 + y^2$; $x = 0$, $y = x$, $y = 1$, $y = 2$
38. $f(x, y) = e^{-y}$; $x = 0$, $x = 4$, $y = 0$, $y = \ln 2$. Sketch the solid.
39. $f(x, y) = 4 - x^2$; $y = 0$, $x = 2$, $y = x$. Sketch the solid.
40. Find the volume of the tetrahedron formed by the planes $3x + 4y + 2z = 12$, $x = 0$, $y = 0$, and $z = 0$.

41. Find the volume of the tetrahedron with vertices $(0, 0, 0)$, $(2, 0, 0)$, $(0, 4, 0)$, and $(0, 0, 4)$.

42. Find the volume inside the paraboloid $z = 4 - x^2 - y^2$ and above the xy-plane.

In Exercises 43–46, evaluate the iterated integral by changing the order of integration.

43. $\displaystyle\int_0^1 \int_x^1 e^{-y^2}\, dy\, dx$

44. $\displaystyle\int_0^1 \int_{\sqrt{y}}^1 \sqrt{1 + x^3}\, dx\, dy$

45. $\displaystyle\int_0^2 \int_y^2 \frac{y}{(1 + x^3)^2}\, dx\, dy$

46. $\displaystyle\int_0^2 \int_{x^2}^4 \frac{1}{1 + y^{3/2}}\, dy\, dx$

In Exercises 47–52, give an equivalent integral with the order of integration reversed.

47. $\displaystyle\int_0^4 \int_{\sqrt{x}}^2 f(x, y)\, dy\, dx$

48. $\displaystyle\int_0^{\pi/2} \int_0^{\sin x} f(x, y)\, dy\, dx$

49. $\displaystyle\int_1^3 \int_{y+1}^4 f(x, y)\, dx\, dy$

50. $\displaystyle\int_0^4 \int_{y^2/4}^{2\sqrt{y}} f(x, y)\, dx\, dy$

51. $\displaystyle\int_0^1 \int_{1-\sqrt{1-x^2}}^{\sqrt{2x-x^2}} f(x, y)\, dy\, dx$

52. $\displaystyle\int_1^2 \int_0^{\ln y} f(x, y)\, dx\, dy$

53. Use an argument similar to that used in the text as a partial justification of Fubini's Theorem to show that if f is continuous and nonnegative on the rectangular region $R = \{(x, y) : a \leq x \leq b,\ c \leq y \leq d\}$, then
$$\iint_R f(x, y)\, dA = \int_c^d \int_a^b f(x, y)\, dx\, dy$$

54. Show that if R is a type I region, then Equation 15.3 for the area of R is consistent with Equation 6.1 for the area between the graphs of two functions.

In Exercises 55–57, evaluate the double integral.

55. $\displaystyle\iint_R x^2 e^y\, dA;\ R = \{(x, y) : 0 \leq x \leq 1,\ -x \leq y \leq x\}$

56. $\displaystyle\iint_R (4xy + 4)\, dA;\ R$ is bounded by $y = 0$ and $y = \sin x$ between $x = 0$ and $x = \pi$.

57. $\displaystyle\iint_R \sqrt{y}\,(x^2 + 1)\, dA;\ R$ is the first-quadrant region bounded by $x^2 y = 4$, $x = 0$, $x = 2$, $y = 0$, and $y = 4$.

In Exercises 58–60, find the area of the region bounded by the given curves, using double integration.

58. Below $y = \sqrt{2x}$ and $x + 2y = 6$, and above $x - 4y = 0$

59. $y = x^2$, $y = x^3 - 2x$

60. $y^2 = x$, $y^2 = 8x$, $x + y = 6$, $5x + y = 48$ (first quadrant)

In Exercises 61–64, find the volume of the solid.

61. Under the surface $z = 2x/y^2$ and above the region bounded by $xy = 6$ and $x^2 + y = 7$

62. Bounded by the surface $z = (1 - \sqrt{x} - \sqrt{y})^2$ and the coordinate planes

63. Bounded by the surface $z = (a^{2/3} - x^{2/3} - y^{2/3})^{3/2}$ and the xy-plane (*Hint:* To evaluate the inner integral use a trigonometric substitution.)

64. Common to the two cylinders $x^2 + y^2 = a^2$ and $x^2 + z^2 = a^2$

CAS *Use a CAS for Exercises 65 through 69.*

65. Find the volume of the region under the paraboloid $f(x, y) = x^2 + 2y^2 + 3$ and above the square in the xy-plane $\{(x, y) : -1 \leq x \leq 1, -1 \leq y \leq 1\}$. Plot the solid in question.

66. Find the volume of the region under the hyperbolic paraboloid $f(x, y) = y^2 - x^2 + 2$ and above the circle in the xy-plane $x^2 + y^2 = 1$. Plot the solid in question.

67. Find the volume of the region under the plane $f(x, y) = 2 - 2(x - 1) + y$ and above the region in the xy-plane bounded by the curves $y = (x - 2)^2 + 1$ and $y = -(x - 2)^2 + 3$. Plot the solid in question.

68. Find the volume of the region under the graph of $f(x, y) = 4 - x^2 - y^2$ and above the graph of $x^2 + 3y^2 - 2$. Plot the region in question.

69. Find the area of the ellipse $\dfrac{x^2}{a^2} + \dfrac{y^2}{b^2} = 1$.

15.3 DOUBLE INTEGRALS IN POLAR COORDINATES

It is frequently more convenient to describe regions in the xy-plane using polar coordinates rather than rectangular coordinates. For example, the region R shown in Figure 15.16 can be expressed in polar coordinates by

$$R = \{(r, \theta) : 0 \leq r \leq 2, \ 0 \leq \theta \leq \pi/4\}$$

Expressing R in rectangular coordinates is more difficult, as you can see by trying it.

The integrand also can often be simplified by changing to polar coordinates. For example, if the integrand were $\sin\sqrt{x^2 + y^2}$, it would become $\sin r$ in polar coordinates.

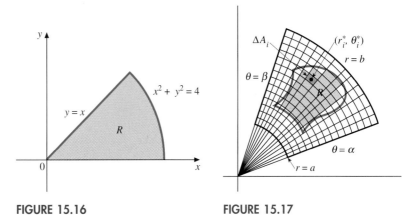

FIGURE 15.16   **FIGURE 15.17**

Suppose, then, that we want to integrate a function f of the polar variables r and θ over a bounded region R that is given in polar coordinates. We need to define the double integral in this case. We follow a procedure similar to that used in Section 15.1. First, we enclose R in a *polar rectangle* of the form $\{(r, \theta) : a \leq r \leq b, \ \alpha \leq \theta \leq \beta\}$, as shown in Figure 15.17. Then we partition this polar rectangle by rays and circular arcs as shown, thus forming finitely many polar subrectangles. We denote the partition by P. The norm of P, again denoted by $\|P\|$, is the length of the longest diagonal of the polar subrectangles formed by P (just as with a rectangle, the diagonal of a polar rectangle is a line segment joining opposite vertices). We number the polar subrectangles that lie entirely in R (shaded in Figure 15.17) from 1 to n. We let ΔA_i denote the area of the ith subrectangle and (r_i^*, θ_i^*) be polar coordinates of an arbitrary point in the ith polar subrectangle. Then we define the double integral of f over R by

$$\iint\limits_R f(r, \theta) \, dA = \lim_{\|P\| \to 0} \sum_{i=1}^n f(r_i^*, \theta_i^*) \, \Delta A_i \tag{15.6}$$

provided this limit exists, independently of the choices of the points (r_i^*, θ_i^*). Again, it can be shown that the integral will exist when f is continuous on R.

To evaluate the double integral of a continuous function by iterated integrals when R is given in polar coordinates, we consider two types of regions only, analogous to type I and type II for rectangular coordinates. Figure 15.18 illustrates these two types. In part (a) of Figure 15.18, R is of the form $\{(r, \theta) : g_1(\theta) \leq r \leq g_2(\theta), \alpha \leq \theta \leq \beta\}$, where g_1 and g_2 are continuous functions on

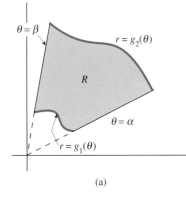

(a)

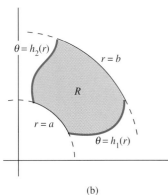

(b)

FIGURE 15.18

$[\alpha, \beta]$. In part (b), R is of the form $\{(r, \theta) : a \leq r \leq b, \; h_1(r) \leq \theta \leq h_2(r)\}$, where h_1 and h_2 are continuous functions on $[a, b]$. Because the first type is the more common, we will concentrate our attention on it.

To help you understand the form of the iterated integral in this case, we consider a typical polar subrectangle formed by a partition P. We have enlarged such a polar subrectangle in Figure 15.19. To simplify our notation we have deleted all subscripts. As shown, let $\Delta\theta$ be the angle between the two rays that form the polar subrectangle, and let Δr be the change in radial distance between the two bounding arcs. In the definition of the double integral, we are free to choose an arbitrary point in each subinterval at which the function is evaluated. We now specify this point (r^*, θ^*) as the center of the subinterval. Then the inner radius is $r^* - \Delta r/2$ and the outer radius is $r^* + \Delta r/2$. The area ΔA is the difference in areas of two circular sectors. Since the area of a circular sector is one-half its angle (in radians) times its radius, we have

$$\Delta A = \frac{1}{2}\left(r^* + \frac{\Delta r}{2}\right)^2 \Delta\theta - \frac{1}{2}\left(r^* - \frac{\Delta r}{2}\right)^2 \Delta\theta$$

Squaring and collecting terms, we get

$$\Delta A = r^* \, \Delta r \, \Delta\theta \tag{15.7}$$

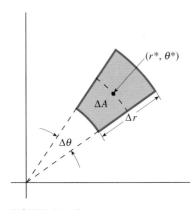

FIGURE 15.19

You might note that $r^*\Delta\theta$ is the length of the arc through the center, which is the average of the lengths of the inner and outer bounding arcs. So ΔA is the average arc length times the radial distance Δr. Notice the similarity with the formula for the area of a trapezoid. The Riemann sum in Equation 15.6 can now be written

$$\sum_{i=1}^{n} f(r_i^*, \theta_i^*) \, \Delta A_i = \sum_{i=1}^{n} f(r_i^*, \theta_i^*) r_i^* (\Delta r \, \Delta\theta)_i$$

where by $(\Delta r \, \Delta\theta)_i$ we mean the product of Δr and $\Delta\theta$ for the ith polar subrectangle.

By taking partitions with norms approaching 0, we are led to the following result, analogous to the stronger form of Fubini's Theorem.

If f is continuous on the polar region

$$R = \{(r, \theta) : g_1(\theta) \leq r \leq g_2(\theta), \; \alpha \leq \theta \leq \beta\}$$

then

$$\iint\limits_R f(r, \theta) \, dA = \int_\alpha^\beta \int_{g_1(\theta)}^{g_2(\theta)} f(r, \theta) r \, dr \, d\theta \tag{15.8}$$

Observe the extra factor r in the integrand of the iterated integral on the right. This discussion is intended to suggest to you where it comes from, but it is not a proof. (In Section 15.8, we will see that this factor is a special case of a more general result having to do with changing variables in double integrals.)

It is useful to write

$$dA = r \, dr \, d\theta$$

and to call this form the *differential of area in polar coordinates*. For rectangular coordinates, the analogous formula is $dA = dy\,dx$. So in going from a double integral in polar coordinates to an iterated integral, dA is replaced by $r\,dr\,d\theta$, whereas in rectangular coordinates dA is replaced by $dy\,dx$. (In each case the order of the differentials may be reversed, depending on the order of integration.)

When the region is of the type in Figure 15.18(b), we have the following result, analogous to that given in Equation 15.8.

If f is continuous on the polar region
$$R = \{(r, \theta) : a \leq r \leq b,\ h_1(r) \leq \theta \leq h_2(r)\}$$
then
$$\iint_R f(r, \theta)\, dA = \int_a^b \int_{h_1(r)}^{h_2(r)} f(r, \theta) r\, d\theta\, dr \tag{15.9}$$

EXAMPLE 15.10 Evaluate the double integral $\iint_R r^2 \sin\theta\, dA$, where R is the region bounded by the polar axis and the upper half of the cardioid $r = 1 + \cos\theta$.

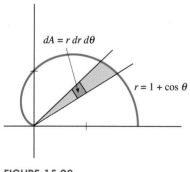

FIGURE 15.20

Solution The region R is pictured in Figure 15.20. We have shown a typical small sector of the type used in a partition along with a typical "differential element" that we imagine to have area $dA = r\,dr\,d\theta$. This shorthand technique is an aid in setting up the integral. The upper half of the cardioid is traced out as θ varies from 0 to π. We can therefore describe the region R by $\{(r, \theta) : 0 \leq r \leq 1 + \cos\theta,\ 0 \leq \theta \leq \pi\}$. From Equation 15.8, we therefore have

$$\iint_R r^2 \sin\theta\, dA = \int_0^\pi \int_0^{1+\cos\theta} (r^2 \sin\theta)\, \overbrace{r\,dr\,d\theta}^{dA}$$

$$= \int_0^\pi \left[\int_0^{1+\cos\theta} r^3 \sin\theta\, dr\right] d\theta$$

$$= \frac{1}{4} \int_0^\pi \left[r^4\right]_0^{1+\cos\theta} \sin\theta\, d\theta$$

$$= \frac{1}{4} \int_0^\pi (1 + \cos\theta)^4 \sin\theta\, d\theta \quad \text{Let } u = 1 + \cos\theta.$$

$$= -\frac{1}{4} \left[\frac{(1 + \cos\theta)^5}{5}\right]_0^\pi = \frac{8}{5} \quad \blacksquare$$

When f is continuous and nonnegative on R, $\iint_R f(r, \theta)\, dA$ again represents the volume above R and under the surface $z = f(r, \theta)$, and in the special case $f(r, \theta) = 1$ for all points (r, θ) in R, as before, the resulting volume is numerically the same as the area of R; that is, $A = \iint_R dA$. The next three examples illustrate these results.

EXAMPLE 15.11 Use polar coordinates to find the volume below the paraboloid $z = 4 - x^2 - y^2$ and above the xy-plane.

Solution In polar coordinates, $x^2 + y^2 = r^2$, so we can write the equation of the paraboloid as $z = 4 - r^2$. Thus, we take $f(r, \theta) = 4 - r^2$. We find the boundary of the region R by setting $z = 0$, giving $r^2 = 4$ or $r = 2$ (see Figure 15.21). So we have

$$V = \iint_R (4 - r^2) \, dA = \int_0^{2\pi} \int_0^2 (4 - r^2) r \, dr \, d\theta$$

$$= \int_0^{2\pi} \left[2r^2 - \frac{r^4}{4} \right]_0^2 d\theta = 4 \int_0^{2\pi} d\theta = 8\pi \qquad \blacksquare$$

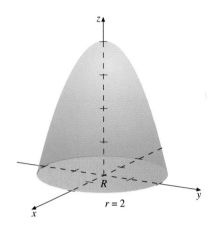

FIGURE 15.21

EXAMPLE 15.12 Find the area inside the circle $r = 5 \cos \theta$ and outside the limaçon $r = 2 + \cos \theta$.

Solution We show the curves in Figure 15.22. To find the points of intersection we solve the equations simultaneously:

$$5 \cos \theta = 2 + \cos \theta$$

$$\cos \theta = \frac{1}{2}$$

$$\theta = \pm \frac{\pi}{3}, \quad r = \frac{5}{2}$$

By symmetry, the total area is twice that for $0 \le \theta \le \pi/3$. By Equation 15.8,

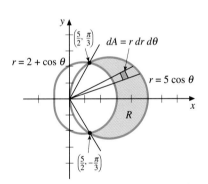

FIGURE 15.22

$$A = \iint_R dA = 2 \int_0^{\pi/3} \int_{2+\cos\theta}^{5\cos\theta} r \, dr \, d\theta = 2 \int_0^{\pi/3} \left[\frac{r^2}{2} \right]_{2+\cos\theta}^{5\cos\theta} d\theta$$

$$= \int_0^{\pi/3} [25 \cos^2 \theta - (4 + 4 \cos \theta + \cos^2 \theta)] \, d\theta$$

$$= 4 \int_0^{\pi/3} (6 \cos^2 \theta - \cos \theta - 1] \, d\theta$$

$$= 4 \int_0^{\pi/3} [3(1 + \cos 2\theta) - \cos \theta - 1] \, d\theta$$

$$= 4 \left[2\theta + \frac{3 \sin 2\theta}{2} - \sin \theta \right]_0^{\pi/3} = \frac{8\pi}{3} + \sqrt{3} \qquad \blacksquare$$

EXAMPLE 15.13 Find the area outside the circle $r = \sqrt{2}$ and inside the lemniscate $r^2 = 4 \cos 2\theta$.

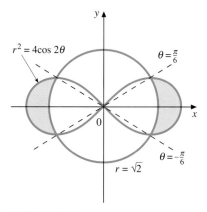

FIGURE 15.23

Solution We show the two curves and the area in question in Figure 15.23. To find the points of intersection, we again solve the two equations simultaneously.

$$4 \cos 2\theta = (\sqrt{2})^2$$

$$\cos 2\theta = \frac{1}{2}$$

$$2\theta = \pm \frac{\pi}{3}$$

$$\theta = \pm \frac{\pi}{6}$$

By symmetry, we can find the area of the first-quadrant region given by

$$\left\{ (r, \theta) : \sqrt{2} \leq r \leq 2\sqrt{\cos 2\theta}, \ 0 \leq \theta \leq \frac{\pi}{6} \right\}$$

and multiply the result by 4. By Equation 15.8, then, we have

$$A = \iint_R dA = 4 \int_0^{\pi/6} \int_{\sqrt{2}}^{2\sqrt{\cos 2\theta}} r \, dr \, d\theta$$

$$= 4 \int_0^{\pi/6} \left[\frac{r^2}{2} \right]_{\sqrt{2}}^{2\sqrt{\cos 2\theta}} d\theta$$

$$= 4 \int_0^{\pi/6} (2 \cos 2\theta - 1) \, d\theta$$

$$= 4 \left[\sin 2\theta - \theta \right]_0^{\pi/6} = 2 \left(\sqrt{3} - \frac{\pi}{3} \right)$$ ∎

Changing from Rectangular to Polar Coordinates

In the next example, we illustrate how an iterated integral in rectangular coordinates can sometimes be evaluated more easily by changing to polar coordinates.

EXAMPLE 15.14 Evaluate the integral

$$\int_0^2 \int_0^{\sqrt{2x-x^2}} \sqrt{x^2 + y^2} \, dy \, dx$$

by changing to polar coordinates.

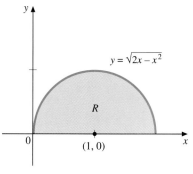

FIGURE 15.24

Solution First, we find the region R determined by the limits of integration. The limits tell us that R is of type I with lower bounding curve $y = 0$ and upper bounding curve $y = \sqrt{2x - x^2}$. Squaring and collecting terms, we find that the upper boundary is that part of the circle $x^2 + y^2 - 2x = 0$ for which $y \geq 0$. So the region R is that shown in Figure 15.24. In polar coordinates, the circle $x^2 + y^2 - 2x = 0$ becomes $r = 2 \cos \theta$. (Verify.) So in polar coordinates the region of integration is

$$\left\{ (r, \theta) : 0 \leq r \leq 2 \cos \theta, \ 0 \leq \theta \leq \frac{\pi}{2} \right\}$$

The integrand, $\sqrt{x^2 + y^2}$, changes to r in polar coordinates, and so the double integral and hence the original iterated integral are both equal to

$$\int_0^{\pi/2} \int_0^{2\cos\theta} r(r\,dr\,d\theta) = \int_0^{\pi/2} \left[\frac{r^3}{3}\right]_0^{2\cos\theta} d\theta$$

$$= \frac{8}{3} \int_0^{\pi/2} \cos^3\theta\,d\theta = \frac{8}{3} \int_0^{\pi/2} (1 - \sin^2\theta)\cos\theta\,d\theta$$

$$= \frac{8}{3} \left[\sin\theta - \frac{\sin^3\theta}{3}\right]_0^{\pi/2} = \frac{8}{3}\left[1 - \frac{1}{3}\right] = \frac{16}{9} \quad \blacksquare$$

The procedure suggested by this example for changing an iterated integral in rectangular coordinates to one in polar coordinates can be summarized as follows:

Procedure for Changing from Rectangular to Polar Coordinates

1. Sketch the region R determined by the limits of integration, and write the equations of the curves that form the boundary of R in polar coordinates.

2. Determine the new limits of integration as in Equation 15.8 or 15.9, depending on whether R is of the type shown in Figure 15.18(a) or (b). (In some cases it may be necessary to express R as the union of two or more such regions.)

3. In the integrand, replace x by $r\cos\theta$ and y by $r\sin\theta$. (Note that $x^2 + y^2$ can be changed immediately to r^2.)

4. Replace $dy\,dx$ or $dx\,dy$, whichever occurs, by $r\,dr\,d\theta$ or by $r\,d\theta\,dr$, depending on the order of integration to be used.

Don't forget the r in step 4. Remember that in rectangular coordinates the differential of area is $dy\,dx$ (or $dx\,dy$), but in polar coordinates it is $r\,dr\,d\theta$ (or $r\,d\theta\,dr$).

If we denote by S the region R described in polar coordinates, then steps 2, 3, and 4 of the procedure can be given by the equation

---

$$\iint_R f(x, y)\,dA = \iint_S f(r\cos\theta, r\sin\theta)\,r\,dr\,d\theta$$

Guidelines for Changing to Polar Coordinates

Changing from rectangular to polar coordinates is desirable if carrying out the integration in rectangular coordinates is more difficult than doing so in polar coordinates. If the curves that make up the boundary of R have simpler polar equations than rectangular equations, a change is indicated. In this connection, be on the lookout especially for bounding curves that are circles centered at the origin ($r = a$), circles centered on an axis and passing through the origin ($r = a\cos\theta$ or $r = a\sin\theta$), and lines through the origin ($\theta = \alpha$). The nature of the integrand also is a factor to be considered. If the original integrand is $f(x, y)$, we see from applying steps 3 and 4 that the new integrand is $rf(r\cos\theta, r\sin\theta)$. Whether the antiderivatives needed in the iterated integration can be found clearly affects the decision. As you practice problems such as Exercises 20–31, you should begin to develop a feel for when a change is worthwhile.

REMARK

■ We investigate changing variables in double integrals more fully in Section 15.8.

Exercise Set 15.3

In Exercises 1–8, evaluate the double integral $\iint_R f(r, \theta) \, dA$, where R is the region described. Draw the region.

1. $f(r, \theta) = r\theta$; $|r| \le 2, 0 \le \theta \le 2\pi$

2. $f(r, \theta) = r(\sin\theta - \cos\theta)$; $\{(r, \theta) : 0 \le r \le 1, \pi/4 \le \theta \le \pi\}$

3. $f(r, \theta) = 2r + 1$; $1 \le r \le 4, 0 \le \theta \le 2\pi$

4. $f(r, \theta) = \sqrt{r}\sin\theta$; $\{(r, \theta) : 0 \le r \le 1 - \cos\theta, 0 \le \theta \le \pi\}$

5. $f(r, \theta) = r^2\cos\theta$; $\{(r, \theta) : 0 \le r \le 2\sin\theta, 0 \le \theta \le \pi/2\}$

6. $f(r, \theta) = r(\sin(\theta/2) - \cos(\theta/2))$; $1 \le r \le 2, 0 \le \theta \le 2\pi$

7. $f(r, \theta) = r^2\sin^2 2\theta$; $\{(r, \theta) : 0 \le r \le 2\sqrt{\cos 2\theta}, 0 \le \theta \le \pi/4\}$

8. $f(r, \theta) = \sqrt{1 + \theta^3}$; $\{(r, \theta) : 0 \le r \le \theta, 0 \le \theta \le 2\}$

In Exercises 9–19, use double integration and polar coordinates to find the area of the region described. Draw the region.

9. Inside $r = 2 + 2\cos\theta$

10. Inside $r^2 = 4\cos 2\theta$

11. Inside $r = 2\cos 2\theta$

12. Inside $r = 2 + \sin\theta$

13. Inside $r = 2(1 + \cos\theta)$ and outside $r = 1$

14. Inside $r = 8\sin\theta$ and outside $r = 3 + 2\sin\theta$

15. Inside $r^2 = 2\cos 2\theta$ and outside $r = 1$

16. Inside $r = 1 + 2\cos\theta$ and to the right of $r\cos\theta = 1$

17. Inside both $r = 2$ and $r = 3 - 2\cos\theta$

18. Inside the small loop of the limaçon $r = 1 + 2\cos\theta$

19. Inside both $r = 6\sin\theta$ and $r = 2(1 + \sin\theta)$

In Exercises 20–31, evaluate the iterated integral by changing to polar coordinates.

20. $\displaystyle\int_0^2 \int_0^{\sqrt{4-x^2}} x^2 y \, dy \, dx$

21. $\displaystyle\int_0^1 \int_x^{\sqrt{2-x^2}} (x^2 + y^2)^2 \, dy \, dx$

22. $\displaystyle\int_{-1}^1 \int_{-\sqrt{1-x^2}}^{\sqrt{1-x^2}} (x^2 + y^2)^{3/2} \, dy \, dx$

23. $\displaystyle\int_0^2 \int_{-\sqrt{8-y^2}}^{-y} (x + y) \, dx \, dy$

24. $\displaystyle\int_0^4 \int_0^{\sqrt{4y-y^2}} y\, dx\, dy$ 25. $\displaystyle\int_0^1 \int_x^{\sqrt{2x-x^2}} x\, dy\, dx$

26. $\displaystyle\int_{-\sqrt{3}}^{\sqrt{3}} \int_1^{\sqrt{4-y^2}} \frac{1}{(x^2+y^2)^{3/2}}\, dx\, dy$

27. $\displaystyle\int_0^{\sqrt{3}} \int_1^{\sqrt{4-x^2}} \frac{x}{y}\, dy\, dx$ 28. $\displaystyle\int_1^{\sqrt{3}} \int_1^x \frac{x^2-y^2}{x^2+y^2}\, dy\, dx$

29. $\displaystyle\int_0^1 \int_y^{1+\sqrt{1-y^2}} \sqrt{4-x^2-y^2}\, dx\, dy$

30. $\displaystyle\int_{-2}^2 \int_2^{2+\sqrt{4-x^2}} \frac{x+y}{y}\, dy\, dx$

31. $\displaystyle\int_{-a}^a \int_{-\sqrt{a^2-y^2}}^{\sqrt{a^2-y^2}} e^{-(x^2+y^2)}\, dx\, dy$

In Exercises 32–39, use polar coordinates to find the indicated volume.

32. Inside the ellipsoid $x^2 + y^2 + 4z^2 = 4$ and above the xy-plane

33. Inside the cone $z = 2 - \sqrt{x^2+y^2}$ and above the xy-plane

34. A sphere of radius a

35. Inside the sphere $x^2 + y^2 + z^2 = a^2$ and outside the cylinder $x^2 + y^2 = b^2$, where $0 < b < a$

36. Inside the cylinder $x^2 + y^2 = a^2$ between the upper and lower sheets of the hyperboloid $z^2 - x^2 - y^2 = a^2$

37. Under the cone $z = \sqrt{x^2+y^2}$ and above the region R in the xy-plane inside the circle $x^2 + y^2 = 2x$

38. Under the cylindrical surface $z = y^2$ and above the region $R = \{(x, y) : 0 \le x \le 1,\ x \le y \le \sqrt{2-x^2}\}$

39. Between the surfaces $z = 6 - x^2 - y^2$ and $z = 2$. (*Hint:* Use the difference of two volumes.)

40. Evaluate by changing to polar coordinates:
$$\int_{2-2\sqrt{2}}^{2+2\sqrt{2}} \int_{-y}^{(4-y^2)/4} (x^2 y + y^3)\, dx\, dy$$

41. Write as a single iterated integral in polar coordinates and evaluate:
$$\int_0^{3/2} \int_{\sqrt{2y-y^2}}^{\sqrt{4y-y^2}} xy\, dx\, dy + \int_{3/2}^3 \int_{y/\sqrt{3}}^{\sqrt{4y-y^2}} xy\, dx\, dy$$

42. Find the area bounded by the curves $r = 1$, $r = 2$, $r = 2\ln\theta$, and $r\theta = 1$.

43. Find the area between the y-axis and the parabola $y^2 = 9 - 6x$ that lies outside the circle $x^2 + y^2 - 4x = 0$.

44. **The Normal Probability Density Function** Carry out the following steps to verify that the area under the normal probability density function is 1:
 (a) Let $I = \int_0^\infty e^{-x^2} dx$. Write I^2 as follows:
$$I^2 = \left(\int_0^\infty e^{-x^2} dx\right)\left(\int_0^\infty e^{-y^2} dy\right)$$
$$= \int_0^\infty \int_0^\infty e^{-(x^2+y^2)}\, dy\, dx$$
 (b) Change the iterated integral in part (a) to polar coordinates and evaluate it. Find the value of I.
 (c) The normal probability density function with mean μ and standard deviation σ is
$$f(x) = \frac{1}{\sqrt{2\pi}\,\sigma} e^{-(x-\mu)^2/2\sigma^2},\quad -\infty < x < \infty$$
Use the result of part (b) to show that $\int_{-\infty}^\infty f(x)\, dx = 1$. (*Hint:* Make a substitution, and use symmetry.)

15.4 MOMENTS, CENTROIDS, AND CENTERS OF MASS USING DOUBLE INTEGRATION

In Chapter 6 we saw how to find the center of mass of a homogeneous lamina (the density is constant) using single integrals. With the aid of double integration we can remove the restriction that the lamina be homogeneous. Suppose the lamina occupies a region R in the xy-plane, and its density at the point (x, y) is given by $\rho(x, y)$, where ρ is a continuous function on R. To explain what is meant by the density at the point (x, y), suppose that ΔA is the area of a small

Section 15.4 Moments, Centroids, and Centers of Mass Using Double Integration 1093

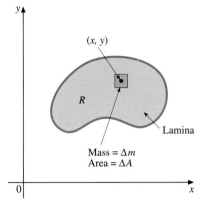

FIGURE 15.25

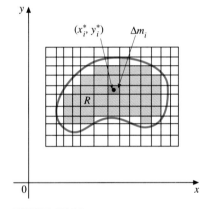

FIGURE 15.26

rectangular element of the lamina, centered at (x, y), as in Figure 15.25. If the mass of this element is Δm, then its average density is $\Delta m/\Delta A$. We define the density $\rho(x, y)$ at the point (x, y) by

$$\rho(x, y) = \lim_{d \to 0} \frac{\Delta m}{\Delta A}$$

where d is the length of the diagonal of the rectangular element.

Now we proceed exactly as in defining the double integral over R, enclosing R in some rectangle and partitioning this rectangle with horizontal and vertical lines, as in Figure 15.26. We number the subrectangles entirely contained within R. Denote the center of the ith subrectangle by (x_i^*, y_i^*). If the norm $\|P\|$ of the partition is small (so that all subrectangles are small), we would expect that the density throughout that part of the lamina in the ith subrectangle would differ very little from $\rho(x_i^*, y_i^*)$. Thus, its mass Δm_i should be approximately $\rho(x_i^*, y_i^*)\Delta A_i$.

We can think of each little rectangular piece of the lamina formed by the partition as if it were a "point mass" concentrated at its center. The total mass m of the lamina is approximately the sum of these point masses:

$$m \approx \sum_{i=1}^{n} \Delta m_i = \sum_{i=1}^{n} \rho(x_i^*, y_i^*)\Delta A_i$$

This last sum is a Riemann sum for the density function ρ over the region R. Thus, as we let $\|P\| \to 0$, we arrive at the following result.

Total Mass of a Lamina of Continuous Density $\rho(x, y)$, Occupying the Region R

$$m = \iint_R \rho(x, y)\, dA \qquad (15.10)$$

Similarly, the moments M_x and M_y about the y-axis and x-axis, respectively, are approximated by

$$M_x \approx \sum_{i=1}^{n} y_i^* \Delta m_i = \sum_{i=1}^{n} y_i^* \rho(x_i^*, y_i^*)\Delta A_i$$

and

$$M_y = \sum_{i=1}^{n} x_i^* \Delta m_i = \sum_{i=1}^{n} x_i^* \rho(x_i^*, y_i^*)\Delta A_i$$

As we take partitions with norm approaching 0 we arrive at the following formulas.

The Moments M_x and M_y About the x- and y-Axes

$$M_x = \iint_R y\rho(x,y)\,dA \qquad (15.11)$$

$$M_y = \iint_R x\rho(x,y)\,dA \qquad (15.12)$$

Recall from Chapter 6 that the center of mass $(\bar{x}, \bar{y})$ satisfies

$$\bar{x} = \frac{M_y}{m} \quad \text{and} \quad \bar{y} = \frac{M_x}{m} \qquad (15.13)$$

If we write $dm = \rho(x,y)\,dA$, we can combine Equations 15.10, 15.11, 15.12, and 15.13 to get the following formulas.

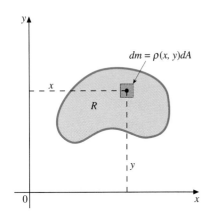

FIGURE 15.27

The Center of Mass

$$\bar{x} = \frac{\iint_R x\,dm}{\iint_R dm} \quad \text{and} \quad \bar{y} = \frac{\iint_R y\,dm}{\iint_R dm} \qquad (15.14)$$

where $dm = \rho(x,y)\,dA$.

REMARK
■ To simplify the process of arriving at Equations 15.14, we can use the shorthand technique of designating a typical element of the lamina by $dm = \rho(x,y)\,dA$, as shown in Figure 15.27. Treating this element as a point mass, we get the moments $x\,dm$ and $y\,dm$ about the y-axis and x-axis, respectively. When we "sum" these moments, in the sense of integration, and divide by the "sum" of the masses, we arrive at Equation 15.14.

EXAMPLE 15.15 A lamina in the shape of the triangular region in Figure 15.28 has density $\rho(x,y) = \sqrt{xy}$. Find the center of mass.

Solution First, let us calculate the mass m of the lamina. By Equation 15.10,

$$m = \iint_R dm = \iint_R \rho(x,y)\,dA = \int_0^2 \int_0^x \sqrt{xy}\,dy\,dx = \frac{16}{9}$$

(The integration here is straightforward, and we have omitted the details.)

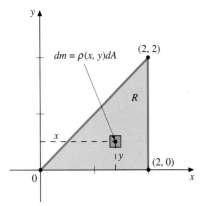

FIGURE 15.28

Next, we calculate the numerators for $\bar{x}$ and $\bar{y}$, which are the moments M_y and M_x about the y-axis and x-axis, respectively. By Equations 15.11 and 15.12,

$$M_y = \iint_R x\,dm = \iint_R x\rho(x,y)\,dA = \int_0^2\int_0^x x\sqrt{xy}\,dy\,dx = \frac{8}{3}$$

$$M_x = \iint_R y\,dm = \iint_R y\rho(x,y)\,dA = \int_0^2\int_0^x y\sqrt{xy}\,dy\,dx = \frac{8}{5}$$

(again, we omit the details of the integration). Thus, by Equation 15.13,

$$\bar{x} = \frac{M_y}{m} = \frac{\frac{8}{3}}{\frac{16}{9}} = \frac{3}{2} \quad \text{and} \quad \bar{y} = \frac{M_x}{m} = \frac{\frac{8}{5}}{\frac{16}{9}} = \frac{9}{10}$$

So the center of mass is the point $(\frac{3}{2}, \frac{9}{10})$. Notice that because of the variable density, the center of mass does not coincide with the centroid of the region, which is two-thirds of the horizontal distance from $(0, 0)$ to $(0, 2)$ and one-third of the vertical distance from $(2, 0)$ to $(2, 2)$—namely, $(\frac{4}{3}, \frac{2}{3})$—as we saw in Section 6.8. ∎

If the density is constant, then the center of mass coincides with the centroid of the region R. In this case, the constant density, say $\rho(x, y) = k$, can be taken outside the integrals for $\bar{x}$ and $\bar{y}$ in Equations 15.14 and then divided out. We thus have the following equations for the coordinates of the centroid.

The Centroid of a Region R in $\mathbb{R}^2$

$$\bar{x} = \frac{\iint_R x\,dA}{\iint_R dA} \quad \bar{y} = \frac{\iint_R y\,dA}{\iint_R dA} \tag{15.15}$$

EXAMPLE 15.16 Find the centroid of the region bounded by the parabola $y = 4 - x^2$ and the x-axis.

Solution We show the region in Figure 15.29. By symmetry, the centroid lies on the y-axis, so $\bar{x} = 0$. (Note that if we were working with a nonhomogeneous lamina, this conclusion would hold only if the mass were symmetrically distributed with respect to the y-axis.) From Equations 15.15, we have

$$\bar{y} = \frac{\iint_R y\,dA}{\iint_R dA} = \frac{\int_{-2}^2 \int_0^{4-x^2} y\,dy\,dx}{\int_{-2}^2 \int_0^{4-x^2} dy\,dx} = \frac{8}{5}$$

(You should verify this result by supplying the details of the integration.) Thus, the centroid is at the point $(0, \frac{8}{5})$. ∎

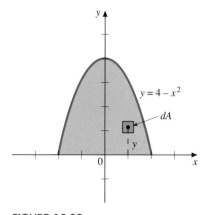

FIGURE 15.29

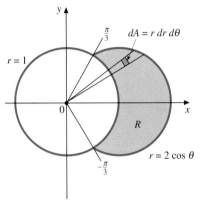

EXAMPLE 15.17 A homogeneous lamina occupies the region outside the circle $x^2 + y^2 = 1$ and inside the circle $x^2 + y^2 - 2x = 0$. Find its center of mass.

Solution The region R in this case is more easily described in polar coordinates (see Figure 15.30). The bounding curves are $r = 1$ and $r = 2\cos\theta$, as you can verify. These intersect at $(1, \pm\frac{\pi}{3})$. By symmetry, $\bar{y} = 0$. For $\bar{x}$, Equations 15.15 give

$$\bar{x} = \frac{\iint_R x \, dA}{\iint_R dA} = \frac{\int_{-\pi/3}^{\pi/3}\int_1^{2\cos\theta}(r\cos\theta)(r\,dr\,d\theta)}{\int_{-\pi/3}^{\pi/3}\int_1^{2\cos\theta} r\,dr\,d\theta} = \frac{8\pi + 3\sqrt{3}}{4\pi + 6\sqrt{3}} \approx 1.321$$

Again, you should supply the missing steps. ∎

FIGURE 15.30

Higher-Order Moments; Moment of Inertia

FIGURE 15.31

Consider again a system of point masses in the xy-plane. Let m be such a point mass and l a line in the plane (see Figure 15.31). Then, as we have seen, the moment of m about l is the product md, where d is the distance from l to m. To distinguish such a moment from those we are about to define, we sometimes refer to md as the **first moment**. Higher-order moments of m about l are defined by

$$md^2, \text{ second moment}$$
$$md^3, \text{ third moment}$$
$$\vdots$$
$$md^k, k\text{th moment}$$

We concentrate on the second moment because it is especially important in physics and engineering. Moments of orders 3 and 4 have applications in fields such as structural design, but the primary use of higher-order moments is in probability theory. We will say more about this use later. In a physical system, the second moment is also called the **moment of inertia**, and the distance d is called the **radius of gyration**. This terminology suggests a rotation, and we will explain shortly the sense in which the moment of inertia is related to a rotating mass.

For a lamina of the type we have been considering, with density function ρ and occupying a region R, we can use the same partition process as before to define the moments of inertia about the x-axis and y-axis, respectively, as follows.

Moments of Inertia About the x- and y-Axes

$$I_x = \lim_{\|P\|\to 0} \sum_{i=1}^{n} (y_i^*)^2 \Delta m_i = \iint_R y^2 \, dm \qquad (15.16)$$

$$I_y = \lim_{\|P\|\to 0} \sum_{i=1}^{n} (x_i^*)^2 \Delta m_i = \iint_R x^2 \, dm \qquad (15.17)$$

where $dm = \rho(x, y) \, dA$.

We also define the *polar moment of inertia*, I_0, as follows.

Polar Moment of Inertia

$$I_0 = \iint_R (x^2 + y^2) \, dm \qquad (15.18)$$

It follows that $I_0 = I_x + I_y$. This polar moment of inertia is sometimes said to be the moment of inertia about the origin. Perhaps a better way to describe it is as the moment of inertia about a line through the origin perpendicular to the xy-plane—that is, about the z-axis (see Figure 15.32).

If a lamina of mass m has moment of inertia I with respect to a line, then the positive number r that satisfies $I = mr^2$ is called the *radius of gyration* of the lamina with respect to that line. This terminology is consistent with the notion of radius of gyration for a point mass if we consider that the entire mass of the lamina is concentrated at a distance r from the line. In particular, if r_x is the radius of gyration with respect to the y-axis, and r_y is that with respect to the x-axis, we have

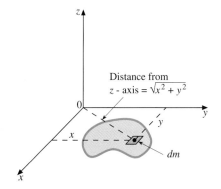

FIGURE 15.32

$$r_x = \sqrt{\frac{I_y}{m}} \quad \text{and} \quad r_y = \sqrt{\frac{I_x}{m}} \qquad (15.19)$$

The point (r_x, r_y) plays a role for second moments analogous to that of $(\bar{x}, \bar{y})$ for first moments. In general, these points are not the same.

The polar moment of inertia is helpful in getting an intuitive feeling for the role moments of inertia play in a dynamic system. Consider a point mass m in the xy-plane located a distance r from the origin. Suppose it is rotating about the z-axis at the constant angular velocity ω (the number of radians it turns through per unit of time). Its kinetic energy is $\frac{1}{2}mv^2$, where v is the linear velocity. By differentiating both sides of the arc length formula $s = r\theta$ (see Appendix 2) with respect to time, we get $v = r\omega$, since $v = ds/dt$ and $\omega = d\theta/dt$. The kinetic energy can thus be written as

$$KE = \frac{1}{2}m(r\omega)^2 = \frac{1}{2}(mr^2)\omega^2 = \frac{1}{2}I_0\omega^2$$

since $I_0 = m(x^2 + y^2) = mr^2$. Comparing this result with the formula $KE = \frac{1}{2}mv^2$, we can see that I_0 plays a role analogous to the mass m. The same result holds if we have a lamina in the xy-plane rotating about the z-axis, as the usual partitioning, summing, and passing to the limit would show. For example, the total kinetic energy of a rotating wheel is $\frac{1}{2}I_0\omega^2$. Now, kinetic energy is equal to the work required to bring the object to rest. So for a constant angular velocity, the moment of inertia is a measure of the work required to bring the rotating wheel to a stop.

EXAMPLE 15.18 A lamina is bounded by the curves $y^2 = x$ and $x - y = 2$, and its density is $\rho(x, y) = 2x$. Find I_x, I_y, r_x, and r_y.

Solution By Equations 15.16 and 15.17, since $dm = \rho(x, y) dA$,

$$I_x = \iint_R y^2 \, dm = \iint_R 2xy^2 \, dA$$

$$I_y = \iint_R x^2 \, dm = \iint_R 2x^3 \, dA$$

The region occupied by the lamina is the type II region pictured in Figure 15.33. The points of intersection $(4, 2)$ and $(1, -1)$ are found by solving the equations simultaneously. Using iterated integrals, we obtain

$$I_x = \int_{-1}^{2} \int_{y^2}^{y+2} 2xy^2 \, dx \, dy \quad \text{and} \quad I_y = \int_{-1}^{2} \int_{y^2}^{y+2} 2x^3 \, dx \, dy$$

The results, found after some computation, are

$$I_x = \frac{531}{35} \quad \text{and} \quad I_y = \frac{369}{5}$$

Thus, $I_0 = I_x + I_y = 3114/35$.

To find r_x and r_y we need the mass m. By Equation 15.10,

$$m = \iint_R \rho \, dA = \int_{-1}^{2} \int_{y^2}^{y+2} 2x \, dx \, dy = \frac{72}{5}$$

So from Equations 15.19 we get, after some simplification,

$$r_x = \sqrt{\frac{I_y}{m}} = \sqrt{\frac{41}{8}} \approx 2.26 \qquad r_y = \sqrt{\frac{I_x}{m}} = \sqrt{\frac{59}{56}} \approx 1.03 \qquad \blacksquare$$

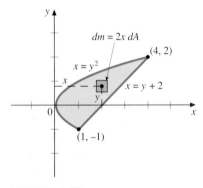

FIGURE 15.33

EXAMPLE 15.19 Consider a lamina that is the annular ring between the circles $x^2 + y^2 = 1$ and $x^2 + y^2 = 4$. The density at any point (x, y) is the reciprocal of its distance from the origin. Find the polar moment of inertia.

Solution The nature of the region (Figure 15.34) suggests that it would be convenient to use polar coordinates. In polar coordinates, Equation 15.18 becomes

$$I_0 = \iint_R r^2 \, dm$$

and $dm = \rho(r, \theta) \, dA = \frac{1}{r}(r \, dr \, d\theta) = dr \, d\theta$. Thus,

$$I_0 = \int_0^{2\pi} \int_1^2 r^2 \, dr \, d\theta = \frac{14\pi}{3} \qquad \blacksquare$$

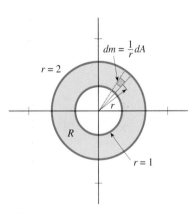

FIGURE 15.34

Moments in Probability Theory

The ideas studied in this section are important in probability theory. Consider first the one-dimensional case in which $f(x)$ is a continuous probability density function on the interval $[a, b]$. The *kth moment about the origin* M_k is

$$M_k = \int_a^b x^k f(x)\, dx$$

This integral is what we referred to in Section 6.8 as the expected value of x^k and denoted by $E[x^k]$. In particular, for $k = 1$, $M_1 = E[x] = \mu$, the mean of the distribution. Since f is a probability density function, $\int_a^b f(x)\, dx = 1$. So we can write

$$\mu = \frac{\int_a^b x f(x)\, dx}{\int_a^b f(x)\, dx}$$

showing that μ is analogous to the center of mass. In fact, we can think of the "probability" as a continuously distributed mass on the interval $[a, b]$, where the "density" is $f(x)$. Indeed, the name *probability density* comes from this analogy.

The *kth moment about the mean* is defined by

$$M_k' = \int_a^b (x - \mu)^k f(x)\, dx$$

which is the same as the expected value $E[(x - \mu)^k]$. In particular, for $k = 2$, we get the variance σ^2:

$$\sigma^2 = E[(x - \mu)^2] = \int_a^b (x - \mu)^2 f(x)\, dx$$

These ideas can be extended to two dimensions. If $f(x, y) \geq 0$ and $\iint_R f(x, y)\, dA = 1$, then f can serve as a probability density function of two random variables. We define the *x-mean* and *y-mean* by

$$\mu_x = \iint_R x f(x, y)\, dA \quad \text{and} \quad \mu_y = \iint_R y f(x, y)\, dA \qquad (15.20)$$

These means are analogous to $\bar{x}$ and $\bar{y}$ for a lamina with density $f(x, y)$. Similarly,

$$\sigma_x^2 = \iint_R (x - \mu_x)^2 f(x, y)\, dA \quad \text{and} \quad \sigma_y^2 = \iint_R (y - \mu_y)^2 f(x, y)\, dA \quad (15.21)$$

Moments of higher order are also sometimes useful, but we will not consider them here.

We illustrate these ideas in the next two examples.

EXAMPLE 15.20   Let $f(x, y) = k(2x + y)$ and $R = \{(x, y) : 0 \leq x \leq 1, 0 \leq y \leq 2x\}$.

(a) Find k so that $f(x, y)$ will be a probability density function on R.
(b) Find μ_x and μ_y.

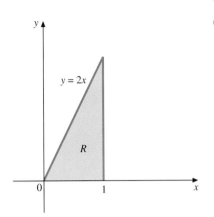

FIGURE 15.35

Solution

(a) We show the region R in Figure 15.35. Note first that in this region both x and y are nonnegative, so $f(x, y) = k(2x + y)$ will be nonnegative if $k > 0$. For $f(x, y)$ to be a probability density function, we must also have $\iint_R f(x, y)\, dA = 1$. So we calculate the integral, set it equal to 1, and solve for k.

$$\iint_R f(x, y)\, dA = k \int_0^1 \int_0^{2x} (2x + y)\, dy\, dx$$

$$= k \int_0^1 \left[2xy + \frac{y^2}{2} \right]_0^{2x} dx$$

$$= k \int_0^1 (4x^2 + 2x^2)\, dx$$

$$= k \left[2x^3 \right]_0^1 = 2k$$

Thus, we must have $2k = 1$, so $k = \frac{1}{2}$.

(b) Using $f(x, y) = \frac{1}{2}(2x + y)$, we have for μ_x, from Equations 15.20,

$$\mu_x = \frac{1}{2} \int_0^1 \int_0^{2x} x(2x + y)\, dy\, dx = \frac{1}{2} \int_0^1 \left[2x^2 y + x \frac{y^2}{2} \right]_0^{2x} dx$$

$$= \frac{1}{2} \int_0^1 (4x^3 + 2x^3)\, dx$$

$$= \frac{1}{2} \left[\frac{6x^4}{4} \right]_0^1 = \frac{3}{4}$$

Similarly, from Equations 15.20, for μ_y, we have

$$\mu_y = \frac{1}{2} \int_0^1 \int_0^{2x} y(2x + y)\, dy\, dx = \frac{1}{2} \int_0^1 \left[y^2 x + \frac{y^3}{3} \right]_0^{2x} dx$$

$$= \frac{1}{2} \int_0^1 \left(4x^3 + \frac{8x^3}{3} \right) dx$$

$$= \frac{1}{2} \int_0^1 \frac{20x^3}{3}\, dx$$

$$= \frac{1}{2} \left[\frac{5x^4}{3} \right]_0^1 = \frac{5}{6} \qquad \blacksquare$$

We can write Equations 15.21 for the variances σ_x^2 and σ_y^2 in a more useful form as follows, making use of Properties 1 and 2 of double integrals given in Section 15.1.

$$\sigma_x^2 = \iint_R (x - \mu_x)^2 f(x, y)\, dA = \iint_R (x^2 - 2x\mu_x + \mu_x^2) f(x, y)\, dA$$

$$= \iint_R x^2 f(x, y)\, dA - 2\mu_x \iint_R x f(x, y)\, dA + \mu_x^2 \iint_R f(x, y)\, dA$$

$$= \iint_R x^2 f(x, y)\, dA - 2\mu_x^2 + \mu_x^2$$

since $\mu_x = \iint_R xf(x, y)\,dA$ and $\iint_R f(x, y)\,dA = 1$. Thus,

$$\sigma_x^2 = \iint_R x^2 f(x, y)\,dA - \mu_x^2 \tag{15.22}$$

A similar calculation shows that

$$\sigma_y^2 = \iint_R y^2 f(x, y)\,dA - \mu_y^2 \tag{15.23}$$

EXAMPLE 15.21 Find σ_x^2 and σ_y^2 for the probability density function in Example 15.20.

Solution In Example 15.20, we found that $f(x, y) = \frac{1}{2}(2x + y)$, $\mu_x = \frac{3}{4}$, and $\mu_y = \frac{5}{6}$. So, from Equation 15.22, we have

$$\sigma_x^2 = \frac{1}{2}\int_0^1\int_0^{2x} x^2(2x + y)\,dy\,dx - \left(\frac{3}{4}\right)^2$$

$$= \frac{1}{2}\int_0^1 \left[2x^3 y + \frac{x^2 y^2}{2}\right]_0^{2x} dx - \frac{9}{16}$$

$$= \frac{1}{2}\int_0^1 (4x^4 + 2x^4)\,dx - \frac{9}{16}$$

$$= \frac{1}{2}\left[\frac{6x^5}{5}\right]_0^1 - \frac{9}{16} = \frac{3}{5} - \frac{9}{16} = \frac{3}{80}$$

Similarly, by Equation 15.23,

$$\sigma_y^2 = \frac{1}{2}\int_0^1\int_0^{2x} y^2(2x + y)\,dy\,dx - \left(\frac{5}{6}\right)^2$$

$$= \frac{1}{2}\int_0^1 \left[\frac{2xy^3}{3} + \frac{y^4}{4}\right]_0^{2x} dx - \frac{25}{36}$$

$$= \frac{1}{2}\int_0^1 \left(\frac{16x^4}{3} + 4x^4\right)\,dx - \frac{25}{36}$$

$$= \frac{1}{2}\left[\frac{28x^5}{15}\right]_0^1 - \frac{25}{36} = \frac{14}{15} - \frac{25}{36} = \frac{43}{180} \blacksquare$$

Exercise Set 15.4

In Exercises 1–8, find the center of mass of the lamina that occupies the region R and has density $\rho(x, y)$.

1. $R = \{(x, y) : 0 \le x \le 1,\ x \le y \le 1\}$; $\rho(x, y) = 2y$

2. $R = \{(x, y) : 0 \le x \le 2,\ 0 \le y \le 1\}$; $\rho(x, y) = x + 2y$

3. $R = \{(x, y) : -y \le x \le y,\ 0 \le y \le 1\}$; $\rho(x, y) = y^2$

4. $R = \{(x, y) : 0 \le x \le 4,\ 0 \le y \le \sqrt{x}\}$; $\rho(x, y) = xy$

5. R is bounded by $y = x$ and $y = 2 - x^2$; $\rho(x, y) = 2$

6. R is bounded by $y = 2x - x^2$ and $y = 0$; $\rho(x, y) = 1$

7. R is bounded by $y = \sqrt{4-x^2}$ and $y = 0$; $\rho(x, y) = \sqrt{x^2 + y^2}$

8. R is the region inside $x^2 + y^2 = 4x$ for which $x \geq 1$; $\rho(x, y) = y^2/x$

In Exercises 9–16, find the centroid of the region R.

9. $R = \{(x, y) : -1 \leq x \leq 1, \ 0 \leq y \leq x^2\}$

10. $R = \{(x, y) : -1 \leq x \leq 1, \ 0 \leq y \leq 2 - x^2\}$

11. $R = \{(r, \theta) : 0 \leq r \leq a, \ 0 \leq \theta \leq \pi\}$

12. $R = \{(r, \theta) : 0 \leq r \leq 2\cos\theta, \ 0 \leq \theta \leq \pi/4\}$

13. R is the region bounded by $y = x^3 - x$ and $y = 7x$ for which $x \geq 0$.

14. $R = \{(x, y) : -\pi/2 \leq x \leq \pi/2, \ 0 \leq y \leq \cos x\}$

15. $R = \{(x, y) : 0 \leq x \leq 2, \ 0 \leq y \leq e^x\}$

16. R is the region above the x-axis and between $x^2 + y^2 = 1$ and $x^2 + y^2 = 16$.

In Exercises 17–24, find I_x, I_y, I_0, r_x, and r_y for the lamina that occupies the region R and has density $\rho(x, y)$.

17. The region in Exercise 1

18. The region in Exercise 2

19. The region in Exercise 3

20. The region in Exercise 4

21. $R = \{(x, y) : 0 \leq x \leq 2, \ 0 \leq y \leq 2-x\}$; $\rho(x, y) = x + y$

22. $R = \{(r, \theta) : 0 \leq r \leq a, \ 0 \leq \theta \leq \pi/2\}$; $\rho(r, \theta) = k$ (constant)

23. R is the region under $y = \sin x$ and above the x-axis, from $x = 0$ to $x = \pi$; $\rho(x, y) = k$ (constant)

24. $R = \{(x, y) : 0 \leq x \leq \sqrt{3}, \ 1 \leq y \leq \sqrt{4-x^2}\}$; $\rho(x, y) = x/y$

25. Prove that the moment of inertia of a homogeneous lamina of density ρ in the shape of a rectangle with base b and altitude h about its base is $I = (bh^3/3)\rho$.

26. Find the centroid of the region bounded by $y^2 = 4x$ and $2x - y = 4$.

27. Suppose a homogeneous lamina of density ρ occupies the region described in Exercise 26. Set up, but do not evaluate, iterated integrals for the moments of inertia about the lines $x + 2 = 0$ and $y - 4 = 0$.

28. Consider the limaçon $r = 3 + 2\cos\theta$ and the circle $r = 3$. Set up, but do not evaluate, iterated integrals for $\bar{x}$ and I_0 for homogeneous laminas of density ρ that occupy each of the following regions:
    (a) outside the circle and inside the limaçon;
    (b) outside the limaçon and inside the circle;
    (c) inside both the limaçon and the circle.

29. Let $f(x, y) = kxy$ on the region $R = \{(x, y): 0 \leq x \leq 1, \ 0 \leq y \leq 1\}$.
    (a) Find k so that $f(x, y)$ will be a probability density function.
    (b) Find μ_x and μ_y.
    (c) Find σ_x^2 and σ_y^2.

30. Let $f(x, y) = k$ on the unit circular disk $\{(x, y) : x^2 + y^2 \leq 1\}$.
    (a) Find k so that $f(x, y)$ is a probability density function.
    (b) Find σ_x^2 and σ_y^2.

31. Let $f(x, y) = k(1 - y)$ on the region $R = \{(x, y): 0 \leq x \leq y \leq 1\}$.
    (a) Find k so that $f(x, y)$ is a probability density function.
    (b) Find μ_x and μ_y.
    (c) Find σ_x^2 and σ_y^2.

32. Prove that the first moment of a lamina about each of the lines $x = \bar{x}$ and $y = \bar{y}$ equals 0.

33. (a) Prove that the moment of inertia $I_{\bar{x}}$ of a lamina about the line $x = \bar{x}$ is
    $$I_{\bar{x}} = I_y - m\bar{x}^2$$
    (b) State and prove an analogous result for the moment of inertia $I_{\bar{y}}$ about the line $y = \bar{y}$.

34. Prove that the medians of a triangle intersect at the centroid of the region enclosed by the triangle.

35. If a homogeneous lamina of density ρ is bounded by a triangle of altitude h and base b, prove that the moment of inertia about its base is
    $$I_b = \frac{bh^3\rho}{12}$$

36. A lamina of density $\rho(x, y) = y^2$ is in the shape of the circular disk $x^2 + y^2 \leq 1$. Find its moment of inertia with respect to the line $x = 2$.

37. Use double integration to prove the following *First Theorem of Pappus:* If R is a plane region and l is a line in the plane of R that does not intersect R, then the volume of the solid formed by revolving R about l is the area of R times the distance traveled by the centroid of R. (See also Exercise 28 in Exercise Set 6.8, where a proof using single integration was called for.)

38. It can be shown that the function

$$f(x, y) = \frac{1}{8}xe^{-(x+y)/2}$$

defined on the infinite region $\{(x, y) : 0 \leq x \leq \infty,\ 0 \leq y \leq \infty\}$ is a probability density function. Assuming that the double integrals over this unbounded region exist whenever the corresponding improper iterated integrals converge, find the following:

(a) μ_x (b) μ_y
(c) σ_x^2 (d) σ_y^2

15.5 SURFACE AREA

We learned in Chapter 6 how to find areas of surfaces of revolution using techniques of single-variable calculus. We now show how to use double integration to find areas of more general surfaces.

Let S be the surface that is the graph of a function f of two variables over a region R, and suppose f has continuous first partial derivatives on R. We partition a rectangle containing R in the usual way and count those subrectangles entirely contained in R. In the ith such subrectangle, whose area is ΔA_i, we select an arbitrary point (x_i^*, y_i^*). Because we have assumed that f_x and f_y are continuous, we know that f is differentiable in R, and so S has a tangent plane at each point. Let T_i denote its tangent plane at $P_i = (x_i^*, y_i^*, f(x_i^*, y_i^*))$. Now project the ith subrectangle vertically. The resulting prism cuts out a patch of the surface S, whose area we denote by $\Delta \sigma_i$, and a corresponding patch of the tangent plane T_i, whose area we denote by ΔT_i. Figure 15.36(a) illustrates this construction.

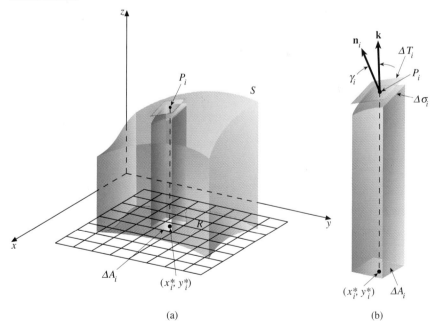

FIGURE 15.36

If the norm $\|P\|$ of the partition is small, we would expect the area of the patch of the surface, $\Delta \sigma_i$, to be approximated by the area of the patch of the tangent plane ΔT_i. Hence, denoting the area of S by $A(S)$, it is reasonable to suppose that

$$A(S) \approx \sum_{i=1}^{n} \Delta \sigma_i \approx \sum_{i=1}^{n} \Delta T_i$$

and that the approximation becomes better and better as we take partitions with norms approaching 0. Our next task is to express ΔT_i in terms of the function f.

As shown in the enlarged view in Figure 15.36(b), let $\mathbf{n}_i$ be a normal vector to the tangent plane at P_i, and let γ_i be the angle between $\mathbf{n}_i$ and the unit vertical vector $\mathbf{k} = \langle 0, 0, 1 \rangle$. Then it can be shown (see Exercise 20, in Exercise Set 15.5) that

$$\Delta T_i |\cos \gamma_i| = \Delta A_i \qquad (15.24)$$

We learned in Section 14.4 that the vector $\langle f_x, f_y, -1 \rangle$ is normal to the surface S, and so we may take

$$\mathbf{n}_i = \langle f_x(x_i^*, y_i^*), f_y(x_i^*, y_i^*), -1 \rangle$$

Thus,

$$\cos \gamma_i = \frac{\mathbf{n}_i \cdot \mathbf{k}}{|\mathbf{n}_i||\mathbf{k}|} = \frac{-1}{\sqrt{f_x^2 + f_y^2 + 1}}$$

where f_x and f_y are evaluated at (x_i^*, y_i^*). Solving for ΔT_i from Equation 15.24 and substituting for $\cos \gamma_i$, we have

$$\Delta T_i = \sqrt{1 + f_x^2 + f_y^2} \, \Delta A_i$$

Thus,

$$A(S) \approx \sum_{i=1}^{n} \sqrt{1 + [f_x(x_i^*, y_i^*)]^2 + [f_y(x_i^*, y_i^*)]^2} \, \Delta A_i$$

We *define* $A(S)$ by the limit of this sum as $\|P\| \to 0$:

$$A(S) = \lim_{\|P\| \to 0} \sum_{i=1}^{n} \sqrt{1 + [f_x(x_i^*, y_i^*)]^2 + [f_y(x_i^*, y_i^*)]^2} \, \Delta A_i$$

By the definition of the double integral, we therefore have the following formula.

Area of the Surface $z = f(x, y)$ Over the Region R

$$A(S) = \iint_R \sqrt{1 + [f_x(x, y)]^2 + [f_y(x, y)]^2} \, dA \qquad (15.25)$$

REMARK
■ In Section 6.4, we developed the formula for arc length

$$L = \int_a^b \sqrt{1 + [f'(x)]^2} \, dx$$

Note the similarity between this formula and Formula 15.25 for surface area.

EXAMPLE 15.22 Find the area of that part of the cylinder $z = \sqrt{4 - x^2}$ above the region R bounded by $y = x$, $x = 0$, and $y = 1$. Sketch the surface.

Solution Figure 15.37 shows the surface. Writing $f(x, y) = \sqrt{4 - x^2}$, we have

$$f_x(x, y) = \frac{-x}{\sqrt{4 - x^2}} \qquad f_y(x, y) = 0$$

So, by Equation 15.25,

$$A(S) = \iint_R \sqrt{1 + \frac{x^2}{4 - x^2}} \, dA = \iint_R \frac{2}{\sqrt{4 - x^2}} \, dA$$

$$= \int_0^1 \int_x^1 \frac{2}{\sqrt{4 - x^2}} \, dy \, dx = \int_0^1 \left(\frac{2}{\sqrt{4 - x^2}} - \frac{2x}{\sqrt{4 - x^2}} \right) dx$$

$$= \left[2 \sin^{-1} \frac{x}{2} + 2\sqrt{4 - x^2} \right]_0^1 = \frac{\pi}{3} + 2\sqrt{3} - 4 \qquad \blacksquare$$

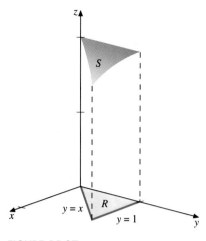

FIGURE 15.37

EXAMPLE 15.23 Find the area of that portion of the paraboloid $z = 2 - x^2 - y^2$ that lies above the xy-plane.

Solution The surface S in question is shown in Figure 15.38. Writing $z = f(x, y)$, we have $f_x = -2x$ and $f_y = -2y$, so by Equation 15.25,

$$A(S) = \iint_R \sqrt{1 + 4x^2 + 4y^2} \, dA$$

The region R is bounded by the xy-trace of the paraboloid—namely, the circle $x^2 + y^2 = 2$. Both the nature of R and the integrand suggest using polar coordinates to evaluate the integral. Since in polar coordinates $r^2 = x^2 + y^2$ and $dA = r \, dr \, d\theta$, we have

$$A(S) = \int_0^{2\pi} \int_0^{\sqrt{2}} \sqrt{1 + 4r^2} \, r \, dr \, d\theta$$

$$= \int_0^{2\pi} \left[\frac{1}{8} \cdot \frac{2}{3} (1 + 4r^2)^{3/2} \right]_0^{\sqrt{2}} d\theta = \frac{13\pi}{3} \qquad \blacksquare$$

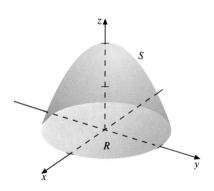

FIGURE 15.38

Surfaces Defined by an Equation of the Form $F(x, y, z) = 0$

If the equation for the surface S has the form $F(x, y, z) = 0$, where F_x, F_y, and F_z are continuous and $F_z \neq 0$, then we know from Section 14.2 that z is implicitly a function of x and y, say $z = f(x, y)$, and

$$f_x = \frac{\partial z}{\partial x} = -\frac{F_x}{F_z} \qquad f_y = \frac{\partial z}{\partial y} = -\frac{F_y}{F_z}$$

When these values are substituted for f_x and f_y in Equation 15.25, we get the alternative formula

$$A(S) = \iint_R \frac{\sqrt{F_x^2 + F_y^2 + F_z^2}}{|F_z|} \, dA \qquad (15.26)$$

By observing that the numerator is the length of the gradient of F, $|\nabla F|$, and that $\nabla F \cdot \mathbf{k} = F_z$, we can put Equation 15.26 in the compact form

$$A(S) = \iint_R \frac{|\nabla F|}{|\nabla F \cdot \mathbf{k}|} \, dA \qquad (15.27)$$

In both Equation 15.26 and Equation 15.27 it must be understood that z is a function of x and y defined implicitly by $F(x, y, z) = 0$.

EXAMPLE 15.24 Find the area of that portion of the sphere $x^2 + y^2 + z^2 = a^2$ that is inside the cylinder $x^2 + y^2 = b^2$, where $0 < b < a$.

Solution Because both surfaces are symmetric with respect to the xy-plane, we can consider the upper hemisphere only and double the result (see Figure 15.39). Taking $F(x, y, z) = x^2 + y^2 + z^2 - a^2$, we have $\nabla F = \langle 2x, 2y, 2z \rangle$, so that

$$|\nabla F| = \sqrt{4x^2 + 4y^2 + 4z^2} = 2\sqrt{x^2 + y^2 + z^2} = 2a$$

The region R in the xy-plane over which the surface lies is bounded by the xy-trace of the cylinder. Thus, $R = \{(x, y) : x^2 + y^2 \leq b^2\}$. So, by Equation 15.26, we have

$$A(S) = 2 \iint_R \frac{2a}{2z} \, dA = 2 \iint_R \frac{a}{\sqrt{a^2 - (x^2 + y^2)}} \, dA$$

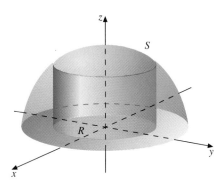

FIGURE 15.39

Using polar coordinates, we obtain

$$A(S) = 2a \int_0^{2\pi} \int_0^b \frac{r \, dr \, d\theta}{\sqrt{a^2 - r^2}} = 2a \int_0^{2\pi} \left[-\sqrt{a^2 - r^2} \right]_0^b \, d\theta$$

$$= 4\pi a \left[a - \sqrt{a^2 - b^2} \right] \qquad \blacksquare$$

DOUBLE INTEGRATION USING COMPUTER ALGEBRA SYSTEMS

As we have seen, computing double integrals can be very messy and tedious. Computer algebra systems can, in many cases, compute double integrals very quickly. In this section, we present several examples that take advantage of the algebraic as well as the graphical capabilities of the CAS in applying double integration.

CAS 58

Find the volume of the region under the paraboloid $z = x^2 + 3y^2 + 4$ and above the unit disk $x^2 + y^2 \leq 1$ in the xy-plane.

The iterated integral that we need to compute for the volume is given by

$$\int_0^1 \int_{-\sqrt{1-x^2}}^{\sqrt{1-x^2}} \left(x^2 + 3y^2 + 2\right) dy\, dx$$

(See Theorem 15.2.)

Maple:

First, we present a plot to get an idea of the region in space. The first plot command plots the paraboloid and the second plots a cylinder with the unit disk as base and that cuts up through the paraboloid. Maple needs several libraries.

with(plots):with(student):
p:=plot3d(x^2+3*y^2+4,x=–2..2,y=–2..2,view=0..10,scaling=
    constrained,axes=boxed):
cylinder:=plot3d([cos(t),sin(t),s],t=0..2*Pi,s=0..27,
    scaling=constrained):
display3d({cylinder,p},orientation=[57,50]);

Mathematica:

P = Plot3D[x^2+3*y^2+4,{x,–2,2},{y,–2,2},
   DisplayFunction–>Identity]

cylinder = ParametricPlot3D[{Cos[t],Sin[t],s},
   {t,0,2*Pi},{s,0,27},DisplayFunction–>Identity]

Show[P,cylinder,DisplayFunction–>$DisplayFunction,
   ViewPoint–>{–2,–2,0},PlotRange–>{0,10}]

Integrate[x^2+3*y^2+4,{x,–1,1},{y,–Sqrt[1–x^2],
   Sqrt[1–x^2]}]

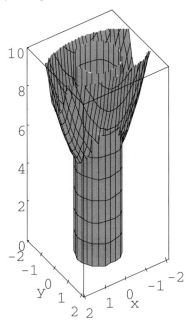

FIGURE 15.5.1

Now, to compute the volume is an easy matter using Maple. You may want to try this integral by hand to appreciate the algebra involved.

volume:=int(int(x^2+3*y^2+4,y=–sqrt(1–x^2)..sqrt(1–x^2)),x=–1..1);

Output: volume := 5π

CAS 59

The formula given for the surface area of $z = f(x, y)$ over some region R in Equation 15.25 is a special case of a more general formula for the surface area when the surface is given in parametric form, say,

$$\mathbf{r}(s, t) = x(s, t)\mathbf{i} + y(s, t)\mathbf{j} + z(s, t)\mathbf{k}$$

where (s, t) are in R. The formula in this case is

$$\iint_R |r_s(s, t) \times r_t(s, t)| \, dA$$

where $\mathbf{r}_s = x_s\mathbf{i} + y_s\mathbf{j} + z_s\mathbf{k}$ and $\mathbf{r}_t = x_t\mathbf{i} + y_t\mathbf{j} + z_t\mathbf{k}$. (Here the parametric surface must be smooth and the surface must be covered only once as (s, t) ranges over R.)

Use this formula to compute the surface area of the torus generated by rotating the circle $(x - 3)^2 + z^2 = 1$ in the xz-plane about the z-axis.

The parametric equations for the circle in the xz-plane are $x = \cos t + 3$ and $z = \sin t$ for $0 \le t \le 2\pi$.

Maple:

To rotate the circle around the z-axis we use:

plot3d([(cos(t)+3)*cos(s),(cos(t)+3)*sin(s),sin(t)], t=0..2*Pi,s=0..2*Pi,scaling=constrained);

Maple will need the linear algebra library:

with(linalg);
X:=(cos(t)+3)*cos(s);
Y:=(cos(t)+3)*sin(s);
Z:=sin(t);
r:=[X,Y,Z];

Mathematica:

ParametricPlot3D[{(Cos[t]+3)*Cos[s],(Cos[t]+3)*Sin[s], Sin[t]},{t,0,2*Pi},{s,0,2*Pi}]

Mathematica needs the library VectorAnalysis.

<<Calculus`VectorAnalysis`
r = {(Cos[t]+3)*Cos[s],(Cos[t]+3)*Sin[s],Sin[t]}

rs = D[r,s]

rt = D[r,t]

cp = CrossProduct[rs,rt]

integrand = Sqrt[cp.cp]

integrand = Simplify[%]

Integrate[integrand,{s,0,2*Pi},{t,0,2*Pi}]

FIGURE 15.5.2

Maple:

Output: $r := [(\cos(t) + 3)\cos(s), (\cos(t) + 3)\sin(s), \sin(t)]$

rs:=diff(r,s);

Output: $rs := [-(\cos(t) + 3)\sin(s), (\cos(t) + 3)\cos(s), 0]$

rt:=diff(r,t);

Output: $rt := [-\sin(t)\cos(s), -\sin(t)\sin(s), \cos(t)]$

cp:=crossprod(rs,rt);

Output:

$cp := [(\cos(t) + 3)\cos(s)\cos(t)(\cos(t) + 3)\sin(s)\cos(t)$
$(\cos(t) + 3)\sin(s)^2\sin(t) + (\cos(t) + 3)\cos(s)^2\sin(t)]$

integrand:=sqrt(dotprod(cp,cp));

Output: $integrand := \Big((\cos(t) + 3)^2\cos(s)^2\cos(t)^2$
$+ (\cos(t) + 3)^2\sin(s)^2\cos(t)^2 + \big((\cos(t) + 3)\sin(s)^2\sin(t)$
$+ (\cos(t) + 3)\cos(s)^2\sin(t)\big)^2\Big)^{1/2}$

We better try to simplify this expression before asking Maple to integrate, because Maple will have trouble with the expression in its current form.

factor(integrand);

Output: $\sqrt{\sin(s)^2 + \cos(s)^2}(\cos(t) + 3)$
$\sqrt{\cos(t)^2 + \sin(t)^2\cos(s)^2 + \sin(s)^2\sin(t)^2}$

integrand:=simplify(");

Output: $integrand := \cos(t) + 3$
This is better and we can now integrate.

int(int(integrand,t=0..2*Pi),s=0..2*Pi);

Output: $12\pi^2$

Exercise Set 15.5

1. Find the area of the portion of the plane $z + 2x + 3y = 6$ that lies above the rectangular region R bounded by $x = 0, y = 0, x = 2,$ and $y = 1$.

2. Find the area of that portion of the plane $4x - 3y - 6z + 12 = 0$ that lies above the triangular region R with vertices $(3, 0), (0, 0),$ and $(0, 4)$.

3. Find the area of the first-octant portion of the cylinder $z = 2 - y^2$ that is cut off by the planes $x = 0, y = x,$ and $z = 0$.

4. Find the area of the portion of the cylinder $z = \sqrt{4 - y^2}$ above the region $R = \{(x, y) : 0 \le x \le 3, \ 0 \le y \le 1\}$.

5. Find the area of the paraboloid $z = x^2 + y^2$ that lies inside the cylinder $x^2 + y^2 = 2$.

6. Find the area of the part of the sphere $x^2 + y^2 + z^2 = 4$ that is inside the cylinder $x^2 + y^2 - 2x = 0$.

7. Find the area of the part of the upper nappe of the cone $z^2 = x^2 + y^2$ that lies inside the cylinder $x^2 + y^2 - 4y = 0$.

8. Find the area of that portion of the cylinder $x^2 + z^2 = a^2$ that lies inside the cylinder $x^2 + y^2 = a^2$.

9. Find the area of the paraboloid $z = 1 + x^2 + y^2$ that is between the planes $z = 2$ and $z = 5$.

10. Find the area of the part of the sphere $x^2 + y^2 + z^2 = 25$ that is between the planes $z = 3$ and $z = 4$.

In Exercises 11–16, find $A(S)$, where S is the portion of the graph $z = f(x, y)$ that lies above the region R.

11. $f(x, y) = 3x + y^2$; $R = \{(x, y) : 0 \leq x \leq y,\ 0 \leq y \leq 2\}$

12. $f(x, y) = \frac{2}{3}(x^{3/2} + y^{3/2})$; $R = \{(x, y) : 0 \leq x \leq 3,\ 0 \leq y \leq 3 - x\}$

13. $f(x, y) = \ln \sec x$; $R = \{(x, y) : -\pi/4 \leq x \leq \pi/4,\ 0 \leq y \leq \sec x\}$

14. $f(x, y) = (1 - y^{2/3})^{3/2}$; $R = \{(x, y) : 0 \leq x \leq 1,\ 0 \leq y \leq (1 - x^{2/3})^{3/2}\}$

15. $f(x, y) = 2 - (x^2 + y^2)/2$; R is the region inside the lemniscate $r^2 = \cos 2\theta$.

16. $f(x, y) = x + y + 2$; R is the region outside $r = 1$ and inside $r = 2\cos\theta$.

17. Use the methods of this section to find the area of the surface of a sphere of radius a.

18. Use the methods of this section to find the lateral surface area of a right circular cone (one nappe) of base radius a and height h.

19. Using the accompanying figure, verify that when a nonnegative smooth function f of one variable, defined on $[a, b]$, is rotated about the x-axis, the equation of the resulting surface of revolution is $y^2 + z^2 = [f(x)]^2$. Using this and the methods of this section, show that its total surface area is

$$2\pi \int_a^b f(x) \sqrt{1 + [f'(x)]^2}\, dx$$

and so is in agreement with Definition 6.3.

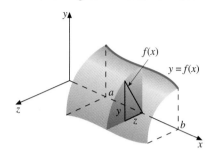

20. In the accompanying figure, $PQRS$ is a parallelogram of area ΔT formed by intersecting a plane with the prism that has the rectangular base $P'Q'R'S'$. The normal $\mathbf{n}$ to the plane makes an angle γ with the unit vertical vector $\mathbf{k}$. If ΔA is the area of the rectangle $P'Q'R'S'$, show that

$$(\vec{PQ} \times \vec{PS}) \cdot \mathbf{k} = (\vec{P'Q'} \times \vec{P'S'}) \cdot \mathbf{k}$$

and then explain how it follows from this that

$$\Delta T |\cos \gamma| = \Delta A$$

(*Hint:* Write $\vec{PQ} = \vec{PP'} + \vec{P'Q'} + \vec{Q'Q}$, and write $\vec{PS}$ in a similar way.)

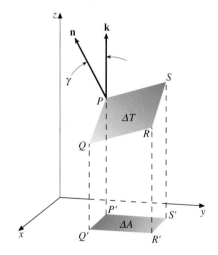

15.6 TRIPLE INTEGRALS

The ideas behind the triple integral of a function of three variables over a region in $\mathbb{R}^3$ are similar to those for a double integral in $\mathbb{R}^2$, so we will be briefer in our treatment. We will designate the three-dimensional region of integration by Q. We begin with the simplest case, in which Q is a rectangular box of the form

$$Q = \{(x, y, z) : a_1 \leq x \leq a_2, b_1 \leq y \leq b_2, c_1 \leq z \leq c_2\}$$

Using planes parallel to each of the coordinate planes, we partition Q into finitely many small boxes that we will call *cells*. We show a typical partition in Figure 15.40. We denote the partition by P and its norm (the length of the longest diagonal of all the cells in P) by $\|P\|$.

We number the n cells of P from 1 to n, in any manner. Let (x_i^*, y_i^*, z_i^*) be any point in the ith cell, and denote the volume of the ith cell by ΔV_i. We define the triple integral of a function f of three variables over the region Q as follows.

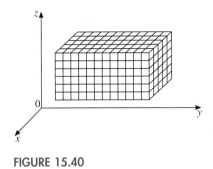

FIGURE 15.40

Definition 15.3
The Triple Integral

$$\iiint_Q f(x, y, z) \, dV = \lim_{\|P\| \to 0} \sum_{i=1}^{n} f(x_i^*, y_i^*, z_i^*) \Delta V_i \qquad (15.28)$$

provided the limit on the right-hand side exists.

It can be shown that the triple integral of a continuous function over a rectangular box Q does exist.

Evaluating Triple Integrals by Iterated Integrals

There is a version of Fubini's Theorem for triple integrals that enables us to evaluate the integral by a triple iterated integral. We can state the theorem as follows.

THEOREM 15.3

Fubini's Theorem for Triple Integrals

If f is continuous on the rectangular box $Q = \{(x, y, z) : a_1 \leq x \leq a_2, b_1 \leq y \leq b_2, c_1 \leq z \leq c_2\}$ then

$$\iiint_Q f(x, y, z) \, dV = \int_{a_1}^{a_2} \int_{b_1}^{b_2} \int_{c_1}^{c_2} f(x, y, z) \, dz \, dy \, dx \qquad (15.29)$$

There are five other iterated integrals that could be used in place of the one on the right-hand side of Equation 15.29, obtained by changing the order of integration.

EXAMPLE 15.25 Evaluate the triple integral

$$\iiint_Q (2xy + 3yz^2) \, dV$$

where Q is the rectangular box $\{(x, y, z) : 0 \leq x \leq 1, -1 \leq y \leq 2, 1 \leq z \leq 3\}$.

Solution By Equation 15.29,

$$\iiint_Q (2xy + 3yz^2) \, dV = \int_0^1 \int_{-1}^2 \int_1^3 (2xy + 3yz^2) \, dz \, dy \, dx$$

$$= \int_0^1 \int_{-1}^2 \left[2xyz + yz^3\right]_1^3 dy \, dx \quad \text{We held } x \text{ and } y \text{ fixed and integrated with respect to } z.$$

$$= \int_0^1 \int_{-1}^2 (6xy + 27y - 2xy - y) \, dy \, dx$$

$$= \int_0^1 \int_{-1}^2 (4xy + 26y) \, dy \, dx$$

$$= \int_0^1 \left[2xy^2 + 13y^2\right]_{-1}^2 dx \quad \text{We held } x \text{ fixed and integrated with respect to } y.$$

$$= \int_0^1 (8x + 52 - 2x - 13) \, dx$$

$$= \int_0^1 (6x + 39) \, dx$$

$$= 3x^2 + 39x \Big]_0^1 = 3 + 39 = 42 \quad \blacksquare$$

In Exercise 7 of Exercise Set 15.6 we will ask you to redo Example 15.25, using two of the other five iterated integrals, to show they give the same result.

Triple Integrals Over General Regions

If the region Q of integration is not a rectangular box but can be enclosed in a rectangular box, then we partition the rectangular box that encloses Q but count only those cells that are completely contained within Q. We show such a partition in Figure 15.41. With the understanding that in Equation 15.28 the n cells are those contained within Q, Definition 15.3 also applies to this more general case.

It is beyond the scope of this book to describe the most general type of three-dimensional region Q for which the triple integral of a continuous function exists. Instead, we will limit consideration to regions of the following type:

$$Q = \{(x, y, z) : (x, y) \in R, \varphi_1(x, y) \leq z \leq \varphi_2(x, y)\}$$

where φ_1 and φ_2 are continuous and the projection R of Q onto the xy-plane is

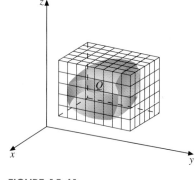

FIGURE 15.41

a two-dimensional region of type I or type II. In Figure 15.42, we have shown such a region Q in which the projection R is a type I region.

It can be shown that if f is continuous on a region Q as described, then the triple integral of f over Q is given by the following formula.

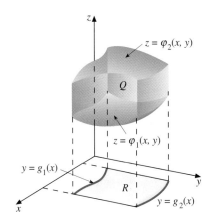

FIGURE 15.42

$$\iiint_Q f(x, y, z)\, dA = \iint_R \left[\int_{\varphi_1(x,y)}^{\varphi_2(x,y)} f(x, y, z)\, dz \right] dA \quad (15.30)$$

To evaluate the integral on the right, first hold x and y fixed and integrate with respect to z. Then the double integral of the result is taken over R.

If R is the type I region shown in Figure 15.42, then Equation 15.30 can be rewritten in the form

$$\iiint_Q f(x, y, z)\, dA = \int_a^b \int_{g_1(x)}^{g_2(x)} \int_{\varphi_1(x,y)}^{\varphi_2(x,y)} f(x, y, z)\, dz\, dy\, dx \quad (15.31)$$

Similarly, if R is the type II region

$$R = \{(x, y) : h_1(y) \leq x \leq h_2(y), c \leq y \leq d\}$$

then Equation 15.30 can be written as

$$\iiint_Q f(x, y, z)\, dA = \int_c^d \int_{h_1(y)}^{h_2(y)} \int_{\varphi_1(x,y)}^{\varphi_2(x,y)} f(x, y, z)\, dz\, dx\, dy \quad (15.32)$$

REMARK
■ Formulas analogous to Equation 15.30 exist for regions of the form

$$Q = \{(x, y, z) : (y, z) \in R, \varphi_1(y, z) \leq x \leq \varphi_2(y, z)\}$$

or

$$Q = \{(x, y, z) : (x, z) \in R, \varphi_1(x, z) \leq y \leq \varphi_2(x, z)\}$$

In the first case, R is the projection of Q onto the yz-plane, and in the second, R is the projection of Q onto the xz-plane. In each case, R can be of type I or type II in the variables in question.

EXAMPLE 15.26 Evaluate the integral $\iiint_Q 2xz\,dV$, where Q is the region enclosed by the planes $x + y + z = 4$, $y = 3x$, $x = 0$, and $z = 0$.

Solution The region Q is shown in Figure 15.43. Its projection onto the xy-plane is a type I region, so we will use Equation 15.31. To find the limits on z, we consider a point (x, y, z) in Q and find the smallest and largest values of z so that the point remains in Q. In this case, z varies from the horizontal plane $z = 0$ to the inclined plane $z = 4 - x - y$ that forms the top of the region. The limits on x and y are determined from the region R that is the xy-projection of Q, just as with double integrals. In this case, the limits on y are the traces $y = 3x$ and $y = 4 - x$. They intersect when $x = 1$, so x varies from 0 to 1. Thus,

$$\iiint_Q 2xz\,dV = \int_0^1 \int_{3x}^{4-x} \int_0^{4-x-y} 2xz\,dz\,dy\,dx$$

$$= \int_0^1 \int_{3x}^{4-x} x(4 - x - y)^2\,dy\,dx$$

$$= \int_0^1 \left[\frac{-x(4-x-y)^3}{3}\right]_{3x}^{4-x} dx = \frac{1}{3}\int_0^1 x(4-4x)^3\,dx$$

$$= \frac{64}{3}\int_0^1 x(1-x)^3\,dx \quad \text{Integrate by parts with } u = x, dv = (1-x)^3\,dx$$

$$= \frac{64}{3}\left[-\frac{x(1-x)^4}{4} - \frac{(1-x)^5}{20}\right]_0^1$$

$$= \frac{64}{3}\left(\frac{1}{20}\right) = \frac{16}{15} \quad \blacksquare$$

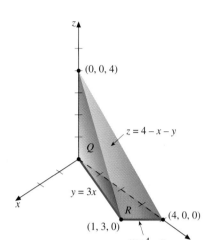

FIGURE 15.43

Applications of Triple Integrals

We can use the triple integral to find volumes and centroids of regions in $\mathbb{R}^3$, masses, centers of mass, and moments of inertia of solids. The ideas parallel those for the two-dimensional case, so we omit the details and simply give the results.

Volume

By taking $f(x, y, z) = 1$, we get the volume of Q:

$$V = \iiint_Q dV \tag{15.33}$$

Mass

If a solid is in the shape of a three-dimensional region Q and has density $\rho(x, y, z)$, then its mass m is

$$m = \iiint_Q dm \tag{15.34}$$

where $dm = \rho(x, y, z)\, dV$.

Moments

The (first) moments with respect to the xy-plane, the xz-plane, and the yz-plane are

$$M_{xy} = \iiint_Q z\, dm, \quad M_{xz} = \iiint_Q y\, dm, \quad M_{yz} = \iiint_Q x\, dm \tag{15.35}$$

Center of Mass

$$\bar{x} = \frac{\iiint_Q x\, dm}{\iiint_Q dm}, \quad \bar{y} = \frac{\iiint_Q y\, dm}{\iiint_Q dm}, \quad \bar{z} = \frac{\iiint_Q z\, dm}{\iiint_Q dm} \tag{15.36}$$

Centroid

The centroid of a region is the same as the center of mass of a homogeneous solid that occupies that region:

$$\bar{x} = \frac{\iiint_Q x\, dV}{V}, \quad \bar{y} = \frac{\iiint_Q y\, dV}{V}, \quad \bar{z} = \frac{\iiint_Q z\, dV}{V} \tag{15.37}$$

Moments of Inertia

With respect to the coordinate axes, the moments of inertia of a region Q are given by

$$I_x = \iiint_Q (y^2 + z^2)\, dm \quad I_y = \iiint_Q (x^2 + z^2)\, dm \quad I_z = \iiint_Q (x^2 + y^2)\, dm \tag{15.38}$$

EXAMPLE 15.27 A solid of density $\rho(x, y, z) = \sqrt{xyz}$ is in the shape of the region Q enclosed by the surfaces $z = 4 - x^2$, $z = x^2 + y^2$, $x = 0$, and $y = 0$. Set up, but do not evaluate, iterated integrals for (a) the mass m, (b) the x-coordinate of the center of mass, $\bar{x}$, and (c) the moment of inertia with respect to the z-axis, I_z.

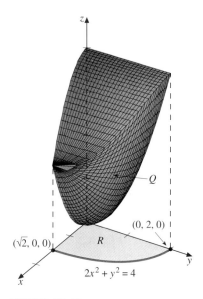

FIGURE 15.44

Solution The region Q is shown in Figure 15.44. To find the projection of Q onto the xy-plane, we solve the equations $z = 4 - x^2$ and $z = x^2 + y^2$ simultaneously. Setting the two values equal to one another yields the equation $2x^2 + y^2 = 4$. Thus, the xy-projection R of Q is given by

$$R = \left\{(x, y) : 0 \leq x \leq \sqrt{2},\ 0 \leq y \leq \sqrt{4 - 2x^2}\right\}$$

The z limits are from the lower bounding surface $z = x^2 + y^2$ to the upper bounding surface $z = 4 - x^2$.

(a) $m = \iiint_Q dm = \int_0^{\sqrt{2}} \int_0^{\sqrt{4-2x^2}} \int_{x^2+y^2}^{4-x^2} \sqrt{xyz}\, dz\, dy\, dx$

(b) $\bar{x} = \dfrac{M_{yz}}{m} = \dfrac{1}{m} \int_0^{\sqrt{2}} \int_0^{\sqrt{4-2x^2}} \int_{x^2+y^2}^{4-x^2} x\sqrt{xyz}\, dz\, dy\, dx$

(c) $I_z = \iiint_Q (x^2 + y^2)\, dm = \int_0^{\sqrt{2}} \int_0^{\sqrt{4-2x^2}} \int_{x^2+y^2}^{4-x^2} (x^2 + y^2)\sqrt{xyz}\, dz\, dy\, dx$

Note that we could have integrated first with respect to y, with limits from 0 to $\sqrt{z - x^2}$, since the xz-projection of Q is the type I region described by $\{(x, z) : 0 \leq x \leq \sqrt{2},\ x^2 \leq z \leq 4 - x^2\}$. Then for part (a),

$$m = \int_0^{\sqrt{2}} \int_{x^2}^{4-x^2} \int_0^{\sqrt{z-x^2}} \sqrt{xyz}\, dy\, dz\, dx$$

and the limits for parts (b) and (c) would have been the same as those for m. ■

Exercise Set 15.6

In Exercises 1–6, evaluate the iterated integrals. Sketch the region Q determined by the limits of integration.

1. $\displaystyle\int_0^1 \int_1^2 \int_{-1}^1 (xy - 2yz)\, dy\, dz\, dx$

2. $\displaystyle\int_2^4 \int_0^1 \int_0^2 \dfrac{2x + y}{z^2}\, dx\, dy\, dz$

3. $\displaystyle\int_0^1 \int_0^{1-x} \int_0^2 x\, dz\, dy\, dx$

4. $\displaystyle\int_0^4 \int_0^{\sqrt{z}} \int_0^2 (2xy^2 - 1)\, dy\, dx\, dz$

5. $\displaystyle\int_0^4 \int_0^2 \int_{\sqrt{y}}^{6-y} 4xz\, dx\, dz\, dy$

6. $\displaystyle\int_0^1 \int_0^{2-2y} \int_2^{4-x-2y} z\, dz\, dx\, dy$

7. Redo Example 15.25 using an iterated integral in which
   (a) the first integration is with respect to y and the second is with respect to x;
   (b) the first integration is with respect to x and the second is with respect to z.

In Exercises 8–11, for the given region Q write an iterated integral whose value equals $\iiint_Q f(x, y, z)\, dV$.

8. Q is bounded by $z = 4 - x^2$, $y = x$, $y = 0$, and $z = 0$.

9. Q is the first-octant portion of the region inside the ellipsoid $4x^2 + y^2 + z^2 = 4$.

10. Q is bounded by $2x + 3y = 6$, $x + z = 3$, and the coordinate planes.

11. $Q = \{(x, y, z) : \sqrt{4 - z^2} \leq x \leq \sqrt{3},\ 0 \leq y \leq 4,\ 1 \leq z \leq 2\}$

12. Evaluate the integral in Exercise 8 for $f(x, y, z) = \sqrt{z}$.

13. Evaluate the integral in Exercise 11 for $f(x, y, z) = x(3 - y)z^2$.

In Exercises 14–17, find the volume of the region Q bounded by the given surfaces.

14. $x + 2y + z = 4$ and the coordinate planes

15. $z = x^2 + y^2$, $z = 2$, $y = x$, $y = 0$, in the first octant

16. $x^2 + z^2 = 4$, $y = x$, $y = 0$, $z = 0$, in the first octant

17. $x + z = 1$, $4x + y + z = 4$, and the coordinate planes

In Exercises 18 and 19, write five different iterated integrals equal to the given integral.

18. $\int_0^1 \int_0^x \int_0^{1-x} f(x, y, z)\, dz\, dy\, dx$

19. $\int_0^4 \int_0^{4-z} \int_0^{\sqrt{y}} f(x, y, z)\, dx\, dy\, dz$

20. Find the centroid of the region bounded by $y^2 = 2x$, $2x + z = 4$, and $z = 0$.

21. For the solid of density $\rho(x, y, z) = 2x$ bounded by the planes $x + y = 1$, $y + z = 1$, and the coordinate planes, find m and $\bar{x}$.

In Exercises 22–28, set up iterated integrals for the specified quantities for the solid that has density $\rho(x, y, z)$ and occupies the region Q.

22. Q is the region above $z = x^2 + y^2$ and below $z = 8 - x^2 - y^2$, $\rho(x, y, z) =$ distance from the xy-plane to the point (x, y, z); m, $\bar{z}$, I_z.

23. The solid of Exercise 21; $\bar{z}$, I_y

24. Q is bounded by $x + 2y + 3z = 6$ and the coordinate planes, $\rho(x, y, z) = x^2 yz$; center of mass.

25. Q is the region inside both $x^2 + y^2 = 4$ and $x^2 + y^2 + z^2 = 16$ above the xy-plane, $\rho(x, y, z) =$ distance from the z-axis to the point (x, y, z); m, $\bar{z}$, I_z.

26. Q is the first-octant portion of the region bounded by $y^2 + z^2 = 4$, $z = x - 2y$, and the coordinate planes, $\rho(x, y, z) = \sqrt{x + y}$; $\bar{x}$, I_x.

27. Q is the first-octant portion of the region inside both $x^2 + y^2 = 1$ and $x^2 + z^2 = 1$, $\rho(x, y, z) = xyz$; center of mass.

28. Q is the first-octant portion of the region inside the ellipsoid $2x^2 + y^2 + z^2 = 4$, $\rho(x, y, z) =$ distance from the y-axis to the point (x, y, z); $\bar{y}$, I_y.

29. Find the volume and the location of the centroid of a pyramid of height h and with a base that is a square of side a.

30. Evaluate the integral
$$\iiint_Q y^2 \sin xy\, dV$$
where Q is bounded by $z = x$, $2xy = \pi$, $y = \pi/2$, $y = \pi$, and $z = 0$. Give two possible physical interpretations of this integral.

31. A solid of density $\rho(x, y, z) = (x+y)^2 e^{(x+y)z}$ occupies the region $Q = \{(x, y, z) : 0 \leq x \leq 2 - y,\ 0 \leq y \leq 1,\ 0 \leq z \leq 1\}$. Find $\bar{z}$.

32. Prove that the first moment of a solid with respect to any plane through its center of mass is 0. (*Hint:* If $\langle a, b, c \rangle$ is a unit normal vector to the plane, the distance d from the plane to a point (x, y, z) is
$$d = \pm[a(x - \bar{x}) + b(y - \bar{y}) + c(z - \bar{z})]$$
which is positive on one side of the plane and negative on the other. (Verify.))

33. The **Parallel Axis Theorem** states that for a solid of mass m, the moment of inertia I_l about any line l is
$$I_l = I_{l'} + md^2$$
where l' is a line through the center of mass parallel to l and at a distance d from it. Prove this theorem.

15.7 TRIPLE INTEGRALS IN CYLINDRICAL AND SPHERICAL COORDINATES

As we saw in Section 15.3, polar coordinates often can be used to simplify double integrals. In a similar way, triple integrals often can be more readily evaluated by using one of two alternatives to rectangular coordinates, called **cylindrical coordinates** and **spherical coordinates**.

Cylindrical Coordinates

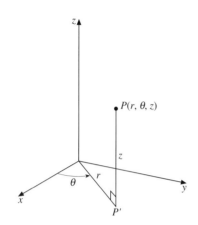

FIGURE 15.45

In a cylindrical coordinate system, polar coordinates are used for two variables and the rectangular coordinate for the third variable. For example, in Figure 15.45 the point P has coordinates (r, θ, z). The projection P' of P onto the xy-plane has polar coordinates r and θ, and z is the usual rectangular z-coordinate. We could equally well use polar coordinates in the xz- or yz-plane, along with the appropriate third rectangular coordinate, but we will concentrate on the situation illustrated in Figure 15.45.

Rectangular and cylindrical coordinates are related by the equations

$$\begin{cases} x = r \cos \theta \\ y = r \sin \theta \\ z = z \end{cases}$$

Some common equations of surfaces in rectangular coordinates, along with the corresponding cylindrical equations, are given below:

	Circular cylinder	Circular cone	Sphere	Paraboloid
Rectangular	$x^2 + y^2 = a^2$	$z^2 = a^2(x^2 + y^2)$	$x^2 + y^2 + z^2 = a^2$	$z = a(x^2 + y^2)$
Cylindrical	$r = a$	$z = ar$	$r^2 + z^2 = a^2$	$z = ar^2$

Triple Integrals in Cylindrical Coordinates

The simplest type of closed region in $\mathbb{R}^3$ to describe in cylindrical coordinates is a set of the form

$$Q = \{(r, \theta, z) : a \leq r \leq b, \ \alpha \leq \theta \leq \beta, \ c \leq z \leq d\}$$

Such a region is shown in Figure 15.46. We will call this region a *cylindrical box*. Notice that its projection onto the xy-plane is a polar rectangle.

Suppose now that f is a function of the cylindrical variables r, θ, and z that is defined on a cylindrical box of the type shown in Figure 15.46. We partition the box by means of horizontal planes, planes containing the z-axis, and circular cylinders centered on the z-axis. These surfaces correspond, respectively, to the equations $z =$ constant, $\theta =$ constant, and $r =$ constant. They divide the box into smaller boxes that we again call *cells*. We denote the partition by P and again call the length of the longest diagonal of all of the cells its norm, which we designate by $\|P\|$. If there are n cells, we number them from 1 to n in any way, and we denote their volumes by $\Delta V_1, \Delta V_2, \ldots, \Delta V_n$. Finally, we choose an arbitrary point $(r_i^*, \theta_i^*, z_i^*)$ in the ith cell, for $i = 1, 2, \ldots, n$. Then we define the triple integral in cylindrical coordinates of f over Q as follows.

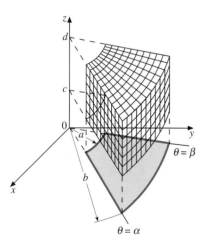

FIGURE 15.46
The cylindrical box $Q = \{(r, \theta, z) : a \leq r \leq b, \alpha \leq \theta \leq \beta, c \leq z \leq d\}$ and a partition of it.

Section 15.7 Triple Integrals in Cylindrical and Spherical Coordinates   1119

Definition 15.4
The Triple Integral in Cylindrical Coordinates

$$\iiint_Q f(r, \theta, z)\, dV = \lim_{\|P\|\to 0} \sum_{i=1}^{n} f(r_i^*, \theta_i^*, z_i^*)\Delta V_i \quad (15.39)$$

provided the limit on the right-hand side exists independently of the choices of the points $(r_i^*, \theta_i^*, z_i^*)$.

Fubini's Theorem takes the following form for a cylindrical box such as that shown in Figure 15.46.

If f is continuous on the cylindrical box $Q = \{(r, \theta, z) : a \leq r \leq b, \alpha \leq \theta \leq \beta, c \leq z \leq d\}$, then

$$\iiint_Q f(r, \theta, z)\, dV = \int_\alpha^\beta \int_a^b \int_c^d f(r, \theta, z)\, r\, dz\, dr\, d\theta \quad (15.40)$$

Equivalently, the iterated integral on the right-hand side may be replaced by any one of the other five iterated integrals obtained by integrating with respect to the three variables in other orders.

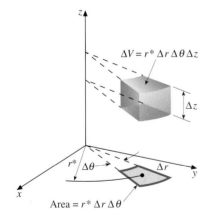

FIGURE 15.47

The factor r in the integrand on the right-hand side of Equation 15.40 is suggested by Figure 15.47, in which we show a typical cell in the partition of Q. The area of the base of this cell is $r^*\Delta r\, \Delta\theta$ (we have omitted subscripts for simplicity), just as in the case of polar coordinates. Its height is Δz. So the volume is $\Delta V = r^*\Delta r\, \Delta\theta\, \Delta z$.

We write

$$dV = r\, dz\, dr\, d\theta$$

(or some permutation of the differentials) and call dV in this case the *differential of volume* in cylindrical coordinates.

EXAMPLE 15.28 Evaluate the integral

$$\iiint_Q (zr \sin\theta)\, dV$$

where $Q = \{(r, \theta, z) : 0 \leq r \leq 2,\ 0 \leq \theta \leq \pi/2,\ 0 \leq z \leq 4\}$.

Solution By Equation 15.40,

$$\iiint_Q (zr \sin\theta)\, dV = \int_0^{\pi/2} \int_0^2 \int_0^4 (zr \sin\theta)\, \overbrace{r\, dz\, dr\, d\theta}^{dV}$$

$$= \int_0^{\pi/2} \int_0^2 \left[\frac{z^2}{2} r^2 \sin\theta \right]_0^4 dr\, d\theta \quad \text{We held } r \text{ and } \theta \text{ constant and integrated with respect to } z.$$

$$= \int_0^{\pi/2} \int_0^2 (8r^2 \sin\theta)\, dr\, d\theta$$

$$= \int_0^{\pi/2} \left[\frac{8r^3}{3}\sin\theta\right]_0^2 d\theta \qquad \text{We held } \theta \text{ constant and integrated with respect to } r.$$

$$= \int_0^{\pi/2} \frac{64}{3}\sin\theta\, d\theta$$

$$= -\frac{64}{3}\cos\theta\Big]_0^{\pi/2} = -\frac{64}{3}(0-1) = \frac{64}{3} \qquad \blacksquare$$

If the region Q is not a cylindrical box but can be enclosed in such a box, we proceed in the usual way by partitioning a cylindrical box that encloses Q and counting only those cells that lie completely inside Q. If there are n such cells in Q, then Definition 15.4 applies to this case also.

If Q is of the form

$$Q = \{(r,\theta,z) : (r,\theta) \in R, \varphi_1(r,\theta) \leq z \leq \varphi_2(r,\theta)\}$$

where R is a polar region of one of the types we considered in Section 15.3, then it can be shown that for f continuous on Q,

$$\iiint_Q f(r,\theta,z)\, dV = \iint_R \left[\int_{\varphi_1(r,\theta)}^{\varphi_2(r,\theta)} f(r,\theta,z)\, dz\right] dA$$

In particular, if $R = \{(r,\theta) : g_1(\theta) \leq r \leq g_2(\theta), \alpha \leq \theta \leq \beta\}$, then

$$\iiint_Q f(r,\theta,z)\, dV = \int_\alpha^\beta \int_{g_1(\theta)}^{g_2(\theta)} \int_{\varphi_1(r,\theta)}^{\varphi_2(r,\theta)} f(r,\theta,z)\, r\, dz\, dr\, d\theta \qquad (15.41)$$

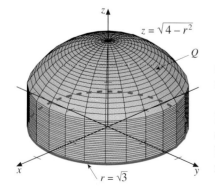

FIGURE 15.48

Analogous formulas exist for the cases in which R is the projection of Q onto the yz-plane or the xz-plane. In these cases, polar coordinates would be used in the yz-plane, or xz-plane, respectively.

EXAMPLE 15.29 A homogeneous solid is bounded laterally by the circular cylinder $x^2 + y^2 = 3$, on the top by the sphere $x^2 + y^2 + z^2 = 4$, and on the bottom by the xy-plane. Find its center of mass and the moment of inertia with respect to the z-axis.

Solution The region Q occupied by the solid is shown in Figure 15.48. In cylindrical coordinates the bounding surfaces are $r = \sqrt{3}$, $z = \sqrt{4-r^2}$, and

$z = 0$. Denote the density by ρ. First, we calculate the mass. By Equations 15.34 and 15.41, we have

$$m = \iiint_Q dm = \iiint_Q \rho \, dV = \int_0^{2\pi} \int_0^{\sqrt{3}} \int_0^{\sqrt{4-r^2}} \rho r \, dz \, dr \, d\theta$$

$$= \rho \int_0^{2\pi} \int_0^{\sqrt{3}} r\sqrt{4-r^2} \, dr \, d\theta = \rho \int_0^{2\pi} \left[-\frac{1}{2} \cdot \frac{2}{3}(4-r^2)^{3/2} \right]_0^{\sqrt{3}} d\theta = \frac{14\pi}{3}\rho$$

By symmetry, we see that $\bar{x} = \bar{y} = 0$, so we need only calculate $\bar{z} = M_{xy}/m$. By Equations 15.35 and 15.41,

$$M_{xy} = \iiint_Q z \, dm = \int_0^{2\pi} \int_0^{\sqrt{3}} \int_0^{\sqrt{4-r^2}} z\rho r \, dz \, dr \, d\theta$$

$$= \rho \int_0^{2\pi} \int_0^{\sqrt{3}} r \left(\frac{4-r^2}{2} \right) dr \, d\theta$$

$$= \rho \int_0^{2\pi} \left[r^2 - \frac{r^4}{8} \right]_0^{\sqrt{3}} d\theta = \frac{15\pi}{4}\rho$$

So

$$\bar{z} = \frac{M_{xy}}{m} = \frac{15\pi\rho}{4} \cdot \frac{3}{14\pi\rho} = \frac{45}{56}$$

From Equations 15.38 and 15.41, we have, for the moment of inertia with respect to the z-axis,

$$I_z = \iiint_Q (x^2 + y^2) \, dm = \iiint_Q r^2 \, dm = \int_0^{2\pi} \int_0^{\sqrt{3}} \int_0^{\sqrt{4-r^2}} \rho r^3 \, dz \, dr \, d\theta$$

$$= \rho \int_0^{2\pi} \int_0^{\sqrt{3}} r^3 \sqrt{4-r^2} \, dr \, d\theta = \frac{94\pi}{15}\rho$$

(You should check the integration on r using either trigonometric substitution, integration by parts, or the substitution $u = \sqrt{4-r^2}$.) ∎

Changing from Rectangular to Cylindrical Coordinates

EXAMPLE 15.30 Write an integral in cylindrical coordinates equivalent to

$$\int_0^1 \int_0^{\sqrt{4-x^2}} \int_{\sqrt{x^2+y^2}}^{6-x^2-y^2} f(x, y, z) \, dz \, dy \, dx$$

Solution The first thing to do is to sketch the region Q determined by the limits of integration. The lower and upper boundaries on z, respectively, are

$$z = \sqrt{x^2 + y^2} \quad \text{and} \quad z = 6 - x^2 - y^2$$

1122   Chapter 15   Multiple Integrals

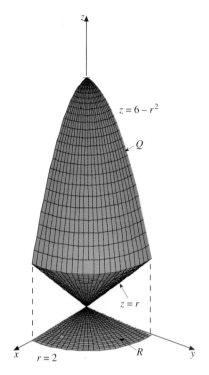

FIGURE 15.49

In cylindrical coordinates these equations become $z = r$ and $z = 6 - r^2$. They intersect when $r = 6 - r^2$, with the positive solution $r = 2$. The limits on x and y show that the projection R of Q onto the xy-plane is one-quarter of the circular disk bounded by $r = 2$, as shown in Figure 15.49. So an equivalent integral in cylindrical coordinates is

$$\int_0^{\pi/2} \int_0^2 \int_r^{6-r^2} f(r\cos\theta, r\sin\theta, z) r \, dz \, dr \, d\theta$$

∎

As we illustrated in Example 15.30, when we are given an iterated integral in rectangular coordinates and wish to convert it to cylindrical coordinates, we first find the region Q of integration determined by the rectangular limits. Then we write Q in cylindrical coordinates in order to determine the limits for r, θ, and z. The original integrand, $f(x, y, z)$, becomes $f(r\cos\theta, r\sin\theta, z)$, and the differential of volume $dV = dz\,dy\,dx$ (or some permutation) is replaced by $r\,dz\,dr\,d\theta$ (or the appropriate permutation).

If we denote by T the region R described in cylindrical coordinates, then the transformation equation can be written as

$$\iiint_Q f(x, y, z)\,dA = \iiint_T f(r\cos\theta, r\sin\theta, z) r \, dz \, dr \, d\theta$$

Spherical Coordinates

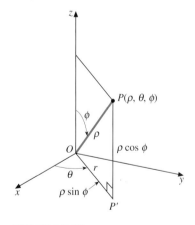

FIGURE 15.50

In the spherical coordinate system, a point P in space is located by its distance ρ from the origin, the polar angle θ from the positive x-axis to the projection OP' of OP onto the xy-plane, and the angle ϕ (phi) from the z-axis to OP. The angles θ and ϕ and the distance ρ are illustrated in Figure 15.50. We restrict ρ and ϕ so that $\rho \geq 0$ and $0 \leq \phi \leq \pi$. From the right triangle $OP'P$, we see that

$$\overline{OP'} = \rho \sin\phi \quad \text{and} \quad \overline{P'P} = \rho \cos\phi$$

REMARK
∎ We have previously used ρ to designate density. Now we are using it as one of the three spherical coordinates. When we are using spherical coordinates, we will designate density by δ.

Thus, since $x = \overline{OP'}\cos\theta$, $y = \overline{OP'}\sin\theta$, $z = \overline{P'P}$, we have

$$\begin{cases} x = \rho \cos\theta \sin\phi \\ y = \rho \sin\theta \sin\phi \\ z = \rho \cos\phi \end{cases} \quad (15.42)$$

as the equations relating rectangular coordinates to spherical coordinates. On squaring each of these and adding, we get (verify it)

$$x^2 + y^2 + z^2 = \rho^2 \quad (15.43)$$

Section 15.7 Triple Integrals in Cylindrical and Spherical Coordinates 1123

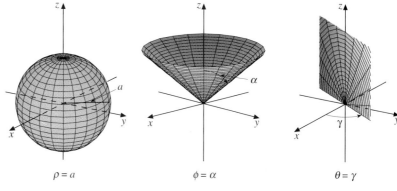

FIGURE 15.51

The simplest surfaces to represent in spherical coordinates are those with equations of the form $\rho = a$, $\phi = \alpha$, and $\theta = \gamma$, where a, α, and γ are constants. These are, respectively, a sphere of radius a centered at the origin, a half-cone with vertex at the origin and axis along the z-axis, and a plane that contains the z-axis. These surfaces are illustrated in Figure 15.51.

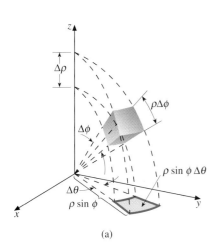

Triple Integrals in Spherical Coordinates

To define the triple integral of a bounded function f over an appropriately restricted closed and bounded region Q, we proceed in a familiar way. Suppose first that Q is of the form $\{(\rho, \theta, \phi) : a \leq \rho \leq b,\ \alpha \leq \theta \leq \beta,\ \gamma \leq \phi \leq \psi\}$. We will call such a region a *spherical box*. Then we partition Q by means of spheres ($\rho = $ constant), half-cones ($\phi = $ constant), and planes ($\theta = $ constant). A typical cell into which this partition subdivides Q is shown in Figure 15.52. If these cells are numbered from 1 to n, we denote their volumes by $\Delta V_1, \Delta V_2, \ldots, \Delta V_n$. In the ith such cell we choose an arbitrary point $(\rho_i^*, \theta_i^*, \phi_i^*)$. Then we define the triple integral of f over Q as follows.

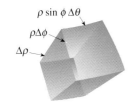

FIGURE 15.52

Definition 15.5
The Triple Integral in Spherical Coordinates

$$\iiint_Q f(\rho, \theta, \phi)\, dV = \lim_{\|P\| \to 0} \sum_{i=1}^{n} f(\rho_i^*, \theta_i^*, \phi_i^*)\, \Delta V_i \qquad (15.44)$$

provided the limit on the right exists independently of the choices of the points $(\rho_i^*, \theta_i^*, \phi_i^*)$.

To help you understand the appropriate form of dV when going to an iterated integral, let us consider the enlarged cell shown in Figure 15.52(b). Its volume ΔV is approximately the same as that of a rectangular box of dimensions $\Delta \rho$ by $\rho \Delta \phi$ by $\rho \sin \phi\, \Delta \theta$. So

$$\Delta V \approx \rho^2 \sin \phi\, \Delta \rho\, \Delta \phi\, \Delta \theta$$

The smaller the norm of the partition—that is, the smaller $\Delta \rho$, $\Delta \theta$, and $\Delta \phi$ are—the better the approximation. It can be proved that when f is continuous over the spherical box Q, the triple integral can be evaluated as the iterated integral in the following equation.

$$\iiint_Q f(\rho, \theta, \phi)dV = \int_\alpha^\beta \int_\gamma^\psi \int_a^b f(r, \theta, \phi)\rho^2 \sin\phi \, d\rho \, d\phi \, d\theta \quad (15.45)$$

We write

$$dV = \rho^2 \sin\phi \, d\rho \, d\phi \, d\theta$$

and call this form the *differential of volume in spherical coordinates*. The factor $\rho^2 \sin\phi$ plays the analogous role to the factor r in the differential of volume in cylindrical coordinates.

If the region Q is not a spherical box but can be enclosed in such a box, we partition a spherical box that encloses Q and count only those cells that lie inside Q, as in previous cases. Then Definition 15.5 continues to hold true, with this restriction on the cells that are counted. The evaluation of the integral of a continuous function f over Q by an iterated integral in this case can be indicated by the following equation.

$$\iiint_Q f(\rho, \theta, \phi)dV = \iiint_{\substack{\text{(appropriate}\\\text{limits)}}} f(r, \theta, \phi)\rho^2 \sin\phi \, d\rho \, d\phi \, d\theta \quad (15.46)$$

The limits of integration depend on the nature of the bounding surfaces. We will illustrate some common types in the examples.

EXAMPLE 15.31 Find the volume and the centroid of the region shaped like an ice cream cone (see Figure 15.53) bounded by the cone $z = \sqrt{3(x^2 + y^2)}$ and the hemisphere $z = \sqrt{4 - x^2 - y^2}$.

Solution For the cone, the spherical equation is

$$\rho \cos\phi = \sqrt{3}\, \rho \sin\phi \quad \text{or} \quad \tan\phi = \frac{1}{\sqrt{3}}$$

So $\phi = \pi/6$. The sphere has the equation $\rho = 2$. Thus,

$$V = \iiint_Q dV = \int_0^{2\pi} \int_0^{\pi/6} \int_0^2 \rho^2 \sin\phi \, d\rho \, d\phi \, d\theta$$

$$= \int_0^{2\pi} \int_0^{\pi/6} \frac{8}{3} \sin\phi \, d\phi \, d\theta = \frac{8}{3} \int_0^{2\pi} \Big[-\cos\phi\Big]_0^{\pi/6} d\theta$$

$$= \frac{8}{3}\left[-\frac{\sqrt{3}}{2} + 1\right] \cdot 2\pi = \frac{8\pi}{3}(2 - \sqrt{3})$$

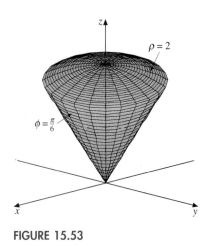

FIGURE 15.53

Section 15.7 Triple Integrals in Cylindrical and Spherical Coordinates 1125

By symmetry, $\bar{x} = \bar{y} = 0$, so we need only calculate $\bar{z}$. First, we calculate M_{xy}:

$$M_{xy} = \iiint_Q z \, dV = \iiint_Q (\rho \cos \phi) \, dV$$

$$= \int_0^{2\pi} \int_0^{\pi/6} \int_0^2 \rho^3 \sin \phi \cos \phi \, d\rho \, d\phi \, d\theta$$

$$= \int_0^{2\pi} \int_0^{\pi/6} 4 \sin \phi \cos \phi \, d\phi \, d\theta = 4 \int_0^{2\pi} \left[\frac{\sin^2 \phi}{2} \right]_0^{\pi/6} d\theta = \pi$$

So

$$\bar{z} = \frac{M_{xy}}{V} = \frac{\pi}{\frac{8\pi}{3}(2 - \sqrt{3})} = \frac{3}{8(2-\sqrt{3})} \cdot \frac{2+\sqrt{3}}{2+\sqrt{3}} = \frac{3(2+\sqrt{3})}{8} \quad \blacksquare$$

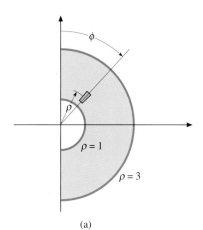

(a)

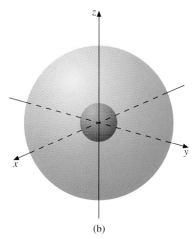

(b)

FIGURE 15.54

EXAMPLE 15.32 Find the mass of the solid between the two spheres $x^2 + y^2 + z^2 = 1$ and $x^2 + y^2 + z^2 = 9$ if the density is inversely proportional to the distance from the origin.

Solution The spheres have equations $\rho = 1$ and $\rho = 3$, and these are the limits of integration on ρ. If we next integrate with respect to ϕ, from $\phi = 0$ to $\phi = \pi$, we will then have integrated over the semiannular region shown in Figure 15.54(a). Then, allowing θ to vary from 0 to 2π in effect rotates this semiannular region around the z-axis to give the entire region shown in Figure 15.54(b). The density at the point (ρ, θ, ϕ) is of the form k/ρ for some constant k, so we have

$$m = \iiint_Q dm = \int_0^{2\pi} \int_0^\pi \int_1^3 \frac{k}{\rho}(\rho^2 \sin \phi \, d\rho \, d\phi \, d\theta)$$

$$= k \int_0^{2\pi} \int_0^\pi \left[\frac{\rho^2}{2} \right]_1^3 \sin \phi \, d\phi \, d\theta = \frac{8k}{2} \int_0^{2\pi} \int_0^\pi \sin \phi \, d\phi \, d\theta = 16k\pi \quad \blacksquare$$

Exercise Set 15.7

In Exercises 1–6, evaluate the iterated integrals.

1. $\displaystyle \int_0^\pi \int_0^1 \int_0^4 rz \sin \theta \, dz \, dr \, d\theta$

2. $\displaystyle \int_0^{\pi/2} \int_1^2 \int_0^{4-r^2} \frac{\cos \theta}{r^2} \, dz \, dr \, d\theta$

3. $\displaystyle \int_0^{2\pi} \int_0^1 \int_0^r zr \, dz \, dr \, d\theta$

4. $\displaystyle \int_0^\pi \int_0^{2 \sin \theta} \int_0^{\sqrt{4-r^2}} r \, dz \, dr \, d\theta$

5. $\displaystyle \int_0^{2\pi} \int_0^{\pi/2} \int_0^1 \rho^2 \sin \phi \, d\rho \, d\phi \, d\theta$

6. $\displaystyle \int_0^{2\pi} \int_{\pi/6}^{\pi/4} \int_0^{1/\cos \phi} \rho^3 \sin \phi \cos \phi \, d\rho \, d\phi \, d\theta$

In Exercises 7–13, use cylindrical coordinates.

7. Find the volume of the region enclosed by the cylinder $x^2 + y^2 = 4$ and the paraboloid $z = 8 - x^2 - y^2$ that lies above the xy-plane.

8. Find the centroid of the region inside the hemisphere $z = \sqrt{4 - x^2 - y^2}$ and outside the cylinder $x^2 + y^2 = 1$.

9. A homogeneous solid of density δ is bounded by $z = \sqrt{x^2 + y^2}$, $x^2 + y^2 = 1$, and $z = 0$. Find I_z.

10. Find the mass of the solid of constant density δ bounded by the surfaces $z = r$, $z = 0$, and $r = 2\cos\theta$.

11. Find the center of mass of the solid in Exercise 10.

12. Find the center of mass of the homogeneous solid of density δ bounded above by $x^2 + y^2 + z^2 = 4$ and below by $3z = x^2 + y^2$.

13. A solid occupies the region

$$Q = \left\{(r, \theta, z) : 1 \leq r \leq 3,\ 0 \leq \theta \leq \frac{\pi}{3},\ 0 \leq z \leq 9 - r^2\right\}$$

Its density at any point is inversely proportional to the distance of the point from the z-axis. Find its center of mass.

In Exercises 14–18, use spherical coordinates.

14. Find the volume of a sphere of radius a.

15. Find the center of mass of a solid that occupies the first-octant region inside a sphere of radius a if the density is proportional to the distance from the z-axis.

16. Find the centroid of the region below the hemisphere $z = \sqrt{4 - x^2 - y^2}$ that lies between the upper nappes of the cones $z^2 = 3(x^2 + y^2)$ and $z^2 = x^2 + y^2$.

17. A solid that occupies the region above the xy-plane and between $z = \sqrt{9 - x^2 - y^2}$ and $z = \sqrt{1 - x^2 - y^2}$ has density proportional to the distance from the origin. Find its mass and its moment of inertia with respect to the z-axis.

18. A solid that occupies the region between the spheres $\rho = 1$ and $\rho = 2$ and inside the cone $\rho = \pi/3$ has density inversely proportional to the distance above the xy-plane. Find its center of mass.

In Exercises 19–23, evaluate the integral by changing to cylindrical or spherical coordinates.

19. $\displaystyle\int_0^2 \int_0^{\sqrt{4-x^2}} \int_0^{\sqrt{x^2+y^2}} \left(z + \sqrt{x^2 + y^2}\right) dz\, dy\, dx$

20. $\displaystyle\int_{-1}^{1} \int_{-\sqrt{1-y^2}}^{\sqrt{1-y^2}} \int_1^{2-x^2-y^2} \frac{1}{z^2}\, dz\, dx\, dy$

21. $\displaystyle\int_0^{\sqrt{2}} \int_0^{\sqrt{2-x^2}} \int_{\sqrt{x^2+y^2}}^{\sqrt{4-x^2-y^2}} \sqrt{x^2 + y^2 + z^2}\, dz\, dy\, dx$

22. $\displaystyle\int_0^1 \int_{-\sqrt{1-z^2}}^{\sqrt{1-z^2}} \int_{-\sqrt{1-y^2-z^2}}^{\sqrt{1-y^2-z^2}} x^2\, dx\, dy\, dz$

23. $\displaystyle\int_0^2 \int_{-\sqrt{2x-x^2}}^{\sqrt{2x-x^2}} \int_0^{\sqrt{4-x^2-y^2}} dz\, dy\, dx$

24. Find the centroid of the region that lies both inside the sphere $x^2 + y^2 + z^2 = 2az$ and outside the sphere $x^2 + y^2 + z^2 = a^2$.

25. A solid spherical ball $x^2 + y^2 + z^2 \leq a^2$ has density equal to the distance from the xy-plane. The ball is cut by planes $z = h_1$ and $z = h_2$, where $0 < h_1 < h_2 < a$. Find the center of mass of the portion of the ball between these two planes. (*Hint:* Use cylindrical coordinates and integrate first with respect to r.)

26. Find the centroid of the region enclosed by the frustum of the cone shown in the accompanying figure.

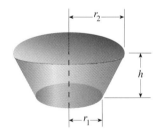

15.8 CHANGING VARIABLES IN MULTIPLE INTEGRALS

We have seen how changing variables from rectangular to polar in double integrals and from rectangular to cylindrical or spherical in triple integrals can often simplify the integration. In this section, we consider the general question of changing variables in double or triple integrals. Changing variables is essentially the same as making a substitution. Just as with substitutions in single integrals, substitutions in multiple integrals require changing not only the integrand, which is usually straightforward, but also the limits of integration and the differential (of area or of volume). These latter two types of changes are generally more difficult, and we concentrate on them.

First, let us review the substitution process in a single integral. Suppose that in the integral $\int_a^b f(x)\,dx$ we make the substitution $x = g(u)$. Then, under suitable restrictions on the functions f and g, we obtain

$$\int_a^b f(x)\,dx = \int_c^d f(g(u))\underbrace{g'(u)\,du}_{dx}$$

where the new limits $u = c$ and $u = d$ satisfy $c = g(a)$ and $d = g(b)$. Notice that the differential dx is replaced by a multiple of du, namely, $g'(u)\,du$.

A similar result holds true when we make the substitution $x = r\cos\theta$, $y = r\sin\theta$ in a double integral. Then, as we have seen, we get

$$\iint_R f(x,y)\,dA = \iint_S f(r\cos\theta, r\sin\theta)\,r\,dr\,d\theta$$

where S denotes the domain of integration described in polar coordinates. Note that the differential of area dA in the variables x and y is replaced by a multiple of $dr\,d\theta$, namely, $r\,dr\,d\theta$. In the case of cylindrical coordinates, the differential of volume dV is replaced again by r times the product $dr\,d\theta\,dz$. In spherical coordinates, the multiple is $\rho^2 \sin\phi$; that is, dV is replaced by $\rho^2 \sin\phi\,d\rho\,d\theta\,d\phi$.

In view of this discussion, it should not be surprising to find that changing from rectangular coordinates to any other set of coordinates causes the differential of area or volume to be replaced by some multiple of the differential of area or volume in the new variables. It turns out that there is a general formula for this multiple, as we now show.

Changing Variables in Double Integrals

Suppose we are given the double integral

$$\iint_R f(x,y)\,dA$$

of the continuous function f over the region R and we want to substitute

$$\begin{cases} x = g(u,v) \\ y = h(u,v) \end{cases} \tag{15.47}$$

Equations 15.47 can be thought of as a function from $\mathbb{R}^2$ to $\mathbb{R}^2$. That is, corresponding to a point (u,v), the equations determine a point (x,y). A function of this type is called a **transformation**, or a **mapping**, from the uv-plane to the xy-plane. Let us denote this transformation by T. Then we can write Equations 15.47 more compactly as

$$T(u, v) = \big(g(u, v), h(u, v)\big)$$

and we can indicate this relationship graphically as in Figure 15.55.

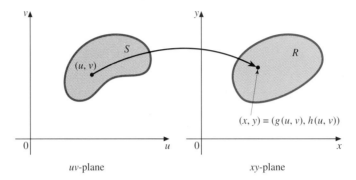

FIGURE 15.55

If (u, v) is a point in the uv-plane, then $T(u, v)$ is called the **image** of (u, v). The **image of a set of points S** in the uv-plane is the set **R** of all images of points in **S**.

EXAMPLE 15.33 Let $T(u, v) = (u^2 - v^2, uv)$. Find the image of the triangle in the uv-plane having vertices $(0, 0)$, $(1, 0)$, and $(1, 1)$.

Solution Along the base of the triangle from $(0, 0)$ to $(1, 0)$, we see that $v = 0$ and u varies from 0 to 1. So its image under T is the set $\{(u^2, 0) : 0 \leq u \leq 1\}$ in the xy-plane. That is, $x = u^2$, $y = 0$ are parametric equations of the image. Since $0 \leq u \leq 1$, it follows also that $0 \leq x \leq 1$. Thus, the image is the line segment from $(0, 0)$ to $(1, 0)$ on the x-axis.

The image of the vertical side of the given triangle from $(1, 0)$ to $(1, 1)$ is the set $\{(1 - v^2, v) : 0 \leq v \leq 1\}$ since u is always 1 on this side. Parametric equations of the image are therefore $x = 1 - v^2$, $y = v$, with $0 \leq v \leq 1$. Eliminating the parameter gives $x = 1 - y^2$, with $0 \leq y \leq 1$. The graph is thus a parabolic arc from $(1, 0)$ to $(0, 1)$.

The image of the third side of the triangle, joining $(0, 0)$ and $(1, 1)$, is the set $\{(0, u^2) : 0 \leq u \leq 1\}$, since on this side $v = u$. The parametric equations $x = 0$, $y = u^2$, with $0 \leq u \leq 1$, define the line segment on the y-axis from $(0, 0)$ to $(0, 1)$.

We show the original triangle in the uv-plane and its image in the xy-plane in Figure 15.56. It is not difficult to show that the entire region S consisting of all points inside and on the triangle is mapped by T onto the region R in the xy-plane. ∎

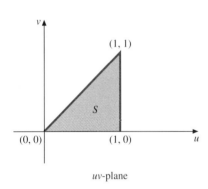

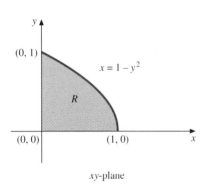

FIGURE 15.56

The transformations that are of most interest to us for our present purposes are one-to-one; that is, they map distinct points in the domain onto distinct points in the range. If T is such a one-to-one transformation given by $T(u, v) = (x, y)$, where $x = g(u, v)$ and $y = h(u, v)$, then there is an inverse transformation T^{-1} such that $T^{-1}(x, y) = (u, v)$. That is, we can solve uniquely for u and v in terms of x and y. We illustrate such a one-to-one mapping and its inverse in the next example.

EXAMPLE 15.34 Let $T(u, v) = (2u - v, u + 3v)$, and show that T is a one-to-one mapping from the entire uv-plane onto the entire xy-plane. Find $T^{-1}(x, y)$.

Solution In order to show that T is one-to-one, we assume that $T(u_1, v_1) = T(u_2, v_2)$ and show as a consequence that $(u_1, v_1) = (u_2, v_2)$. It will follow, then, that no two distinct points in the uv-plane have the same image, or equivalently, if two points (u_1, v_1) and (u_2, v_2) are distinct, their images are distinct.

Assume, then, that $T(u_1, v_1) = T(u_2, v_2)$. Then

$$(2u_1 - v_1, u_1 + 3v_1) = (2u_2 - v_2, u_2 + 3v_2)$$

Thus,

$$\begin{cases} 2u_1 - v_1 = 2u_2 - v_2 \\ u_1 + 3v_1 = u_2 + 3v_2 \end{cases}$$

If we multiply both sides of the top equation by 3 and add the result to the bottom equation, we get $7u_1 = 7u_2$, so $u_1 = u_2$. By substituting u_1 for u_2 in the second equation and simplifying, we get $3v_1 = 3v_2$, so $v_1 = v_2$. Thus, whenever $T(u_1, v_1) = T(u_2, v_2)$, we have $(u_1, v_1) = (u_2, v_2)$, proving that T is one-to-one.

To find T^{-1}, let us set $(x, y) = T(u, v)$. Then

$$x = 2u - v \quad \text{and} \quad y = u + 3v$$

What we want to do is to solve this system of two equations for u and v in terms of x and y. We find that

$$u = \frac{3x + y}{7} \quad \text{and} \quad v = -\frac{x - 2y}{7}$$

or we can write

$$T^{-1}(x, y) = \left(\frac{3x + y}{7}, -\frac{x - 2y}{7} \right)$$

No restriction was placed on either (u, v) or (x, y), so the domain of T is the entire uv-plane, and its range is the entire xy-plane. Similarly, the domain of T^{-1} is the entire xy-plane, and its range is the entire uv-plane. ∎

Another requirement we will place on the transformations we consider is that the component functions have continuous first partial derivatives. That is, if $T(u, v) = (g(u, v), h(u, v))$, we want g_u, g_v, h_u, h_v all to be continuous functions. If we write $x = g(u, v)$ and $y = h(u, v)$, these partials can be denoted by $\partial x/\partial u$, $\partial x/\partial v$, $\partial y/\partial u$, and $\partial y/\partial v$. A transformation having this continuity property is called a C^1 **transformation** (here C denotes "continuous" and the superscript 1 denotes first partial derivatives; a C^2 transformation has continuous second-order partials).

Let $T(u, v) = (g(u, v), h(u, v))$ denote a one-to-one C^1 transformation. We want to see how it transforms a small rectangular element of area, Δu units by Δv units, in the uv-plane. Such an element is of the type that would occur in a partition of a region. As we show in Figure 15.57(a), let (u_0, v_0) denote the lower left-hand corner of the rectangle. On the lower boundary of the rectangle,

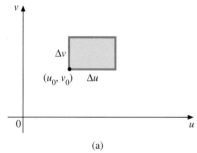

(a)

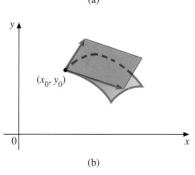

(b)

FIGURE 15.57

v is constant, namely, $v = v_0$. So its image is the set $\{(g(u, v_0), h(u, v_0)) : u_0 \leq u \leq u_0 + \Delta u\}$. Parametric equations of this image are

$$\begin{cases} x = g(u, v_0) \\ y = h(u, v_0) \end{cases} \quad u_0 \leq u \leq u_0 + \Delta u$$

Equivalently, we can write its vector equation as

$$\mathbf{r}_1(u) = g(u, v_0)\mathbf{i} + h(u, v_0)\mathbf{k}, \quad u_0 \leq u \leq u_0 + \Delta u$$

Similarly, the image of the left boundary of the rectangle can be given by the vector equation

$$\mathbf{r}_2(v) = g(u_0, v)\mathbf{i} + h(u_0, v)\mathbf{j}, \quad v_0 \leq v \leq v_0 + \Delta v$$

We show these images, along with those of the other two sides of the rectangle, in Figure 15.57(b), where we have denoted by (x_0, y_0) the image of (u_0, v_0).

The images of the left and right endpoints of the lower boundary of the rectangle are $\mathbf{r}_1(u_0)$ and $\mathbf{r}_1(u_0 + \Delta u)$. If Δu is small, the secant vector $\mathbf{r}_1(u_0 + \Delta u) - \mathbf{r}_1(u_0)$ approximates the curved image of this lower boundary. Since

$$\mathbf{r}'_1(u_0) = \lim_{\Delta u \to 0} \frac{\mathbf{r}_1(u_0 + \Delta u) - \mathbf{r}_1(u_0)}{\Delta u}$$

it follows that

$$\mathbf{r}_1(u_0 + \Delta u) - \mathbf{r}_1(u_0) \approx \mathbf{r}'_1(u_0) \Delta u$$

if Δu is small. Consequently, the lower boundary image is approximated by the vector $\mathbf{r}'_1(u_0)\Delta u$. Similarly, the image of the left side of the rectangle is approximated by the vector $\mathbf{r}'_2(v_0)\Delta v$. We show these two vectors in red, drawn from the point (x_0, y_0) in Figure 15.57(b).

Let ΔA denote the area of the image of the rectangular element. Then ΔA is approximately equal to the area of the parallelogram having the two tangent vectors $\mathbf{r}'_1(u_0)\Delta u$ and $\mathbf{r}'_2(v_0)\Delta v$ as adjacent sides. We learned in Section 11.4 that this area is the magnitude of the cross product of the two vectors:

$$\Delta A \approx |\mathbf{r}'_1(u_0)\Delta u \times \mathbf{r}'_2(v_0)\Delta v| = |\mathbf{r}'_1(u_0) \times \mathbf{r}'_2(v_0)|\Delta u \Delta v$$

(Here, we must treat the vectors as three-dimensional, with third component 0, in order for the cross product to be defined.) If we write $x = g(u, v)$ and $y = h(u, v)$, we can calculate the cross product as follows:

$$\mathbf{r}'_1(u_0) \times \mathbf{r}'_2(v_0) = \begin{vmatrix} \mathbf{i} & \mathbf{j} & \mathbf{k} \\ \frac{\partial x}{\partial u} & \frac{\partial y}{\partial u} & 0 \\ \frac{\partial x}{\partial v} & \frac{\partial y}{\partial v} & 0 \end{vmatrix} = \begin{vmatrix} \frac{\partial x}{\partial u} & \frac{\partial y}{\partial u} \\ \frac{\partial x}{\partial v} & \frac{\partial y}{\partial v} \end{vmatrix} \mathbf{k}$$

where it is understood that the partial derivatives are evaluated at the point (u_0, v_0). Since $|\mathbf{k}| = 1$, we conclude that

$$\Delta A \approx \left| \frac{\partial x}{\partial u} \frac{\partial y}{\partial v} - \frac{\partial x}{\partial v} \frac{\partial y}{\partial u} \right| \Delta u \Delta v$$

The second-order determinant involved here is given a special name and symbol, after the German mathematician Carl Jacobi (1804–1851).

Definition 15.6
The Jacobian of a Transformation

The **Jacobian** of the C^1 transformation $T(u, v) = (x, y)$, where $x = g(u, v)$ and $y = h(u, v)$, is denoted by $J(u, v)$ and has the value

$$J(u, v) = \begin{vmatrix} \dfrac{\partial x}{\partial u} & \dfrac{\partial y}{\partial u} \\ \dfrac{\partial x}{\partial v} & \dfrac{\partial y}{\partial v} \end{vmatrix} = \dfrac{\partial x}{\partial u}\dfrac{\partial y}{\partial v} - \dfrac{\partial x}{\partial v}\dfrac{\partial y}{\partial u} \qquad (15.48)$$

REMARKS

- The Jacobian is also frequently designated by $\partial(x, y)/\partial(u, v)$. Thus,

$$\dfrac{\partial(x, y)}{\partial(u, v)} = J(u, v)$$

- Note also that $J(u, v)$ can be written as

$$J(u, v) = \begin{vmatrix} \dfrac{\partial x}{\partial u} & \dfrac{\partial x}{\partial v} \\ \dfrac{\partial y}{\partial u} & \dfrac{\partial y}{\partial v} \end{vmatrix} \qquad (15.49)$$

since the value of this determinant is the same as the value given by Equation 15.48. Using the alternative notation

$$\dfrac{\partial(x, y)}{\partial(u, v)}$$

suggests writing the Jacobian with partials of x on the first row and partials of y on the second, as above.

EXAMPLE 15.35 Find the Jacobian of the transformation given in Example 15.33.

Solution The transformation is $T(u, v) = (u^2 - v^2, uv)$. That is,

$$x = u^2 - v^2 \quad \text{and} \quad y = uv$$

Thus, by Equation 15.48,

$$J(u, v) = \begin{vmatrix} \dfrac{\partial x}{\partial u} & \dfrac{\partial y}{\partial u} \\ \dfrac{\partial x}{\partial v} & \dfrac{\partial y}{\partial v} \end{vmatrix} = \begin{vmatrix} 2u & v \\ -2v & u \end{vmatrix} = 2u^2 + 2v^2 \qquad \blacksquare$$

We have seen that under the change of variables $(x, y) = T(u, v)$, a small element of area ΔA in the xy-plane is related to the area $\Delta u \Delta v$ of a rectangular element of area in the uv-plane by the approximation

$$\Delta A \approx |J(u, v)| \Delta u \Delta v$$

It should come as no surprise, then, that when we make this change of variables in a double integral, the differential of area dA in the original integral is replaced by $|J(u, v)| du\, dv$ in the new integral.

Our discussion should make the following theorem seem plausible, but it is not a proof. A complete proof can be found in most advanced calculus textbooks. We assume that each of the regions R and S consists of all points inside or on a piecewise-smooth, simple, closed curve.

THEOREM 15.4

Change of Variables in a Double Integral

Let $T(u, v) = (g(u, v), h(u, v))$ be a one-to-one C^1 transformation, with nonzero Jacobian, that maps the region S in the uv-plane onto the region R in the xy-plane. If f is continuous on R, then

$$\iint_R f(x, y) \, dA = \iint_S f(g(u, v), h(u, v)) |J(u, v)| \, du \, dv \quad (15.50)$$

REMARK

■ We are using $du \, dv$ here for the differential of area in the uv coordinates instead of the usual dA. We are reserving dA for the differential of area in the xy coordinates. So in the context of the double integral on the right-hand side of Equation 15.50, the order of writing the differentials du and dv is not significant. The order becomes significant only when we evaluate the integral by means of iterated integrals.

EXAMPLE 15.36 Find the result of changing from rectangular to polar coordinates in the integral $\iint_R f(x, y) \, dA$.

Solution The transformation equations are

$$\begin{cases} x = r \cos \theta \\ y = r \sin \theta \end{cases}$$

or equivalently,

$$T(r, \theta) = (r \cos \theta, r \sin \theta)$$

Then (replacing u and v by r and θ, respectively), we have, by Equation 15.49,

$$J(r, \theta) = \begin{vmatrix} \cos \theta & \sin \theta \\ -r \sin \theta & r \cos \theta \end{vmatrix} = r \cos^2 \theta + r \sin^2 \theta = r$$

By restricting r to be the nonnegative value $r = \sqrt{x^2 + y^2}$ and θ to satisfy $0 \leq \theta \leq 2\pi$, the transformation is one-to-one and $J(r, \theta) \neq 0$ except at the origin, where $r = 0$. (It can be shown that Theorem 15.4 continues to hold true if the Jacobian is zero at finitely many points.) Let S be the region in the (r, θ)-plane having image R in the xy-plane. Then, by Equation 15.50

$$\iint_R f(x, y) \, dA = \iint_S f(r \cos \theta, r \sin \theta) r \, dr \, d\theta \quad ■$$

Example 15.36 agrees with the result arrived at in Section 15.3, which we have used repeatedly.

Sometimes the nature of the region R suggests a suitable transformation, as the next example shows.

EXAMPLE 15.37 Make a suitable change of variables to evaluate the integral

$$\iint_R (3x - y)\,dA$$

where R is the parallelogram with vertices $(1, 2)$, $(3, 4)$, $(4, 3)$, and $(6, 5)$.

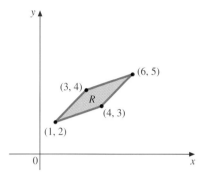

FIGURE 15.58

Solution We show the region R in Figure 15.58. To evaluate the integral without changing variables, we would need to divide the region into three parts in order to have each part either type I or type II. The equations of the sides of the parallelogram are

$$x - y = -1, \quad x - y = 1, \quad x - 3y = -5, \quad \text{and} \quad x - 3y = -9$$

Suppose we make the substitution

$$u = x - y \quad \text{and} \quad v = x - 3y$$

Then, the region S in the uv-plane corresponding to R is the rectangle with sides $u = -1$, $u = 1$, $v = -5$, and $v = -9$, shown in Figure 15.59. Our change of variables has therefore simplified the region of integration.

In substituting for u and v in terms of x and y, we have given the inverse of the transformation T described in Theorem 15.4. To find T, we solve for x and y in terms of u and v:

$$\begin{array}{r} x - y = u \\ x - 3y = v \\ \hline 2y = u - v \quad \text{Subtract.} \end{array}$$

$$y = \frac{u - v}{2}$$

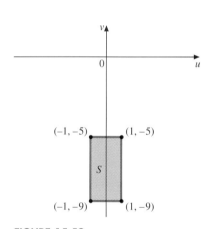

FIGURE 15.59

Thus,

$$x = u + y = u + \frac{u - v}{2} = \frac{3u - v}{2}$$

We can therefore write the transformation T as

$$T(u, v) = \left(\frac{3u - v}{2}, \frac{u - v}{2} \right)$$

The Jacobian of the transformation is

$$J(u, v) = \begin{vmatrix} \frac{3}{2} & \frac{1}{2} \\ -\frac{1}{2} & -\frac{1}{2} \end{vmatrix} = -\frac{3}{4} + \frac{1}{4} = -\frac{2}{4} = -\frac{1}{2}$$

Under the transformation, the original integrand becomes

$$3x - y = 3\left(\frac{3u - v}{2}\right) - \left(\frac{u - v}{2}\right) = \frac{9u - 3v - u + v}{2}$$

$$= \frac{8u - 2v}{2} = 4u - v$$

All of the hypotheses of Theorem 15.4 are readily seen to be satisfied. Thus, by that theorem, we have

$$\iint_R (3x - y) \, dA = \iint_S (4u - v) \left| -\frac{1}{2} \right| du \, dv$$

$$= \frac{1}{2} \int_{-9}^{-5} \int_{-1}^{1} (4u - v) du \, dv$$

$$= \frac{1}{2} \int_{-9}^{-5} \left[2u^2 - uv \right]_{-1}^{1} dv$$

$$= \frac{1}{2} \int_{-9}^{-5} [(2 - v) - (2 + v)] \, dv$$

$$= \frac{1}{2} \int_{-9}^{-5} (-2v) \, dv = -\frac{1}{2} \left[v^2 \right]_{-9}^{-5} = -\frac{25}{2} + \frac{81}{2} = 28 \quad \blacksquare$$

In the next example, the substitution is motivated by the form of the integrand.

EXAMPLE 15.38 Evaluate the integral

$$\iint_R (x + y) e^{x^2 - y^2} dA$$

by substituting $u = x - y$ and $v = x + y$, where R is the region shown in Figure 15.60.

Solution Notice that as it stands, there is no elementary antiderivative for the integrand regardless of the order of integration. By making the specified change of variables, the integrand becomes

$$(x + y)e^{x^2 - y^2} = (x + y)e^{(x-y)(x+y)} = ve^{uv}$$

To find the region S in the uv-plane corresponding to R, we note that since $x^2 - y^2 = uv$, the boundaries $x^2 - y^2 = 1$ and $x^2 - y^2 = -1$ transform to $uv = 1$ and $uv = -1$, respectively. The boundaries $x + y = 1$ and $x + y = 2$ transform to $v = 1$ and $v = 2$, respectively. The region S is therefore as shown in Figure 15.61.

By solving for x and y in terms of u and v, we find that

$$(x, y) = T(u, v) = \left(\frac{u + v}{2}, \frac{v - u}{2} \right)$$

So

$$J(u, v) = \begin{vmatrix} \frac{1}{2} & -\frac{1}{2} \\ \frac{1}{2} & \frac{1}{2} \end{vmatrix} = \frac{1}{2}$$

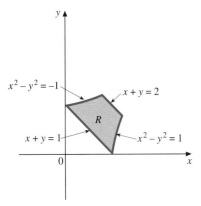

FIGURE 15.60

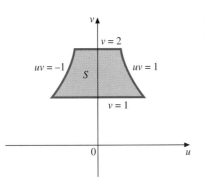

FIGURE 15.61

Consequently, T is a one-to-one C^1 transformation, and $J(u, v) \neq 0$. So by Theorem 15.4,

$$\iint_R (x+y)e^{x^2-y^2} dA = \iint_S ve^{uv} \left|\frac{1}{2}\right| du\,dv$$

$$= \frac{1}{2} \int_1^2 \int_{-1/v}^{1/v} ve^{uv} du\,dv$$

$$= \frac{1}{2} \int_1^2 \left[e^{uv}\right]_{-1/v}^{1/v} dv$$

$$= \frac{1}{2} \int_1^2 (e - e^{-1})\,dv = \frac{e - e^{-1}}{2} = \sinh 1 \quad \blacksquare$$

Changing Variables in Triple Integrals

A result analogous to Theorem 15.4 holds true for triple integrals. The transformation T in this case maps a point (u, v, w) onto a point (x, y, z) by equations of the form

$$x = g(u, v, w), \quad y = h(u, v, w), \quad z = k(u, v, w)$$

We again assume that T is a C^1 transformation—that is, the first partial derivatives of g, h, and k are all continuous on the domain in question—and that T is one-to-one. The Jacobian is the third-order determinant

$$J(u, v, w) = \begin{vmatrix} \dfrac{\partial x}{\partial u} & \dfrac{\partial y}{\partial u} & \dfrac{\partial z}{\partial u} \\ \dfrac{\partial x}{\partial v} & \dfrac{\partial y}{\partial v} & \dfrac{\partial z}{\partial v} \\ \dfrac{\partial x}{\partial w} & \dfrac{\partial y}{\partial w} & \dfrac{\partial z}{\partial w} \end{vmatrix}$$

which can also be denoted by

$$\frac{\partial(x, y, z)}{\partial(u, v, w)}$$

If T has nonzero Jacobian and maps the region G onto the region Q, and f is continuous on Q, then the following equation holds true:

$$\iiint_Q f(x, y, z)\,dV$$

$$= \iiint_G f(g(u, v, w),\ h(u, v, w),\ k(u, v, w))\,|J(u, v, w)|\,du\,dv\,dw$$

(15.51)

We leave it as an exercise (Exercise 11 in Exercise Set 15.8) for you to show that the Jacobian of the transformation $T(\rho, \theta, \phi) = (x, y, z)$, where

$$x = \rho \sin\theta \cos\phi, \quad y = \rho \sin\theta \sin\phi, \quad z = \rho \cos\phi$$

is

$$J(\rho, \theta, \phi) = \rho^2 \sin\phi$$

This transformation effects a change of variables from rectangular to spherical coordinates.

Exercise Set 15.8

In Exercises 1–4, find the image R of the region S under the given transformation.

1. $S = \{(u, v) : 1 \le u \le 2, \ 0 \le v \le 3\}$; $T(u, v) = (2u - 3v, \ 3u + 2v)$

2. $S = \{(u, v) : 0 \le u \le 2, \ 0 \le v \le u\}$; $T(u, v) = (2uv, \ v^2 - u^2)$

3. $S = \{(u, v) : 0 \le u \le 1, \ -u^2 \le v \le u^2\}$; $T(u, v) = (u^2 + v, \ u^2 - v)$

4. $S = \{(u, v) : |u| + |v| \le 1\}$; $T(u, v) = (u - v, u + v)$

In Exercises 5–10, find the Jacobian of the transformation.

5. $x = u + 3v, \ y = 3u - v$

6. $x = u/v, \ y = uv$

7. $x = e^u \cos v, \ y = e^u \sin v$

8. $x = u^2 + v^2, \ y = 2uv$

9. $x = 2u - v + w, \ y = u + v - 2w, \ z = u - 2v + 3w$

10. $x = uv, \ y = vw, \ z = uw$

11. Verify that the Jacobian of the transformation $x = \rho \cos\theta \sin\phi, \ y = \rho \sin\theta \sin\phi, \ z = \rho \cos\phi$ is $\rho^2 \sin\phi$.

In Exercises 12–16, evaluate the given integral by making the indicated change of variables.

12. $\iint_R (x + 2y)\, dA$, where R is the square with vertices $(1, 0), (0, 1), (1, 2),$ and $(2, 1)$; $x = (u + v)/2, \ y = (u - v)/2$

13. $\iint_R (2x - y)\, dA$, where R is the triangle with vertices $(0, 0), (1, 2),$ and $(3, 3)$; $x = u - v, \ y = u - 2v$

14. $\iint_R y e^{xy}\, dA$, where R is the region bounded by the hyperbolas $xy = 1$ and $xy = 3$ and the lines $y = 1$ and $y = 3$; $x = u/v, \ y = v$

15. $\iint_R y \sin(y^2 - x)\, dA$, where R is the region bounded by $y = \sqrt{x}, \ x = 2,$ and $y = 0$; $x = v, \ y = \sqrt{u + v}$

16. $\iint_R xy\, dA$, where R is the region bounded by the ellipse $x^2 + 4y^2 = 4$; $x = 2u, \ y = v$

17. Set up a triple integral for the volume enclosed by the ellipsoid

$$\frac{x^2}{a^2} + \frac{y^2}{b^2} + \frac{z^2}{c^2} = 1$$

and evaluate it by making the change of variables $x = au, \ y = bv, \ z = cw$.

In Exercises 18–22, evaluate the integral by making an appropriate change of variables.

18. $\iint_R \cos(x - 2y)\, dA$, where R is the region bounded by the lines $y = 2x, \ y = 2x - 4, \ 2y - x = 0,$ and $2y - x = 3$

19. $\iint_R \frac{4}{(x - y)^2}\, dA$, where R is the region bounded by the lines $x - y = 2, \ x - y = 4, \ x = 0,$ and $y = 0$

20. $\iint_R (x - 2y)^2\, dA$, where R is the region enclosed by the triangle with vertices $(0, 0), (1, -1),$ and $(2, 1)$

21. $\iint_R \dfrac{1}{x}\, dA$, where R is the region bounded by the curves $xy = 1$, $xy = 4$, $y = 1$, and $y = 2$

22. $\iint_R 6xy\, dA$, where R is the region enclosed by the ellipse $x^2/9 + y^2/4 = 1$

In Exercises 23 and 24, evaluate the iterated integral by making a suitable change of variables.

23. $\displaystyle\int_0^{1/2}\int_y^{1-y} \dfrac{\sin(x-y)}{\cos(x+y)}\, dx\, dy$

24. $\displaystyle\int_1^2 \int_0^{x-1} (x-y)^{-3} e^{(x+y)/(x-y)}\, dy\, dx$

Chapter 15 Review Exercises

In Exercises 1 and 2, evaluate the integrals. Sketch the region of integration.

1. $\displaystyle\int_1^5 \int_x^{2x+1} \dfrac{6x}{(x+y)^2}\, dy\, dx$

2. $\displaystyle\int_0^1 \int_{-\sqrt{y}}^{2-y} xy\sqrt[3]{x^2+4y}\, dx\, dy$

In Exercises 3 and 4, evaluate $\iint_R f(x, y)\, dA$.

3. $f(x, y) = (x+y)/y^2$; $R = \{(x, y) : y^2 \le x \le y + 2,\ 1 \le y \le 2\}$

4. $f(x, y) = xe^y$; R is the region bounded by $y = x^2$ and $x - y + 2 = 0$.

In Exercises 5–9, use double integration.

5. Find the area of the region bounded by $y = \ln x$ and $y = (x-1)/(e-1)$.

6. Find the area outside the circle $r = 1$ and inside the limaçon $r = 3 + 4\cos\theta$.

7. Find the volume of the region under the graph of $f(x, y) = x + y$ that lies above $R = \{(x, y) : 0 \le x \le 2,\ 0 \le y \le \sqrt{4-x^2}\}$.

8. Find the volume of the region under the graph of
$$f(x, y) = \dfrac{xy}{x^2+y^2}$$
above the region inside $r = 3\cos\theta$ that is outside $r = 1 + \cos\theta$ and in the first quadrant.

9. Find the volume enclosed by the tetrahedron with vertices $(0, 0, 0)$, $(0, 2, 0)$, $(0, 2, 4)$, and $(1, 2, 0)$.

10. Give an equivalent integral with the order of integration reversed:
$$\int_0^4 \int_1^{\sqrt{2x+1}} f(x, y)\, dy\, dx$$

11. Evaluate the integral
$$\int_0^1 \int_{x^2}^1 x\cos^2(y^2)\, dy\, dx$$

12. Change to polar coordinates and evaluate:
$$\int_0^9 \int_{y/3}^{\sqrt{10y-y^2}} (x+y)\, dx\, dy$$

13. A lamina occupies the region bounded by $2y = x^2$, $x = 2$, and $y = 0$. Its density is $\rho(x, y) = 5(x+y)$. Find $(\bar{x}, \bar{y})$ and (r_x, r_y).

14. Find the centroid of the leaf of the rose curve $r = 4\sin 2\theta$ that lies in the first quadrant.

15. A lamina occupies the region $R = \{(x, y) : 0 \le x \le \pi,\ 0 \le y \le \sin x\}$ and has density $\rho(x, y) = y$. Find $(\bar{x}, \bar{y})$, I_x, and r_y.

In Exercises 16–21, find the volume described using triple integration in rectangular, cylindrical, or spherical coordinates, whichever seems most appropriate.

16. Inside $z = 4 - \sqrt{x^2+y^2}$ and outside $x^2 + y^2 = 1$, in the first octant

17. Bounded by $z = \sqrt{x}$, $x + z = 2$, $y = 0$, and $y = 4$, in the first octant

18. Inside $x^2 + y^2 = 4$, above $z = 0$, and below $2x + y + z = 8$

19. Bounded by $z = 12 - x^2 - y^2$, $y = x^2$, and $x = 0$, in the first octant

20. Inside $x^2 + y^2 + z^2 = a^2$, between $z^2 = x^2 + y^2$ and $z^2 = 3(x^2 + y^2)$

21. Bounded by $y = x^2 + z^2$, $z = \sqrt{3}x$, $z = 0$, and $y = 4$, in the first octant. (*Hint:* Use cylindrical coordinates having the polar variables in the xz-plane.)

22. A solid occupies the first-octant region bounded by the surfaces $y = x^2$, $y = z$, $y = 1$, $y = 4$, and $z = 0$. If its density is $\rho(x, y) = (x + z)/\sqrt{y}$, find its center of mass.

23. For the region that is common to the two half-cones $z = \sqrt{x^2 + y^2}$ and $z = 3 - 2\sqrt{x^2 + y^2}$, find (a) the centroid and (b) I_z.

24. For the solid described in Example 15.32, find the moment of inertia with respect to the z-axis.

25. Find the moment of inertia of the homogeneous solid inside $z = \sqrt{2 - x^2 - y^2}$, below $z = \sqrt{x^2 + y^2}$, and above $z = 0$, with respect to the line $\mathbf{r}(t) = 2\mathbf{j} + t\mathbf{k}$. (*Hint:* Use the law of cosines.)

In Exercises 26–28, find the area of the surface described.

26. The portion of the paraboloid $z = 4 - x^2 - y^2$ above the plane $z = 4$

27. The portion of the surface $z = 2e^{x/2} \sin(y/2)$ that lies above the region
$$R = \{(x, y) : 0 \leq x \leq \ln 3, \ 0 \leq y \leq e^x\}$$

28. The band on the sphere $x^2 + y^2 + z^2 = a^2$ cut off by the planes $y = b_1$ and $y = b_2$, where $0 < b_1 < b_2 < a$.

29. Let $f(x, y) = (kx)/y$ on the region bounded by $x - 2y = 0$, $y = 1$, $y = 3$, and $x = 0$. Find the following:
    (a) k so that f will be a probability density function;
    (b) $P(x > 2)$;
    (c) μ_x and μ_y;
    (d) σ_x and σ_y.

30. Rewrite each integral in either cylindrical or spherical coordinates, as seems most appropriate, but do not evaluate.

    (a) $\displaystyle\int_0^2 \int_0^{\sqrt{2y - y^2}} \int_0^{2 - \sqrt{x^2 + y^2}} (xy - y^2) \, dz \, dx \, dy$

    (b) $\displaystyle\int_{-1}^1 \int_{-\sqrt{1-x^2}}^{\sqrt{1-x^2}} \int_{\sqrt{x^2+y^2}}^{\sqrt{2 - x^2 - y^2}} \frac{1}{x^2 + y^2 + z^2} \, dz \, dy \, dx$

    (c) $\displaystyle\int_0^2 \int_0^{\sqrt{2x - x^2}} \int_0^{x^2 + y^2} \frac{1}{1 + x^2 + y^2} \, dz \, dy \, dx$

    (d) $\displaystyle\int_{-1}^1 \int_{-\sqrt{1-z^2}}^{\sqrt{1-z^2}} \int_{-\sqrt{5 - x^2 - z^2}}^{\sqrt{5 - x^2 - z^2}} \frac{z^2 - x^2}{z^2 + x^2} \, dy \, dx \, dz$

(*Hint:* Permute the variables in the transformation equations.)

In Exercises 31 and 32, find the Jacobian of the given transformation.

31. (a) $T(u, v) = (u^2 - v, u + 2v^2)$
    (b) $x = ue^v$, $y = ve^u$

32. (a) $T(u, v, w) = (2u - 3v + w, \ u - 2w, \ 4u + v - w)$
    (b) $x = uv + w$, $y = 2vw$, $z = uvw$

In Exercise 33–35, evaluate the double integral by making the given change of variables.

33. $\displaystyle\iint_R (x - 4y) \sin \frac{\pi(x - y)}{2} \, dA$, where R is the region enclosed by the parallelogram with vertices $(-2, -2)$, $(2, -1)$, $(2, 2)$, and $(6, 3)$; $x = 4u - v$, $y = u - v$

34. $\displaystyle\iint_R \frac{(x + y)^2}{\sqrt{xy}} \, dA$, where R is the region between the curves $y = \sqrt{x}$ and $y = x^2$; $x = u^2$, $y = v^2$

35. $\iint_R xy(\ln x)^2 \, dA$, where R is the region bounded by the graphs of $y = 1/x$, $x = 1$, $x = 2$, and $y = 0$; $x = v$, $y = u/v$

In Exercises 36–38, evaluate the integral by making a suitable change of variables.

36. $\iint_R (2x - 3y) \, dA$, where R is the region enclosed by the triangle with vertices $(0, 0)$, $(-1, 1)$, and $(2, 2)$

37. $\iint_R (x - y)e^{x+y} \, dA$, where R is the trapezoid bounded by the lines $y = x + 2$, $y = x + 4$, $x = 0$, and $y = 0$

38. $\int_0^2 \int_0^{2-y} (x^2 - y^2) \sin(x + y)^2 \, dx \, dy$

Chapter 15 Concept Quiz

1. Define each of the following:
   (a) The double integral of a bounded function f over a region R in $\mathbb{R}^2$
   (b) A type I region
   (c) A type II region
   (d) The triple integral of a bounded function f over a region Q in $\mathbb{R}^3$
   (e) The Jacobian of the transformation $x = g(u, v)$, $y = h(u, v)$
   (f) A C^1 transformation

2. In each of the following cases, express the double integral $\iint_R f(x, y)\, dA$ as an iterated integral.
   (a) $R = \{(x, y) : a \leq x \leq b,\ g_1(x) \leq y \leq g_2(x)\}$
   (b) $R = \{(x, y) : h_1(y) \leq x \leq h_2(y),\ c \leq y \leq d\}$

3. (a) Give equations relating rectangular coordinates x and y to polar coordinates r and θ. Also give the formula for dA in polar coordinates.
   (b) Give equations relating rectangular coordinates $x, y,$ and z to cylindrical coordinates $r, \theta,$ and z. Also give the formula for dV in terms of cylindrical coordinates.
   (c) Repeat part (b), replacing cylindrical coordinates with spherical coordinates $\rho, \theta,$ and ϕ.

4. For a lamina of density $\rho(x, y)$ occupying a region R, give formulas in terms of double integrals for each of the following:
   (a) the mass of the lamina
   (b) the coordinates $\bar{x}$ and $\bar{y}$ of the center of mass;
   (c) the moments of inertia I_x, I_y, and I_0.

5. Give a formula for the area of a surface S over a region R in the xy-plane if the equation of S is of the form
   (a) $z = f(x, y)$
   (b) $F(x, y, z) = 0$

6. Suppose the change of variables $x = g(u, v)$, $y = h(u, v)$ is made in the double integral
$$\iint_R f(x, y)\, dA.$$
   Make appropriate assumptions about the transformation, and write the equivalent double integral in terms of the new variables.

APPLYING CALCULUS

1. Epidemiologists often model the spread of a disease in a population by assuming that the probability an infected individual will spread the disease to a healthy individual is a function of the distance between them. Consider a circular city, of radius 10 mi. Assume that the population is uniformly distributed and that the individuals with a certain disease are also uniformly distributed throughout the city, with k such individuals per square mile. Suppose that the probability, $f(P)$, that an infected individual at location $P(x, y)$ will infect a healthy person at location $P_0(x_0, y_0)$ is given by

$$f(P) = 0.05[20 - d(P, P_0)]$$

where $d(P, P_0)$ is the distance between P and P_0. Define the total exposure, $E(P_0)$, of a healthy person at P_0 to be the sum of all probabilities of catching the disease from all infected persons.

   (a) Write a double integral for $E(P_0)$.

   (b) Evaluate the integral in part (a) for P_0 at the center of the city.

   (c) Evaluate $E(P_0)$ for P_0 on the edge of the city.

   (d) For minimum exposure, is it better to live at the center or on the edge?

2. Assume that the population density (number of people per unit of area) for a certain city is given by

$$N(r) = N_0 e^{-kr}$$

where N_0 is the population density at the center of the city, r is the radial distance from the center, and k is a constant. Let D be a circular disk of radius R, centered at the city center.

   (a) Find an expression for the total number of people living in the region.

   (b) Evaluate the quotient

$$\frac{\iint_D r N(r) \, dA}{\iint_D N(r) \, dA}$$

   and explain what its value gives in practical terms.

   (c) In parts (a) and (b), let $R \to \infty$ and explain the meaning of the result in each case.

3. A lamina of constant density $\rho = k$ occupies the circular region described in polar coordinates by $R = \{(r, \theta) : 0 \leq r \leq a, 0 \leq \theta \leq 2\pi\}$. A particle of mass m_1 is located on the z-axis at the point $(0, 0, h)$, where $h > 0$.

   (a) Show that the magnitude of the force exerted by the laminar mass on the point mass m_1 is given by

$$\iint_R \frac{G m_1 h k}{(r^2 + h^2)^{3/2}} \, dA$$

   and evaluate the integral. (*Hint:* Use Newton's Law of Universal Gravitation and the fact that the relevant component of the force is the z component.)

   (b) Suppose the lamina occupies the entire xy-plane. By letting $a \to \infty$ in part (a) show that the total force is independent of the height h.

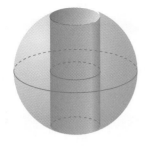

4. A cylindrical hole of radius b is drilled through the center of a solid sphere of radius a, where $0 < b < a$.
   (a) Determine the volume of the material that is removed. (*Hint:* Set up a double integral and use polar coordinates.)
   (b) What is the volume of the "ring" that remains after the hole has been drilled?
   (c) Assume that the sphere has uniform density k. Find the moment of inertia of the "ring" about its axis of symmetry.

5. Schrödinger's Equation in two dimensions has the form

$$-\frac{h^2}{8\pi^2 m}\left(\frac{\partial^2 \psi(x, y)}{\partial x^2} + \frac{\partial^2 \psi(x, y)}{\partial y^2}\right) + V(x, y)\psi(x, y) = E\psi(x, y)$$

where h, m, ψ, V, and E have the meanings noted in Exercise 6 of the Applying Calculus Exercises for Chapter 13. Consider an atomic particle in the two-dimensional "box" $B = \{(x, y) : 0 \leq x \leq L, 0 \leq y \leq L\}$, and suppose that the potential energy V is zero inside the box. Since ψ^2 is the probability density function for the position of the particle, and the particle is confined to the box, we must have

$$\iint_B \psi^2(x, y)\, dA = 1$$

and outside the box $\psi^2(x, y) = 0$, reflecting the fact that the probability of the particle's being outside the box is zero. It follows also that $\psi(x, y) = 0$ for points outside the box.

Assume that $\psi(x, y)$ can be written in the form $f(x)g(y)$ where

$$f(x) = A\sin k_x x + B\cos k_x x$$

and

$$g(y) = C\sin k_y y + D\cos k_y y$$

Impose the continuity requirement on ψ on the boundary of the box to conclude that $B = 0$ and $D = 0$. Also show that k_x and k_y must be integral multiples of π/L. That is, $k_x = m_x \pi/L$ and $k_y = m_y \pi/L$, for integers m_x and m_y. Then use the condition

$$\iint_B \psi^2(x, y)\, dA = 1$$

to find the constant $C_1 = AC$ for which

$$\psi(x, y) = C_1 \sin\frac{m_x \pi x}{L}\sin\frac{m_y \pi y}{L}$$

6. A spherical planet of radius R has an atmosphere in which the density decreases exponentially with the altitude above the surface. That is, the density δ at height h above the surface is $\delta = \delta_0 e^{-kh}$, where δ_0 is the density at the surface and k is the constant of proportionality.
   (a) Determine the total mass of the planet's atmosphere.
   (b) For the planet Earth, half of the atmosphere lies below 5,500 m. Determine the mass of the Earth's atmosphere if the density is 1.20 kg/m³ at sea level. Use 6,370 km for the radius of the Earth.
   (c) Determine the average density of the Earth's atmosphere for that part that lies between 0 and 5,500 m.

A PERSONAL VIEW

Alfred L. Goldberg

Do you use calculus in your medical research?

We use the concepts that we learned in calculus in our analyses of physiological and biochemical processes all the time. Even though in our type of research we seldom use precise mathematical formulations, the concepts of the differential and integral calculus, of differential equations, and of multivariate calculus are nevertheless implicit in our thinking about biochemical and physiological phenomena. For example, if one analyzes a biochemical process, which in living cells is catalyzed by enzymes, the rates of product formation will depend on the concentrations of each of the precursor molecules and on the concentrations and properties of the enzyme. Therefore, we must analyze these processes experimentally as if we were solving multivariate equations. In our laboratory research, we handle logarithmic functions all the time in our analysis of bacterial growth. Bacteria grow exponentially at rates that depend on nutrient supply, temperature, and their genetic makeup, and the analysis of exponential curves can be very informative. Similarly, we use radioactive tracers in analyzing metabolism of cells and tissues, and therefore have to take into account the exponential decay of the radioisotopes when we design such experiments.

In my own research, we are very interested in how cells adapt to new conditions or to hormonal signals by increasing the levels of a specific enzyme. These responses follow a classic differential equation commonly seen in elementary calculus and commonly used in physics or engineering to describe the charging and discharging of an electrical capacitor. The levels of an enzyme depend on the rate at which the cell synthesizes it, which is a linear function of time, and the rate at which the cells destroy that enzyme, which turns out to be an exponential process proportional to the enzyme's concentration. Therefore, the rate of biochemical adaptation depends on an exponential rate constant, and one can predict how quickly the system adapts by knowing the intracellular half-life of the protein.

Is multivariable calculus of use in your field?

We very often encounter multivariable problems, and the researcher has to define the critical variable; that is, to identify the rate-limiting parameter in a process or on an experimental system. Our approach in the laboratory amounts to the experimental solution of a multivariate differential equation in which we control independently each parameter and define how the outcome depends on different variables. We are using calculus subconsciously.

Some areas of medical research are much more quantitative and really involve explicit solutions of multivariate differential equations. For example, in the development of new drugs, pharmacologists have to do research called *pharmacokinetics*, in order to understand the dynamic properties and the stability of a drug in the body. If you were trying to develop a new medication for the treatment of arthritis (disease of the joints), you would want to know how rapidly after ingestion the pill goes into solution in bodily fluids, how rapidly the resulting soluble drug is transported across the intestines into the blood stream, how rapidly it reaches the painful joints, how rapidly it acts once there, and how rapidly the body destroys the drug in the liver or secretes it through the kidneys. Each of these processes is described by rate equations. Moreover, each of these functions may differ with sex, age, or disease, and can determine the important result: how long the drug will be maintained in the blood in an active form. It is then the task of the chemist to synthesize different variant forms of the drug that optimize its pharmacokinetic properties.

These examples from our types of research involve only limited use of the calculus. However, there are very important areas of biochemistry and molecular biology that depend totally on advanced calculus and computer technology for rapid solutions of complex equations. Specifically, the recent revolutionary advances in our knowledge about DNA, enzyme, and protein function have come about from the elucidation of their molecular architectures by X ray diffraction analysis, in which crystals of these large molecules are bombarded with X rays. From the pattern of scattering of the X rays, it is possible through a complex Fourier analysis to determine how the protein or DNA is folded and where individual atoms are located. By this Fourier analysis, very exciting and enlightening images can be obtained of the structure of a virus or of how big protein molecules fold in space, interact with drugs, or are altered in disease states. For example, present efforts to develop drugs to combat AIDS rely on such a biophysical analysis of the critical enzymes of the virus. On this basis, small molecular drugs are being synthesized, and their interactions with targets in the virus are being analyzed by similar biochemical and mathematical approaches.

A somewhat similar mathematical analysis, which also involves Fourier transforms, is extremely useful in studies of large cellular structure. From tomographic analysis of electron-microscopic images, it is possible to obtain precise structural information and to construct three-dimensional images. At the very gross anatomical level, it is now routine in medical diagnosis to use similar mathematical approaches to construct images from nuclear magnetic resonance spectroscopy of the human body. In this case, temporary changes in the physical state of nuclei of the atoms within the body are induced by short exposures to magnetic fields. This perturbation leads to quantitative data that, if analyzed from all angles surrounding the body, can be converted with computers to useful images that indicate whether the structure is normal or affected by disease.

Do you always realize when you are using the concepts of calculus?

No, in fact, when you asked me initially, "Do I use calculus in my work?" my reaction was to deny it, because I always associated calculus with college courses, painful homework, and challenging exams. In fact, although I have forgotten most of the details, those couple of years of calculus have subconsciously permeated my thinking, so that nowadays I practice calculus without a license. In other words, learning how to describe variables precisely, to understand complex functions, focus on one parameter at a time (as occurs in multivariate calculus) was really an exercise in clear, rigorous thinking, from which I have greatly benefited and subconsciously use all the time in research. Also, I have often had the impression that some of my colleagues have contributed so much to our knowledge of biology because of their extensive studies in physics or mathematics, which trained them to think in rigorous, quantitative terms.

Calculus is a requirement for premedical students. Why?

Our medical school requires all entering students to have had a year of calculus and college-level physics, because many of the most common physiological phenomena that doctors have to think about involve the principles of rates or forces, integration over time, electrical events, and so on. For example, the performance of the heart is analyzed in terms of cardiac output, which is the amount of blood pumped per minute, and depends on the heart rate, the force generation during each contraction, the time the valves of the heart are open, and the back pressure and capacitance of the arteries. All these variables are taken into account by the cardiologist in analyzing the nature of cardiac disease and choosing the appropriate therapy. Thus, the cardiologist, like us researchers, is thinking in the rigorous terms of the calculus, though not explicitly formulating or solving mathematical equations. However, the machinery being used in diagnosis, or even in our medical student physiological laboratory exercises, automatically integrates or differentiates the measured parameters to convert them into medically important variables.

Do you use concepts of gradients and flows, which are discussed in multivariate calculus courses?

Absolutely. The concepts of gradients and flows permeates all our analyses of living systems. In evolution, when biological systems could not function adequately simply through diffusion, then active-transport systems or bulk-flow systems evolved, such as the cardiovascular and respiratory systems. Many of these different physiological systems have been analyzed rigorously by multivariate equations. A specific area of physiology and biophysics, where concepts of fields and potentials dominate our thinking, is in understanding the electrical events underlying communication in the nervous system. Signals along nerves and muscles travel as action potentials, which are electrical events whose generation and termination depend on the electrical potential, the ionic fluxes across the cell membrane, and the permeability of the cell membrane to charged ions, such as sodium, potassium, or chloride. Furthermore, these possibilities in turn are nonlinear functions of the electrical potential.

The analysis of these events by Hodgkin and Huxley in the 1950s was a classic achievement that led to our modern understanding of the mechanisms of nerve conduction. At present, many scientists are studying the integration of these ionic events at the synapses between neurons in order to understand how drugs or disease processes alter the actual potentials across the membrane, the molecular channels for different ions, and the resulting fluxes of ions across membranes. We now understand various disease processes in man and animals that lead to changes in ionic flux across membranes and influence whether signals are carried along neurons and muscles. None of these phenomena could be analyzed without knowledge of the mathematics of fields and potentials.

What advice do you have for people in biological studies?

I continue personally to find a life in scientific research exciting and fulfilling, as do my colleagues. My own ability to think rigorously about biological phenomena and to analyze data, I am sure, would be greater were I more adept in use of sophisticated mathematical tools, such as calculus and statistics. Also, I would hope that young people who are excellent in mathematics, physics, and engineering would greatly become more knowledgeable about biology and medicine, where their training and quantitative strengths would enable them to make valuable practical contributions.

Alfred L. Goldberg is professor of cell biology at the Harvard Medical School. He teaches human physiology to medical students and cell biology to graduate students, and trains postdoctoral fellows. A native of Providence, Rhode Island, Prof. Goldberg graduated from Harvard College, was a Churchill Scholar at the University of Cambridge, attended Harvard Medical School, and received his Ph.D. in physiology.

CHAPTER 16

VECTOR FIELD THEORY

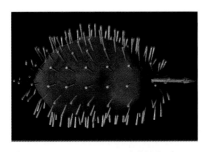

The vector gravitational field on Mars's moon Phobos is shown as blue spikes. The direction to Mars itself is shown with a long, blue arrow to the right.

We introduce the concept of a vector field in this chapter, and we study the calculus associated with vector fields. We will give a formal definition in the first section, but we can indicate the idea of a vector field by considering air currents in some region of space, say in a football stadium. At each point think of a vector giving the magnitude and direction of the air velocity at that point. The totality of such vectors is a vector field that is called a *velocity field*. Or consider the electric force exerted by an electric charge at a fixed point on other charges at various distances from it. At a given location of another charge, there is a force, which can be represented by a vector. The force is attractive if the charges are unlike (one positive and one negative) and repulsive if they are like charges (both positive or both negative). Again, the totality of such vectors is a vector field. It is called an electric *force field*. Another example of a force field is the gravitational field of the Earth.

Vector field theory is of fundamental importance in the physical sciences. For example, it is used in the study of ocean currents, in meteorology, in aerodynamics (such as the flow of air around an airplane or a car), in electricity and magnetism, in heat flow, and in the analysis of effects on objects moving through force fields (such as gravitational forces on a space vehicle).

The two most important tools of calculus used in the study of vector fields are *line integrals* and *surface integrals*. We define both of these types of integrals in this chapter and investigate their properties. We also relate them to integrals we have previously studied. We show applications to vector fields in calculating such things as *work*, *circulation*, and *flux* (all of which we will define). Along the way, we will give some important theorems concerning the various types of integrals we have previously encountered and those we introduce in this chapter.

16.1 VECTOR FIELDS

In Chapter 12 we studied vector-valued functions, which assign either two-dimensional or three-dimensional vectors to real numbers in their domains. Now we consider vector-valued functions whose domains are subsets of $\mathbb{R}^2$ or $\mathbb{R}^3$. We give such functions a special name.

Definition 16.1
A Vector Field

If D is a subset of $\mathbb{R}^2$, a **vector field** on D is a function **F** that assigns a two-dimensional vector $\mathbf{F}(x, y)$ to each point (x, y) in D. If D is a subset of $\mathbb{R}^3$, a vector field **F** on D assigns a three-dimensional vector $\mathbf{F}(x, y, z)$ to each point (x, y, z) in D.

FIGURE 16.1

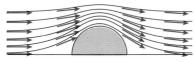

FIGURE 16.2

As one illustration of a vector field, consider the flow of air over a moving car. We have shown a photograph simulating such airflow in Figure 16.1, with several vectors in the vector field of velocities indicated. Each vector represents the direction and magnitude of the wind velocity at the initial point of the vector. We call the vector field a **velocity field** in this case.

In Figure 16.2 we show the velocity field of the flow of water around a submerged object (such as a half-buried pipe). The lines shown in each of these figures are called *streamlines*. A streamline is the path followed by a particle of fluid (air in Figure 16.1, water in Figure 16.2). The velocity vectors are tangent to the streamlines.

If **F** is a two-dimensional vector field, we can express $\mathbf{F}(x, y)$ in terms of its component functions, say f and g, as

$$\mathbf{F}(x, y) = \langle f(x, y), g(x, y) \rangle$$

or, alternatively, as

$$\mathbf{F}(x, y) = f(x, y)\mathbf{i} + g(x, y)\mathbf{j}$$

The component functions f and g are scalar functions; that is, their values are real numbers, rather than vectors.

Similarly, if **F** is three-dimensional, with component functions f, g, and h, we can write

$$\mathbf{F}(x, y, z) = \langle f(x, y, z), g(x, y, z), h(x, y, z) \rangle$$
$$= f(x, y, z)\mathbf{i} + g(x, y, z)\mathbf{j} + h(x, y, z)\mathbf{k}$$

EXAMPLE 16.1 Let $\mathbf{F}(x, y) = \langle y, -x \rangle$. Sketch several members of the vector field defined by **F**.

Solution In Figure 16.3, we show vectors $\mathbf{F}(x, y)$ drawn at selected points (x, y). Notice that this vector field suggests a clockwise rotation, with increasing magnitudes as the distance from the origin increases.

We can show that $\mathbf{F}(x, y)$ is tangent to the circle about the origin that passes through the point (x, y) as follows. Let $\mathbf{r} = \langle x, y \rangle$ be the position vector of this point. Then

$$\mathbf{r} \cdot \mathbf{F}(x, y) = \langle x, y \rangle \cdot \langle y, -x \rangle = xy - xy = 0$$

So $\mathbf{F}(x, y)$ is perpendicular to the position vector $\mathbf{r}$, and since $\mathbf{r}$ is a radius vector of the circle, $\mathbf{F}(x, y)$ is tangent to the circle at (x, y). ∎

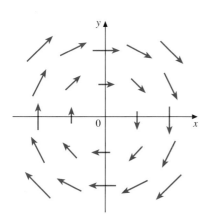

FIGURE 16.3
The vector field $\mathbf{F}(x, y) = \langle y, -x \rangle$

EXAMPLE 16.2 Sketch several vectors in the three-dimensional vector field $\mathbf{F}(x, y, z) = \langle 0, y, 0 \rangle$.

Solution We show several vectors in Figure 16.4. Note that all the vectors are parallel to the y-axis. For points to the right of the xz-plane, the vectors point in the direction of the positive y-axis; for points to the left of the xz-plane, they point in the opposite direction. The farther a point is from the xz-plane, the greater the magnitude of the corresponding vector. ∎

The next example illustrates the gravitational field associated with a body such as the earth.

EXAMPLE 16.3 Newton's Law of Gravitation states that the magnitude of the gravitational force between two bodies is

$$|\mathbf{F}| = \frac{GmM}{|\mathbf{r}|^2}$$

where G is the universal gravitational constant (approximately 6.67×10^{-11} N·m²/kg²), m and M are the respective masses of the two bodies, and $\mathbf{r}$ is the vector from the center of mass of one of the bodies to the other. Suppose the body of mass M has its center of mass at the origin. Find the gravitational force field produced by the mass M on a mass m located at a point (x, y, z) in space.

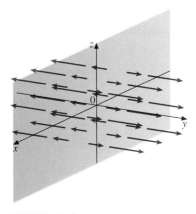

FIGURE 16.4
The vector field $\mathbf{F}(x, y, z) = \langle 0, y, 0 \rangle$

Solution Let $\mathbf{r}$ denote the position vector $\langle x, y, z \rangle$ of mass m. Then the gravitational force of mass M acting on mass m is in the direction from m toward M, that is, in the direction $-\mathbf{r}$. To find the force $\mathbf{F}$, we multiply its magnitude by the unit vector $-\mathbf{r}/|\mathbf{r}|$ in the direction of the force. Thus,

$$\mathbf{F}(\mathbf{r}) = \frac{-GmM}{|\mathbf{r}|^3}\mathbf{r} \tag{16.1}$$

In terms of components, since $\mathbf{r} = \langle x, y, z \rangle$, Equation 16.1 can be written in the form

$$\mathbf{F}(x, y, z) = \frac{-GmM}{\left(x^2 + y^2 + z^2\right)^{3/2}} (x\mathbf{i} + y\mathbf{j} + z\mathbf{k})$$

We show some of the vectors in this gravitational field in Figure 16.5. Notice that all vectors point toward the origin and increase in magnitude as points come closer to the origin. ∎

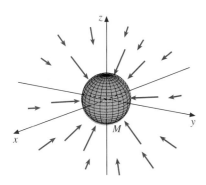

FIGURE 16.5
The gravitational field
$\mathbf{F}(\mathbf{r}) = \dfrac{-GmM}{|\mathbf{r}|^3}\mathbf{r}$

The electric force field between an electric charge Q and other electric charges is similar to the gravitational field in Example 16.3. According to Coulomb's Law, if Q is located at the origin and q is another electric charge, located at the point (x, y, z) whose position vector is $\mathbf{r}$, then the force $\mathbf{F}(\mathbf{r})$ between the two charges is

$$\mathbf{F}(\mathbf{r}) = \frac{EqQ}{|\mathbf{r}|^3}\mathbf{r} \tag{16.2}$$

where E is a constant. When Q and q are like charges, $qQ > 0$, and the force is repulsive, and when Q and q are unlike, $qQ < 0$, and the force is attractive. Equations 16.1 and 16.2 have essentially the same form.

Gradient Fields

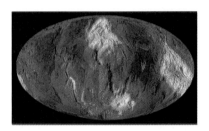

A false-color map of the gravitational potential of the Earth, which is known as the *geoid*.

If f is a scalar function of two variables, then its gradient

$$\nabla f(x, y) = \langle f_x(x, y), f_y(x, y) \rangle$$

is a vector field, called a **gradient field**. The function f is called a **potential function** for the gradient field.

Similarly, if f is a scalar function of three variables, then it is a potential function for the three-dimensional gradient field

$$\nabla f(x, y, z) = \langle f_x(x, y, z), f_y(x, y, z), f_z(x, y, z) \rangle$$

Finding the gradient field for a given potential function is straightforward, as the next example illustrates.

EXAMPLE 16.4 Find the gradient field of the potential function

$$f(x, y) = \tan^{-1} \frac{y}{x}$$

Sketch several vectors in the field.

Solution First, we calculate the partial derivatives of f.

$$\frac{\partial f}{\partial x} = \frac{1}{1 + \frac{y^2}{x^2}} \cdot \left(-\frac{y}{x^2}\right) = \frac{-y}{x^2 + y^2}$$

$$\frac{\partial f}{\partial y} = \frac{1}{1 + \frac{y^2}{x^2}} \cdot \left(\frac{1}{x}\right) = \frac{x}{x^2 + y^2}$$

Thus, the gradient field is

$$\nabla f(x, y) = \left\langle \frac{-y}{x^2 + y^2}, \frac{x}{x^2 + y^2} \right\rangle$$

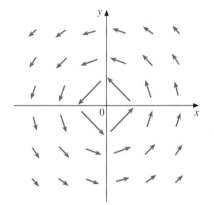

FIGURE 16.6
The gradient field
$\nabla f = \left\langle -\dfrac{y}{x^2 + y^2}, \dfrac{x}{x^2 + y^2} \right\rangle$

We show some of the vectors in Figure 16.6. As in Example 16.1, the vector $\nabla f(x, y)$ is tangent to the circle centered at the origin and passes through the point (x, y). ∎

Gradient fields are special types of vector fields that have many interesting and useful properties, as we will show in Section 16.3. Such fields are said to be **conservative**. Finding the gradient field of a given potential is relatively easy, as we have seen. Determining whether a given vector field is conservative and finding a potential function for it when it is conservative is, in general, more difficult. We will take up this question in Section 16.3 also.

Exercise Set 16.1

In Exercises 1–10, draw several members of the vector field **F**.

1. $\mathbf{F}(x, y) = \langle -y, x \rangle$
2. $\mathbf{F}(x, y) = \langle 2x, y \rangle$
3. $\mathbf{F}(x, y) = \dfrac{x\mathbf{i} + y\mathbf{j}}{\sqrt{x^2 + y^2}}$
4. $\mathbf{F}(x, y) = -x\mathbf{i} - y\mathbf{j}$
5. $\mathbf{F}(x, y) = \dfrac{y\mathbf{i}}{|y|}$
6. $\mathbf{F}(x, y) = \langle x + y, y - x \rangle$
7. $\mathbf{F}(x, y, z) = \langle 1, 1, 0 \rangle$
8. $\mathbf{F}(x, y, z) = \dfrac{x\mathbf{i} + y\mathbf{j} + z\mathbf{k}}{\sqrt{x^2 + y^2 + z^2}}$
9. $\mathbf{F}(x, y, z) = z\mathbf{k}$
10. $\mathbf{F}(x, y, z) = \langle 0, -z, y \rangle$

In Exercises 11–14, find the gradient field of the given potential function, and sketch several of its vectors.

11. $f(x, y) = x^2 - 2y^2$
12. $f(x, y) = \sqrt{x^2 + y^2}$
13. $f(x, y) = \ln\sqrt{x^2 + y^2}$
14. $f(x, y) = \tan^{-1}\dfrac{x}{y}$

In Exercises 15–20, find the gradient field for the given potential function f.

15. $f(x, y) = e^{-x} \sin 2y$
16. $f(x, y) = \ln(\cos xy)$
17. $f(x, y, z) = xy - xz + yz$
18. $f(x, y, z) = \sqrt{x^2 + y^2 + z^2}$
19. $f(x, y, z) = x^2 y^3 z$
20. $f(x, y, z) = \ln(x^2 + y^2 + z^2)$

21. Show that the gravitational field in Example 16.3 is conservative by showing it is the gradient field of the potential function

$$f(x, y, z) = \dfrac{GMm}{\sqrt{x^2 + y^2 + z^2}}$$

22. Use a CAS to draw the vector field **F**. (See page 1158.)
    (a) $\mathbf{F}(x, y) = \dfrac{(y\mathbf{i} - x\mathbf{j})}{\sqrt{x^2 + y^2}}$
    (b) $\mathbf{F}(x, y) = 2x\mathbf{i} + y\mathbf{j}$
    (c) $\mathbf{F}(x, y) = -y\mathbf{i} + x\mathbf{j}$
    (d) $\mathbf{F}(x, y, z) = x\mathbf{i} + y\mathbf{j} + z\mathbf{k}$

16.2 LINE INTEGRALS AND WORK

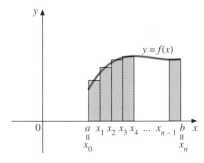

FIGURE 16.7

$$\int_a^b f(x)\,dx = \lim_{\max \Delta x_k \to 0} \sum_{k=1}^n f(c_k)\Delta x_k$$

In this section we introduce a new type of integral, called a **line integral**, that is especially useful in studying properties of vector fields. As one application, we will show how a line integral can be used to find the work done by a force field in moving a particle along a curve.

In Chapter 5, we defined the integral $\int_a^b f(x)\,dx$ of a function of one variable as a limit of Riemann sums formed by partitioning the interval $[a, b]$ on the x-axis. (See Figure 16.7.) For the line integral, we replace the interval $[a, b]$ by a curve C in the plane, and we replace the function $f(x)$ of one variable by a function $f(x, y)$ of two variables.

We begin with a curve C in the xy-plane given by the parametric equations

$$x = x(t), \; y = y(t), \quad a \le t \le b$$

and we assume that C is rectifiable (has finite length). We partition the parameter interval $[a, b]$ by points

$$a = t_0 < t_1 < t_2 < \cdots < t_n = b$$

which induces a partition of the curve C by the points

$$P_0 = (x(t_0), y(t_0)), P_1 = (x(t_1), y(t_1)), \ldots, P_n = (x(t_n), y(t_n))$$

as shown in Figure 16.8(a). Let Δs_k denote the length of the arc of C from P_{k-1} to P_k and let $P_k^* = (x_k^*, y_k^*)$ denote any point on C between P_{k-1} and P_k, for $k = 1, 2, \ldots, n$. In Figure 16.8(b) we show an enlarged view of a typical subarc from P_{k-1} to P_k, along with the point P_k^*. Now we form the sum

$$\sum_{k=1}^{n} f(x_k^*, y_k^*) \Delta s_k$$

where f is a function of two variables whose domain contains the curve C. Note the similarity between this sum and a Riemann sum for the integral $\int_a^b f(x)\,dx$. If this sum approaches a finite limit as we take partitions such that the lengths Δs_k of the subarcs all approach zero, then this limit is the line integral of f along C.

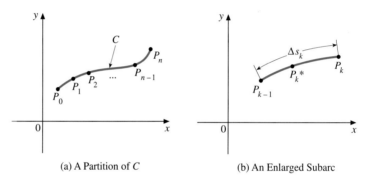

(a) A Partition of C (b) An Enlarged Subarc

FIGURE 16.8

Definition 16.2
The Line Integral

If f is a function of two variables defined on the rectifiable curve C, then the **line integral of f along C** is

$$\int_C f(x, y)\,ds = \lim_{\max \Delta s_k \to 0} \sum_{k=1}^{n} f(x_k^*, y_k^*) \Delta s_k \qquad (16.3)$$

provided this limit exists independently of the choices of the points (x_k^*, y_k^*).

REMARK
■ If f is continuous on C, and C is rectifiable, then it can be shown that the line integral of f along C exists.

Note that if f is the constant function $f(x, y) = 1$, then Equation 16.3 gives the length of C, since the sum on the right-hand side simply adds up the lengths of the subarcs.

Section 16.2 Line Integrals and Work 1151

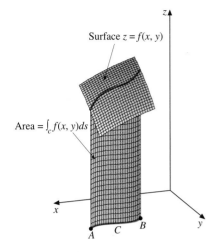

FIGURE 16.9

$$\text{Length of } C = \int_C ds$$

The name *line integral* is unfortunate, since we normally interpret "line" to mean "straight line." *Curvilinear integral* would be a better name. We will use *line integral*, however, since it occurs so widely in the literature.

Geometric Interpretation

When f is a nonnegative continuous function on C, we can give a geometric interpretation to $\int_C f(x, y)\, ds$ as the area of the cylindrical surface obtained by projecting C upward until it meets the surface $z = f(x, y)$. (See Figure 16.9.) To see why, observe that each term $f(x_k^*, y_k^*)\,\Delta s_k$ of the sum on the right-hand side of Equation 16.3 gives the area of a rectangle that approximates the area of a vertical strip with curved base Δs_i. Figure 16.10(a) shows one such strip. In Figure 16.10(b) we show a rectangle whose area approximates the area of the strip. In the limit we get the exact area. Note the similarity with the interpretation of $\int_a^b f(x)\, dx$ as the area above the interval $[a, b]$ and below the graph of f, when f is nonnegative.

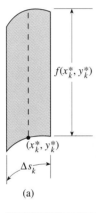

(a)

The Area of a Cylindrical Surface

If f is nonnegative and continuous on the rectifiable curve C, then the area A of the cylindrical surface formed by projecting C upward until it meets the surface $z = f(x, y)$ is

$$A = \int_C f(x, y)\, ds$$

A Computational Formula for the Line Integral

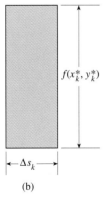

(b)

FIGURE 16.10

THEOREM 16.1

Definition 16.2 does not provide a very useful means of calculating the values of line integrals. The following theorem, which we state without proof, provides a more efficient computational method. Recall that an arc C is *smooth* if it has a parameterization $x = x(t)$, $y = y(t)$, for $a \le t \le b$, where x' and y' are continuous on $[a, b]$ and not simultaneously 0 there. We saw in Chapter 9 that smooth curves are rectifiable.

If f is continuous on the smooth curve C, with parameterization $x = x(t)$, $y = y(t)$ for $a \le t \le b$, then

$$\int_C f(x, y)\, ds = \int_a^b f(x(t), y(t)) \sqrt{\left(\frac{dx}{dt}\right)^2 + \left(\frac{dy}{dt}\right)^2}\, dt \qquad (16.4)$$

The result seems plausible since we saw in Chapter 9 that

$$ds = \sqrt{\left(\frac{dx}{dt}\right)^2 + \left(\frac{dy}{dt}\right)^2}\, dt$$

According to Theorem 16.1 we can evaluate a line integral as an ordinary integral over the parameter interval $[a, b]$ by writing both the integrand and the differential in terms of the parameter.

REMARK

■ Recall that a curve C can have more than one parameterization. It can be shown that the value of the integral is independent of the parameterization used. This fact is important because some choices of parameterization simplify computation of the integral.

EXAMPLE 16.5 Find the area of the portion of the parabolic cylinder $y = x^2/2$ from $x = 0$ to $x = 2$ that lies in the first octant and is bounded above by the cone $2z = \sqrt{x^2 + 4y^2}$.

Solution The surfaces in question are shown in Figure 16.11. The desired area is given by the line integral

$$\text{Area} = \int_C f(x, y)\, ds$$

where C is the arc of the parabola $y = x^2/2$ from $(0, 0)$ to $(2, 2)$ and $f(x, y) = \sqrt{x^2 + 4y^2}/2$. A parameterization of C can be obtained by setting $x = t$:

$$\begin{cases} x = t \\ y = \dfrac{t^2}{2} \end{cases} \quad 0 \le t \le 2$$

With this parameterization, we have

$$ds = \sqrt{\left(\frac{dx}{dt}\right)^2 + \left(\frac{dy}{dt}\right)^2}\, dt = \sqrt{1 + t^2}\, dt$$

By Equation 16.4, we therefore have

$$\text{Area} = \int_C \frac{\sqrt{x^2 + 4y^2}}{2}\, ds = \int_0^2 \frac{\sqrt{t^2 + t^4}}{2} \sqrt{1 + t^2}\, dt$$

$$= \frac{1}{2} \int_0^2 t(1 + t^2)\, dt = 3$$

(You should supply the details of the integration.) ■

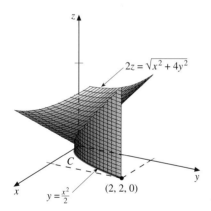

FIGURE 16.11

Another parameterization of the curve C in Example 16.5 is $x = e^t$, $y = e^{2t}/2$, for $-\infty < t < \ln 2$, as you can verify by eliminating the parameter. In the exercises (Exercise 11 in Exercise Set 16.2) we will ask you to show that with this parameterization, the result found in the example is unchanged.

Sectionally Smooth Curves

If C is composed of finitely many smooth curves joined end-to-end, then C is said to be **sectionally smooth**. Suppose, for example, that the smooth curves $C_1, C_2, \ldots, C_n$ are joined end-to-end to form the curve C. Then we write $C = C_1 \cup C_2 \cup C_3 \cup \cdots \cup C_n$ and define the line integral of f over C by

$$\int_C f(x,y)\,ds = \int_{C_1} f(x,y)\,ds + \int_{C_2} f(x,y)\,ds + \cdots + \int_{C_n} f(x,y)\,ds$$

EXAMPLE 16.6 Find

$$\int_C xy^2 \, ds$$

where C is the sectionally smooth closed curve shown in Figure 16.12, consisting of the line segment $y = x$ from $(0, 0)$ to $\left(\sqrt{2}, \sqrt{2}\right)$, the circular arc $x^2 + y^2 = 4$ from $\left(\sqrt{2}, \sqrt{2}\right)$ to $(0, 2)$, and the y-axis from $(0, 2)$ to $(0, 0)$.

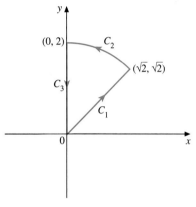

FIGURE 16.12

Solution Denote the three parts of C by $C_1, C_2,$ and C_3, respectively, as shown. We can parameterize these curves as follows:

$$C_1 : \begin{cases} x = t \\ y = t \end{cases} \quad 0 \le t \le \sqrt{2} \qquad C_2 : \begin{cases} x = 2\cos t \\ y = 2\sin t \end{cases} \quad \pi/4 \le t \le \pi/2$$

$$C_3 : \begin{cases} x = 0 \\ y = 2 - t \end{cases} \quad 0 \le t \le 2$$

Then on C_1, $ds = \sqrt{1+1}\,dt = \sqrt{2}\,dt$. So by Equation 16.4,

$$\int_{C_1} xy^2\,ds = \int_0^{\sqrt{2}} t^3 \left(\sqrt{2}\,dt\right) = \sqrt{2}\left[\frac{t^4}{4}\right]_0^{\sqrt{2}} = \sqrt{2}$$

On C_2, $ds = \sqrt{(-2\sin t)^2 + (2\cos t)^2}\,dt = \sqrt{4(\sin^2 t + \cos^2 t)}\,dt = 2\,dt$, so

$$\int_{C_2} xy^2\,ds = \int_{\pi/4}^{\pi/2} (2\cos t)(4\sin^2 t)(2\,dt) = 16 \int_{\pi/4}^{\pi/2} \sin^2 t \cos t\,dt$$

$$= \frac{16}{3} \sin^3 t \Big]_{\pi/4}^{\pi/2}$$

$$= \frac{16}{3} - \frac{4}{3}\sqrt{2}$$

On C_3 the integrand $xy^2 = (0)(2-t)^2 = 0$, so the integral is also 0.
Combining the results, we have

$$\int_C xy^2\,ds = \int_{C_1} xy^2\,ds + \int_{C_2} xy^2\,ds + \int_{C_3} xy^2\,ds$$

$$= \sqrt{2} + \frac{16}{3} - \frac{4\sqrt{2}}{3} = \frac{16}{3} - \frac{\sqrt{2}}{3}$$ ∎

Parameterizing Curves

We have seen that if a curve is given in the form $y = f(x)$ from (x_1, y_1) to (x_2, y_2), then we can take x as the parameter and write

$$\begin{cases} x = t \\ y = f(t) \end{cases} \quad x_1 \leq t \leq x_2$$

(If $x_2 < x_1$, then the t limits are from x_2 to x_1.) Similarly, if $x = g(y)$ from (x_1, y_1) to (x_2, y_2), we can use y as the parameter to get

$$\begin{cases} x = g(t) \\ y = t \end{cases} \quad y_1 \leq t \leq y_2$$

(where, again, if $y_2 < y_1$, the lower t limit is y_2 and the upper limit is y_1).

For a circle $x^2 + y^2 = a^2$, using the polar angle θ (see Figure 16.13) as the parameter gives $x = a \cos \theta$ and $y = a \sin \theta$, or on replacing θ by t,

$$\begin{cases} x = a \cos t \\ y = a \sin t \end{cases}$$

The parameter interval in this case depends on the orientation. Starting from $(a, 0)$, if the orientation is counterclockwise, then t goes from 0 to 2π. If only a part of the circle is to be used, the limits have to be adjusted accordingly.

To parameterize a straight line segment from $P_1(x_1, y_1)$ to $P_2(x_2, y_2)$, recall that a direction vector for the line through P_1 and P_2 is $\langle x_2 - x_1, y_2 - y_1 \rangle$. So a vector equation of the entire line is

$$\mathbf{r}(t) = \langle x_1, y_1 \rangle + \langle x_2 - x_1, y_2 - y_1 \rangle t \tag{16.5}$$

By limiting t to the interval $0 \leq t \leq 1$, we get the line segment from P_1 to P_2. Thus, the parametric equations are

$$\begin{cases} x = x_1 + (x_2 - x_1)t \\ y = y_1 + (y_2 - y_1)t \end{cases} \quad 0 \leq t \leq 1 \tag{16.6}$$

Although Equations 16.6 always work, certain line segments, such as parts of horizontal or vertical lines, can usually be parameterized more easily. For example, we could use $x = t, y = b$ for a horizontal line segment along $y = b$. The t limits are determined by the x-coordinates of the endpoints (taking direction into account) of the line segment. Similarly, $x = a, y = t$ is a parameterization of a vertical line segment along $x = a$, with the t limits determined by the endpoint y values.

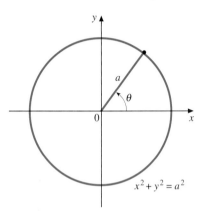

FIGURE 16.13
$x = a \cos \theta, y = a \sin \theta$

Work Done by a Variable Force

An important application of line integrals is in calculating the work done by a variable force $\mathbf{F}$ that acts on a particle as it moves along a curve C. This application extends the definition given in Chapter 11 of the work done by a variable force as the particle moves along a straight line. The force $\mathbf{F}$ varies from point to point and so is a vector-valued function of x and y. That is, $\mathbf{F}$ is a force field, which as we saw in Section 16.1 is a particular kind of vector field.

To motivate the definition, let $\mathbf{F}(x, y) = \langle f(x, y), g(x, y) \rangle$ be a continuous force field whose domain includes the smooth curve C with vector equation $\mathbf{r}(t) = \langle x(t), y(t) \rangle$ for $a \leq t \leq b$. Partition the interval $[a, b]$, thereby obtaining a partition of the curve C, as in Definition 16.2, by the points $P_k = (x_k, y_k)$,

Section 16.2 Line Integrals and Work 1155

$k = 0, 1, 2, \ldots, n$. If the subarc of length Δs_k is sufficiently small, the force can be considered approximately equal to the constant value $\mathbf{F}(x_k^*, y_k^*)$ on that subarc, where (x_k^*, y_k^*) is any point on the subarc. Thus, the work ΔW_k in moving the particle along the kth subarc is approximately equal to the tangential component of $\mathbf{F}(x_k^*, y_k^*)$ multiplied by the length of the subarc, Δs_k (see Figure 16.14). That is,

$$\Delta W_k \approx \text{Comp}_\mathbf{T} \mathbf{F}(x_k^*, y_k^*) \Delta s_k \quad \text{See Section 11.2}$$
$$= \mathbf{F}(x_k^*, y_k^*) \cdot \mathbf{T}(x_k^*, y_k^*) \Delta s_k \quad \text{By Equation 11.9}$$

where $\mathbf{T}(x_k^*, y_k^*)$ is the unit tangent vector to the curve C at (x_k^*, y_k^*). (See Equation 11.7.) The total work W is therefore

$$W = \sum_{k=1}^{n} \Delta W_k \approx \sum_{k=1}^{n} \mathbf{F}(x_k^*, y_k^*) \cdot \mathbf{T}(x_k^*, y_k^*) \Delta s_k$$

By taking the limit as max $\Delta s_k \to 0$, we are led to the following definition.

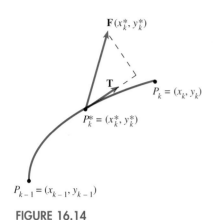

FIGURE 16.14

Definition 16.3
Work Done by a Variable Force Along a Curve

The work W done by a continuous force $\mathbf{F}$ in moving a particle along a smooth curve C in the domain of $\mathbf{F}$ is

$$W = \int_C \mathbf{F} \cdot \mathbf{T} \, ds \tag{16.7}$$

where $\mathbf{T}$ is the unit tangent vector to the curve C.

Before giving an example, we obtain a convenient alternative formulation of the integral in Equation 16.7. Using the vector equation $\mathbf{r}(t) = \langle x(t), y(t) \rangle$ of C, we can write the unit tangent vector $\mathbf{T}(t)$ as

$$\mathbf{T}(t) = \frac{\mathbf{r}'(t)}{|\mathbf{r}'(t)|} \quad \text{By Equation 12.10}$$

So from Equation 16.7 and Equation 16.4,

$$W = \int_a^b \left[\mathbf{F}(\mathbf{r}(t)) \cdot \frac{\mathbf{r}'(t)}{|\mathbf{r}'(t)|} \right] |\mathbf{r}'(t)| \, dt \quad \text{Since } ds = |\mathbf{r}'(t)| \, dt$$

That is,

$$W = \int_a^b \mathbf{F}(\mathbf{r}(t)) \cdot \mathbf{r}'(t) \, dt \tag{16.8}$$

We can also write the unit tangent vector $\mathbf{T}$ as

$$\mathbf{T} = \frac{d\mathbf{r}}{ds} \quad \text{By Equation 12.6}$$

Formally, then

$$\mathbf{F} \cdot \mathbf{T} \, ds = \mathbf{F} \cdot \frac{d\mathbf{r}}{ds} \, ds = \mathbf{F} \cdot d\mathbf{r}$$

So, another way to write Equation 16.7 is

$$W = \int_C \mathbf{F} \cdot d\mathbf{r} \tag{16.9}$$

Combining Equations 16.7 and 16.9, we have the following alternative ways of calculating work.

> The work W done by the continuous force field $\mathbf{F}$ in moving a particle along the smooth curve C given by the vector-valued function $\mathbf{r}(t)$, $a \leq t \leq b$, is
> $$W = \int_C \mathbf{F} \cdot d\mathbf{r} = \int_a^b \mathbf{F}(\mathbf{r}(t)) \cdot \mathbf{r}'(t)\, dt \qquad (16.10)$$

EXAMPLE 16.7 Find the work done by the force field $\mathbf{F}(x, y) = x^2 \mathbf{i} + 2xy \mathbf{j}$ on a particle as it moves along the curve C defined by $\mathbf{r}(t) = (\cos t)\mathbf{i} + (\sin t)\mathbf{j}$, for $0 \leq t \leq \pi$.

Solution Note that C is the upper half of the circle $x^2 + y^2 = 1$, with a counterclockwise orientation. Since $\mathbf{r}'(t) = (-\sin t)\mathbf{i} + (\cos t)\mathbf{j}$, we have, by Equation 16.10,

$$W = \int_C \mathbf{F} \cdot d\mathbf{r} = \int_0^\pi [(\cos^2 t)\mathbf{i} + 2(\cos t \sin t)\mathbf{j}] \cdot [(-\sin t)\mathbf{i} + (\cos t)\mathbf{j}]\, dt$$

$$= \int_0^\pi (-\cos^2 t \sin t + 2\cos^2 t \sin t)\, dt$$

$$= \int_0^\pi \cos^2 t \sin t\, dt = -\left.\frac{\cos^3 t}{3}\right]_0^\pi = \frac{2}{3}$$

The units depend on those for force and distance. If, as is usual, distance is in meters and force is in newtons, then work is in joules. ∎

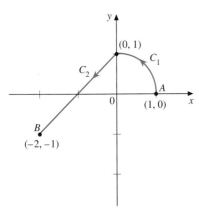

FIGURE 16.15

EXAMPLE 16.8 Find the work done by the force $\mathbf{F}(x, y) = \langle x - y, xy \rangle$ that acts on a particle in moving it from point A to point B along the curve $C = C_1 \cup C_2$ shown in Figure 16.15. Assume force is measured in newtons and distance in meters.

Solution First, we obtain parameterizations for C_1 and C_2. For the circular arc C_1 we have $\mathbf{r}(t) = \langle \cos t, \sin t \rangle$, where $0 \leq t \leq \pi/2$. For C_2 we use the vector equation of a line segment given in Equation 16.5:

$$\mathbf{r}(t) = \langle 0, 1 \rangle + t\langle -2, -2 \rangle = \langle -2t, 1 - 2t \rangle, \qquad 0 \leq t \leq 1$$

The work done is given by

$$W = \int_C \mathbf{F} \cdot d\mathbf{r} = \int_{C_1} \mathbf{F} \cdot d\mathbf{r} + \int_{C_2} \mathbf{F} \cdot d\mathbf{r}$$

We evaluate each integral on the right separately and then add the results:

$$\int_{C_1} \mathbf{F} \cdot d\mathbf{r} = \int_0^{\pi/2} \langle \cos t - \sin t, \cos t \sin t \rangle \cdot \langle -\sin t, \cos t \rangle\, dt$$

$$= \int_0^{\pi/2} (-\sin t \cos t + \sin^2 t + \cos^2 t \sin t)\, dt = \frac{\pi}{4} - \frac{1}{6}$$

(Supply the missing steps.)

$$\int_{C_2} \mathbf{F} \cdot d\mathbf{r} = \int_0^1 \langle -2t - (1-2t), -2t(1-2t) \rangle \cdot \langle -2, -2 \rangle \, dt$$

$$= \int_0^1 (2 + 4t - 8t^2) \, dt = \frac{4}{3}$$

So $W = (\pi/4 - 1/6) + 4/3 = (\pi/4 + 7/6)$ joules. ∎

The integral $\int_C \mathbf{F} \cdot d\mathbf{r}$ can be expressed in yet another way by setting $d\mathbf{r} = \langle dx, dy \rangle$ and evaluating the dot product $\mathbf{F} \cdot d\mathbf{r}$. If $\mathbf{F}(x, y) = \langle f(x, y), g(x, y) \rangle$, then $\mathbf{F} \cdot d\mathbf{r} = f(x, y)dx + g(x, y)dy$. So we can write

$$\int_C \mathbf{F} \cdot d\mathbf{r} = \int_C f(x, y) \, dx + g(x, y) \, dy \tag{16.11}$$

The form on the right is sometimes called the *differential form* of the line integral $\int_C \mathbf{F} \cdot d\mathbf{r}$.

EXAMPLE 16.9 Evaluate the line integral

$$\int_C (2x + y) \, dx + x^2 \, dy$$

where C is the curve

$$\begin{cases} x = t \\ y = 2t^2 \end{cases} \quad 0 \leq t \leq 2$$

Solution The given integral is the differential form of $\int_C \mathbf{F} \cdot d\mathbf{r}$, where $\mathbf{F}(x, y) = \langle 2x + y, x^2 \rangle$ and $\mathbf{r}(t) = \langle t, 2t^2 \rangle$. We can evaluate it by writing everything in terms of the parameter t. We have $dx = dt$ and $dy = 4t \, dt$. So, on substituting for x, y, dx, and dy, we get

$$\int_C (2x + y) \, dx + x^2 \, dy = \int_0^2 \left[\left(2t + 2t^2 \right) dt + t^2 \left(4t \, dt \right) \right]$$

$$= 2 \int_0^2 \left(t + t^2 + 2t^3 \right) dt = \frac{76}{3}$$
∎

REMARK

■ Although the line integral $\int_C f(x, y) \, ds$ is unchanged if C is given a new parameterization, the integral $\int_C \mathbf{F} \cdot \mathbf{T} \, ds$, or equivalently, $\int_C \mathbf{F} \cdot d\mathbf{r}$, changes in sign if the orientation of C is reversed. The sign changes because the unit tangent vector $\mathbf{T}$ reverses direction when the orientation is reversed. So the function $\mathbf{F} \cdot \mathbf{T}$ changes in sign. Thus, if $-C$ denotes the curve C with orientation reversed, we have

$$\int_{-C} \mathbf{F} \cdot d\mathbf{r} = - \int_C \mathbf{F} \cdot d\mathbf{r}$$

Extension to Three Dimensions

The concept of the line integral extends in a natural way to functions of three variables defined on space curves, and we will use such integrals freely from now on. For example, if the smooth curve C has vector representation $\mathbf{r}(t) = \langle x(t), y(t), z(t) \rangle$, and if $\mathbf{F}(x, y, z) = \langle f(x, y, z), g(x, y, z), h(x, y, z) \rangle$ is continuous on C, then

$$\int_C \mathbf{F} \cdot \mathbf{T}\, ds = \int_C \mathbf{F} \cdot d\mathbf{r} = \int_C \mathbf{F}(x(t), y(t), z(t)) \cdot \mathbf{r}'(t)\, dt$$

and in differential form

$$\int_C \mathbf{F} \cdot d\mathbf{r} = \int_C f(x, y, z)\, dx + g(x, y, z)\, dy + h(x, y, z)\, dz$$

VECTOR FIELDS AND WORK USING COMPUTER ALGEBRA SYSTEMS

A vector field is a function $\mathbf{F}$ that maps points (x, y) in the plane (or space) to a vector $\mathbf{F}(x, y)$ in the plane (or space). One way of describing a vector field is to draw a collection of arrows in the plane (or in space), representing the vector starting at the point (x, y). Sketching vector fields can be done easily and quickly using a CAS. The example in this section examines the work done by a force field in moving an object along a curve in the plane.

CAS 60

Find the work done by the force field $\mathbf{F}(x, y) = -2x\mathbf{i} + (y - 1)\mathbf{j}$ in moving an object along one arch of the cycloid $\mathbf{r}(t) = (t - \sin t)\mathbf{i} + (1 - \cos t)\mathbf{j}$, for $0 \leq t \leq 2\pi$. The formula we will need, given in Equation 16.8, is

$$W = \int_0^{2\pi} \mathbf{F}(\mathbf{r}(t)) \cdot \mathbf{r}'(t)\, dt$$

Maple:

First we show a sketch of the cycloid and the vector field $\mathbf{F}$. See Figures 16.2.1 and 16.2.2.

with(plots):with(linalg):
plot([t–sin(t),1–cos(t),t=–4*Pi..4*Pi],scaling=constrained);
fieldplot([–2*x,y–1],x=–4..4,y=–4..4);

Mathematica:

<<Graphics`Graphics`

ParametricPlot[{t–Sin[t],1–Cos[t]},{t,–4*Pi,4*Pi},
 AspectRatio–>Automatic]

PlotVectorField[{–2*x,y–1},{x=–4,4},{y,–4,4}]

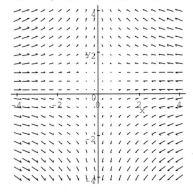

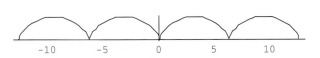

FIGURE 16.2.1

FIGURE 16.2.2

Maple:

Now we prepare to compute the integral.

F:=(x,y)->[-2*x,y-1];
R:=[t-sin(t),1-cos(t)];
F(R[1],R[2]);

Output: $[-2t + 2\sin(t), -\cos(t)]$

diff(R,t);

Output: $[1 - \cos(t), \sin(t)]$

integrand:=dotprod(F(R[1],R[2]),diff(R,t));

Output: $integrand := (-2t+2\sin(t))(1-\cos(t))-\cos(t)\sin(t)$

int(integrand,t=0..2*Pi);

Output: $-4\pi^2$

Mathematica:

F[x_,y_] = {-2*x,y-1}

R = {t-Sin[t],1-Cos[t]}

D[R,t]

integrand = F[t-Sin[t],1-Cos[t]].D[R,t]

Integrate[integrand,{t,0,2*Pi}]

Exercise Set 16.2

In Exercises 1–6, evaluate the integral $\int_C f(x, y)\, ds$, where C has the given parameterization.

1. $f(x, y) = x^2 y$; $x = 2t - 1$, $y = 3t$, $-1 \le t \le 2$

2. $f(x, y) = x - 2y$; $x = \cos 2t$, $y = \sin 2t$, $0 \le t \le \pi/2$

3. $f(x, y) = x/y$; $x = 2t$, $y = t^2 + 1$, $0 \le t \le 2$

4. $f(x, y) = ye^x$; $x = t$, $y = 2t - 3$, $0 \le t \le 1$

5. $f(x, y) = x - y$; $x = 2\cos^2 t$, $y = \sin 2t$, $0 \le t \le \pi/2$

6. $f(x, y) = (x^3 + 9x)/y$; $x = 3t$, $y = t^2 + 1$, $0 \le t \le 2$

In Exercises 7–10, find the area of the vertical cylindrical surface that has xy-trace C and is bounded above by $z = f(x, y)$ and below by the xy-plane.

7. C is the circle $x^2 + y^2 = 1$; $f(x, y) = 4 - x - 2y$.

8. C is the line segment from $(1, 0)$ to $(0, 2)$; $f(x, y) = x^2 + y^2$.

9. C is the curve $y = \sin x$ from $x = 0$ to $x = \pi/2$; $f(x, y) = \sin 2x$.

10. C is the parabolic arc $y = x^2$ from $(0, 0)$ to $(1, 1)$; $f(x, y) = \sqrt{1 + 4x^2}$.

11. Redo Example 16.5 using the parameterization $x = e^t$, $y = e^{2t}/2$, for $-\infty < t < \ln 2$, and show that the answer is unchanged.

12. Redo Exercise 10 using the parameterization $x = \sin t$, $y = \sin^2 t$, for $0 \le t \le \pi/2$, and show that the answer is unchanged.

In Exercises 13–16, find the work done by the force $\mathbf{F}$ acting on a particle moving in the positive direction along C.

13. $\mathbf{F}(x, y) = 16xy^2\mathbf{i} - (y/x^2)\mathbf{j}$; C is given by $x = \cosh t$, $y = \sinh t$, $0 \le t \le \ln 2$.

14. $\mathbf{F}(x, y) = x\mathbf{i} + y\mathbf{j}$; C is the arc of the parabola $y = 2x^2 - 1$ from $(0, -1)$ to $(1, 1)$.

15. $\mathbf{F}(x, y, z) = \left\langle \dfrac{x+y}{z}, xyz, \dfrac{1}{z^2} \right\rangle$; C has the vector equation $\mathbf{r}(t) = \left\langle t, t^2, \dfrac{1}{t} \right\rangle$, $1 \le t \le 3$.

16. $\mathbf{F}(x, y, z) = xy\mathbf{i} + xz\mathbf{j} + yz\mathbf{k}$; C is the arc of the circular helix, $x = \cos t$, $y = \sin t$, $z = t$ for $0 \le t \le \pi/2$.

In Exercises 17–22, evaluate the line integral $\int_C \mathbf{F} \cdot d\mathbf{r}$, where C is the graph of $\mathbf{r}(t)$.

17. $\mathbf{F}(x, y) = \langle xy, x - y \rangle$; $\mathbf{r}(t) = \langle 2t, 1 - t \rangle$, $1 \le t \le 2$

18. $\mathbf{F}(x, y) = \langle x^2 - y^2, 2xy \rangle$; $\mathbf{r}(t) = \langle \cos t, \sin t \rangle$, $0 \le t \le \pi$

19. $\mathbf{F}(x, y) = (x + y)\mathbf{i} + (3x - 2y)\mathbf{j}$; $\mathbf{r}(t) = t^2\mathbf{i} - 2t\mathbf{j}$, $-1 \le t \le 1$

20. $\mathbf{F}(x, y) = xy\mathbf{i} - x^2\mathbf{j}$; $\mathbf{r}(t) = e^t\mathbf{i} + e^{-t}\mathbf{j}$, $0 \le t \le 1$

21. $\mathbf{F}(x, y, z) = \langle x+y-z, x-2z, y+z \rangle$; $\mathbf{r}(t) = \langle t+1, t-1, t \rangle$, $1 \le t \le 2$

22. $\mathbf{F}(x, y, z) = xyz^2\mathbf{i} + (x/y)\mathbf{j} + xz\mathbf{k}$; $\mathbf{r}(t) = t\mathbf{i} + (1/t^2)\mathbf{j} + \sqrt{t}\mathbf{k}$, $1 \le t \le 4$

In Exercises 23–28, evaluate the line integral along the given curve C.

23. $\int_C y\,dx - x\,dy$; C is the curve $y = e^x$, $0 \le x \le \ln 2$.

24. $\int_C 2xy\,dx + x^2\,dy$; C is the arc of the parabola $y = x^2$ from $(-1, 1)$ to $(2, 4)$.

25. $\int_C (x^2 - y^2)\,dx + xy\,dy$; C is the line segment from $(0, 1)$ to $(1, 2)$.

26. $\int_C (2x - 3y)\,dx + (4x + 2y)\,dy$; $C = C_1 \cup C_2$, where C_1 is the line segment from $(0, 0)$ to $(2, 0)$ and C_2 is the line segment from $(2, 0)$ to $(2, 4)$.

27. $\int_C x\,dx + y\,dy$; $C = C_1 \cup C_2 \cup C_3$, where C_1 is the arc of the parabola $y = x^2$ from $(0, 0)$ to $(1, 1)$, C_2 is the line segment from $(1, 1)$ to $(0, 1)$, and C_3 is the line segment from $(0, 1)$ to $(0, 0)$.

28. $\int_C (x^2 + y^2)\,dx + (1 - x^2)\,dy$; $C = C_1 \cup C_2$, where C_1 is the semicircle $y = \sqrt{1 - x^2}$ from $(1, 0)$ to $(-1, 0)$, and C_2 is the line segment from $(-1, 0)$ to $(1, 0)$.

29. For a certain vector field $\mathbf{F}$, defined on all of $\mathbb{R}^2$ except the origin, the vector at each point has length inversely proportional to the distance of that point from the origin, and it is directed toward the origin.
    (a) Draw several typical vectors of the field $\mathbf{F}$.
    (b) Find, without integration, the work done by $\mathbf{F}$ on a particle moving on a circle $x^2 + y^2 = a^2$. Explain your reasoning.
    (c) Find a formula for $\mathbf{F}(x, y)$.

30. Let $\mathbf{F}(x, y) = -y\mathbf{i} + x\mathbf{j}$. Find and show geometrically the vectors in this field at $45°$ intervals around the circle C: $x^2 + y^2 = a^2$ directed in a counterclockwise direction. Show that for this curve C, $\mathbf{F} \cdot \mathbf{T} = |\mathbf{F}| = a$, and use this to evaluate $\int_C \mathbf{F} \cdot \mathbf{T}\,ds$ by inspection.

31. Evaluate $\int_C xyz\,ds$ along the circular helix $x = 2\cos t$, $y = 2\sin t$, $z = t$, for $0 \le t \le \pi$.

32. Evaluate $\int_C (x + z)\,dx + (x - y)\,dy + (2y - z)\,dz$, where C is the line segment from $(2, -1, 3)$ to $(3, 0, 4)$.

33. Evaluate $\int_C (2x + y)\,dx + (x - y)\,dy$ where C is the path from A to B shown in the accompanying figure.

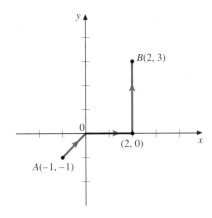

34. Evaluate $\int_C (x^2 + y^2)\,dx + 2xy\,dy$, where C is
    (a) the path shown in the figure for Exercise 33
    (b) the straight line segment from A to B.

35. Evaluate $\int_C x^2 y\,dx + (x - 2y)\,dy$, where C is the closed curve shown in the accompanying figure.

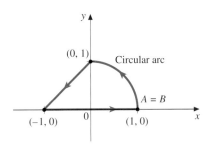

36. Show that

$$\int_C (2x + 3y)\,dx + (3x - 4y)\,dy = 0$$

for the curve C in the figure for Exercise 35.

37. Evaluate

$$\int_C (x+z)\,dx + (y-z)\,dy + (x-y)\,dz$$

where C is the path from A to B shown in the accompanying figure.

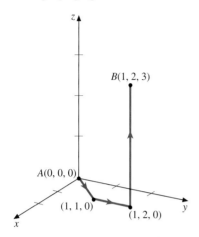

38. Evaluate $\int_C (x^2 - 4y)\,ds$, where C is defined by $x = 2t$, $y = t^2/2 - \ln t$, with $1 \le t \le e$.

39. Evaluate $\int_C \mathbf{F} \cdot d\mathbf{r}$ if C is given by $\mathbf{r}(t) = (e^t \cos t)\mathbf{i} + (e^t \sin t)\mathbf{j}$ for $0 \le t \le \pi$, and

$$\mathbf{F}(x, y) = \frac{-x}{\sqrt{x^2 + y^2}}\mathbf{i} + \frac{y}{\sqrt{x^2 + y^2}}\mathbf{j}$$

40. Evaluate $\int_C (x + y - z)\,ds$, where C is defined by $x = 4\cos^3 t$, $y = 4\sin^3 t$, $z = 3\sin 2t$, with $0 \le t \le \pi/2$.

41. Find the work done by the force field $\mathbf{F}(x, y, z) = \langle xz, xy^2, -yz \rangle$ on a unit mass moving along the elliptical helix $\mathbf{r}(t) = \langle 2\cos t, \sin t, 3t \rangle$, where $0 \le t \le 2\pi$.

Exercises 42 and 43 refer to a thin wire of density $\rho(x, y)$ (mass per unit length) in the shape of a smooth, plane curve C.

42. By partitioning C, explain the plausibility of defining the mass m of the wire by $m = \int_C dm$, where $dm = \rho(x, y)\,ds$.

43. (a) By using ideas from Section 15.4, show that the natural definition of the center of mass $(\bar{x}, \bar{y})$ of the wire is

$$\bar{x} = \frac{\int_C x\,dm}{m} \qquad \bar{y} = \frac{\int_C y\,dm}{m}$$

(b) Similarly, show that the natural definition of the moments of inertia I_x and I_y are

$$I_x = \int_C y^2\,dm \qquad I_y = \int_C x^2\,dm$$

44. A wire is in the shape of the catenary $y = a\cosh(x/a)$, where $-a \le x \le a$, and its density at any point is inversely proportional to the distance from the x-axis to the point. Use the results of Exercises 42 and 43 to find (a) its center of mass and (b) I_x and I_y.

45. Using results analogous to those in Exercises 42 and 43, find the center of mass of a homogeneous wire of density ρ in the shape of the circular helix

$$\mathbf{r}(t) = (2\cos t)\mathbf{i} + (2\sin t)\mathbf{j} + (3t)\mathbf{k}$$

from $t = 0$ to $t = 3\pi/2$. Also find I_x, I_y, and I_z.

46. An *inverse-square force field* is of the form

$$\mathbf{F}(x, y, z) = \frac{k}{|\mathbf{r}|^2}\mathbf{u}$$

where $\mathbf{u} = \mathbf{r}/|\mathbf{r}|$. Find the work done by such a force field in moving a particle of unit mass from $(1, 1, 1)$ to $(2, 1, 3)$.

47. Use Newton's Second Law of Motion, $\mathbf{F} = m\mathbf{a}$, to prove that the work done by a force $\mathbf{F}$ in moving a particle of mass m along a curve C: $\mathbf{r}(t) = \langle x(t), y(t), z(t) \rangle$, from $A = (x(a), y(a), z(a))$ to $B = (x(b), y(b), z(b))$, equals the change in kinetic energy $K(B) - K(A)$, where $K = \frac{1}{2}mv^2$. *Hint:* Show that

$$\int_C \mathbf{F} \cdot d\mathbf{r} = \int_a^b m\mathbf{r}'' \cdot \mathbf{r}'\,dt = \frac{m}{2}\int_a^b \frac{d}{dt}(\mathbf{r}' \cdot \mathbf{r}')\,dt$$

$$= \frac{m}{2}\int_a^b \frac{d}{dt}|\mathbf{r}'|^2\,dt$$

16.3 GRADIENT FIELDS AND PATH INDEPENDENCE

Recall from Section 16.1 that the gradient of a scalar function is a particular type of vector field, called a **gradient field**. For example, if ϕ is a scalar function of two variables, then $\nabla\phi = \langle \phi_x, \phi_y \rangle$ is a gradient field. The function ϕ is called a **potential function** for this gradient field. Note also that for any constant k, $\phi + k$ is another potential function for this field since the gradient is still $\langle \phi_x, \phi_y \rangle$.

If we begin with a scalar function (whose partial derivatives exist), it is easy to find the gradient field $\nabla \phi$. A more difficult problem in general is to determine if a given vector field $\mathbf{F}$ is a gradient field and if so, to find a potential function for it. In certain special cases this determination may not be difficult. For example, if $\mathbf{F}(x, y) = \langle 2x, 2y \rangle$, then it is easy to see that $\mathbf{F}(x, y) = \nabla \phi(x, y)$, where $\phi(x, y) = x^2 + y^2$. Or we could take $\phi(x, y) = x^2 + y^2 + 2$ or $x^2 + y^2 - 3$, or $x^2 + y^2 + k$ for any constant k. In each case, we would have $\nabla \phi(x, y) = \langle 2x, 2y \rangle = \mathbf{F}(x, y)$.

Later in this section, we will show a general procedure for determining whether a vector field is a gradient field and for finding a potential function for it when it is. For now we concentrate on some special properties of gradient fields not shared by other vector fields. The most important property is given in the next theorem, sometimes called the *fundamental theorem for line integrals*. The theorem refers to a function defined on an **open** region in $\mathbb{R}^2$, which means a set consisting only of interior points (no boundary points). That is, for every point (x_0, y_0) in the set, there is some circle centered at (x_0, y_0) enclosing only points of the set. For example, the region enclosed by a rectangle (excluding the rectangle itself) is an open region.

THEOREM 16.2

Fundamental Theorem for Line Integrals

Let $\mathbf{F}$ be a continuous gradient field in an open region R of $\mathbb{R}^2$, and let ϕ be any potential function for $\mathbf{F}$. If $A = (x_1, y_1)$ and $B = (x_2, y_2)$ are any two points in R and C is any piecewise-smooth curve from A to B lying entirely in R, then

$$\int_C \mathbf{F} \cdot d\mathbf{r} = \int_C \nabla \phi \cdot d\mathbf{r} = \phi(x_2, y_2) - \phi(x_1, y_1) \qquad (16.12)$$

Proof We prove the theorem when C is a smooth curve only. The result for piecewise-smooth curves can be obtained by adding the results for each of the smooth component curves of C. (See Exercise 38 of Exercise Set 16.3.)

Let $\mathbf{F}(x, y) = \langle f(x, y), g(x, y) \rangle$. Then, since $\mathbf{F} = \nabla \phi = \langle \phi_x, \phi_y \rangle$, it follows that $f(x, y) = \phi_x(x, y)$ and $g(x, y) = \phi_y(x, y)$. So

$$\int_C \mathbf{F} \cdot d\mathbf{r} = \int_C f(x, y)\,dx + g(x, y)\,dy = \int_C \phi_x(x, y)\,dx + \phi_y(x, y)\,dy$$

If C has the vector equation $\mathbf{r}(t) = \langle x(t), y(t) \rangle$ for $a \leq t \leq b$, such that $\mathbf{r}(a) = (x_1, y_1)$ and $\mathbf{r}(b) = (x_2, y_2)$, we can evaluate the last integral in terms of the parameter t:

$$\int_C \phi_x(x, y)\,dx + \phi_y(x, y)\,dy$$

$$= \int_a^b \left[\phi_x(x(t), y(t))\,x'(t) + \phi_y(x(t), y(t))\,y'(t) \right] dt$$

$$= \int_a^b \left[\frac{\partial \phi}{\partial x} \frac{dx}{dt} + \frac{\partial \phi}{\partial y} \frac{dy}{dt} \right] dt \qquad \text{Using Leibniz notation}$$

By the Chain Rule given in Equation 14.10, the integrand is just $d\phi/dt$. So we have

$$\int_C \mathbf{F} \cdot d\mathbf{r} = \int_a^b \left[\frac{d}{dt} \phi(x(t), y(t)) \right] dt$$

$$= \phi(x(b), y(b)) - \phi(x(a), y(a)) \quad \text{By the Second Fundamental Theorem of Calculus (Theorem 5.3)}$$

$$= \phi(x_2, y_2) - \phi(x_1, y_1) \quad \blacksquare$$

The striking feature of Theorem 16.2 is that the value of the line integral $\int_C \mathbf{F} \cdot d\mathbf{r}$ depends only on the initial and terminal points of C and not on C itself. Thus, *any* piecewise-smooth curve C from (x_1, y_1) to (x_2, y_2) in the domain of $\mathbf{F}$ will give the same value of the line integral. Because of this fact, the line integral $\int_C \mathbf{F} \cdot d\mathbf{r}$ is said to be *independent of the path*. The term *path* is used here to mean a piecewise-smooth curve (which may, in particular, be simply a smooth curve). Our theorem can therefore be restated as follows:

Path Independence

If $\mathbf{F}$ is a continuous gradient field in an open region R, then $\int_C \mathbf{F} \cdot d\mathbf{r}$ is independent of the path C in R.

Because of this path independence, the integral $\int_C \mathbf{F} \cdot d\mathbf{r}$ is sometimes written as $\int_{(x_1, y_1)}^{(x_2, y_2)} \mathbf{F} \cdot d\mathbf{r}$ or even as $\int_A^B \mathbf{F} \cdot d\mathbf{r}$, where $A = (x_1, y_1)$ and $B = (x_2, y_2)$. Using the latter notation, the result of Theorem 16.2 assumes the form

$$\int_A^B \mathbf{F} \cdot d\mathbf{r} = \phi(B) - \phi(A)$$

which looks very much like the result of the Second Fundamental Theorem of Calculus.

Another result is immediately apparent. If C is a *closed* curve, then the initial point $A = (x_1, y_1)$ and the terminal point $B = (x_2, y_2)$ coincide. Thus,

$$\int_C \mathbf{F} \cdot d\mathbf{r} = \phi(x_2, y_2) - \phi(x_1, y_1) = 0$$

So we have the following additional result.

COROLLARY 16.2

If $\mathbf{F}$ is a continuous gradient field in an open region R, and C is any piecewise-smooth *closed* curve in R, then

$$\int_C \mathbf{F} \cdot d\mathbf{r} = 0$$

EXAMPLE 16.10 Evaluate the integral $\int_C 2x\,dx + 2y\,dy$, where C is a piecewise-smooth curve from $A = (1, -2)$ to $B = (3, 6)$.

Solution As we observed earlier, the vector field $\mathbf{F} = \langle 2x, 2y \rangle$ is a gradient field with potential $\phi(x, y) = x^2 + y^2$, since $\nabla\phi = \langle 2x, 2y \rangle = \mathbf{F}$, valid throughout all of $\mathbb{R}^2$. Thus, by Theorem 16.2, the integral

$$\int_C \mathbf{F} \cdot d\mathbf{r} = \int_C 2x\,dx + 2y\,dy$$

is independent of the path from A to B, and

$$\int_C 2x\,dx + 2y\,dy = \phi(3, 6) - \phi(1, -2)$$
$$= (9 + 36) - (1 + 4) = 40 \qquad \blacksquare$$

In Exercise 29 of Exercise Set 16.3, you will be asked to evaluate the integral in Example 16.10 along various paths from A to B by parameterizing the curves to verify that the answer is always the same.

EXAMPLE 16.11 Evaluate the integral $\int_C 2x\,dx + 2y\,dy$, where C is the path shown in Figure 16.16.

Solution The hard way to do this problem is to parameterize each part of C and evaluate the integral on each part. The easy way is to use the fact that $\mathbf{F}(x, y) = \langle 2x, 2y \rangle$ is a gradient field, and so by Corollary 16.2, $\int_C \mathbf{F} \cdot d\mathbf{r} = 0$ for every piecewise-smooth closed path. Thus, $\int_C 2x\,dx + 2y\,dy = 0$. $\blacksquare$

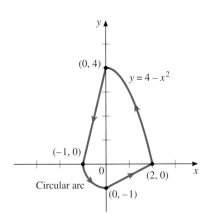

FIGURE 16.16

Recall that a force field that is a gradient field is called *conservative*. This term comes from the law of conservation of energy in physics, stating that the sum of the kinetic energy and potential energy of a particle moving through a conservative force field is constant.

EXAMPLE 16.12 Let $\mathbf{F}$ be the conservative force field $\mathbf{F}(x, y) = \nabla \phi(x, y)$, where $\phi(x, y) = x^2 - 2xy + y^3$. Find the work required to move a particle through this force field from the point $(-1, 1)$ to $(3, -2)$.

Solution By Equation 16.10, the work W is

$$W = \int_C \mathbf{F} \cdot d\mathbf{r}$$

where the curve C joins the two given points. But since $\mathbf{F}$ is conservative, the integral is independent of the path, and its value is, by Theorem 16.2,

$$W = \int_C \nabla\phi \cdot d\mathbf{r} = \phi(3, -2) - \phi(-1, 1)$$
$$= \left[3^2 - 2(3)(-2) + (-2)^3\right] - \left[(-1)^2 - 2(-1)(1) + 1^3\right]$$
$$= 13 - 4 = 9$$

Assuming the magnitude of the force $\mathbf{F}$ is in meters, the work required is 9 joules. $\blacksquare$

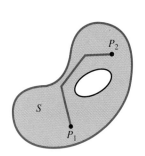

(a) S is connected

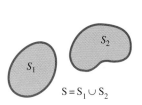

(b) S is disconnected

FIGURE 16.17

The converse of Theorem 16.2 is also true, provided we assume that R is **connected**, meaning that any two points in R can be joined by a sectionally smooth curve lying entirely in R. (See Figure 16.17.)

Section 16.3 Gradient Fields and Path Independence 1165

THEOREM 16.3

> If $\mathbf{F}(x, y) = \langle f(x, y), g(x, y) \rangle$ is a continuous vector field in an open connected region R of $\mathbb{R}^2$ and $\int_C \mathbf{F} \cdot d\mathbf{r}$ is independent of the path C in R, then $\mathbf{F}$ is a gradient field.

Proof To prove the theorem, it is sufficient to construct a function ϕ such that $\mathbf{F} = \nabla \phi$. To do so, we begin with an arbitrary fixed point (a, b) in R. Define $\phi(x, y)$ by

$$\phi(x, y) = \int_C \mathbf{F} \cdot d\mathbf{r} = \int_{(a,b)}^{(x,y)} \mathbf{F} \cdot d\mathbf{r}$$

where C is any piecewise-smooth curve in R from (a, b) to (x, y). Since by hypothesis $\int_C \mathbf{F} \cdot d\mathbf{r}$ is independent of the path, $\phi(x, y)$ depends only on the point (x, y) and not on the curve C connecting (a, b) to (x, y). Consequently, ϕ is a valid scalar-valued function of x and y. We now show that $\phi_x(x, y) = f(x, y)$.

We fix (x, y) temporarily. By Definition 13.7,

$$\phi_x(x, y) = \lim_{h \to 0} \frac{\phi(x + h, y) - \phi(x, y)}{h}$$

Since R is open, $|h|$ can be chosen small enough that the line segment from (x, y) to $(x + h, y)$ lies in R. Let C_1 be any path from (a, b) to (x, y), and let C_2 be the line segment from (x, y) to $(x + h, y)$ (see Figure 16.18). Then we have

$$\phi(x + h, y) - \phi(x, y) = \int_{C_1 \cup C_2} \mathbf{F} \cdot d\mathbf{r} - \int_{C_1} \mathbf{F} \cdot d\mathbf{r} = \int_{C_2} \mathbf{F} \cdot d\mathbf{r}$$

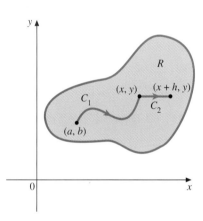

FIGURE 16.18

A parameterization of C_2 is $\mathbf{r}(t) = \langle t, y \rangle$, where $x \leq t \leq x + h$, so that $d\mathbf{r} = \langle dt, 0 \rangle$. We now have

$$\int_{C_2} \mathbf{F} \cdot d\mathbf{r} = \int_x^{x+h} \mathbf{F}(\mathbf{r}(t)) \cdot \mathbf{r}'(t) \, dt = \int_x^{x+h} \langle f(t, y), g(t, y) \rangle \cdot \langle 1, 0 \rangle \, dt$$

$$= \int_x^{x+h} f(t, y) \, dt$$

By the Mean-Value Theorem for Integrals (Theorem 5.1), the last integral can be written as

$$\int_x^{x+h} f(t, y) \, dt = f(c, y) h$$

where c is between x and $x + h$. Finally, then,

$$\phi_x(x, y) = \lim_{h \to 0} \frac{\phi(x + h, y) - \phi(x, y)}{h} = \lim_{h \to 0} \frac{f(c, y) h}{h} = f(x, y)$$

The last equality follows from the continuity of f, since as $h \to 0, c \to x$, so that $f(c, y) \to f(x, y)$.

A similar argument in which C_2 is the vertical segment from (x, y) to $(x, y + k)$ would show that $\phi_y(x, y) = g(x, y)$ (see Exercise 40 in Exercise Set 16.3). It follows that $\mathbf{F} = \nabla \phi$, and so $\mathbf{F} is a gradient field. ∎

REMARK
■ The way the function ϕ was defined in the proof of this theorem provides one means of obtaining a potential function for $\mathbf{F}$, once we know that $\int_C \mathbf{F} \cdot d\mathbf{r}$

is independent of the path in a region R. We define ϕ by

$$\phi(x, y) = \int_{(a,b)}^{(x,y)} \mathbf{F} \cdot d\mathbf{r}$$

and evaluate the integral by selecting a convenient initial point (a, b) and a convenient path from (a, b) to (x, y). Often a path that consists of vertical and horizontal line segments is useful, or the line segment that goes from (a, b) to (x, y) might work well. The complete path, however, must lie in R. We will illustrate this technique after we have developed a test for determining path independence (see Example 16.14).

We have seen that when $\mathbf{F}$ is a continuous gradient field in an open region R, and C is any piecewise-smooth *closed* curve in R, $\int_C \mathbf{F} \cdot d\mathbf{r} = 0$. We now use Theorem 16.3 to show that the converse of this result is also true, providing another way to show when $\mathbf{F}$ is a gradient field.

THEOREM 16.4

If $\mathbf{F}$ is a continuous vector field in an open connected region R, with the property that $\int_C \mathbf{F} \cdot d\mathbf{r} = 0$ for *every* piecewise-smooth closed curve C in R, then $\mathbf{F}$ is a gradient field.

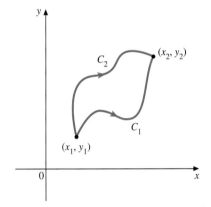

FIGURE 16.19

Proof Let (x_1, y_1) and (x_2, y_2) be any two points in R, and consider any two piecewise-smooth curves C_1 and C_2 in R from (x_1, y_1) to (x_2, y_2). Then $C_1 \cup (-C_2)$ is a closed path in R (see Figure 16.19), and so by hypothesis, $\int_{C_1 \cup (-C_2)} \mathbf{F} \cdot d\mathbf{r} = 0$. But then we have

$$\int_{C_1 \cup (-C_2)} \mathbf{F} \cdot d\mathbf{r} = \int_{C_1} \mathbf{F} \cdot d\mathbf{r} + \int_{-C_2} \mathbf{F} \cdot d\mathbf{r} = \int_{C_1} \mathbf{F} \cdot d\mathbf{r} - \int_{C_2} \mathbf{F} \cdot d\mathbf{r} = 0$$

so that

$$\int_{C_1} \mathbf{F} \cdot d\mathbf{r} = \int_{C_2} \mathbf{F} \cdot d\mathbf{r}$$

This proves that $\int_C \mathbf{F} \cdot d\mathbf{r}$ is independent of the path. Thus, by Theorem 16.3, $\mathbf{F}$ is a gradient field. ∎

REMARK

■ Recall from Section 14.1 that the differential of a differentiable function ϕ of two variables is given by

$$d\phi = \frac{\partial \phi}{\partial x} dx + \frac{\partial \phi}{\partial y} dy$$

When ϕ is a potential function for the continuous gradient field $\mathbf{F} = \langle f, g \rangle$, then $\phi_x = f$ and $\phi_y = g$, so that

$$d\phi = f(x, y) dx + g(x, y) dy$$

For this reason we call $f(x, y) dx + g(x, y) dy$ an *exact differential*, since it is exactly equal to the differential of ϕ. Using this terminology, we can say that *the continuous vector field $\mathbf{F} = \langle f, g \rangle$ in an open connected region R is a gradient field if and only if $f\, dx + g\, dy$ is an exact differential.*

Testing for Gradient Fields and Finding Potential Functions

In the study of gradient fields two crucial questions remain: (1) How can one tell whether a vector field is a gradient field? and (2) When it is a gradient field, how can a potential function for it be found? We discuss these questions now.

A simple test for determining whether **F** is a gradient field is given in the next theorem. First, though, we must distinguish between two types of open connected regions. We say that R is **simply connected** if, for every simple closed curve C in R, all points inside C are also in R. If R is not simply connected, it is said to be **multiply connected**. Figure 16.20 illustrates both types. Intuitively you can think of simple connectedness as meaning "no holes."

THEOREM 16.5 If f and g have continuous first partial derivatives in the simply connected open region R, then $\mathbf{F} = \langle f, g \rangle$ is a gradient field if and only if

$$\frac{\partial f}{\partial y} = \frac{\partial g}{\partial x} \tag{16.13}$$

Partial Proof We will prove only that when **F** is a gradient field, Equation 16.13 is true. The converse is more difficult and will be omitted. Suppose then that **F** is a gradient field with potential function ϕ. Then $\mathbf{F} = \nabla \phi$ and so by definition

$$f(x, y) = \frac{\partial \phi}{\partial x} \quad \text{and} \quad g(x, y) = \frac{\partial \phi}{\partial y}$$

Thus,

$$\frac{\partial f}{\partial y} = \frac{\partial^2 \phi}{\partial y\, \partial x} \quad \text{and} \quad \frac{\partial g}{\partial x} = \frac{\partial^2 \phi}{\partial x\, \partial y}$$

In view of the continuity of f_y and g_x, it follows from Theorem 13.1 that the two second-order mixed partials of ϕ are equal; that is,

$$\frac{\partial^2 \phi}{\partial y\, \partial x} = \frac{\partial^2 \phi}{\partial x\, \partial y}$$

Consequently, we have the result that

$$\frac{\partial f}{\partial y} = \frac{\partial g}{\partial x}$$

This part of the proof did not require simple connectedness. It is only in proving the converse that this property comes into play. ∎

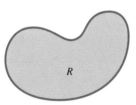

Simply connected
(a)

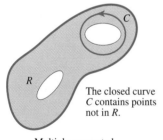

The closed curve C contains points not in R.

Multiply connected
(b)

FIGURE 16.20

Theorem 16.5 provides the key for determining when a vector field is a gradient field, at least in a simply connected region. In the next example we use this test and then illustrate a general technique for finding a potential function.

EXAMPLE 16.13 Show that

$$\mathbf{F}(x, y) = \langle e^y - 2xy, xe^y - x^2 + 2y \rangle$$

is a gradient field throughout $\mathbb{R}^2$, and find a potential function for **F**.

Solution The functions $f(x, y) = e^y - 2xy$ and $g(x, y) = xe^y - x^2 + 2y$ are each continuous, and they have continuous partial derivatives in all of $\mathbb{R}^2$. The set $\mathbb{R}^2$, which is the entire xy-plane, is a simply connected open region. Furthermore,

$$\frac{\partial f}{\partial y} = e^y - 2x \quad \text{and} \quad \frac{\partial g}{\partial x} = e^y - 2x$$

so that the condition in Equation 16.13 is met. Thus, **F** is a gradient field.

We know, then, that a function ϕ exists for which

$$\frac{\partial \phi}{\partial x} = e^y - 2xy \quad \text{and} \quad \frac{\partial \phi}{\partial y} = xe^y - x^2 + 2y$$

To find ϕ, we can proceed in either of two ways: integrate $e^y - 2xy$ with respect to x while holding y fixed, or integrate $xe^y - x^2 + 2y$ with respect to y while holding x fixed. Let us choose the first:

$$\phi(x, y) = \int (e^y - 2xy) \, dx = xe^y - x^2 y + C(y) \qquad \text{y is held fixed.}$$

Note that we have written the general antiderivative with the "constant" of integration as $C(y)$ to allow for the fact that it may be a function of y. Since $\partial C(y)/\partial x = 0$, we have

$$\frac{\partial}{\partial x}\left[xe^y - x^2 y + C(y)\right] = e^y - 2xy$$

as required. To find $C(y)$, we force $\partial \phi/\partial y$ to equal $xe^y - x^2 + 2y$:

$$\frac{\partial \phi}{\partial y} = xe^y - x^2 + C'(y) = xe^y - x^2 + 2y$$

$$C'(y) = 2y$$

$$C(y) = y^2 + C_1$$

where C_1 is an arbitrary (numerical) constant. Since we are seeking any potential function for **F**, we might as well let $C_1 = 0$. Then

$$\phi(x, y) = xe^y - x^2 y + y^2$$

You can verify that $\mathbf{F} = \nabla \phi$. ∎

As we indicated in this example, once we have determined that the function $\mathbf{F} = \langle f, g \rangle$ is a gradient field, we know it has a potential function ϕ for which $\phi_x = f$ and $\phi_y = g$. So we can obtain $\phi(x, y)$ by integrating $f(x, y)$ with respect to x or $g(x, y)$ with respect to y. In the first case, the integration constant is a function $C(y)$ of y whose value can be determined by forcing $\phi_y = g$, as we did in the example. If the second approach is used, the integration constant is of the form $C(x)$, and its value is determined by forcing $\phi_x = f$. Although either way will work, sometimes one of the integrations may be simpler than the other, and so you should be on the lookout for this possibility.

An alternative way of finding a potential function when you know that **F** is a gradient field was suggested in the remark following Theorem 16.3. We illustrate this method in the next example.

EXAMPLE 16.14 Find a potential function for the gradient field **F** of Example 16.13 using the method suggested in the remark following Theorem 16.3.

Solution The function **F** is

$$\mathbf{F}(x, y) = \langle e^y - 2xy, xe^y - x^2 + 2y \rangle$$

which was shown in Example 16.13 to be a gradient field throughout $\mathbb{R}^2$. Define ϕ by $\phi(x, y) = \int_{(0,0)}^{(x,y)} \mathbf{F} \cdot d\mathbf{r}$, where the path taken from $(0, 0)$ to (x, y) is the polygonal path $C_1 \cup C_2$ shown in Figure 16.21. We can parameterize C_1 by $\mathbf{r}(t) = \langle t, 0 \rangle$, $0 \le t \le x$, and C_2 by $\mathbf{r}(t) = \langle x, t \rangle$, $0 \le t \le y$. So

$$\phi(x, y) = \int_{C_1} \mathbf{F} \cdot d\mathbf{r} + \int_{C_2} \mathbf{F} \cdot d\mathbf{r} = \int_0^x \mathbf{F}(t, 0) \cdot \langle dt, 0 \rangle + \int_0^y \mathbf{F}(x, t) \cdot \langle 0, dt \rangle$$

$$= \int_0^x dt + \int_0^y (xe^t - x^2 + 2t) \, dt = x + \left[xe^t - x^2 t + t^2 \right]_0^y$$

$$= x + [(xe^y - x^2 y + y^2) - x] = xe^y - x^2 y + y^2 \quad \blacksquare$$

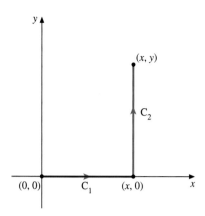

FIGURE 16.21

Summary of Results

The main results of this section can be summarized by the following list of equivalent statements for a continuous vector field $\mathbf{F} = \langle f, g \rangle$ in an open connected region R. They are equivalent in the sense that if any one of the statements is true, each of the others is also true.

1. **F** is a gradient field.
2. $\int_C \mathbf{F} \cdot d\mathbf{r}$ is independent of the path.
3. $f \, dx + g \, dy$ is an exact differential.
4. $\int_C \mathbf{F} \cdot d\mathbf{r} = 0$ for every piecewise-smooth closed curve C in R.

Figure 16.22 illustrates these equivalencies. If, in addition, R is simply connected and f and g have continuous first partial derivatives in R, we can add a fifth statement equivalent to each of these four:

5. $\dfrac{\partial f}{\partial y} = \dfrac{\partial g}{\partial x}$

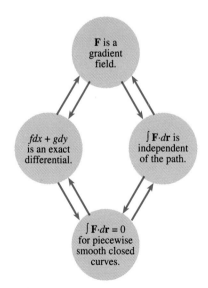

FIGURE 16.22

Extension to Three Dimensions

Most of the ideas of this section have natural extensions to three dimensions. In particular, the equivalence of the first four conditions given above remains true, where $\mathbf{F} = \langle f, g, h \rangle$. In Condition 3, the differential form becomes $f \, dx + g \, dy + h \, dz$. We will show in Section 16.7 that the appropriate form of Condition 5 in the three-dimensional case is the following:

$$\frac{\partial f}{\partial y} = \frac{\partial g}{\partial x} \qquad \frac{\partial g}{\partial z} = \frac{\partial h}{\partial y} \qquad \frac{\partial h}{\partial x} = \frac{\partial f}{\partial z} \qquad (16.14)$$

Conservation of Energy

We conclude this section by verifying the *Law of Conservation of Energy*. Let $\mathbf{F}$ be a three-dimensional conservative force field with potential function ϕ, so that $\mathbf{F} = \nabla \phi$. If the position vector of a particle of mass m moving through the field is $\mathbf{r}(t)$, then the **kinetic energy** of the particle is defined by

$$K(\mathbf{r}(t)) = \frac{1}{2} m \left| \mathbf{r}'(t) \right|^2$$

(Recall that $\left| \mathbf{r}'(t) \right|$ is the speed of the particle.) Its **potential energy** is defined by

$$P(\mathbf{r}(t)) = -\phi(\mathbf{r}(t))$$

Thus, $\mathbf{F} = -\nabla P$. Note that both kinetic energy and potential energy are scalars. The Conservation Law can be stated as follows.

Law of Conservation of Energy

In a conservative force field, the sum of the potential energy and kinetic energy remains constant as the object moves between any two points in the field.

We prove this law by making use of the ideas of this section. Our approach is to find the work done in moving the particle between two points in two different ways and then to equate the results. Let $A = \mathbf{r}(a)$ and $B = \mathbf{r}(b)$ be any two points in the vector field $\mathbf{F}$. By the three-dimensional analogue of Theorem 16.2,

$$W = \int_C \mathbf{F} \cdot d\mathbf{r} = -\int_C \nabla P \cdot d\mathbf{r} = -[P(B) - P(A)] = P(A) - P(B)$$

We also can find the work by using Newton's Second Law of Motion, which states that $\mathbf{F} = m\mathbf{a}$, where $\mathbf{a}$ is the acceleration of the particle. Since $\mathbf{a} = \mathbf{r}''$, we have

$$\begin{aligned}
W = \int_C \mathbf{F} \cdot d\mathbf{r} &= \int_a^b m \mathbf{r}''(t) \cdot \mathbf{r}'(t) \, dt \\
&= \frac{m}{2} \int_a^b \frac{d}{dt} \left[\mathbf{r}'(t) \cdot \mathbf{r}'(t) \right] dt \\
&= \frac{m}{2} \int_a^b \frac{d}{dt} \left| \mathbf{r}'(t) \right|^2 dt \\
&= \frac{m}{2} \left[\left| \mathbf{r}'(t) \right|^2 \right]_a^b \\
&= \frac{m}{2} \left[\left| \mathbf{r}'(b) \right|^2 - \left| \mathbf{r}'(a) \right|^2 \right] \\
&= \frac{1}{2} m \left| \mathbf{r}'(b) \right|^2 - \frac{1}{2} m \left| \mathbf{r}'(a) \right|^2 \\
&= K(B) - K(A)
\end{aligned}$$

Section 16.3 Gradient Fields and Path Independence   1171

The two values we have found for the work must be equal:
$$P(A) - P(B) = K(B) - K(A)$$
That is,
$$P(A) + K(A) = P(B) + K(B)$$
So the sum of the potential energy and kinetic energy remains constant from one point to another.

Exercise Set 16.3

In Exercises 1–12, determine whether $\mathbf{F}$ is a gradient field. If so, find a potential function for $\mathbf{F}$ using the method of Example 16.13. Assume the domain is $\mathbb{R}^2$ unless otherwise specified.

1. $\mathbf{F}(x, y) = \langle 2x + 3y, 3x - 2y + 1 \rangle$

2. $\mathbf{F}(x, y) = (3x^2 - 4xy)\mathbf{i} + (6y^2 - 2x^2)\mathbf{j}$

3. $\mathbf{F}(x, y) = \langle x^2 - y^2, 2xy \rangle$

4. $\mathbf{F}(x, y) = \langle y \sin x, \cos x + y^2 \rangle$

5. $\mathbf{F}(x, y) = x^2(3y^2 + 2)\mathbf{i} + 2(x^3y - 1)\mathbf{j}$

6. $\mathbf{F}(x, y) = \left\langle \dfrac{xy}{\sqrt{x^2+1}}, \sqrt{x^2+1} \right\rangle$

7. $\mathbf{F}(x, y) = \langle xe^{xy}(xy+2), x^3 e^{xy} \rangle$

8. $\mathbf{F}(x, y) = \dfrac{x\mathbf{i} + y\mathbf{j}}{x^2 + y^2}$; $\mathbb{R}^2 - \{(0,0)\}$

9. $\mathbf{F}(x, y) = \left(\dfrac{y}{\sqrt{x^2+y^2}} \right)\mathbf{i} + \left(\tan^{-1} \dfrac{y}{x} \right)\mathbf{j}$

10. $\mathbf{F}(x, y) = \langle \sin^2 xy, 1 - \cos xy \rangle$

11. $\mathbf{F}(x, y) = \langle y^2 \cos xy, xy \cos xy + \sin xy - 2y \rangle$

12. $\mathbf{F}(x, y) = (2xy \sec^2 xy + 2 \tan xy + 1)\mathbf{i} + (2x^2 \sec^2 xy + 4y)\mathbf{j}$; in $\{(x, y) : 0 < x < \infty, -\pi/(2x) < y < \pi/(2x)\}$

In Exercises 13–20, use results from Exercises 1–12 to evaluate the given integrals.

13. $\displaystyle\int_{(-1,-2)}^{(3,5)} (2x + 3y)\,dx + (3x - 2y + 1)\,dy$

14. $\displaystyle\int_{(0,2)}^{(1,4)} (3x^2 - 4xy)\,dx + (6y^2 - 2x^2)\,dy$

15. $\displaystyle\int_{(-1,2)}^{(3,1)} (3x^2 y^2 + 2x^2)\,dx + (2x^3 y - 2)\,dy$

16. $\displaystyle\int_{(0,0)}^{(2,5)} \dfrac{xy\,dx}{\sqrt{x^2+1}} + \sqrt{x^2+1}\,dy$

17. $\displaystyle\int_{(1,0)}^{(2,\ln 2)} xe^{xy}(xy + 2)\,dx + x^3 e^{xy}\,dy$

18. $\displaystyle\int_{(1,1)}^{(3,-4)} \dfrac{x\,dx + y\,dy}{x^2 + y^2}$

19. $\displaystyle\int_{(1,0)}^{(5\pi/9,\,3/2)} \mathbf{F} \cdot d\mathbf{r}$, where $\mathbf{F}$ is given in Exercise 11

20. $\displaystyle\int_{(0,0)}^{(\pi/2,\,1/2)} \mathbf{F} \cdot d\mathbf{r}$, where $\mathbf{F}$ is given in Exercise 12

In Exercises 21–24, show that $\mathbf{F}$ is independent of the path in the given region, and use the method shown in Example 16.14 to find a potential function for $\mathbf{F}$.

21. $\mathbf{F}(x, y) = \langle 2x - 2y^2, 3y^2 - 4xy \rangle$; all of $\mathbb{R}^2$

22. $\mathbf{F}(x, y) = \left\langle \dfrac{1}{x-y}, \dfrac{-1}{x-y} \right\rangle$; $\{(x, y) : x > y\}$ [*Hint:* Use the straight-line path from $(1, 0)$ to (x, y).]

23. $\mathbf{F}(x, y) = \langle \sin y + y \sin x, x \cos y - \cos x \rangle$; all of $\mathbb{R}^2$

24. $\mathbf{F}(x, y) = \left\langle \ln \sqrt{x^2 + y^2}, \tan^{-1} xy \right\rangle$; $\mathbb{R}^2 - \{(0, 0)\}$. [*Hint:* Use $C_1 \cup C_2$, where C_1 is the vertical line segment from $(1, 0)$ to $(1, y)$ and C_2 is the horizontal line segment from $(1, y)$ to (x, y).]

In Exercises 25–28, prove that the given integral is independent of the path and then evaluate it in two ways: (a) using Theorem 16.2, where C is any piecewise-smooth curve from A to B, and (b) integrating along the specified curve.

25. $\int_C (2x - 2y)\,dx + (6y - 2x)\,dy$ from $A = (-1, 2)$ to $B = (3, -2)$; C is the straight line segment from A to B.

26. $\int_C \dfrac{x\,dx + y\,dy}{\sqrt{x^2 + y^2}}$ from $A = (-3, 4)$ to $B = (5, 12)$, where C does not pass through the origin; C is the horizontal line segment from A to $(5, 4)$ followed by the vertical line segment from $(5, 4)$ to B.

27. $\int_C \dfrac{-y\,dx + x\,dy}{x^2 + y^2}$ from $A = (\sqrt{6}, -\sqrt{6})$ to $B = (3, \sqrt{3})$, where C does not pass through the origin; C is the smaller arc of the circle $x^2 + y^2 = 12$ from A to B.

28. $\int_C e^{xy}(y \cos x - \sin x)\,dx + xe^{xy}\cos x\,dy$ from $A = (0, 4)$ to $B = (\pi/3, 0)$; C is the vertical line segment from A to $(0, 0)$ followed by the horizontal line segment from $(0, 0)$ to B.

29. Evaluate the integral in Example 16.10 by integrating along each of the following paths:
    (a) C is the line segment from A to B.
    (b) C is the horizontal line segment from A to $(3, -2)$ followed by the vertical line segment from $(3, -2)$ to B.
    (c) C is the arc of the parabola $y = 2x^2 - 4x$ from A to B.

30. Evaluate the integral in Example 16.11 by integrating along each of the component curves of C.

31. Show that the force field
$$\mathbf{F}(x, y) = (2x - 3y)\mathbf{i} + 3(y^2 - x)\mathbf{j}$$
is conservative throughout $\mathbb{R}^2$, and find the work done by $\mathbf{F}$ on a particle moving from $(2, 1)$ to $(5, -3)$.

32. Show that the force field
$$\mathbf{F}(x, y) = \dfrac{y}{x}\mathbf{i} + \left(\ln x - \dfrac{2}{y}\right)\mathbf{j}$$
is conservative in the region
$$R = \{(x, y) : x > 0, y > 0\}$$
and find the work done by $\mathbf{F}$ on a particle as it goes once around the ellipse
$$4x^2 + y^2 - 16x - 6y + 21 = 0$$

33. Evaluate the integral
$$\int_C \cos x \cos y\,dx - \sin x \sin y\,dy$$
along each path in the accompanying figure. (*Hint:* There is an easy way and a hard way.)

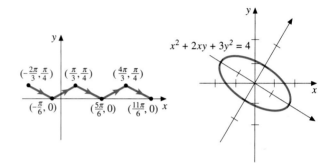

In Exercises 34–37, show that the given expression is an exact differential, and find ϕ so that the expression equals $d\phi$.

34. $(2x - 3y)\,dx + (8y - 3x)\,dy$

35. $(2x + y^2 \cos xy)\,dx + (\sin xy + xy \cos xy)\,dy$

36. $\dfrac{xy^2}{(1 + x^2)^2}\,dx + \dfrac{x^2 y}{1 + x^2}\,dy$

37. $(\ln \cosh y)\,dx + (x \tanh y - 1)\,dy$

38. Using the proof of Theorem 16.2 for a smooth curve C, show that the theorem is also true for any piecewise-smooth curve.

39. By inspection determine a potential function for $\mathbf{F}(x, y, z) = \langle 2x, 2y, 2z \rangle$ and use the result to evaluate
$$\int_{(1, -1, 2)}^{(3, 2, -1)} 2x\,dx + 2y\,dy + 2z\,dz$$

40. Complete the proof of Theorem 16.3 by showing that $\phi_y(x, y) = g(x, y)$, taking C_1 as any path from (a, b) to (x, y) and C_2 as the vertical line segment from (x, y) to $(x, y + k)$ for $|k|$ sufficiently small.

41. Using Equations 16.14, show that

$$\mathbf{F}(x, y, z) = \langle 2xyz - 3z^2, x^2z + 8yz^3, x^2y - 6xz + 12y^2z^2 \rangle$$

is a gradient field in $\mathbb{R}^3$, and follow a procedure similar to that used in Example 16.7 to find a potential function for $\mathbf{F}$.

42. Prove that the gravitational field

$$\mathbf{F} = -\frac{C}{|\mathbf{r}|^3}\mathbf{r}$$

of a point mass is conservative, where $\mathbf{r}$ is the vector $\overrightarrow{OP}$ from the point located at O to a point P in space. Find the work done by $\mathbf{F}$ on an object as it moves from A to B, where $|\overrightarrow{OA}| = r_1$ and $|\overrightarrow{OB}| = r_2$.

16.4 GREEN'S THEOREM

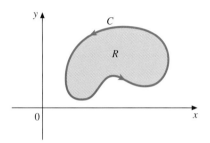

FIGURE 16.23

We now introduce Green's Theorem, which has far-reaching consequences in both pure and applied mathematics. The theorem provides a relationship between a line integral around a simple closed curve C and a double integral over the region R enclosed by C. (See Figure 16.23.) Recall that the direction, or *orientation* of a curve is determined by the parameterization, with the positive direction corresponding to increasing parameter values. In Green's Theorem the positive direction of the closed curve C is counterclockwise, as indicated in Figure 16.23.

THEOREM 16.6

Green's Theorem in the Plane

Let the simple closed curve C be oriented in the counterclockwise direction and let R be the region enclosed by C. If f and g have continuous first partial derivatives in an open set containing R, then

$$\int_C f(x, y)\, dx + g(x, y)\, dy = \iint_R \left(\frac{\partial g}{\partial x} - \frac{\partial f}{\partial y} \right) dA \qquad (16.15)$$

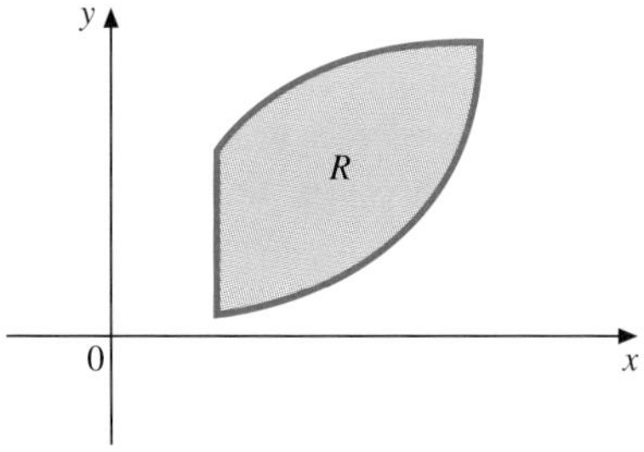

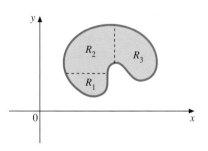

FIGURE 16.24
A simple region R

Recall that the left-hand side of Equation 16.15 is equivalent to $\int \mathbf{F} \cdot \mathbf{T}\, ds$ and also to $\int \mathbf{F} \cdot d\mathbf{r}$. We will prove Green's Theorem only for a region R that is simultaneously of type I and type II, as defined in Section 15.2. We call a region of this type a **simple region**. Figure 16.24 shows an example of a simple region. In Exercise 37 of Exercise Set 16.4 you will be asked to show that Green's Theorem is also true when R can be divided by horizontal and vertical line segments into a finite number of simple regions, as in Figure 16.25.

Partial Proof of Green's Theorem Let R be a simple region with boundary C oriented counterclockwise. Figure 16.26(a) shows such a region. In Figure 16.26(b) it is viewed as type I and in Figure 16.26(c) as type II. The proof of Equation 16.15 is accomplished by showing that

$$\int_C f(x, y)\, dx = -\iint_R \frac{\partial f}{\partial y}\, dA \qquad (16.16)$$

and

$$\int_C g(x, y)\, dy = \iint_R \frac{\partial g}{\partial x}\, dA \qquad (16.17)$$

FIGURE 16.25
$R = R_1 \cup R_2 \cup R_3$ is the union of three simple regions

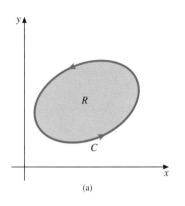
(a)

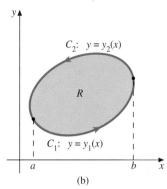
(b)

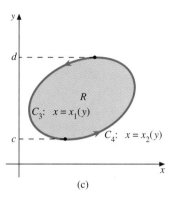
(c)

FIGURE 16.26

George Green (1793–1841) was the leading English mathematician at a time when most mathematical advances were taking place on the European continent. Green had little formal education and worked as a miller in Nottingham until he was more than 30 years old. At the age of 35, without prior history of publication, he wrote his "Essay on the Application of Mathematical Analysis to the Theories of Electricity and Magnetism." What we now call Green's Theorem was included. But the essay was not widely known. Seven years later, when William Thompson (who later became Lord Kelvin) was a senior at Cambridge University, he saw a reference to "the ingenious Essay by Mr. Green of Nottingham" and wound up introducing the work to a wider audience. Recently a memorial was dedicated to Green in Westminster Abbey, as was a stained glass window (photo below) in his old Cambridge college, where he began studying for a mathematics degree at age 40.

By adding the corresponding sides of these two equations, we get the desired result.

To prove Equation 16.16, we express R as the type I region

$$R = \{(x, y) : a \leq x \leq b, y_1(x) \leq y \leq y_2(x)\}$$

(See Figure 16.26(b).) Then the right-hand side of Equation 16.16 becomes

$$-\iint_R \frac{\partial f}{\partial y} dA = \int_a^b \int_{y_1(x)}^{y_2(x)} \frac{\partial f(x, y)}{\partial y} dy\, dx$$

$$= -\int_a^b \left[f(x, y_2(x)) - f(x, y_1(x)) \right] dx \quad \text{By the Second Fundamental Theorem} \quad (16.18)$$

In Figure 16.26(b), the lower boundary curve for the region R is denoted by C_1, with equation $y = y_1(x)$, and the upper boundary curve by C_2, with equation $y = y_2(x)$. To compute the left-hand side of Equation 16.16, we consider the curve C oriented in the counterclockwise direction as the union of the curves C_1 and C_2. To parameterize C_1, we can use x as the parameter to get $x = x$, $y = y_1(x)$, with $a \leq x \leq b$. For C_2, the parameter x will vary from b to a, since we want to traverse C_2 from right to left. Thus, $-C_2$ is traversed from left to right, and its parameterization is $x = x$, $y = y_2(x)$, with $a \leq x \leq b$. We can therefore write

$$\int_{C_2} f(x, y)\, dx = -\int_{-C_2} f(x, y)\, dx$$

$$= -\int_a^b f(x, y_2(x))\, dx$$

Since $C = C_1 \cup C_2$, we have

$$\int_C f(x, y)\, dx = \int_{C_1} f(x, y)\, dx + \int_{C_2} f(x, y)\, dx$$

$$= \int_a^b f(x, y_1(x))\, dx - \int_a^b f(x, y_2(x))\, dx$$

$$= -\int_a^b \left[f(x, y_2(x)) - f(x, y_1(x)) \right] dx$$

When we compare this result with Equation 16.18, we see that Equation 16.16 is true. The proof of Equation 16.17 is similar, using Figure 16.26(c). You will be asked to carry out the details in Exercise 36 of Exercise Set 16.4. Equation 16.15 follows, as previously indicated, by adding the results of Equations 16.16 and 16.17. ∎

REMARK
■ There is a version of Green's Theorem for three dimensions that we will consider in Section 16.7, which explains the phrase "in the plane" in the name of the theorem as stated here.

We can view Equation 16.15 in either of two ways: (a) as a means of evaluating a line integral using a double integral, or (b) as a means of evaluating a double integral using a line integral. Both points of view are important, and we illustrate them in the examples that follow.

EXAMPLE 16.15 Use Green's Theorem to evaluate the line integral

$$\int_C \left(e^{-x^2} + xy^2\right) dx + \left(x^3 + \sqrt{1+y^3}\right) dy$$

where C is the path shown in Figure 16.27.

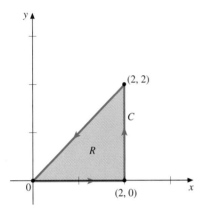

FIGURE 16.27

Solution Observe first that evaluating the integral by parameterizing the component curves of C would lead to integrals that cannot be evaluated by elementary means. (Try it.) Let us see, then, if we can use Green's Theorem to replace the line integral with a double integral that we can evaluate. Note that the region R bounded by C is a simple region (both type I and type II). To apply Green's Theorem, we take

$$f(x, y) = e^{-x^2} + xy^2 \quad \text{and} \quad g(x, y) = x^3 + \sqrt{1+y^3}$$

Since

$$\frac{\partial g}{\partial x} = 3x^2 \quad \text{and} \quad \frac{\partial f}{\partial y} = 2xy$$

we have, by Green's Theorem,

$$\int_C \left(e^{-x^2} + xy^2\right) dx + \left(x^3 + \sqrt{1+y^3}\right) dy = \iint_R (3x^2 - 2xy) \, dA$$

$$= \int_0^2 \int_0^x (3x^2 - 2xy) \, dy \, dx$$

$$= \int_0^2 \left[3x^2 y - xy^2\right]_0^x dx$$

$$= \int_0^2 2x^3 \, dx$$

$$= 8 \qquad \blacksquare$$

In Example 16.15, Green's Theorem was useful in evaluating a difficult line integral by means of a double integral. In the next example, the situation is reversed.

EXAMPLE 16.16 Use Green's Theorem to evaluate the integral $\iint_R y^2 \, dA$, where R is the elliptical region shown in Figure 16.28.

Solution Direct evaluation by iterated integrals results in the integral

$$\int_{-3}^{3} \int_{(-2/3)\sqrt{9-x^2}}^{(2/3)\sqrt{9-x^2}} y^2 \, dy \, dx$$

which is difficult to evaluate. By Green's Theorem, we can replace this integral with a line integral that may be easier to evaluate. To use Green's Theorem, we must find functions f and g that satisfy the continuity requirements over R for which

$$\iint_R y^2 \, dA = \iint_R \left(\frac{\partial g}{\partial x} - \frac{\partial f}{\partial y} \right) dA$$

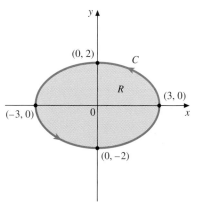

FIGURE 16.28

Many choices are possible, but a particularly simple one is $g(x, y) = xy^2$ and $f(x, y) = 0$. A parameterization of the ellipse C that gives a counterclockwise orientation is $x = 3\cos t$, $y = 2\sin t$, where $0 \leq t \leq 2\pi$. So we have, by Equation 16.15,

$$\iint_R y^2 \, dA = \iint_R \left(\frac{\partial g}{\partial x} - \frac{\partial f}{\partial y} \right) dA = \int_C f(x, y) \, dx + g(x, y) \, dy$$

$$= \int_C xy^2 \, dy = \int_0^{2\pi} (3\cos t)(4\sin^2 t)(2\cos t \, dt)$$

$$= 6 \int_0^{2\pi} \sin^2 2t \, dt = 3 \int_0^{2\pi} (1 - \cos 4t) \, dt \quad \text{Since } \sin^2 \theta = (1 - \cos 2\theta)/2$$

$$= 6\pi \quad \blacksquare$$

In the next example, we use Green's Theorem to evaluate a line integral around the boundary of a region that is the union of two simple regions.

EXAMPLE 16.17 Evaluate the line integral

$$\int_C (e^x - x^2 y) \, dx + (xy^2 + y^3) \, dy$$

using Green's Theorem, where C is the boundary of the semiannular region shown in Figure 16.29, oriented in a counterclockwise direction.

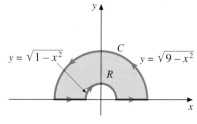

FIGURE 16.29

Solution Let R denote the region enclosed by C. Although R is not simple, it is the union of the two simple regions into which the y-axis divides R. To apply Green's Theorem, we let $f(x, y) = e^x - x^2 y$ and $g(x, y) = xy^2 + y^3$. Then

$$\frac{\partial g}{\partial x} - \frac{\partial f}{\partial y} = y^2 - (-x^2) = x^2 + y^2$$

So we have, by Equation 16.15,

$$\int_C (e^x - x^2 y)\, dx + (xy^2 + y^3)\, dy = \iint_R (x^2 + y^2)\, dA$$

It is easier to use polar coordinates in this case. The bounding curves are $r = 1$ and $r = 3$, with θ going from 0 to π. Thus,

$$\iint_R (x^2 + y^2)\, dA = \int_0^\pi \int_1^3 r^2(r\, dr\, d\theta) = \int_0^\pi \int_1^3 r^3\, dr\, d\theta$$

$$= \int_0^\pi \frac{r^4}{4}\Big]_1^3 d\theta = \left(\frac{81}{4} - \frac{1}{4}\right)\pi = 20\pi \qquad \blacksquare$$

Areas by Line Integration

We can use Green's Theorem to find the area of a region as a line integral around its boundary, as we show in the following theorem.

THEOREM 16.7

Let R be the region enclosed by a piecewise-smooth, simple, closed curve C, oriented in a counterclockwise direction. Then the area A of the region R is given by any one of the following formulas:

$$A = \int_C x\, dy \qquad (16.19)$$

$$A = -\int_C y\, dx \qquad (16.20)$$

$$A = \frac{1}{2}\int_C x\, dy - y\, dx \qquad (16.21)$$

Proof To prove Equation 16.19, we let $f(x, y) = 0$ and $g(x, y) = x$. Then, by Green's Theorem, we have

$$\int_C x\, dy = \iint_R 1\, dA = A$$

To prove Equation 16.20, we let $f(x, y) = y$ and $g(x, y) = 0$. By Green's Theorem

$$\int_C y\, dx = \iint_R (-1)\, dA = -A$$

So $A = -\int_C y\, dx$. To prove Equation 16.21, we add corresponding sides of Equations 16.19 and 16.20, getting $2A = \int_C x\, dy - \int_C y\, dx$. Then we divide by 2 and combine the two integrals. $\blacksquare$

EXAMPLE 16.18 Find the area of the region R enclosed by the ellipse

$$\frac{x^2}{a^2} + \frac{y^2}{b^2} = 1$$

(See Figure 16.30.)

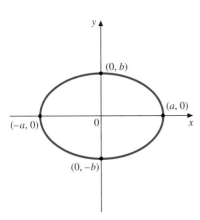

FIGURE 16.30

The ellipse $\dfrac{x^2}{a^2} + \dfrac{y^2}{b^2} = 1$

Solution Parametric equations of the ellipse that give it the correct orientation are $x = a\cos t$, $y = b\sin t$, where $0 \le t \le 2\pi$. Although each of the equations in Theorem 16.7 will give the same result, we choose to use Equation 16.21. Using it, we get

$$A = \frac{1}{2}\int_C x\,dy - y\,dx = \frac{1}{2}\int_0^{2\pi}[(a\cos t)(b\cos t\,dt) - (b\sin t)(-a\sin t\,dt)]$$
$$= \frac{1}{2}\int_0^{2\pi} ab(\cos^2 t + \sin^2 t)\,dt$$
$$= \pi ab \qquad \blacksquare$$

Multiply-Connected Regions

We now discuss how to extend Green's Theorem to certain multiply-connected regions (regions that are not simply connected). In particular, suppose R is a closed and bounded region with a boundary that consists of finitely many simple closed curves that are sectionally smooth and do not intersect one another. We illustrate such a region in Figure 16.31, with boundary curves C_1, C_2, C_3, and C_4. Let each of these boundary curves be oriented so that when it is traversed in its positive direction, the region R lies on the *left*, and let C be the union of all of these boundary curves. When the component curves of C are oriented according to this left-hand rule, we say that C is *positively oriented with respect to R*.

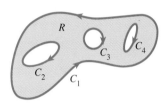

FIGURE 16.31
$C = C_1 \cup C_2 \cup C_3 \cup C_4$ is positively oriented with respect to R.

Green's Theorem holds for regions of the type described; that is, if f and g have continuous first partial derivatives in an open set that contains R, then

$$\int_C f\,dx + g\,dy = \iint_R \left(\frac{\partial g}{\partial x} - \frac{\partial f}{\partial y}\right) dA$$

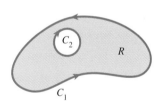

FIGURE 16.32

We indicate the idea of the proof for the simple case shown in Figure 16.32 in which the multiply-connected region R has one hole. The outer boundary of R is C_1 and the inner boundary C_2, both positively oriented with respect to R. We denote the total boundary of R by C. So $C = C_1 \cup C_2$. In Figure 16.33, we show R cut into two simply-connected regions, R_1 and R_2. Let us denote the boundary of R_1 by B_1 and the boundary of R_2 by B_2. Since Green's Theorem applies to R_1 and R_2 individually, we have

$$\iint_R \left(\frac{\partial g}{\partial x} - \frac{\partial f}{\partial y}\right) dA = \iint_{R_1}\left(\frac{\partial g}{\partial x} - \frac{\partial f}{\partial y}\right) dA + \iint_{R_2}\left(\frac{\partial g}{\partial x} - \frac{\partial f}{\partial y}\right) dA$$

$$= \int_{B_1} f\,dx + g\,dy + \int_{B_2} f\,dx + g\,dy \qquad \text{By Equation 16.10 applied to both } R_1 \text{ and } R_2$$

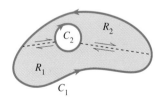

FIGURE 16.33

The boundaries B_1 and B_2 have the two line segments that we introduced in common, but oriented in opposite directions. So when we add the integrals along these line segments, they cancel, and we are left with the integrals around C_1 and C_2:

$$\iint_R \left(\frac{\partial g}{\partial x} - \frac{\partial f}{\partial y}\right) dA = \int_{C_1} f\,dx + g\,dy + \int_{C_2} f\,dx + g\,dy = \int_C f\,dx + g\,dy$$

Thus, Green's Theorem holds true for the multiply-connected region R.

A very useful consequence of Green's Theorem for multiply connected regions is the following. It shows that in certain cases line integrals around different closed paths are equal.

THEOREM 16.8 Let C_1 and C_2 be any two nonintersecting piecewise-smooth simple closed curves, and let R be the closed annular region bounded by C_1 and C_2. If f and g have continuous partial derivatives and

$$\frac{\partial f}{\partial y} = \frac{\partial g}{\partial x}$$

throughout some open set that contains R, then

$$\int_{C_1} f\,dx + g\,dy = \int_{C_2} f\,dx + g\,dy$$

provided C_1 and C_2 are similarly oriented.

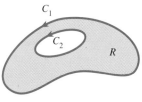

FIGURE 16.34

Proof Suppose C_1 and C_2 are both oriented in the counterclockwise direction and C_2 is interior to C_1, as shown in Figure 16.34. Then Green's Theorem for the multiply-connected region R applies, where the boundary of R is $C = C_1 \cup (-C_2)$, since C is positively oriented with respect to R. Thus,

$$\int_C f\,dx + g\,dy = \iint_R \left(\frac{\partial g}{\partial x} - \frac{\partial f}{\partial y}\right) dA = 0 \quad \text{Since } \frac{\partial f}{\partial y} = \frac{\partial g}{\partial x}$$

But

$$\int_C f\,dx + g\,dy = \int_{C_1} f\,dx + g\,dy + \int_{-C_2} f\,dx + g\,dy$$

$$= \int_{C_1} f\,dx + g\,dy - \int_{C_2} f\,dx + g\,dy$$

Since we just showed that this difference equals 0, the result follows. If C_1 and C_2 are oriented in the clockwise direction, we let $C = (-C_1) \cup C_2$, and the reasoning follows the same lines. ∎

This theorem sometimes enables us to replace a complicated closed path with a simpler one. The next example illustrates this approach.

EXAMPLE 16.19 Evaluate the integral

$$\int_C \frac{-y}{x^2+y^2}\,dx + \frac{x}{x^2+y^2}\,dy$$

where C is the ellipse $x^2 - 2xy + 3y^2 = 4$.

Solution The ellipse is pictured in Figure 16.35. Let

$$f(x,y) = \frac{-y}{x^2+y^2} \quad \text{and} \quad g(x,y) = \frac{x}{x^2+y^2}$$

It is easily verified that for all points except $(0, 0)$, $\partial g/\partial x = \partial f/\partial y$. If it were not for the exceptional point $(0, 0)$, we could conclude that $\langle f, g \rangle$ is a gradient field (why?) and therefore that the integral around the closed path C is 0. But this reasoning breaks down, since f and g fail even to exist at the origin.

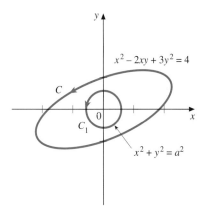

FIGURE 16.35

It is rather difficult to parameterize this ellipse, so we introduce a new curve C_1 inside C and with the origin in its interior. Because of its simplicity, we might as well take C_1 to be a circle centered at the origin. Let its radius be a. It does not matter what a is, as long as it is small enough that C_1 lies inside C. The conditions of Theorem 16.8 are now met, so

$$\int_C \frac{-y}{x^2+y^2}\,dx + \frac{x}{x^2+y^2}\,dy = \int_{C_1} \frac{-y}{x^2+y^2}\,dx + \frac{x}{x^2+y^2}\,dy$$

A parameterization of C_1 is $x = a\cos t$, $y = a\sin t$, where $0 \leq t \leq 2\pi$. So $dx = -a\sin t\,dt$ and $dy = a\cos t\,dt$. Also, $x^2 + y^2 = a^2\cos^2 t + a^2\sin^2 t = a^2$. So we have

$$\int_{C_1} \frac{-y}{x^2+y^2}\,dx + \frac{x}{x^2+y^2}\,dy = \int_0^{2\pi} \frac{-a\sin t}{a^2}(-a\sin t\,dt) + \frac{a\cos t}{a^2}(a\cos t\,dt)$$

$$= \int_0^{2\pi} (\sin^2 t + \cos^2 t)\,dt = \int_0^{2\pi} dt = 2\pi$$

The fact that the integral around the closed path is nonzero again shows that $\langle f, g \rangle$ is not a gradient field in the region enclosed by C. ∎

REMARK
■ The ellipse in Example 16.19 could have been replaced by any other sectionally smooth simple closed curve with the origin inside it. The answer would have been the same.

Green's Theorem and Two-Dimensional Fluid Flow

We conclude this section with a brief discussion of how Green's Theorem can be used to determine certain characteristics of the flow of a fluid. We consider here a two-dimensional "laminar" flow (think of a thin sheet of water flowing over a flat surface), and in Sections 16.6 and 16.7 we will extend the ideas to three dimensions.

Let $\mathbf{F} = \langle f, g \rangle$ be the velocity field of the fluid, and let a region R bounded by the curve C, as in Green's Theorem, be in the field of flow. We assume the flow is in *steady state*, meaning that $\mathbf{F}$ does not change with time. At each point P on C, we wish to consider the component $\mathbf{F} \cdot \mathbf{T}$ of $\mathbf{F}$ in the direction of the unit tangent vector $\mathbf{T}$ and the component $\mathbf{F} \cdot \mathbf{n}$ in the direction of the *outer* unit normal $\mathbf{n}$, as shown in Figure 16.36. The integral $\int_C \mathbf{F} \cdot \mathbf{T}\,ds$ gives the amount of fluid per unit of time flowing tangentially around C, whereas $\int_C \mathbf{F} \cdot \mathbf{n}\,ds$ gives the amount of fluid per unit of time flowing out of R orthogonally across C. These integrals are given names as follows:

$$\int_C \mathbf{F} \cdot \mathbf{T}\,ds = \text{The \textbf{Circulation} of } \mathbf{F} \text{ Around } C$$

$$\int_C \mathbf{F} \cdot \mathbf{n}\,ds = \text{The \textbf{Flux} of } \mathbf{F} \text{ Through } C$$

Although the circulation and flux can be evaluated directly as line integrals, we wish to show how Green's Theorem can be used to find them. For the

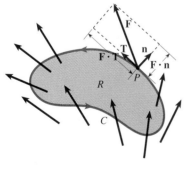

FIGURE 16.36

circulation, this result is immediate, since $\mathbf{F} \cdot \mathbf{T}\,ds = \mathbf{F} \cdot d\mathbf{r} = f\,dx + g\,dy$, so that Green's Theorem gives

$$\text{Circulation of } \mathbf{F} \text{ Around } C = \int_C \mathbf{F} \cdot \mathbf{T}\,ds = \iint_R \left(\frac{\partial g}{\partial x} - \frac{\partial f}{\partial y}\right) dA \quad (16.22)$$

To obtain a similar representation of the flux, we can show (see Exercise 33 of Section 16.4) that the outer unit normal vector $\mathbf{n}$ is given by

$$\mathbf{n} = \left\langle \frac{dy}{ds}, \frac{-dx}{ds} \right\rangle$$

We can therefore write

$$\mathbf{F} \cdot \mathbf{n}\,ds = \langle f, g \rangle \cdot \left\langle \frac{dy}{ds}, \frac{-dx}{ds} \right\rangle ds = -g\,dx + f\,dy$$

Now we can apply Green's Theorem, replacing f with $-g$ and g with f in Equation 16.15, to get

$$\text{Flux of } \mathbf{F} \text{ Through } C = \int_C \mathbf{F} \cdot \mathbf{n}\,ds = \iint_R \left(\frac{\partial f}{\partial x} + \frac{\partial g}{\partial y}\right) dA \quad (16.23)$$

At each point P of R, the integrands of the double integrals in Equations 16.22 and 16.23 are called, respectively, the **rotation of F** at P and the **divergence of F** at P. These are abbreviated rot $\mathbf{F}$ and div $\mathbf{F}$, respectively. So we have

$$\text{rot } \mathbf{F} = \frac{\partial g}{\partial x} - \frac{\partial f}{\partial y}$$

$$\text{div } \mathbf{F} = \frac{\partial f}{\partial x} + \frac{\partial g}{\partial y}$$

Each of these is a scalar function. Some authors use *scalar curl* or *two-dimensional curl* for rotation. (We will study the curl of a vector field in Section 16.7 and see how it is related to rotation.) With these definitions, we can rewrite Equations 16.22 and 16.23 as

$$\int_C \mathbf{F} \cdot \mathbf{T}\,ds = \iint_R (\text{rot } \mathbf{F})\,dA \quad (16.24)$$

$$\int_C \mathbf{F} \cdot \mathbf{n}\,ds = \iint_R (\text{div } \mathbf{F})\,dA \quad (16.25)$$

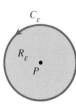

FIGURE 16.37

To get an intuitive understanding of the physical significance of the rotation and circulation of $\mathbf{F}$ at a point P in the field of flow, consider a small circle C_ε, of radius ε, centered at P, and denote the region it encloses by R_ε (see Figure 16.37). Then, by Equation 16.24, the circulation around C_ε is

$$\int_{C_\varepsilon} \mathbf{F} \cdot \mathbf{T}\,ds = \iint_{R_\varepsilon} (\text{rot } \mathbf{F})\,dA$$

Now there is a mean-value theorem for double integrals analogous to the one for single integrals that enables us to write (when rot $\mathbf{F}$ is continuous)

$$\iint_{R_\varepsilon} (\text{rot } \mathbf{F})\,dA = [\text{rot } \mathbf{F}(Q)](\text{Area of } R_\varepsilon)$$

where Q is some point in R_ε. Solving for rot $\mathbf{F}(Q)$, we have

$$\text{rot}\,\mathbf{F}(Q) = \frac{\iint_{R_\varepsilon}(\text{rot}\,\mathbf{F})\,dA}{\text{Area of }R_\varepsilon} = \frac{\int_{C_\varepsilon}\mathbf{F}\cdot\mathbf{T}\,ds}{\text{Area of }R_\varepsilon}$$

If we let $\varepsilon \to 0$, then $Q \to P$, and we get

$$\text{rot}\,\mathbf{F}(P) = \lim_{\varepsilon \to 0} \frac{\text{Circulation of }\mathbf{F}\text{ Around }C_\varepsilon}{\text{Area of }R_\varepsilon}$$

So rot $\mathbf{F}$ at a point P is the *circulation per unit area at P*. If rot $\mathbf{F} \neq 0$ at P, the fluid forms a whirlpool, called a **vortex**, at P. If rot $\mathbf{F} = 0$ for all points of a region, then $\mathbf{F}$ is said to be **irrotational** in that region.

In an exactly analogous way, we can show that

$$\text{div}\,\mathbf{F}(P) = \lim_{\varepsilon \to 0} \frac{\text{Flux of }\mathbf{F}\text{ Through }C_\varepsilon}{\text{Area of }R_\varepsilon}$$

so that div $\mathbf{F}$ at a point P is the *flux per unit area* at that point. If div $\mathbf{F}(P) > 0$, fluid is emerging from P, and we say that P is a **source**. If div $\mathbf{F}(P) < 0$, fluid is flowing into P, and we say that P is a **sink**. If div $\mathbf{F} = 0$ for all points of a region, we say the fluid is **incompressible**.

REMARK

■ Although the concepts of circulation, flux, rotation, and divergence were introduced for fluid flow, the terms are frequently used for other types of vector fields as well.

EXAMPLE 16.20   Let $\mathbf{F} = -2xy\mathbf{i} + x^2\mathbf{j}$ be the velocity field of a two-dimensional fluid flow, and let R be the region enclosed by the triangle C that has vertices $(0,0)$, $(2,0)$, and $(2,4)$, oriented counterclockwise. Use Green's Theorem to find the circulation of $\mathbf{F}$ around C and the flux of $\mathbf{F}$ through C.

Solution   The region R is shown in Figure 16.38. First we calculate rot $\mathbf{F}$ and div $\mathbf{F}$. We write $f(x,y) = -2xy$ and $g(x,y) = x^2$, so that $\mathbf{F} = \langle f, g\rangle$. So

$$\text{rot}\,\mathbf{F} = \frac{\partial g}{\partial x} - \frac{\partial f}{\partial y} = 2x - (-2x) = 4x$$

and

$$\text{div}\,\mathbf{F} = \frac{\partial f}{\partial x} + \frac{\partial g}{\partial y} = -2y + 0 = -2y$$

Thus, by Equations 16.24 and 16.25,

$$\text{Circulation of }\mathbf{F}\text{ Around }C = \int_C \mathbf{F}\cdot\mathbf{T}\,ds = \iint_R (\text{rot}\,\mathbf{F})\,dA$$

$$= \int_0^2 \int_0^{2x} 4x\,dy\,dx = \int_0^2 \Big[4xy\Big]_0^{2x}\,dx$$

$$= \int_0^2 8x^2\,dx = \frac{64}{3}$$

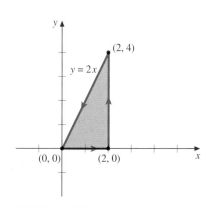

FIGURE 16.38

Flux of **F** Through $C = \int_C \mathbf{F} \cdot \mathbf{n}\,ds = \iint_R (\text{div } \mathbf{F})\,dA$

$$= \int_0^2 \int_0^{2x} (-2y)\,dy\,dx = -\int_0^2 \left[y^2\right]_0^{2x} dx$$

$$= -\int_0^2 4x^2\,dx = -\frac{32}{3}$$

Since the circulation is positive, the net flow of fluid around C is in the counterclockwise direction, and since the flux is negative, there is a net inflow of fluid into R. ∎

REMARK

■ Since we are dealing with two-dimensional flow, the "amount" of fluid is given by area rather than volume. Thus, for example, if $|\mathbf{F}|$ is in centimeters per second and distance is in centimeters, both circulation and flux will be in square centimeters per second.

Exercise Set 16.4

In Exercises 1–14, use Green's Theorem to evaluate the line integral. In each case, C has a counterclockwise orientation. For Exercises 1–4, take C to be the rectangle with vertices $(0, 0)$, $(2, 0)$, $(2, 1)$, and $(0, 1)$.

1. $\int_C \left(x^2 y - 2y^2\right) dx + \left(x^3 - 2y^2\right) dy$

2. $\int_C \left(\ln\sqrt{x+1} + xy\right) dx + \left(x^2 y + e^{y^2}\right) dy$

3. $\int_C y\cos\frac{\pi x}{2}\,dx - x\sin\frac{\pi y}{2}\,dy$

4. $\int_C \frac{x}{1+y^2}\,dx + \frac{y}{1+x^2}\,dy$

5. $\int_C e^{x+2y}\,dx$, C is the triangle with vertices $(0, 0)$, $(1, 1)$, and $(0, 1)$.

6. $\int_C (\tan^{-1} x)\,dy$, C is the boundary of the region between $y = 2 - x^2$ and $y = x$.

7. $\int_C y\sin 2x\,dx + \sin^2 x\,dy$, C is the ellipse $2x^2 + 3y^2 = 6$.

8. $\int_C (x^2 + y^2)\,dx + (x^2 - y^2)\,dy$, C is the boundary of the region determined by $y = x^2$, $y = 0$, and $x = 1$.

9. $\int_C \left(x^2 y + \frac{y^3}{3}\right) dx + (2x - y^5)\,dy$, C is the circle $x^2 + y^2 = 4$.

10. $\int_C (e^x + y^3)\,dx + (x^2 - \sqrt{y})\,dy$, C is the boundary of the region determined by $y = \sqrt{x}$, $y = 0$, and $x - y = 2$.

11. $\int_C y^2\,dx + x^2\,dy$, C is the boundary of the region determined by $x = -\sqrt{9 - y^2}$, $x + y = 3$, $y = 0$.

12. $\int_C (2y^3 - 3x^2)\,dx + (2x^3 + 5y^2)\,dy$, C is the boundary of the region determined by $y = \sqrt{4 - x^2}$ and $y = 0$.

13. $\int_C y^2\,dx + 3xy\,dy$, C is the cardioid $r = 1 + \cos\theta$.

14. $\int_C \sqrt{x^2 + 1}\,dx + x(1 + y)\,dy$, C is the boundary of the region outside the circle $r = 1$ and inside the cardioid $r = 2(1 + \cos\theta)$.

In Exercises 15–18, make use of Green's Theorem to evaluate each double integral by means of a line integral.

15. $\iint_R x\,dA$, R is the triangle with vertices $(0, 0)$, $(1, 1)$, and $(-1, 2)$.

16. $\iint_R [2(x - 1) - 2y]\,dA$, R is the circle $x^2 + y^2 = 2x$.

17. $\iint_R \left(x\sqrt{1 - y^2} - 4x^2 y\right) dA$, R is the ellipse $x^2 + 4y^2 = 4$.

18. $\iint_R y\, dA$, R is the parallelogram with vertices $(0, 0)$, $(4, 0)$, $(5, 2)$, and $(1, 2)$.

In Exercises 19–25, find the area of the specified region using line integration.

19. Bounded by the parallelogram with vertices $(0, 0)$, $(3, 1)$, $(4, 3)$, and $(1, 2)$

20. Between $y = x^2$ and $y = 2x$

21. Bounded by the triangle with vertices $(0, 2)$, $(1, 1)$, and $(2, 3)$

22. Inside the loop of $x = t^2 - 1$, $y = t^3 - t$, where $-\infty < t < \infty$

23. Inside the four-cusp hypocycloid $x = a\cos^3 t$, $y = a\sin^3 t$, where $0 \le t \le 2\pi$

24. Bounded by $x^{1/2} + y^{1/2} = 1$, $x = 0$, and $y = 0$. (*Hint:* Take $t = x^{1/2}$.)

25. Above the x-axis and under one arch of the cycloid $x = a(t - \sin t)$, $y = a(1 - \cos t)$

26. Let R be the region inside the circle C, having equation $x^2 + y^2 = a^2$, oriented in a counterclockwise direction. For each of the following vector fields find the circulation of $\mathbf{F}$ around C and the flux of $\mathbf{F}$ through C:
(a) $\mathbf{F}(x, y) = x\mathbf{i} + y\mathbf{j}$ (b) $\mathbf{F}(x, y) = -y\mathbf{i} + x\mathbf{j}$

In Exercises 27–30, $\mathbf{F}$ is the velocity field of two-dimensional fluid flow and R is the region enclosed by the curve C oriented in a counterclockwise direction. Find the circulation of $\mathbf{F}$ around C and the flux of $\mathbf{F}$ through C.

27. $\mathbf{F}(x, y) = \langle x^2 - y^2, 2xy\rangle$; R is the region in the first quadrant bounded by $y = \sqrt{1 - x^2}$, $x = 0$, and $y = 0$.

28. $\mathbf{F}(x, y) = \langle xy^2 - 3y, 2x + x^2 y\rangle$; C is the circle $x^2 + y^2 = 9$.

29. $\mathbf{F}(x, y) = -y^3\mathbf{i} + x^3\mathbf{j}$; C is the circle $x^2 + y^2 = 4$.

30. $\mathbf{F}(x, y) = (2x^2 - 3y^2)\mathbf{i} + (4y^2 - x^2)\mathbf{j}$; R is the region enclosed by the lines $y = x$, $y = 2 - x$, and $y = 0$.

In Exercises 31 and 32, C is the ellipse $x = 3\cos t$, $y = 2\sin t$, where $0 \le t \le 2\pi$.

31. Let $\mathbf{F} = \langle \sqrt{1 + x^4} - 4xy, x^3 - e^{y^2}\rangle$. Find the circulation of $\mathbf{F}$ around C.

32. Let $\mathbf{F} = \langle 2x + \cosh y^2, 4y - \sinh x^2\rangle$. Find the flux of $\mathbf{F}$ through C.

33. Let C be a smooth simple closed curve, oriented in a counterclockwise direction, and let C be parameterized by arc length. Prove that the outer unit normal $\mathbf{n}$ is given by

$$\mathbf{n} = \left\langle \frac{dy}{ds}, -\frac{dx}{ds}\right\rangle$$

(*Hint:* Let θ denote the smallest positive angle from the unit vector $\mathbf{i}$ to the unit tangent vector $\mathbf{T}$. Then

$$\mathbf{T} = \left\langle \frac{dx}{ds}, \frac{dy}{ds}\right\rangle = \langle \cos\theta, \sin\theta\rangle$$

Now show that $\mathbf{n} = \langle \cos(\theta - \pi/2), \sin(\theta - \pi/2)\rangle$.)

34. Make use of Theorem 16.8 to evaluate the integral

$$\int_C \ln\sqrt{x^2 + y^2}\, dx - \left(\tan^{-1}\frac{y}{x}\right) dy$$

where C is the limaçon $r = 4 + 2\cos\theta$ (see Section 9.5). Explain why Green's Theorem is not applicable in this case.

35. Prove that for

$$F(x, y) = \frac{x\mathbf{i} + y\mathbf{j}}{\sqrt{x^2 + y^2}}$$

and for any piecewise-smooth simple closed curve C that does not pass through the origin,

$$\int_C \mathbf{F} \cdot d\mathbf{r} = 0$$

Consider separately the cases in which the origin is interior to C and exterior to C.

36. Complete the proof of Theorem 16.6 for a simple region by showing that Equation 16.17 is true.

37. Prove Green's Theorem for a region that can be divided by a horizontal or a vertical line segment into two simple regions. Extend this result by mathematical induction to a region that can be divided in this manner into finitely many simple regions.

16.5 SURFACE INTEGRALS

In this section we define a type of integral called a **surface integral** that is particularly useful in the study of three-dimensional vector fields. For example, surface integrals are used in the study of fluid dynamics, heat transfer, and electric and magnetic field theory. Surface integrals can also be used to find the mass, center of gravity, and moments of inertia of such objects as sheets of metal in various shapes.

The term *surface integral* comes from the fact that the domain of integration is a surface S in space, in contrast to a double integral, which involves integration over a region R in the xy-plane. In what follows, we will make use of results obtained in Section 15.5 on areas of surfaces, so you may need to review that section.

Let S be the surface defined by $z = f(x, y)$, where the domain of f is a region R in the xy-plane. Figure 16.39 illustrates such a surface. We will refer to the region R as the *xy projection* of S. Our primary concern will be with regions R of type I or type II, or that can be divided into a finite number of subregions of these two types. We assume that the function f and its first partial derivatives f_x and f_y are continuous on R. This assumption ensures that S has a tangent plane at each of its points.

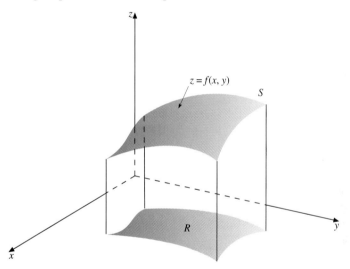

FIGURE 16.39

Following exactly the same procedure as in Section 15.5, we partition a rectangle containing R and count only the subrectangles completely contained in R. When we project the ith subrectangle vertically, the resulting prism cuts out a patch $\Delta\sigma_i$ (called *a cell*) on the surface S that is approximated by the patch ΔT_i on the tangent plane, drawn at a point (x_i^*, y_i^*, z_i^*) on S that is the vertical projection on S of a point (x_i^*, y_i^*) in the subrectangle. We denote the area of the ith subrectangle by ΔA_i. The relationships among $\Delta\sigma_i$, ΔT_i, and ΔA_i are shown in Figure 16.40. We define the **surface integral of g over S** by

$$\iint\limits_S g(x, y, z)\, d\sigma = \lim_{\|P\| \to 0} \sum_{i=1}^n g(x_i^*, y_i^*, z_i^*) \Delta\sigma_i \qquad (16.26)$$

where as usual, $\|P\|$ is the norm of the partition of the region R, that is, the length of the longest diagonal of all the subrectangles in the partition. It can be shown that when g is continuous on S, and the function f and the region R satisfy the conditions stated, then the limit in Equation 16.26 exists, so the surface integral exists.

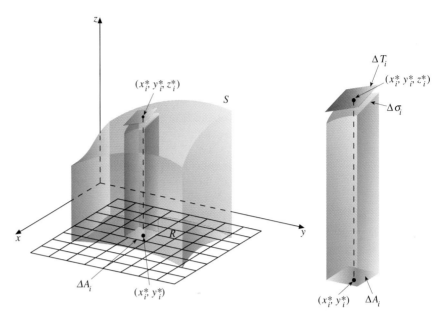

FIGURE 16.40

In Section 15.5, we found that the area $\Delta\sigma_i$ of the ith cell on the surface is approximated by ΔT_i, given by the following formula:

$$\Delta T_i \approx \sqrt{1 + \left[f_x(x_i^*, y_i^*)\right]^2 + \left[f_y(x_i^*, y_i^*)\right]^2}\, \Delta A_i$$

Thus, we can obtain the computational formula for the surface integral, given in the following theorem.

THEOREM 16.9

Let the surface S be the graph of $z = f(x, y)$, and let R be the xy projection of S. If f_x and f_y are continuous on R, and g is a function of three variables that is continuous on S, then

$$\iint\limits_S g(x, y, z)\, d\sigma$$

$$= \iint\limits_R g(x, y, f(x, y)) \sqrt{1 + \left[f_x(x, y)\right]^2 + \left[f_y(x, y)\right]^2}\, dA \qquad (16.27)$$

REMARK

■ Equation 16.27 enables us to evaluate a surface integral as an ordinary double integral. Compare this result with that of Equation 16.4 in Theorem 16.1, in which a line integral is given as an ordinary integral of a function of one variable.

If the surface S is defined by the equation $x = f(y, z)$ and its yz projection is R, then by permuting variables in Equation 16.27, we obtain

$$\iint_S g(x, y, z)\, d\sigma = \iint_R g(f(y, z), y, z) \sqrt{1 + [f_y(y, z)]^2 + [f_z(y, z)]^2}\, dA$$
(16.28)

Similarly, if S is defined by $y = f(x, z)$ and R is the xz projection of S, we have

$$\iint_S g(x, y, z)\, d\sigma = \iint_R g(x, f(x, z), z) \sqrt{1 + [f_x(x, z)]^2 + [f_z(x, z)]^2}\, dA$$
(16.29)

In Equation 16.27, if we let $g(x, y, z) = 1$ for all points (x, y, z) on the surface S, then we see that the double integral on the right-hand side is exactly the area of the surface S, as given by Equation 15.25. Thus, we have the following result:

Let $A(S)$ denote the area of the surface S. Then

$$A(S) = \iint_S d\sigma \qquad (16.30)$$

EXAMPLE 16.21 Evaluate the surface integral $\iint_S (2xy + xz)\,d\sigma$, where S is the portion of the plane $3x + 2y + z = 6$ in the first octant.

Solution We will do this problem in two ways, first using R as the xy projection of S and second as the yz projection, to illustrate that we get the same result.

In the first method, $z = f(x, y)$ (xy projection). Solving for z, we get $f(x, y) = 6 - 3x - 2y$. The region R is the triangular region in the xy-plane bounded by the x- and y-axes and the xy trace of S, $3x + 2y = 6$, as shown in Figure 16.41. Thus, $f_x = -3$ and $f_y = -2$, so $\sqrt{1 + f_x^2 + f_y^2} = \sqrt{14}$. Using Equation 16.27, we get (omitting some details)

$$\iint_S (2xy + xz)\,d\sigma = \iint_R [2xy + x(6 - 3x - 2y)]\sqrt{14}\,dA$$

$$= \sqrt{14} \iint_R (6x - 3x^2)\,dA$$

$$= 3\sqrt{14} \int_0^2 \int_0^{(6-3x)/2} (2x - x^2)\,dy\,dx = 6\sqrt{14}$$

In the second method, $x = f(y, z)$ (yz projection). Solving for x, we get $f(y, z) = (6 - 2y - z)/3$. This time, the region R is the yz projection of S bounded by the y- and z-axes and the line $2y + z = 6$. Since $f_y = -2/3$ and

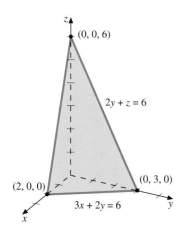

FIGURE 16.41

$f_z = -1/3$, we get $\sqrt{1 + f_y^2 + f_z^2} = \sqrt{14}/3$. Thus, by Equation 16.28

$$\iint_S (2xy + xz)\,d\sigma = \iint_R \left[2\left(\frac{6-2y-z}{3}\right)y + \left(\frac{6-2y-z}{3}\right)z\right]\frac{\sqrt{14}}{3}\,dA$$

$$= \frac{\sqrt{14}}{9}\int_0^3\int_0^{6-2y}(12y - 4y^2 - 4yz + 6z - z^2)\,dz\,dy$$

$$= 6\sqrt{14}$$

Note that in this problem the integration is simpler using the first method. When there is a choice of methods, it pays to look ahead to anticipate which method may result in the easiest integration. ∎

Moments, Mass, and Center of Mass of a Lamina

Suppose a thin sheet (a lamina) of some material (such as metal) is in the shape of a surface S. If the density is $\rho(x, y, z)$ at the point (x, y, z) on S, then by the usual reasoning we can obtain the following formulas for the mass m, the center of mass $(\bar{x}, \bar{y}, \bar{z})$, and the moments of inertia I_x, I_y, and I_z with respect to the coordinate axes.

Mass, Center of Mass, and Moments of Inertia of a Lamina in the Shape of a Surface S

$$m = \iint_S \rho(x, y, z)\,d\sigma \tag{16.31}$$

$$\bar{x} = \frac{1}{m}\iint_S x\,dm, \qquad \bar{y} = \frac{1}{m}\iint_S y\,dm, \qquad \bar{z} = \frac{1}{m}\iint_S z\,dm \tag{16.32}$$

$$I_x = \iint_S (y^2 + z^2)\,dm, \qquad I_y = \iint_S (x^2 + z^2)\,dm, \qquad I_z = \iint_S (x^2 + y^2)\,dm$$
$$\tag{16.33}$$

where $dm = \rho(x, y, z)\,d\sigma$.

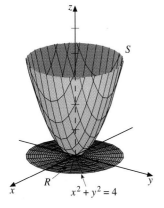

FIGURE 16.42

EXAMPLE 16.22 A homogeneous lamina of density ρ is in the shape of the portion of the paraboloid $z = x^2 + y^2$ between $z = 0$ and $z = 4$. Find the center of mass.

Solution The xy projection of S is the circular region bounded by $x^2 + y^2 = 4$, as shown in Figure 16.42. With $f(x, y) = x^2 + y^2$, we get

$$\sqrt{1 + f_x^2 + f_y^2} = \sqrt{1 + 4(x^2 + y^2)}$$

We first calculate the mass. By Equation 16.31,

$$m = \iint_S \rho\,d\sigma = \iint_R \rho\sqrt{1 + 4(x^2 + y^2)}\,dA = \rho\int_0^{2\pi}\int_0^2 \sqrt{1 + 4r^2}\,r\,dr\,d\theta$$

$$= \frac{\pi\rho}{6}\left[(17)^{3/2} - 1\right]$$

where again we have omitted details of integration. By symmetry, $\bar{x} = \bar{y} = 0$. For $\bar{z}$ we have, by the third of Equations 16.32,

$$m\bar{z} = \iint_S z\, dm = \iint_S z\rho\, d\sigma = \rho \iint_R (x^2 + y^2)\sqrt{1 + 4(x^2 + y^2)}\, dA$$

$$= \rho \int_0^{2\pi} \int_0^2 r^3 \sqrt{1 + 4r^2}\, dr\, d\theta = \frac{\pi \rho}{60}\left[23(17)^{3/2} + 1\right]$$

(The integration with respect to r can be accomplished by the substitution $u = \sqrt{1 + 4r^2}$. You should supply the details.) Thus,

$$\bar{z} = \frac{1}{10}\left[\frac{23(17)^{3/2} + 1}{(17)^{3/2} - 1}\right] \approx 2.335$$

∎

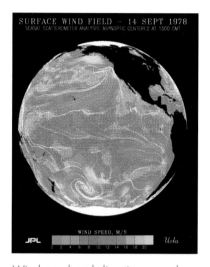

Windspeed and direction over the Pacific Ocean

Three-Dimensional Fluid Flow

Suppose now that $\mathbf{F}(x, y, z)$ is the velocity field of some fluid in steady state. The flow of water in a stream, ocean currents, wind flow, radiation in a star, and the flow of blood in the vascular system can be assumed to have an approximate steady-state motion, at least over relatively short periods of time. Let S be a surface in the given vector field through which the fluid can flow unimpeded. You can think of S as being a screen or netting (or even an imaginary surface). Let $\Delta \sigma$ be the area of one of the cells on S that results from a partition of S, and let $\mathbf{n}$ be a unit normal vector to S at an arbitrary point in this cell. If the cell is small, we can assume the velocity is approximately constant throughout the cell. The dot product $\mathbf{F} \cdot \mathbf{n}$ is the component of $\mathbf{F}$ in the direction of $\mathbf{n}$, and if we multiply this component by $\Delta \sigma$ we get the approximate volume of fluid that flows orthogonally through this cell per unit of time. This volume is represented in Figure 16.43 by the prism with height $\mathbf{F} \cdot \mathbf{n}$. By summing over all such cells and passing to the limit as the norms of partitions approach 0, we

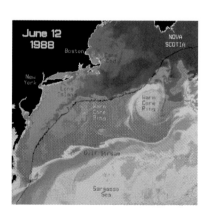

Ocean currents

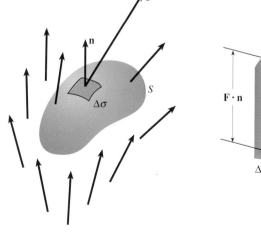

FIGURE 16.43

obtain the *flux of* **F** *through S*:

$$\text{Flux of } \mathbf{F} \text{ Through } S = \iint_S \mathbf{F} \cdot \mathbf{n}\, d\sigma \qquad (16.34)$$

The flux is the total net volume of the fluid that flows through S per unit of time. This volume flux is analogous to flux in two dimensions. If S is a closed surface (such as an ellipsoid), we typically take **n** to be directed toward the exterior to S, called the *outer unit normal*. Then if the flux is positive, there is a net outflow of fluid through S, and we say there is a *source* inside S. If the flux is negative, there is a net inflow of fluid, and we say there is a *sink* inside S. If the flux is 0, the flow is said to be *incompressible*. As we have defined it, flux is the net volume that passes through S per unit of time. If the fluid has density $\rho(x, y, z)$, then the integral $\iint_S \rho \mathbf{F} \cdot \mathbf{n}\, d\sigma$ is the *mass flux*—that is, the net mass of fluid that passes through S per unit of time.

EXAMPLE 16.23 Find the flux of a fluid that has velocity field $\mathbf{F} = 4x\mathbf{i} + 4y\mathbf{j} + 3z\mathbf{k}$ through the parabolic surface S defined by $z = 4 - x^2 - y^2$ for $z \geq 0$ in the direction of outer unit normals.

Solution As shown in Figure 16.44, the outer normals are also the upward normals. In Section 14.4, we showed that a normal vector to the surface $z = f(x, y)$ is $f_x\mathbf{i} + f_y\mathbf{j} - \mathbf{k}$. But we take the negative of this normal so that the z component is positive (upward-directed). We make it into a unit vector by dividing by its length. Thus,

$$\mathbf{n} = \frac{-f_x\mathbf{i} - f_y\mathbf{j} + \mathbf{k}}{\sqrt{1 + f_x^2 + f_y^2}} = \frac{2x\mathbf{i} + 2y\mathbf{j} + \mathbf{k}}{\sqrt{1 + 4(x^2 + y^2)}}$$

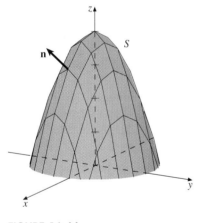

FIGURE 16.44

The flux is therefore

$$\iint_S \mathbf{F} \cdot \mathbf{n}\, d\sigma = \iint_R \frac{8x^2 + 8y^2 + 3(4 - x^2 - y^2)}{\sqrt{1 + 4(x^2 + y^2)}} \sqrt{1 + 4(x^2 + y^2)}\, dA$$

$$= \int_0^{2\pi} \int_0^2 (5r^2 + 12)r\, dr\, d\theta = 88\pi$$

If velocity is in meters per second, for example, and area is in square meters, then the net amount of fluid that flows out of the surface each second is 88π cubic meters. ∎

Although we have illustrated the idea of flux using fluid dynamics, we still call the integral $\iint_S \mathbf{F} \cdot \mathbf{n}\, d\sigma$ the flux of **F** through S for any vector field **F**, whether or not it represents velocity. Some other areas in which this notion is useful are heat flow, electricity, magnetism, and gravitational fields.

Exercise Set 16.5

In Exercises 1–8, evaluate the surface integral $\iint_S g(x, y, z)\, d\sigma$.

1. $g(x, y, z) = 2x - y + z$; S is the first-octant portion of the plane $x + y + z = 2$.

2. $g(x, y, z) = x^2 y - 2z$; S is the first-octant portion of the plane $z = x + 2y$ that lies below $z = 4$.

3. $g(x, y, z) = xz$; S is the portion of the plane $z = 2x - 3y$ inside the cylinder $x^2 + y^2 = 9$.

4. $g(x, y, z) = xy$; S is the first-octant portion of the cylinder $x^2 + z^2 = 4$ between $y = 0$ and $y = 4$ and above $z = 1$.

5. $g(x, y, z) = x^2 + y^2 - z$; S is the portion of the paraboloid $z = 2 - x^2 - y^2$ above the xy-plane.

6. $g(x, y, z) = xyz$; S is the portion of the cone $z^2 = x^2 + y^2$ between $z = 1$ and $z = 2$.

7. $g(x, y, z) = 8/z^2$; S is the portion of the sphere $x^2 + y^2 + z^2 = 25$ above $z = 3$.

8. $g(x, y, z) = xz^2$; S is the portion of the parabolic cylinder $y = x^2$ in the first octant bounded by $y = 2$, $y = 6$, $z = 0$, and $z = 4$. (*Hint:* Use a yz projection.)

In Exercises 9–12, find the mass and center of mass of the lamina in the shape of the surface S with density $\rho(x, y, z)$.

9. S is the first-octant portion of the plane $x + 2y + 4z = 8$; $\rho(x, y, z) = z$.

10. S is the portion of the cylinder $3z = x^2$ lying above the region $R = \{(x, y) : 0 \le x \le 2, 0 \le y \le 4\}$; $\rho = $ constant.

11. S is the portion of the paraboloid $z = 4 - x^2 - y^2$ that is inside the cylinder $x^2 + y^2 = 2$; $\rho = $ constant.

12. S is the upper portion of the sphere $x^2 + y^2 + z^2 = 16$ that is inside the cylinder $x^2 + y^2 = 8$; $\rho = 1/\sqrt{z}$.

In Exercises 13 and 14, set up, but do not evaluate, iterated integrals for I_x, I_y, and I_z for the specified laminas.

13. The lamina of Exercise 10

14. The lamina of Exercise 12

In Exercises 15 and 16, set up, but do not evaluate, iterated integrals for calculating the given surface integrals using (a) xy projections, (b) yz projections, and (c) xz projections.

15. $\iint_S x^2 y z^3\, d\sigma$; S is the first-octant portion of the plane $2x + y - z = 0$ that is below the plane $z = 4$

16. $\iint_S (x + 2yz)\, d\sigma$; S is the first-octant portion of the elliptic paraboloid $z = 12 - 3x^2 - 4y^2$

In Exercises 17 and 18, set up, but do not evaluate, two iterated integrals for calculating the given surface integrals using projections on two different planes.

17. $\iint_S xyz\, d\sigma$; S is the portion of the cylinder $z = 2 - x^2$ in the first octant below $z = 1$ and between $y = 0$ and $y = 5$.

18. $\iint_S (xz/y)\, d\sigma$; S is the first-octant portion of the cylinder $y = e^x$ bounded by the planes $y = 2$ and $z = 3$.

In Exercises 19–26, find the flux of $\mathbf{F}$ through S. Take $\mathbf{n}$ as the upward normal unless otherwise specified.

19. $\mathbf{F} = \langle 3xy, yz, 2z \rangle$; S is the portion of the plane $x + 3y + z = 5$ lying above the region $R = \{(x, y) : 0 \le x \le 2, 0 \le y \le 1\}$.

20. $\mathbf{F} = \langle x, y, z \rangle$; S is the upper portion of the sphere $x^2 + y^2 + z^2 = 25$ that is inside the cylinder $x^2 + y^2 = 16$.

21. $\mathbf{F} = 2x\mathbf{i} + 2y\mathbf{j} + 3z\mathbf{k}$; S is the portion of the paraboloid $z = 4 - x^2 - y^2$ above the xy-plane.

22. $\mathbf{F} = yz\mathbf{i} + (x^3/z)\mathbf{j} + 2z^2\mathbf{k}$; S is the first-octant portion of the cylinder $z = e^x$ bounded above by $z = 2$ and on the right by $y = 2$. Use downward-directed normals.

23. $\mathbf{F} = \langle -x^3, y^3, -z \rangle$; S is the portion of the cone $z^2 = x^2 + y^2$ between $z = 1$ and $z = 2$. Use downward-directed normals.

24. $\mathbf{F} = 3x\mathbf{i} + 3y\mathbf{j} - z\mathbf{k}$; S is the portion of the hemisphere $z = \sqrt{8 - x^2 - y^2}$ that is inside the cone $z^2 = x^2 + y^2$.

25. $\mathbf{F} = (x^3 yz)\mathbf{i} + (x - y^2)\mathbf{j} + z^2\mathbf{k}$; S is the surface $z = 1 - |y|$ above the xy-plane and between $x = 0$ and $x = 4$. (*Hint:* Divide S into two parts and add the integrals over the separate parts.)

26. $\mathbf{F} = \langle xyz, x^2 - y^2, xz - yz \rangle$; S is the cube that is one unit on an edge that has one vertex at the origin and three of its edges on the positive coordinate axes. Use outward-directed normals. Is there a source or a sink within S? (*Hint:* Find the flux through each face and add the results.)

27. Find the mass and center of mass of a homogeneous hemispherical shell of radius a and density $\rho = k$. (*Hint:* To evaluate the improper integral, first integrate over the circular region R_b with radius $b < a$, and after integration let $b \to a^-$.)

28. The velocity field for a certain liquid is $\mathbf{F} = x\mathbf{i} - y\mathbf{j} + z\mathbf{k}$. In it is submerged a closed surface that forms the boundary of the region below the hemisphere $z = \sqrt{2a^2 - x^2 - y^2}$ and above the paraboloid $az = x^2 + y^2$. Find the total flux through S, using outward-directed normals.

29. Let $\mathbf{F}$ be the inverse-square field

$$\mathbf{F} = \frac{k\mathbf{u}}{|\mathbf{r}|^2}$$

where $\mathbf{u}$ is the unit vector in the direction of $\mathbf{r} = x\mathbf{i} + y\mathbf{j} + z\mathbf{k}$. Show that the flux $\mathbf{F}$ through a sphere S centered at the origin is independent of the radius of S.

30. If $\mathbf{F} = \langle L, M, N \rangle$ and S is the graph of $z = f(x, y)$ that has projection R in the xy-plane, then

$$\iint_S \mathbf{F} \cdot \mathbf{n}\, d\sigma = \iint_R (-Lf_x - Mf_y + N)\, dA$$

assuming the appropriate hypotheses. Prove this result.

31. Give a derivation to justify Equations 16.31, 16.32, and 16.33.

16.6 THE DIVERGENCE THEOREM

FIGURE 16.45
A Möbius strip

In this section and the next, we will be making use of surface integrals of vector fields, and for this purpose we limit consideration to surfaces that have two distinct sides, called **orientable surfaces**. It may surprise you to learn that some surfaces have only one side. The best known example is the *Möbius strip*, shown in Figure 16.45. You can easily construct such a surface by taking a strip of paper, giving one end a half twist, and pasting the ends together. To convince yourself that this surface is one-sided, take a crayon and start coloring the "top" side and continue until you return to the starting point. You will discover you have colored the entire surface without ever lifting the crayon from the paper!

Another way to describe an orientable surface S is by means of its unit normal vectors. If a direction for a unit normal can be chosen in such a way that, starting from any point on S and going around any closed curve C on S, the unit normal returns to its original direction at the starting point, then S is orientable. In Figure 16.46, we indicate how the Möbius strip fails this test. We call whichever unit normal we have selected positive, and then we say that S is *oriented* with respect to that normal. In effect, we have designated a positive side to the surface. For example, a surface may be oriented by upward normals. Then we are calling the top side positive.

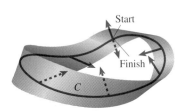

FIGURE 16.46
Normals along C do not return to their original position.

Green's Theorem enabled us to express a line integral around a closed path C as a double integral over the region enclosed by C. The theorem we consider in this section enables us to express a surface integral over a closed surface S as a triple integral over the region enclosed by S. Before stating the theorem, we introduce the notion of the *divergence* of a vector field in $\mathbb{R}^3$, which is the natural extension of divergence in $\mathbb{R}^2$.

Definition 16.4
The Divergence of a Vector Field

Let $\mathbf{F} = \langle f, g, h \rangle$ be a vector field in $\mathbb{R}^3$ for which the partial derivatives of f, g, and h exist. The **divergence** of $\mathbf{F}$, written $\text{div}\,\mathbf{F}$, is the scalar field defined by

$$\text{div}\,\mathbf{F} = \frac{\partial f}{\partial x} + \frac{\partial g}{\partial y} + \frac{\partial h}{\partial z}$$

The next theorem is known as both the *Divergence Theorem* and *Gauss's Theorem*. A precise formulation of the hypotheses of the theorem would require a deeper background in three-dimensional regions and their boundaries than we have presented. Our formulation is sufficient for most applications.

THEOREM 16.10

The Divergence Theorem (Gauss's Theorem)

Let G be a closed and bounded three-dimensional region with boundary S that is a piecewise-smooth closed surface. If $\mathbf{F}$ is a continuously differentiable vector field on some open set that contains G, then

$$\iint_S \mathbf{F} \cdot \mathbf{n}\, d\sigma = \iiint_G \operatorname{div} \mathbf{F}\, dV \tag{16.35}$$

where $\mathbf{n}$ is the outer unit normal to S.

REMARK

■ Observe the similarity between Equations 16.35 and 16.25 for two dimensions.

Proof We will give a proof of the theorem when G is a special (but commonly occurring) type of region that we call a **simple** three-dimensional region. By a simple region, we mean that G is bounded above and below by the graphs of two smooth functions $z = \phi_1(x, y)$ and $z = \phi_2(x, y)$ with $\phi_1(x, y) \leq \phi_2(x, y)$ for all points (x, y) in the projection R of G onto the xy-plane. Similarly, for G to be simple, we require that the lateral bounding surfaces of G be graphs of smooth functions of y and z in the x direction and of x and z in the y direction. A rectangular box is an example of a simple region, as is an ellipsoid.

Let $\mathbf{F}(x, y, z) = f(x, y, z)\mathbf{i} + g(x, y, z)\mathbf{j} + h(x, y, z)\mathbf{k}$. Then

$$\iint_S \mathbf{F} \cdot \mathbf{n}\, d\sigma = \iint_S f\mathbf{i} \cdot \mathbf{n}\, d\sigma + \iint_S g\mathbf{j} \cdot \mathbf{n}\, d\sigma + \iint_S h\mathbf{k} \cdot \mathbf{n}\, d\sigma$$

Also,

$$\iiint_G \operatorname{div} \mathbf{F}\, dV = \iiint_G \frac{\partial f}{\partial x}\, dV + \iiint_G \frac{\partial g}{\partial y}\, dV + \iiint_G \frac{\partial h}{\partial z}\, dV$$

Thus, to prove Equation 16.35, it is sufficient to prove that

$$\iint_S f\mathbf{i} \cdot \mathbf{n}\, d\sigma = \iiint_G \frac{\partial f}{\partial x}\, dV \tag{16.36}$$

$$\iint_S g\mathbf{j} \cdot \mathbf{n}\, d\sigma = \iiint_G \frac{\partial g}{\partial y}\, dV \tag{16.37}$$

$$\iint_S h\mathbf{k} \cdot \mathbf{n}\, d\sigma = \iiint_G \frac{\partial h}{\partial z}\, dV \tag{16.38}$$

We will prove Equation 16.38 only and leave the proofs of the other two equations as an exercise. (See Exercise 19 in Exercise Set 16.6.)

We are assuming G is a simple region. So we can write

$$G = \{(x, y, z): (x, y) \in R, \phi_1(x, y) \leq z \leq \phi_2(x, y)\}$$

where R is the xy projection of G. As we show in Figure 16.47, let S_1 be the lower bounding surface (the graph of $z = \phi_1(x, y)$) and S_2 the upper bounding surface (the graph of $z = \phi_2(x, y)$). Also, let S_3, S_4, S_5, and S_6 be the vertical lateral bounding surfaces, as shown. (In some cases, such as an ellipsoid, there may not be any vertical surfaces.)

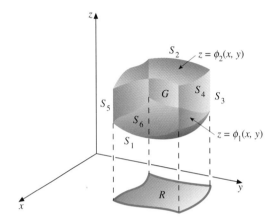

FIGURE 16.47

On each of the lateral surfaces S_3, S_4, S_5, and S_6, the unit normal vector $\mathbf{n}$ is horizontal. Thus, $\mathbf{k} \cdot \mathbf{n} = 0$ on these lateral surfaces. It follows that

$$\iint_S h\mathbf{k} \cdot \mathbf{n} \, d\sigma = \iint_{S_1} h\mathbf{k} \cdot \mathbf{n} \, d\sigma + \iint_{S_2} h\mathbf{k} \cdot \mathbf{n} \, d\sigma$$

The outer unit normal on S_1 is directed downward. So on S_1

$$\mathbf{n} = \frac{\frac{\partial \phi_1}{\partial x}\mathbf{i} + \frac{\partial \phi_1}{\partial y}\mathbf{j} - \mathbf{k}}{\sqrt{1 + \left(\frac{\partial \phi_1}{\partial x}\right)^2 + \left(\frac{\partial \phi_1}{\partial y}\right)^2}}$$

Thus,

$$\mathbf{k} \cdot \mathbf{n} = \frac{-1}{\sqrt{1 + \left(\frac{\partial \phi_1}{\partial x}\right)^2 + \left(\frac{\partial \phi_1}{\partial y}\right)^2}}$$

Similarly, on the upper boundary, S_2,

$$\mathbf{n} = \frac{-\frac{\partial \phi_2}{\partial x}\mathbf{i} - \frac{\partial \phi_2}{\partial y}\mathbf{j} + \mathbf{k}}{\sqrt{1 + \left(\frac{\partial \phi_2}{\partial x}\right)^2 + \left(\frac{\partial \phi_2}{\partial y}\right)^2}}$$

so that

$$\mathbf{k} \cdot \mathbf{n} = \frac{1}{\sqrt{1 + \left(\dfrac{\partial \phi_2}{\partial x}\right)^2 + \left(\dfrac{\partial \phi_2}{\partial y}\right)^2}}$$

We therefore have

$$\iint_S h\mathbf{k} \cdot \mathbf{n}\, d\sigma = \iint_{S_1} h\mathbf{k} \cdot \mathbf{n}\, d\sigma + \iint_{S_2} h\mathbf{k} \cdot \mathbf{n}\, d\sigma$$

$$= \iint_{S_1} \frac{-h(x, y, z)\, d\sigma}{\sqrt{1 + \left(\dfrac{\partial \phi_1}{\partial x}\right)^2 + \left(\dfrac{\partial \phi_1}{\partial y}\right)^2}}$$

$$+ \iint_{S_2} \frac{h(x, y, z)\, d\sigma}{\sqrt{1 + \left(\dfrac{\partial \phi_2}{\partial x}\right)^2 + \left(\dfrac{\partial \phi_2}{\partial y}\right)^2}}$$

We evaluate the two surface integrals on the right, by Equation 16.27, as double integrals over the xy projection R:

$$\iint_{S_1} \frac{-h(x, y, z)\, d\sigma}{\sqrt{1 + \left(\dfrac{\partial \phi_1}{\partial x}\right)^2 + \left(\dfrac{\partial \phi_1}{\partial y}\right)^2}}$$

$$= \iint_R \frac{-h(x, y, \phi_1(x, y))}{\sqrt{1 + \left(\dfrac{\partial \phi_1}{\partial x}\right)^2 + \left(\dfrac{\partial \phi_1}{\partial y}\right)^2}} \sqrt{1 + \left(\dfrac{\partial \phi_1}{\partial x}\right)^2 + \left(\dfrac{\partial \phi_1}{\partial y}\right)^2}\, dA$$

$$= \iint_R -h(x, y, \phi_1(x, y))\, dA$$

Similarly,

$$\iint_{S_2} \frac{h(x, y, z)\, d\sigma}{\sqrt{1 + \left(\dfrac{\partial \phi_2}{\partial x}\right)^2 + \left(\dfrac{\partial \phi_2}{\partial y}\right)^2}}$$

$$= \iint_R \frac{h(x, y, \phi_2(x, y))}{\sqrt{1 + \left(\dfrac{\partial \phi_2}{\partial x}\right)^2 + \left(\dfrac{\partial \phi_2}{\partial y}\right)^2}} \sqrt{1 + \left(\dfrac{\partial \phi_2}{\partial x}\right)^2 + \left(\dfrac{\partial \phi_2}{\partial y}\right)^2}\, dA$$

$$= \iint_R h(x, y, \phi_2(x, y))\, dA$$

Combining the integrals over S_1 and S_2 gives

$$\iint_S h\mathbf{k} \cdot \mathbf{n}\, d\sigma = \iint_R \left[h(x, y, \phi_2(x, y)) - h(x, y, \phi_1(x, y))\right] dA \qquad (16.39)$$

The right-hand side of Equation 16.38 can be written as

$$\iiint_G \frac{\partial h}{\partial z} dV = \iint_R \left[\int_{\phi_1(x,y)}^{\phi_2(x,y)} \frac{\partial h}{\partial z} dz \right] dA$$

$$= \iint_R \left[h(x, y, \phi_2(x, y)) - h(x, y, \phi_1(x, y)) \right] dA \quad \text{By the Second Fundamental Theorem of Calculus}$$

The equality of this last integral and the integral on the right-hand side of Equation 16.39 establishes the result given in Equation 16.38:

$$\iint_S h\mathbf{k} \cdot \mathbf{n} \, d\sigma = \iiint_G \frac{\partial h}{\partial z} dV$$

Equations 16.36 and 16.37 are proved in a similar manner. Thus, Equation 16.35 is true when G is simple. ∎

The proof we have given can be extended to a region $G = G_1 \cup G_2$, where G_1 and G_2 are simple regions with an intersection that is a sectionally smooth surface, and by induction to any finite union of such simple regions. You will be asked to show this extension in Exercise 20 of Exercise Set 16.6.

EXAMPLE 16.24 Use the Divergence Theorem to evaluate $\iint_S \mathbf{F} \cdot \mathbf{n} \, d\sigma$, where S is the sphere $x^2 + y^2 + z^2 = a^2$ and $\mathbf{F} = xy^2\mathbf{i} + yz^2\mathbf{j} + x^2z\mathbf{k}$. Assume $\mathbf{n}$ is the outer unit normal.

Solution Let G be the sphere together with its interior. Then, by Equation 16.35,

$$\iint_S \mathbf{F} \cdot \mathbf{n} \, d\sigma = \iiint_G \text{div } \mathbf{F} \, dV$$

$$= \iiint_G (y^2 + z^2 + x^2) \, dV$$

Because G is a sphere and because of the nature of the integrand, it is convenient to change to spherical coordinates. The sphere has equation $\rho = a$, and the integrand becomes ρ^2. Thus,

$$\iint_S \mathbf{F} \cdot \mathbf{n} \, d\sigma = \int_0^\pi \int_0^{2\pi} \int_0^a \rho^2(\rho^2 \sin \phi \, d\rho \, d\theta \, d\phi) = \frac{4\pi a^5}{5}$$

(You may wish to try evaluating the surface integral in this problem *without* using the Divergence Theorem, to compare the difficulty.) ∎

EXAMPLE 16.25 Evaluate the integral $\iint_S \mathbf{F} \cdot \mathbf{n} \, d\sigma$, where S is the boundary of the region G below the paraboloid $z = 4 - x^2 - y^2$, inside the cylinder $x^2 + y^2 = 1$, and above the xy-plane, and where

$$\mathbf{F} = \left\langle 2x + \sqrt{z^3}, 3y - e^{z^2}, (x^3 + y^3)^{4/3} \right\rangle$$

Use outer unit normals. (See Figure 16.48.)

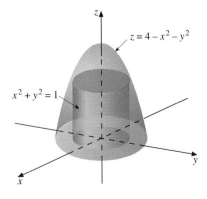

FIGURE 16.48

Solution Since div $\mathbf{F} = 2+3+0 = 5$, we have, by the Divergence Theorem,

$$\iint\limits_{S} \mathbf{F} \cdot \mathbf{n}\, d\sigma = \iiint\limits_{G} 5\, dV$$

This time we will use cylindrical coordinates. The paraboloid has the cylindrical equation $z = 4 - r^2$, and the cylinder has the equation $r = 1$. So we have

$$\iint\limits_{S} \mathbf{F} \cdot \mathbf{n}\, d\sigma = 5 \int_{0}^{2\pi} \int_{0}^{1} \int_{0}^{4-r^2} r\, dz\, dr\, d\theta$$

$$= 5 \int_{0}^{2\pi} \int_{0}^{1} (4r - r^3)\, dr\, d\theta = \frac{35\pi}{2}$$

Without the Divergence Theorem this problem would be virtually impossible to solve. (Try it!) ∎

Relationship Between Flux and Divergence

If $\mathbf{F}$ is the velocity field of a fluid, we know that $\iint_S \mathbf{F} \cdot \mathbf{n}\, d\sigma$ is the flux of $\mathbf{F}$ through S. So, assuming appropriate conditions on $\mathbf{F}$, S, and G, the Divergence Theorem says that

$$\text{Flux of } \mathbf{F} \text{ Through } S = \iiint\limits_{G} \text{div}\, \mathbf{F}\, dV$$

The physical interpretation of div $\mathbf{F}$ at a point P is analogous to that in two dimensions. We let S_ε be a sphere of radius ε centered at P, and let G_ε be the region enclosed by S_ε. Then, using a Mean-Value Theorem for triple integrals, we can write

$$\iiint\limits_{G_\varepsilon} \text{div}\, \mathbf{F}\, dV = [\text{div}\, \mathbf{F}(Q)]\, (\text{Volume of } G_\varepsilon)$$

where Q is some point in G_ε. Thus,

$$\text{div}\, \mathbf{F}(Q) = \frac{\text{Flux of } \mathbf{F} \text{ Through } S_\varepsilon}{\text{Volume of } G_\varepsilon}$$

Now we let $\varepsilon \to 0$, so that $Q \to P$, and if div $\mathbf{F}$ is continuous, div $\mathbf{F}(Q) \to$ div $\mathbf{F}(P)$. Thus,

$$\text{div}\, \mathbf{F}(P) = \lim_{\varepsilon \to 0} \frac{\text{Flux of } \mathbf{F} \text{ Through } S_\varepsilon}{\text{Volume of } G_\varepsilon}$$

The limit on the right is the flux per unit volume at P, called the *flux density* of $\mathbf{F}$ at P. So the divergence of $\mathbf{F}$ at a point is the flux density at that point. Just as with two-dimensional flow, we call P a *source* if div $P > 0$ and a *sink* if div $P < 0$.

The Divergence Theorem says that we can obtain the flux of $\mathbf{F}$ through the closed surface S by integrating the flux density over the volume G enclosed by S:

$$\text{Flux of } \mathbf{F} \text{ Through } S = \iiint\limits_{G} (\text{Flux Density of } \mathbf{F})\, dV$$

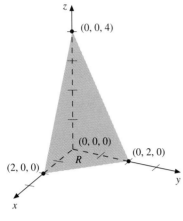

FIGURE 16.49

EXAMPLE 16.26   Use the Divergence Theorem to find the flux of the velocity field $\mathbf{F}(x, y, z) = 2xy\mathbf{i} + 3yz\mathbf{j} + xz\mathbf{k}$ through the tetrahedron with vertices $(0, 0, 0)$, $(2, 0, 0)$, $(0, 2, 0)$, and $(0, 0, 4)$.

Solution   Let S denote the surface of the tetrahedron. We show it in Figure 16.49. Let G denote the region enclosed by S. Its xy projection is the triangular region R, as shown, bounded by the lines $x = 0$, $y = 0$, and $x + y = 2$. The inclined plane that forms the upper bounding surface of G has the equation $2x + 2y + z = 4$, so $z = 4 - 2x - 2y$. Applying the Divergence Theorem, we have

$$\text{Flux of } \mathbf{F} \text{ Through } S = \iiint_G (\text{div } \mathbf{F}) \, dV$$

$$= \int_0^2 \int_0^{2-x} \int_0^{4-2x-2y} (2y + 3z + x) \, dz \, dy \, dx = 12 \quad \blacksquare$$

Exercise Set 16.6

In Exercises 1–6, use the Divergence Theorem to evaluate $\iint_S \mathbf{F} \cdot \mathbf{n} \, d\sigma$, where $\mathbf{n}$ is the outer unit normal.

1. $\mathbf{F}(x, y, z) = x^2\mathbf{i} + y^2\mathbf{j} + z^2\mathbf{k}$; S is the rectangular parallelepiped formed by the planes $x = 1$, $x = 3$, $y = 2$, $y = 6$, $z = 0$, and $z = 4$.

2. $\mathbf{F}(x, y, z) = (2x + y)\mathbf{i} + (x - y)\mathbf{j} + x^2 y^3 \mathbf{k}$; S is the tetrahedron formed by the planes $3x + 2y + z = 6$, $x = 0$, $y = 0$, and $z = 0$.

3. $\mathbf{F}(x, y, z) = \langle x + z, y^2, yz \rangle$; S is the boundary of the first-octant region enclosed by the cylinder $z = 4 - x^2$ and the planes $y = 0$, $z = 0$, and $y = x$.

4. $\mathbf{F}(x, y, z) = \langle xz, 2xy, 4yz \rangle$; S is the tetrahedron formed by the planes $x + y = 2$, $y + z = 2$, $x = 0$, $y = 0$, and $z = 0$.

5. $\mathbf{F}(x, y, z) = (e^z \sin y)\mathbf{i} + (e^z \cos x)\mathbf{j} + z\mathbf{k}$; S is the boundary of the region inside the cylinder $x^2 + y^2 = 4$ between the planes $z = 0$ and $z = 6 + x + 2y$.

6. $\mathbf{F}(x, y, z) = \langle x^3, y^3, \cosh x^3 \rangle$; S is the boundary of the region inside the cone $z = \sqrt{x^2 + y^2}$ between $z = 1$ and $z = 2$.

In Exercises 7–12, use the Divergence Theorem to find the flux of $\mathbf{F}$ through S in the direction of outer unit normals.

7. $\mathbf{F}(x, y, z) = (2x - 3y)\mathbf{i} + (4y + 2z)\mathbf{j} + (x + z)\mathbf{k}$; S is the sphere $x^2 + y^2 + z^2 = 16$.

8. $\mathbf{F}(x, y, z) = (e^y \cos z)\mathbf{i} + (e^z \sin x)\mathbf{j} + (e^{x^2+y^2})\mathbf{k}$; S is the ellipsoid $3x^2 + 7y^2 + 12z^2 = 84$.

9. $\mathbf{F}(x, y, z) = \langle y/x, x/y, 1/z^2 \rangle$; S is the cube formed by the planes $x = 1$, $x = 2$, $y = 1$, $y = 2$, $z = 2$, and $z = 3$.

10. $\mathbf{F}(x, y, z) = \langle xy^2, yz^2, zx^2 \rangle$; S is the "ice cream cone" formed by the cone $z = \sqrt{x^2 + y^2}$ and the hemisphere $z = \sqrt{4 - x^2 - y^2}$.

11. $\mathbf{F}(x, y, z) = x^2 y\mathbf{i} + xy^2\mathbf{j} + z^2\mathbf{k}$; S is the boundary of the region inside the paraboloid $z = x^2 + y^2$ and below the hemisphere $z = \sqrt{2 - x^2 - y^2}$.

12. $\mathbf{F}(x, y, z) = xy\mathbf{i} + y^2\mathbf{j} + yz\mathbf{k}$; S is the boundary of the first-octant region inside the cylinder $x^2 + z^2 = 4$ between the planes $y = 0$ and $y = 2x$.

In Exercises 13–16, verify the Divergence Theorem by calculating $\iint_S \mathbf{F} \cdot \mathbf{n} \, d\sigma$ and $\iiint_G \text{div } \mathbf{F} \, dV$.

13. $\mathbf{F}(x, y, z) = 2x\mathbf{i} - 3y\mathbf{j} + 4z\mathbf{k}$; G is the region inside the paraboloid $z = x^2 + y^2$ and below the plane $z = 4$.

14. $\mathbf{F}(x, y, z) = \langle x - y, x + y, 2x \rangle$; G is the region enclosed by the tetrahedron formed by the coordinate planes and the plane $x + y + z = 2$.

15. $\mathbf{F}(x, y, z) = \langle 2x, 3y, z \rangle$; G is the first-octant region inside both of the cylinders $x^2 + y^2 = a^2$ and $x^2 + z^2 = a^2$.

16. $\mathbf{F}(x, y, z) = x^{3/2}\mathbf{i} + y^{3/2}\mathbf{j} + z^{3/2}\mathbf{k}$; G is the region enclosed by the planes $x + y = 4$, $z = 4$, and the coordinate planes.

17. Show that if a region G and its boundary S satisfy the conditions of the Divergence Theorem, then for $\mathbf{F}(x, y, z) = x\mathbf{i} + y\mathbf{j} + z\mathbf{k}$,
$$\iint_S \mathbf{F} \cdot \mathbf{n}\, d\sigma = 3V$$
where V is the volume of G.

18. Let G satisfy the Divergence Theorem hypotheses. Prove that the flux of any constant vector field through S is 0.

19. Verify Equations 16.36 and 16.37 for a simple region.

20. Let $G = G_1 \cup G_2$, where G_1 and G_2 are simple regions with an intersection that is a sectionally smooth surface T. Prove that the Divergence Theorem holds true for G. Extend the result by induction to a finite union of simple regions. (*Hint:* The outer unit normals for G_1 and G_2 across their common boundary T are oppositely directed.)

21. Let S be the sphere $x^2 + y^2 + (z - a)^2 = a^2$ and $\mathbf{F} = \langle x^2, y^2, z^2 \rangle$. Show that $\iint_S \mathbf{F} \cdot \mathbf{n}\, d\sigma = 8\pi a^4/3$. (*Hint:* Use the Divergence Theorem and spherical coordinates.)

22. By Coulomb's Law the force field $\mathbf{F}$ of a point charge of q coulombs located at the origin is
$$\mathbf{F} = \frac{cq\mathbf{u}}{|\mathbf{r}|^2}$$
where $\mathbf{r}$ is the position vector of a point P in space, $\mathbf{u}$ is a unit vector in the direction of $\mathbf{r}$, and c is a constant. Prove that the flux of $\mathbf{F}$ through any sectionally smooth closed surface S with the origin in its interior is $4\pi qc$. (*Hint:* Use the Divergence Theorem to show that if S_1 is a sphere centered at the origin lying inside S, then $\iint_S \mathbf{F} \cdot \mathbf{n}\, d\sigma = -\iint_{S_1} \mathbf{F} \cdot \mathbf{n}\, d\sigma$ with $\mathbf{n}$ directed toward the origin for S_1.)

Exercises 23–26 are to be done in sequence. The formulas in Exercises 23 and 24 are known as **Green's Identities**.

23. If f is a scalar function whose second partials exist, the **Laplacian** of f, denoted by $\nabla^2 f$, is defined by
$$\nabla^2 f = \frac{\partial^2 f}{\partial x^2} + \frac{\partial^2 f}{\partial y^2} + \frac{\partial^2 f}{\partial z^2}$$
Prove that if u and v are scalar functions that satisfy appropriate continuity requirements and G and S are as in the Divergence Theorem, then
$$\iiint_G (u\nabla^2 v + \nabla u \cdot \nabla v)\, dV = \iint_S u\nabla v \cdot \mathbf{n}\, d\sigma$$
Hint: Take $\mathbf{F} = \left\langle u\dfrac{\partial v}{\partial x}, u\dfrac{\partial v}{\partial y}, u\dfrac{\partial v}{\partial z} \right\rangle$.

24. Using Exercise 23, prove that
$$\iiint_G (u\nabla^2 v - v\nabla^2 u)\, dV = \iint_S (u\nabla v - v\nabla u) \cdot \mathbf{n}\, d\sigma$$
(*Hint:* Make use of Exercise 23 twice, once as it stands and once with u and v interchanged.)

25. Prove that
$$\iiint_G \nabla^2 u\, dV = \iint_S \nabla u \cdot \mathbf{n}\, d\sigma$$

26. Let $\mathbf{F}$ be a gradient field with potential function ϕ. Prove that if div $\mathbf{F} = 0$,
$$\iiint_G |\mathbf{F}|^2\, dV = \iint_S \phi\mathbf{F} \cdot \mathbf{n}\, d\sigma$$

16.7 STOKES'S THEOREM

The main result of this section is a generalization of Green's Theorem to three dimensions, called Stokes's Theorem, named for the English mathematical physicist George Stokes (1819–1903). Green's Theorem relates a line integral around a closed curve C in the plane to the double integral over the region enclosed by C. In a similar way, Stokes's Theorem relates the line integral around a closed curve C in space to the surface integral over a surface that has C as its boundary.

Definition 16.5
Curl of a Vector Field

Let $\mathbf{F} = \langle f, g, h \rangle$ be a vector field for which the first partial derivatives of f, g, and h exist in some open region of $\mathbb{R}^3$. Then the **curl of F** is the vector

$$\operatorname{curl} \mathbf{F} = \left\langle \frac{\partial h}{\partial y} - \frac{\partial g}{\partial z}, \frac{\partial f}{\partial z} - \frac{\partial h}{\partial x}, \frac{\partial g}{\partial x} - \frac{\partial f}{\partial y} \right\rangle \qquad (16.40)$$

We can compute curl **F** using the symbolic determinant

$$\operatorname{curl} \mathbf{F} = \begin{vmatrix} \mathbf{i} & \mathbf{j} & \mathbf{k} \\ \dfrac{\partial}{\partial x} & \dfrac{\partial}{\partial y} & \dfrac{\partial}{\partial z} \\ f & g & h \end{vmatrix} \qquad (16.41)$$

EXAMPLE 16.27 Find curl **F**, where

$$\mathbf{F}(x, y, z) = x^2 y \mathbf{i} + (2y - z)\mathbf{j} + xyz\mathbf{k}$$

Solution Using Equation 16.41, we get

$$\operatorname{curl} \mathbf{F} = \begin{vmatrix} \mathbf{i} & \mathbf{j} & \mathbf{k} \\ \dfrac{\partial}{\partial x} & \dfrac{\partial}{\partial y} & \dfrac{\partial}{\partial z} \\ x^2 y & 2y - z & xyz \end{vmatrix} = (xz + 1)\mathbf{i} - yz\mathbf{j} - x^2 \mathbf{k} \qquad \blacksquare$$

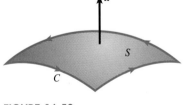

FIGURE 16.50

Now let S be a smooth oriented surface, and let its boundary be a sectionally smooth simple closed curve C. We will say that C is *positively oriented with respect to S* if, when viewed from the tip of a positive unit normal **n** to S, C is oriented in a counterclockwise direction. This orientation is illustrated in Figure 16.50. The direction of C then is such that if you walked around C in its positive direction, with your head in the direction of **n**, the surface S would always be on your left.

THEOREM 16.11

Stokes's Theorem

Let C be a piecewise-smooth simple closed curve that forms the boundary of a smooth oriented surface S, and let C be positively oriented with respect to S. Then if **F** is a continuously differentiable vector field on some open set that contains both S and C,

$$\int_C \mathbf{F} \cdot d\mathbf{r} = \iint_S (\operatorname{curl} \mathbf{F}) \cdot \mathbf{n}\, d\sigma \qquad (16.42)$$

We will not give a proof here, but we illustrate the result and see some of its consequences. First, let us show that the theorem does generalize Green's Theorem in the plane. We can think of the two-dimensional vector field

$\mathbf{F} = \langle f, g \rangle$ as being three-dimensional, with $h = 0$; that is, we can write $\mathbf{F} = \langle f, g, 0 \rangle$. Then it is easy to verify that

$$\text{curl } \mathbf{F} = \left(\frac{\partial g}{\partial x} - \frac{\partial f}{\partial y}\right) \mathbf{k}$$

A region R in the plane with boundary C oriented counterclockwise is a smooth surface in $\mathbb{R}^3$ oriented by upward unit normals; that is, $\mathbf{n} = \mathbf{k}$. Hence,

$$(\text{curl } \mathbf{F}) \cdot \mathbf{n} = \frac{\partial g}{\partial x} - \frac{\partial f}{\partial y}$$

and so Equation 16.42 reduces to

$$\int_C f\, dx + g\, dy = \iint_R \left(\frac{\partial g}{\partial x} - \frac{\partial f}{\partial y}\right) dA$$

which is the conclusion in Green's Theorem.

EXAMPLE 16.28 Use Stokes's Theorem to evaluate the integral

$$\int_C (x^2 + y^2)\, dx + xy^2\, dy + xyz\, dz$$

where C is the boundary of the surface S consisting of the first-octant portion of the cylinder $z = 4 - x^2$ between the planes $y = 0$ and $y = 2x$. Orient S with upward unit normals and orient C positively with respect to S.

Solution In Figure 16.51, we show the surface S and its boundary $C = C_1 \cup C_2 \cup C_3$. We could parameterize each of these component curves and evaluate the integral directly, but it is easier to use Stokes's Theorem. Let $\mathbf{F} = (x^2 + y^2)\mathbf{i} + xy^2\mathbf{j} + xyz\mathbf{k}$. By Equation 16.40 or 16.41, we find that curl $\mathbf{F} = xz\mathbf{i} - yz\mathbf{j} + (y^2 - 2y)\mathbf{k}$. The upward unit normal $\mathbf{n}$ to $z = 4 - x^2$ is

$$\mathbf{n} = \frac{-\frac{\partial z}{\partial x}\mathbf{i} - \frac{\partial z}{\partial y}\mathbf{j} + \mathbf{k}}{\sqrt{1 + \left(\frac{\partial z}{\partial x}\right)^2 + \left(\frac{\partial z}{\partial y}\right)^2}} = \frac{2x\mathbf{i} + \mathbf{k}}{\sqrt{1 + 4x^2}}$$

So we have

$$\int_C \mathbf{F} \cdot d\mathbf{r} = \iint_S (\text{curl } \mathbf{F}) \cdot \mathbf{n}\, d\sigma = \iint_S \frac{2x^2 z + y^2 - 2y}{\sqrt{1 + 4x^2}}\, d\sigma$$

Using Equation 16.27, this surface integral can be evaluated as the following double integral over the xy projection R of the surface S.

$$\iint_R \frac{2x^2(4 - x^2) + y^2 - 2y}{\sqrt{1 + 4x^2}} \sqrt{1 + 4x^2}\, dA$$

$$= \int_0^2 \int_0^{2x} (8x^2 - 2x^4 + y^2 - 2y)\, dy\, dx$$

$$= \frac{64}{3} \qquad \blacksquare$$

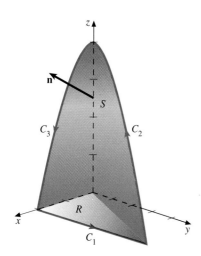

FIGURE 16.51

George Stokes (1819–1903) was a British mathematician and physicist. Following publication of his papers on fluid motion, he was appointed to the Lucasian Professorship of Mathematics of Cambridge at the age of 30. He originated the word *fluorescence* and used the property to study ultraviolet light. He argued on the side of those who believed that light is made of waves, not particles, and that light traveled through an ether, an idea later overturned through the experiments of Michelson and Morley. Sir George, as he became, was the first person since Newton to achieve the positions of Lucasian Professor and then secretary and president of the Royal Society.

EXAMPLE 16.29 Verify Stokes's Theorem for $\mathbf{F}(x, y, z) = \langle x+y, y-x, z \rangle$, where S is the portion of the paraboloid $z = x^2 + y^2$ below $z = 4$, oriented by downward unit normals.

Solution The surface S and its boundary C are shown in Figure 16.52. Note that with the unit normal $\mathbf{n}$ directed downward, the positive orientation for C is clockwise, as shown. We will verify Equation 16.42 by calculating each side separately.

The curve $\check{C}$ can be parameterized by

$$\begin{cases} x = 2\cos(-t) = 2\cos t \\ y = 2\sin(-t) = -2\sin t, \\ z = 4 \end{cases} \quad 0 \le t \le 2\pi$$

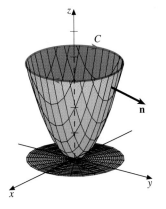

FIGURE 16.52

So $dx = -2\sin t\, dt$, $dy = -2\cos t\, dt$, and $dz = 0$. Thus,

$$\int_C \mathbf{F} \cdot d\mathbf{r} = \int_C (x+y)\, dx + (y-x)\, dy + z\, dz$$

$$= \int_0^{2\pi} [(2\cos t - 2\sin t)(-2\sin t)$$

$$+ (-2\sin t - 2\cos t)(-2\cos t)]\, dt$$

$$= 4\int_0^{2\pi} (\sin^2 t + \cos^2 t)\, dt = 8\pi$$

Next, we calculate the integral $\iint_S (\operatorname{curl} \mathbf{F}) \cdot \mathbf{n}\, d\sigma$ on the right-hand side of Equation 16.42. We find that $\operatorname{curl} \mathbf{F} = -2\mathbf{k}$. For $z = x^2 + y^2$, the downward unit normal is

$$\mathbf{n} = \frac{\dfrac{\partial z}{\partial x}\mathbf{i} + \dfrac{\partial z}{\partial y}\mathbf{j} - \mathbf{k}}{\sqrt{1 + \left(\dfrac{\partial z}{\partial x}\right)^2 + \left(\dfrac{\partial z}{\partial y}\right)^2}} = \frac{2x\mathbf{i} + 2y\mathbf{j} - \mathbf{k}}{\sqrt{1 + 4x^2 + 4y^2}}$$

So, by Equation 16.27,

$$\iint_S (\operatorname{curl} \mathbf{F}) \cdot \mathbf{n}\, d\sigma = \iint_S \frac{2}{\sqrt{1 + 4x^2 + 4y^2}}\, d\sigma$$

$$= \iint_R \frac{2}{\sqrt{1 + 4x^2 + 4y^2}} \sqrt{1 + 4x^2 + 4y^2}\, dA$$

$$= 2\iint_R dA = 2(\text{Area of } R) = 2(4\pi) = 8\pi$$

Our answers for the integrals on the two sides of Equation 16.42 agree, so we have verified the truth of Stokes's Theorem in this case. ∎

REMARK

■ Using ideas similar to those for extending Green's Theorem to multiply connected regions, we can also extend Stokes's Theorem to surfaces with holes, whose boundaries therefore consist of unions of two or more disjoint closed curves. We will not pursue these ideas here, however.

Application of Stokes's Theorem to Fluid Flow

To gain some insight into the physical interpretation of Stokes's Theorem, we return to the notion of fluid flow in which **F** is the velocity field of the fluid. For a closed curve C in the region of flow, then, just as in two-dimensional flow, the integral

$$\int_C \mathbf{F} \cdot d\mathbf{r} = \int_C \mathbf{F} \cdot \mathbf{T}\, ds \qquad \text{Where } \mathbf{T} \text{ is the unit tangent vector to } C$$

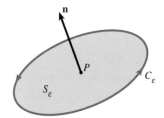

FIGURE 16.53

is called the **circulation of F around** C. Suppose now that S_ε is a small disk of radius ε, centered at a point P in the velocity field. Let **n** be the upward unit normal to S_ε, and let C_ε be the positively oriented boundary of S_ε, as in Figure 16.53. Then, by Stokes's Theorem,

$$\int_{C_\varepsilon} \mathbf{F} \cdot d\mathbf{r} = \iint_{S_\varepsilon} (\text{curl } \mathbf{F}) \cdot \mathbf{n}\, d\sigma$$

A mean-value theorem for surface integrals enables us to write

$$\iint_{S_\varepsilon} (\text{curl } \mathbf{F}) \cdot \mathbf{n}\, d\sigma = [\text{curl } \mathbf{F}(Q) \cdot \mathbf{n}](\text{Area of } S_\varepsilon)$$

where Q is some point in S_ε. Thus,

$$\text{curl } \mathbf{F}(Q) \cdot \mathbf{n} = \frac{\iint_{S_\varepsilon} (\text{curl } \mathbf{F}) \cdot \mathbf{n}\, d\sigma}{\text{Area of } S_\varepsilon}$$

$$= \frac{\int_{C_\varepsilon} \mathbf{F} \cdot d\mathbf{r}}{\text{Area of } S_\varepsilon}$$

$$= \frac{\text{Circulation of } \mathbf{F} \text{ Around } C_\varepsilon}{\text{Area of } S_\varepsilon}$$

Now we let $\varepsilon \to 0$, so that $Q \to P$, and assuming the continuity of curl **F**, curl $\mathbf{F}(Q) \to$ curl $\mathbf{F}(P)$. So

$$\text{curl } \mathbf{F}(P) \cdot \mathbf{n} = \lim_{\varepsilon \to 0} \frac{\text{Circulation of } \mathbf{F} \text{ Around } C_\varepsilon}{\text{Area of } S_\varepsilon}$$

The limit on the right is called the **rotation of F around n** at P. So we can write

$$\text{curl } \mathbf{F}(P) \cdot \mathbf{n} = \text{Rotation of } \mathbf{F} \text{ around } \mathbf{n} \text{ at } P \qquad (16.43)$$

As we saw earlier in this section, when we interpret the two-dimensional vector field $\mathbf{F} = \langle f, g \rangle$ as being the same as the three-dimensional field $\mathbf{F} = \langle f, g, 0 \rangle$,

$$(\text{curl } \mathbf{F}) \cdot \mathbf{n} = \left(\frac{\partial g}{\partial x} - \frac{\partial f}{\partial y} \right) \mathbf{k} \cdot \mathbf{k} = \frac{\partial g}{\partial x} - \frac{\partial f}{\partial y}$$

and this scalar function is what we called the rotation of $\mathbf{F}$ at P in Section 16.4. So, in view of Equation 16.43, we see that rot $\mathbf{F}$ for two dimensions is the same as the rotation of $\mathbf{F}$ about $\mathbf{k}$ in three dimensions.

From Equation 16.43, if curl $\mathbf{F}(P) = 0$ or if $\mathbf{n}$ is perpendicular to curl $\mathbf{F}(P)$, then the rotation of $\mathbf{F}$ around $\mathbf{n}$ will be 0 at P. If curl $\mathbf{F} \neq \mathbf{0}$ at P, there is a circular motion **vortex** at P. The flow is said to be **irrotational** in a region if curl $\mathbf{F} = \mathbf{0}$ for all points in that region.

From Equation 16.43, we can see that the rotation at a point will be a maximum when $\mathbf{n}$ is in the direction of curl $\mathbf{F}$. The maximum value is $|\text{curl } \mathbf{F}|$ evaluated at the point. Suppose, for example, that curl $\mathbf{F}(P) \neq \mathbf{0}$ and a paddle wheel is submerged in the fluid at P, as in Figure 16.54. Then the paddle wheel will rotate as long as its axis is not perpendicular to curl $\mathbf{F}(P)$ (in which case curl $\mathbf{F} \cdot \mathbf{n} = 0$). It will rotate most rapidly when its axis is in the same direction as curl $\mathbf{F}(P)$.

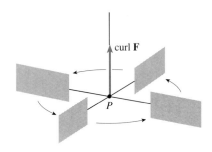

FIGURE 16.54

Three-Dimensional Conservative Vector Fields

The following list of equivalent conditions for conservative vector fields is analogous to those in Section 16.3 for two-dimensional conservative fields. We omit the proofs. We assume $\mathbf{F} = \langle f, g, h \rangle$ is a continuous vector field in an open region Q of $\mathbb{R}^3$.

1. $\mathbf{F}$ is a gradient field.
2. $\int_C \mathbf{F} \cdot d\mathbf{r}$ is independent of the path.
3. $f\, dx + g\, dy + h\, dz$ is an exact differential.
4. $\int_C \mathbf{F} \cdot d\mathbf{r} = 0$ for every piecewise-smooth simple closed curve C in Q.

If Q is simply connected and $\mathbf{F}$ is continuously differentiable, we can add a fifth condition equivalent to these four:

5. curl $\mathbf{F} = \mathbf{0}$ in Q.

Section 16.7 Stokes's Theorem 1205

Summary of Main Theorems

We conclude with a brief summary of the main theorems of this chapter.

CHART 16.1

Fundamental Theorem for Line Integrals

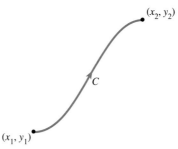

C is any path from (x_1, y_1) to (x_2, y_2),

$$\int_{(x_1,y_1)}^{(x_2,y_2)} \nabla \phi \cdot d\mathbf{r} = \phi(x_2, y_2) - \phi(x_1, y_1)$$

Green's Theorem

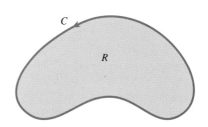

R is a plane region bounded by the plane curve C.

$$\int_C \mathbf{F} \cdot d\mathbf{r} = \iint_R \left(\frac{\partial g}{\partial x} - \frac{\partial f}{\partial y} \right) dA$$

where $\mathbf{F} = \langle f, g \rangle$

Divergence Theorem (Gauss's Theorem)

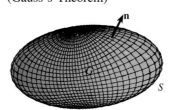

G is a three-dimensional region bounded by the surface S.

$$\iint_S \mathbf{F} \cdot \mathbf{n} \, d\sigma = \iiint_G \text{div } \mathbf{F} \, dV$$

Stokes's Theorem

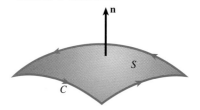

S is a three-dimensional surface bounded by the space curve C.

$$\int_C \mathbf{F} \cdot d\mathbf{r} = \iint_S (\text{curl } \mathbf{F}) \cdot \mathbf{n} \, d\sigma$$

Exercise Set 16.7

In Exercises 1–6, find curl **F**.

1. $\mathbf{F}(x, y, z) = \langle 2xyz, x - y, y + 2z \rangle$

2. $\mathbf{F}(x, y, z) = \langle x^2 y^2, x^2 z^2, y^2 z^2 \rangle$

3. $\mathbf{F}(x, y, z) = e^x yz\mathbf{i} + e^y xz\mathbf{j} + e^z xy\mathbf{k}$

4. $\mathbf{F}(x, y, z) = (y \ln xz)\mathbf{i} + (x \ln yz)\mathbf{j} + (z \ln xy)\mathbf{k}$

5. $\mathbf{F}(x, y, z) = \langle \cos xy, \sin xy, e^{-z^2} \rangle$

6. $\mathbf{F}(x, y, z) = \left\langle \ln \sqrt{x^2 + y^2}, \tan^{-1} \frac{y}{x}, \frac{z}{xy} \right\rangle$

In Exercises 7–10, use Stokes's Theorem to evaluate the line integral, where C is oriented in a counterclockwise direction when viewed from above.

7. $\int_C xy\,dx + (y + z)\,dy + (x - yz)\,dz$; C is the triangle formed by the traces of the plane $3x + 2y + z = 6$ on the coordinate planes.

8. $\int_C x^2 yz\,dx + xy^2 z^3\,dy + x^4 y^3 z^2\,dz$; C is the curve $\mathbf{r}(t) = \langle \cos t, \sin t, 2 \rangle$, where $0 \le t \le 2\pi$.

9. $\int_C \mathbf{F} \cdot d\mathbf{r}$; $\mathbf{F}(x, y, z) = \langle x + yz, 2yz, x - y \rangle$; C is the intersection of the cylinder $x^2 + y^2 = 4$ and the plane $x + y + z = 1$.

10. $\int_C \mathbf{F} \cdot d\mathbf{r}$; $\mathbf{F}(x, y, z) = (x + y)\mathbf{i} + (x - z)\mathbf{j} + (y + z)\mathbf{k}$; C is the intersection of the hemisphere $z = \sqrt{1 - x^2 - y^2}$ and the cylinder $x^2 + y^2 = x$.

*In Exercises 11–16, verify Stokes's Theorem for the given surface S and vector field **F**. Assume S is oriented by upward unit normals unless otherwise specified.*

11. S is the part of the surface $z = 4 - x^2 - y^2$ above the xy-plane; $\mathbf{F}(x, y, z) = \langle y - z, x - z, x - y \rangle$.

12. S is the part of the plane $x + z = 2$ in the first octant between $y = 0$ and $y = 4$; $\mathbf{F}(x, y, z) = \langle x^2 z, yz^2, x^2 + z^2 \rangle$.

13. S is the triangular surface with vertices $(2, 0, 0)$, $(0, 1, 0)$, and $(0, 0, 3)$; $\mathbf{F} = (1 - xy)\mathbf{i} + (y + z)\mathbf{j} + (2x + 3z)\mathbf{k}$.

14. S is the hemisphere $z = \sqrt{4 - x^2 - y^2}$; $\mathbf{F}(x, y, z) = xyz\mathbf{i} + (x + 1)\mathbf{j} + xz^2\mathbf{k}$.

15. S is the part of the cone $z = \sqrt{x^2 + y^2}$ below $z = 1$, oriented by downward unit normals; $\mathbf{F}(x, y, z) = \langle 2x - 3z, xy + 2z, y - xz \rangle$.

16. S is the first-octant portion of the surface $z = e^x$ between $y = 0$ and $y = 3$, below $z = 2$, oriented by downward unit normals; $\mathbf{F}(x, y, z) = \langle ye^x, ze^x, e^x \rangle$.

*In Exercises 17 and 18, show that **F** is a gradient field, and find a potential function for **F**.*

17. $\mathbf{F}(x, y, z) = \langle 2xyz, x^2 z, x^2 y \rangle$

18. $\mathbf{F}(x, y, z) = \langle 2y - 4z, 2x + 5z, 5y - 4x \rangle$

*In Exercises 19 and 20, show that the given force field **F** is conservative, and find the work done by **F** on a particle moving from A to B.*

19. $\mathbf{F}(x, y, z) = (y - 2z)e^x \mathbf{i} + e^x \mathbf{j} - 2e^x \mathbf{k}$; $A = (0, 0, 0)$, $B = (0, -3, -5)$

20. $\mathbf{F}(x, y, z) = 3xz\sqrt{x^2 + y^2}\mathbf{i} + 3yz\sqrt{x^2 + y^2}\mathbf{j} + [(x^2 + y^2)^{3/2} + 2]\mathbf{k}$; $A = (-2, 0, 3)$, $B = (3, 4, -1)$

In Exercises 21 and 22, show that the given expression is an exact differential and find a function for which it is the differential.

21. $e^z \cos x \cos y\,dx - e^z \sin x \sin y\,dy + e^z \sin x \cos y\,dz$

22. $(\ln y + zx)\,dx + (xy - \ln z)\,dy + (\ln x - yz)\,dz$

23. Show that div(curl **F**) = 0.

24. Show that curl($\nabla \phi$) = 0.

25. Let S be a closed surface that satisfies the hypotheses of the Divergence Theorem, and let **F** be a continuously differentiable vector field on some open region that contains S. Prove that

$$\iint_S (\text{curl } \mathbf{F}) \cdot \mathbf{n}\,d\sigma = 0$$

(*Hint:* Use the result of Exercise 23.)

26. A fluid has velocity field $\mathbf{F}(x, y, z) = y^2\mathbf{i} + z^2\mathbf{j} + x^2\mathbf{k}$. Find its circulation around the curve of intersection of the surfaces $z = x^2 + y^2$ and $z = 2(x + y + 1)$, oriented counterclockwise when viewed from above.

27. Let $\mathbf{a} = \langle a_1, a_2, a_3 \rangle$ be any constant vector and $\mathbf{r} = \langle x, y, z \rangle$ be the position vector of a point in $\mathbb{R}^3$.
(a) Prove that $\text{curl}(\mathbf{a} \times \mathbf{r}) = 2\mathbf{a}$.
(b) Show that $\int_C (\mathbf{a} \times \mathbf{r}) \cdot d\mathbf{r} = 2 \iint_S \mathbf{a} \cdot \mathbf{n}\, d\sigma$, where S and C satisfy the hypotheses of Stokes's Theorem.

28. Prove that if u has continuous first partials and v has continuous second partials in an open region Q of $\mathbb{R}^3$, then
$$\text{curl}(u\nabla v) = \nabla u \times \nabla v$$

29. Under the assumptions of Exercise 28, prove the following:
(a) $\iint_S (\nabla u \times \nabla v) \cdot \mathbf{n}\, d\sigma = \int_C u \nabla v \cdot d\mathbf{r}$
(b) $\int_C (u\nabla v - v\nabla u) \cdot d\mathbf{r} = 2 \iint_S (\nabla u \times \nabla v) \cdot \mathbf{n}\, d\sigma$, where S and C are in Q and satisfy the hypotheses of Stokes's Theorem.

Chapter 16 Review Exercises

1. Let C_1 be the quarter of the unit circle from $(1, 0)$ to $(0, 1)$, let C_2 be the line segment from $(0, 1)$ to $(-1, 0)$, and let $C = C_1 \cup C_2$. Find $\int_C f(x, y)\, ds$ for $f(x, y) = (x + y)^2$.

2. Let $\mathbf{F}(x, y, z) = \langle xy + z, x - y, z^2 - y^2 \rangle$ and C be the arc of the helix $\mathbf{r}(t) = \langle 2\cos t, \sin t, 3t \rangle$ from $(2, 0, 0)$ to $(0, 1, 3\pi/2)$. Find $\int_C \mathbf{F} \cdot d\mathbf{r}$.

3. Find the work done by the force field $\mathbf{F}(x, y, z) = x^2\mathbf{i} - y^2\mathbf{j} + xyz\mathbf{k}$ acting on a unit mass along the curve $\mathbf{r}(t) = \sqrt{t}\mathbf{i} + t^{3/2}\mathbf{j} + (\ln t)\mathbf{k}$ from $(1, 1, 0)$ to $(\sqrt{e}, e^{3/2}, 1)$.

4. Evaluate $\int_C e^{-x} \cos y\, dx + e^{-x} \sin y\, dy$ for each path in the accompanying figure. (*Hint:* There is an easy way.)

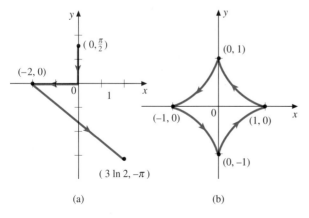

5. For each of the following, determine whether $\mathbf{F}$ is a gradient field in the specified domain. If so, find a potential function for $\mathbf{F}$.
(a) $\mathbf{F}(x, y) = \left\langle \dfrac{y^2}{(x^2 + y^2)^{3/2}} - 2, \dfrac{-xy}{(x^2 + y^2)^{3/2}} + 3 \right\rangle$; $\mathbb{R}^2 - \{(0, 0)\}$

(b) $\mathbf{F}(x, y) = \left\langle \dfrac{y}{x(x + y)}, \dfrac{x}{y(x + y)} \right\rangle$; $x > 0$, $y > 0$
(c) $\mathbf{F}(x, y) = y(2x + \tan xy)\mathbf{i} + x(x + \sec^2 xy)\mathbf{j}$; $\mathbb{R}^2$
(d) $\mathbf{F}(x, y) = \langle 2x \tanh x^2 + \tanh y, x \,\text{sech}^2 y - 2y \rangle$; $\mathbb{R}^2$

In Exercises 6 and 7, use Green's Theorem to evaluate $\int_C \mathbf{F} \cdot d\mathbf{r}$, where C has a counterclockwise orientation.

6. $\mathbf{F}(x, y) = \langle x^2y, x/y^2 \rangle$; C is the boundary of the region enclosed by $xy = 2$, $y = x + 1$, and $y = 1$.

7. $\mathbf{F}(x, y) = (y^2 - \cos x^3)\mathbf{i} + (2xy + e^{y^2})\mathbf{j}$; $C = C_1 \cup C_2$, where C_1 is the arc of the parabola $y = x^2$ from $(-1, 1)$ to $(2, 4)$ and C_2 is the line segment from $(2, 4)$ to $(-1, 1)$.

In Exercises 8 and 9, evaluate the double integral using Green's Theorem.

8. $\iint_R (3x^2 - 2y)\, dA$, where R is the region bounded by the triangle with vertices $(0, 0)$, $(1, 0)$, and $(1, 1)$.

9. $\iint_R (y^2/4 - x^3/3)\, dA$, where R is the region enclosed by the ellipse $4x^2 + 9y^2 = 36$.

In Exercises 10–12, use Green's Theorem to find the area of R.

10. R is the region bounded by the quadrilateral with vertices $(0, 0)$, $(1, 2)$, $(3, 4)$, and $(-1, 3)$.

11. R is the region inside the loop of the *strophoid* $x = \cos 2t$, $y = \tan t \cos 2t$. (See the figure.)

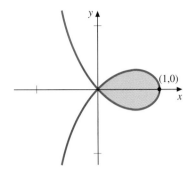

12. R is the region inside $\mathbf{r}(t) = \langle \cos t, \sin 2t \rangle$.

13. Evaluate the integral
$$\int_C \frac{xy^2\, dy - y^3\, dx}{x^2 + y^2}$$
where C is the star-shaped path that has vertices $(\pm 3, 0)$, $(\pm 1, 1)$, $(\pm 1, -1)$, and $(0, \pm 3)$ with a counterclockwise orientation. (*Hint:* Use Theorem 16.8.)

In Exercises 14 and 15, evaluate the surface integral $\iint_S g(x, y, z)\, d\sigma$.

14. $g(x, y, z) = 3y - x^2 - z$; S is the first-octant portion of the plane $z = 2x + y$ below $z = 4$.

15. $g(x, y, z) = 2x^2 + y^2 + z^2$; S is the portion of the sphere $x^2 + y^2 + z^2 = 5$ above $z = 1$.

16. A homogeneous lamina is in the shape of that portion of the hemisphere $z = \sqrt{4 - x^2 - y^2}$ that is inside the cylinder $(x - 1)^2 + y^2 = 1$.
    (a) Find its mass and center of mass.
    (b) Set up, but do not evaluate, iterated integrals for I_x, I_y, and I_z.

17. Set up, but do not evaluate, two iterated integrals for calculating $\iint_S x^2 yz^3\, d\sigma$, using projections on two different planes, where S is the first-octant portion of the cylinder $z = 4 - x^2$ bounded by $y = 0$, $y = x$, and $z = 0$.

18. Let S be the portion of the upper half of the hyperboloid $x^2 + y^2 - z^2 = 3$ below $z = 2$, oriented by downward normals. Find the flux of
$$\mathbf{F}(x, y, z) = \langle xy^2, -x^2 y, -2 \rangle$$
through S.

19. Use the Divergence Theorem to find the outward flux of $\mathbf{F}(x, y, z) = 3x\mathbf{i} + 2y\mathbf{j} + z\mathbf{k}$ through the boundary of the region G between the paraboloids $z = 2 - x^2 - y^2$ and $z = x^2 + y^2$.

In Exercises 20 and 21, use the Divergence Theorem to evaluate $\iint_S \mathbf{F} \cdot \mathbf{n}\, d\sigma$, *where* $\mathbf{n}$ *is the outer unit normal to S.*

20. $\mathbf{F}(x, y, z) = \langle x + \sqrt{1 + y^3}, e^{x^3} + \ln(1 + z^2), xz \rangle$; S is the boundary of the region G between $z = 0$, $z = y$, and $y = 2x - x^2$.

21. $\mathbf{F}(x, y, z) = 2xz\mathbf{i} - 2yz\mathbf{j} + z^2\mathbf{k}$; S is the boundary of the region G enclosed by the hemisphere $z = \sqrt{4 - x^2 - y^2}$, the plane $z = 0$, and the cylinder $x^2 + y^2 = x + \sqrt{x^2 + y^2}$. (*Hint:* You will recognize the xy trace of the cylinder when you change to cylindrical coordinates.)

22. A function $f(x, y, z)$ is said to be **harmonic** in a region G if it satisfies the **Laplace Equation**
$$\frac{\partial^2 f}{\partial x^2} + \frac{\partial^2 f}{\partial y^2} + \frac{\partial^2 f}{\partial z^2} = 0$$
for all (x, y, z) in G. Show that if u and v are harmonic in a bounded region G, with boundary S, where G and S satisfy the conditions of the Divergence Theorem, then
$$\iint_S u \nabla v \cdot \mathbf{n}\, d\sigma = \iint_S v \nabla u \cdot \mathbf{n}\, d\sigma$$
(See Exercise 24 in Exercise Set 16.6.)

In Exercises 23–25, use Stokes's Theorem. In each case C is oriented counterclockwise when viewed from above.

23. Evaluate the integral $\int_C 2yz\, dx + 2xz\, dy + 3xy\, dz$, where C is the triangle with vertices $(3, 0, 0)$, $(0, 2, 0)$, and $(0, 0, 6)$.

24. Evaluate $\int_C \mathbf{F} \cdot d\mathbf{r}$, where $\mathbf{F}(x, y, z) = \langle xz, y + 1, y^2 \rangle$ and C is the union of the traces of the ellipsoid $2x^2 + 4y^2 + z^2 = 8$ on the first-octant parts of the coordinate planes.

25. Find the circulation of the velocity field $\mathbf{F}(x, y, z) = yz\mathbf{i} + 8z\mathbf{j} + xy\mathbf{k}$ around the curve of the intersection of the cylinder $x^2 + y^2 = 4y$ and the plane $3x + 2y + z = 12$.

26. Verify Stokes's Theorem for $\mathbf{F}(x, y, z) = \langle 2yz, y^2, xy \rangle$ and S the portion of the hemisphere $z = \sqrt{5 - x^2 - y^2}$ inside the cylinder $x^2 + y^2 = 1$, oriented by upward normals.

27. Make use of the Divergence Theorem to find the volume of the region bounded by the elliptic paraboloids $z = 2x^2 + 3y^2$ and $z = 4 - 2x^2 - y^2$. (*Hint*: Choose **F** so that div **F** = 1.)

28. Show that the force field $\mathbf{F}(x, y, z) = (2xy - 1)\mathbf{i} + (x^2 + 2z^2)\mathbf{j} + (4yz + 3x^2)\mathbf{k}$ is conservative in $\mathbb{R}^3$ and find the work done by **F** on a particle moving from $(0, 1, -1)$ to $(2, 1, 3)$.

29. Let u and v have continuous second partial derivatives in an open region G of $\mathbb{R}^3$, and let C be a sectionally smooth simple closed curve in G. Prove that
$$\int_C (u \nabla v + v \nabla u) \cdot d\mathbf{r} = 0$$
(*Hint*: Let S be a surface in G with C as its boundary such that S and C satisfy the conditions of Stokes's Theorem. Then use Exercises 28 and 29 in Exercise Set 16.7.)

Chapter 16 Concept Quiz

1. Define each of the following.
   (a) A vector field
   (b) A gradient field
   (c) A potential function for a gradient field
   (d) A conservative field
   (e) The line integral of a function f along a curve C
   (f) The work done by a force field in moving a particle along a curve C
   (g) The divergence of a vector field $\mathbf{F} = \langle f, g, h \rangle$
   (h) The curl of a vector field $\mathbf{F} = \langle f, g, h \rangle$
   (i) The flux of a vector field through a surface S
   (j) The circulation of a vector field around a closed curve C
   (k) A simply-connected region

2. State the following.
   (a) The Fundamental Theorem for Line Integrals
   (b) Green's Theorem
   (c) The Divergence Theorem
   (d) Stokes's Theorem

3. (a) Give a formula for evaluating the line integral
   $$\int_C f(x, y) ds,$$
   where $C = \{(x, y) : x = x(t), y = y(t) \text{ for } a \leq t \leq b\}$.

   (b) Give a formula for evaluating the surface integral $\iint_S f(x, y, z) d\sigma$, where S is the graph of a smooth function $z = \phi(x, y)$ and the xy projection of S is R.

4. State a test for determining whether a vector field $\mathbf{F} = f\mathbf{i} + g\mathbf{j}$ is conservative. If it is conservative, explain how to find a potential function for **F**.

5. Let R be a simply-connected region in $\mathbb{R}^2$. State four conditions that are equivalent to the condition that $\mathbf{F} = \langle f, g \rangle$ is a gradient field in R.

6. Indicate which of the following statements are true and which are false.
   (a) In a conservative force field, the work done in moving a particle along a straight line from (x_1, y_1), to (x_2, y_2) is the same as moving it along the arc of a curve between these points.
   (b) If $\int_C \mathbf{F} \cdot d\mathbf{r} \neq 0$ for a closed path C, then there is no function ϕ for which $\nabla \phi = \mathbf{F}$.
   (c) If $\int_C (\mathbf{F} - \mathbf{G}) \cdot d\mathbf{r} = 0$ for every simple closed curve C in the common domain of **F** and **G**, then $\mathbf{F} = \mathbf{G}$.
   (d) The value of $\int_C \mathbf{F} \cdot d\mathbf{r}$ depends on the parameterization of C.
   (e) If f is a scalar function with continuous partial derivatives in a region R, then $\int_C \nabla f \cdot d\mathbf{r}$ is independent of the path in R.

APPLYING CALCULUS

The Hubble Space Telescope after its 1993 repair

1. When the space shuttle went into orbit to service the Hubble Space Telescope, they had a combined mass of 20,000 kg and orbited the Earth in a circular orbit with radius 8,850 km from the center of the Earth.
   (a) Find the work W done by gravity on the space shuttle during one-half a revolution.
   (b) Find the work done by gravity on the shuttle during one full revolution.
   (c) In general, suppose that a continuous force acts on an object in a direction normal to its path. Show that the work done by the force on the object is zero. How does this result compare with your answers in parts (a) and (b)?

2. A vector field $\mathbf{F}$ in space (or in the plane) is a *central force field* if $\mathbf{F}(P)$ is parallel to the position vector $\overrightarrow{OP}$ from the origin to P for all points P for which $\mathbf{F}$ is defined. A central force field is *radially symmetric* if $|\mathbf{F}(P)| = |\mathbf{F}(Q)|$ whenever P and Q are points on the same circle centered at the origin.
   (a) Show that a radially symmetric vector field $\mathbf{F}$ has the form
   $$\mathbf{F}(P) = f(|\mathbf{r}|)\mathbf{r}$$
   where $\mathbf{r} = \overrightarrow{OP}$ and f is a scalar function.
   (b) Show that a radially symmetric field is conservative.
   (c) Assume that $\mathbf{F}$ is a radially symmetric vector field that is defined everywhere in space except at the origin and that $\mathbf{F}$ is differentiable. Show that if div $\mathbf{F} = 0$, then $f(|\mathbf{r}|) = k/|\mathbf{r}|^3$. Thus, a gravitational field is a radially symmetric vector field with divergence equal to 0.

3. A particle of mass m moves along a plane curve C, with position vector $\mathbf{r}(t)$ for $t_1 \leq t \leq t_2$. It is acted on by a force field given by $\mathbf{F}(\mathbf{r}) = -k\mathbf{r}$, where k is a positive constant.
   (a) Show that $\mathbf{F}$ is a conservative force field, and find a potential function for it.
   (b) Find the work done by $\mathbf{F}$ in moving the particle from $\mathbf{r}(t_1)$ to $\mathbf{r}(t_2)$.
   (c) Suppose that the force field is modified to include a damping force proportional to the velocity of the particle and opposite to the direction of motion. Then $\mathbf{F}(\mathbf{r}) = -k\mathbf{r} - c\mathbf{v}$, where $\mathbf{v} = \mathbf{r}'(t)$ and c is a positive constant. Give an integral in terms of the parameter t for the work done in this case in moving the particle from $\mathbf{r}(t_1)$ to $\mathbf{r}(t_2)$. Show that the force field in this case is not conservative. (*Hint:* By considering two different paths from $\mathbf{r}(t_1)$ to $\mathbf{r}(t_2)$, show that the integral is not independent of the path.)

4. In 1864, the Scottish physicist James Clerk Maxwell unified electricity and magnetism with the following set of equations, now known as **Maxwell's Equations**:

   (1) div $\mathbf{E} = \dfrac{\rho}{\varepsilon}$ (3) curl $\mathbf{E} = -\dfrac{\partial \mathbf{B}}{\partial t}$

   (2) div $\mathbf{B} = 0$ (4) curl $\mathbf{B} = \varepsilon\mu\dfrac{\partial \mathbf{E}}{\partial t} + \mu\mathbf{J}$

   In these equations $\mathbf{E}$ represents an electric field and $\mathbf{B}$ a magnetic field. The constant ε is an electrical constant called the *permittivity* and μ is a magnetic constant called the *permeability*. The charge density is ρ and the current density is $\mathbf{J}$, where $\mathbf{J} = \rho\mathbf{v}$, for charge velocity $\mathbf{v}$. Note that a changing electric field results in a magnetic field, and vice versa. Maxwell's Equations govern all electromagnetic waves, including light, and among many other things led to the discoveries of radio and television.

(a) Show that Equation 3 implies Equation 2.

(b) Show that Equation 3 implies **Faraday's Induction Law**, that the electromotive force around any closed contour C that is sectionally smooth and forms the boundary of an orientable surface S in the electric field $\mathbf{E}$ is the negative of the time rate of change of the magnetic flux through S:

$$\int_C \mathbf{E} \cdot \mathbf{T}\, ds = -\frac{\partial}{\partial t} \iint_S \mathbf{B} \cdot \mathbf{n}\, d\sigma$$

(c) Use the special case of Equation 4 in which $\partial \mathbf{E}/\partial t = 0$ to prove **Ampere's Law** for a steady current, that the circulation of the magnetic field $\mathbf{B}$ induced by $\mathbf{J}$ around the boundary C of an orientable surface S is proportional to the total current flowing through S:

$$\int_C \mathbf{B} \cdot \mathbf{T}\, ds = \mu \iint_S \mathbf{J} \cdot \mathbf{n}\, d\sigma$$

5. Refer to Maxwell's Equations in Exercise 4 to derive two forms of **Gauss's Law** as follows:

(a) Use Equation 1 to derive Gauss's Law for the electric field, that the net outward flux is equal to the electric charge q inside, divided by the permittivity:

$$\iint_S \mathbf{E} \cdot \mathbf{n}\, d\sigma = \frac{q}{\varepsilon}$$

(b) Use Equation 2 to derive Gauss's Law for the magnetic field, that the net outward flux is equal to 0:

$$\iint_S \mathbf{B} \cdot \mathbf{n}\, d\sigma = 0$$

6. Consider an incompressible fluid (no sources or sinks) flowing in space with variable density $\rho = \rho(x, y, z, t)$ and velocity $\mathbf{v} = \mathbf{v}(x, y, z, t)$. Let S_ε be a sphere of radius ε centered at a point (x, y, z) in the velocity field, and let G_ε be the spherical region enclosed by S_ε.

(a) Write a triple integral for the mass $m(t)$ of fluid inside S_ε at time t, and show that

$$m'(t) = \iiint_{G_\varepsilon} \frac{\partial \rho}{\partial t}\, dV$$

(Assume differentiation inside the integral is valid.)

(b) Explain why $m'(t)$ is also given by

$$m'(t) = -\iint_{S_\varepsilon} \rho \mathbf{v} \cdot \mathbf{n}\, d\sigma$$

where $\mathbf{n}$ is the outer unit normal to S_ε.

(c) Use the Divergence Theorem to write the result in part (b) as a triple integral.

(d) Equate the triple integrals for $m'(t)$ in parts (a) and (c), and combine the integrals. By using a mean-value theorem for triple integrals (as in Section 16.6) and then letting $\varepsilon \to 0$, obtain the **continuity equation**

$$\frac{\partial \rho}{\partial t} + \text{div}(\rho \mathbf{v}) = 0$$

A PERSONAL VIEW

Frank McGrath

Your Ph.D. is in partial differential equations. How have you found the mathematics you learned in university useful in your professional life?

It's impossible to solve most real-world problems without mathematics. When I was a student solving problems like "how long a board can be carried around this corner?" I sometimes wondered whether calculus had any real-world use. It turned out that many of the problems people were willing to pay me to solve were much more interesting calculus problems than those I solved in class. So market forces are not necessarily bad when it comes to science. Partial differential equations is calculus and most of the engineering problems that one runs into involve some form of differential equations.

What role do computers play in these kinds of problems?

What computers are doing is numerically integrating differential equations, which involve derivatives. Often one has to numerically integrate. That is, you have to solve most of the problems that are of practical value on a computer. Very few real-world problems are amenable to closed-form solution and so you have to numerically integrate them—which, I contend, can often require more understanding of the nature of the differential equations than merely working with or solving the analytical forms of the equations.

Can you give us an example?

Sure. There are the problems of computing an orbit to a planet, which is relatively straightforward on the surface—just applying Newton's laws. They make a very simple differential equation, force equals mass times acceleration, except you need to account for solar wind, gravitational fields, and thrust that burns fuel which changes mass and the center of gravity. Force and mass become complex functions in the differential equation. It's a hard problem to arrive at a planet that itself is a moving target.

I was involved in tracking and intercepting satellites. Here there are a multiplicity of problems, all of which involve differential equations. Imagine that you have a satellite in orbit and you want to know its position with great precision. If you're going to intercept something and you can't see it until you are almost there, it's kind of nice to know where it is in advance. If you're going to arrive at a point in space at the same time the satellite arrives at the same point when the closing velocity may be 20,000 feet per second or more, you have to know the satellite's orbit with great precision. So where's the satellite? It's found by measuring the changing position of the satellite at earlier times, usually with radars, and then using mathematics to project forward in time both the position of the satellite and the interceptor's position to the point of rendezvous. Precise determination of position as a function of time is very, very important in this kind of problem. You have to consider things like solar wind and irregularities in the gravitational field. You might not think that the solar wind pushing on a satellite in an orbit of 300-mile altitude would have much effect on the orbit. In fact, its effect on the position in orbit must be taken into account whenever high accuracy is required, especially in applications such as navigation and measurement of gravitational fields. Also, you have to predict the gravitational field with great precision because, unlike rocket thrust, the acceleration due to gravity cannot be measured during powered flight.

To intercept your target, you have to know where you are at all times while you are still controlling the vehicle with rocket thrust. The cycle is: Where am I now? Exactly what velocity vector do I need to reach my target from this position? Do I have this velocity? If not, in what direction do I need to point the rocket thrust to get this velocity? When you arrive at the velocity that will create the trajectory to hit your target you shut the motor off. Every step in this cycle involves calculus. When you use polar and Cartesian coordinates in calculus you will see that some problems are better solved in polar coordinates and others in Cartesian. It turns out that the best coordinate system for the differential equations that are used in the cycle is a coordinate system called *inertial coordinates*. Devices that measure forces in inertial coordinates are based on gyroscopes, and the design of these gyros is a problem solved with mathematics based on calculus. When you shut the rocket motors off because you have reached the desired velocity, you still have to be able to predict the effect of terminating thrust. This involves the noninstantaneous termination of thrust and the relaxation of the metal of the final stage of the booster, which can add additional thrust. The mathematics needed to solve the problems of each of these steps involves differential equations, which are studied in advanced courses like applied mechanics, control theory, and continuum mechanics.

Is this is a problem for the space shuttle as well?

Yes. It is one very large mathematical problem after another to be able to fly the shuttle from Cape Kennedy to intercept the Hubble Space Telescope. But these mathematical problems are also relevant to the design of the shuttle. For instance, Thiokol, the company that builds the solid fuel rocket booster that unfortunately blew up the Challenger space shuttle because of a design flaw, had already solved a very difficult problem for solid rocket motors. When you ignite the solid fuel, huge pressures build up inside the rocket. If not properly controlled, they can oscillate back and forth and you can get a resonance that can destroy the rocket. So how do you prevent this resonance? The first step is to write down the applicable laws of physics in mathematical notation. You end up with equations that involve the same integrals and differentials that you learn about in calculus class. In some sense, this is the easy part of the problem. Then some person has the very interesting challenge of solving these equations. The difference being that if your solution is not correct the rocket might blow up, which is much more serious than getting a "C" in your calculus class. All of this very complex engineering gets back to calculus.

What other mathematics courses are particularly important for practical applications after you take your calculus?

I've already said that if you want to work on engineering problems in industry, differential equations are extremely important, both ordinary and partial. Real and complex variable analysis is also important, as is linear algebra. You need to understand how all this fits together. Another very useful mathematical area in industry that applied math majors often miss is that of Fourier transforms and spectral analysis. These too involve integrals—you just can't seem to get away from calculus.

For example, you may have noticed that there are several antenna towers at a radio or television transmitter. These antennas combine into something called a phased array. Basically, the engineering problem is to receive or transmit waves, sound or electromagnetic, on each of a number of omnidirectional receivers or transmitters and then combine all of these omnidirectional waves in such a way that the result is a beam that points in one direction. The more omnidirectional receivers or transmitters you have, the narrower you can make the beam. When the phased array is an antenna receiving sound, a narrow beam excludes sound from all directions except in the direction of the beam. The sound in one beam is then decomposed into its spectral components, or tones, by numerically integrating the integral, that is, the Fourier transform. Each sound source has a unique signature of tones that is used to distinguish the sound source from all other sound sources.

There is not only the Fourier transform involved in forming a beam and then producing the spectrum, but if your real-world problem is to simultaneously form a large number of beams, say 500, and then break the sound in each beam into its spectral components, you have one whale of a computation problem. You might be able to write the analytical equations and then find out that computers large enough to solve the equations within the necessary time *do not exist*! You didn't give enough thought to computational efficiency. In these types of problems, you have to worry about saving a microsecond or less in certain intermediate computations. For instance, a method of numerically integrating the Fourier Transform integral, called the Fast Fourier Transform, makes an enormous improvement in computational efficiency for beam forming and spectral decomposition by replacing very computationally expensive multiplications by right or left shifts of the string of 1s and 0s in the binary representation of numbers. Shifts are extremely efficient on a computer. So you not only need to know the mathematics, you also need to understand computation both at the level of computer functions and numerical algorithms. Be wary of computing a lot of trigonometric functions. Don't divide when you can multiply. Addition is better yet and shifts can be even better.

What do you look for in terms of mathematical aptitude when you hire employees?

Many people who think they're trained in mathematics have learned how to turn the crank and do computation but don't really understand mathematics. There is an enormous value for people to understand mathematics as opposed to merely learning that the derivative of x^2 is $2x$. I really need people who understand a derivative and what a differential is, and the difference between the two, and how does this apply to solving a practical problem? Similarly, if you don't understand the physics behind an engineering problem you're going to be very limited in what you can do. You just can't work on real-world engineering problems without a broad base of knowledge to supplement your mathematics. You've got to understand the physics; you've got to understand the math; you've got to understand how they relate to each other. This is necessary, but not sufficient, for you to have the two most in-demand skills: understanding thought processes of mathematics and having mathematical creativity. You can't just be a worker of algorithms because computers can work the algorithms better than you can. You have to truly understand the problems and this involves the ability to reason quantitatively.

Mathematicians working in industry also have excellent opportunities to be part of a team developing large-scale software. The creative thought process of mathematics is the key skill you can bring to the team. One of the problems of great economic importance is how to improve productivity and quality in very large computer programs, say half a million instructions or more. The central problem is how to decompose this problem that can't be understood by one person into a large number of problems, each of which can be understood. The best current and future (potential) technology for doing this is called object-oriented design, which decomposes into levels of abstraction. Does this sound like mathematics? It may not at your stage of undergraduate calculus, but it will if you do a Ph.D. in applied mathematics. You are currently moving up the levels of abstraction in mathematics. Calculus generalized many of the more concrete properties of numbers. When you complete your calculus and then eventually get up the ladder of abstraction to abstract algebra and point set topology, you will appreciate why a mathematics education gives the thought process necessary to pull off the design of large-scale software.

Frank McGrath is a native of Ohio and is Vice-President for Information Technology at Logicon, Inc., in Arlington, VA. He received his B.A. in Mathematics from Iowa State University, and his Ph.D. in Applied Mathematics from the University of California at Berkeley. His speciality—both as a research and occupational interest—is developing extremely large computer systems involving both software and hardware. These projects have tackled a wide variety of problems involving direct applications of mathematics and physics.

CHAPTER 17

DIFFERENTIAL EQUATIONS

We introduced the notion of a differential equation in Chapter 4, during our discussion of antiderivatives. At that stage, we considered only simple differential equations of the form

$$\frac{dy}{dx} = f(x)$$

having the general solution

$$y = F(x) + C$$

where F is an antiderivative of f. In Chapter 7, in our study of exponential growth and decay, we described the solution technique called *separation of variables*. This technique enabled us to solve differential equations that were more complex than those we had previously studied.

In this chapter we will review and expand on the basic idea of a differential equation and the nature of solutions. We will review the technique of separation of variables and introduce other methods for solving first-order differential equations (those involving only first derivatives). We will also study two techniques for solving second-order equations. We will conclude the chapter by indicating how to use power series to solve equations that otherwise would be difficult or impossible to solve.

The subject of differential equations is almost as old as calculus itself; both Newton and Leibniz originated many of the concepts involved. A tremendous body of knowledge on the subject exists, but much is still not known, and the subject continues to be an area of intensive research. In this single chapter we can only scratch the surface of this vast field, but the techniques we will present are nevertheless important. We leave to a later course the deeper theoretical aspects, in particular those having to do with existence and uniqueness of solutions.

The importance of differential equations in applications lies in their use as mathematical models to describe a wide variety of phenomena having to do with rates of change. Such phenomena occur in biology, chemistry, economics, engineering, physics, and psychology, to name only a few areas of study. For example, velocities, accelerations, chemical reaction rates, rates of spread of diseases, learning rates, and growth rates of investments all enter into models that are expressed by differential equations. We will illustrate a variety of such applications in this chapter.

17.1 BASIC CONCEPTS; GENERAL SOLUTIONS AND PARTICULAR SOLUTIONS

Recall that a differential equation is an equation that involves a function and one or more of its derivatives. If there is only one independent variable, so that ordinary derivatives occur, the equation is called an **ordinary differential equation**. If there are two or more independent variables, then partial derivatives occur, and the equation is called a **partial differential equation**. For example, the equation

$$Q'(t) = kQ(t)$$

of exponential growth or decay that we studied in Chapter 7 is an ordinary differential equation. The equation

$$\frac{\partial u}{\partial t} = c\frac{\partial^2 u}{\partial x^2}$$

is a partial differentiation equation, called the *heat equation*, that we discussed briefly in Chapter 13 in connection with our study of partial derivatives. In this chapter we limit our consideration to ordinary differential equations. When we use the term *differential equation*, we will be referring to an ordinary differential equation.

One way of classifying differential equations is by their order. The **order** of a differential equation is defined to be the order of the highest derivative that occurs. For example, the differential equations

$$\frac{dy}{dx} = 3x^2$$

$$y'' - 2y' - 3y = e^x$$

$$\frac{d^4x}{dt^4} - (\sin t)\left(\frac{dx}{dt}\right)^2 = t^3 + 1$$

have orders 1, 2, and 4, respectively.

REMARK
■ When we write y', it will be understood to mean $y'(x)$, or equivalently, dy/dx, unless a different independent variable is specified.

A function $y = f(x)$ is a **solution** of a differential equation if when y and its derivatives are substituted into the equation, the equation becomes an identity. For example, we know that $y = x^3$ is a solution of the differential equation $dy/dx = 3x^2$. Also, $y = x^3 + C$, where C is any constant, is a solution.

We call $x^3 + C$ the **general solution**. All solutions to the equation can be obtained from the general solution by appropriate choices of the constant C. This general solution is really a **family** of solutions. We refer to it as a **one-parameter** family, since there is only one arbitrary constant (the parameter C) involved. Several members of the family are sketched in Figure 17.1.

In solving differential equations, we often seek such a general solution from which all solutions can be obtained.* For first-order equations, the general solution will always be a one-parameter family. This result seems plausible, since you would expect to have to carry out only one integration, giving rise to one arbitrary constant. Similarly, a second-order equation will have a two-parameter family as its general solution, and in general, an nth-order equation will have an n-parameter family as its general solution.

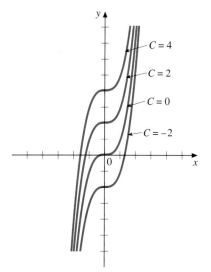

FIGURE 17.1
The family of solutions $y = x^3 + C$

REMARK
■ It is important to point out that, when speaking of the number of parameters in the general solution, we mean the number of *essential* constants. For example, the equation $y + C_1 = x^2 + C_2$ may appear to define a two-parameter family, but by writing it as $y = x^2 + (C_2 - C_1)$ and defining $C = C_2 - C_1$, we see that it is really the one-parameter family $y = x^2 + C$.

When specific values are given to the arbitrary constants in the general solution, we obtain a **particular** solution. In a first-order equation, specifying the value of y for one value of x is sufficient to determine the constant. For example, suppose we want the particular solution to $y' = 3x^2$ that satisfies $y(1) = 3$. Graphically, we want the curve of the family that passes through the point $(1, 3)$. To find it, we substitute $x = 1$ and $y = 3$ into the general solution $y = x^3 + C$. This substitution gives $3 = (1)^3 + C$, or $C = 2$. Thus, $y = x^3 + 2$ is the desired particular solution.

More generally, for a first-order differential equation, if we want the solution $y = y(x)$ that satisfies $y(x_0) = y_0$, we substitute $x = x_0$ and $y = y_0$ into the general solution to find C. The condition $y(x_0) = y_0$ is called an **initial condition**, and the differential equation with such a specified initial condition is called an **initial-value problem**. For a second-order equation, two initial conditions are needed to determine a particular solution, as the next example illustrates.

EXAMPLE 17.1 Verify that $y = C_1 e^{3x} + C_2 e^{-x}$ is a two-parameter family of solutions of the differential equation

$$y'' - 2y' - 3y = 0$$

and find the particular solution in this family that satisfies $y(0) = 2$ and $y'(0) = -1$. (*Note:* In Section 17.5 we will see that the given family is the general solution of this equation and see how it was obtained.)

Solution We first calculate y' and y'' for the given family:

$$y' = 3C_1 e^{3x} - C_2 e^{-x}$$
$$y'' = 9C_1 e^{3x} + C_2 e^{-x}$$

---

*Sometimes there are additional solutions not obtainable from the general solution. These are called **singular** solutions, but we will not consider them here.

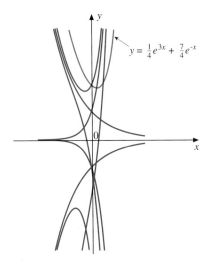

FIGURE 17.2

So
$$y'' - 2y' - 3y = 9C_1e^{3x} + C_2e^{-x} - 2(3C_1e^{3x} - C_2e^{-x}) - 3(C_1e^{3x} + C_2e^{-x})$$
$$= (9C_1 - 6C_1 - 3C_1)e^{3x} + (C_2 + 2C_2 - 3C_2)e^{-x} = 0$$

Thus, the differential equation is identically satisfied by the given function, so we have a two-parameter family of solutions. To find C_1 and C_2 that satisfy the given initial conditions, we impose those conditions on y and y' to get

$$C_1 + C_2 = 2$$
$$3C_1 - C_2 = -1$$

The simultaneous solution is easily found to be $C_1 = \frac{1}{4}$, $C_2 = \frac{7}{4}$. So the particular solution in question is

$$y = \frac{1}{4}e^{3x} + \frac{7}{4}e^{-x}$$

In Figure 17.2 we show several members of the family given by the general solution, along with the particular solution. ∎

In the next example, a particular solution of a second-order equation is obtained from its general solution by specifying values of y at two different values of x. Such values of y are called **boundary values** and are of the general form $y(x_0) = y_0$ and $y(x_1) = y_1$. The terminology reflects the fact that in applications the two known values of y occur at the endpoints (that is, on the boundary) of the interval under consideration. A differential equation to be solved, subject to such boundary conditions, is known as a **boundary-value problem**.

EXAMPLE 17.2 Show that for all choices of the parameters C_1 and C_2, the function $y = C_1 \cos x + C_2 \sin x$ is a solution (it is the general solution) of the equation $y'' + y = 0$, and determine C_1 and C_2 such that the boundary conditions $y(0) = 2$ and $y(\pi/2) = 1$ are satisfied.

Solution Since for the given family $y' = -C_1 \sin x + C_2 \cos x$, and $y'' = -C_1 \cos x - C_2 \sin x$, we see that

$$y'' + y = (-C_1 \cos x - C_2 \sin x) + (C_1 \cos x + C_2 \sin x) = 0$$

So the equation is satisfied. From $y(0) = 2$, we get $C_1 = 2$, and from $y(\frac{\pi}{2}) = 1$, we get $C_2 = 1$. Thus, the solution to the boundary-value problem is

$$y = 2\cos x + \sin x$$

We show its graph in Figure 17.3. ∎

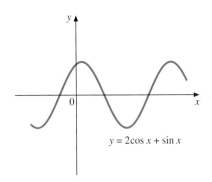

FIGURE 17.3

It is desirable to express a solution to a differential equation as an explicit function, $y = f(x)$, but sometimes doing so is difficult or even impossible and we settle for an **implicit solution** in the form $F(x, y) = 0$. The next example illustrates this situation.

EXAMPLE 17.3 Show that if y is defined as a function of x by the equation
$$x^3 - 2xy^2 + y^3 - 5 = C$$
then y is a solution to the differential equation
$$\frac{dy}{dx} = \frac{3x^2 - 2y^2}{4xy - 3y^2}$$

Solution   Let $F(x, y) = x^3 - 2xy^2 + y^3 - 5 - C$. Then from Chapter 14 we know that

$$\frac{dy}{dx} = -\frac{F_x}{F_y} = -\frac{3x^2 - 2y^2}{-4xy + 3y^2} = \frac{3x^2 - 2y^2}{4xy - 3y^2}$$

as long as the denominator is nonzero. So the function y defined implicitly by $F(x, y) = 0$ satisfies the given differential equation. ∎

Sometimes we are interested in finding the differential equation of a given family. That is, we want to find the differential equation, given its general solution. We illustrate this procedure in the next example.

EXAMPLE 17.4   In each of the following, find the differential equation having the given equation as its general solution.
(a) $x^2 + y^2 = Cx$ (b) $y = C_1 e^{-x} + C_2 e^{2x}$

Solution

(a) Since there is one arbitrary constant, we expect the differential equation to be of first order. By differentiating both sides with respect to x, we obtain

$$2x + 2yy' = C$$

The differential equation cannot contain the constant of integration, C, so we want to eliminate this constant. We can do so by replacing C in the original equation by the value just found for C, namely, $2x + 2yy'$. Thus,

$$x^2 + y^2 = (2x + 2yy')x$$

or, on simplification,

$$2xyy' = y^2 - x^2$$

which is the desired differential equation.

(b) This time we differentiate twice, since there are two arbitrary constants:

$$y' = -C_1 e^{-x} + 2C_2 e^{2x}$$
$$y'' = C_1 e^{-x} + 4C_2 e^{2x}$$

By using these two equations, together with the original equation, we can eliminate the two constants. Adding the equations for y and y' gives

$$y + y' = 3C_2 e^{2x}$$

Similarly, adding y' and y'' gives

$$y' + y'' = 6C_2 e^{2x}$$

Thus,

$$2(y + y') = y' + y''$$

or equivalently,

$$y'' - y' - 2y = 0$$

This equation is the one we are seeking. ∎

Slope Fields

The techniques we will study in subsequent sections will make it possible to solve a wide variety of differential equations, but just as there are functions that cannot be integrated in an exact form, there are differential equations for which exact solutions cannot be found. There are various approximation techniques studied in more advanced courses, but we will not consider them here. We will, however, mention one geometric approach that is sometimes useful in obtaining approximate graphs of the solution curves. Suppose the differential equation is of first order. By solving for dy/dx, we can write the equation in the form

$$\frac{dy}{dx} = F(x, y)$$

Since dy/dx is the slope of a given solution curve at a given point on the graph, we can obtain slopes at various points in the plane (in the domain of F). If at a large number of points we draw short line segments with slope $F(x, y)$, we call the result a **slope field** (also called a **direction field**) for the given differential equation. We illustrate such a slope field in Figure 17.4. A slope field is similar to a vector field except that instead of vectors we draw short line segments (parts of tangent lines). Once we have a slope field, we can sketch solution curves by following the tangent line segments. By starting at a particular point, we obtain a sketch of the particular solution passing through that point. We illustrate this technique in the next example.

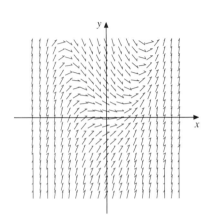

FIGURE 17.4
A slope field

EXAMPLE 17.5 Sketch the slope field for the differential equation

$$\frac{dy}{dx} = 2x - y$$

Use the result to sketch the graph of the particular solution that passes through the point $(1, 2)$.

Solution We obtain the slope field in Figure 17.5 by substituting for x and y to calculate the slope $2x - y$ at various points (x, y). For example, the slopes at $(-1, 2)$, $(0, 1)$, and $(1, 2)$ are -4, -1, and 0, respectively. So at these points we draw short segments with the given slopes.

To obtain the particular solution passing through the point $(1, 2)$, we start our sketch at that point and then move to the right and left, using the tangent line segments as a guide. (See Figure 17.5) ∎

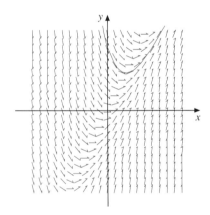

FIGURE 17.5

SOLVING DIFFERENTIAL EQUATIONS USING COMPUTER ALGEBRA SYSTEMS

The theory of differential equations is one of the oldest and richest in mathematics. There exist many techniques for finding explicit solutions for particular classes of differential equations as well as a multitude of approximation techniques for those that cannot be solved explicitly. In practice, the latter occur more frequently. A CAS can find explicit solutions for only a small collection of types of differential equations and can be used effectively in finding numerical approximations to solutions. In this section we give two examples to describe several ways a CAS can be useful in analyzing differential equations. The methods for solving the equations considered will be developed in the chapter.

CAS 61

Consider the first-order linear differential equation

$$\frac{dy}{dx} = x - y$$

Use Maple and Mathematica to first sketch a slope field for the differential equation that will describe the nature of the solutions. Then use the CAS to find the general solution and sketch several specific solutions along with the slope field.

Maple:

with(DEtools):with(plots):
dfieldplot(x−y,[x,y],x=−3..3,y=−3..3,scaling=constrained);
dsolve(diff(y(x),x)+y(x)−x=0,y(x));

Output: $y(x) = x - 1 + e^{-x} \_C1$

In returning the general solution, Maple indicates an arbitrary constant by _C1. Next, we set in turn the constant to 1, 0, and $4/e^3$ to get three particular solutions from the family of solutions and at the same time create plots that can then be generated together.

sol1:=plot(x−1+exp(−x),x=−3..3,y=−3..3):
sol2:=plot(x−1,x=−3..3,y=−3..3):
sol3:=plot(x−1+4/E^3*exp(−x),x=−3..3,y=−3..3):
slopefield:=dfieldplot(x−y,[x,y],x=−3..3,y=−3..3,scaling=constrained):
display({slopefield,sol1,sol2,sol3});

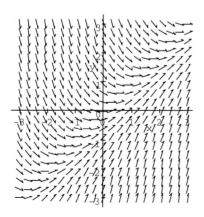

FIGURE 17.1.1
Slope field

Mathematica:

DSolve[y′[x]==x−y[x],y[x],x]

Plot[{x−1+Exp[−x],x−1,x−1+4/E^3*Exp[−x]},{x,−3,3}, PlotRange−>{−3,3}]

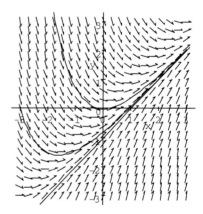

FIGURE 17.1.2
Slope field and several solutions

CAS 62

Solve the initial-value problem

$$y'' + y = e^x, \; y(0) = 1, \text{ and } y'(0) = 1$$

Maple:*

de:=diff(y(x),x$2)+y−exp(x);

Output: $de := \left(\dfrac{\partial^2}{\partial x^2} y(x)\right) + y - e^x$

The particular solution to the initial-value problem can be solved by:

dsolve({de=0,y(0)=1,D(y)(0)=1},y(x));

Output: $y(x) = \dfrac{1}{2}e^x + \dfrac{1}{2}\sin(x) + \dfrac{1}{2}\cos(x)$

plot(1/2*exp(x)+1/2*sin(x)+1/2*cos(x),x=−10..10,y=−10..10, scaling=constrained);

Mathematica:

DSolve[y''[x]+y[x]==Exp[x],y[x],x]

y[x_] = Exp[x]/2+A*Cos[x]+B*Sin[x]

Solve[{y[0]==0,y'[0]==1},{A,B}]

Plot[Exp[x]/2−1/2*Cos[x]+1/2*Sin[x],{x,−10,10}, PlotRange−>{−10,10}]

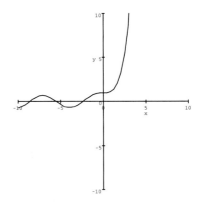

FIGURE 17.1.3

---

*In the Maple command **diff**, the use of x$2 means the second derivative with respect to x.

Exercise Set 17.1

In Exercises 1–6, solve the given initial-value problem.

1. $\dfrac{dy}{dx} = x - 2; \; y(1) = 4$

2. $\dfrac{dy}{dx} = xe^{-x}; \; y(0) = -2$

3. $y' = \tan x$ on $[0, \pi/2); \; y(0) = 3$

4. $y' - x/\sqrt{9 + x^2} = 0; \; y(-4) = 3$

5. $y'' = 2; \; y(1) = 2, \; y'(1) = 3$

6. $d^2x/dt^2 = 3t; \; x(0) = 0, \; x'(0) = 2$

In Exercises 7–12, verify that for all choices of C_1 and C_2 the given function is a solution of the differential equation.

7. $y = C_1 e^{-x} + C_2 e^{-2x}; \; d^2y/dx^2 + 3\,dy/dx + 2y = 0$

8. $y = C_1 e^{-2x} + C_2 e^{4x}$; $y'' - 2y' - 8y = 0$

9. $y = C_1 \cos ax + C_2 \sin ax$; $y'' + a^2 y = 0$

10. $y = C_1 \cosh ax + C_2 \sinh ax$; $y'' - a^2 y = 0$

11. $x = e^{-t}(C_1 \cos t + C_2 \sin t)$; $d^2x/dt^2 + 2\,dx/dt + 2x = 0$

12. $y = C_1 x^{-1} + C_2 x^{-2}$; $x^2 y'' + 4xy' + 2y = 0$ $(x > 0)$

In Exercises 13–18, find the particular solution of the problem specified that satisfies the given initial conditions or boundary conditions.

13. Exercise 7; $y(0) = 0$, $y'(0) = -2$

14. Exercise 8; $y(0) = 1$, $y'(0) = 0$

15. Exercise 9; $y(0) = 2$, $y\left(\dfrac{\pi}{2a}\right) = 1$

16. Exercise 10; $y(0) = 1$, $y'(0) = 4a$

17. Exercise 11; $x(0) = 0$, $x'(0) = 1$

18. Exercise 12; $y(1) = 2$, $y(2) = 3$

In Exercises 19–21, verify that any differentiable function $y = y(x)$ defined implicitly by the given equation is a solution of the differential equation.

19. $2x^3 y - 3xy^2 + y^4 = 7$; $\dfrac{dy}{dx} = \dfrac{3y(y - 2x^2)}{2x^3 - 6xy + 4y^3}$

20. $y \tan xy - 2x = 1$; $\dfrac{dy}{dx} = \dfrac{2 - y^2 \sec^2 xy}{xy \sec^2 xy + \tan xy}$

21. $\tan^{-1} \dfrac{y}{x} + \ln \sqrt{x^2 + y^2} = 1$; $\dfrac{dy}{dx} = \dfrac{y - x}{y + x}$

In Exercises 22 and 23, solve the initial-value problem by first showing that the given family is a solution and then determining the proper constants.

22. $y''' + 3y'' - 4y = e^{-2x}$; $y(0) = 1$, $y'(0) = -2$, $y''(0) = 3$;
$y = C_1 e^x + C_2 e^{-2x} + C_3 x e^{-2x} - (1/6) x^2 e^{-2x}$

23. $y''' + y' - 10y = 10x^2 - 2x + 5$; $y(0) = 2$, $y'(0) = 1$, $y''(0) = 0$; $y = e^{-x}(C_1 \cos 2x + C_2 \sin 2x) + C_3 e^{2x} - x^2 - \tfrac{1}{2}$

In Exercises 24–29, find the differential equation that has the given family as a general solution.

24. $y = Cx^2$

25. $xy - y^2 = Cx$

26. $y = C_1 e^x + C_2 e^{-x}$

27. $y = C_1 \cos x + C_2 \sin x$

28. $y = C_1 x + C_2 x^2$

29. $y = C_1 x + C_2 x \ln x$

30. The general solutions of the equations
$$\dfrac{dy}{dx} = F(x, y) \quad \text{and} \quad \dfrac{dy}{dx} = -\dfrac{1}{F(x, y)}$$
are called **orthogonal trajectories** of each other. Explain the graphical significance of such orthogonal trajectories. Find the family of orthogonal trajectories of the family $y = x^3 + C$, and draw several members of each family. (*Hint*: First find the differential equation that has $y = x^3 + C$ as its general solution.)

In Exercises 31–34, sketch the slope field of the given differential equations. Then use your result to sketch the graph of the solution curve passing through the given point.

31. $\dfrac{dy}{dx} = x + y$; $(-1, 0)$

32. $\dfrac{dy}{dx} = x^2 - y$; $(0, 1)$

33. $\dfrac{dy}{dx} = xy$; $(1, 1)$

34. $\dfrac{dy}{dx} = x^2 + y^2$; $(2, 0)$

35. If $y' = F(x, y)$, then on each of the level curves $F(x, y) = c$, the slopes of solution curves all have the value c. These level curves are called **isoclines**. For each of the following, draw several isoclines and use them as an aid in making a sketch of the slope field.
   (a) $y' = x - y^2$
   (b) $y' = x^2/y$

Use a CAS for Exercises 36 through 38.

36. In each of the following, first sketch a slope field for the differential equation. Then solve the equation and plot several solution curves on the same plot as the slope field.
   (a) $dy/dx = x - y$
   (b) $dy/dx - 4y = e^x$
   (c) $dy/dx - \sin(x) y = \sin x$
   (d) $dy/dx + x^2 - y = 0$

37. In each of the following, solve the initial-value problem and plot the solution curve.
   (a) $y'' - y' - 6x = 0$, $y(0) = 1$, $y'(0) = 0$
   (b) $(x + y) y' = x - y$, $y(0) = 1$
   (c) $y'' + y' = \sin x$, $y(0) = 0$, $y'(0) = 1$

38. Solve the differential equation $y' + x^2 y - y^2 = 0$.

17.2 FIRST-ORDER SEPARABLE AND FIRST-ORDER HOMOGENEOUS DIFFERENTIAL EQUATIONS

Separable Equations

A first-order differential equation is one that can be written in the form

$$\frac{dy}{dx} = F(x, y)$$

In this section and the next, we introduce solution techniques for cases in which $F(x, y)$ is of a special form. In the first case, $F(x, y)$ is the quotient of a function of x only and a function of y only.

Definition 17.1
Separable Equations

A first-order differential equation that can be written in the form

$$\frac{dy}{dx} = \frac{f(x)}{g(y)} \quad (g(y) \neq 0) \tag{17.1}$$

is said to be *separable*.

In order to explain the meaning of *separable*, let $y = y(x)$ be a solution of Equation 17.1, and write the equation in the form

$$g(y(x)) \frac{dy}{dx} = f(x) \tag{17.2}$$

If G is an antiderivative of g, by the Chain Rule, we have

$$\frac{d}{dx} G(y(x)) = G'(y(x)) \frac{dy}{dx} = g(y(x)) \frac{dy}{dx}$$

Thus, we can write Equation 17.2 as

$$\frac{d}{dx} G(y(x)) = \frac{d}{dx} F(x)$$

where F is an antiderivative of f. The equality of these two derivatives implies that the functions F and G differ by a constant:

$$G(y(x)) = F(x) + C$$

Equivalently, we have

$$\int g(y) \, dy = \int f(x) \, dx \tag{17.3}$$

where we understand that y is a function of x.

In Equation 17.3, the variables x and y are separated. We integrate the left-hand side with respect to y and the right-hand side with respect to x.

In practice, we usually shorten the process leading to Equation 17.3 by the following steps. First, we write Equation 17.1 in the *differential form*

$$g(y) \, dy = f(x) \, dx$$

(thus separating the variables). Then we integrate both sides to get Equation 17.3.

EXAMPLE 17.6
Find the general solution of the differential equation
$$\frac{dy}{dx} = \frac{2x}{y-1}, \quad y > 1$$

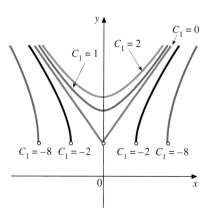

FIGURE 17.6
The family $y = 1 + \sqrt{2x^2 + C_1}$

Solution Separating variables, we get $(y-1)\,dy = 2x\,dx$. Thus, integrating both sides, we have
$$\int (y-1)\,dy = \int 2x\,dx$$
$$\frac{(y-1)^2}{2} = x^2 + C$$

Since $y > 1$, we can solve explicitly for y by taking the positive square root:
$$y - 1 = \sqrt{2(x^2 + C)}$$
or, on writing $C_1 = 2C$,
$$y = 1 + \sqrt{2x^2 + C_1}$$

We show several members of this family in Figure 17.6. ∎

EXAMPLE 17.7
Find the general solution of the differential equation
$$(x^2 + 1)\,dy = xy\,dx$$

Solution Assume for the moment that $y \neq 0$. Then we can separate variables by dividing both sides of the equation by $(x^2 + 1)y$:
$$\frac{dy}{y} = \frac{x}{x^2 + 1}\,dx$$

Thus,
$$\int \frac{dy}{y} = \int \frac{x}{x^2 + 1}\,dx \quad \text{Substitute } u = x^2 + 1 \text{ on the right-hand side.}$$
$$\ln|y| = \frac{1}{2}\ln(x^2 + 1) + C$$

We can solve for y more readily by writing the constant C in the form $\ln|C_1|$, where $C_1 \neq 0$. Then we have
$$\ln|y| = \ln\sqrt{x^2 + 1} + \ln|C_1|$$
$$= \ln\left(|C_1|\sqrt{x^2 + 1}\right)$$
$$|y| = |C_1|\sqrt{x^2 + 1}$$

or equivalently, since C_1 can be either positive or negative,
$$y = C_1\sqrt{x^2 + 1}$$

Now let us return to the original equation and consider the case $y = 0$. The equation will be satisfied in this case if $dy = 0$ also—that is, if y is the zero function. But if we permit C_1 to be 0, the solution $y = C_1\sqrt{x^2 + 1}$ includes $y = 0$ as a particular case. Thus, $y = C_1\sqrt{x^2 + 1}$ is the general solution. (See Figure 17.7.) ∎

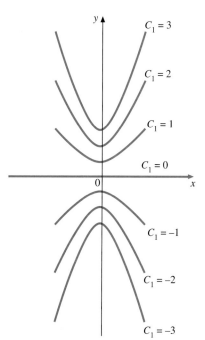

FIGURE 17.7
The family $y = C_1\sqrt{x^2 + 1}$

REMARK
■ The device of writing the constant of integration in the form $\ln|C|$ is frequently useful when other natural logarithm terms are involved. We will frequently do this in subsequent examples.

EXAMPLE 17.8 Solve the initial-value problem
$$y' = xy - y + x - 1; \quad y(1) = 3$$

Solution We factor the right-hand side by grouping, and then we separate variables:

$$\frac{dy}{dx} = y(x-1) + (x-1)$$

$$\frac{dy}{dx} = (y+1)(x-1)$$

$$\frac{dy}{y+1} = (x-1)\,dx \quad (y \neq -1)$$

$$\ln|y+1| = \frac{(x-1)^2}{2} + \ln|C| \quad \text{Write the constant as } \ln|C|.$$

$$\ln\left|\frac{y+1}{C}\right| = \frac{(x-1)^2}{2} \quad \text{Since } \ln|y+1| - \ln|C| = \ln\left|\frac{y+1}{C}\right|$$

$$\frac{y+1}{C} = e^{(x-1)^2/2}$$

$$y = Ce^{(x-1)^2/2} - 1$$

Now we substitute $x = 1$ and $y = 3$ to get $3 = C - 1$, or $C = 4$. The desired particular solution is therefore

$$y = 4e^{(x-1)^2/2} - 1$$

We show the graph in Figure 17.8. ∎

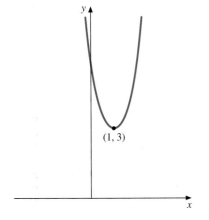

FIGURE 17.8

Homogeneous Equations

Sometimes a differential equation in which the variables are not separable can be transformed into one in which they are. This transformation is especially applicable to so-called **homogeneous** first-order differential equations. To explain the procedure, we first define what is meant by a homogeneous function. Then we use this definition to define a first-order homogeneous differential equation.

Definition 17.2
A Homogeneous Function

A function $f(x, y)$ is said to be **homogeneous of degree** n if, for all $t > 0$ for which (tx, ty) is in the domain of f, the equation

$$f(tx, ty) = t^n f(x, y)$$

holds true.

EXAMPLE 17.9 Show that each of the following functions is homogeneous, and give the degree of homogeneity:

(a) $f(x, y) = x^3 - 2x^2y + xy^2$

(b) $g(x, y) = \dfrac{\sqrt{x^2 + y^2}}{x + y}$

Solution

(a) $f(tx, ty) = t^3x^3 - 2(t^2x^2)(ty) + (tx)(t^2y^2)$
$= t^3(x^3 - 2x^2y + xy^2) = t^3 f(x, y)$

Thus, f is homogeneous of degree 3.

(b) $g(tx, ty) = \dfrac{\sqrt{t^2x^2 + t^2y^2}}{tx + ty} = \dfrac{t\sqrt{x^2 + y^2}}{t(x + y)} = \dfrac{\sqrt{x^2 + y^2}}{x + y} = t^0 g(x, y)$

So g is homogeneous of degree 0. Note that we used the fact that $t > 0$ when we wrote $\sqrt{t^2x^2 + t^2y^2} = t\sqrt{x^2 + y^2}$. ∎

Definition 17.3
Homogeneous Differential Equation

An equation that can be expressed in the form

$$\dfrac{dy}{dx} = F(x, y)$$

where F is homogeneous of degree 0 is said to be a first-order **homogeneous differential equation**.

EXAMPLE 17.10 Show that each of the following differential equations is homogeneous:

(a) $\dfrac{dy}{dx} = \dfrac{\sqrt{x^2 + y^2}}{x + y}$

(b) $\left(y + x \sin \dfrac{x}{y}\right) dx + \left(y \cos \dfrac{x}{y} - 2x\right) dy = 0$

Solution

(a) We saw in Example 17.9, part (b), that $\sqrt{x^2 + y^2}/(x + y)$ is homogeneous of degree 0. So, by Definition 17.3, the given differential equation is homogeneous.

(b) Solving for dy/dx, we get

$$\dfrac{dy}{dx} = \dfrac{y + x \sin \dfrac{x}{y}}{2x - y \cos \dfrac{x}{y}}$$

Denote the function on the right by $F(x, y)$. Then

$$F(tx, ty) = \frac{ty + tx \sin \frac{tx}{ty}}{2tx - ty \cos \frac{tx}{ty}} = \frac{t\left(y + x \sin \frac{x}{y}\right)}{t\left(2x - y \cos \frac{x}{y}\right)} = F(x, y)$$

So F is homogeneous of degree 0, which means that the differential equation is homogeneous. ∎

If a differential equation is written in the differential form

$$f(x, y)\,dx + g(x, y)\,dy = 0$$

then the homogeneity requirement is satisfied provided f and g are homogeneous functions of the *same* degree. We can see this result by writing

$$\frac{dy}{dx} = -\frac{f(x, y)}{g(x, y)}$$

and observing that $F = -f/g$ will be homogeneous of degree 0 when f and g are homogeneous of the same degree. In Example 17.10(b), we see this result, where both $f(x, y) = y + x \sin(x/y)$ and $g(x, y) = y \cos(x/y) - 2x$ are homogeneous functions of degree 1.

The important and useful property common to all first-order homogeneous differential equations is that either of the substitutions

$$v = \frac{y}{x} \quad \text{or} \quad v = \frac{x}{y}$$

will invariably transform the equation to one in which the variables are separable. We will illustrate this result using $v = y/x$, but you should keep in mind that the second substitution $v = x/y$ also works and in some cases may involve simpler calculations.

Suppose, then, that $dy/dx = F(x, y)$ is a homogeneous differential equation. Setting $v = y/x$, we obtain

$$y = vx \quad \text{and} \quad \frac{dy}{dx} = v + x \frac{dv}{dx} \qquad (17.4)$$

On substitution in the original differential equation, we get

$$v + x \frac{dv}{dx} = F(x, vx) \qquad (17.5)$$

By our assumption that the original differential equation is homogeneous, we know by Definition 17.3 that F is a homogeneous function of degree 0. Thus, by Definition 17.2 (with x replacing t and v replacing y),

$$F(x, vx) = x^0 F(1, v) = F(1, v)$$

From Equation 17.5, we therefore have

$$v + x \frac{dv}{dx} = F(1, v)$$

Solving for dv/dx, we get

$$\frac{dv}{dx} = \frac{F(1, v) - v}{x}$$

Section 17.2 First-Order Separable and First-Order Homogeneous Differential Equations   1229

The variables x and v can now be separated:

$$\frac{dv}{F(1,v) - v} = \frac{dx}{x} \tag{17.6}$$

If F were known, we could solve this equation (assuming we can integrate the left-hand side) and then replace v by y/x to get the final result.

REMARK
- There is no need to memorize Equation 17.6. The important thing to remember is that once you have identified a differential equation as being homogeneous, the substitution in Equations 17.4 should be made (or the similar substitution corresponding to $v = x/y$). Then in each individual case the variables can be separated and the resulting equation solved.

EXAMPLE 17.11   Find the general solution of the differential equation

$$\frac{dy}{dx} = \frac{x+y}{x-y}$$

Solution   If we write

$$F(x,y) = \frac{x+y}{x-y}$$

we see that F is homogeneous of degree 0, since

$$F(tx, ty) = \frac{tx + ty}{tx - ty} = \frac{x+y}{x-y} = F(x,y)$$

Letting $y = vx$, we have

$$F(x, vx) = \frac{x + vx}{x - vx} = \frac{1+v}{1-v}$$

By Equation 17.5,

$$v + x\frac{dv}{dx} = \frac{1+v}{1-v}$$

Simplifying, we obtain

$$x\frac{dv}{dx} = \frac{1+v}{1-v} - v$$

$$= \frac{1 + v - v(1-v)}{1-v}$$

$$= \frac{1+v^2}{1-v}$$

Now we can separate variables to get

$$\frac{1-v}{1+v^2}dv = \frac{dx}{x}$$

Thus,
$$\int \frac{1-v}{1+v^2} \, dv = \int \frac{dx}{x}$$

We divide the integral on the left into two integrals, obtaining
$$\int \frac{dv}{1+v^2} - \int \frac{v \, dv}{1+v^2} = \int \frac{dx}{x}$$

Integrating, we get
$$\tan^{-1} v - \frac{1}{2} \ln(1+v^2) = \ln|x| + C$$

Finally, we replace v by y/x:
$$\tan^{-1} \frac{y}{x} - \frac{1}{2} \ln\left(1 + \frac{y^2}{x^2}\right) = \ln|x| + C$$

Since $\ln[1 + (y^2/x^2)] = \ln[(x^2+y^2)/x^2] = \ln(x^2+y^2) - \ln x^2$ and $\ln x^2 = 2 \ln|x|$, we can simplify our answer to get
$$\tan^{-1} \frac{y}{x} - \frac{1}{2} \ln(x^2 + y^2) = C$$

Solving for y would be difficult, so we leave the answer in implicit form. ∎

When you make the substitution $y = vx$ to solve a homogeneous differential equation, you will first get an answer in terms of x and v. Don't forget then to replace v by y/x to get the final answer. A similar remark applies when you substitute $x = vy$.

EXAMPLE 17.12 Solve the initial-value problem
$$(2x - 3y) \, dx + y \, dy = 0; \qquad y(1) = 3$$

Solution The coefficients of dx and dy are both homogeneous functions of degree 1, so the differential equation is homogeneous. Setting $v = y/x$ gives $y = vx$ and $dy = v \, dx + x \, dv$. Thus, we get
$$(2x - 3vx) \, dx + vx(v \, dx + x \, dv) = 0$$

On dividing the preceding equation by x and simplifying, we get
$$(v^2 - 3v + 2) \, dx + xv \, dv = 0$$

We now separate variables and use partial fractions:
$$\frac{v \, dv}{(v-2)(v-1)} = -\frac{dx}{x}$$
$$\left(\frac{2}{v-2} - \frac{1}{v-1}\right) dv = -\frac{dx}{x}$$

Thus, on integrating both sides, we get

$$2\ln|v-2| - \ln|v-1| = -\ln|x| + \ln|C|$$

Write $\ln|C|$ for the constant of integration.

$$\ln\frac{(v-2)^2}{|v-1|} = \ln\left|\frac{C}{x}\right|$$

$$(v-2)^2 = (v-1)\frac{C}{x}$$

Now we replace v with y/x and simplify to get $(y-2x)^2 = C(y-x)$. Imposing the initial condition $y(1) = 3$ gives $C = \frac{1}{2}$ (found by substituting $x = 1$ and $y = 3$). Thus, the solution in question is given implicitly by

$$y = x + 2(y - 2x)^2$$

We show its graph in Figure 17.9.

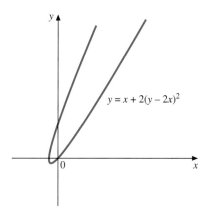

FIGURE 17.9

Exercise Set 17.2

In Exercises 1–5, use separation of variables to find the general solution.

1. $\dfrac{dy}{dx} = \dfrac{y}{x+1}$

2. $xy\,dy = (y^2 + 1)\,dx$

3. $y' = \dfrac{x+1}{y-2}$

4. $y' = \dfrac{y}{x^2y + x^2}$

5. $e^{-x}\,dy - (1 + y^2)\,dx = 0$

In Exercises 6–10, show that the differential equation is homogeneous, and find its general solution.

6. $y' = \dfrac{x+y}{x}$

7. $\dfrac{dy}{dx} = \dfrac{y(x+y)}{x(x-y)}$

8. $2x^2\,dx = (x^2 + y^2)\,dy$

9. $\dfrac{dy}{dx} = \dfrac{x}{x+2y}$

10. $xy\,dy = y\left(y + x\cos^2\dfrac{y}{x}\right)dx$

In Exercises 11–20, use whatever method is appropriate to find the general solution. If an initial value is given, find the particular solution that satisfies it.

11. $y' = x\sqrt{4 - y^2}$

12. $y' = \dfrac{\ln x}{2xy - 3x}; \; y(1) = 1$

13. $x\,dy = \left(\sqrt{x^2 - y^2} + y\right)dx$

14. $\tan x \sec y\,dx = \sin^2 y\,dy$

15. $y' = \dfrac{2xy}{x^2 - y^2}; \; y(1) = 2$

16. $\sqrt{y^2 + 4}\,dx + (2y - xy)\,dy = 0$

17. $ye^{x/y}\,dx - (y + xe^{x/y})\,dy = 0; \; y(0) = e$

18. $\dfrac{dy}{dx} = \dfrac{3x - y}{x + y}$

19. $(xy - x + y - 1)\,dy = (xy + x)\,dx$

20. $xy' = y + x\tan y/x; \; y(2) = \pi/3$

21. Show that if $y' = F(x, y)$ is homogeneous, then

$$x = C\exp\left(\int\frac{dv}{F(1,v) - v}\right)$$

where $v = y/x$. (Recall that $\exp x$ means e^x.)

22. Show that with the substitution $v = x/y$, the homogeneous differential equation $dy/dx = F(x, y)$ is transformed into the separable equation

$$y\frac{dv}{dy} = \frac{1}{F(v, 1)} - v$$

23. Use the method of Exercise 22 to solve the differential equation

$$\frac{dy}{dx} = \frac{y^3 + x^2y}{x^3 + y^3}$$

24. Find the general solution of the equation
$$3(1 + y^2)\,dx + (y^2 - x^3 y^2 - 2x^3 y + 2y)\,dy = 0$$

25. Solve the initial-value problem
$$y' = \frac{5xy + y - 5x - 1}{x^3 - 3x + 2}; \quad y(0) = 2$$

26. Use the idea of Exercise 30 in Exercise Set 17.1 to find the family of orthogonal trajectories of the family $x^2 - xy + y^2 = C$.

27. Prove that $dy/dx = F(x, y)$ is homogeneous if and only if there is a function G of one variable such that
$$F(x, y) = G\left(\frac{y}{x}\right)$$

17.3 FIRST-ORDER EXACT AND FIRST-ORDER LINEAR DIFFERENTIAL EQUATIONS

Exact Equations

Consider a first-order differential equation in differential form
$$f(x, y)\,dx + g(x, y)\,dy = 0$$
with f and g having continuous first partial derivatives in a region R. Suppose further that there is a function $\phi(x, y)$ such that
$$\frac{\partial \phi(x, y)}{\partial x} = f(x, y) \quad \text{and} \quad \frac{\partial \phi(x, y)}{\partial y} = g(x, y)$$
For example, the differential equation
$$(3x^2 - y)\,dx + (2y - x)\,dy = 0$$
can be written as
$$\frac{\partial \phi(x, y)}{\partial x}\,dx + \frac{\partial \phi(x, y)}{\partial y}\,dy = 0 \tag{17.7}$$
with $\phi(x, y) = x^3 - xy + y^2$, as you can readily verify. In Chapter 16, the expression on the left-hand side of Equation 17.7 was defined as the *exact differential* of ϕ. Differential equations that can be written in the form of Equation 17.7 are called *exact differential equations*.

Definition 17.4
Exact Differential Equation

The first-order differential equation
$$f(x, y)\,dx + g(x, y)\,dy = 0 \tag{17.8}$$
is said to be **exact** in a region R if there exists a function ϕ such that
$$f(x, y) = \frac{\partial \phi(x, y)}{\partial x} \quad \text{and} \quad g(x, y) = \frac{\partial \phi(x, y)}{\partial y}$$
in R.

If Equation 17.8 is exact, then by Definition 14.2, we know that $f(x, y)\,dx + g(x, y)\,dy$ is the exact differential of ϕ. That is, $d\phi(x, y) = f(x, y)\,dx + g(x, y)\,dy$.

Section 17.3 First-Order Exact and First-Order Linear Differential Equations 1233

From Section 16.2, we know that Equation 17.8 is exact if

$$\frac{\partial g}{\partial x} = \frac{\partial f}{\partial y} \quad (17.9)$$

throughout R. In this case, Equation 17.8 is equivalent to

$$d\phi(x, y) = 0$$

whose solution is

$$\phi(x, y) = C \quad (17.10)$$

(See Exercise 35 in Exercise Set 17.3.) Thus, Equation 17.10 is an implicit solution to Equation 17.8.

In the next example, we review the procedure for finding the function ϕ when Equation 17.9 is satisfied.

EXAMPLE 17.13 Show that the differential equation

$$(3x^2 - y)\,dx + (2y - x)\,dy = 0$$

is exact, and find its general solution.

Solution We have already seen that the left-hand side of this equation is the exact differential of the function $\phi(x) = x^3 - xy + y^2$, but for purposes of this example, we will assume this result is not known.

Equation 17.9 is satisfied, since with $f(x, y) = 3x^2 - y$ and $g(x, y) = 2y - x$, we have

$$\frac{\partial g}{\partial x} = -1 \quad \text{and} \quad \frac{\partial f}{\partial y} = -1$$

So we know the equation is exact. Let ϕ denote a function for which

$$\frac{\partial \phi}{\partial x} = 3x^2 - y \quad \text{and} \quad \frac{\partial \phi}{\partial y} = 2y - x$$

Integrating both sides of the first of these equations with respect to x gives

$$\phi(x, y) = \int (3x^2 - y)\,dx = x^3 - xy + C_1(y) \qquad \text{The "constant" of integration can be a function of } y.$$

Now, imposing the condition $\partial \phi / \partial y = 2y - x$ gives

$$-x + C_1'(y) = 2y - x$$
$$C_1'(y) = 2y$$
$$C_1(y) = y^2$$

(We do not need to include a constant of integration at this stage.) Thus, we have verified that

$$\phi(x, y) = x^3 - xy + y^2$$

The original differential equation is $d\phi(x, y) = 0$, so its solution is $\phi(x, y) = C$. That is,

$$x^3 - xy + y^2 = C \qquad \blacksquare$$

EXAMPLE 17.14 Solve the initial-value problem
$$\frac{dy}{dx} = \frac{xe^{-x} + \cos y}{x \sin y - \cos y}, \qquad y(0) = \pi$$

Solution The equation can be written in the differential form
$$(xe^{-x} + \cos y)\,dx + (\cos y - x \sin y)\,dy = 0$$

The variables cannot be separated, and the equation is not homogeneous. Testing for exactness, we have
$$\frac{\partial}{\partial y}(xe^{-x} + \cos y) = -\sin y, \qquad \frac{\partial}{\partial x}(\cos y - x \sin y) = -\sin y$$

Thus, the equation is exact. So we know that a function ϕ exists for which $d\phi = (xe^{-x} + \cos y)\,dx + (\cos y - x \sin y)\,dy$. To find ϕ, we choose this time to begin by integrating the coefficient of dy with respect to y, since it looks relatively easy to do.

$$\phi(x, y) = \int (\cos y - x \sin y)\,dy = \sin y + x \cos y + C_1(x)$$

Now we must also have $\partial\phi/\partial x = xe^{-x} + \cos y$. So
$$\cos y + C_1'(x) = xe^{-x} + \cos y$$
$$C_1'(x) = xe^{-x}$$
$$C_1(x) = -xe^{-x} - e^{-x} \qquad \text{Obtained by integrating by parts}$$

Thus
$$\phi(x, y) = \sin y + x \cos y - xe^{-x} - e^{-x}$$

and since the original equation can be written as $d\phi = 0$, its general solution is $\phi(x, y) = C$; that is,
$$\sin y + x \cos y - xe^{-x} - e^{-x} = C$$

Imposing the initial condition $y(0) = \pi$ gives $-1 = C$. The particular solution in question can therefore be written in the form
$$\sin y + x \cos y - e^{-x}(x + 1) + 1 = 0 \qquad \blacksquare$$

Linear Equations

Another class of differential equations that in many cases can be solved explicitly consists of equations involving y and its derivatives to the first power only, with coefficients that are functions of x only. Such equations are called *linear* differential equations. The formal definition is as follows:

Definition 17.5
Linear Differential Equation

An nth-order differential equation is said to be **linear** if it can be written in the form
$$a_n(x)y^{(n)} + a_{n-1}(x)y^{(n-1)} + \cdots + a_1(x)y' + a_0(x)y = g(x) \qquad (17.11)$$
with $a_n(x) \neq 0$.

The term *linear* is used here because each term on the left-hand side of Equation 17.11 involves y or one of its derivatives raised to the first power only. To emphasize this reference to y and its derivatives, we sometimes say that the equation is *linear in y*. The coefficient functions $a_0(x), a_1(x), \ldots, a_n(x)$ may be highly nonlinear. For example, both

$$\frac{dy}{dx} + (\tan x)y = \sin x \quad \text{and} \quad x^2 y'' + 2xy' - 3y = \ln x$$

are linear in y, whereas

$$\left(\frac{dy}{dx}\right)^2 + 2xy = 0 \quad \text{and} \quad y'y'' - 3y = x$$

are nonlinear.

If $n = 1$ in Equation 17.11, we have the first-order linear equation

$$a_1(x)y' + a_0(x)y = g(x), \quad a_1(x) \neq 0$$

If we divide both sides by $a_1(x)$ and write $P(x) = a_0(x)/a_1(x)$ and $Q(x) = g(x)/a_1(x)$, we obtain the following, called the **standard form** of a first-order linear differential equation.

Standard Form of a First-Order Linear Differential Equation

$$y' + P(x)y = Q(x) \tag{17.12}$$

In order to develop a general procedure for solving first-order linear differential equations, let us rewrite Equation 17.12 in the differential form

$$[P(x)y - Q(x)]\,dx + dy = 0 \tag{17.13}$$

Our goal is to transform this equation into an exact differential equation, which we can then solve by the technique we have already developed.

Sometimes equations that are not exact can be transformed to exact equations by multiplying both sides by a function of x. For example, the linear equation

$$(5x + 2y)\,dx + x\,dy = 0$$

is not exact, but if we multiply both sides by x, we get

$$(5x^2 + 2xy)\,dx + x^2\,dy = 0$$

which is exact, since $\partial(x^2)/\partial x = 2x$ and $\partial(5x^2 + 2xy)/\partial y = 2x$ are equal. The factor x that we multiplied by is called an **integrating factor** for the differential equation.

We now generalize the idea illustrated in this example by attempting to find an integrating factor, which we denote by $\mu(x)$, for Equation 17.13. That is, we try to find a function $\mu(x)$ such that

$$[\mu(x)P(x)y - \mu(x)Q(x)]\,dx + \mu(x)\,dy = 0 \tag{17.14}$$

is an exact equation. The requirement for exactness is

$$\frac{\partial}{\partial y}[\mu(x)P(x)y - \mu(x)Q(x)] = \frac{\partial}{\partial x}[\mu(x)]$$

That is,
$$\mu'(x) = \mu(x)P(x) \tag{17.15}$$

We can separate variables in Equation 17.15 to get
$$\frac{d\mu(x)}{\mu(x)} = P(x)\,dx$$

When we integrate both sides, we obtain
$$\ln \mu(x) = \int P(x)\,dx$$

from which we get the following formula for $\mu(x)$.

An Integrating Factor for Linear First-Order Differential Equations
$$\mu(x) = e^{\int P(x)\,dx} \tag{17.16}$$

Note that $\mu(x)$ is positive, since e raised to any power is positive. We can take the constant of integration as 0, since we want *any* integrating factor, so we might as well use the simplest one.

We could proceed now to find a function ϕ for which the left-hand side of Equation 17.14 is the exact differential, using the technique we have developed. However, in this case there is an easier way. If we multiply both sides of Equation 17.12 by $\mu(x)$, we have
$$\mu(x)y' + \mu(x)P(x)y = \mu(x)Q(x)$$

But by Equation 17.15 this equation can be written as (suppressing the parentheses and the independent variable x)
$$\mu y' + \mu' y = \mu Q$$

The left-hand side is precisely the derivative of the product μy. Thus,
$$\frac{d}{dx}[\mu y] = \mu Q \tag{17.17}$$

So, on integrating both sides,
$$\mu y = \int \mu Q\,dx + C$$

and
$$y = \frac{1}{\mu}\int \mu Q\,dx + \frac{C}{\mu} \tag{17.18}$$

Direct substitution will show that the value of y given by Equation 17.18 is the solution we were seeking. There is no need to memorize the solution given by Equation 17.18. You can in each individual case carry out the following procedure.

Procedure for Solving a First-Order Linear Differential Equation

1. If the equation is linear, write it in the standard form
$$y' + P(x)y = Q(x)$$

2. Find the integrating factor
$$\mu(x) = e^{\int P(x)\,dx}$$

3. Multiply both sides of the standard form by $\mu(x)$, thereby obtaining
$$\frac{d}{dx}[\mu y] = \mu Q$$
(It is a good idea to check at this stage that the left-hand side is indeed the derivative of the product of μ and y.)

4. Take antiderivatives and solve for y.

EXAMPLE 17.15 Solve the differential equation
$$xy' + 2y = 2x^2 + 3x, \qquad x > 0$$

Solution The equation is a first-order linear differential equation in y. To put it in standard form, we divide by x:
$$y' + \frac{2}{x}y = 2x + 3$$

So $P(x) = \dfrac{2}{x}$ and $Q(x) = 2x + 3$. The integrating factor is therefore
$$\mu(x) = e^{\int P(x)\,dx} = e^{\int (2/x)\,dx} = e^{2\ln x} = e^{\ln x^2} = x^2$$

(Note the use of the properties of logarithms and exponentials and the stipulation that $x > 0$.) We next multiply both sides of the standard form by $\mu(x)$:
$$x^2 y' + 2xy = 2x^3 + 3x^2$$

If our theory is correct, the left-hand side should be the derivative of μy—that is, of $x^2 y$—and a quick check using the Product Rule shows that this is correct. So we have
$$\frac{d}{dx}(x^2 y) = 2x^3 + 3x^2$$

Integrating both sides gives
$$x^2 y = \frac{x^4}{2} + x^3 + C$$
$$y = \frac{x^2}{2} + x + \frac{C}{x^2}$$

We show several solution curves in Figure 17.10. ∎

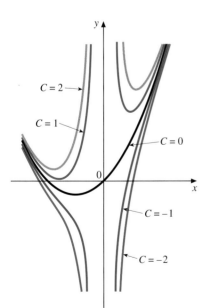

FIGURE 17.10
The family $y = \dfrac{x^2}{2} + x + \dfrac{C}{x^2}$

A common mistake in this procedure is to forget the constant of integration in the next-to-last step. The constant is essential, however; otherwise we have not found the general solution.

EXAMPLE 17.16 Solve the initial-value problem
$$\frac{dy}{dx} = \frac{x+y}{x-1}; \quad y(2) = 3$$

Solution We see that the equation is linear in y, since both y and y' appear to the first degree. In standard form the equation is
$$y' + \left(\frac{1}{1-x}\right)y = \frac{x}{x-1} \quad \text{Since } \frac{-y}{x-1} = \frac{y}{1-x}$$

Since we want the solution passing through (2, 3), we restrict x so that $x > 1$. Thus,
$$\mu(x) = e^{\int dx/(1-x)} = e^{-\ln|1-x|} = e^{\ln(1/|1-x|)} = \frac{1}{|1-x|} = \frac{1}{x-1} \quad \begin{array}{l}\text{Since } x > 1,\\ |1-x| = x-1.\end{array}$$

We multiply both sides by $\mu(x)$ to get
$$\frac{d}{dx}\left[\left(\frac{1}{x-1}\right)y\right] = \frac{x}{(x-1)^2} \quad \text{Verify.}$$

Hence,
$$\frac{1}{x-1}y = \int \frac{x}{(x-1)^2}\,dx$$

Using partial fractions, we can write the integral on the right-hand side as
$$\int\left[\frac{1}{x-1} - \frac{1}{(x-1)^2}\right]dx$$

Thus,
$$\frac{1}{x-1}y = \ln(x-1) - \frac{1}{x-1} + C$$
$$y = (x-1)\ln(x-1) - 1 + C(x-1)$$

This is the general solution. Now we set $x = 2$ and $y = 3$, getting $C = 4$. Substituting this constant and simplifying gives the particular solution
$$y = (x-1)\ln(x-1) + 4x - 5$$

We show its graph in Figure 17.11.

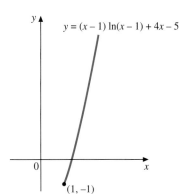

$y = (x-1)\ln(x-1) + 4x - 5$

(1, −1)

FIGURE 17.11

Exercise Set 17.3

In Exercises 1–6, show that the given equation is exact, and find its general solution. If an initial condition is given, find the corresponding particular solution.

1. $\dfrac{dy}{dx} = \dfrac{x-y+1}{x-2}$

2. $(e^y + 2x)\,dx + (xe^y - 2y)\,dy = 0$

3. $2xyy' = 3x^2 - y^2; \quad y(1) = 2$

4. $\dfrac{dy}{dx} = \dfrac{y\sin x - \cos y}{\cos x - x\sin y}$

5. $\dfrac{y}{x}\,dx + (y + \ln x)\,dy = 0; \quad y(1) = 4$

6. $y' = \dfrac{x\ln\cos^2 y}{x^2\tan y - 3y^2}; \quad y(0) = 2$

In Exercises 7–12, show that the given equation is linear, and find its general solution. If an initial condition is given, find the corresponding particular solution.

7. $y' + y = x$

8. $y' - 2xy = x$

9. $x\dfrac{dy}{dx} = x^2 - y;\ y(2) = 1$

10. $dy = (\sin x + y \tan x)\,dx \quad \left(-\dfrac{\pi}{2} < x < \dfrac{\pi}{2}\right)$

11. $\dfrac{dy}{dx} + \dfrac{y}{1-x} = 1 - x^2;\ y(0) = 0$

12. $xy' = x^4 e^x + 2y;\ y(1) = 0$

Solve Exercises 13–27 by any method.

13. $(x - 2y)\,dx + (3y - 2x)\,dy = 0$

14. $y' = x - xy$

15. $(y + 1)\,dx + (x - 1)\,dy = 0;\ y(0) = 2$

16. $y' + y \cot x = 0 \quad (0 < x < \pi)$

17. $y' - \dfrac{2xy}{x^2 + 1} = 1;\ y(1) = \pi$

18. $\dfrac{dy}{dx} = \dfrac{x^2 - y^2}{y(2x - y)}$

19. $\left(\dfrac{y}{x} - e^x + 2\right)dx + (\ln xy + 3y^2 - 1)\,dy = 0 \quad (x > 0)$

20. $(\cosh x)\,dy = (\cosh^3 x + y \sinh x)\,dx$

21. $\left(\dfrac{x}{\sqrt{x^2 + y^2}} + 1\right)dx + \left(\dfrac{y}{\sqrt{x^2 + y^2}} - 2\right)dy = 0;$
$y(3) = 4$

22. $x\,dy + y(x + 1)\,dx = x^2\,dx \quad (x > 0)$

23. $\left(\dfrac{x - y}{x}\right)dx + (y - \ln x)\,dy = 0 \quad (x > 0)$

24. $(e^x + 1)\,dy = e^x(x - y)\,dx$

25. $\dfrac{dy}{dx} = \dfrac{1 + y^2}{y(1 + x + y^2)}$ (Hint: Show it is linear in x.)

26. $y' + y \sin x = \sin 2x;\ y\left(\dfrac{\pi}{2}\right) = 3$

27. $[\ln(x^2 + y^2)]\,dx = 2\left(\tan^{-1}\dfrac{y}{x}\right)dy$

28. An equation of the form

$$y' + P(x)y = y^n Q(x) \quad (n \neq 1)$$

is called a **Bernoulli** equation. Show that the substitution $v = y^{1-n}$ transforms it into an equation that is linear in v. (Hint: First divide both sides by y^n.)

In Exercises 29 and 30, use the method of Exercise 28 to find the solution.

29. $y' + 2xy = xy^2$

30. $y' + \dfrac{2y}{x} = \sqrt{y} \sin x;\ y\left(\dfrac{\pi}{2}\right) = 4$

31. Show that a differential equation of the form

$$f(x, y)\,dx + g(x, y)\,dy = 0$$

if not already exact, can be made into an exact equation by multiplying both sides by an integrating factor of the form $\mu(x)$ provided $(f_y - g_x)/g$ is a function of x only. What is $\mu(x)$ in this case?

32. Find a condition analogous to that in Exercise 31 that will ensure the existence of an integrating factor of the form $\mu(y)$.

In Exercises 33 and 34, use the method of Exercise 31 or 32 to find an integrating factor, and solve the equation.

33. $(y^2 + 4x)\,dx + y\,dy = 0$

34. $2 \sin x \cos^2 y\,dx + \cos x \sin 2y\,dy = 0$

35. Prove that if ϕ is differentiable in a region R of P^2, then $y = y(x)$ is a solution to the differential equation $d\phi(x, y) = 0$ if and only if $\phi(x, y(x)) = C$.

17.4 APPLICATIONS OF FIRST-ORDER DIFFERENTIAL EQUATIONS

Differential equations are particularly useful in modeling real-world phenomena involving rates of change (derivatives). The following diagram illustrates the usual stages involved in mathematical modeling.

1240 Chapter 17 Differential Equations

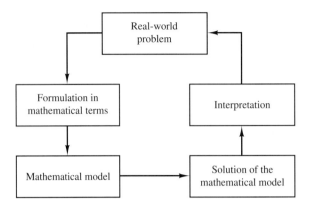

In this section we show how first-order differential equations can be used as models in a variety of applications. The first example is familiar from Section 7.5.

Radioactive Decay

Mushroom cloud from the explosion of a plutonium bomb

EXAMPLE 17.17 The isotope plutonium-241 produced in a nuclear explosion has a half-life of approximately 13.2 yr. Assume that the substance decays at a rate proportional to the amount present. Of a given initial amount, how much will remain after 5 yr? How long will it take for 90% of the original amount to decay?

Solution Let $Q(t)$ denote the quantity present at time t (in years). Then since the rate of change dQ/dt is proportional to Q, we have $dQ/dt = kQ$. Denote the initial amount $Q(0)$ by Q_0. Since the half-life is 13.2 yr, we have

$$Q(13.2) = 0.5 Q_0$$

Our problem is thus modeled by the differential equation $dQ/dt = kQ$, together with the known value $Q(13.2) = 0.5 Q_0$. We can solve the differential equation by separating variables (note that it can also be solved as a linear equation).

$$\frac{dQ}{Q} = k\, dt$$

Integrating both sides gives

$$\ln Q = kt + \ln C \quad \text{Write the constant of integration as } \ln C.$$

$$\ln Q - \ln C = kt$$

$$\ln \frac{Q}{C} = kt$$

$$Q(t) = C e^{kt}$$

Letting $t = 0$ gives $Q_0 = C$. So the solution is of the form

$$Q(t) = Q_0 e^{kt}$$

To find k, we use the fact that $Q(13.2) = 0.5Q_0$. Thus,

$$0.5Q_0 = Q_0 e^{13.2k}$$

from which we find k by dividing by Q_0 and taking the natural logarithm of each side:

$$k = \frac{\ln 0.5}{13.2} \approx -0.0525$$

and the solution becomes

$$Q(t) = Q_0 e^{-0.0525t}$$

We show the graph of Q in Figure 17.12.

For $t = 5$, we have

$$Q(5) = Q_0 e^{(-0.0525)5} \approx 0.769 Q_0$$

So after 5 yr, about 77% of the original amount remains.

To find how long it takes for 90% to decay, we set $Q(t) = 0.1 Q_0$ and solve for t, since if 90% has decayed, 10% remains:

$$0.1 Q_0 = Q_0 e^{-0.0525t}$$

$$t = \frac{\ln 0.1}{-0.0525} \approx 43.7 \text{ yr} \qquad \blacksquare$$

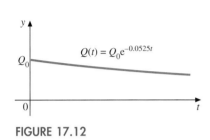

FIGURE 17.12

Population Growth

As our second example, we solve a model for population growth that in general gives a more realistic estimate than the Malthus Model that we considered in Section 7.4. In most models of population growth, the instantaneous rate of change of the size Q of the population is assumed to be some function of Q:

$$\frac{dQ}{dt} = f(Q) \qquad (17.19)$$

For the Malthus Model, $f(Q)$ is taken to be kQ for $k > 0$, and just as in the previous example, the solution is $Q(t) = Q_0 e^{kt}$. In 1837, the Belgian mathematician Pierre-François Verhulst (1804–1849) proposed modifying the Malthus Model by taking $f(Q) = \alpha Q - \beta Q^2$, in Equation 17.19, where α and β are positive constants. Subtracting the term βQ^2 was Verhulst's way of accounting for competition among members of the population. The next example shows one way to solve Verhulst's equation.

EXAMPLE 17.18 Solve the Verhulst equation for population growth,

$$\frac{dQ}{dt} = \alpha Q - \beta Q^2 \qquad (\alpha > 0, \beta > 0)$$

where the initial size $Q(0)$ of the population is Q_0. Analyze the result.

Solution One way to solve the problem is by separating variables, making use of partial fractions in the integration. We choose to illustrate a method suggested in Exercise 28 of Exercise Set 17.3 whereby we make a substitution that transforms the equation into a linear one. Equations of this type are called **Bernoulli equations**. We first rewrite the equation as

---

The Bernoullis were the greatest family of mathematicians in history. Jakob Bernoulli (1654–1705) was born and educated in Switzerland. After extending Descartes's book on geometry, he showed how to divide a triangle into four equal parts with two perpendicular straight lines.

Jakob elaborated on Leibniz's ideas and considered the problem of a body descending in a curve under the force of gravity. In this analysis the word *integral* first came to be used. He also worked out the parabolic and logarithmic spirals.

Jakob taught mathematics to his younger brother, Johann I (1667–1748), who became professor of mathematics in Basel, and extended the ideas of infinitesimals. He developed important ideas about series, about large numbers in probability theory, and on the method we call mathematical induction.

(continued next page)

Jakob and Johann together became the first to fully understand the differential calculus as expressed by Leibniz. In 1696, Johann proposed the problem of determining the curve by which a bead would descend most swiftly, the famous Brachistochrone Problem bested by Newton.

Johann taught calculus to L'Hôpital, the greatest French mathematician of the time. Johann's subsequent letters, after he left Paris, were the basis of the first textbook on differential calculus, published under the name of L'Hôpital.

Nikolaus I (1647–1759), the nephew of Jakob and Johann, studied with his uncles. Among his best known results is the proof that the binomial expansion $(1+x)^n$ diverges for $x > 1$.

Daniel (1700–1782), one of Johann I's sons worked on differential equations and series. He is also noted for his work in physics. In fact, we credit "Bernoulli's Principle" for explaining why airplanes fly—that pressure is lowest where speed is highest.

$$\frac{dQ}{dt} - \alpha Q = -\beta Q^2$$

and then divide through by Q^2:

$$Q^{-2}\frac{dQ}{dt} - \alpha Q^{-1} = -\beta$$

Now let $v = Q^{-1}$ and note that the derivative of v with respect to time t is

$$\frac{dv}{dt} = -Q^{-2}\frac{dQ}{dt}$$

When we substitute and multiply both sides by -1, we get

$$\frac{dv}{dt} + \alpha v = \beta$$

This equation is linear in v. The integrating factor is $\mu(t) = e^{\alpha t}$, and on multiplying by this value of $\mu(t)$ and simplifying (by the Product Rule), we get

$$\frac{d}{dt}[ve^{\alpha t}] = \beta e^{\alpha t}$$

Thus, on integrating both sides, we get

$$ve^{\alpha t} = \frac{\beta}{\alpha}e^{\alpha t} + C$$

The constant C can be found by setting $t = 0$, observing that $v(0) = \frac{1}{Q(0)} = \frac{1}{Q_0}$. Thus,

$$C = \frac{1}{Q_0} - \frac{\beta}{\alpha} = \frac{\alpha - \beta Q_0}{\alpha Q_0}$$

Substituting this constant, we can solve for v and then invert to obtain Q, since $Q = 1/v$. The details of the algebra are omitted, but the result is

$$Q(t) = \frac{\alpha Q_0}{\beta Q_0 + (\alpha - \beta Q_0)e^{-\alpha t}} \tag{17.20}$$

We make several observations about this solution. First, if β is small relative to α, the solution differs little from the Malthus solution $Q(t) = Q_0 e^{\alpha t}$, as you can see by neglecting the terms involving β. Second, if the denominator is written in the form

$$\beta Q_0(1 - e^{-\alpha t}) + \alpha e^{-\alpha t}$$

we see that for t near 0, the factor $1 - e^{-\alpha t} \approx 0$, so that again the solution is approximately $Q_0 e^{\alpha t}$. Finally, let us see what happens as $t \to \infty$:

$$\lim_{t \to \infty} Q(t) = \frac{\alpha Q_0}{\beta Q_0 + 0} = \frac{\alpha}{\beta}$$

Here, the result is strikingly different from that of Malthus. There is a limit to the size of the population, whereas in the Malthus model the growth is unlimited.

The Verhulst differential equation is sometimes referred to as a **logistic equation**, and its solution as given in Equation 17.20 is called the **Law of Logistic Growth**. The graph of Q is shown in Figure 17.13. ∎

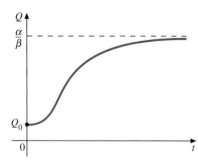

FIGURE 17.13
The logistic curve
$$Q(t) = \frac{\alpha Q_0}{\beta Q_0 + (\alpha - \beta Q_0)e^{-\alpha t}}$$

REMARK

■ If we put $m = \alpha/\beta$, the Verhulst equation can be written in the form

$$\frac{dQ}{dt} = \beta Q(m - Q)$$

and since m is the limiting maximum size of the population, this equation can be interpreted as saying that the rate of change of the population at any time is jointly proportional to the population at that time and the remaining capacity to expand. Equation 17.20 can then be written as

$$Q(t) = \frac{mQ_0}{Q_0 + (m - Q_0)e^{-m\beta t}}$$

(Compare this with Exercise 24 in Exercise Set 7.5.)

Mixing Problems

The third example is typical of *mixing problems*.

EXAMPLE 17.19   A fish tank with a capacity of 12,000 L has 6,000 L of pure water in it initially, and holds tilapia, which can live in both fresh and salt water. To convert the tank to salt water, brine containing $\frac{1}{4}$ kg of salt per liter is then fed into the tank at the rate of 40 L/min, and the well-stirred mixture is allowed to drain out at the rate of 20 L/min. Find how much salt is in the tank just as it begins to overflow.

Solution   Let $Q(t)$ be the number of kilograms of salt in the tank t minutes after the brine begins to enter. Because initially there was fresh water in the tank, we see that $Q(0) = 0$. The basic idea of this and other similar mixture problems is that the rate of change of Q is the rate at which salt (or some other substance) enters the tank minus the rate at which it leaves:

$$\frac{dQ}{dt} = (\text{Rate Salt Comes In}) - (\text{Rate Salt Goes Out})$$

Since 40 L of brine enters the tank each minute, and each liter contains $\frac{1}{4}$ kg of salt, it follows that salt enters the tank at the rate of 10 kg/min. To determine how much salt leaves the tank each minute, it is necessary to find the *concentration* (number of kilograms per liter) of salt in the tank at any given time. The net total increase in liquid in the tank is 20 L/min (40 L comes in and 20 goes out). So after t minutes it contains $6{,}000 + 20t$ liters of solution (see Figure 17.14). Contained within this solution are $Q(t)$ kg of salt. So we have at time t

$$\text{Concentration of Salt in Tank} = \frac{Q(t)}{6{,}000 + 20t} \text{ kg/L}$$

Since 20 L of solution leaves the tank each minute, the amount of salt leaving is

$$(20 \text{ L/min})\left(\frac{Q}{6{,}000 + 20t} \text{ kg/L}\right) = \frac{Q}{300 + t} \text{ kg/min}$$

The differential equation is

$$\frac{dQ}{dt} = 10 - \frac{Q}{300 + t} \quad \text{or} \quad \frac{dQ}{dt} + \left(\frac{1}{300 + t}\right)Q = 10$$

A fish tank containing tilapia

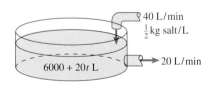

FIGURE 17.14

This equation is linear in Q with integrating factor

$$\mu(t) = e^{\int dt/(300+t)} = e^{\ln(300+t)} = 300 + t$$

Multiplying the equation by this integrating factor, we get

$$\frac{d}{dt}[(300+t)Q] = 10(300+t)$$

$$Q(t) = 5(300+t) + \frac{C}{300+t} \qquad \text{Before the tank overflows.}$$

Since the initial amount of salt in the tank is $Q(0) = 0$, we find that $C = -5(300)^2$. The tank will overflow when $6{,}000 + 20t = 12{,}000$, or $t = 300$. The amount of salt in the tank at that instant is

$$Q(300) = 5(600) - \frac{5(300)^2}{2(300)}$$

$$= 3{,}000 - 750 = 2{,}250 \text{ kg} \qquad \blacksquare$$

Falling Body Subject to Air Resistance

In the next example we consider a falling body problem similar to ones we studied in Chapter 4 (see Example 4.26) but with the important difference that we now take air resistance into account.

EXAMPLE 17.20 From a height of s_0 meters above the ground an object of mass m is given an upward initial velocity of v_0 meters per second. Assume that air resistance has magnitude proportional to the speed. Find the height above the ground and the velocity of the object at time t. If the object could continue indefinitely without hitting the ground, what would be its limiting velocity?

Solution As shown in Figure 17.15, let the distance $s(t)$ be measured positively upward from the ground, and let the velocity be $v(t)$. Then $s(0) = s_0$ and $v(0) = v_0$. The forces that act on the body are the pull of gravity, with magnitude mg, acting downward, and air resistance acting opposite to the direction of motion. (See Figure 17.15.)

Let $k > 0$ be the proportionality constant for the air resistance. When the object is moving upward, $v > 0$, and the resisting force is downward and so equals $-kv$. When the object is moving downward, $v < 0$, and the resisting force is upward, but again $-kv$ is the correct value, since this product is positive when v is negative. Thus, in all cases the resisting force is $-kv$.

Now we use Newton's Second Law of Motion, which for bodies of constant mass can be written as $m(dv/dt) = F$, where F is the magnitude of the net force acting on the body. In this case, then, we have

$$m\frac{dv}{dt} = -kv - mg$$

This differential equation can be written in the standard linear form

$$\frac{dv}{dt} + \frac{k}{m}v = -g$$

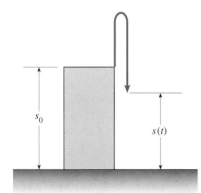

FIGURE 17.15

An integrating factor is $\mu(t) = e^{kt/m}$, and if we multiply both sides of this equation by $\mu(t)$ and simplify, we get

$$\frac{d}{dt}[ve^{kt/m}] = -ge^{kt/m}$$

Integrating both sides, we obtain

$$ve^{kt/m} = -\frac{gm}{k}e^{kt/m} + C_1$$

Setting $t = 0$ and $v = v_0$ gives

$$C_1 = v_0 + \frac{gm}{k}$$

Thus,

$$v(t) = \left(v_0 + \frac{gm}{k}\right)e^{-kt/m} - \frac{gm}{k}$$

To determine the limiting velocity (also called the *terminal* velocity), if the object could continue indefinitely, we let $t \to \infty$ to obtain

$$\lim_{t \to \infty} v(t) = -\frac{gm}{k}$$

Observe that this value is independent of the initial velocity.

To find $s(t)$, we integrate $v(t)$:

$$s(t) = -\frac{m}{k}\left(v_0 + \frac{gm}{k}\right)e^{-kt/m} - \frac{gmt}{k} + C_2$$

Since $s(0) = s_0$, we find that

$$C_2 = s_0 + \frac{m}{k}\left(v_0 + \frac{gm}{k}\right)$$

and $s(t)$ can be written as

$$s(t) = s_0 + \frac{m}{k}\left(v_0 + \frac{gm}{k}\right)(1 - e^{-kt/m}) - \frac{gmt}{k} \qquad \blacksquare$$

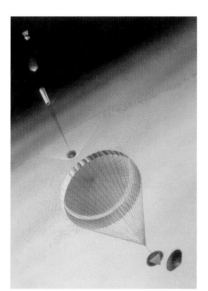

NASA's Pathfinder mission to Mars uses a parachute to slow the spacecraft during its descent through the Martian atmosphere.

Chemical Reactions

In certain types of chemical reactions, called *second-order reactions*, two substances, say A and B, react with each other to form a third substance C. This reaction is written symbolically as A + B → C. Suppose that initially the concentration of substance A is a (usually given in moles per liter) and of substance B is b. If, after t seconds, x moles per liter of A and of B have decomposed, the concentrations of what is left of A and of B are $a - x$ and $b - x$, respectively, and the concentration of C is x. For a second-order reaction, the rate of change of x is jointly proportional to $a - x$ and $b - x$:

$$\frac{dx}{dt} = k(a - x)(b - x)$$

In the next example, we solve this equation.

EXAMPLE 17.21 For the second-order chemical reaction described above, find the concentration $x(t)$ of the substance C after time t.

Solution   The initial-value problem to be solved is

$$\frac{dx}{dt} = k(a-x)(b-x), \qquad x(0) = 0$$

We separate variables and make use of partial fractions to perform the integration:

$$\frac{dx}{(a-x)(b-x)} = k\,dt$$

$$\frac{1}{a-b} \int \left[\frac{1}{b-x} - \frac{1}{a-x} \right] dx = kt + C$$

$$\frac{1}{a-b} \ln\left(\frac{a-x}{b-x}\right) = kt + C$$

When $t = 0$, $x = 0$. So

$$C = \frac{1}{a-b} \ln \frac{a}{b}$$

and we obtain

$$\frac{1}{a-b}\left[\ln\frac{a-x}{b-x} - \ln\frac{a}{b}\right] = kt$$

$$\ln\frac{b(a-x)}{a(b-x)} = (a-b)kt$$

$$\frac{b(a-x)}{a(b-x)} = e^{(a-b)kt}$$

Now we solve for x. You should verify that the answer can be put in the form

$$x(t) = \frac{ab(e^{akt} - e^{bkt})}{ae^{akt} - be^{bkt}}$$ ∎

Electrical Circuits

In the theory of the flow of electricity through an electrical circuit, it is known that the algebraic sum of all voltage drops is 0. That is, a voltage must be supplied by a battery or a generator to compensate for voltage losses elsewhere in the circuit. This result is known as **Kirchhoff's Second Law**. We will apply this to a circuit of the type shown in Figure 17.16, known as an **RL series circuit** since it involves resistance (R) and inductance (L). The customary units used are $R =$ ohms, $L =$ henrys, $I(t) =$ amperes, and $E(t) =$ volts. It can be shown that the voltage drops are RI across the resistor and $L(dI/dt)$ across

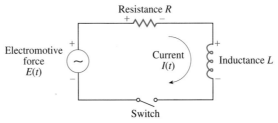

FIGURE 17.16

Section 17.4 Applications of First-Order Differential Equations 1247

the inductor. The electromotive force (for example, a generator) provides the only voltage increase. Thus, by Kirchhoff's Second Law, we have

$$L\frac{dI}{dt} + RI = E(t) \tag{17.21}$$

We make use of this equation in the next example.

EXAMPLE 17.22 Find the current $I = I(t)$ in an RL circuit in which the electromotive force $E(t) = V$, a constant. Assume $I(0) = 0$. Draw the graph of I.

Solution From Equation 17.21, we have

$$L\frac{dI}{dt} + RI = V$$

In standard form, this linear equation is

$$\frac{dI}{dt} + \frac{R}{L}I = \frac{V}{L}$$

and an integrating factor is $e^{(R/L)t}$. So we have

$$\frac{d}{dt}\left[Ie^{(R/L)t}\right] = \frac{V}{L}e^{(R/L)t}$$

$$Ie^{(R/L)t} = \frac{V}{R}e^{(R/L)t} + C$$

From the initial condition $I(0) = 0$, we get $C = -V/R$. Thus,

$$I(t) = \frac{V}{R}\left(1 - e^{-(R/L)t}\right)$$

The graph is shown in Figure 17.17. ∎

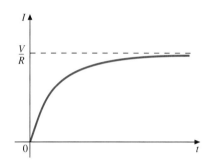

FIGURE 17.17
Graph of $I(t) = \frac{V}{R}\left(1 - e^{-(R/L)t}\right)$

Orthogonal Trajectories

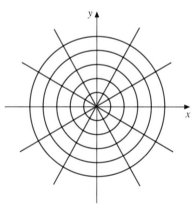

FIGURE 17.18
Orthogonal trajectories of circles $x^2 + y^2 = C_1$ are the lines $y = C_2 x$.

For our final example, we give a geometric application. Suppose two one-parameter families of curves $F(x, y) = C_1$ and $G(x, y) = C_2$ have the property that whenever a curve of one family intersects a curve of the second family, it does so orthogonally (that is, the tangent lines are perpendicular). Then we say that the families are **orthogonal trajectories** of each other (see Exercise 30 in Exercise Set 17.1 and Exercise 26 in Exercise Set 17.2). For example, the family of circles $x^2 + y^2 = C_1$ and the family of lines $y = C_2 x$ are easily seen to be orthogonal trajectories of each other (see Figure 17.18).

If the differential equation of one family is

$$\frac{dy}{dx} = \frac{f(x, y)}{g(x, y)}$$

then, since slopes of orthogonal curves are negative reciprocals of one another, the differential equation of the family of orthogonal trajectories is

$$\frac{dy}{dx} = -\frac{g(x, y)}{f(x, y)}$$

This relation is the key to finding orthogonal trajectories of a given family, as we show in the example that follows.

EXAMPLE 17.23 Find the family of orthogonal trajectories of the family of curves $3x^2 + y^2 = Cx$.

Solution To find the differential equation that has the given family as its general solution, we proceed, as in Example 17.4(a), by first differentiating both sides of the given equation:
$$6x + 2yy' = C$$
Now we substitute $C = 6x + 2yy'$ into the original equation:
$$3x^2 + y^2 = 6x^2 + 2xyy'$$
So, if $x \neq 0$ and $y \neq 0$,
$$y' = \frac{y^2 - 3x^2}{2xy}$$
Since at each point of intersection of a curve of the original family with any curve of the family of orthogonal trajectories, the tangents to the two curves are perpendicular, so that their slopes are negative reciprocals, the differential equation satisfied by the orthogonal trajectories is
$$y' = \frac{2xy}{3x^2 - y^2}$$
To solve this differential equation, we observe that it is homogeneous, so we substitute $y = vx$:
$$v + x\frac{dv}{dx} = \frac{2x^2 v}{3x^2 - v^2 x^2}$$
$$x\frac{dv}{dx} = \frac{2v}{3 - v^2} - v$$
$$x\frac{dv}{dx} = \frac{v^3 - v}{3 - v^2}$$
We now separate variables and make use of partial fractions to perform the integration:
$$\int \left[\frac{(3 - v^2)}{v(v+1)(v-1)}\right] dv = \int \frac{dx}{x}$$
$$\int \left(-\frac{3}{v} + \frac{1}{v+1} + \frac{1}{v-1}\right) dv = \int \frac{dx}{x}$$
$$-3\ln|v| + \ln|v+1| + \ln|v-1| = \ln|x| + \ln|C_1|$$
_{Write the constant of integration as $\ln|C_1|$.}
$$\ln\left|\frac{v^2 - 1}{v^3}\right| = \ln|C_1 x|$$

Thus,
$$\frac{v^2 - 1}{v^3} = C_1 x$$

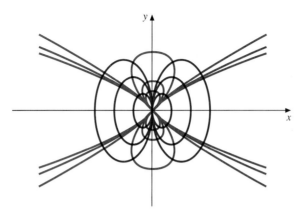

FIGURE 17.19

On replacing v by y/x and simplifying, we get

$$x^2 = y^2(1 - C_1 y)$$

as the equation of the orthogonal trajectories. Figure 17.19 shows several curves of the two families. ∎

Exercise Set 17.4

1. Assume that the rate of growth of a culture of bacteria is proportional to the number present. If a culture of 100 bacteria grows to a size of 150 after 2 hr, how long will it take for the size to double? How many will be present after 5 hr?

2. When interest on an investment is compounded continuously, the rate of growth of the amount of money in the account is proportional to the amount present, where the proportionality constant is the annual interest rate (expressed as a decimal). If an initial amount of P dollars is invested at $r\%$ compounded continuously, find a formula for the amount in the account at time t yr after it is invested. What value of r would cause the amount to double after 7 yr?

3. Use the result of Exercise 2 to determine the amount after 30 yr that results from an initial amount of $1000 invested at 6% compounded continuously. How many years will it take for the amount in the account to be $5000?

4. If 10% of a radioactive substance decays after 33 yr, how much will remain after 200 yr? What is the half-life of the substance?

5. Radium-226 has a half-life of 1620 yr. How long will it take for 80% of a given amount to decay?

6. If 75% of a quantity of the radioactive isotope uranium-232 remains after 30 yr, how much will be present after 100 yr? What is its half-life?

7. The half-life of thorium-228 is approximately 1.913 yr. If 100 g are on hand, how much will remain after 3 yr? How long will it take for the amount left to be 10 g?

8. Newton's Law of Cooling states that the surface temperature of an object changes at a rate proportional to the difference between the temperature of the object and that of the surrounding medium. Let $T(t)$ be the temperature of the object at time t, and let T_m be the temperature of the surrounding medium. If $T(0) = T_0$ (where $T_0 > T_m$), find the formula for $T(t)$.

9. Use the result of Exercise 8 to find the temperature of a body that was initially at 30°C, 30 min after it is placed in a medium of constant temperature 5°C if it cools to 20°C after 5 min. According to the model, will the object ever cool to 5°C? What does your answer tell about the model?

10. A thermometer registering 70°F is taken outside where the temperature is 28°F. After 5 min the thermometer registers 55°F. When will it register 30°F? (See Exercise 8.)

11. A vat contains 200 L of a 20% dye solution. A 40% solution of the same dye is then fed into the tank at the rate of 10 L/min, and the well-mixed solution is drained off at the same rate. Find an expression for the amount of pure dye in the tank at any time t. What is the concentration of dye in the tank after 30 min?

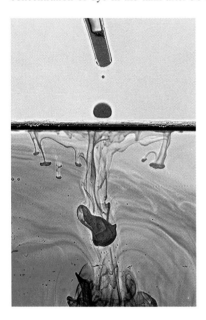

12. A tank with a 500-L capacity initially contains 100 L of brine with a salt concentration of 0.2 kg/L. Fresh water is allowed to enter the tank at the rate of 3 L/min, and the well-stirred mixture is drained from the tank at the rate of 1 L/min. How much salt does the tank contain after 20 min? How long will it take for the concentration to be reduced to 0.1 kg/L? Will this happen before the tank overflows? What is the salt concentration just as the tank overflows?

13. Solve the Verhulst model (Example 17.18) by separating variables.

14. Supply all the missing details in the solution of the Verhulst model in Example 17.18. Show that for $Q_0 < \alpha/\beta$, $Q(t)$ is an increasing function, and for $Q_0 > \alpha/\beta$, it is decreasing.

15. A body of mass m is dropped from a balloon high above the earth. Assume air resistance is proportional to the square of the velocity. Use Newton's Second Law of Motion to find a formula for the velocity at time t. What is the theoretical limiting velocity as $t \to \infty$? (Take the downward direction as positive.)

16. In Example 17.21, if the concentrations a and b are the same, then the solution given is not valid. (Why?) Solve the equation under this assumption, where $x(0) = 0$.

17. A 12-V battery is connected to an RL series circuit with a 6-ohm resistance and an inductance of 1 henry. If $I(0) = 0$, find $I(t)$.

18. Solve Equation 17.21 if $E(t) = E_0 \cos \omega t$ and $I(0) = I_0$.

19. A reasonably accurate model for the rate of dissemination of a drug injected into the bloodstream is given by

$$\frac{dy}{dt} = a - by, \qquad a > 0, \ b > 0$$

where $y = y(t)$ is the concentration of the drug at time t. Find $y(t)$ if the initial concentration of the drug is y_0.

20. Find the orthogonal trajectories of the family $x^2 + y^2 = Cx$. Sketch several members of each family.

21. Third-order chemical reactions are rare in the gaseous state, and those that do occur are almost always of the form $2A + B \to C$. For example, $2NO + O_2 \to 2NO_2$, a reaction that occurs in combustion of fossil fuels. If x is the concentration of C at time t, and a and b are the initial concentrations of A and B, respectively, then the rate of change of x is given by

$$\frac{dx}{dt} = k(a - 2x)^2(b - x)$$

Solve this equation with the initial condition $x(0) = 0$.

22. One model for the spread of an infectious disease in a community of N individuals is that the rate of change of the number $x(t)$ infected is jointly proportional to the number infected and the number uninfected. Set up and solve the relevant differential equation. Suppose $x(0) = 1$. Determine the limiting value of $x(t)$ as $t \to \infty$. (Observe that this model is mathematically the same as the Verhulst population growth model.)

23. The air in a room with dimensions $12 \times 20 \times 8$ ft initially contains 1% carbon dioxide. Air that contains 0.02% carbon dioxide is then forced into the room at the rate of 200 cu ft/min, and the well-circulated air leaves the room at the same rate. What will be the concentration of carbon dioxide after 5 min? After how long a time will the concentration be reduced to 0.05%?

24. A problem that is important in ecology has to do with the rise and decline of two species, where one species is a predator and the other is its prey. This problem is called the **predator–prey problem**. The following model is known as the **Lotka–Volterra model**, proposed by the American mathematician and biologist A. J. Lotka (1880–1949) and the Italian mathematician Vito Volterra (1860–1940) in the study of the interaction between the Alaskan snowshoe hare and the lynx:

$$\begin{cases} \dfrac{dx}{dt} = x(a - by) \\ \dfrac{dy}{dt} = y(-c + dx) \end{cases}$$

where a, b, c, and d are positive constants. Here, $x(t)$ is the population of the predator at time t and $y(t)$ is the population of the prey. By dividing the second equation by the first, find y as a function of x.

In Exercises 25–27, different models for population growth are given by specifying $f(Q)$ in Equation 17.19. Solve each model under the assumption that $Q(0) = Q_0$. The constants α and β are positive unless otherwise indicated.

25. $f(Q) = \alpha Q \cos \beta t$. This model is useful in describing seasonal growth, in which the population periodically increases and decreases. Sketch the graph of the solution.

26. $f(Q) = \alpha Q - \beta Q \ln Q$. This equation, called the **Gompertz model**, has applications in economic theory as well as population growth. For $\alpha > \beta \ln Q_0$, analyze and draw the graph of Q under each of the following circumstances:
    (a) $\alpha > 0, \beta > 0$ (b) $\alpha > 0, \beta < 0$

27. $f(Q) = (Q - m)(\alpha - \beta Q)$. This equation is an appropriate model for the growth of a population that becomes extinct when its numbers are too small. Show that when $Q_0 < m$, $Q(t_1) = 0$ for some t_1. Find t_1.

17.5 SECOND-ORDER LINEAR DIFFERENTIAL EQUATIONS WITH CONSTANT COEFFICIENTS: THE HOMOGENEOUS CASE

We have seen a variety of applications of first-order differential equations. Second-order equations also are important in applications because they can serve as models for physical problems involving second derivatives, such as those dealing with acceleration. Oscillatory motion, such as that of a vibrating spring, can be modeled by a second-order differential equation, as we will show in Section 17.7. Many applications involve only second-order linear equations, and we will concentrate on these.

From Definition 17.5, we know that a second-order linear differential equation is of the form

$$a_2(x)y'' + a_1(x)y' + a_0(x)y = g(x)$$

with $a_2(x) \neq 0$. We limit our consideration to the case where the coefficient functions are constant, say $a_2(x) = a$, $a_1(x) = b$, and $a_0(x) = c$. When $g(x) = 0$ for all x on some interval, the equation is said to be **homogeneous**. Here we are using *homogeneous* in a different sense from that in Section 17.2. We will study the homogeneous case in this section and the nonhomogeneous case in the next section.

The homogeneous equation with constant coefficients has the form

$$ay'' + by' + cy = 0 \quad (a \neq 0) \tag{17.22}$$

Our main objective in this section is to find the general solution of the differential equation given by Equation 17.22. It will simplify our notation if we introduce the symbol $L[y]$ for the left-hand side of Equation 17.22. That is,

$$L[y] = ay'' + by' + cy$$

Then Equation 17.22 is equivalent to $L[y] = 0$. We call L a **linear differential operator**. The word *linear* is used here to mean that

$$L[ky] = kL[y] \quad \text{for any constant } k \tag{17.23}$$

and

$$L[y_1 + y_2] = L[y_1] + L[y_2] \tag{17.24}$$

We will ask you to prove these two properties in Exercise 37 of Exercise Set 17.5. (These properties are similar to properties of the linear *function* $f(x) = mx + b$.)

Suppose we have found two solutions, say $y_1 = y_1(x)$ and $y_2 = y_2(x)$, of Equation 17.22. Then $L[y_1] = 0$ and $L[y_2] = 0$. We can now show that any **linear combination**

$$C_1 y_1 + C_2 y_2$$

(C_1 and C_2 are constants) is also a solution. To see why, note that by Equation 17.23,

$$L[C_1 y_1] = C_1 L[y_1] = 0 \quad \text{and} \quad L[C_2 y_2] = C_2 L[y_2] = 0$$

Then, by Equation 17.24,

$$L[C_1 y_1 + C_2 y_2] = L[C_1 y_1] + L[C_2 y_2] = 0$$

Clearly, if C_1 and C_2 are both 0, then $C_1 y_1 + C_2 y_2 = 0$. If the *only* way that the linear combination $C_1 y_1 + C_2 y_2$ can be identically 0 (0 for all x) is for both C_1 and C_2 to be 0, then the functions y_1 and y_2 are said to be **linearly independent**. Otherwise, they are **linearly dependent**. For example, $y_1 = e^x$ and $y_2 = e^{-2x}$ are linearly independent, since no linear combination $C_1 e^x + C_2 e^{-2x}$ can be identically 0 except when both C_1 and C_2 are 0. On the other hand, $y_1 = 2x - 1$ and $y_2 = \frac{1}{2} - x$ are linearly dependent, since with $C_1 = 1$ and $C_2 = 2$,

$$C_1 y_1 + C_2 y_2 = 1(2x - 1) + 2\left(\frac{1}{2} - x\right) = 0$$

The importance of linear independence is that when y_1 and y_2 are linearly independent solutions of Equation 17.22, then $y = C_1 y_1 + C_2 y_2$ is the *general* solution of Equation 17.22. We leave a proof of this result to a course in differential equations.

> **REMARK**
> ■ The definition we have given for the linear independence of two functions extends in a natural way to more than two. In the particular case of two functions, however, it is readily seen (see Exercise 38 of this section) that they are linearly independent if and only if *neither function is a multiple of the other*.

The Auxiliary Equation

With this background, we can obtain the general solution of Equation 17.22 if we can find two linearly independent solutions. For the remainder of this section we will concentrate on how to find two such solutions. To find a solution to Equation 17.22, we need to find a function y so that a constant times its second derivative y'', plus a constant times its first derivative y', plus a constant times the function y itself, add to give 0. Since an exponential function of the form $y = e^{mx}$ has the property that each successive derivative is a multiple of the function itself, this function seems as though it would be a natural candidate for a solution. So let us try $y = e^{mx}$. We have $y' = me^{mx}$ and $y'' = m^2 e^{mx}$, so that

$$L[y] = ay'' + by' + cy = am^2 e^{mx} + bme^{mx} + ce^{mx}$$
$$= (am^2 + bm + c)e^{mx}$$

Since e^{mx} is always positive, it follows that $L[y]$ will be 0 if and only if

$$am^2 + bm + c = 0 \qquad (17.25)$$

Equation 17.25 is called the **auxiliary equation** for Equation 17.22. (It is also referred to as the **characteristic equation**.) Let m_1 and m_2 denote the roots of Equation 17.25. Then by the Quadratic Formula, we have

$$m_1 = \frac{-b + \sqrt{b^2 - 4ac}}{2a} \quad \text{and} \quad m_2 = \frac{-b - \sqrt{b^2 - 4ac}}{2a}$$

We distinguish three cases, according to the nature of the discriminant $b^2 - 4ac$.

Case 1: $b^2 - 4ac > 0$

In this case m_1 and m_2 are distinct real roots. Since $e^{m_1 x}$ and $e^{m_2 x}$ are linearly independent, the general solution is $y = C_1 e^{m_1 x} + C_2 e^{m_2 x}$.

Distinct Real Roots

If the auxiliary equation $am^2 + bm + c = 0$ has two distinct real roots m_1 and m_2, then the general solution of the equation

$$ay'' + by' + cy = 0$$

is

$$y = C_1 e^{m_1 x} + C_2 e^{m_2 x}$$

EXAMPLE 17.24 Find the general solution of the equation $y'' - 3y' - 4y = 0$.

Solution The auxiliary equation is $m^2 - 3m - 4 = 0$. (Notice how the pattern of the auxiliary equation resembles the pattern of the differential equation.) By factoring, we have

$$(m-4)(m+1) = 0$$

So the two solutions are $m_1 = 4$ and $m_2 = -1$. These roots are real and distinct, so the general solution is

$$y = C_1 e^{4x} + C_2 e^{-x}$$ ∎

Case 2: $b^2 - 4ac = 0$

In this case, we have

$$m_1 = \frac{-b + \sqrt{0}}{2a} \quad \text{and} \quad m_2 = \frac{-b - \sqrt{0}}{2a}$$

That is, $m_1 = m_2 = -b/2a$. So there is only one solution of the form e^{mx}, namely,

$$y_1 = e^{-bx/2a}$$

By direct substitution into Equation 17.22, it can be shown that a second solution, independent from y_1, is

$$y_2 = xe^{-bx/2a}$$

(See Exercise 39 in Exercise Set 17.5.) Thus, writing $m = -b/2a$, the general solution is

$$y = C_1 e^{mx} + C_2 x e^{mx}$$

Equal Real Roots

If the roots of the auxiliary equation $am^2 + bm + c = 0$ are equal, with the common value $m = -b/2a$, then the general solution of the equation

$$ay'' + by' + cy = 0$$

is

$$y = C_1 e^{mx} + C_2 x e^{mx}$$

EXAMPLE 17.25 Find the general solution of the equation $y'' - 4y' + 4y = 0$.

Solution The auxiliary equation is $m^2 - 4m - 4 = 0$, or equivalently $(m - 2)^2 = 0$, which has the double root $m = 2$. So the general solution is

$$y = C_1 e^{2x} + C_2 x e^{2x}$$ ∎

Case 3: $b^2 - 4ac < 0$

In this case, the roots m_1 and m_2 of the auxiliary equation are the *complex conjugates*

$$m_1 = \alpha + i\beta \quad \text{and} \quad m_2 = \alpha - i\beta$$

where

$$\alpha = -\frac{b}{2a} \quad \text{and} \quad \beta = \frac{\sqrt{4ac - b^2}}{2a}$$

The complex number i satisfies $i^2 = -1$. Although the general solution can be written as

$$y = C_1 e^{(\alpha + i\beta)x} + C_2 e^{(\alpha - i\beta)x}$$

it is possible to rewrite the general solution in a form involving functions of real variables only, as we now show.

We will make use of the following formula, known as **Euler's Formula**:

$$e^{i\theta} = \cos\theta + i\sin\theta \tag{17.26}$$

To see why this formula is true, we begin by defining $e^{i\theta}$ as the result of replacing x with $i\theta$ in the Maclaurin series for e^x. (See Section 10.8.) When we do so and use $i^2 = -1$, so that $i^3 = -i$, $i^4 = 1$, and so on, we obtain

$$\begin{aligned}e^{i\theta} &= 1 + (i\theta) + \frac{(i\theta)^2}{2!} + \frac{(i\theta)^3}{3!} + \frac{(i\theta)^4}{4!} + \frac{(i\theta)^5}{5!} + \cdots \\ &= \left[1 - \frac{\theta^2}{2!} + \frac{\theta^4}{4!} - \cdots\right] + i\left[\theta - \frac{\theta^3}{3!} + \frac{\theta^5}{5!} - \cdots\right]\end{aligned}$$

where we have rearranged the terms into the so-called real and imaginary parts. But from Section 10.8 we recognize that the bracketed series are the Maclaurin series for $\cos\theta$ and $\sin\theta$, respectively. Thus, $e^{i\theta} = \cos\theta + i\sin\theta$.

REMARK
■ If we set $\theta = \pi$ in Euler's Formula, we get $e^{i\pi} = -1$, or equivalently

$$e^{i\pi} + 1 = 0 \tag{17.27}$$

Equation 17.27 is one of the most remarkable equations in all of mathematics. Here, in one equation, appear the five special constants $0, 1, \pi, e$, and i.

Returning now to the solution of Equation 17.22, we write

$$y = C_1 e^{(\alpha+i\beta)x} + C_2 e^{(\alpha-i\beta)x}$$
$$= e^{\alpha x}[C_1 e^{i\beta x} + C_2 e^{-i\beta x}]$$
$$= e^{\alpha x}[C_1(\cos \beta x + i \sin \beta x) + C_2(\cos \beta x - i \sin \beta x)]$$
$$= e^{\alpha x}[(C_1 + C_2)\cos \beta x + i(C_1 - C_2)\sin \beta x]$$
$$= e^{\alpha x}[C_3 \cos \beta x + C_4 \sin \beta x]$$

where C_3 and C_4 are new arbitrary constants, namely, $C_3 = C_1 + C_2$ and $C_4 = i(C_1 - C_2)$. Note that in applying Euler's Formula to $e^{-i\beta x}$, we used $\cos(-\beta x) = \cos \beta x$ and $\sin(-\beta x) = -\sin \beta x$. When the roots of the auxiliary equation are imaginary, we will use the result just obtained to write the general solution; that is, if the roots are $\alpha + i\beta$ and $\alpha - i\beta$, then we will write the general solution as

$$y = e^{\alpha x}(C_1 \cos \beta x + C_2 \sin \beta x) \qquad (17.28)$$

(There is no longer any need to call the constants C_3 and C_4.)

Imaginary Roots

If the roots of the auxiliary equation $am^2 + bm + c = 0$ are the complex conjugates $m_1 = \alpha + i\beta$ and $m_2 = \alpha - i\beta$, then the general solution of the equation

$$ay'' + by' + cy = 0$$

is

$$y = e^{\alpha x}(C_1 \cos \beta x + C_2 \sin \beta x)$$

EXAMPLE 17.26   Find the general solution of each of the following:

(a) $y'' + 4y = 0$ (b) $y'' - 2y' + 3y = 0$

Solution

(a) The auxiliary equation is $m^2 + 4 = 0$, with roots $m_1 = 2i$ and $m_2 = -2i$. So we have $\alpha = 0$ and $\beta = 2$. Thus, since $e^{\alpha x} = e^0 = 1$, the general solution given by Equation 17.28 is

$$y = C_1 \cos 2x + C_2 \sin 2x$$

(You can check to see that this value of y does satisfy the differential equation.)

(b) The auxiliary equation is $m^2 - 2m + 3 = 0$, with the solutions

$$m = \frac{2 \pm \sqrt{4 - 12}}{2} = 1 \pm i\sqrt{2}$$

Thus, $\alpha = 1$ and $\beta = \sqrt{2}$. So, by Equation 17.28, we can write the general solution in the form

$$y = e^x(C_1 \cos \sqrt{2}x + C_2 \sin \sqrt{2}x) \qquad \blacksquare$$

Initial-Value and Boundary-Value Problems

The next two examples illustrate an initial-value problem and a boundary-value problem that involve second-order equations.

EXAMPLE 17.27 Solve the initial-value problem

$$y'' - 2y' - 8y = 0; \quad y(0) = 1, \ y'(0) = -4$$

Solution The auxiliary equation is

$$m^2 - 2m - 8 = 0$$
$$(m-4)(m+2) = 0$$
$$m = 4, \ -2$$

Thus, the general solution is

$$y = C_1 e^{4x} + C_2 e^{-2x}$$

From $y(0) = 1$, we get $1 = C_1 + C_2$. Now we calculate y' and then apply the second initial condition:

$$y' = 4C_1 e^{4x} - 2C_2 e^{-2x}$$
$$-4 = 4C_1 - 2C_2 \quad \text{or} \quad 2C_1 - C_2 = -2$$

To find C_1 and C_2, we solve the system

$$\begin{cases} C_1 + C_2 = 1 \\ 2C_1 - C_2 = -2 \end{cases}$$

simultaneously. The solution is found to be $C_1 = -\frac{1}{3}$, $C_2 = \frac{4}{3}$. Thus, the desired particular solution is

$$y = -\frac{1}{3} e^{4x} + \frac{4}{3} e^{-2x} \qquad \blacksquare$$

EXAMPLE 17.28 Find the solution of the equation $x''(t) + x(t) = 0$ that satisfies the boundary conditions $x(0) = 3$ and $x(\pi/2) = 5$.

Solution The auxiliary equation $m^2 + 1 = 0$ has roots $\pm i$, and so by Equation 17.28 the general solution is

$$x(t) = C_1 \cos t + C_2 \sin t$$

Since $x(0) = C_1$ and $x(\pi/2) = C_2$, we see immediately that $C_1 = 3$ and $C_2 = 5$. Thus, the particular solution in question is

$$x(t) = 3 \cos t + 5 \sin t \qquad \blacksquare$$

The General Solution of a Second-Order Linear Homogeneous Differential Equation with Constant Coefficients

To find the general solution of

$$ay'' + by' + cy = 0$$

first solve the auxiliary equation

$$am^2 + bm + c = 0$$

Denote the roots by m_1 and m_2.

Case 1. If m_1 and m_2 are real and unequal, write the general solution as

$$y = C_1 e^{m_1 x} + C_2 e^{m_2 x}$$

Case 2. If m_1 and m_2 are real and equal, with common value m, write the general solution as

$$y = C_1 e^{mx} + C_2 x e^{mx}$$

Case 3. If m_1 and m_2 are the complex conjugates $m_1 = \alpha + i\beta$ and $m_2 = \alpha - i\beta$, with $\beta \neq 0$, write the general solution as

$$y = e^{\alpha x}(C_1 \cos \beta x + C_2 \sin \beta x)$$

Exercise Set 17.5

In Exercises 1–16, find the general solution of the differential equation. Unless otherwise indicated, the independent variable is x.

1. $y'' + 3y' + 2y = 0$
2. $y'' - y' - 2y = 0$
3. $y'' + 8y' + 16y = 0$
4. $y'' - 2y' + y = 0$
5. $y'' + 9y = 0$
6. $y'' + 9y' = 0$
7. $y'' - 2y' + 5y = 0$
8. $y'' + y' + 2y = 0$
9. $2y'' - 3y' - 5y = 0$
10. $4y'' + 5y' - 6y = 0$
11. $\dfrac{d^2 y}{dt^2} - 3\dfrac{dy}{dt} - 4y = 0$
12. $2x''(t) + 6x'(t) + 5x(t) = 0$
13. $2\dfrac{d^2 u}{dx^2} - 3\dfrac{du}{dx} = 0$
14. $\dfrac{d^2 v}{dx^2} - 9v = 0$
15. $4s''(t) - 12s'(t) + 9s(t) = 0$
16. $\dfrac{d^2 s}{dt^2} + 3\dfrac{ds}{dt} + 4s = 0$

In Exercises 17–24, solve the given initial-value problem or boundary-value problem.

17. $y'' + 2y' - 15y = 0;\ y(0) = 2,\ y'(0) = 3$
18. $2y'' - 3y' - 9y = 0;\ y(0) = -2,\ y'(0) = -6$
19. $y'' - 3y' = 0;\ y(0) = 2,\ y'(0) = 9$
20. $y'' - 9y = 0;\ y(0) = 0,\ y'(0) = 1$
21. $y'' - 4y' + 4y = 0;\ y(0) = 2,\ y(1) = e^2$
22. $y'' + 4y = 0;\ y(0) = 5,\ y(\pi/4) = 3$

23. $y'' + y = 0$; $y(\pi/6) = 0, y(\pi/3) = 1$

24. $y'' - 4y' + 5y = 0$; $y(0) = \frac{2}{3}$, $y'(0) = -\frac{1}{3}$

25. Show that when the roots m_1 and m_2 of the auxiliary equation are real and unequal, the general solution of Equation 17.22 can be written in the form

$$y = e^{ux}(C_1 \cosh vx + C_2 \sinh vx)$$

where $u = -b/2a$ and $v = \sqrt{b^2 - 4ac}/2a$.

26. Find the general solution of $y'' + 3y' + y = 0$ in the form given in Exercise 25. Find the particular solution that satisfies $y(0) = 4$, $y'(0) = -3$.

27. An equation of the form

$$ax^2y'' + bxy' + cy = 0 \quad (a \neq 0, \ x > 0)$$

is called an **Euler Equation**. Show that the substitution $t = \ln x$ changes it into a linear equation with constant coefficients.

In Exercises 28–30, use the result of Exercise 27 to find the general solution of the differential equation.

28. $x^2 y'' + 3xy' + 3y = 0$, $x > 0$

29. $\dfrac{d^2 s}{dt^2} + \dfrac{1}{t}\dfrac{ds}{dt} = 0$, $t > 0$ (*Hint: Multiply by t^2.*)

30. $v\dfrac{d^2 \psi}{dv^2} - \dfrac{1}{v}\psi = 0$, $v > 0$ (*Hint: Multiply by v.*)

In Exercises 31–36, assume that the ideas of this section extend to higher-order linear homogeneous equations with constant coefficients, and find the general solution.

31. $y''' - 2y'' - 3y' = 0$

32. $y^{(4)} - 16y = 0$

33. $y^{(4)} + 5y'' + 4y = 0$

34. $y''' - 3y' + 2y = 0$

35. $y''' - y'' + 4y' - 4y = 0$. Also find the particular solution that satisfies $y(0) = 0$, $y'(0) = 1$, $y''(0) = 2$.

36. $y''' - 3y'' + 3y' - y = 0$

37. Prove that $L[ky] = kL[y]$ and that $L[y_1 + y_2] = L[y_1] + L[y_2]$.

38. Prove that two functions $y_1(x)$ and $y_2(x)$ are linearly independent on an interval if and only if neither is a multiple of the other.

39. Prove that if the solutions of the auxiliary equation of $ay'' + by' + cy = 0$ are equal, then $y_2 = xe^{-bx/2a}$ is a solution of the differential equation.

17.6 SECOND-ORDER LINEAR DIFFERENTIAL EQUATIONS WITH CONSTANT COEFFICIENTS: THE NONHOMOGENEOUS CASE

In this section, we give methods for solving second-order linear differential equations of the form

$$ay'' + by' + cy = g(x) \tag{17.29}$$

where g is a continuous function on some interval I and is not identically 0 on I. We refer to $g(x)$ as the *nonhomogeneous* term. As we will show, the corresponding homogeneous equation

$$ay'' + by' + cy = 0 \tag{17.30}$$

plays an important role in obtaining the general solution of Equation 17.29. We call Equation 17.30 the **complementary equation** to Equation 17.29, and its general solution, which we denote by y_c, is called the **complementary solution**.

The next theorem shows the relationship between the general solution of the nonhomogeneous equation (Equation 17.29) and the general solution y_c of the complementary equation (Equation 17.30).

THEOREM 17.1

If y_p is any particular solution of the nonhomogeneous equation

$$ay'' + by' + cy = g(x)$$

then its general solution is

$$y = y_c + y_p \qquad (17.31)$$

where y_c is the general solution of the homogeneous complementary equation $ay'' + by' + cy = 0$.

Proof We must show that every solution of Equation 17.29 can be put in the form of Equation 17.31 for appropriate choices of the constants that occur in the complementary solution y_c. To simplify the proof, we make use of the linear differential operator, $L[y] = ay'' + by' + c$, that we introduced in Section 17.5. With this notation, Equation 17.29 can be written as $L[y] = g(x)$ and the complementary equation (Equation 17.30) as $L[y] = 0$. Now by hypothesis, $L[y_p] = g(x)$ and $L[y_c] = 0$. Suppose $y = Y(x)$ is any other solution of Equation 17.29; that is, $L[Y] = g(x)$. Then, by the properties of L,

$$L[Y - y_p] = L[Y] - L[y_p] = g(x) - g(x) = 0$$

Thus, $Y - y_p$ is a solution of the complementary equation $L[y] = 0$, whose general solution we know is y_c. This means, then, that for appropriate choices of the constants in y_c, $Y - y_p = y_c$ or $Y = y_c + y_p$. We have therefore shown that every solution of Equation 17.29 is a member of the family $y_c + y_p$. Thus, $y_c + y_p$ is the general solution of Equation 17.29. ∎

REMARK

■ The remarkable thing about this theorem is that y_p can be *any* particular solution of Equation 17.29. So, for example, if y_{p_1} and y_{p_2} are two different particular solutions, then, even though the solutions $y_c + y_{p_1}$ and $y_c + y_{p_2}$ would *look* different, they would, in fact, describe the same family.

EXAMPLE 17.29 Show that $y_p = x + 2$ is a particular solution of

$$y'' - 3y' + 2y = 2x + 1$$

and find the general solution.

Solution For $y_p = x + 2$, we have $y'_p = 1$ and $y''_p = 0$. So

$$y''_p - 3y'_p + 2y_p = 0 - 3(1) + 2(x + 2) = 2x + 1$$

Thus, $y_p = x + 2$ is a particular solution. According to Theorem 17.1, the general solution is $y_c + y_p$, where y_c is the complementary solution; that is, y_c is the general solution of the homogeneous equation

$$y'' - 3y' + 2y = 0$$

The auxiliary equation is $m^2 - 3m + 2 = 0$ and has roots $m_1 = 1$, $m_2 = 2$.

Section 17.6 Second-Order Linear Differential Equations: Nonhomogeneous Case

Thus,
$$y_c = C_1 e^x + C_2 e^{2x}$$
The general solution of the nonhomogeneous equation is therefore
$$y = \underbrace{C_1 e^x + C_2 e^{2x}}_{y_c} + \underbrace{x + 2}_{y_p}$$

The crucial question, clearly, is how to find y_p. Sometimes we can find a solution by inspection. For example, the equation $y'' + 2y' = 4$ has $y_p = 2x$ as a particular solution, since $y_p' = 2$ and $y_p'' = 0$. Usually, however, more work is involved. We will describe two methods. The first, called the **method of undetermined coefficients**, works when the nonhomogeneous term belongs to a certain class of functions. Although the restriction to this class limits the applicability of the method, it turns out that many of the functions that occur in common applications fall in this class. The second method, called **variation of parameters**, is more general in its applicability but is often more difficult to apply.

Method of Undetermined Coefficients

Before describing the method of undetermined coefficients in general, let us consider an example.

EXAMPLE 17.30 Find a particular solution of the equation
$$y'' + 2y' - 8y = g(x)$$
where

(a) $g(x) = 5e^{3x}$ (b) $g(x) = 5e^{2x}$

Solution

(a) Since $g(x)$ is an exponential function, it is reasonable to suppose that a solution of the form
$$y_p = Ae^{3x}$$
exists, since derivatives of y_p will all be multiples of e^{3x}. The coefficient A is yet to be determined (hence the name *undetermined* coefficients for this method). Trying this solution, we have $y_p' = 3Ae^{3x}$ and $y_p'' = 9Ae^{3x}$. Substituting into the original equation gives
$$9Ae^{3x} + 2(3Ae^{3x}) - 8(Ae^{3x}) = 5e^{3x}$$
$$7Ae^{3x} = 5e^{3x}$$
$$A = \frac{5}{7}$$

So our trial solution works, with $A = \frac{5}{7}$; that is, a particular solution is $y_p = \frac{5}{7}e^{3x}$.

(b) Proceeding as in part (a), we try
$$y_p = Ae^{2x}$$

Then $y_p' = 2Ae^{2x}$ and $y_p'' = 4Ae^{2x}$. So, on substitution, we get
$$4Ae^{2x} + 2(2Ae^{2x}) - 8(Ae^{2x}) = 5e^{2x}$$
$$0 = 5e^{2x}$$

Clearly, something has gone wrong, since we have arrived at an impossibility. A look at the complementary solution reveals the problem. You can verify that
$$y_c = C_1 e^{2x} + C_2 e^{-4x}$$

Taking $C_2 = 0$, we see that any function of the form $C_1 e^{2x}$ satisfies the homogeneous equation. Since our trial solution $y_p = Ae^{2x}$ is of this form, it has no chance of satisfying the nonhomogeneous equation.

In this situation we alter our initial trial solution by multiplying it by x to obtain a new trial solution. Multiplying by x is analogous to the repeated root situation for the auxiliary equation. Thus, we try
$$y_p = Axe^{2x}$$

Then
$$y_p' = Ae^{2x} + 2Axe^{2x}$$
$$y_p'' = 2Ae^{2x} + 2Ae^{2x} + 4Axe^{2x}$$
$$= 4Ae^{2x} + 4Axe^{2x}$$

Now we substitute into the original differential equation:
$$4Ae^{2x} + 4Axe^{2x} + 2(Ae^{2x} + 2Axe^{2x}) - 8Axe^{2x} = 5e^{2x}$$

After collecting terms, we get $6Ae^{2x} = 5e^{2x}$, which is true if $A = \frac{5}{6}$. Our desired particular solution is therefore
$$y_p = \frac{5}{6} x e^{2x}$$ ■

The method of undetermined coefficients works when the nonhomogeneous term $g(x)$ is one of the following types:

1. an exponential function
2. a polynomial function
3. a sine or cosine function

or else is a finite product of functions of one of these three types. In each case, our initial trial for y_p is a generalized function of the same type as $g(x)$, where unknown coefficients are used and where all *derived* terms (that is, terms obtained by differentiation) are included. Here are some examples to help guide your strategy:

$g(x)$	Initial Trial for y_p
$3x^2$	$Ax^2 + Bx + C$
$5 \sin 2x$	$A \sin 2x + B \cos 2x$
$2e^{-x} \cos 3x$	$e^{-x}(A \cos 3x + B \sin 3x)$
$(3x + 4)e^{2x}$	$(Ax + B)e^{2x}$
$x^2 \cos 3x$	$(Ax^2 + Bx + C) \cos 3x + (Dx^2 + Ex + F) \sin 3x$

Finally, suppose $g(x)$ is the sum of two functions of the type described above—say, $g(x) = g_1(x) + g_2(x)$. Then we find particular solutions y_{p_1} and y_{p_2} that satisfy

$$L[y_{p_1}] = g_1(x) \quad \text{and} \quad L[y_{p_2}] = g_2(x)$$

and set $y_p = y_{p_1} + y_{p_2}$. Since

$$L[y_p] = L[y_{p_1} + y_{p_2}] = L[y_{p_1}] + L[y_{p_2}] = g_1(x) + g_2(x)$$

it follows that y_p is a particular solution of the original equation. This procedure can be extended in a natural way to any finite sum.

Example 17.30 suggests the following algorithm for finding a particular solution by the method of undetermined coefficients:

An Algorithm for Finding y_p by the Method of Undetermined Coefficients

1. Find y_c.
2. Determine a trial solution that has the same general form as the nonhomogeneous term $g(x)$, together with terms of the form obtained from it by differentiation.
3. If the trial solution has no term in common with y_c, substitute it into the original equation to find the unknown coefficients.
4. If the trial solution does have a term in common with y_c, multiply the trial solution by x. Use this product as a new trial solution, and return to step 3.

Notice that when step 4 applies, the new trial solution may again have a term in common with y_c, in which case we have to multiply by x again. For example, in the equation

$$y'' - 4y' + 4y = 5e^{2x}$$

we would determine $y_c = C_1 e^{2x} + C_2 x e^{2x}$. (Verify.) Since both trial solutions Ae^{2x} and Axe^{2x} occur in y_c, we would use

$$y_p = Ax^2 e^{2x}$$

EXAMPLE 17.31 Solve the initial-value problem

$$y'' - 2y' = 3x + 2e^{2x} \cos x; \quad y(0) = 0, \ y'(0) = 1$$

Solution We find from the auxiliary equation $m^2 - 2m = 0$ that

$$y_c = C_1 + C_2 e^{2x}$$

Since $g(x)$ is the sum of two functions of the types we have described, we write $g(x) = g_1(x) + g_2(x)$, where $g_1(x) = 3x$ and $g_2(x) = 2e^{2x} \cos x$. Our initial trial solution for $L[y] = g_1(x)$ is $y_{p_1} = Ax + B$, but this function contains a constant term (B) that duplicates a term (C_1) in y_c. Thus, we modify our initial trial and use

$$y_{p_1} = Ax + Bx^2$$

Substituting this value of y_{p_1} into $L[y] = 3x$ enables us to find A and B. You should verify the results: $A = -\frac{3}{4}$, $B = -\frac{3}{4}$.

For the equation $L[y] = g_2(x)$, the initial trial is

$$y_{p_2} = e^{2x}(C \cos x + D \sin x)$$

and since this form does not duplicate any term in y_c, we use it. (Even though e^{2x} occurs, it occurs only in combination with a sine or cosine and so does not duplicate the term $C_2 e^{2x}$ that occurs in y_c.) Calculating y'_{p_2} and y''_{p_2} and substituting into $L[y] = 2e^{2x} \cos x$ give

$$e^{2x}[(-C + 2D) \cos x + (-2C - D) \sin x] = 2e^{2x} \cos x$$

(Supply the missing steps.) For this equation to be an identity (like terms on left and right have the same coefficient), we must have

$$\begin{cases} -C + 2D = 2 \\ -2C - D = 0 \end{cases}$$

The simultaneous solution is $C = -\frac{2}{5}$, $D = \frac{4}{5}$.

A particular solution of the original differential equation is therefore

$$y_p = y_{p_1} + y_{p_2} = -\frac{3}{4}x - \frac{3}{4}x^2 + e^{2x}\left(-\frac{2}{5} \cos x + \frac{4}{5} \sin x\right)$$

and the general solution $y_c + y_p$, is

$$y = C_1 + C_2 e^{2x} - \frac{3}{4}x - \frac{3}{4}x^2 + e^{2x}\left(-\frac{2}{5} \cos x + \frac{4}{5} \sin x\right)$$

The determination of the constants C_1 and C_2 so that the two initial conditions are satisfied is a bit messy. The result is

$$y = -\frac{19}{40} + \frac{7}{8}e^{2x} - \frac{3}{4}x - \frac{3}{4}x^2 + e^{2x}\left(-\frac{2}{5} \cos x + \frac{4}{5} \sin x\right) \quad \blacksquare$$

Variation of Parameters

Suppose we have already solved the homogeneous equation $ay'' + by' + cy = 0$ to obtain the complementary solution

$$y_c = C_1 y_1 + C_2 y_2.$$

In the method of **variation of parameters**, we replace the constants C_1 and C_2 (the parameters) in this complementary solution by variables (hence the name *variation* of parameters). That is, we try a particular solution y_p of the form

$$y_p(x) = u_1(x) y_1(x) + u_2(x) y_2(x)$$

where u_1 and u_2 are functions yet to be determined. We will be able to find u_1 and u_2 if we can obtain two equations relating them that we can solve simultaneously. One equation results from the requirement that y_p must satisfy the original differential equation. We are free to choose a second equation however we want. We do so in a way that simplifies the derivative y_p'. First, we calculate this derivative:

$$y_p' = (u_1'y_1 + u_1 y_1') + (u_2' y_2 + u_2 y_2')$$
$$= (u_1 y_1' + u_2 y_2') + (u_1' y_1 + u_2' y_2)$$

Since we are free to impose one condition on u_1 and u_2, we require them to satisfy

$$u_1' y_1 + u_2' y_2 = 0 \tag{17.32}$$

Then y_p' becomes

$$y_p' = u_1 y_1' + u_2 y_2'$$

So

$$y_p'' = (u_1' y_1' + u_1 y_1'') + (u_2' y_2' + u_2 y_2'')$$

Now we substitute y_p and its derivatives into the original differential equation

$$ay'' + by' + cy = g(x)$$

After rearranging terms, we get (you should verify it)

$$u_1 \left(ay_1'' + by_1' + cy_1\right) + u_2 \left(ay_2'' + by_2' + cy_2\right) + a\left(u_1' y_1' + u_2' y_2'\right) = g(x)$$

But y_1 and y_2 are solutions of the complementary equation, so $L[y_1]$ and $L[y_2]$ both are 0. That is, the quantities inside the first two sets of parentheses vanish, and we have

$$a(u_1' y_1' + u_2' y_2') = g(x) \tag{17.33}$$

Equations 17.32 and 17.33 can now be solved simultaneously for u_1' and u_2', from which u_1 and u_2 can be obtained by integration. (Of course, it is not always easy or even always possible to carry out this last step.)

We summarize our procedure below.

An Algorithm for Finding y_p by the Method of Variation of Parameters

1. Find the complementary solution $y_c = C_1 y_1 + C_2 y_2$.
2. Solve the system of equations

$$\begin{cases} u_1' y_1 + u_2' y_2 = 0 \\ a(u_1' y_1' + u_2' y_2') = g(x) \end{cases}$$

simultaneously for u_1' and u_2'.

3. Integrate to find u_1 and u_2.
4. Determine y_p from

$$y_p = u_1 y_1 + u_2 y_2$$

REMARK
■ In carrying out step 3, the constants of integration can be omitted, since we want *any* two functions u_1 and u_2 that work, and we might as well choose the simplest ones.

EXAMPLE 17.32 Solve the differential equation
$$y'' + y = \sec^3 x \qquad 0 < x < \frac{\pi}{2}$$
by the method of variation of parameters.

Solution From the auxiliary equation $m^2 + 1 = 0$, whose solutions are $m_1 = i$ and $m_2 = -i$, we know that the complementary solution can be written in the form
$$y_c = C_1 \cos x + C_2 \sin x$$
Thus, $y_1 = \cos x$ and $y_2 = \sin x$. To carry out step 2 of our algorithm, we solve the system
$$\begin{cases} u_1' \cos x + u_2' \sin x = 0 \\ -u_1' \sin x + u_2' \cos x = \sec^3 x \end{cases}$$
for u_1' and u_2'. The result is (which you will be asked to verify in Exercise 19 of Exercise Set 17.6)
$$u_1' = -\tan x \sec^2 x \qquad \text{and} \qquad u_2' = \sec^2 x$$
Thus, on integration, we obtain
$$u_1 = -\frac{1}{2} \tan^2 x \qquad \text{and} \qquad u_2 = \tan x$$
A particular solution is therefore
$$y_p = \left(-\frac{1}{2} \tan^2 x\right) \cos x + \tan x \sin x$$
The general solution is $y = y_c + y_p$:
$$y = C_1 \cos x + C_2 \sin x - \frac{1}{2} \tan^2 x \cos x + \tan x \sin x \qquad \blacksquare$$

REMARK
■ Although the method of variation of parameters will work for a broader class of functions $g(x)$ than the method of undetermined coefficients, it is often the more difficult of the two methods to apply since it involves integration, whereas undetermined coefficients involves only differentiation. So when there is a choice, you will probably find it easier to use the method of undetermined coefficients.

Exercise Set 17.6

In Exercises 1–4, find a particular solution by inspection. Then give the general solution.

1. $y'' + y = 2x$
2. $y'' - 3y' = 1$
3. $y'' - y' - 2y = 3$
4. $y'' - 4y = 3x + 5$

In Exercises 5–18, find a particular solution using the method of undetermined coefficients.

5. $y'' + y' - 6y = 3e^x$
6. $2y'' - 3y' - 5y = 3x - 1$
7. $y'' + 4y = \sin x$
8. $y'' - 4y = 2x^2$
9. $y'' - y' - 2y = e^{2x}$
10. $y'' - 2y' = x - 2$
11. $y'' + y = 3 \cos x$
12. $y'' - y = 3e^x$

13. $y'' - 2y' + y = e^x$
14. $y'' + 4y' + 4y = 2e^{-2x}$
15. $y'' - 2y' = xe^{2x}$
16. $y'' + y = e^x \sin x$
17. $2y'' + y' + y = 2e^{-x} \cos x$
18. $3y'' - 2y' + 5y = (2x - 1)e^x$

19. Supply the missing details in Example 17.32 for calculating u_1' and u_2'.

20. Show that the simultaneous solution of Equations 17.32 and 17.33 is
$$u_1' = -\frac{1}{a}\frac{y_2 g(x)}{W}, \qquad u_2' = \frac{1}{a}\frac{y_1 g(x)}{W}$$
where
$$W = \begin{vmatrix} y_1 & y_2 \\ y_1' & y_2' \end{vmatrix}$$

[Note: W is called the *Wronskian* of y_1 and y_2, after the Polish mathematician Josef Wronski (1778–1853).]

In Exercises 21–30, use the method of variation of parameters to find a particular solution. Then write the general solution.

21. $y'' + y = \sec x \tan x, \quad 0 < x < \pi/2$
22. $y'' + y = \tan x, \quad 0 < x < \pi/2$
23. $\dfrac{d^2 x}{dt^2} + x = \sec t, \quad 0 < t < \pi/2$
24. $\dfrac{d^2 y}{dx^2} + y = \sec^2 x, \quad 0 < x < \pi/2$
25. $y'' - y = xe^x$
26. $y'' + y' - 2y = e^x$
27. $y'' + 3y' + 2y = \cos(e^x)$
28. $y'' - 2y' - 3y = \cosh 2x$
29. $y'' + 2y' + 2y = e^x \sin x$
30. $y'' - y' = \dfrac{e^x}{1 + e^x}$

In Exercises 31–38, determine an appropriate form for a particular solution using undetermined coefficients, after finding the complementary solution. Do not calculate the coefficients.

31. $y'' + 3y' - 4y = x^3 e^x$
32. $y'' - 4y' + 4y = 3e^{2x} + x \cos x$
33. $y'' - 2y' = 2x^2 - 1 + xe^{2x}$
34. $y'' - 2y' + 5y = 3e^x \sin 2x$
35. $y'' - 6y' + 25y = e^{3x} \cos 4x - 1$
36. $3y'' + 4y' - 7y = x^2 \sin 2x$
37. $2y'' + 3y' - 9y = xe^{-x} \sin x$
38. $y'' + y' + 2y = 2 \sin^2 x$

In Exercises 39–46, solve the initial-value problem or boundary-value problem. Use either undetermined coefficients or variation of parameters to find a particular solution.

39. $y'' - 4y = x + 3; \; y(0) = 0, \; y'(0) = 1$
40. $y'' + 2y' = 2e^{-x}; \; y(0) = 1, \; y'(0) = -1$
41. $y'' + y = \cos 2x; \; y(0) = 2, \; y(\pi/2) = 3$
42. $y'' + 3y' - 10y = x^2 + 2x; \; y(0) = 1, \; y'(0) = 0$
43. $y'' - 4y' + 5y = 2e^{-2x}; \; y(0) = -1, \; y'(0) = 2$
44. $y'' + 4y = 3 \sin x; \; y(0) = 4, \; y(\frac{\pi}{4}) = 0$
45. $y'' - y' - 2y = e^{2x}; \; y(0) = 2, \; y'(0) = 3$
46. $y'' + 4y' - 5y = xe^x; \; y(0) = 0, \; y'(0) = 1$

In Exercises 47–51, find the general solution.

47. $y'' - y = \cosh x$
48. $y'' + 4y = 2 \cos^2 x - 1$
49. $y'' + 4y = 2 \sin x \cos x$
50. $d^2 x/dt^2 + \omega_0^2 x = k \cos \omega t$, where (a) $\omega \neq \omega_0$ and (b) $\omega = \omega_0$
51. $d^2 x/dt^2 - 6(dx/dt) + 9x = 2e^{3t} + t^2$. Also find the particular solution that satisfies $x(0) = 4, x'(0) = -2$.

52. Prove that
$$ay'' + by' + cy = x^n \qquad (a \neq 0, \; n \geq 0)$$
has a particular solution of the form
$$y_p = A_n x^n + A_{n-1} x^{n-1} + \cdots + A_1 x + A_0$$
if and only if $c \neq 0$.

In Exercises 53–56, use the same undetermined coefficient procedure as for second-order equations to find a particular solution. Also give the general solution.

53. $y''' - 3y'' = x^2$

54. $y''' - y'' + 4y' - 4y = \sin 2x$

55. $y''' - 3y' + 2y = 2e^x + 3x$

56. $d^4y/dx^4 - 16y = \cosh 2x + \cos 2x$

17.7 THE VIBRATING SPRING

An important application of second-order differential equations is in modeling problems dealing with oscillatory motion. We restrict our attention here to a discussion of a vibrating spring, although the ideas presented are applicable to a broad range of problems involving oscillatory motion, extending into quantum mechanics.

Consider a spring of natural length l attached to a support, as shown in Figure 17.20(a). Suppose a weight w is attached, causing the spring to stretch a distance Δl as it comes to an equilibrium position, as shown in Figure 17.20(b). We assume the weight is not great enough to stretch the spring beyond its elastic limit. According to Hooke's Law, the force exerted by the spring is proportional to the elongation, within this elastic limit. When the spring and weight are in equilibrium and the force exerted by the spring is w, we have

$$w = k\,\Delta l,$$

where k is the constant of proportionality (the *spring constant*). Since $w = mg$, where m is the mass of the spring, we have $mg = k\,\Delta l$, or

$$mg - k\,\Delta l = 0 \qquad (17.34)$$

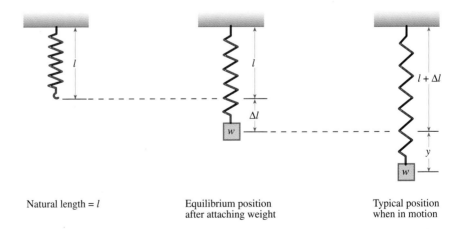

Natural length = l

(a)

Equilibrium position after attaching weight

(b)

Typical position when in motion

(c)

FIGURE 17.20

Now suppose the spring with attached weight is set in motion in some way. For example, it might be pulled below the equilibrium position and then released. As shown in Figure 17.20(c), we let $y = y(t)$ be the distance of the weight from the equilibrium position at time t. We take y as positive downward and try to determine y as a function of t. By Newton's Second Law of Motion,

$$m\frac{d^2y}{dt^2} = F$$

where F is the summation of all the forces acting on the weight. The force of gravity is mg, acting downward. The force exerted by the spring is of magnitude $k(\Delta l + y)$, since $\Delta l + y$ is the total displacement from the spring's natural length. When $\Delta l + y > 0$, the force of the spring is upward and so equals $-k(\Delta l + y)$. But this value is also correct when $\Delta l + y < 0$, since the spring is then compressed and its force is downward, in agreement with $-k(\Delta l + y) > 0$.

Two other forces may need to be considered. If the action takes place in air, we might reasonably neglect air resistance, but often vibrations occur in some viscous medium, such as oil, in which case the resistance of the medium cannot be neglected. Experimentally, it can be shown that for relatively small velocities, it is reasonable to assume that the resisting force is proportional to the velocity. Since this force is always opposite to the direction of motion, it is of the form $-c(dy/dt)$ for $c > 0$. Finally, we allow for some external force $f(t)$. For example, $f(t)$ might be a force applied to the entire support system.

We can now sum all the forces and obtain the following result from Newton's Second Law:

$$m\frac{d^2y}{dt^2} = \underbrace{mg}_{\substack{\text{Force} \\ \text{of} \\ \text{Gravity}}} - \underbrace{k(\Delta l + y)}_{\substack{\text{Force of} \\ \text{Spring}}} - \underbrace{c\frac{dy}{dt}}_{\substack{\text{Resisting} \\ \text{Force}}} + \underbrace{f(t)}_{\substack{\text{External} \\ \text{Force}}}$$

Since by Equation 17.34, $mg - k\Delta l = 0$, we can collect terms to obtain

---

Differential Equation for the Vibrating Spring

$$m\frac{d^2y}{dt^2} + c\frac{dy}{dt} + ky = f(t) \qquad (17.35)$$

This equation is the basic differential equation for the vibrating spring. We now consider certain special cases.

Undamped Free Vibrations

The simplest case is the one in which the resisting force is so small that it can be neglected, and for which there is no external force. We describe this situation by saying the motion is **undamped** (no resisting force) and **free** (no external force). Equation 17.35 then reduces to

$$m\frac{d^2y}{dt^2} + ky = 0$$

or, on dividing by m and writing $\omega^2 = k/m$ (ω is the Greek letter omega), we have the following formula.

Undamped Free Vibrations

$$\frac{d^2y}{dt^2} + \omega^2 y = 0 \tag{17.36}$$

Equation 17.36 is a second-order linear homogeneous equation that has auxiliary equation $m^2 + \omega^2 = 0$, with roots $m = \pm i\omega$. So the general solution is

$$y = C_1 \cos \omega t + C_2 \sin \omega t \tag{17.37}$$

The motion described by Equation 17.37 is called **simple harmonic motion**. Its **period** T, the time required to go through one complete cycle and return to the original position, is $2\pi/\omega$. **Frequency**, the number of cycles completed per unit of time, is $1/T = \omega/2\pi$. In Exercise 15 of this section you will be asked to show that Equation 17.37 can be written in the form

$$y = C \sin(\omega t + \alpha) \tag{17.38}$$

where $C = \sqrt{C_1^2 + C_2^2}$ and α satisfies $\sin \alpha = C_1/\sqrt{C_1^2 + C_2^2}$, $\cos \alpha = C_2/\sqrt{C_1^2 + C_2^2}$. In this form we see that the motion is sinusoidal, with amplitude C. Its graph is shown in Figure 17.21.

EXAMPLE 17.33 A spring is attached to a rigid overhead support. A 4-lb weight is attached to the end of the spring, causing it to stretch 6 in. Then the weight is pulled down 3 in. below the equilibrium position and released. Air resistance is negligible. Describe the resulting motion.

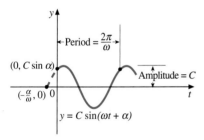

FIGURE 17.21
Simple harmonic motion

Solution For consistency, we will use feet instead of inches as the unit of distance. Then we have $y(0) = \frac{1}{4}$ and $y'(0) = 0$, since the weight was released without imparting any velocity. Since the 4-lb weight stretched the spring $\frac{1}{2}$ ft, we get the spring constant k from Hooke's Law:

$$4 = k \cdot \frac{1}{2}$$
$$k = 8 \text{ lb/ft}$$

Approximating g as 32 ft/s^2, we have

$$m = \frac{w}{g} = \frac{4}{32} = \frac{1}{8}$$

So using Equation 17.36 we have the initial-value problem

$$\frac{d^2y}{dt^2} + 64y = 0; \quad y(0) = \frac{1}{4}, \ y'(0) = 0$$

Here we have used the fact that $\omega^2 = k/m = 8 \cdot 8 = 64$. The general solution is

$$y = C_1 \cos 8t + C_2 \sin 8t$$

From the initial conditions, we get $C_1 = \frac{1}{4}$ and $8C_2 = 0$, so that $C_2 = 0$. Thus, the equation of motion is

$$y = \frac{1}{4} \cos 8t$$

The period is $2\pi/\omega = \pi/4$, and the frequency is $4/\pi$ cycles per second. The amplitude is 1/4 ft. We show the graph of the solution in Figure 17.22. ■

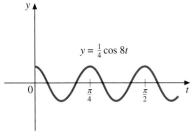

FIGURE 17.22

Damped Free Vibrations

According to the undamped model, the motion of the spring continues forever without diminishing. This solution does not accurately describe reality, although it does give reasonable accuracy over a relatively short time span when the medium in which the motion takes place offers little resistance. When this resistance is taken into consideration and there is no external force, Equation 17.35 takes the following form.

Damped Free Vibrations

$$m\frac{d^2y}{dt^2} + c\frac{dy}{dt} + ky = 0 \qquad (17.39)$$

It is important to remember in Equation 17.39 that all the constants m, c, and k are positive. The nature of the motion in this case depends on the relative sizes of the resistance constant c and the spring constant k. On solving the auxiliary equation, we get the roots

$$r_1 = \frac{-c + \sqrt{c^2 - 4km}}{2m} \quad \text{and} \quad r_2 = \frac{-c - \sqrt{c^2 - 4km}}{2m}$$

There are three cases to consider.

1. $c^2 > 4km$: the roots are real and unequal, so the solution is of the form
$$y = C_1 e^{r_1 t} + C_2 e^{r_2 t}$$
Observe that since $\sqrt{c^2 - 4km} < c$, it follows that $r_1 < 0$ and $r_2 < 0$.

2. $c^2 = 4km$: the roots are real and equal. Denote the common value by r; that is, $r = -c/2m < 0$. The solution is therefore of the form
$$y = (C_1 + C_2 t)e^{rt}$$

3. $c^2 < 4km$: the roots are imaginary—say, $r_1 = \alpha + i\beta$ and $r_2 = \alpha - i\beta$, where $\alpha = -c/2m < 0$ and $\beta = \sqrt{4km - c^2}/2m > 0$. The solution then is of the form
$$y = e^{\alpha t}(C_1 \cos \beta t + C_2 \sin \beta t)$$

REMARK
■ In Case 3 we wrote
$$\sqrt{c^2 - 4km} = \sqrt{(-1)(4km - c^2)}$$
$$= i\sqrt{4km - c^2}$$

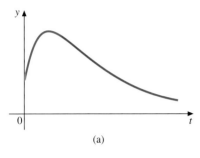

FIGURE 17.23
Overdamped motion $c^2 > 4km$

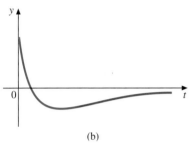

FIGURE 17.24
Critically damped motion $c^2 = 4km$

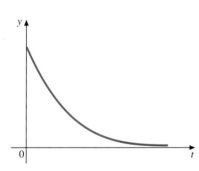

FIGURE 17.25
Underdamped motion $c^2 < 4km$

We analyze the motion for each case. For Case 1, $c^2 > 4km$, the solution is of the form

$$y = C_1 e^{r_1 t} + C_2 e^{r_2 t}$$

where both r_1 and r_2 are negative. As t increases, y approaches 0 and there is no oscillation. This type of motion is said to be **overdamped**. The resistance is so strong that the motion rapidly dies out. The form of the graph of y versus t depends on the initial conditions that determine C_1 and C_2. Figure 17.23 shows two possibilities.

In Case 2, $c^2 = 4km$, the solution is of the form

$$y = (C_1 + C_2 t)e^{rt}, \quad r < 0$$

Again there is no oscillation, and the motion tends to die out as t increases because of the factor e^{rt}, where $r < 0$. The slightest change in the resisting force changes the situation to either Case 1 or Case 3. We describe this as **critically damped** motion. The graph is similar to that for overdamping, with the initial conditions determining the exact form. Figure 17.24 shows one possibility.

For Case 3, $c^2 < 4km$, the solution is of the form

$$y = e^{\alpha t}(C_1 \cos \beta t + C_2 \sin \beta t), \quad \alpha < 0, \ \beta > 0$$

The motion in this case is oscillatory but with decreasing "amplitude" of the oscillations because of the $e^{\alpha t}$ factor (where $\alpha < 0$). The motion is said to be **underdamped**. A typical situation is shown in Figure 17.25.

EXAMPLE 17.34 An 8-lb weight stretches a spring 1.6 ft beyond its natural length. A damping force equal in magnitude to that of the velocity is present. If the spring is pushed upward 4 in. above the equilibrium position and then given a downward velocity of 6 ft/sec, find the equation of motion.

Solution Using $g \approx 32$, we get $m = w/g = 1/4$, and by Hooke's Law $8 = k(1.6)$, so that $k = 5$. The resistance constant is $c = 1$. So the differential equation is

$$\frac{1}{4} \cdot \frac{d^2 y}{dt^2} + \frac{dy}{dt} + 5y = 0$$

or equivalently,

$$\frac{d^2 y}{dt^2} + 4\frac{dy}{dt} + 20y = 0$$

and the initial conditions are $y(0) = -\frac{1}{3}$, $y'(0) = 6$.

The auxiliary equation has roots $-2 \pm 4i$, so the general solution is

$$y = e^{-2t}(C_1 \cos 4t + C_2 \sin 4t)$$

To find C_1 and C_2, we impose the two initial conditions. From $y(0) = -\frac{1}{3}$, we have immediately that $C_1 = -\frac{1}{3}$. We find y' to be

$$y' = e^{-2t}[(4C_2 - 2C_1)\cos 4t + (-4C_1 - 2C_2)\sin 4t] \quad \text{Verify.}$$

So from $y'(0) = 6$, we get

$$4C_2 - 2C_1 = 6$$

Substituting for C_1, we find $C_2 = \frac{4}{3}$. Thus, the solution is

$$y = -\frac{e^{-2t}}{3}(\cos 4t - 4\sin 4t)$$

The motion is underdamped in this case. (See Figure 17.26.)

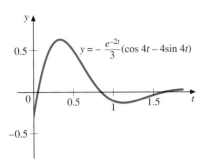

FIGURE 17.26

Forced Vibrations

When an external force $f(t)$ is present, we describe the motion as having **forced vibrations**, and $f(t)$ is called the **forcing function**. Frequently the forcing function is sinusoidal, as in the example that follows.

EXAMPLE 17.35 A 4-lb weight stretches a spring 2 ft beyond its natural length. Air resistance is negligible, but the system is subjected to an external force $f(t) = 2\cos \lambda t$. The weight is pulled down 6 in. and released. Describe the motion for each of the following values of λ: (a) $\lambda = 3$ and (b) $\lambda = 4$.

Solution In the usual way we find $m = \frac{1}{8}$ and $k = 2$. We assume $c = 0$. Thus, Equation 17.35 becomes

$$\frac{1}{8} \cdot \frac{d^2y}{dt^2} + 2y = 2\cos \lambda t$$

or equivalently,

$$\frac{d^2y}{dt^2} + 16y = 16\cos \lambda t$$

and the initial conditions are $y(0) = \frac{1}{2}$, $y'(0) = 0$. The complementary solution y_c is

$$y_c = C_1 \cos 4t + C_2 \sin 4t$$

(a) For $\lambda = 3$, we expect a particular solution of the form

$$y_p = A\cos 3t + B\sin 3t$$

By substituting this value into the equation and comparing coefficients, we get $A = \frac{16}{7}$, $B = 0$. The general solution is therefore

$$y = C_1 \cos 4t + C_2 \sin 4t + \frac{16}{7}\cos 3t$$

Imposing the initial conditions, we find $C_1 = -\frac{25}{14}$ and $C_2 = 0$, giving the equation of motion as

$$y = -\frac{25}{14}\cos 4t + \frac{16}{7}\cos 3t$$

(b) For $\lambda = 4$, the forcing function $f(t) = \cos 4t$ is a solution of the homogeneous equation, and so our trial solution y_p is of the form

$$y_p = t(A\cos 4t + B\sin 4t)$$

A straightforward calculation shows that $A = 0$, $B = 2$. Then applying the initial conditions to the general solution $y = y_c + y_p$, we obtain the equation of motion

$$y = \frac{1}{2}\cos 4t + 2t\sin 4t$$

We show its graph in Figure 17.27.

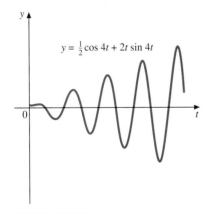

FIGURE 17.27

It is instructive to analyze the motion in part (b) of the preceding example. The presence of the factor t in the term $2t \sin 4t$ means that as t increases, the values of y become unbounded. For example, when $t = (2n+1)\pi/8$, we get

$$y\left(\frac{2n+1}{8}\pi\right) = \frac{(2n+1)\pi}{4}(-1)^n$$

so that as $n \to \infty$, $|y| \to \infty$, with y alternating between positive and negative values. This phenomenon is known as **resonance**. It occurs when the forcing function is periodic and has the same period as the complementary solution (called the natural period of the system). The consequences of resonance to a physical system can be catastrophic. The presence of the factor t indicates large oscillations. Physical structures usually break, or the system changes so that the differential equation no longer describes it.

Tacoma Narrows Bridge as it collapsed

Exercise Set 17.7

In Exercises 1–10, find the equation of motion. Neglect the resisting force unless otherwise indicated. When the resisting force is considered, identify the motion as underdamped, overdamped, or critically damped.

1. A 4-lb weight stretches a spring 6 in. beyond its natural length. After the system comes to equilibrium, the weight is pulled down 6 more inches and released. Give the amplitude and period.

2. In Exercise 1, instead of the weight being pulled down, it is given an upward push from the equilibrium position of 4 ft/s.

3. A 16-lb weight stretches a spring 2 ft beyond its natural length. The weight is then pulled down 8 in. below the equilibrium position and given an upward velocity of 12 ft/s.

4. Repeat Exercise 3 under the assumption that there is a resisting force equal to 4 times the velocity.

5. A mass of 50 g stretches a spring 5 cm beyond its natural length. The mass is pushed upward 10 cm and released. (Take $g = 980$ cm/s^2.)

6. A 32-lb weight is attached to a spring with a natural length of 3 ft. When the system comes to equilibrium, the spring is 4 ft long. The system is in a viscous medium that produces a resisting force equal to the velocity and opposite in direction. The weight is started in motion with a downward velocity from the equilibrium position of 6 ft/s.

7. A 20-g mass attached to a spring stretches it 10 cm beyond its natural length. The mass is attached to a mechanism that produces a resisting force of magnitude $420|v|$ dynes, where v is the velocity. The mass is pulled down 4 cm and also given an initial downward velocity of 7 cm/s.

8. An 8-lb weight stretches a spring from its natural length of 20 in. to a length of 28 in. The weight is then pulled down an additional 4 in. and given a downward velocity of 4 ft/s. Find (a) the time when the weight first passes through the equilibrium position and (b) its maximum displacement from the equilibrium position.

9. Repeat Exercise 1 if there is an external force $f(t) = 3 \cos 4t$ pounds acting on the system.

10. An 8-lb weight stretches a spring 1 ft beyond its natural length. The system is subjected to an external force of $f(t) = 2\cos t + 3\sin t$ pounds.

11. A weight of 8 lb is attached to a spring with a spring constant of 5 lb/ft. Assume a resisting force equal to $-c(dy/dt)$, where $c > 0$. Find c such that the motion is (a) overdamped, (b) critically damped, and (c) underdamped.

12. The angle θ from the vertical to a pendulum (see the figure) can be shown to satisfy the nonlinear second-order differential equation

$$\frac{d^2\theta}{dt^2} + \frac{g}{l}\sin\theta = 0$$

where l is the length of the pendulum rod. For small values of θ, $\sin\theta \approx \theta$. Using this approximation, determine the equation of motion of a pendulum 2 ft long, where θ is initially $\frac{1}{4}$ radian (with the positive direction for θ taken as counterclockwise, measured from the vertical), if the initial angular velocity is $d\theta/dt = \sqrt{3}$ radians per second.

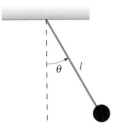

13. For the pendulum of Exercise 12, find the following:
    (a) the maximum angle from the vertical through which the pendulum will swing;
    (b) the time to complete one back-and-forth swing;
    (c) the instant when the pendulum will be vertical for the first time;
    (d) the magnitude of its velocity when it is in the vertical position.

14. Prove that for free damped vibrations of a spring, in the overdamped and critically damped cases, the weight will pass through the equilibrium position at most one time.

15. Show that the equation

$$y = C_1 \cos\omega t + C_2 \sin\omega t$$

of simple harmonic motion can be expressed in either of the following ways:
(a) $y = C\sin(\omega t + \alpha)$ (b) $y = C\cos(\omega t - \beta)$
where $C = \sqrt{C_1^2 + C_2^2}$. Describe the angles α and β.

16. Consider the following equation for damped forced vibrations:

$$m\frac{d^2y}{dt^2} + c\frac{dy}{dt} + ky = F_0\cos\lambda t$$

Find the general solution. Show that if $c \neq 0$, there can be no resonance. Find $\lim_{t\to\infty} y(t)$. (This is called the *steady-state* solution.)

17. Consider a simple series circuit that contains a resistance of R ohms, an inductance of L henrys, a capacitance of C farads, and an electromotive force of $E(t)$ volts (see the figure). By one of Kirchhoff's laws, the current I, in amperes, satisfies

$$L\frac{dI}{dt} + RI + \frac{Q}{C} = E(t)$$

where Q is the electric charge, related to I by $I = dQ/dt$. By differentiating both sides of the differential equation, we obtain

$$L\frac{d^2I}{dt^2} + R\frac{dI}{dt} + \frac{I}{C} = E'(t)$$

Find I if $L = 10$, $R = 20$, $C = 0.02$, and $E(t) = 200\sin t$. What is the steady-state current? (Let $t \to \infty$.)

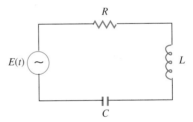

17.8 SERIES SOLUTIONS

Using methods we have studied, we can solve a large number of differential equations, but there are many others for which none of the methods works. For example, we have not yet shown a means of finding the general solution of

$$y'' + xy' + y = 0$$

This equation is a homogeneous linear equation, but one of the coefficients is a variable, so the theory of Section 17.5 is not applicable. For this problem and others it is often possible to find a solution in the form of a power series,

$$y = \sum_{n=0}^{\infty} a_n x^n \qquad (17.40)$$

Recall from Chapter 10 that every such power series defines a function within its interval of convergence, $|x| < R$. We assume $R > 0$. The series can be differentiated term by term, and the resulting series converges to y' in the same interval, $|x| < R$. Similarly, differentiating again, we obtain a series that converges to y'' for $|x| < R$, and so on. Thus, if y is given by Equation 17.40, then

$$y' = \sum_{n=1}^{\infty} n a_n x^{n-1}$$

and

$$y'' = \sum_{n=2}^{\infty} n(n-1) a_n x^{n-2}$$

Notice we started the index of summation with 1 for y' and with 2 for y'', since the constant term in each case drops out on differentiation.

It is useful to shift the index of summation at times. For example, we can see by expanding the terms that

$$\sum_{n=0}^{\infty} a_n x^n = \sum_{n=1}^{\infty} a_{n-1} x^{n-1}$$

The key to handling such shifts is that when the initial value of n on the summation sign is increased, n in the summand must be decreased by the same amount, and vice versa. (More formally, in the example shown, we could make the change of index $n = m - 1$ and write the second sum in terms of m, but the way we have suggested is faster and is easy to apply.) Test yourself by seeing whether you agree that

$$\sum_{n=2}^{\infty} n(n-1) a_n x^{n-2} = \sum_{n=1}^{\infty} (n+1)(n) a_{n+1} x^{n-1} = \sum_{n=0}^{\infty} (n+2)(n+1) a_{n+2} x^n$$

We give two examples to illustrate solution by series. The first problem could be done more easily by separating variables, but we include it to illustrate the technique first on a simple problem and also to provide a means of comparing our answer with that obtained by separating variables.

EXAMPLE 17.36 Find the general solution of the equation $y' = 2xy$ using infinite series.

Solution We try a solution of the form given in Equation 17.40: $y = \sum_{n=0}^{\infty} a_n x^n$. The equation can then be written as

$$\sum_{n=1}^{\infty} n a_n x^{n-1} - 2 \sum_{n=0}^{\infty} a_n x^{n+1} = 0 \quad \text{Since } x \sum_{n=0}^{\infty} a_n x^n = \sum_{n=0}^{\infty} a_n x^{n+1}$$

To have the same power of x appearing in the two summations, we shift the first index down by 1 and the second one up by 1:

$$\sum_{n=0}^{\infty} (n+1) a_{n+1} x^n - 2 \sum_{n=1}^{\infty} a_{n-1} x^n = 0$$

After separating the $n = 0$ term of the first summation, both sums will begin with $n = 1$, and so they can be brought together, giving

$$a_1 + \sum_{n=1}^{\infty} [(n+1) a_{n+1} - 2 a_{n-1}] x^n = 0$$

Section 17.8 Series Solutions 1277

We want this equation to be an identity in x. So we must have $a_1 = 0$, and every coefficient of x^n for $n = 1, 2, 3, \ldots$ equal to 0:

$$(n+1)a_{n+1} - 2a_{n-1} = 0, \quad n \geq 1$$

The last equation can be written in the form

$$a_{n+1} = \frac{2a_{n-1}}{n+1}, \quad n \geq 1$$

This equation is a **recursion formula** that enables us to find the coefficients $a_2, a_3, \ldots$ in sequence:

$\underline{n=1}: \quad a_2 = \dfrac{2a_0}{2} = a_0$

$\underline{n=2}: \quad a_3 = \dfrac{2a_1}{3} = 0 \quad \text{Since } a_1 = 0$

$\underline{n=3}: \quad a_4 = \dfrac{2a_2}{4} = \dfrac{a_2}{2} = \dfrac{a_0}{2} \quad \text{Since } a_2 = a_0$

$\underline{n=4}: \quad a_5 = \dfrac{2a_3}{5} = 0 \quad \text{Since } a_3 = 0$

$\underline{n=5}: \quad a_6 = \dfrac{2a_4}{6} = \dfrac{a_4}{3} = \dfrac{a_0}{3 \cdot 2} \quad \text{Since } a_4 = \dfrac{a_0}{2}$

$\vdots \qquad \qquad \vdots$

Continuing in this way, we see that

$$\begin{cases} a_{2k-1} = 0 & \text{for } k = 1, 2, 3, \ldots \\ a_{2k} = \dfrac{a_0}{k!} & \text{for } k = 1, 2, 3, \ldots \end{cases}$$

Since no condition is placed on a_0, it is arbitrary. Thus, our solution $y = \sum_{n=0}^{\infty} a_n x^n$ becomes

$$y = a_0 + \sum_{k=1}^{\infty} a_{2k} x^{2k}$$

$$= a_0 + a_0 \sum_{k=1}^{\infty} \frac{x^{2k}}{k!}$$

Since $0! = 1$, we can combine the first term with the summation by starting k with 0:

$$y = a_0 \sum_{k=0}^{\infty} \frac{x^{2k}}{k!}$$

By the Ratio Test, we can determine that the series converges for all values of x.

Now let us compare this solution with that obtained by separating variables. We have

$$\frac{dy}{y} = 2x \, dx$$

$$\ln|y| = x^2 + \ln|C| \quad \text{Write the constant as } \ln|C|.$$

$$\ln\left|\frac{y}{C}\right| = x^2$$

$$y = Ce^{x^2}$$

To see that our series solution agrees with this result, recall from Section 10.8 that the Maclaurin series for e^{x^2} is

$$e^{x^2} = \sum_{k=0}^{\infty} \frac{x^{2k}}{k!}$$

Thus, the solution found by separating variables agrees with the series solution, with the arbitrary constant called C in one case and a_0 in the other. ∎

EXAMPLE 17.37 Find the general solution of the equation $y'' + xy' + y = 0$ using infinite series.

Solution This problem is the one we posed at the beginning of this section. Substituting $y = \sum_{n=0}^{\infty} a_n x^n$, we get

$$\sum_{n=2}^{\infty} n(n-1)a_n x^{n-2} + \sum_{n=1}^{\infty} n a_n x^n + \sum_{n=0}^{\infty} a_n x^n = 0$$

We shift the index down by 2 on the first sum, and by observing that the second sum is unaltered by starting with $n = 0$ (since the term corresponding to $n = 0$ will be 0), we obtain

$$\sum_{n=0}^{\infty} (n+2)(n+1)a_{n+2} x^n + \sum_{n=0}^{\infty} n a_n x^n + \sum_{n=0}^{\infty} a_n x^n = 0$$

The terms can be brought together now in one summation:

$$\sum_{n=0}^{\infty} [(n+2)(n+1)a_{n+2} + (n+1)a_n] x^n = 0$$

To be an identity in x, every coefficient must be 0. Thus,

$$(n+2)(n+1)a_{n+2} + (n+1)a_n = 0$$

or

$$a_{n+2} = -\frac{a_n}{n+2}, \qquad n \geq 0$$

We consider even subscripts and odd subscripts separately:

n Even	n Odd
$a_2 = -\dfrac{a_0}{2}$	$a_3 = -\dfrac{a_1}{3}$
$a_4 = -\dfrac{a_2}{4} = \dfrac{a_0}{2 \cdot 4}$	$a_5 = -\dfrac{a_3}{5} = \dfrac{a_1}{3 \cdot 5}$
$a_6 = -\dfrac{a_4}{6} = -\dfrac{a_0}{2 \cdot 4 \cdot 6}$	$a_7 = -\dfrac{a_5}{7} = -\dfrac{a_1}{3 \cdot 5 \cdot 7}$
$\vdots$	$\vdots$
$a_{2k} = \dfrac{(-1)^k a_0}{2 \cdot 4 \cdot 6 \cdot \ldots \cdot (2k)}$	$a_{2k+1} = \dfrac{(-1)^k a_1}{3 \cdot 5 \cdot 7 \cdot \ldots \cdot (2k+1)}$

Since a_0 and a_1 have no restrictions, they are arbitrary. The solution can now be written as

$$y = a_0 \left[1 + \sum_{k=1}^{\infty} \frac{(-1)^k}{2 \cdot 4 \cdot 6 \cdot \cdots \cdot (2k)} x^{2k} \right]$$

$$+ a_1 \left[x + \sum_{k=1}^{\infty} \frac{(-1)^k}{3 \cdot 5 \cdot 7 \cdot \cdots \cdot (2k+1)} x^{2k+1} \right]$$

The expression in the first bracket can be simplified somewhat by observing that $2 \cdot 4 \cdot 6 \cdot \cdots \cdot (2k) = 2^k \cdot k!$, and since $0! = 1$, we can combine the first term with the summation by starting k with 0. Similarly, we can include the first term of the second bracket with the summation by starting with $n = 0$. We can therefore write

$$y = a_0 \sum_{k=0}^{\infty} \frac{(-1)^k x^{2k}}{2^k k!} + a_1 \sum_{k=0}^{\infty} \frac{(-1)^k}{1 \cdot 3 \cdot 5 \cdot \cdots \cdot (2k+1)} x^{2k+1}$$

Both series can be shown to converge for all x. If we let y_1 be the first sum and y_2 the second, and write $C_1 = a_0$ and $C_2 = a_1$, then the solution is in the familiar form $y = C_1 y_1 + C_2 y_2$. ∎

REMARK
■ In the two preceding examples, the recursion formulas were such that we could find explicit formulas for the coefficients. In some cases finding such explicit formulas is not possible, and the best we can do is calculate as many coefficients as we need in sequential order.

The procedure we have illustrated will also work for nonhomogeneous equations provided that the nonhomogeneous term can be expanded in a power series. In particular, if the nonhomogeneous term is a polynomial, it is already a (finite) power series. By comparing coefficients of like powers of x, we can find the unknown coefficients a_n in the series $\sum a_n x^n$.

As you might expect, there is a good deal more to solution by power series than we have gone into here, but the deeper aspects will have to be deferred to a course on differential equations.

Exercise Set 17.8

In Exercises 1–13, find the general solution in terms of power series in x. Where possible, solve the equation by other means and compare answers. If initial values are given, find the particular solution.

1. $y' = x^2 y$
2. $y' - 2xy = x$
3. $y'' = y$
4. $y'' - xy' - y = 0$
5. $y'' = xy$
6. $y' - y = 2x$; $y(0) = 3$
7. $y'' + y = 0$; $y(0) = 1$, $y'(0) = 0$
8. $(1 - x)y'' = y'$
9. $y'' + x^2 y = 0$
10. $(1 + x^2)y'' + 2xy' = 0$
11. $(1 - x^2)y'' - 5xy' - 3y = 0$
12. $(1 + x^2)y'' - 3xy' - 5y = 0$
13. $y'' + xy' + 2y = x^2 + 1$; $y(0) = 2$, $y'(0) = 1$
14. Obtain the general solution of the equation

$$y'' + (x - 2)y = 0$$

as a power series about $x = 2$—that is, in the form $\sum a_n (x - 2)^n$.

15. Find two linearly independent solutions in the form of power series for the equation

$$y'' + (1 - x)y = 0$$

Then write the general solution. (*Hint:* For one solution take $a_0 = 1$ and $a_1 = 0$; then do the opposite.)

16. Obtain the first 10 nonzero terms of the Maclaurin series solution

$$y(x) = y(0) + y'(0)x + \frac{y''(0)}{2!}x^2 + \frac{y'''(0)}{3!}x^3 + \cdots$$

for the initial-value problem

$$y'' - xy' + 2y = 0; \quad y(0) = 1, \; y'(0) = -1$$

by carrying out the following steps:

(a) Solve the equation for y''.
(b) Put $x = 0$ and substitute the given values of $y(0)$ and $y'(0)$ to get $y''(0)$.
(c) Differentiate the equation in step (a) to get y'''.
(d) Put $x = 0$ and substitute known values for $y(0)$, $y'(0)$, and $y''(0)$ to get $y'''(0)$.
(e) Differentiate the equation in step (c) and continue in this manner.

17. Follow the procedure in Exercise 16 to obtain the first 10 nonzero terms of the Maclaurin series solution of the following initial-value problem:

$$y'' + (x - 1)y = 0; \quad y(0) = 2, \; y'(0) = 1$$

Chapter 17 Review Exercises

In Exercises 1–18, find the general solution of the differential equation. If initial or boundary conditions are given, find the particular solution that satisfies them.

1. $(2 + e^{-x} \cos y) \, dx + (3 + e^{-x} \sin y) \, dy = 0$

2. $xy' = y + x \cot \frac{y}{x}; \; y(1) = 0$

3. $y' = x\sqrt{1 - y^2}; \; y(0) = 1$

4. $(\cosh x)y' + (\sinh x)y = \cosh^2 x$

5. $\dfrac{dy}{dx} = \dfrac{y(2x + y)}{x(x + y)}$

6. $xy' - 1 = e^{-y}; \; y(1) = 0$

7. $(y - 3x) \, dy + (3y + 4x) \, dx = 0; \; y(1) = 0$

8. $\left(\dfrac{2x^2 + y^2}{\sqrt{x^2 + y^2}} \right) dx + \left(\dfrac{xy}{\sqrt{x^2 + y^2}} - 3 \right) dy = 0; \; y(3) = 4$

9. $\dfrac{dy}{dx} = 1 + \dfrac{y(1 - x)}{x(2 - x)}; \; y(1) = 2$

10. $\dfrac{dy}{\tan y} = \dfrac{\sin 2x}{1 + \sin^2 x} \, dx$

11. $(\cos xy - xy \sin xy) \, dx = (x^2 \sin xy) \, dy$

12. $y' + \dfrac{3y}{x} = \dfrac{x}{y^2}$ (See Exercise 28 in Exercise Set 17.3.)

13. $e^{2x^2} dy + x(2ye^{2x^2} - 1) \, dx = 0; \; y(0) = 1$

14. $2y'' + 5y' - 3y = 0; \; y(0) = 1, \; y'(0) = 0$

15. $y'' + 6y' + 9y = 0; \; y(0) = 3, \; y'(0) = -4$

16. $y'' + 4y' + 5y = 0$

17. $\dfrac{d^2s}{dt^2} + 2\dfrac{ds}{dt} = 0; \; s(0) = 0, \; s'(0) = 4$

18. $y'' + 9y = 0; \; y(0) = 3, \; y(\pi/2) = -2$

In Exercises 19–26, find a particular solution by: (a) undetermined coefficients, (b) variation of parameters.

19. $\dfrac{d^2x}{dt^2} - \dfrac{dx}{dt} - 6x = 2e^t$

20. $y'' + 2y' - 3y = 2e^{-x} - 1$

21. $y'' - y' = x + 3e^x$

22. $y'' + 4y = 2\cos^2 x$

23. $2y'' + 3y' - 5y = e^x \cos x$

24. $y'' + y = xe^{-x}$

25. $y'' - y' = \sinh x; \; y(0) = 0, \; y'(0) = 1$

26. $y'' - 4y' + 4y = 8x^2; \; y(0) = 1, \; y'(0) = -1$

In Exercises 27 and 28, give the appropriate form for a particular solution using undetermined coefficients. Do not calculate the coefficients.

27. $y'' - 3y' - 4y = xe^{-x} + \cos 2x$

28. $y'' + 2y' + 2y = x^2 e^{-x} \sin x$

29. If 10% of a certain radioactive substance decays in 2 yr, find its half-life. How long will it take for 90% to decay?

30. A 50% dye solution is fed at the rate of 8 L/min into a vat that originally contained 200 L of pure water, and the well-mixed solution is drained off at the same rate. How long will it take for the mixture to become a 25% dye solution?

31. A modification of the Malthus model that accounts for the culling of a population $Q(t)$ at a constant rate H (for example, the culling of a deer population by hunters or of some other animal species by predators) is given by

$$Q'(t) = \alpha Q(t) - H \quad (\alpha > 0)$$

with $Q(0) = Q_0$. Solve this equation and show that the model predicts three possible outcomes, depending on the relative sizes of H and α: (1) the population dies out in a finite time, (2) the population grows without limit, or (3) the population size stays constant. Determine the relationship between H and α that produces each result.

32. In an *autocatalytic* chemical reaction, a substance A is converted to a substance B in such a way that the reaction is stimulated by the substance being produced. If the original concentration of A is a and at time t the concentration of B is $x = x(t)$, then the reaction rate is modeled by

$$\frac{dx}{dt} = kx(a - x)$$

Find the solution if $x(0) = x_0$.

33. A 4-lb weight stretches a spring 0.64 ft beyond its natural length. The system is pushed up 4 in. from equilibrium and then given a downward velocity of 5 ft/sec. There is a damping force present of $0.25\, v$. Find the equation of motion.

34. In the equation

$$L\frac{d^2 I}{dt^2} + R\frac{dI}{dt} + \frac{I}{C} = E'(t)$$

obtained for the RLC circuit in Exercise 17 in Exercise Set 17.7, suppose $E(t) = E \sin \omega t$ and $R^2 C = 4L$. Find $I(t)$. To simplify notation, write $\alpha = (R/2L)$.

In Exercises 35 and 36, find the general solution using power series. Find the largest interval in which the solution is valid.

35. $y'' - 2xy' - 4y = 0$

36. $(1 - x^2)y'' - 6xy' - 4y = 0$

Chapter 17 Concept Quiz

1. Define each of the following:
   (a) The order of a differential equation
   (b) A first-order separable differential equation
   (c) An exact differential equation
   (d) A homogeneous first-order differential equation
   (e) A linear differential equation of order n
   (f) A homogeneous function of degree k in x and y
   (g) Linearly independent functions $f_1, f_2, \ldots, f_n$

2. Explain in your own words how to solve a first-order differential equation of each of the following types:
   (a) Separable   (b) Homogeneous
   (c) Exact   (d) Linear

3. Consider the equation $ay'' + by' + cy = 0$.
   (a) Write the auxiliary equation and express its solutions m_1 and m_2 in terms of a, b, and c.
   (b) Give the general solution in each of the following cases:
       (i) $b^2 - 4ac > 0$
       (ii) $b^2 - 4ac = 0$
       (iii) $b^2 - 4ac < 0$

4. Explain how to use the method of variation of parameters.

5. Consider the equation $ay'' + by' + cy = g(x)$. For each of the following values of $g(x)$, tell what your initial trial would be for a particular solution, y_p. Also, tell under what circumstances you would modify your initial trial and in what way you would modify it.
   (a) $g(x) = e^{kx}$ (b) $g(x) = kx^2$
   (c) $g(x) = \cos kx$ (d) $g(x) = xe^{kx}$

6. Consider the equation

$$m\frac{d^2y}{dt^2} + c\frac{dy}{dt} + ky = 0$$

for the motion of a mass m attached to a spring having spring constant k, if there is a damping force proportional to velocity in which the constant of proportionality is c. Describe the motion for each of the following cases:
(a) $c^2 - 4km > 0$, (b) $c^2 - 4km = 0$, (c) $c^2 - 4km < 0$

7. Fill in the blanks.
   (a) A test for exactness of the differential equation $f(x, y)dx + g(x, y)dy$ is that _____ = _____.
   (b) An integrating factor, μ, for the equation $y' + P(x)y = Q(x)$ is _____.
   (c) If $f(x, y)dx + g(x, y)dy = 0$ and f and g are homogeneous functions of the same degree, then either of the substitutions _____ or _____ will result in a _____ equation.
   (d) If $dy/dx = f(x, y)$, the differential equation of the orthogonal trajectories of its family of solutions is _____.

   (e) In forced, undamped vibrations, if the forcing function is periodic and has the same period as the complementary solution, the phenomenon of _____ occurs.
   (f) If y_1 and y_2 are solutions of the homogeneous differential equation $ay'' + by' + cy = 0$, then _____ is also a solution.

8. Which of the following statements are true and which are false?
   (a) It is impossible to have three linearly independent solutions of the equation $ay'' + by' + cy = 0$.
   (b) In the general solution $y = y_c + y_p$ of a nonhomogeneous linear differential equation with constant coefficients, both y_c and y_p are uniquely determined.
   (c) If a first-order differential equation is homogeneous, it cannot be exact.
   (d) If y_1, y_2, and y_3 are all solutions of $ay'' + by' + cy = 0$, and y_2 is not a multiple of y_1, then there must be constants C_1 and C_2 such that $y_3 = C_1y_1 + C_2y_2$.
   (e) The sum of two solutions of a nonhomogeneous equation $ay'' + by' + cy = g(x)$ is also a solution.

APPLYING CALCULUS

1. A skydiver and his parachute weigh 175 pounds. He free-falls from rest (i.e., his initial velocity is 0) from a height of 10,000 feet for 10 seconds at which time his parachute opens. Assume that the air resistance during the free fall is $(1/5)v$ and that the air resistance after his chute opens is $5v$, where $v = v(t)$ is his velocity at time t.
   (a) Determine expressions for the skydiver's velocity v and position s at any time t. (*Hint*: the motion is described by two differential equations.)
   (b) How long will it take for the skydiver to hit the ground?
   (c) What is the skydiver's terminal velocity and how does this compare with the velocity with which he hits the ground?

2. A dog sees a rabbit running in a straight line across an open field and immediately begins to chase it. In a rectangular coordinate system, assume:

   (i) The rabbit is at the origin and the dog is at the point $(L, 0)$ at the instant the dog sees the rabbit.

   (ii) The rabbit runs up the y-axis and the dog always runs toward the rabbit.

   (iii) The rabbit runs with speed u and the dog runs with speed v.

   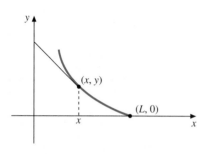

   Suppose that the dog's path is the graph of a function $y = y(x)$.
   (a) Show that the function y must satisfy the differential equation
   $$x \frac{d^2 y}{dx^2} = \frac{u}{v} \sqrt{1 + \left(\frac{dy}{dx}\right)^2}$$
   (b) Determine $y(t)$ such that $y(L) = 0$ and $y'(L) = 0$.
   (c) Show that the dog will catch the rabbit only if $v > u$. When this condition is satisfied, where will the dog catch the rabbit?

3.
   (a) If an object of mass m is projected vertically upward from the ground with an initial velocity v_0, and if air resistance is neglected, then it follows from Newton's Second Law that
   $$m \frac{d^2 y}{dt^2} = -mg.$$
   where $y = y(t)$ represents the position of the object at time t, g is the acceleration due to gravity and $y(0) = 0$, $y'(0) = v_0$. Let t_a denote the time for the object to reach its highest point, called the *ascent time*, and let t_d denote the time that it takes for the object to descend from its highest point to the ground—the *descent time*. Let v_f denote the velocity with which the object hits the ground. Show that $t_a = t_d$ and $|v_0| = |v_f|$.

   (b) Now suppose that the object is projected vertically upward from the ground with initial velocity v_0 and that air resistance exerts a force proportional to the velocity. Then
   $$m \frac{d^2 y}{dt^2} = -k \frac{dy}{dt} - mg$$
   subject to the initial conditions $y(0) = 0$, $y'(0) = v_0$.

   (i) Determine the solution of this initial-value problem.

(ii) Determine the ascent time t_a. (*Note*: It can be shown, with some difficulty, that $t_a < t_d$.)

(iii) Let $E(t) = (1/2)mv^2 + mgy$ be the total energy of the object at time t. Show that $E(t)$ is strictly decreasing on the interval $[0, T]$, where $T = t_a + t_d$ is the total time of flight. Conclude that $|v_0| > |v_f|$.

4. If glucose is given intravenously at a constant rate, the change in the overall concentration $c(t)$ of glucose in the blood with respect to time may be described by the differential equation

$$\frac{dc}{dt} = \frac{G}{100V} - kc$$

where G, V, and k are positive constants, G being the rate at which glucose is administered, in milligrams per minute, and V the volume of blood in the body, in liters (around 5 L for an adult). The concentration $c(t)$ is measured in milligrams per centiliter. The term $-kc$ is included because the glucose is assumed to be changing continually into other molecules at a rate proportional to its concentration.

(a) Solve the equation above for $c(t)$, using c_0 to denote $c(0)$.

(b) Find the so-called steady-state concentration, $\lim_{t \to \infty} c(t)$.

5.
(a) The volume of Lake Michigan is approximately constant at 4,900 km³ (water enters and leaves the lake at approximately the same rate). If the concentration of pollutants in the lake is now 0.20%, and water with a concentration of 0.05% pollutants enters the lake at 158 km³/yr, how long will it take for the concentration of pollutants in the lake to be reduced to 0.10%?

(b) Lake Superior has a volume of approximately 12,200 km³, and the rate of inflow and outflow is about 65.2 km³/yr. Redo part (a) for Lake Superior, assuming the same concentrations of pollutants.

6.
(a) Show that if a radioactive substance A decomposes into a second radioactive substance B, then the amount $y(t)$ of substance B present at time t satisfies the differential equation

$$\frac{dy}{dt} = -k_2 y + k_1 x_0 e^{-k_1 t}$$

where x_0 is the initial amount of substance A. Solve for $y(t)$, with the initial condition $y(0) = 0$.

(b) A serious problem in many homes worldwide is that radioactive radon-222 enters the basements through cracks. Radon results from the decomposition of radium-226 in the soil. The half-life of radon-222 is 3.82 days, and the half-life of radium-226 is 1620 yr. Suppose a new home is built in an area with a radon problem and that cracks in the basement cause infiltration of radon. Use the result of part (a) to determine how long it will take for the amount of radon in the basement to be a maximum.

7. For a series circuit containing a resistance of R ohms, an inductance of L henrys, a capacitance of C farads, and an electromotive force of $E(t)$ volts, one of Kirchhoff's laws yields the differential equation

$$L\frac{dI}{dt} + RI + \frac{Q}{C} = E(t) \tag{1}$$

where $I = I(t)$ is the current, in amperes, and $Q = Q(t)$ is the electric charge. The current and charge are related by the equation $I = dQ/dt$. Thus, from Equation 1, Q satisfies the second-order differential equation

$$L\frac{d^2Q}{dt^2} + R\frac{dQ}{dt} + \frac{Q}{C} = E(t) \tag{2}$$

It is common practice in electrical engineering to express $E(t)$ in the complex exponential form

$$E(t) = E_0 e^{i\omega t} \tag{3}$$

Recall that by Euler's Formula (Equation 17.26)

$$e^{i\omega t} = \cos \omega t + i \sin \omega t$$

(a) Find the general solution of Equation 2, with $E(t)$ given by Equation 3. Assume that $R^2 C < 4L$, and to simplify notation, let $\alpha = R/(2L)$ and $\beta = \sqrt{1/(LC) - \alpha^2}$. (*Hint*: For a particular solution try $Q_p = A e^{i\omega t}$.)

(b) Find the steady-state current. (*Hint*: Let $t \to \infty$ and use the fact that $I = dQ/dt$.)

(c) By finding the real and imaginary parts of the solution in part (b), determine the steady-state current when the impressed voltage $E(t)$ is of the form $E_0 \cos \omega t$ and when it is of the form $E_0 \sin \omega t$.

ANSWERS

CHAPTER 10

Exercise Set 10.1

1. $1 - x + \frac{x^2}{2!} - \frac{x^3}{3!} + \frac{x^4}{4!} - \frac{x^5}{5!}$ **3.** $1 - x + x^2 - x^3 + x^4 - x^5 + x^6$ **5.** $1 + \frac{x^2}{2!} + \frac{x^4}{4!} + \frac{x^6}{6!} + \frac{x^8}{8!}$
7. $1 - (x-1) + (x-1)^2 - (x-1)^3 + (x-1)^4 - \ldots + (-1)^n(x-1)^n$. **9.** $x - \frac{x^3}{3!} + \frac{x^5}{5!} - \ldots + \frac{(-1)^{k+1}x^{2k-1}}{(2k-1)!}$, $n = 2k-1$ or $n = 2k$
$k = 1, 2, 3, \ldots$ **11.** $-(x-1) - \frac{(x-1)^2}{2} - \frac{(x-1)^3}{3} - \frac{(x-1)^4}{4} - \ldots - \frac{(x-1)^n}{n}$ **13.** $R_n(x) = \frac{(-1)^{n+1}(x-1)^{n+1}}{c^{n+2}}$; c is between 1 and x,
where $x \in (0,2)$ **15.** $R_{2k}(x) = \frac{(-1)^k \cos c}{(2k+1)!} x^{2k+1}$, $k = 1, 2, \ldots$; c is between 0 and x; $x \in (-\infty, \infty)$
17. $R_n(x) = \frac{-(x-1)^{n+1}}{(2-c)^{n+1}(n+1)}$, c is between 0 and x; $x \in (0, 2)$ **19.** $P_5(1) \approx 0.36667$, $|R_5(1)| \leq \frac{1}{6!} \approx 0.0013889$, $P_5(1)$ has at least a
2-decimal place accuracy. The most we can say, then, is $e^{-1} \approx 0.37$. **21.** $P_7(1.5) \approx 0.99739$, $|R_8(1.5)| \leq \frac{(1.5)^9}{9!} \approx 0.0001059$,
$\sin 1.5 \approx 0.997$, correct to 3 places. **23.** $P_8(2) \approx -0.415873$, $|R_9(2)| \leq \frac{2^{10}}{(10)!} \approx 0.0002822$, $\cos 2 \approx -0.416$, correct to 3 places.
25. $P_5(5) \approx 2.236076$, $|R_5(5)| \leq \frac{1 \cdot 3 \cdot 5 \cdot 7 \cdot 9}{2^6 \cdot 2^{11} \cdot 6!} \approx 0.000010014$. $\sqrt{5} \approx 2.2361$ to 4 places.
27. $P_{12}(3) \approx 20.0852$; $|R_{12}(3)| < \frac{3^{16}}{(13)!} \approx 0.0069$, $e^3 \approx 20.1$, with accuracy assured only to 1 place. **29.** $P_6(1) \approx 1.54306$,
$R_7(1) < \frac{2}{8!} \approx 0.0000496$, $\cosh 1 \approx 1.5431$ to 4 places. **31.** $P_4\left(\frac{\pi}{3} - \frac{\pi}{90}\right) \approx 0.5299193$; $\left|R_4\left(\frac{\pi}{3} - \frac{\pi}{90}\right)\right| \leq \frac{\left(\frac{\pi}{90}\right)^5}{5!} \approx 4.3 \times 10^{-10}$,
$\cos 58° = 0.5299193$ is accurate to 7 places, and we would have accuracy to 9 places if the calculator went that far.
33. $k = 4$, $\cos 72° = \cos \frac{2\pi}{5} \approx 0.309017$. By calculator, $\cos 72° = 0.309016994$. **35.** Use $k = 7$. $\sinh 2 \approx 3.626860$
37. (a) $P_n(x) = 1 + \alpha x + \frac{\alpha(\alpha-1)}{2!}x^2 + \frac{\alpha(\alpha-1)(\alpha-2)}{3!}x^3 + \ldots + \frac{\alpha(\alpha-1)(\alpha-2)\ldots(\alpha-n+1)}{n!}x^n$; $R_n(x) = \frac{\alpha(\alpha-1)(\alpha-2)\ldots(\alpha-n)}{(n+1)!(1+c)^{n+1-\alpha}}x^{n+1}$; c is between 0 and x
(b) If $\alpha = m$, then $R_m(x) = 0$, since $(\alpha - m) = m - m = 0$. Then, $f(x) = P_m(x) + R_m(x) = P_m(x)$ **(c)** $f^{(n+1)}(x)$ fails
to exist for $x = -1$; the neighborhood of 0 in which $f(x) = P_n(x) + R_n(x)$ is $(-1, 1)$. **(d)** Let $x = \frac{1}{2}$, $P_4\left(\frac{1}{2}\right) \approx 1.1440$;
$\left|R_4\left(\frac{1}{2}\right)\right| \leq 0.000943$; $\sqrt[3]{1.5} \approx 1.14$, is correct to 2 places **39. (a)** If $\frac{1+x}{1-x} = u$, then $x = \frac{u-1}{u+1} = 1 - \frac{2}{u+1} < 1$. Also
$1 - \frac{2}{u+1} > -1$. For any positive number u, there is an x in $(-1, 1)$ such that $\ln u = \ln \frac{1+x}{1-x}$. **(b)** $P_{2k-1}(x) = P_{2k}(x) =$
$2\left[x + \frac{x^3}{3} + \frac{x^5}{5} + \frac{x^7}{7} + \ldots + \frac{x^{2k-1}}{(2k-1)}\right]$, $k = 1, 2, 3, \ldots$, $R_{2k}(x) = \left[\frac{1}{(1+c)^{2k+1}} + \frac{1}{(1-c)^{2k+1}}\right]\frac{x^{2k+1}}{2k+1}$, $k = 1, 2, \ldots$ **(c)** Let $u = 2$.
Then, $x = \frac{1}{3}$, $\left|R_8\left(\frac{1}{3}\right)\right| \leq 0.000223$, $P_8\left(\frac{1}{3}\right) = 2\left[\frac{1}{3} + \frac{1}{3 \cdot 3^3} + \frac{1}{5 \cdot 3^5} + \frac{1}{7 \cdot 3^7}\right] \approx 0.693$.

41. (a) $P_n(x) = x - \frac{x^2}{2} + \frac{x^3}{3} - \frac{x^4}{4} + \ldots + (-1)^{n-1}\frac{x^n}{n}$ (b) 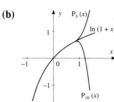 (c) The graphs of $P_n(x)$ approach the graph of $\ln(1+x)$

on $(-1, 1)$, but diverge outside that interval. (d) Conjecture: $R_n(x) \to 0$ as $n \to \infty$ for $x \in (-1, 1)$. The actual interval is $(-1, 1]$. **43.** $-0.33098 \le x \le 0.33098$. **45.** (a) $P_n(x) = P_{2k}(x) = x - \frac{x^3}{3!} + \frac{x^5}{5!} + \ldots + (-1)^k \frac{x^{2k+1}}{(2k+1)!}$ (b)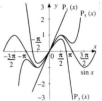

(c) $P_3(x): (-1.04, 1.04)$.
$P_5(x): (-1.75, 1.75)$.
$P_{10}(x): (-3.25, 3.25)$.

47. (a) $P_n(x) = 1 - \frac{1}{3}x - \frac{2}{9}\frac{x^2}{2!} - \frac{10}{27}\frac{x^3}{3!} - \ldots - \frac{1 \cdot 2 \cdot 5 \ldots (3n-4)}{3^n}\frac{x^n}{n!}$ (b)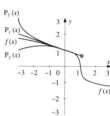

(c) $P_3(x): (-0.78, 0.60)$.
$P_5(x): (-0.94, 0.74)$.
$P_{10}(x): (-0.995, 0.875)$.

49. (a) About $x = 0$: $P_n(x) = 1 + \frac{1}{2}x - \frac{1}{4}\frac{x^2}{2!} + \frac{3}{8}\frac{x^3}{3!} - \frac{15}{16}\frac{x^4}{4!} + \ldots + (-1)^{n+1}\frac{1 \cdot 3 \cdot 5 \ldots (2n-3)}{2^n}\frac{x^n}{n!}$ (b)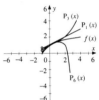

(c) $P_3(x): (-0.611, 0.792)$.
$P_5(x): (-0.755, 0.974)$.
$P_{10}(x): (-0.883, 1.083)$.

(a) About $x = 2$: $P_n(x) = \sqrt{3} + \frac{1}{2\sqrt{3}}(x-2) - \frac{1}{4 \cdot 3^{3/2}}\frac{(x-2)^2}{2!} + \frac{3}{8 \cdot 3^{5/2}}\frac{(x-2)^3}{3!} + \ldots + (-1)^{n+1}\frac{1 \cdot 3 \cdot 5 \ldots (2n-3)}{2^n 3^{(2n-1)/2}}\frac{(x-2)^n}{n!}$ (b)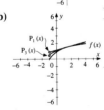

(c) $P_3(x): (0.361, 4.046)$.
$P_5(x): (-0.107, 4.647)$.
$P_{10}(x): (-0.554, 5.084)$.

Exercise Set 10.2

1. (a) $1, \frac{-2}{3}, \frac{3}{5}, \frac{-4}{7}, \frac{5}{9}$ (b) $2, \frac{4}{2!}, \frac{8}{3!}, \frac{16}{4!}, \frac{32}{5!}$ **3.** (a) $1, -2, 6, -24, 120$ (b) $1, 2, 2, 1, \frac{1}{2}$ **5.** (a) $a_n = \frac{2^n - 1}{2^n}$ (b) $a_n = \frac{(-1)^{n-1}(2n)}{2n+3}$ **7.** $\frac{1}{3}$ **9.** diverges **11.** 0 **13.** Does not exist. $a_{2k} = 0, a_{2k+1} = 1$ **15.** 1 **17.** 0 **19.** 0 **21.** 0 **23.** Does not exist. **25.** 0 **27.** 0 **29.** 0 **31.** monotone increasing. **33.** monotone decreasing for $n \ge n_0 = 2$. **35.** monotone decreasing. **37.** monotone decreasing. **39.** monotone increasing.
41. $\frac{a_{n+1}+1}{a_n} = \frac{e}{n+1} < 1$ for $n \ge 2$ $\therefore \{a_n\}$ is decreasing. Also, $a_n \ge 0$. $\therefore$ convergent. **43.** Let $y = x^{1/x}$.
$y' < 0$ for $x \ge e$. $\therefore x^{1/x}$ is decreasing for $x \ge e$, and $\{n^{1/n}\}$ is decreasing. Since it is bounded below by 0, the sequence converges.
45. Show that $\{a_n\}$ is monotone increasing and bounded above by 2. **47.** Suppose that L_1 and L_2 are both limits of $\{a_n\}$, $L_1 \ne L_2$, say $L_2 > L_1$. Let $\varepsilon = \frac{L_2 - L_1}{2}$. Then for all n large enough, $|a_n - L_1| < \varepsilon$, so that $a_n < L_1 + \varepsilon = \frac{L_1 + L_2}{2}$. Similarly, $a_n > L_2 - \varepsilon = \frac{L_1 + L_2}{2}$. The two inequalities cannot both be true. **49.** Since $|a_n - 0| = ||a_n| - 0|$, it follows that for any $\varepsilon > 0$, there is an N such that, for all $n > N$, $|a_n - 0| < \varepsilon$ if, and only if, there is an N such that, for all $n > N$, $||a_n| - 0| < \varepsilon$. Or $\lim_{n \to \infty} a_n = 0$ if, and only if, $\lim_{n \to \infty} |a_n| = 0$. **51.** $a_{n+1} - a_n > 0$. $a_n < n\left(\frac{1}{n+1}\right) < 1$. $\therefore \{a_n\}$ is monotone increasing and

bounded above; it is convergent. **53.** $\frac{1+\sqrt{5}}{2}$ **55. (a)** The limit is 1 if $0 < x_1 < \infty$ and 0 if $x_1 = 0$. Proof: Let $\lim_{n\to\infty} x_n = L$. Then, $L = \sqrt{L}$; $L^2 - L = 0$; $L(L-1) = 0$; $L = 0, 1$. The answer depends on x_1 if x_1 may be zero; otherwise, it does not. **(b)** $x_{x+1} = x_n^2 - 2$. Let $\lim_{x\to\infty} x_n = L$. Then $L = L^2 - 2$; $L^2 - L - 2 = 0$; $L = -1, 2$. The numbers -1 and 2 are fixed points of the function $f(x) = x^2 - 2$. If some other number is substituted into the function, the interation may or may not converge. If we let $x_1 = -1 + \delta$ or $x_1 = 2 + \delta$, both sequences diverge.

57. **(a)** upper bound of 2 **(b)** Assume that $a_1 = 1$, which is ≤ 2. Assume that $a_n \leq 2$. Then $a_{n+1} = \sqrt{2 + a_n} = \sqrt{b}$, where $b \leq 4$. Hence, $a_{n+1} \leq 2$. By induction, $a_n \leq 2$ for all n. **(c)** $a_{n+1}^2 - a_n^2 = 2 + a_n - (2 + a_{n+1}) = 2 + a_n - a_n^2 = (2 - a_n)(1 + a_n) > 0$, since $2 - a_n$ and $1 + a_n$ are > 0. Then, $a_{n+1}^2 - a_n^2 > 0$, which shows that $\{a_n\}$ is monotone increasing $(a_n > 0)$. **(d)** $L = 2$.

59. (a) $\left\{11 + \frac{n-1}{n}\right\}$; **(b)** $\{n\}$ **(c)** $\{\sin n\}$ **(d)** $\left\{\frac{\sin n}{n}\right\}$ **(e)** $\{n\}$ **(f)** $\left\{7\left(1 + \frac{\sin(n\pi/2)}{n}\right)\right\}$ **(g)** Such a sequence is not possible because if it converges to 7, there exists a finite number N such that $|a_n - 7| < \epsilon$ for $n > N$. **(h)** Not possible, since every convergent sequence is bounded. **(i)** $\left\{1 + \frac{1}{n}\right\}$ and $\left\{1 - \frac{1}{n}\right\}$ **(j)** Not possible, by the same reasoning as in part **(g)**, with 7 replaced by 2.

Exercise Set 10.3

1. $\frac{1}{2} + \frac{2}{5} + \frac{3}{10} + \frac{4}{17} + \ldots + \frac{n}{n^2+1} + \ldots$ $S_1 = \frac{1}{2}$, $S_2 = \frac{9}{10}$, $S_3 = \frac{6}{5}$, $S_4 = \frac{122}{85}$ **3.** $\frac{\ln 2}{1!} + \frac{\ln 3}{2!} + \frac{\ln 4}{3!} + \frac{\ln 5}{4!} + \ldots + \frac{\ln(n+1)}{n!} + \ldots$ $S_1 = \ln 2$, $S_2 = \ln 2 + \frac{\ln 3}{2}$, $S_3 = \ln 2 + \frac{\ln 3}{2} + \frac{\ln 4}{6}$, $S_4 = \ln 2 + \frac{\ln 3}{2} + \frac{\ln 4}{6} + \frac{\ln 5}{24}$ **5.** $\sum_{n=1}^{\infty} \frac{(-1)^{n-1}}{3^{n-1}}$; $S_n = \sum_{k=1}^{n} \frac{(-1)^{k-1}}{3^{k-1}}$

7. $\sum_{k=1}^{\infty} \frac{1}{(k+1)\ln(k+1)}$; $S_n = \sum_{k=1}^{n} \frac{1}{(k+1)\ln(k+1)}$ **9.** converges to 1. **11.** converges to 2. **13.** $a = 2$, $r = -\frac{1}{2}$; $S = \frac{4}{3}$

15. $a = 3$, $r = \frac{1}{2}$; $S = 6$ **17.** $a = \frac{2}{e}$, $r = \frac{-2}{e}$; $S = \frac{2}{e+2}$ **19.** $a = 1, r = 0.99$; $S = 100$ **21.** $\frac{5}{33}$ **23.** $\frac{4}{27}$

25. $\frac{1}{2}$ **27.** 1 **29.** 2 **31.** $-\frac{1}{2}$ **33.** 50 m **35. (a)** $\lim_{n\to\infty} \frac{n}{100n+1} = \frac{1}{100} \neq 0$ **(b)** $\lim_{x\to\infty} \frac{x}{(\ln x)^2} = \infty \neq 0$

37. (a) False **(b)** False **(c)** True **(d)** False **(e)** True **39. (a)** Suppose that $\sum (a_n - b_n)$ converges. Then, $-\sum b_n = \sum -b_n = \sum [(a_n - b_n) - a_n] = \sum (a_n - b_n) - \sum a_n$ converges also, contrary to the statement that $\sum b_n$ diverges. Thus, $\sum (a_n - b_n)$ diverges. **(b)** Let $A = \sum_{n=1}^{\infty} \frac{1}{2m-1}$ and $B = \sum_{n=1}^{\infty} \frac{1}{2n}$; both diverge but $A - B$ converges. **41.** Since $a_n \geq 0$ and $S_{n+1} = S_n + a_n$, it follows that $\{S_n\}$ is monotone increasing. Also, $S_n = a_1 + a_2 + \ldots + a_n \leq k$ for all n. Thus $\{S_n\}$ is bounded above. $\therefore$ the sequence $\{S_n\}$ converges. Call its limit S. Then, by definition, $\sum_{n=1}^{\infty} a_n = S$. **43. (a)** $x < -5$ or $x > 1$ **(b)** $e^{-1} < x < e$; $\frac{1}{1-\ln x}$; $\frac{1}{x+5}$ **45.** For a convergent series, $|S_n - L| < \epsilon$ whenever $n > N$, a finite number. The partial sum S_n is the sum of the first N terms, and since addition is associative, these terms may be arranged in any order whatsoever. **47.** ≈ 5.1873775

49. (a) $s_n = 3 \cdot 4^{n-1}$, $l_n = \frac{a}{3^{n-1}}$, $P_n = 3a\left(\frac{4}{3}\right)^{n-1}$ **(b)** $\lim_{n\to\infty} P_n = \infty$, since $\lim_{n\to\infty} r^n = \infty$ if $r > 1$, and $r = \frac{4}{3}$ **(c)** After the first term the series is geometric, with $r = \frac{4}{9}$. Area $= \frac{2\sqrt{3}}{5}a^2$

Exercise Set 10.4

1. diverges **3.** converges **5.** diverges **7.** diverges **9.** diverges **11.** converges **13.** diverges **15.** diverges
17. converges **19.** converges **21.** diverges **23.** converges **25.** converges **27.** diverges **29.** converges
31. converges **33.** converges **35.** converges **37.** converges **39.** converges **41.** diverges **43.** converges
45. diverges **47.** converges **49.** converges if, and only if, $p > 1$ **51. (a)** diverges **(b)** diverges **(c)** converges **53.** Let $\lim_{n\to\infty} na_n = L > 0$. Then, there is a natural number N such that, for all $n > N$, $na_n > \frac{L}{2}$; that is, $a_n > \frac{L}{2} \cdot \frac{1}{n}$ for all $n > N$. But $\sum_{n=N+1}^{\infty} \frac{L}{2} \cdot \frac{1}{n} = \frac{L}{2} \sum_{n=N+1}^{\infty} \frac{1}{n}$ diverges, since $\sum_{n=1}^{\infty} \frac{1}{n}$ diverges. $\therefore \sum a_n$ diverges. **55. (a)** The first series behaves like $\sum \frac{1}{n}$ and diverges **(b)** The second series behaves like $\sum \frac{1}{n^2}$ and converges

Exercise Set 10.5

1. converges 3. diverges 5. converges 7. diverges 9. converges 11. converges 13. converges 15. converges 17. converges 19. converges 21. diverges 23. converges 25. diverges 27. (a) $\lim_{n\to\infty} \frac{a_{n+1}}{a_n} = \lim_{n\to\infty} \frac{n}{n+1} = 1$, No conclusion (b) $\lim_{n\to\infty} \frac{n^2}{(n+1)^2} = 1$, No conclusion 29. If $0 \leq L < 1$, choose r such that $L < r < 1$. Then, $\sqrt[n]{a_n} < r$, $a_n < r^n$; $\sum_{k=N}^{\infty} a_k$ converges as a geometric series. If $1 < L$, choose r such that $1 < r < L$ and $\sum_{k=N}^{\infty} a_k$ diverges. If $L = 1$, no conclusion can be drawn. 31. (a) $\lim_{n\to\infty} \sqrt[n]{a_n} = \frac{1}{2} < 1$ (b) n even: 2^{-3}; n odd: 2^1 $\lim_{n\to\infty} \frac{a_{n+1}}{a_n}$ does not exist 33. It appears that $n^{1/n}$ passes through a maximum at $n = 3$ and then decreases toward its limit of 1.

Exercise Set 10.6

1. converges 3. converges 5. converges 7. diverges 9. converges 11. Error $\leq a_{10} = \frac{1}{10^3} \approx 0.001$ 13. Error $\leq a_6 = \frac{1}{6^6} \approx 0.0000214$. 15. $n \geq 39,998$ 17. $n \geq 101$ 19. ≈ 0.531 21. converges conditionally 23. converges absolutely 25. converges absolutely 27. converges absolutely 29. converges absolutely 31. diverges. 33. Alternate the 2's from 2^n with the elements of $n!$; group factors to get form for a_n; compare to $\sum_{n=0}^{\infty} \frac{1}{2n+1}$ which diverges. $0 < a_n < \frac{1}{\sqrt{n+1}}$ implies $\lim_{n\to\infty} a_n = 0$ and the alternating series converges conditionally. 35. $S_{2n+1} = (a_1 - a_2) + (a_3 - a_4) + \ldots + (a_{2n-1} - a_{2n}) + a_{2n+1} \geq 0$ for all n. Also, $S_{2n+1} - S_{2n-1} = a_{2n+1} - a_{2n} \leq 0$, since $a_{2n+1} \leq a_{2n}$. Hence, $\{S_{2n+1}\}$ is a decreasing sequence and is bounded below by 0. 37. (a) $p_n - q_n = a_n$, since when $a_n \geq 0$, $p_n = a_n$ and $q_n = 0$, and when $a_n < 0$, $-q_n = a_n$ and $p_n = 0$. Also, $p_n + q_n = \begin{cases} a_n & \text{if } a_n \geq 0 \\ -a_n & \text{if } a_n < 0 \end{cases}$ or $p_n + q_n = |a_n|$. (b) The series could not both converge, since if they did, $\sum (p_n + q_n) = \sum |a_n|$ would also converge and $\sum a_n$ would be absolutely convergent. Assume one series converges and compare S_n of each to get a contradiction. 39. If the process is continued, the sum can be made arbitrarily close to 2. At any stage the maximum difference between the partial sum and 2 is q_k, for the last q_k subtracted. Since $\lim_{k\to\infty} Q_k = 0$, the partial sums approach 2. 41. $1 - \frac{1}{2} + \left(\frac{1}{3} + \frac{1}{5} + \frac{1}{7}\right) - \frac{1}{4} + \left(\frac{1}{9} + \frac{1}{11} + \ldots + \frac{1}{19}\right) - \frac{1}{6} + \ldots$ 43. Add terms of $\sum p_k$ until the sum first exceeds S. Then subtract terms of $\sum q_k$ until the sum less than S.

Exercise Set 10.7

1. $-1 < x < 1$ 3. $-\infty < x < \infty$ 5. $-1 \leq x \leq 1$ 7. $0 \leq x < 2$ 9. $-5 \leq x < 1$ 11. $-1 < x \leq 1$ 13. $-\frac{1}{2} \leq x \leq \frac{1}{2}$ 15. $-1 \leq x < 1$ 17. $-\infty < x < \infty$ 19. $\frac{5}{2} \leq x \leq \frac{7}{2}$ 21. $-\frac{7}{3} < x < \frac{1}{3}$ 23. $-\frac{1}{3} < x < \frac{1}{3}$ 25. $4 \leq x \leq 6$ 27. $-\infty < x < \infty$ 29. $\frac{1}{3} \leq x < 1$ 31. converges only if $x = -4$. 33. $-1 \leq x < 1$ 35. $-2 < x < 2$ 37. $-\infty < x < \infty$ 39. $R = 1$ 41. (a) Choose x_1 such that $|x_0| < |x_1| \leq R$ and such that $\sum \frac{|x_1|^n}{a^n}$ converges. If no such x_1 existed, then $|x_0|$ would be an upper bound to the set for which the series converges absolutely, contradicting the definition of R. (b) Thus, by Theorem 10.18, $\sum a_n x^n$ converges absolutely for all x such that $|x| < |x_0|$. (c) If $|x| > R$ and $\sum a_n x^n$ converged, then we could find an x_1 such that $R < |x_1| < |x|$ and conclude by Theorem 10.18 that $\sum a_n x_1^n$ converges absolutely, contrary to the definition of R. Hence, $\sum a_n x^n$ diverges.

Exercise Set 10.8

1. $\sum_{n=1}^{\infty} x^{n-1}$, $|x| < 1$. 3. $\sum_{n=1}^{\infty} \frac{(-1)^n x^{2n-1}}{(2n-1)!}$, $|x| < \infty$. 5. $\sum_{n=1}^{\infty} \frac{nx^{n-1}}{2^{n+1}}$, $|x| < 2$ 7. $\sum_{n=0}^{\infty} \frac{x^{n+1}}{(n+1)^2} + C$, $|x| < 1$. 9. $\sum_{n=0}^{\infty} \frac{(-1)^n x^{2n+1}}{(2n+1)!} + C$, $|x| < \infty$. 11. $\sum_{n=0}^{\infty} \frac{(-1)^n x^{2n+1}}{2^n} + C$, $|x| < \sqrt{2}$. 13. $\sum_{n=0}^{\infty} \left(\frac{1}{2}\right)^{n+1} = 1$ 15. $\sum_{n=0}^{\infty} \frac{2^{n+1}}{(n+1)!} - \sum_{n=0}^{\infty} \frac{1}{(n+1)!}$, $e^2 - e$. 17. $\sum_{n=0}^{\infty} \frac{(-1)^n}{2^n(2n+1)^2}$ 19. $\sum_{n=0}^{\infty} (-1)^{n-1} nx^{n-1}$, $R = 1$. 21. $-\sum_{n=0}^{\infty} \frac{x^{n+1}}{n+1}$, $R = 1$ 23. $2\sum_{n=1}^{\infty} nx^{2n-1}$, $R = 1$, 25. $\frac{1}{2}\sum_{n=0}^{\infty} \frac{(-1)^n x^{n+1}}{n+1}$, $R = 1$. 27. $\sum_{n=0}^{\infty} \frac{(-1)^n x^{2n+1}}{2^{2n+1}(2n+1)}$, $R = \sqrt{2}$

29. $\frac{1}{2}\sum_{n=2}^{\infty} n(n-1)x^{n-2}$, $R = 1$ **31.** $\sum_{k=1}^{\infty} kx^{k+1}$, $R = 1$ **33.** $\int_0^1 f(x)dx = \frac{1}{2}(e^2-1) = \sum_{n=0}^{\infty} \frac{2^n}{(n+1)!}$ **35.** $\frac{x^2+x}{(1-x)^3}$

37. $\frac{3}{2}$ **39.** $\sum_{n=1}^{\infty} \frac{x^{n-1}}{n!}$ **41.** ≈ 0.74682413 **43. (a)** $\frac{10^{50}}{50!} \approx 3.288 \times 10^{-15}$ **(b)** $\frac{20^{100}}{100!} \approx 1.3583 \times 10^{-28}$

(c) $\frac{100^{1000}}{1000!} \approx 10^{-567.6}$

Exercise Set 10.9

1. $\sum_{k=0}^{\infty} \frac{(-1)^k x^{2k+1}}{(2k+1)!}$, $-\infty < x < \infty$.   **3.** $\sum_{k=0}^{\infty} \frac{x^{2k}}{(2k)!}$, $-\infty < x < \infty$ **5.** $\sum_{k=0}^{\infty} \frac{x^{2k+1}}{2k+1}$, $-1 < x < 1$

7. $\sum_{k=1}^{\infty} \frac{(-1)^{k-1} 2^{2k-1} x^{2k}}{(2k)!}$, $-\infty < x < \infty$.   **9.** $\sum_{k=0}^{\infty} \left(\binom{-3}{k}\right)(-1)^k x^k$, $-1 < x < 1$ **11.** $\sum_{k=0}^{\infty} \frac{(\ln 2)^k x^k}{k!}$, $-\infty < x < \infty$.

13. $\sum_{k=1}^{\infty} \frac{(-1)^{k+1} x^{2k-2}}{(2k)!}$, $-\infty < x < \infty$ **15.** $\sum_{k=0}^{\infty} (-1)^k \left(\binom{1/3}{k}\right) \frac{x^k}{2^{3k+1}}$, $-8 < x < 8$ **17.** $\sum_{k=0}^{\infty} \frac{(-1)^k x^k}{(2k)!}$, $-\infty < x < \infty$

19. $e \sum_{k=0}^{\infty} \frac{(x-1)^k}{k!}$, $-\infty < x < \infty$ **21.** $-\sum_{k=0}^{\infty} (x+1)^k$, $-2 < x < 0$.

23. $\frac{1}{2}\left[1 + \sqrt{3}\left(\frac{\pi}{6}\right)(x-1) - \left(\frac{\pi}{6}\right)^2 \frac{(x-1)^2}{2!} - \sqrt{3}\left(\frac{\pi}{6}\right)^3 \frac{(x-1)^3}{3!} + \left(\frac{\pi}{6}\right)^4 \frac{(x-1)^4}{4!} + - - + + \ldots\right]$, $-\infty < x < \infty$

25. (a) $|R_{2k+1}(x)| = |R_{2k+2}(x)| = \frac{|\cos c| |x|^{2k+2}}{(2k+2)!} \le \frac{|x|^{2k+2}}{(2k+2)!}$, and $\lim_{k \to \infty} |R_{2k+2}(x)| = 0$ **(b)** $|R_{2k+1}(x)| = |R_{2k+2}(x)| = \frac{(\cosh c) |x|^{2k+2}}{(2k+2)!}$,

and $\lim_{k \to \infty} R_{2k+2}(x) = 0$.   **(c)** $|R_{2k}(x)| = |R_{2k+1}(x)| = \frac{(\cosh c) |x|^{2k+1}}{(2k+1)!} \le \frac{(\cosh x) |x|^{2k+1}}{(2k+1)!}$ and $\lim_{k \to \infty} R_{2k+1}(x) = 0$.

27. $R = \lim_{n \to \infty} \left|\frac{a_n}{a_{n+1}}\right| = \lim_{n \to \infty} \left|\frac{\alpha(\alpha-1)(\alpha-2)\ldots(\alpha-n+1)}{n!} \cdot \frac{(n+1)!}{\alpha(\alpha-1)(\alpha-2)\ldots(\alpha-n)}\right| = \lim_{n \to \infty} \left|\frac{n+1}{\alpha-n}\right| = 1$ **29.** The series converges to 0 for all x

and so represents $f(x)$ only at $x = 0$. By using the definition of the derivative $f'(0) = 0$. Also $f^{(n)}(0) = 0$ for all n.

31. ≈ 0.5077. error ≤ 0.000006.   **33.** ≈ 0.19737, error ≤ 0.0000003.   **35.** $1 + x^2 + \frac{2}{3}x^4$

37. $\frac{2}{\sqrt{\pi}} \sum_{n=0}^{\infty} \frac{(-1)^n}{(2n+1)n!}$; erf $(1) \approx 0.843$ **39. (a)** $|\sin^{-1}(\sin x)| = |x|$; but with a period of π **(b)** The graph of the sum of the first three terms is

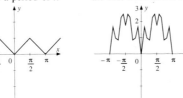

Exercise Set 10.10

1. $\frac{a_0}{2} = \frac{\int_{-\pi}^{\pi} f(x)dx}{\pi - (-\pi)} = f_{ave}$ on $[-\pi, \pi]$ **3.** $\int_{-\pi}^{\pi} f(x) \sin mx \, dx = \frac{a_0}{2} \int_{-\pi}^{\pi} \sin mx \, dx + \sum_{n=1}^{\infty} \left[a_n \int_{-\pi}^{\pi} \cos nx \sin mx \, dx\right.$

$\left. + b_n \int_{-\pi}^{\pi} \sin nx \sin mx \, dx\right]$. Apply equations 10.37, 10.34 and 10.35. $\int_{-\pi}^{\pi} f(x) \sin mx \, dx = b_m \int_{-\pi}^{\pi} \sin^2 mx \, dx = b_m \pi$ (equation

10.26). Hence, $b_m = \frac{1}{\pi} \int_{-\pi}^{\pi} f(x) \sin mx \, dx$.   **5.** $\frac{\pi}{4} - \frac{2}{\pi}\left(\cos x + \frac{\cos 3x}{9} + \frac{\cos 5x}{25} + \ldots\right) + \sin x - \frac{\sin 2x}{2} + \frac{\sin 3x}{3} - \ldots$

7. $a_n = 0$, except for $a_2 = 1$, and $b_n = 0$.   **9.** $a_0 = 0$, $a_n = \frac{1}{n}\sin\frac{n\pi}{2}$, $b_n = 0$, So $f(x) = \cos x - \frac{\cos 3x}{3} + \frac{\cos 5x}{5} - \ldots$

11. For 10 terms   **13.** Graphs for $n = 1$ and 3   The graph for $n = 7$ **15.** The graphs are identical for $n = 2, 5, 8, 11$.

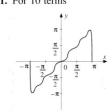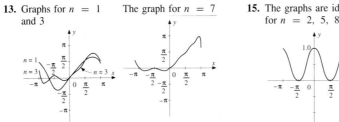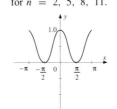

Chapter Review

1. $P_5(x) = -x - \frac{x^2}{2} - \frac{x^3}{3} - \frac{x^4}{4} - \frac{x^5}{5}$; error ≤ 0.0000407; actual error ≈ 0.000013

3. $P_5(x) = 1 + 2\left(x - \frac{\pi}{4}\right) + 2\left(x - \frac{\pi}{4}\right)^2 + \frac{8}{3}\left(x - \frac{\pi}{4}\right)^3 + \frac{10}{3}\left(x - \frac{\pi}{4}\right)^4 + \frac{64}{15}\left(x - \frac{\pi}{4}\right)^5$

5. $\sqrt{2}\sin\left(x - \frac{\pi}{4}\right) = \sqrt{2}\left[x - \frac{\pi}{4} - \frac{(x-\pi/4)^3}{3!} + \frac{(x-\pi/4)^5}{5!} - \cdots + (-1)^{2n+1}\frac{(x-\pi/4)^{2n+1}}{(2n+1)!} + \cdots\right]$

7. $P_3(x) = \frac{\pi}{4} + \frac{1}{2}(x-1) - \frac{1}{4}(x-1)^2 + \frac{1}{12}(x-1)^3$, $R_3(x) = \frac{c(1-c)^2}{(1+c^2)^4}(x-1)^4$

9.

x	$P_1(x)$	$P_2(x)$	$P_3(x)$	$P_4(x)$	e^x
0.25	1.25	1.2813	1.2839	1.2840	1.2840
-0.25	0.75	0.7813	0.7786	0.7788	0.7788
0.50	1.50	1.6250	1.6458	1.6484	1.6487
-0.50	0.50	0.6250	0.6042	0.6068	0.6065
0.75	1.75	2.0313	2.1016	2.1147	2.1170
-0.75	0.25	0.5313	0.4609	0.4741	0.4724
1.00	2.00	2.5000	2.6667	2.7083	2.7183
-1.00	0.00	0.5000	0.3333	0.3750	0.3779

11. **(a)** 2   **(b)** 0

13. **(a)** Increasing, bounded above by e, limit is e.   **(b)** Increasing, 2   15. For large n, $|a_n^2 - L^2| = |a_n + L||a_n - L| \leq (|a_n| + |L|)|a_n - L| < (2|L| + 1)|a_n - L|$. For $\epsilon > 0$ and large n, $|a_n - L| < \frac{\epsilon}{2|L|+1}$, and $|a_n^2 - L^2| < \epsilon$. The converse is not true. Consider $a_n = \frac{(-1)^n n}{n+1}$.

17. $S_n = 2 - \frac{2}{n+1}$; $\lim_{n \to \infty} S_n = 2$ 19. **(a)** $\frac{3}{10}$ **(b)** $\frac{3}{4}$ 21. diverges   23. converges   25. diverges.

27. converges.   29. converges conditionally   31. converges.   33. converges   35. converges   37. conditionally convergent.

39. converges.   41. diverges   43. converges conditionally.   45. $-\frac{3}{2} < x < \frac{3}{2}$ 47. $1 \leq x < 3$.

49. $-1 < x < 1$ 51. $-\frac{1}{2} < x \leq \frac{3}{2}$.   53. $-2\sum_{n=0}^{\infty}\frac{x^{2n+2}}{2n+2}$, $R = 1$.   55. $\sum_{n=0}^{\infty}\binom{-1/2}{n}\frac{x^{2n+1}}{2n+1}$, $R = 1$.

57. **(a)** $\lim_{x \to 0}\left(1 - \frac{x^2}{3!} + \frac{x^4}{5!} - \cdots\right) = 1$ **(b)** $\lim_{x \to 0}\left(\frac{1}{2!} - \frac{x^2}{4!} + \frac{x^4}{6!} - \cdots\right) = \frac{1}{2}$ **(c)** $\lim_{x \to 0}\left(1 + \frac{x}{2!} + \frac{x^2}{3!} + \cdots\right) = 1$

59. $\frac{1}{2}\sum_{n=1}^{\infty}\frac{(-1)^{n-1}(x+1)^n}{n}$ 61. ≈ 0.444, |error| ≤ 0.000052 63. $\cos^2 x - \sin^2 x = \cos 2x$; $\lim_{n \to \infty} R_n(x) = 0$; $\sum_{k=0}^{\infty}\frac{(-1)^{k+1}2^{2k+1}(x-\pi/4)^{2k+1}}{(2k+1)!}$

CHAPTER 11

Exercise Set 11.1

1. $\langle 2, 5 \rangle$ 3. $\langle -3, 1 \rangle$ 5. $\langle -6, 1 \rangle$ 7. $\langle 1, -5 \rangle$ 9. $\langle 0, 7 \rangle$

11. $\langle 11, 2 \rangle$ 13. $\langle -4, -2 \rangle$ 15. $\langle -3, -4 \rangle$ 17. $(3, -2)$ 19. $(-2, 0)$

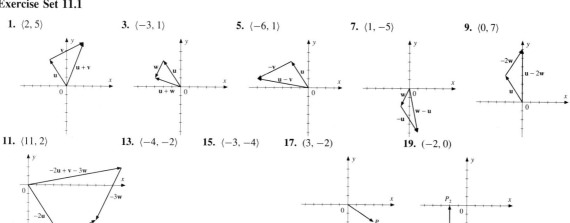

21. 5   **23.** 17   **25.** $\sqrt{5}$ **27.** (a) $\sqrt{5}$ (b) $\sqrt{41}$ (c) 7 (d) $5\sqrt{10}$ **29.** $\frac{5}{13}\mathbf{i} + \frac{12}{13}\mathbf{j}$ **31.** $\left\langle -\frac{2}{\sqrt{13}}, -\frac{3}{\sqrt{13}} \right\rangle$ **33.** $\langle 2\sqrt{5}, -4\sqrt{5} \rangle$

35. $\frac{12}{5}\mathbf{i} + \frac{16}{5}\mathbf{j}$ **37.** $3\sqrt{5}\mathbf{i} - 6\sqrt{5}\mathbf{j}$ **39.** Inclination of **R** ≈ 46.6°   **41.** Speed ≈ 294.62 kph. Heading of plane ≈ 240.16°.

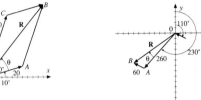

43. Let $\mathbf{u} = \langle a, b \rangle$ and $\mathbf{v} = \langle c, d \rangle$, and consider geometric representatives of **u** and **v** with initial points at the origin.

45. (1) $\mathbf{u} + \mathbf{0} = \langle a, b \rangle + \langle 0, 0 \rangle = \langle a+0, b+0 \rangle = \langle a, b \rangle = \mathbf{u}$ (2) $\mathbf{u} + (-\mathbf{u}) = \langle a, b \rangle + \langle -a, -b \rangle = \langle 0, 0 \rangle = \mathbf{0}$
(3) $(1)\mathbf{u} = (1)\langle a, b \rangle = \langle a, b \rangle = \mathbf{u}$ (4) $(-1)\mathbf{u} = (-1)\langle a, b \rangle = \langle -a, -b \rangle = -\mathbf{u}$ (5) $\mathbf{u} = 0\langle a, b \rangle = \langle 0, 0 \rangle = \mathbf{0}$
(6) $k\mathbf{0} = k\langle 0, 0 \rangle = \langle 0, 0 \rangle = \mathbf{0}$

47. $a = 2, b = -1$ **49.** Let $\mathbf{w} = \langle w_1, w_2 \rangle = a\langle u_1, u_2 \rangle + b\langle v_1, v_2 \rangle = \langle au_1 + bv_1, au_2 + bv_2 \rangle$ and solve the simultaneous equations $u_1 a + v_1 b = w_1; u_2 a + v_2 b = w_2$ for $a = (v_2 w_1 - v_1 w_2)/(u_1 v_2 - u_2 v_1)$ and $b = (u_1 w_2 - u_2 w_1)/(u_1 v_2 - u_2 v_1)$, provided $u_1 v_2 - u_2 v_1 \neq 0$, which is equivalent to $\mathbf{u} \neq k\mathbf{v}$ for all scalars k.   **51.** Let M_1 and M_2 be the midpoints of $\overrightarrow{AC}$ and $\overrightarrow{BD}$, respectively. Then, $\overrightarrow{AM}_1 = \frac{1}{2}(\overrightarrow{AB} + \overrightarrow{AD}))$ and $\overrightarrow{AM}_2 = \overrightarrow{AD}\frac{1}{2}(\overrightarrow{DB})\overrightarrow{AD} + \frac{1}{2}(\overrightarrow{AB} - \overrightarrow{AD}) = \frac{1}{2}(\overrightarrow{AB} + \overrightarrow{AD})$. Hence, $\overrightarrow{AM}_1 = \overrightarrow{AM}_2$; that is, the midpoints coincide.   **55.** Heading = 348.43°. On the return trip, heading = 191.57°. Total time = 3.50 hrs

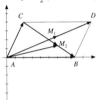

Exercise Set 11.2

1. −6   **3.** −4   **5.** $\mathbf{u} \cdot \mathbf{v} = 0$ **7.** $\mathbf{u} \cdot \mathbf{v} = 0$ **9.** $-\frac{3}{5}$ **11.** $-\frac{1}{5\sqrt{2}}$ **13.** $\frac{7}{25}$ **15.** $\frac{1}{2}; \theta = \frac{\pi}{3}$ **17.** $-\frac{1}{7}, 7$

19. $-\frac{5}{\sqrt{2}}$ **21.** $-\frac{5}{\sqrt{2}}$ **23.** 28 joules   **25.** 30 joules   **27.** $\text{Proj}_\mathbf{v}\mathbf{u} = \left\langle -\frac{1}{5}, -\frac{2}{5} \right\rangle; \text{Proj}_\mathbf{v}^\perp\mathbf{u} = \left\langle \frac{16}{5}, -\frac{8}{5} \right\rangle$

29. $\text{Proj}_\mathbf{v}\mathbf{u} = \frac{6}{5}\mathbf{i} - \frac{8}{5}\mathbf{j}; \text{Proj}_\mathbf{v}^\perp\mathbf{u} = \frac{24}{5}\mathbf{i} + \frac{18}{5}\mathbf{j}$ **31.** 500 lb   **33.** Let $A = P(2, 1), B = P(6, 9), C = P(-2, 3)$ Let $\mathbf{u} = \overrightarrow{AB} = \langle 4, 8 \rangle, \mathbf{v} = \overrightarrow{AC} = \langle -4, 2 \rangle$ $\mathbf{u} \cdot \mathbf{v} = -16 + 16 = 0$. ∴ $\angle BAC = 90°$; Area = 20 square units

35. Let $P_1 = (-5, 2), P_2 = (-3, -2), P_3 = (6, 1), P_4 = (4, 5)$ $\overrightarrow{P_1P_2} = \langle 2, -4 \rangle, \overrightarrow{P_3P_4} = \langle -2, 4 \rangle$. ∴ $\overrightarrow{P_1P_2} \| \overrightarrow{P_3P_4}$.
$\overrightarrow{P_1P_4} = \langle 9, 3 \rangle, \overrightarrow{P_2P_3} = \langle 9, 3 \rangle$. ∴ $\overrightarrow{P_1P_4} \| \overrightarrow{P_2P_3}$; angle at $P_1 \approx 81.9°$; angle at $P_2 \approx 98.1°$

37. (a) $|\mathbf{u} + \mathbf{v}|^2 + |\mathbf{u} - \mathbf{v}|^2 = (\mathbf{u} + \mathbf{v}) \cdot (\mathbf{u} + \mathbf{v}) + (\mathbf{u} - \mathbf{v}) \cdot (\mathbf{u} - \mathbf{v})$ (Theorem 11.4)
$= |\mathbf{u}|^2 + 2\mathbf{u} \cdot \mathbf{v} + |\mathbf{v}|^2 + |\mathbf{u}|^2 - 2\mathbf{u} \cdot \mathbf{v} + |\mathbf{v}|^2 = 2\left(|\mathbf{u}|^2 + |\mathbf{v}|^2\right)$ (b) $|\mathbf{u} + \mathbf{v}|^2 - |\mathbf{u} - \mathbf{v}|^2 = 2\mathbf{u} \cdot \mathbf{v} - (-2\mathbf{u} \cdot \mathbf{v}) = 4\mathbf{u} \cdot \mathbf{v}$ (following the method used in part a)

39. $(\text{Proj}_\mathbf{v}\mathbf{u}) \cdot (\text{Proj}_\mathbf{v}^\perp\mathbf{u}) = \left(\frac{\mathbf{u} \cdot \mathbf{v}}{|\mathbf{v}|^2}\right) \mathbf{v} \cdot \left[\mathbf{u} - \left(\frac{\mathbf{u} \cdot \mathbf{v}}{|\mathbf{v}|^2}\right)\mathbf{v}\right] = \frac{(\mathbf{u} \cdot \mathbf{v})(\mathbf{v} \cdot \mathbf{u})}{|\mathbf{v}|^2} - \frac{(\mathbf{u} \cdot \mathbf{v})^2(\mathbf{v} \cdot \mathbf{v})}{|\mathbf{v}|^4} = \frac{(\mathbf{u} \cdot \mathbf{v})^2}{|\mathbf{v}|^2}\left[1 - \frac{|\mathbf{v}|^2}{|\mathbf{v}|^2}\right] = 0$

41. From Exercise 40, $\mathbf{n} = \langle a, b \rangle$ is perpendicular to the line $ax + by + c = 0$. Also, $\overrightarrow{P_0P_1} = \langle x_1 - x_0, y_1 - y_0 \rangle$. Then,
$d = |\text{Comp}_\mathbf{n} \overrightarrow{P_0P_1}| = \frac{|\langle x_1 - x_0, y_1 - y_0 \rangle \cdot \langle a, b \rangle|}{|\langle a, b \rangle|} = \frac{|ax_1 + by_1 - ax_0 - by_0|}{\sqrt{a^2 + b^2}} = \frac{|ax_1 + by_1 + c|}{\sqrt{a^2 + b^2}}$

Exercise Set 11.3

1. (a) 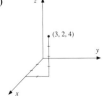   (b) 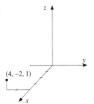   (c)   (d) 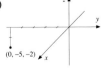   (e)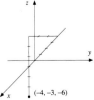

3. (a) xy-plane   (b) yz-plane   (c) xz-plane   (d) x-axis   (e) a point on a coordinate axis   **5.** (a) $\overrightarrow{PQ} = \langle -2, -4, 3 \rangle$

Answers 1293

(b) $\overrightarrow{PQ} = \langle -3, 1, 4 \rangle$ (c) $\overrightarrow{PQ} = \langle 3, 2, -2 \rangle$ (d) $\overrightarrow{PQ} = \langle 6, -2, -2 \rangle$ **7.** (a) -14 (b) -27 (c) $3\sqrt{14}$ (d) $\langle 3, 2, -3 \rangle$
9. (a) $\langle \frac{2}{3}, -\frac{1}{3}, \frac{2}{3} \rangle$ (b) $\langle \frac{4}{5\sqrt{2}}, \frac{3}{5\sqrt{2}}, -\frac{1}{\sqrt{2}} \rangle$ (c) $\frac{1}{\sqrt{3}}\mathbf{i} - \frac{1}{\sqrt{3}}\mathbf{j} + \frac{1}{\sqrt{3}}\mathbf{k}$ (d) $\frac{2}{3\sqrt{5}}\mathbf{i} - \frac{4}{3\sqrt{5}}\mathbf{j} + \frac{5}{3\sqrt{5}}\mathbf{k}$
11. (a) $\mathbf{u} \cdot \mathbf{v} = 8 - 6 - 2 = 0$ (b) $x = \frac{1}{2}$ **13.** (a) $\cos\alpha = \frac{2}{3}$, $\cos\beta = -\frac{1}{3}$, $\cos\gamma = \frac{2}{3}$ (b) $\cos\alpha = \frac{3}{5\sqrt{2}}$, $\cos\beta = -\frac{1}{\sqrt{2}}$, $\cos\gamma = \frac{4}{5\sqrt{2}}$
15. $\langle -\frac{1}{2}, \frac{1}{2}, \frac{1}{\sqrt{2}} \rangle$ **17.** $\cos\alpha \leq \frac{1}{2}$; $\cos\beta \leq \frac{1}{2}$; $\cos^2\gamma = 1 - \cos^2\alpha - \cos^2\beta \geq \frac{1}{2} \Rightarrow \cos\gamma \geq \frac{1}{\sqrt{2}}$, and $\gamma \leq \frac{\pi}{4}$. **19.** $\frac{31}{5}$ **21.** $-\frac{2}{\sqrt{3}}$
23. 184 ergs **25.** $\text{Proj}_\mathbf{v}\mathbf{u} = \langle -\frac{2}{7}, -\frac{6}{7}, \frac{4}{7} \rangle$; $\text{Proj}_\mathbf{v}^\perp\mathbf{u} = \langle \frac{37}{7}, -\frac{1}{7}, \frac{17}{7} \rangle$ **27.** $\text{Proj}_\mathbf{v}\mathbf{u} = \left(\frac{11}{14}\mathbf{i} + \frac{11}{7}\mathbf{j} - \frac{33}{14}\mathbf{k} \right)$; $\text{Proj}_\mathbf{v}^\perp\mathbf{u} = \left(\frac{17}{14}\mathbf{i} - \frac{32}{7}\mathbf{j} - \frac{37}{14}\mathbf{k} \right)$
29. Applying the rules for the addition and multiplication of real numbers, we have: (1) $\mathbf{u} + \mathbf{v} = \langle u_1, u_2, u_3 \rangle + \langle v_1, v_2, v_3 \rangle = \langle u_1 + v_1, u_2 + v_2, u_3 + v_3 \rangle = \langle v_1 + u_1, v_2 + u_2, v_3 + u_3 \rangle = \langle v_1, v_2, v_3 \rangle + \langle u_1, u_2, u_3 \rangle = \mathbf{v} + \mathbf{u}$ (2) $\mathbf{u} + (\mathbf{v} + \mathbf{w}) = \langle u_1, u_2, u_3 \rangle + (\langle v_1, v_2, v_3 \rangle + \langle w_1, w_2, w_3 \rangle) = \langle u_1 + v_1 + w_1, u_2 + v_2 + w_2, u_3 + v_3 + w_3 \rangle = \langle u_1 + v_1, u_2 + v_2, u_3 + v_3 \rangle + \langle w_1, w_2, w_3 \rangle = (\mathbf{u} + \mathbf{v}) + \mathbf{w}$
(3) $k(\mathbf{u} + \mathbf{v}) = k\langle u_1 + v_1, u_2 + v_2, u_3 + v_3 \rangle = \langle ku_1 + kv_1, ku_2 + kv_2, ku_3 + kv_3 \rangle = \langle ku_1, ku_2, ku_3 \rangle + \langle kv_1, kv_2, kv_3 \rangle = k\mathbf{u} + k\mathbf{v}$
(4) $(k+l)\mathbf{u} = (k+l)\langle u_1, u_2, u_3 \rangle = \langle (k+l)u_1, (k+l)u_2, (k+l)u_3 \rangle = \langle ku_1 + lu_1, ku_2 + lu_2, ku_3 + lu_3 \rangle = \langle ku_1, ku_2, ku_3 \rangle + \langle lu_1, lu_2, lu_3 \rangle = k\mathbf{u} + l\mathbf{u}$
(5) $k(l\mathbf{u}) = k(l\langle u_1, u_2, u_3 \rangle) = k\langle lu_1, lu_2, lu_3 \rangle = \langle klu_1, klu_2, klu_3 \rangle = (kl)\langle u_1, u_2, u_3 \rangle = (kl)\mathbf{u}$
31. (1) $|\mathbf{u}| = \sqrt{u_1^2 + u_2^2 + u_3^2} \geq 0$; $|\mathbf{u}| = 0$ if and only if u_1, u_2, and u_3 are all zero and $\mathbf{u} = \mathbf{0}$.
(2) $|-\mathbf{u}| = |\langle -u_1, -u_2, -u_3 \rangle| = \sqrt{u_1^2 + u_2^2 + u_3^2} = |\mathbf{u}|$ (3) $|k\mathbf{u}| = |\langle ku_1, ku_2, ku_3 \rangle| = |k\langle u_1, u_2, u_3 \rangle| = |k||\mathbf{u}|$
(4) $(\mathbf{u} + \mathbf{v}) \cdot (\mathbf{u} + \mathbf{v}) = |\mathbf{u} + \mathbf{v}|^2 = |\mathbf{u}|^2 + 2\mathbf{u} \cdot \mathbf{v} + |\mathbf{v}|^2$; $|\mathbf{u}|^2 + 2\mathbf{u} \cdot \mathbf{v} + |\mathbf{v}|^2 \leq |\mathbf{u}|^2 + 2|\mathbf{v}||\mathbf{v}| + |\mathbf{v}|^2 = (|\mathbf{u}| + |\mathbf{v}|)^2$; Take the square root of both sides of the equation to get the desired result $|\mathbf{u} + \mathbf{v}| \leq |\mathbf{u}| + |\mathbf{v}|$. **35.** $\frac{2}{\sqrt{93}}\mathbf{i} - \frac{8}{\sqrt{93}}\mathbf{j} - \frac{5}{\sqrt{93}}\mathbf{k}$ **37.** $a = \frac{5}{3}, b = \frac{7}{3}, c = \frac{4}{3}$
39. Suppose that $\alpha + \beta < \frac{\pi}{2} \Rightarrow \alpha < \left(\frac{\pi}{2} - \beta \right)$. Then, $\cos^2\alpha + \cos^2\beta > (\cos\pi/2 \cos\beta + \sin\pi/2 \sin\beta)^2 + \cos^2\beta = \sin^2\beta + \cos^2\beta = 1 \Rightarrow \cos^2\alpha + \cos^2\beta > 1$. Now, since $\cos^2\alpha + \cos^2\beta + \cos^2\gamma = 1$, we have $\cos^2\gamma = 1 - (\cos^2\alpha + \cos^2\beta) < 0$, which is impossible. $\alpha + \beta \geq \frac{\pi}{2}$. Since α and β were chosen arbitrarily from the set $\{\alpha, \beta, \gamma\}$, the conclusion that the sum of any two of the three direction angles must be $> \frac{\pi}{2}$ follows. **41.** $\cos\alpha = .486, \cos\beta = .741, \cos\gamma = .463$; $\alpha = 60.904°, \beta = 42.184°, \gamma = 117.588°$; $\alpha = 1.063$ radians, $\beta = .736$ radians, $\gamma = 2.052$ radians **43.** $\cos\alpha = \frac{1}{165}\sqrt{165}, \cos\beta = -\frac{8}{165}\sqrt{165}, \cos\gamma = \frac{2}{33}\sqrt{165}$; $\alpha = 85.535°, \beta = 128.521°, \gamma = 38.877°$; $\alpha = 1.493$ radians, $\beta = 2.243$ radians, $\gamma = .679$ radians **45.** The two vectors are $\langle -\frac{3}{4}, -\frac{9}{8}, 1 \rangle$ and $\langle -\frac{3}{2}, -\frac{9}{4}, 2 \rangle$. **51.** (a) $\mathbf{u} + \mathbf{v} = \langle -2, 1, 12 \rangle$; $\mathbf{u} - \mathbf{v} = \langle 6, -3, -2 \rangle$; $-2\mathbf{u} = \langle -4, 2, -10 \rangle$ (b) $\sqrt{69}$
(c) $\langle -\frac{4}{69}\sqrt{69}, \frac{2}{69}\sqrt{69}, \frac{7}{69}\sqrt{69} \rangle$ (d) 25 (e) .989 radians or 56.668° (f) $25\sqrt{69}$ (g) $\langle -\frac{100}{69}, \frac{50}{69}, \frac{175}{69} \rangle$

Exercise Set 11.4

1. $\langle 3, -1, 4 \rangle$ **3.** $3\mathbf{i} + 9\mathbf{j} + 6\mathbf{k}$ **5.** $\langle -9, -1, -21 \rangle$ **7.** $19\mathbf{i} + 13\mathbf{j} + 5\mathbf{k}$ **9.** $\langle 11, -6, -2 \rangle$ **11.** $14\mathbf{i} - 14\mathbf{j} + 14\mathbf{k}$ or $\mathbf{i} - \mathbf{j} + \mathbf{k}$
13. 12 **15.** $12\mathbf{i} + 12\mathbf{j} - 12\mathbf{k}$ **17.** 69 **19.** $\mathbf{v} \cdot (\mathbf{u} \times \mathbf{v}) = \langle v_1, v_2, v_3 \rangle \cdot \langle (u_2v_3 - u_3v_2), (u_3v_1 - u_1v_3), (u_1v_2 - u_2v_1) \rangle = v_1(u_2v_3 - u_3v_2) + v_2(u_3v_1 - u_1v_3) + v_3(u_1v_2 - u_2v_1) = u_2v_1v_3 - u_3v_1v_2 + u_3v_1v_2 - u_1v_2v_3 + u_1v_2v_3 - u_2v_1v_3 = 0$, or $\mathbf{v} \perp (\mathbf{u} \times \mathbf{v})$
21. $\sqrt{146}$ **23.** $3\sqrt{78}$ **25.** $\frac{3\sqrt{5}}{2}$ **27.** $\langle \frac{4}{\sqrt{89}}, -\frac{3}{\sqrt{89}}, -\frac{8}{\sqrt{89}} \rangle$ **29.** 23 **31.** $\overrightarrow{AB} \cdot (\overrightarrow{AC} \times \overrightarrow{AD}) = \langle 2, -3, -1 \rangle \cdot (\langle -1, 2, 0 \rangle \times \langle 0, 1, -1 \rangle) = 0$
33. Assume that the second and third rows of the determinant are identical. If this is not the case, then interchange row 1 with either row 2 or row 3, as needed. This will change the sign of the determinant only.

$$\begin{vmatrix} a_1 & a_2 & a_3 \\ b_1 & b_2 & b_3 \\ b_1 & b_2 & b_3 \end{vmatrix} = a_1 \begin{vmatrix} b_2 & b_3 \\ b_2 & b_3 \end{vmatrix} - a_2 \begin{vmatrix} b_1 & b_3 \\ b_1 & b_3 \end{vmatrix} + a_3 \begin{vmatrix} b_1 & b_2 \\ b_1 & b_2 \end{vmatrix} = 0$$

35. Property 1. $\mathbf{u} \times \mathbf{v} = \langle u_1, u_2, u_3 \rangle \times \langle v_1, v_2, v_3 \rangle = \langle u_2v_3 - u_3v_2, u_3v_1 - u_1v_3, u_1v_2 - u_2v_1 \rangle = -(v_2u_3 - v_3u_2, v_3u_1 - v_1u_3, v_1u_2 - v_2u_1) = -\mathbf{v} \times \mathbf{u}$
Property 2. $k(\mathbf{u} \times \mathbf{v}) = k\langle u_2v_3 - u_3v_2, u_3v_1 - u_1v_3, u_1v_2 - u_2v_1 \rangle = \langle (ku_2)v_3 - (ku_3)v_2, (ku_3)v_1 - (ku_1)v_3, (ku_1)v_2 - (ku_2)v_1 \rangle = (k\mathbf{u}) \times \mathbf{v}$
Property 3. $\mathbf{u} \times \mathbf{0} = \langle u_1, u_2, u_3 \rangle \times \langle 0, 0, 0 \rangle = \langle 0, 0, 0 \rangle = \mathbf{0}$ **37.** Use components to show in each case that the two sides are equal.
39. $(\mathbf{u}+\mathbf{v}) \times (\mathbf{u}-\mathbf{v}) = \mathbf{u} \times (\mathbf{u}-\mathbf{v}) + \mathbf{v} \times (\mathbf{u}-\mathbf{v})$ (Property 6) $= \mathbf{u} \times \mathbf{u} - \mathbf{u} \times \mathbf{v} + \mathbf{v} \times \mathbf{u} - \mathbf{v} \times \mathbf{v}$ (Property 5) $= \mathbf{0} - \mathbf{u} \times \mathbf{v} - \mathbf{u} \times \mathbf{v} - \mathbf{0} = -2(\mathbf{u} \times \mathbf{v}) = 2(\mathbf{v} \times \mathbf{u})$
41. $\mathbf{u} \times (\mathbf{v} + \mathbf{w}) + \mathbf{v} \times (\mathbf{w} + \mathbf{u}) + \mathbf{w} \times (\mathbf{u} + \mathbf{v}) = \mathbf{u} \times \mathbf{v} + \mathbf{u} \times \mathbf{w} + \mathbf{v} \times \mathbf{w} + \mathbf{v} \times \mathbf{u} + \mathbf{w} \times \mathbf{u} + \mathbf{w} \times \mathbf{v}$ (Property 5) $= \mathbf{u} \times \mathbf{v} + \mathbf{u} \times \mathbf{w} + \mathbf{v} \times \mathbf{w} - \mathbf{u} \times \mathbf{v} - \mathbf{u} \times \mathbf{w} - \mathbf{v} \times \mathbf{w}$ (Property 1) $= 0$
43. (a) The area of the given triangle is one-half the area of the parallelogram with adjacent sides $\overrightarrow{P_1P_2}$ and $\overrightarrow{P_1P_3}$. ∴ the area of the triangle $= \frac{1}{2} \left| \overrightarrow{P_1P_2} \times \overrightarrow{P_1P_3} \right|$ (b) Evaluate the cross product in part (a). **45.** $\langle 17, -4, -5 \rangle$ **47.** Volume of the parallelepiped = 44 cubic units

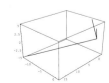

Exercise Set 11.5

1. $\langle 2, 5, -1 \rangle + t\langle -3, 1, 2 \rangle$; $x = 2 - 3t, y = 5 + t, z = -1 + 2t$
3. $\langle 5, 8, -6 \rangle + t\langle 2, -3, 4 \rangle$; $x = 5 + 2t, y = 8 - 3t, z = -6 + 4t$
5. $\langle 4, -1, 8 \rangle + t\langle -1, 3, -3 \rangle$ **7.** $\langle 7, -2, -4 \rangle + t\langle -4, 3, 2 \rangle$ **9.** $\theta \approx 52.5°$ **11.** $\theta \approx 92.8°$ **13.** $\langle 3, -2, 1 \rangle \cdot \langle 7, 8, -5 \rangle = 0$
15. $x = 3 - 3t, y = -1 + t, z = 2 + 5t$ **17.** $\langle 5 + 2t, 2 + t, -3 \rangle$ **19.** The point of intersection is $(-1, 2, 1)$. A vector orthogonal is $\langle -1 - 7t, 2 - 9t, 1 + 3t \rangle$ or $x = -1 - 7t, y = 2 - 9t, z = 1 + 3t$ **21.** The lines are skew.
23. The point of intersection is $(-1, 3, 2)$ **25.** Let $x = x_0 + at \Rightarrow t = \frac{x-x_0}{a}, a \neq 0$; $y = y_0 + bt \Rightarrow t = \frac{y-y_0}{b}, b \neq 0$; $z = z_0 + ct \Rightarrow t = \frac{z-z_0}{c}, c \neq 0$. Equate the expressions for t: $\frac{x-x_0}{a} = \frac{y-y_0}{b} = \frac{z-z_0}{c}$ **27.** (a) $\frac{x-3}{-2} = \frac{y+1}{4} = \frac{z-4}{3}$ (b) $x = 3 - 2t, y = -1 + 4t, z = 4 + 3t$
(c) $\langle 3, -1, 4 \rangle + t\langle -2, 4, 3 \rangle$ (d) $\left(0, 5, \frac{17}{2}\right); \left(\frac{5}{2}, 0, \frac{19}{4}\right); \left(\frac{17}{3}, -\frac{19}{3}, 0\right)$
29. $t = -1 : \mathbf{r} = \langle -1 + 3s, 2 + 4s, 8 - 5s \rangle; t = \frac{7}{3} : \mathbf{r} = \langle -1 + 19s, 2 - 18s, 8 + 5s \rangle$ **31.** $|\overrightarrow{PQ} \times \mathbf{v}| = |\overrightarrow{PQ}| \cdot |\mathbf{v}| \cdot \sin\theta$;
$d = |\overrightarrow{PQ}| \cdot \sin\theta = \frac{|\overrightarrow{PQ} \times \mathbf{v}|}{|\overrightarrow{PQ}| \cdot |\mathbf{v}|} \cdot |\overrightarrow{PQ}| = \frac{|\overrightarrow{PQ} \times \mathbf{v}|}{|\mathbf{v}|}$ **33.** The point of intersection is $(3, 0, 1)$; $10.025°$ **35.** Solution of parametric equations is $x = 1 - 3t, y = -2 + 7t, z = 4 + 2t$

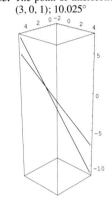

Exercise Set 11.6

1. (a) $\langle x - 4, y - 2, z - 6 \rangle \cdot \langle 3, 2, -1 \rangle = 0$ (b) $3(x - 4) + 2(y - 2) - (z - 6) = 0$ (c) $3x + 2y - z - 10 = 0$
3. (a) $\langle x - 5, y + 3, z + 4 \rangle \cdot \langle 2, 3, -4 \rangle = 0$ (b) $2(x - 5) + 3(y + 3) - 4(z + 4) = 0$ (c) $2x + 3y - 4z - 17 = 0$ **5.** $x - 2y - 4z + 8 = 0$
7. $3x - 2y + 4z + 25 = 0$ **9.** $z = 4$ **11.** $3x - 4y + 5z = 0$ **13.** $x - 4y - 3z - 1 = 0$ **15.** $5x + y - 4z - 9 = 0$
17. $\theta \approx 78.55°$ **19.** $\frac{\mathbf{n}_1 \cdot \mathbf{n}_2}{|\mathbf{n}_1||\mathbf{n}_2|} = \frac{\langle 3, -4, 2 \rangle \cdot \langle 2, 3, 3 \rangle}{|\mathbf{n}_1||\mathbf{n}_2|} = 0$ **21.** $3y - z - 2 = 0$ **23.** $17x + 5y + 19z - 55 = 0$

25. 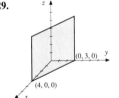 **27.** 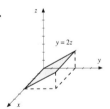 **29.** **31.**

33. **35.** $x = t, y = -13t, z = -2 + 8t$. **37.** $\frac{22}{5\sqrt{5}}$ **39.** $\mathbf{n}_2 = \langle 3, -6, 6 \rangle = 3\mathbf{n}_1; D = \frac{14}{9}$ **41.** The intersection occurs at $(-1, -4, 3)$; $x - y - z = 0$.

43. $(-1, 9, 4)$ **45.** Parametric equations for the line of intersection are $x = 1 + 5t, y = -1 + 10t, z = 5t$ **47.** Direction vectors for the lines are $\mathbf{u} = \langle 1, -1, 1 \rangle$ and $\mathbf{v} = \langle 2, 3, 1 \rangle$ respectively, and $\mathbf{u} \cdot \mathbf{v} = 0$.

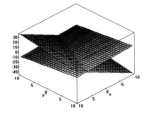
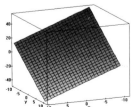

Answers   1295

49. $\frac{1}{6}\sqrt{6}$ **51.** 72 units

Chapter Review

1. (a) 201   **(b)** $\sqrt{5}\mathbf{i} + 2\sqrt{5}\mathbf{j}$ **3. (a)** -82 **(b)** $\theta = 45°$ **5. (a)** $3\sqrt{30}$ **(b)** $\left\langle -\frac{20}{9}, \frac{10}{9}, \frac{20}{9} \right\rangle$

7. If two sides of the triangle are the vectors $\mathbf{u}$ and $\mathbf{v}$, then the third side is $\mathbf{u} - \mathbf{v}$. In the smaller triangle, the sides are $\mathbf{v}/2$ and $(\mathbf{u} - \mathbf{v})/2$ and $\mathbf{x}$; we have $\mathbf{x} = \frac{\mathbf{v}}{2} + \frac{\mathbf{u}-\mathbf{v}}{2} = \frac{\mathbf{u}}{2}$. Hence, $\mathbf{x}$ is parallel to $\mathbf{u}$, and its length is one-half that of $\mathbf{u}$.

9. 253.74 mph; Heading $= 186.79°$

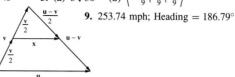

11. $\langle 3, 1, -2 \rangle + t\langle -1, 2, 5 \rangle$; $x = 3 - t$, $y = 1 + 2t$, $z = -2 + 5t$ **13.** $\langle 4, 2, -1 \rangle + t\langle 5, -3, 4 \rangle$; $x = 4 + 5t$, $y = 2 - 3t$, $z = -1 + 4t$
15. Lines do not intersect.   **17.** $x - y + z - 1 = 0$ **19. (a)** $\alpha \approx \alpha 107.3°$, $\beta \approx 53.4°$, $\gamma \approx 41.8°$ **(b)** $\langle -\frac{16}{3}, \frac{28}{3}, \frac{16}{3} \rangle$
21. (a) $\frac{16}{5}$ **(b)** $\frac{5}{3}$ **23.** Let $a\mathbf{u} + b\mathbf{v} + c\mathbf{w} = 0$. Then $a(\mathbf{u} \cdot \mathbf{u}) + b(\mathbf{u} \cdot \mathbf{v}) + c(\mathbf{u} \cdot \mathbf{w}) = a|\mathbf{u}|^2 = 0 \Rightarrow a = 0$, since $|\mathbf{u}| \neq 0$;
$a(\mathbf{u} \cdot \mathbf{v}) + b(\mathbf{v} \cdot \mathbf{v}) + c(\mathbf{v} \cdot \mathbf{w}) = b|\mathbf{v}|^2 = 0 \Rightarrow b = 0$, since $|\mathbf{v}| \neq 0$; $a(\mathbf{u} \cdot \mathbf{w}) + b(\mathbf{v} \cdot \mathbf{w}) + c(\mathbf{w} \cdot \mathbf{w}) = a|\mathbf{w}|^2 = 0 \Rightarrow c = 0$, since $|\mathbf{w}| \neq 0$

CHAPTER 12

Exercise Set 12.1

1. $\{t : t < 1\}$, or $-\infty < t < 1$ **3.** $\{t : 1 \leq t < \infty\}$ **5.** $\{t : t > 0, t \neq \frac{n\pi}{2}, n = 1, 2, 3, \ldots\}$ **7.**

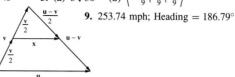

9.

11.

13.

15.

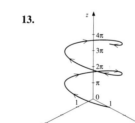
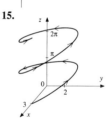

17.

19.

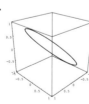
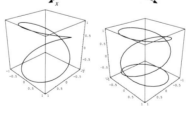

21.

Exercise Set 12.2

1. $\langle 1, 2, 3 \rangle$ **3.** $\mathbf{i} + 2\mathbf{j}$ **5.** $\{t : t < 1\}$ **7.** $\{t : -1 < t < 1\}$ **9. (a)** $\langle 3t - 1, t - 2, 3 - t \rangle$ **(b)** $\langle t + 3, 2t + 6, 3t + 6 \rangle$
11. (a) $\langle t^2 - 6, 3 - 3t - 2t^2, 3t + t^2 \rangle$ **(b)** $\langle t, 2, 1 - t \rangle$ **13. (a)** $(2t \cos t)\mathbf{i} + (2t \sin t)\mathbf{j} + (2t \sin t)\mathbf{k}$
(b) $\cos t \sin t + \sin t \cos t + \sin t = \sin 2t + \sin t$ **15.** $\left\langle 2t, 2e^{2t}, -\frac{1}{t^2} \right\rangle$ **17.** $\langle t \cos t + \sin t, 1 + \ln t \rangle$ **19.** $e^{-t}\mathbf{i} + e^{2t}(2t + 1)\mathbf{j}$
21. $\frac{1}{(1-t^2)^{3/2}}\mathbf{i} + \frac{1}{\sqrt{1-t^2}}\mathbf{j} - \frac{t}{\sqrt{1-t^2}}\mathbf{k}$ **23.** $x = 2 + 2u$, $y = 3 + 6u$, $z = 1 + 3u$ **25.** $x = 4 - 16u$, $y = \frac{1}{2} + u$; $z = \frac{1}{4} + 2u$

27. $x = -1 + u, y = 4 + \frac{5}{2}u, z = 1 - u$ **29.** $\mathbf{r}'\left(\frac{\pi}{3}\right) = \langle -2\sqrt{3}, 1 \rangle$ **31.** $\mathbf{r}'(-1) = -2\mathbf{i} + 3\mathbf{j}$ **33.** $\mathbf{r}'(1) = \langle 2, -3, 3 \rangle$

35. $\left\langle -\cos t, t - \sin t, \frac{t^2}{2} \right\rangle + \langle C_1, C_2, C_3 \rangle$ **37.** $\left(\frac{\pi}{4}, \frac{2}{3}, \frac{1}{3}\right)$ **39.** $\frac{\pi}{4}\mathbf{i} + \frac{1}{3}\mathbf{j}$ **41.** $\frac{-3\sqrt{3}}{8}\mathbf{i} + \left(\sqrt{3} - \frac{\pi}{3}\right)\mathbf{j} + \frac{\pi}{3}\mathbf{k}$

43. $\frac{1}{3}(32 - \sqrt{2}) \approx 10.195$ **45.** $\left| \int_a^b \mathbf{r}(t)dt \right| = |\langle \cos a - \cos b, \sin b - \sin a, b - a \rangle| = [2 - 2\cos(a-b) + (b-a)^2]^{1/2}$ and
$\int_a^b |\mathbf{r}(t)|dt = \sqrt{2}(b - a)$. Let $(b - a) = \alpha \geq 0$. $\sqrt{\alpha^2 - 2\cos\alpha + 2} \leq \sqrt{2}\alpha$ since $\alpha^2 + 2\cos\alpha - 2$ is nondecreasing.

47. $\frac{d}{dt}[c\mathbf{u}(t)] = \frac{d}{dt}[\langle cu_1(t), cu_2(t), \ldots, cu_n(t) \rangle] = \langle cu_1'(t), cu_2'(t), \ldots, cu_n'(t) \rangle = c\mathbf{u}'(t)$

49. $\frac{d}{dt}[\mathbf{u}(t) \cdot \mathbf{v}(t)] = \frac{d}{dt}[u_1(t)v_1(t) + u_2(t)v_2(t) + \ldots + u_n(t)v_n(t)] = u_1(t)v_1'(t) + u_1'(t)v_1(t) + u_2(t)v_2'(t) + u_2'(t)v_2(t) + \ldots + u_n(t)v_n'(t) + u_n'(t)v_n(t) =$
$u_1(t)v_1'(t) + u_2(t)v_2'(t) + \ldots + u_n(t)v_n'(t) + u_1'(t)v_1(t) + u_2'(t)v_2(t) + \ldots + u_n'(t)v_n(t) = \mathbf{u}(t) \cdot \mathbf{v}'(t) + \mathbf{u}'(t) \cdot \mathbf{v}(t)$

51. $\frac{d}{dt}[\mathbf{u}(f(t))] = \frac{d}{dt}[\langle u_1(f(t)), u_2(f(t)), \ldots, u_n(f(t)) \rangle] = \langle u_1'(f(t))f'(t), u_2'(f(t))f'(t), \ldots, u_n'(f(t))f'(t) \rangle = \mathbf{u}'(f(t))f'(t)$

53. $c\int_a^b \mathbf{u}(t)dt = c\int_a^b \langle u_1(t), u_2(t), \ldots, u_n(t) \rangle dt = \int_a^b \langle cu_1(t), cu_2(t), \ldots, cu_n(t) \rangle dt = \int_a^b c\mathbf{u}(t)dt$ **55.** $\mathbf{K} \cdot \int_a^b \mathbf{u}(t)dt =$
$\langle k_1, k_2, \ldots, k_n \rangle \cdot \langle \int_a^b u_1(t)dt, \int_a^b u_2(t)dt, \ldots, \int_a^b u_n(t)dt \rangle = k_1 \int_a^b u_1(t)dt + k_2 \int_a^b u_2(t)dt + \ldots + k_n \int_a^b u_n(t)dt = \int_a^b \mathbf{K} \cdot \mathbf{u}(t)dt$

57. $\mathbf{u}'(t) = \mathbf{v}'(t) \Rightarrow u_i'(t) = v_i'(t), i = 1, 2, \ldots, n$ Since the derivatives of $u_i(t)$ and $v_i(t)$ are equal, $u_i(t)$ and $v_i(t)$ can differ only by a constant, say C_i. Thus, $\langle u_1(t), u_2(t), \ldots, u_n(t) \rangle = \langle v_1(t) + C_1, v_2(t) + C_2, \ldots, v_n(t) + C_n \rangle$. $\mathbf{u}(t) = \mathbf{v}(t) + \mathbf{C}$.

59. Parametric equations of the tangent line are $x = t + 2$, $y = 4t + 4$, $y = 12t + 8$ **61.** $\left(\frac{3}{2}, 2\sqrt{3}, 0\right)$

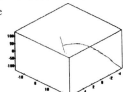

Exercise Set 12.3

1. $2\sqrt{14}$ **3.** $4(3\sqrt{6} - \sqrt{2})$ **5.** $\sqrt{5}\pi$ **7.** $\sqrt{65}(t + 1)$ **9.** $\frac{3}{2}(t^2 + 2t)$ **11.** $\frac{1}{27}[(9t^2 + 4)^{3/2} - 8]$

13. $\mathbf{R}(u) = \mathbf{r}(\alpha(u)) = \left\langle \sqrt{u}, 2\sqrt{u}, \frac{1}{\sqrt{u}} \right\rangle$, $u > 0$; The orientation is unchanged.   **15.** $\mathbf{R}(u) = (1 - u)\mathbf{i} + u\mathbf{j} + (1 + u)\mathbf{k}$, $\frac{1}{2} \leq u \leq 1$; The orientation is reversed.

17. $\mathbf{R}(s) = \mathbf{r}\left(\frac{s}{a}\right) = \langle a\cos(s/a), a\sin(s/a) \rangle$ **19.** $\mathbf{R}(s) = \mathbf{r}\left(\frac{s}{5}\right) = \left(4\cos\frac{s}{5}\right)\mathbf{i} + 3\frac{s}{5}\mathbf{j} + \left(4\sin\frac{s}{5}\right)\mathbf{k}$

21. $L = \frac{1}{27}(44^{3/2} - 17^{3/2}) \approx 8.2137$; domain of $\mathbf{R}$ is $0 \leq u \leq \ln 2$ **23.** $\frac{2}{27}(343 - 80\sqrt{10}) \approx 6.6680$ **25.** $s(t) = \frac{5}{2}(t^2 - 1) = u$; so
$\mathbf{R}(u) = \left\langle \frac{3u}{5} + \frac{3}{2}, 6 + \frac{4u}{5}, 3 \right\rangle$ and $\mathbf{R}'(u) = \left\langle \frac{3}{5}, \frac{4}{5}, 0 \right\rangle$ **27.** $\mathbf{R}'(u) = \mathbf{r}'(\alpha(u))\alpha'(u)$ Assume that $\alpha'(u) > 0$. Let $I = [a, b] = [\alpha(c), \alpha(d)]$,
where $J = [c, d]$. $L = \int_c^d |\mathbf{r}'(\alpha(u))||\alpha'(u)|du =$ length of graph of $\mathbf{R}$ over J. Let $t = \alpha(u), dt = \alpha'(u)du$. Then, $L = \int_a^b |\mathbf{r}'(t)|dt =$ length of graph of $\mathbf{r}$ over I. If $\alpha'(u) < 0$, the orientation of the curve is reversed. Then $I = [\alpha(d), \alpha(c)] = [a, b]$, and the conclusion follows.

29. 8.151998483

Exercise Set 12.4

1. $\mathbf{T}(2) = \left\langle \frac{1}{\sqrt{17}}, \frac{4}{\sqrt{17}} \right\rangle$; $\mathbf{N}(2) = \left\langle -\frac{4}{\sqrt{17}}, \frac{1}{\sqrt{17}} \right\rangle$ **3.** $\mathbf{T}\left(\frac{\pi}{4}\right) = -\frac{1}{\sqrt{2}}\mathbf{i} + \frac{1}{\sqrt{2}}\mathbf{j}$; $\mathbf{N}\left(\frac{\pi}{4}\right) = -\frac{1}{\sqrt{2}}\mathbf{i} - \frac{1}{\sqrt{2}}\mathbf{j}$

5. $\mathbf{T}\left(\frac{3\pi}{4}\right) = \left\langle -\frac{1}{\sqrt{10}}, -\frac{1}{\sqrt{10}}, \frac{2}{\sqrt{5}} \right\rangle$; $\mathbf{N}\left(\frac{3\pi}{4}\right) = \left\langle \frac{1}{\sqrt{2}}, -\frac{1}{\sqrt{2}}, 0 \right\rangle$ **7.** $\mathbf{T}(1) = \frac{4}{5}\mathbf{i} + \frac{3}{5}\mathbf{j} + 0\mathbf{k}$; $\mathbf{N}(1) = -\frac{3}{5}\mathbf{i} + \frac{4}{5}\mathbf{j} + 0\mathbf{k}$

9. $\mathbf{T}(0) = \left\langle \frac{1}{3}, -\frac{2}{3}, \frac{2}{3} \right\rangle$; $\mathbf{N}(0) = \left\langle \frac{2}{3}, \frac{2}{3}, \frac{1}{3} \right\rangle$ **11.** $\frac{6}{13\sqrt{13}}$ **13.** $\frac{3}{2\sqrt{2}}$ **15.** $\frac{32}{343}$ **17.** $\frac{2}{5\sqrt{5}}$ **19.** $\frac{4}{5\sqrt{5}}$ **21.** $\frac{14}{17\sqrt{17}}$ **23.** $\frac{2}{9}$

Answers   1297

25. $\frac{\sqrt{2}}{3}$ **27.** $\left(\frac{7}{4}, 0\right)$ **29.** $Q = (-2, 3)$ **31. (a)** $\mathbf{T} = \langle |\mathbf{T}|\cos\theta, |\mathbf{T}|\sin\theta\rangle = \langle\cos\theta, \sin\theta\rangle$, since $|\mathbf{T}| = 1$

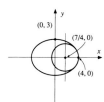

(b) $\kappa = \left|\frac{d\mathbf{T}}{ds}\right| = \left|\frac{d\mathbf{T}}{d\theta}\right|\left|\frac{d\theta}{ds}\right| = |\langle-\sin\theta, \cos\theta\rangle|\left|\frac{d\theta}{ds}\right| = \left|\frac{d\theta}{ds}\right|$ **(c)** $\mathbf{N} = \frac{1}{\kappa}\frac{d\mathbf{T}}{ds} = \frac{1}{|d\theta/ds|}\langle-\sin\theta, \cos\theta\rangle\frac{d\theta}{ds} = \langle-\sin\theta, \cos\theta\rangle\frac{d\theta}{ds}\bigg/\left|\frac{d\theta}{ds}\right| =$
$\begin{cases}\langle-\sin\theta, \cos\theta\rangle & \text{if } d\theta/ds > 0 \\ \langle\sin\theta, -\cos\theta\rangle & \text{if } d\theta/ds < 0\end{cases}$; If $d\theta/ds > 0$, C is concave toward the left, and the x-component of $\mathbf{N}$ is negative $(0 < \theta < \pi)$. If $d\theta/ds < 0$, C is concave toward the right, and again the normal is directed toward the concave side of C.

33. $\mathbf{T}(t) = \left\langle\frac{-ab\sin bt}{\sqrt{a^2b^2+c^2}}, \frac{ab\cos bt}{\sqrt{a^2b^2+c^2}}, \frac{c}{\sqrt{a^2b^2+c^2}}\right\rangle$; $\mathbf{N}(t) = \frac{\mathbf{T}'(t)}{|\mathbf{T}'(t)|} = \langle-\cos bt, -\sin bt, 0\rangle$; $\mathbf{B} = \left\langle\frac{c\sin bt}{\sqrt{a^2b^2+c^2}}, \frac{-c\cos bt}{\sqrt{a^2b^2+c^2}}, \frac{ab}{\sqrt{a^2b^2+c^2}}\right\rangle$; $\kappa = \frac{|a|b^2}{a^2b^2+c^2}$

35. $\mathbf{r}' \times \mathbf{r}'' = |\mathbf{r}'|\kappa\left(\frac{ds}{dt}\right)^2 \mathbf{B}$; $\mathbf{r}' = |\mathbf{r}'|\mathbf{T}\mathbf{j}$; So $(\mathbf{r}' \times \mathbf{r}'') \times \mathbf{r}' = |\mathbf{r}'|^2\left(\frac{ds}{dt}\right)^2\kappa\mathbf{N}$; (since $\mathbf{B} \times \mathbf{T} = \mathbf{N}) = |\mathbf{r}'|^4\kappa\mathbf{N}$ ∴ $\mathbf{N} = \frac{(\mathbf{r}' \times \mathbf{r}'') \times \mathbf{r}'}{\kappa|\mathbf{r}'|^4}$

37. $\mathbf{B} \times \mathbf{T} = (\mathbf{T} \times \mathbf{N}) \times \mathbf{T} = -\mathbf{T} \times (\mathbf{T} \times \mathbf{N}) = -(\mathbf{T} \cdot \mathbf{N})\mathbf{T} + (\mathbf{T} \cdot \mathbf{T})\mathbf{N} = 0\mathbf{T} + \mathbf{N} = \mathbf{N}$ (Theorem 11.8).∴ $\mathbf{B} \times \frac{d\mathbf{T}}{ds} + \frac{d\mathbf{B}}{ds} \times \mathbf{T} = \frac{d\mathbf{N}}{ds}$ Since $\mathbf{N} = \frac{1}{\kappa}\frac{d\mathbf{T}}{ds}, \frac{d\mathbf{T}}{ds} = \kappa\mathbf{N}$. Also, $\frac{d\mathbf{B}}{ds} = -\tau\mathbf{N}$ Thus, $\frac{d\mathbf{N}}{ds} = \kappa(\mathbf{B} \times \mathbf{N}) - \tau(\mathbf{N} \times \mathbf{T}) = -\kappa\mathbf{T} - \tau(-\mathbf{B}) = \tau\mathbf{B} - \kappa\mathbf{T}$

39. (a) 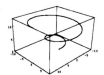   **(b)** velocity $= \langle\cos(t) - t\sin(t), \sin(t) + t\cos(t), 1\rangle$

speed $= \sqrt{(\cos(t) - t\sin(t))^2 + (\sin(t) + t\cos(t))^2 + 1}$; accel $= \langle-2\sin(t) - t\cos(t), 2\cos(t) - t\sin(t), 0\rangle$

(c) $\kappa = \frac{\sqrt{5t^2+8+t^4}}{(2+t^2)^{3/2}}$ **(d)** $\mathbf{T} = \frac{\langle\cos(t)-t\sin(t),\sin(t)+t\cos(t),1\rangle}{\sqrt{2+t^2}}$;
$\mathbf{N} = (2\langle-2\sin(t) - t\cos(t), 2\cos(t) - t\sin(t), 0\rangle + \langle-2\sin(t) - t\cos(t),$
$2\cos(t) - t\sin(t), 0\rangle t^2 - \langle\cos(t) - t\sin(t), \sin(t) + t\cos(t), 1\rangle t)/\left(\sqrt{2+t^2}\sqrt{t^4 + 5t^2 + 8}\right)$ **(e)** $T_a = \frac{t}{\sqrt{4+t^2}}$; $N_a = \frac{\sqrt{5t^2+8+t^4}}{\sqrt{2+t^2}}$

41. $\mathbf{T} = 2\frac{\left\langle\frac{3}{2}t^2, 2t\right\rangle}{t\sqrt{9t^2+6}}$; $\mathbf{N} = -\frac{1}{6}\frac{-9\langle 3t, 2\rangle t^3 - 16\langle 3t, 2\rangle t + 18\left\langle\frac{3}{2}t^2, 2t\right\rangle t^2 + 16\left\langle\frac{3}{2}t^2, 2t\right\rangle}{t^2\sqrt{9t^2+16}}$; $\kappa = 3\frac{t^2}{\left(\frac{9}{4}t^4+4t^2\right)^{3/2}} \approx .0320019$ at $t = 2$

Exercise Set 12.5

1. $\mathbf{v}\left(\frac{\pi}{4}\right) = \langle-\sqrt{2}, \frac{3\sqrt{2}}{2}\rangle$;
$\mathbf{a}\left(\frac{\pi}{4}\right) = \langle-\sqrt{2}, -\frac{3\sqrt{2}}{2}\rangle$;
$\left|\mathbf{v}\left(\frac{\pi}{4}\right)\right| = \frac{\sqrt{26}}{2}$

3. $\mathbf{v}(0) = 2\mathbf{i} - 3\mathbf{j}$; $\mathbf{a}(0) = 2\mathbf{i} + 3\mathbf{j}$; $|\mathbf{v}(0)| = \sqrt{13}$

5. $\mathbf{v}(2) = \langle 2, 4\rangle$; $\mathbf{a}(2) = \langle 0, 2\rangle$; $|\mathbf{v}(2)| = 2\sqrt{5}$

7. $\mathbf{v}(1) = 2\mathbf{i} + 2\mathbf{j} + 3\mathbf{k}$; $\mathbf{a}(1) = 0\mathbf{i} + 2\mathbf{j} + 6\mathbf{k}$; $|\mathbf{v}(1)| = \sqrt{17}$

9. $\mathbf{v}(0) = \langle 1, 2, -1\rangle$; $\mathbf{a}(0) = \langle 1, 0, 1\rangle$; $|\mathbf{v}(0)| = \sqrt{6}$

11. $\mathbf{v}(t) = -3(5-2t)^{1/2}\mathbf{i} + (t+2)\mathbf{j}$;
$\mathbf{a}(t) = 3(5-2t)^{-1/2}\mathbf{i} + \mathbf{j}$; $|\mathbf{v}(t)| = 7-t$

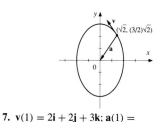
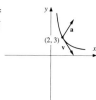
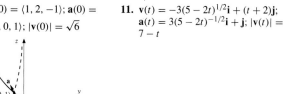

13. $v(t) = \langle -t\sin t + \cos t, t\cos t + \sin t, (2t)^{1/2}\rangle$; $a(t) = \langle -t\cos t - 2\sin t, -t\sin t + 2\cos t, (2t)^{-1/2}\rangle$; $|v(t)| = t+1$

15. $v(t) = \langle -e^t\sin t + e^t\cos t, e^t\cos t + e^t\sin t, e^t\rangle$; $a(t) = \langle -2e^t\sin t, 2e^t\cos t, e^t\rangle$; $|v(t)| = \sqrt{3}e^t$ **17. (a)** Force will become $\sqrt{840} \approx 29.0$ N   **(b)** Force will become $\sqrt{612} \approx 24.7$ N   **19.** $h \approx 284$ km   **21.** $y_{max} \approx 4592$ m $= 4.60$ km; $x_{max} \approx 31{,}800$ m $= 31.8$ km

23. $v_0 \approx 256$ m/s   **25.** $v \cdot a + a \cdot v = 0 \Rightarrow 2a \cdot v = 0 \Rightarrow a \cdot v = 0 \Rightarrow a \perp v$ **27.** Suppose that $a(t) = 0$. Then, $v(t) = C_1$, and $r(t) = C_1 t + C_2$ (let $C_1 = A$ and $C_2 = B$). Now suppose that $r(t) = At + B$. Then $v(t) = A$, and $a(t) = 0$. The motion is along a line with direction vector A, and the speed is constant $= |A|$.   **29.** $x_{max} = (500)(\cos 25°)(47.4) \approx 21{,}500$ m; $|r'(t)| \approx 519$ m/s at impact

31. $v(t) = \langle 2, -3, 1 - \frac{t^2}{2}\rangle$, or $2i - 3j + \left(1 - \frac{t^2}{2}\right)k$; $r(t) = \langle 2t + 4, -3t + 2, t - \frac{t^3}{6}\rangle$, or $(2t+4)i + (-3t+2)j + \left(t - \frac{t^3}{6}\right)k$

33. $\omega \times r(t) = \begin{vmatrix} i & j & k \\ 0 & 0 & \omega \\ r\cos\omega t & r\sin\omega t & 0 \end{vmatrix} = \langle -r\omega\sin\omega t, r\omega\cos\omega t, 0\rangle$
$= r'(t) = v(t)$
The same result holds if $r(t) = \langle r\cos\omega t, r\sin\omega t, C\rangle$

35. $a_T(1) = \frac{22}{13}$; $a_N = \frac{6}{\sqrt{13}}$

37. $a_T(\ln 2) = \frac{15}{2\sqrt{34}}$; $a_N = \frac{4}{\sqrt{34}}$

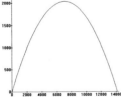

39. $a_T(0) = \frac{9}{\sqrt{5}}$; $a_N = \frac{2}{\sqrt{5}}$

41. $\frac{6}{13\sqrt{13}}$ **43.** $\frac{32}{17\sqrt{34}}$ **45.** $\frac{2}{5\sqrt{5}}$ **47.** $a_T = -\frac{2}{t^3}$; $a_N = \frac{2}{t^2}$

49. $a_T = 3$; $a_N = 3\sqrt{t^2 + 3 + \frac{1}{2t}}$ **51.** $a_T = \sqrt{5}$; $a_N = t$ **53.** $T = \left\langle \frac{2t}{1+2t^2}, \frac{1}{1+2t^2}, \frac{2t^2}{1+2t^2}\right\rangle$; $N = \left\langle \frac{1-2t^2}{1+2t^2}, \frac{-2t}{1+2t^2}, \frac{2t}{1+2t^2}\right\rangle$; $\kappa = \frac{2t^2}{(1+2t^2)^2}$

55. $T = \frac{t\cos t + \sin t}{t+1}i + \frac{\cos t - t\sin t}{t+1}j + \frac{(2t)^{1/2}}{t+1}k$; $N = \frac{-t^2\sin t + t(\cos t - \sin t) + 2\cos t - \sin t}{d}i - \frac{t^2\cos t + t(\sin t + \cos t) + 2\sin t + \cos t}{d}j + \frac{(1-t)/(2t)^{1/2}}{d}k$, where $d = (t+1)\sqrt{t^2 + 3 + 1/(2t)}$; $\kappa = \frac{\sqrt{2t^3 + 6t + 1}}{3(t+1)^2\sqrt{2t}}$ **57.** $T = \left\langle \frac{\sin t}{\sqrt{5}}, \frac{\cos t}{\sqrt{5}}, \frac{2}{\sqrt{5}}\right\rangle$; $N = \langle \cos t, -\sin t, 0\rangle$; $\kappa = \frac{1}{5t}$

59. $a_N = \kappa\left(\frac{ds}{dt}\right)^2 = \frac{|v \times a|}{|v|}$. Then, $\kappa = \frac{a_N}{(ds/dt)^2} = \frac{|v \times a|}{|v||v|^2} = \frac{|v \times a|}{|v|^3}$ **61.** $F = kN = ma$, or $a = \frac{k}{m}N$ ∴ $a_T = \frac{d^2s}{dt^2} = 0$, and $a_N = \frac{k}{m}$; $\frac{d^2s}{dt^2} = 0 \Rightarrow \frac{ds}{dt} = v = C$ **63. (a)** $v = \frac{dr}{dt} = r\frac{du_r}{dt} + \frac{dr}{dt}u_r = ru_\theta\frac{d\theta}{dt} + \frac{dr}{dt}u_r$ **(b)** $|v| = \left|\frac{dr}{dt}\right| = \sqrt{r^2\left(\frac{d\theta}{dt}\right)^2 + \left(\frac{dr}{dt}\right)^2}$, since $|u_\theta| = |u_r| = 1$ **65. (a)** $\text{Comp}_{u_r} a = \frac{a \cdot u_r}{|u_r|} = a \cdot u_r$ From Exercise 64, since $u_r \cdot u_r = 1$ and $u_r \cdot u_\theta = 0$, $a \cdot u_r = \frac{d^2r}{dt^2} - r\left(\frac{d\theta}{dt}\right)^2$

(b) Similarly, $\text{Comp}_{u_\theta} a = a \cdot u_\theta = r\frac{d^2\theta}{dt^2} + 2\frac{dr}{dt}\frac{d\theta}{dt}$

67. $v = (-2t\sin t^2)u_r + (2t + 2t\cos t^2)u_\theta$; $a = -2(\sin t^2 + 2t^2 + 4t^2\cos t^2)u_r + 2(1 + \cos t^2 - 4t^2\sin t^2)u_\theta$

69. Maximum height = 2040.816327; Range = 14139.19027 The speed at impact is equal to the initial speed.

71. Maximum height = 7485.001135; Range = 25122.64675

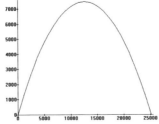

Exercise Set 12.6

1. $\dfrac{\left(x+\frac{pe^2}{1-e^2}\right)^2}{\left[\frac{e^2p^2}{(1-e^2)^2}\right]} + \dfrac{y^2}{\left(\frac{e^2p^2}{1-e^2}\right)} = 1$ with $a^2 = \dfrac{e^2p^2}{(1-e^2)^2}$ and $b^2 = \dfrac{e^2p^2}{1-e^2}$. 3. (a) $\mathbf{u}_r \cdot \mathbf{u}_\theta = \langle\cos\theta, \sin\theta, 0\rangle \cdot \langle-\sin\theta, \cos\theta, 0\rangle = 0$

(b) $\mathbf{u}_r \times \mathbf{u}_\theta = \langle 0, 0, 1\rangle = \mathbf{k}$ (c) $\mathbf{r} = r\mathbf{u}_r$; $\mathbf{v} = \dfrac{d\mathbf{r}}{dt} = r\dfrac{d\mathbf{u}_r}{dt} + \dfrac{dr}{dt}\mathbf{u}_r$; $\dfrac{d\mathbf{u}_r}{dt} = \mathbf{u}_\theta \dfrac{d\theta}{dt}$ and $\dfrac{d\mathbf{u}_\theta}{dt} = -\mathbf{u}_r \dfrac{d\theta}{dt}$ $\therefore \mathbf{v} = \dfrac{dr}{dt}\mathbf{u}_r + r\dfrac{d\theta}{dt}\mathbf{u}_\theta$

5. (a) We know that $\mathbf{c} = \mathbf{r} \times \mathbf{v}$ is directed along the positive z-axis (by choice). Hence, $\mathbf{v} \times \mathbf{c}$ is perpendicular to the z-axis. $\therefore \mathbf{v} \times \mathbf{c}$ lies in the xy-plane. (b) Since $\mathbf{d} = \mathbf{v} \times \mathbf{c} - GM\mathbf{u}_r$ (by Equation 12.37) and both $\mathbf{v} \times \mathbf{c}$ and $\mathbf{u}_r$ lie in the xy-plane, so also does $\mathbf{d}$.

7. (a) $\dfrac{dA}{dt} = \dfrac{c}{2}$; $\therefore A(t) = \dfrac{c}{2}t + C_1$. Since $A(0) = 0, C_1 = 0$. Let $A = A(T)$ be the area swept out in one revolution. Then, $A = \dfrac{1}{2}cT$. (b) $b = a\sqrt{1-e^2}$; hence, $\pi a^2\sqrt{1-e^2} = \dfrac{1}{2}cT$. $T = \dfrac{2\pi a^2}{c}\sqrt{1-e^2}$. (c) $a = \dfrac{ep}{1-e^2} \Rightarrow 1 - e^2 = \dfrac{ep}{a}$; thus, $T^2 = \dfrac{4\pi^2 a^4}{c^2}\left(\dfrac{ep}{a}\right) = \dfrac{4\pi^2 ep}{c^2} a^3$. Since $ep = \dfrac{c^2}{GM}$, we can also write $T^2 = \dfrac{4\pi^2}{GM}a^3$ 9. (a) $r_0 = $ minimum distance of planet from the sun when $\theta = 0$; $r_0 = \dfrac{ep}{1+e\cos 0} = \dfrac{ep}{1+e}$ (b) $|\mathbf{v} \times \mathbf{c}| = |\mathbf{v}||\mathbf{c}|\sin\dfrac{\pi}{2} = |\mathbf{v}||\mathbf{c}| = vc$, since the angle between $\mathbf{v}$ and $\mathbf{c}$ is $\dfrac{\pi}{2}$. Also, $\mathbf{v} \times \mathbf{c} = GM\mathbf{u}_r + \mathbf{d}$. At $\theta = 0$, $\mathbf{u}_r$ and $\mathbf{d}$ are collinear, and $|GM\mathbf{u}_r + \mathbf{d}| = GM + d$. Thus, $v_0 = \dfrac{(GM+d)}{c}$. (c) Since $d = GMe$, we can write $v_0 = \dfrac{GM(1+e)}{c} = c\left(\dfrac{GM}{c^2}\right)(1+e) = \dfrac{c(1+e)}{ep} = \dfrac{c}{r_0}$ (since $ep = \dfrac{c^2}{GM}$)

Chapter Review

1. (a) $t^3 \ln t - 2t^3 + 3 - 2t - t^2$ (b) $\langle e^{2t}, e^{3t}, 1 - e^t\rangle$ (c) $\langle t^2 - t\ln t, t^3 + 2, -2 - 2t\rangle$
(d) $\langle 0, -2, 4\rangle$ (e) $\langle 3, -3, -3\rangle$ 3. (a) $\left\langle \sin^{-1} t, -\sqrt{1-t^2}, \ln\dfrac{\sqrt{1+t}}{1-t}\right\rangle + \langle C_1, C_2, C_3\rangle$
(b) $\left(\dfrac{\pi}{6} + \dfrac{\sqrt{3}}{8}\right)\mathbf{i} - \dfrac{5}{24}\mathbf{j} + \left(\sqrt{3} - \dfrac{\pi}{3}\right)\mathbf{k}$ 5. (a) $4(\ln 2)\mathbf{i} + \mathbf{j} + 4\mathbf{k}$ (b) 5 7. 30 9. (a) 6
(b) $\mathbf{R}(u) = \left\langle \ln\dfrac{1+u^2}{2u}, 2\tan^{-1}\dfrac{1}{u}, -\sqrt{3}\ln u\right\rangle$; $e \geq u \geq e^{-2}$ (c) 6 11. $\mathbf{T} = \left\langle \dfrac{2}{3}[\cos u(s) + \sin u(s)], \dfrac{2}{3}[-\sin u(s) + \cos u(s)], \dfrac{1}{3}\right\rangle$ with $u(s) = \ln\left(\dfrac{1+s}{3}\right)$; $\mathbf{N} = \left\langle \dfrac{\cos u(s) - \sin u(s)}{\sqrt{2}}, \dfrac{-\cos u(s) - \sin u(s)}{\sqrt{2}}, 0\right\rangle$; $\kappa = \dfrac{2\sqrt{2}}{3(s+3)}$; $\mathbf{T}(0) = \left\langle\dfrac{2}{3}, \dfrac{2}{3}, \dfrac{1}{3}\right\rangle$, $\mathbf{N}(0) = \left\langle\dfrac{1}{\sqrt{2}}, -\dfrac{1}{\sqrt{2}}, 0\right\rangle$, $\kappa(0) = \dfrac{2\sqrt{2}}{9}$

13. $\mathbf{v}(t) = \langle e^{-t}(2\cos 2t - \sin 2t), e^{-t}(-2\sin 2t - \cos 2t), -2e^{-t}\rangle$; $\mathbf{a}(t) = \langle e^{-t}(-3\sin 2t - 4\cos 2t), e^{-t}(-3\cos 2t + 4\sin 2t), 2e^{-t}\rangle$; $\dfrac{ds}{dt} = 3e^{-t}$

15. (a) $v \approx 28,100$ km/hr (b) $h \approx 1670$ km; $v = \dfrac{2\pi(8110)}{2} \approx 25,500$ km/hr

17. (a) $\kappa = |\sin x|$ (b) $\kappa = \dfrac{8}{5\sqrt{10}}$ 19. $\mathbf{a}(t) = \left\langle\dfrac{-1}{(t+1)^2}, 0, 2\right\rangle$; $a_\mathbf{T} = \dfrac{-1}{(t+1)^2} + 2$; $a_\mathbf{N} = \dfrac{2}{t+1}$ 21. The greatest distance occurs when $\theta = \pi$. Then, $\mathbf{u}_r = -\mathbf{i}$ and $\mathbf{d} = d\mathbf{i}$. $r_m = \dfrac{ep}{1-e}$; $v_m = \dfrac{|\mathbf{v}\times\mathbf{c}|}{c} = \dfrac{|GM\mathbf{u}_r+\mathbf{d}|}{c} = \dfrac{|(-GM+d)\mathbf{i}|}{c} = \dfrac{GM-d}{c} = \dfrac{GM-GMe}{c} = \dfrac{GM}{c}(1-e)$; $c = \sqrt{GMep} = \sqrt{GMa(1-e^2)}$; and $v_m = \dfrac{GM(1-e)}{\sqrt{GMa(1-e^2)}} = \sqrt{\dfrac{GM}{a}}\sqrt{\dfrac{1-e}{1+e}} = \sqrt{\dfrac{4\pi^2 a^3}{T^2 a}}\sqrt{\dfrac{1-e}{1+e}} = \dfrac{2\pi a}{T}\sqrt{\dfrac{1-e}{1+e}}$; $v_m \approx 65,620$ mi/hr.

CHAPTER 13

Exercise Set 13.1

1. (a) $\dfrac{3}{2}$ 3. (a) 1 5. (a) 0 (b) a 7. (a) $\dfrac{2}{3(-3+h)}$ 9. $1 + \tan 2t$ 11. (a) domain $= \{(x, y) : y \neq x^2\}$
(b) 0 (b) 3 (c) $\dfrac{x^2 + 2x\Delta x + (\Delta x)^2 - y}{x + \Delta x - y^2}$ (b) $\dfrac{-4}{3(3+k)}$
(c) $\dfrac{3}{22}$ (c) 2 (d) $\dfrac{x^2 - y - \Delta y}{x - y^2 - 2y\Delta y - (\Delta y)^2}$
(d) $\dfrac{1}{4}$ (d) 0

(b) domain 13. domain 15. domain $= \{(x, y) :$ 17. domain $= \{(x, y) :$
$= \{(x, y) : y \neq \pm x\}$ $= \{(x, y) : xy \geq 0\}$ $x^2 + y^2 \neq 0\}$ $x^2 > y^2\}$

  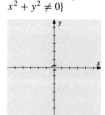 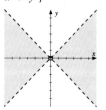

19. domain = $\{(x, y) : x^2 > 2y\}$

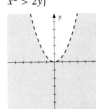

21. domain = $\{(x, y) : xy > 0\}$

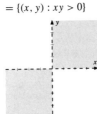

23. domain = $\{(x, y, z) : x \neq 1, y \neq -2, z \neq 3\}$ = all of $\mathbb{R}^3$ except the planes $x = 1$, $y = -2$, and $z = 3$

25. domain = $\{(x, y, z); x + y - z > 0\}$ = all points in $\mathbb{R}^3$ below the plane $z = x + y$

27. $x^2 + 2x\Delta x + (\Delta x)^2 - 3(xy + x\Delta y + y\Delta x + \Delta x \Delta y) + 2(y^2 + 2y\Delta y + (\Delta y)^2) = x^2 - 3xy + 2y^2 + (2x - 3y)\Delta x + (4y - 3x)\Delta y + (\Delta x - 3\Delta y)\Delta x + (2\Delta y)\Delta y$
Then, $\varepsilon_1 = \Delta x - 3\Delta y$ and $\varepsilon_2 = 2\Delta y$. We can take $g(x, y) = x - 3y$, $h(x, y) = 2y$

29. domain = $\{(x, y) : |x| + |y| < 1\}$

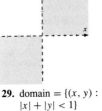

31. $C(r, h, p) = \pi r p (5r + 2h)$

33.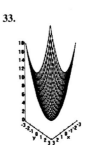

35.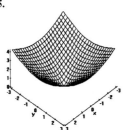

37.

39.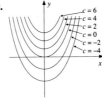

41.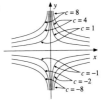

43.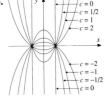

45.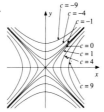

47.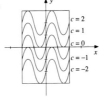

49. Circles of radius $\frac{k}{c}$. $x^2 + y^2 = \frac{625}{9}$ or circles of radius $\frac{25}{3}$.

51. $p = \frac{dv^2}{k}$.

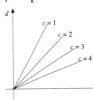

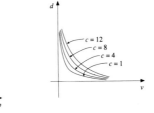

53.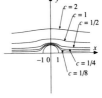

55. (a)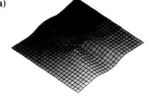

(b)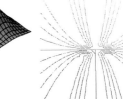

Answers   1301

(c) 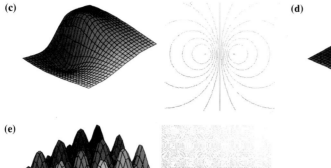   **(d)**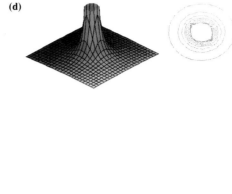

(e)

Exercise Set 13.2

1. 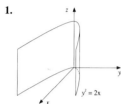 **3.** 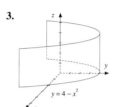 **5.** 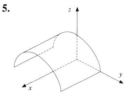 **7.**

9.   **11.** Sphere, center (1,3,4), radius 4 **13.** Degenerate sphere: the point $(-5, 1, -3)$ **15.** Sphere, center $(3, -4, 4)$, radius $2\sqrt{2}$

 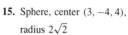

17. Elliptic hyperboloid of one sheet **19.** Elliptic cone **21.** Elliptic paraboloid **23.** One nappe of an elliptic cone

25. domain $= \mathbb{R}^2$, range $= \mathbb{R}$; plane **27.** domain $= \mathbb{R}^2$, range $= \mathbb{R}$; plane **29.** domain $= \{(x, y) : x \geq 0\}$, range $= \{z : z \geq 0\}$; half a parabolic cylinder **31.** domain $= \mathbb{R}^2$, range $= \{z : z \leq 9\}$; parabolic cylinder

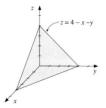 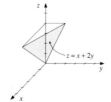 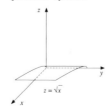 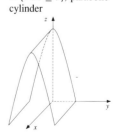

33. domain = $\mathbb{R}^2$, range = $\{z : z \geq 0\}$; paraboloid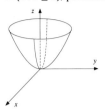

35. domain = $\{(x, y) : 4x^2 - 3y^2 \leq 12\}$, range = $\{z : z \geq 2\sqrt{3}\}$; upper half of a hyperboloid of one sheet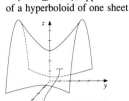

37. domain = $\mathbb{R}^2$, range = $\{z : z \geq 1\}$; paraboloid of revolution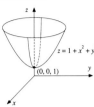

39. domain = $\{(x, y) : x^2 + y^2 \geq 4\}$, range = $\{z : z \geq 0\}$; upper half of a hyperboloid of one sheet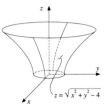

41. domain = $\{(x, y) : x^2 \leq 4y\}$, range = $\{z : z \geq 0\}$; elliptic paraboloid, upper half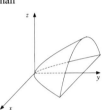

43. domain = $\mathbb{R}^2$, range = $\mathbb{R}$; hyperbolic paraboloid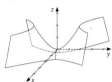

45. $\frac{(z+2)^2}{4} - \frac{(x-1)^2}{2} - \frac{(y-2)^2}{8} = 1$; hyperboloid of two sheets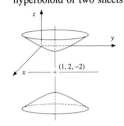

47. $4(x-3)^2 + (y-2)^2 = 4(z+5)$; elliptic paraboloid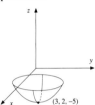

49. $z - 4 = \frac{(y+2)^2}{4} - \frac{(x-2)^2}{6}$; hyperbolic paraboloid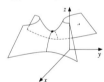

51.

53.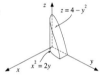

55.

Exercise Set 13.3

1. 13   **3.** −8   **5.** 3   **7.** 1   **9.** 0   **11.** 0   **13.** If $y = x$, the limit is $-\frac{1}{2}$, but if $y = 2x$, the limit is -2 **15.** Along the x-axis $f(x, y) = f(x, 0) = \frac{x}{x^2} = \frac{1}{x}$ if $x \neq 0$, and $\lim_{x \to 0} f(x, 0)$ does not exist.   **17.** Consider $x = 0$ and $x = y^2$. The respective limits are 0 and $\frac{1}{2}$.   **19.** Continuous on all of $\mathbb{R}^2$ **21.** Continuous on $\mathbb{R}^2 - \{(x, y) : y = x\}$ **23.** Continuous on $\mathbb{R}^2 - \{(x, y) : y = -x\}$ **25.** Continuous on $\mathbb{R}^2 - \{(x, y) : |y| = |x|\}$ **27.** Continuous on $\mathbb{R}^2$ **29.** Continuous on $\mathbb{R}^2 - \{(0, 0)\}$ **31.** f is continuous at $(0, 0)$ **33.** f is continuous at $(0, 0)$ **35.** f is continuous at $(0, 0)$ **37.** Let $\varepsilon > 0$. Let $P = (x, y)$. Choose δ_1 and δ_2 such that $|f(P) - L| < \frac{\varepsilon}{2}$ if $0 < |\overrightarrow{P_0 P}| < \delta_1$ and $|g(P) - M| < \frac{\varepsilon}{2}$ if $0 < |\overrightarrow{P_0 P}| < \delta_2$. Let $\delta = \min(\delta_1, \delta_2)$. Then, if $0 < |\overrightarrow{P_0 P}| < \delta$, $|(f + g)(P) - (L + M)| = |(f(P) - L) - (g(P) - M)| \leq |f(P) - L| + |g(P) - M| < \frac{\varepsilon}{2} + \frac{\varepsilon}{2} = \varepsilon$. Thus, $\lim_{P \to P_0} (f + g)(P) = L + M$.   **39.** No, the limit does not exist.

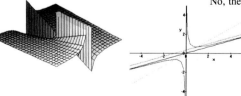

Exercise Set 13.4

1. $f_x = 2x$, $f_y = 2y$ **3.** $f_x = y \cos xy$, $f_y = x \cos xy$ **5.** $f_x = \frac{-2y}{(x-y)^2}$, $f_y = \frac{2x}{(x-y)^2}$ **7.** $f_x = \ln y - \frac{y}{x}$, $f_y = \frac{x}{y} - \ln x$ **9.** $f_x = \frac{-y^3}{(x^2-y^2)^{3/2}}$, $f_y = \frac{x^3}{(x^2-y^2)^{3/2}}$ **11.** $\frac{\partial z}{\partial x} = \frac{x}{(1-x^2-y^2)^{3/2}}$, $\frac{\partial z}{\partial y} = \frac{y}{(1-x^2-y^2)^{3/2}}$ **13.** $\frac{\partial z}{\partial x} = -\tan(x - y)$, $\frac{\partial z}{\partial y} = \tan(x - y)$ **15.** $\frac{\partial z}{\partial x} = \frac{-y}{\sqrt{1-x^2 y^2}}$, $\frac{\partial z}{\partial y} = \frac{-x}{\sqrt{1-x^2 y^2}}$ **17.** $f_x(3, -2) = -12$, $f_y(3, -2) = 17$ **19.** $f_r\left(0, \frac{\pi}{3}\right) = \frac{1}{2}$, $f_\theta\left(0, \frac{\pi}{3}\right) = -\frac{\sqrt{3}}{2}$ **21.** $\left.\frac{\partial w}{\partial s}\right|_{(3,-2)} = -\frac{1}{2}$, $\left.\frac{\partial w}{\partial t}\right|_{(3,-2)} = -1$ **23.** $\frac{\partial z}{\partial x} = \frac{-y}{x^2+y^2}$, $\frac{\partial z}{\partial y} = \frac{x}{x^2+y^2}$; $x\frac{\partial z}{\partial y} - y\frac{\partial z}{\partial x} = \frac{x^2+y^2}{x^2+y^2} = 1$ **25.** $f_x = 4x - 3y^2$, $f_y = -6xy + 9y^2$;

$f_x(3, 2) = 12 - 12 = 0 =$ slope of tangent to C_1 at $(3, 2, 6)$; $f_y(3, 2) = -6(6) + 9(4) = 0 =$ slope of tangent to C_2 at $(3, 2, 6)$

27. $f_{xx} = \frac{2(3y^2-x^2)}{(x^2+3y^2)^2}$, $f_{yy} = \frac{6(x^2-3y^2)}{(x^2+3y^2)^2}$, $f_{xy} = f_{yx} = \frac{-12xy}{(x^2+3y^2)^2}$ **29.** $f_{xx} = 2ye^{x^2y}(2x^2y + 1)$, $f_{yy} = x^4e^{x^2y}$, $f_{xy} = f_{yx} = 2xe^{x^2y}(x^2y + 1)$

31. $f_{xx} = e^{-x}\sin y - e^{-y}\cos x$, $f_{yy} = -e^{-x}\sin y + e^{-y}\cos x$, $f_{xy} = f_{yx} = -e^{-x}\cos y + e^{-y}\sin x$

33. $u_{xx} = -18(\cos 3x)e^{-9kt}$, $u_t = -18k(\cos 3x)e^{-9kt} = ku_{xx}$ **35.** $y_{xx} = -\sin x \cos at$, $y_{tt} = -a^2 \sin x \cos at = a^2 y_{xx}$

37. $u_{xx} = -\cos x \cosh y$, $u_{yy} = \cos x \cosh y$, $u_{xx} + u_{yy} = 0$, u is harmonic   **39.** $u_{xx} = \frac{2xy}{(x^2+y^2)^2}$, $u_{yy} = \frac{-2xy}{(x^2+y^2)^2}$, $u_{xx} + u_{yy} = 0 \Rightarrow u$
harmonic   **41.** v is harmonic since $v_{xx} = \sin x \sinh y$, $v_{yy} = -\sin x \sinh y$; $v_{xx} + v_{yy} = 0$. $u_x = -\sin x \cosh y = v_y$ and
$u_y = \cos x \sinh y = -v_x$; u and v are harmonic conjugates.   **43.** In Exercise 42, $\tan^{-1}\frac{y}{x}$ and $\ln\sqrt{x^2+y^2}$ were proved to be harmonic
conjugates. Here, the sign of one of the functions has been changed. Suppose that $u_x = v_y$ and $u_y = -v_x$, with the original function v.
If v is replaced by $w = -v$, we have $u_x = v_y = -w_y$ and $u_y = -v_x = w_x$. But, these equations can be written as $w_x = u_y$ and
$w_y = -u_x$. $\therefore -v$ and u are harmonic conjugates if u and v are.

45. (a) $u_{xx} = 2$; $u_{yy} = -2$; $u_{xx} + u_{yy} = 0 \Rightarrow u$ harmonic.   (d)
$v_{xx} = 0$; $v_{yy} = 0$; $v_{xx} + v_{yy} = 0 \Rightarrow v$ harmonic.
(b) $u_x = v_y$ and $u_y = -v_x \Rightarrow u$ and v are harmonic conjugates
(c) Level curves for u: $x^2 - y^2 + 2x + 1 = c_1$;
Level curves for v: $2xy + 2y = c_2$;
slope of $u(x, y) = m_1 = \frac{x+1}{y}(y \neq 0)$,
slope of $v(x, y) = m_2 = \frac{-y}{x+1}(x \neq -1)$
$m_1m_2 = -1 \Rightarrow$ level curves of $u \perp$ level curves of v (excluding
the lines $x = -1$ and $y = 0$)

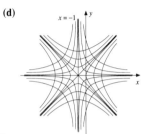

47. (a) $f_{xxy} = \frac{-z}{x^2} = f_{xyx} = f_{yxx}$ (b) $f_{xyy} = f_{yxy} = f_{yyx} = \frac{-z}{y^2}$ (c) $f_{xzz} = f_{zxz} = f_{zzx} = \frac{-y}{z^2}$

(d) $f_{xyz} = f_{yxz} = f_{xzy} = f_{zxy} = f_{yzx} = f_{zyx} = \frac{1}{z} + \frac{1}{y} + \frac{1}{x}$ **49.** (a) \$20 (b) \$30 (c) \$2

51. $Q_{KK} = \alpha(\alpha-1)AK^{\alpha-2}L^{1-\alpha}$; $Q_{LL} = (1-\alpha)(-\alpha)AK^\alpha L^{-\alpha-1}$; $K^2 Q_{KK} = \alpha(\alpha-1)AK^\alpha L^{1-\alpha} = L^2 Q_{LL}$

53. $f_x(0, 0) = \lim\limits_{h \to 0} \frac{f(h,0)-f(0,0)}{h} = \lim\limits_{h \to 0} \frac{h^3/h^2-0}{h} = 1$; $f_y(0, 0) = \lim\limits_{k \to 0} \frac{f(0,k)-f(0,0)}{k} = \lim\limits_{k \to 0} \frac{-k^3/k^2-0}{k} = -1$

55. $f_x = \lim\limits_{h \to 0} \frac{\frac{hy(h^2-y^2)}{h^2+y^2}}{h} = \lim\limits_{h \to 0} \frac{y(h^2-y^2)}{h^2+y^2} = -y$; $f_{xy}(0, 0) = -1$; $f_y = \lim\limits_{k \to 0} \frac{\frac{xk(x^2-k^2)}{x^2+k^2}}{k} = \lim\limits_{k \to 0} \frac{x(x^2-k^2)}{x^2+k^2} = x$; $f_{yx}(0, 0) = 1$

57. (a) slope is 2   (b) slope is 2

Chapter Review

1. (a) domain $= \{(x, y) : x^2 + y^2 > 1\}$ (b) domain $= \{(x, y): x^2 > y^2\}$ **3.** (a) parabolic cylinder   (b) sphere

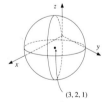

5. (a) hyperbolic paraboloid   (b) cylinder   **7.** (a) domain $= \mathbb{R}^2$, range $= \{z : z \geq 2\}$, half a hyperboloid of two sheets   (b) domain $= \mathbb{R}^2$, range $= \{z : z \leq 4\}$, paraboloid of revolution

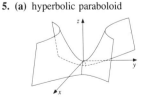
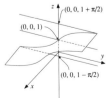

9. $T_y(1, 3) = -300e^{-5}$, decreasing

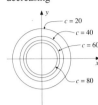

11. (a) $f_x = \frac{1}{x}\sinh\frac{2x}{y} - \frac{y}{x^2}\cosh^2\frac{x}{y}$,
$f_y = -\frac{1}{y}\sinh\frac{2x}{y} + \frac{1}{x}\cosh^2\frac{x}{y}$
(b) $f_x = \frac{1}{\sqrt{2x(y^2-2x)}}$,
$f_y = \frac{2x}{y\sqrt{2x(y^2-2x)}}$

13. $u_{xx} = -\sin x \cosh y$, $u_{yy} = \sin x \cosh y$, and $u_{xx} + u_{yy} = 0 \Rightarrow u$ harmonic
$v_{xx} = -\cos x \sinh y$, $v_{yy} = \cos x \sinh y$, $v_{xx} + v_{yy} = 0 \Rightarrow v$ harmonic $u_x = \cos x \cosh y = v_y$ and $u_y = \sin x \sinh y = -v_x \Rightarrow u$ and v harmonic conjugates

15. (a) $\frac{\partial^2 w}{\partial z^2} = \frac{-x^2-y^2}{(z^2-x^2-y^2)^{3/2}}$ **(b)** $\frac{\partial^2 w}{\partial x \partial z} = \frac{xz}{(z^2-x^2-y^2)^{3/2}}$ **(c)** $\frac{\partial^2 w}{\partial y \partial z} = \frac{yz}{(z^2-x^2-y^2)^{3/2}}$ **(d)** $\frac{\partial^2 w}{\partial y \partial x} = \frac{-xy}{(z^2-x^2-y^2)^{3/2}}$

17. $|f(x, y) - 0| \le \frac{x^2|x+3y|}{x^2+2y^2} \le \frac{(x^2+y^2)(|x|+3|y|)}{(x^2+y^2)} \le 4\sqrt{x^2+y^2} < \varepsilon$ if $\sqrt{x^2+y^2} < \frac{\varepsilon}{4}$. Thus, $\lim_{(x,y)\to(0,0)} f(x, y) = 0$. Assigning 0 to $f(0, 0)$.

19. (a) $f(x, 0) = \frac{x^3}{3x^3} = \frac{1}{3}$ if $x \ne 0$; $f(0, y) = \frac{-2y^3}{4y^3} = -\frac{1}{2}$ if $y \ne 0$ **(b)** $f(x, 0) = 0$, $f(y^2, y) = \frac{3}{7}$ if $y \ne 0$

21. $\frac{\partial P}{\partial V} = -\frac{nRT}{V^2}$; $\frac{\partial V}{\partial T} = \frac{nR}{P}$; $\frac{\partial T}{\partial P} = \frac{V}{nR}$; $\frac{\partial P}{\partial V}\frac{\partial V}{\partial T}\frac{\partial T}{\partial P} = -1$ **23. (a)** $y_{xx} = -\frac{2a^2t^2+2x^2}{(x^2-a^2t^2)^2}$, $y_{tt} = \frac{-2a^2(x^2-a^2t^2)-4a^4t^2}{(x^2-a^2t^2)^2} = a^2 y_{xx}$

(b) $y_{xx} = Ae^x e^{at} + Be^x e^{-at}$; $y_{tt} = Aa^2 e^x e^{at} + Ba^2 e^x e^{-at} = a^2 y_{xx}$ **25.** $Q_K(240, 150) = 18\left(\frac{150}{240}\right)^{0.4} \approx 14.92$, $Q_L(240, 150) = 12\left(\frac{240}{150}\right)^{0.6} \approx 15.91$; Q_K represents the marginal change in Q for a unit change in K. Q_L represents the marginal change in Q for a unit change in L. **27. (a)** $f_x(0, 0) = \lim_{h\to 0}\frac{2h^3/h^2}{h} = 2$ **(b)** $f_x(x, y) = \frac{(x^2+2y^2)(6x^2)-(2x^3-y^3)(2x)}{(x^2+2y^2)^2} = \frac{2x^4+12x^2y^2+2xy^3}{(x^2+2y^2)^2}$ for $(x, y) \ne (0, 0)$ **(c)** $\lim_{(x,0)\to(0,0)} f_x(x, 0) = 2$, but $\lim_{(x,x)\to(0,0)} f_x(x, x) = \frac{16}{9}$ ∴ $\lim_{(x,y)\to(0,0)} f_x(x, y)$ does not exist.

(d) $f_y(0, 0) = \lim_{k\to 0}\frac{-k^3/(2k^2)}{k} = -\frac{1}{2}$; $f_y(x, y) = \frac{(x^2+2y^2)(-3y^2)-(2x^3-y^3)(4y)}{(x^2+2y^2)^2} = \frac{-2y^4-3x^2y^2-8x^3y}{(x^2+2y^2)^2}$; $\lim_{(0,y)\to(0,0)} f_y(0, y) = -\frac{1}{2}$ but $\lim_{(y,y)\to(0,0)} f_y(y, y) = \frac{-13}{9}$ ∴ $\lim_{(x,y)\to(0,0)} f_y(x, y)$ does not exist. **29.** Suppose that S is open, and let $P \in S$. Then P is not a boundary point of S, and so some neighborhood of P contains only points of S. Conversely, let S be any nonempty set with the given property. Then no point of S is a boundary point of S, and so S is open.

CHAPTER 14

Exercise Set 14.1

1. $f_x = 3x^2 + 6xy$; $f_y = 3x^2 - 8y^3$; these are continuous on $\mathbb{R}^2$ **3.** $f_x = e^x \sin y$; $f_y = e^x \cos y$; these are continuous on all of $\mathbb{R}^2$
5. $f_x = \frac{1}{x} - \frac{2x}{x^2+y^2}$; $f_y = \frac{1}{y} - \frac{2y}{x^2+y^2}$. For $xy > 0$, both f_x and f_y are continuous. **7.** $(4x - 3y)dx + (3y^3 - 3x)dy$
9. $ye^{xy}dx + xe^{xy}dy$ **11.** $\frac{2y}{(x+y)^2}dx - \frac{2x}{(x+y)^2}dy$ **13.** $\frac{y}{x^2+y^2}dx - \frac{x}{x^2+y^2}dy$ **15.** $\frac{x}{\sqrt{x^2+y^2+z^2}}dx + \frac{y}{\sqrt{x^2+y^2+z^2}}dy + \frac{z}{\sqrt{x^2+y^2+z^2}}dz$
17. $df = 0.58$ **19.** $df = 0.012$ **21.** $df = -0.0044$ **23.** $\Delta V \approx dV = 51.6$ cu ft; $\approx 2.56\%$ **25.** $\Delta T \approx dT = 0.20$; Actual change = 0.2059; Approximate percentage change $\approx 25.7\%$ **27.** $\Delta C \approx dC = \$54.10$; Approximate percentage increase in cost $\approx 8.78\%$ **29.** Approximate percentage change $\approx 1.60\%$ **31.** $\Delta f = y\Delta x + x\Delta y + \Delta x \Delta y = f_x \Delta x + f_y \Delta y + E_1 \Delta x + E_2 \Delta y$, where $E_1 = \Delta y$, $E_2 = 0$ Then, $E_1 \to 0$ and $E_2 \to 0$ as $(\Delta x, \Delta y) \to (0, 0)$ **33.** $\Delta f = [(x + \Delta x) - 3y(y + \Delta y)]^2 - (x - 3y)^2 = 2(x - 3y)\Delta x - 6(x - 3y)\Delta y + (\Delta x - 6\Delta y)\Delta x + 9\Delta y(\Delta y) = f_x \Delta x + f_y \Delta y + E_1 \Delta x + E_2 \Delta y$, where $E_1 = \Delta x - 6\Delta y$, $E_2 = 9\Delta y$ Then, $E_1 \to 0$ and $E_2 \to 0$ as $(\Delta x, \Delta y) \to (0, 0)$ **35.** $f_x(0, 0) = \lim_{h\to 0}\frac{f(h,0)-f(0,0)}{h} = \lim_{h\to 0}\frac{0}{h} = 0$ Similarly, $f_y(0, 0) = 0$. For $(x, y) \ne (0, 0)$, $f_x(x, y) = \frac{2xy^4}{(x^2+y^2)^2}$ and $f_y(x, y) = \frac{2x^4y}{(x^2+y^2)^2}$. If $(x, y) \ne (0, 0)$, then f_x and f_y are continuous. To show continuity at $(0, 0)$, let $\varepsilon > 0$ be given. $|f_x(x, y) - f_x(0, 0)| = \frac{|2xy^4|}{(x^2+y^2)^2} \le \frac{2\sqrt{x^2+y^2}(x^2+y^2)^2}{(x^2+y^2)^2} < \varepsilon$ if $\sqrt{x^2+y^2} < \frac{\varepsilon}{2}$. Hence, f_x is continuous at $(0, 0)$; so is f_y. Thus, f_x and f_y are continuous on all of $\mathbb{R}^2$, and so, by Theorem 14.2, f is differentiable on $\mathbb{R}^2$.

37. (a) $\lim_{(x,y)\to(0,0)} f(x, y) = \lim_{(x,y)\to(0,0)} |\sqrt{|xy|} - 0| = \sqrt{|x||y|} \le \sqrt{x^2+y^2} < \varepsilon$ if $\sqrt{x^2+y^2} < \delta = \varepsilon$. Since $\lim_{(x,y)\to(0,0)} f(x, y)$ exists and equals $f(0, 0)$, f is continuous at $(0, 0)$. **(b)** $f_x(0, 0) = \lim_{h\to 0}\frac{f(h,0)-f(0,0)}{h} = \lim_{h\to 0}\frac{0-0}{h} = 0$ By symmetry $f_y(0, 0) = 0$ **(c)** $\Delta f = \sqrt{|\Delta x \Delta y|} - 0$. Take $\Delta x = \Delta y$, to get $\Delta f = |\Delta x| = 0 + 0 + (E_1 + E_2)\Delta x$. If $\Delta x > 0$, $E_1 + E_2 = 1$; if $\Delta x < 0$, $E_1 + E_2 = -1$. Since E_1 and E_2 do not go to zero as $(\Delta x, \Delta y) \to (0, 0)$, f is not differentiable at $(0, 0)$.

Exercise Set 14.2

1. 56   **3.** 2   **5.** $-(\tan t + \cot t)$ **7.** $1 - \cos t^2 + 2t^2 \sin t^2$ **9.** $\frac{\partial z}{\partial u}\Big|_{(2,-1)} = -4$; $\frac{\partial z}{\partial v}\Big|_{(2,-1)} = 8$

11. $\frac{\partial z}{\partial u}\Big|_{(1,1/2)} = 2e^2$; $\frac{\partial z}{\partial v}\Big|_{(1,1/2)} = 0$ **13.** $\frac{\partial z}{\partial u} = 2v \tan 2uv$; $\frac{\partial z}{\partial v} = 2u \tan 2uv$ **15.** $\frac{\partial z}{\partial u} = -4u$; $\frac{\partial z}{\partial v} = 6v$ **17.** $\frac{3x^2 - 2y^2}{4y(x-y^2)}$

19. $\frac{\sin y - y \cos x}{\sin x - x \cos y}$ **21.** $\frac{2x - x^2 y + y^3}{2y - xy^2 + x^3}$ **23.** $\frac{\partial z}{\partial x} = \frac{-2xyz}{x^2 y + 6z}$; $\frac{\partial z}{\partial y} = \frac{4y - x^2 z}{6z + x^2 y}$ **25.** $\frac{\partial z}{\partial x} = \frac{2z(1 - 2xyz - 2y^2 z)}{(x+y)(1+4xyz)}$; $\frac{\partial z}{\partial y} = \frac{2z(1 - 2x^2 z - 2xyz)}{(x+y)(1+4xyz)}$

27. $\frac{\partial z}{\partial x} = \frac{-z \cos xz}{x \cos xz - y \tan xz}$; $\frac{\partial z}{\partial y} = \frac{z \tan yz}{x \cos xz - y \tan xz}$ **29.** $\frac{\partial w}{\partial x} = \frac{2yz(y^2 + z^2)}{(x^2 + y^2 + z^2)^{3/2}}$; $\frac{\partial w}{\partial y} = \frac{2xz(x^2 + z^2)}{(x^2 + y^2 + z^2)^{3/2}}$; $\frac{\partial w}{\partial z} = \frac{2xy(x^2 + y^2)}{(x^2 + y^2 + z^2)^{3/2}}$

31. $\frac{\partial z}{\partial r} = \frac{\partial z}{\partial x} \cos \theta + \frac{\partial z}{\partial y} \sin \theta$; $\frac{\partial z}{\partial \theta} = \frac{\partial z}{\partial x}(-r \sin \theta) + \frac{\partial z}{\partial y}(r \cos \theta)$; $\left(\frac{\partial z}{\partial r}\right)^2 + \frac{1}{r^2}\left(\frac{\partial z}{\partial \theta}\right)^2 = \left(\frac{\partial z}{\partial x}\right)^2 \cos^2 \theta + 2 \frac{\partial z}{\partial x} \frac{\partial z}{\partial y} \sin \theta \cos \theta$
$+ \left(\frac{\partial z}{\partial y}\right)^2 \sin^2 \theta + \left(\frac{\partial z}{\partial x}\right)^2 \sin^2 \theta - 2 \frac{\partial z}{\partial x} \frac{\partial z}{\partial y} \sin \theta \cos \theta + \left(\frac{\partial z}{\partial y}\right)^2 \cos^2 \theta = \left(\frac{\partial z}{\partial x}\right)^2 + \left(\frac{\partial z}{\partial y}\right)^2$ **33.** 0.68 km²/hr   **35.** Let

$u = \frac{x}{y}$; $\frac{\partial z}{\partial x} = f'(u)\left(\frac{1}{y}\right)$, $\frac{\partial z}{\partial y} = f'(u)\left(\frac{-x}{y^2}\right)$, and $x \frac{\partial z}{\partial x} + y \frac{\partial z}{\partial y} = \frac{x}{y} f'(u) - \frac{x}{y} f'(u) = 0$

37. $\frac{\partial z}{\partial u} = f_x g_u + f_y h_u$; $\frac{\partial z}{\partial v} = f_x g_v + f_y h_v$; $\frac{\partial^2 z}{\partial u^2} = g_u(f_{xx} g_u + f_{xy} h_u) + f_x g_{uu} + h_u(f_{yx} g_u + f_{yy} h_u) + f_y h_{uu}$
$= f_{xx} g_u^2 + 2 f_{xy} g_u h_u + f_{yy} h_u^2 + f_x g_{uu} + f_y h_{uu}$ (assuming that $f_{xy} = f_{yx}$). Similarly, $\frac{\partial^2 z}{\partial v^2} = f_{xx} g_v^2 + 2 f_{xy} g_v h_v + f_{yy} h_v^2 + f_x g_{vv} + f_y h_{vv}$

39. $\frac{\partial^2 z}{\partial \theta \partial r} = \left(\frac{\partial^2 z}{\partial y^2} - \frac{\partial^2 z}{\partial x^2}\right)(r \sin \theta \cos \theta) + \frac{\partial^2 z}{\partial y \partial x}(r \cos^2 \theta - r \sin^2 \theta) - \frac{\partial z}{\partial x} \sin \theta + \frac{\partial z}{\partial y} \cos \theta$ **41.** $\frac{dh}{dt} = \frac{9}{10\pi}$ ft/min

43. Let $w = f(x, y, z)$, where $x = x(t), y = y(t)$, and $z = z(t)$ Then the chain rule can be written as
$D_1(f)(x(t), y(t), z(t)) \left(\frac{\partial}{\partial t} x(t)\right) + D_2(f)(x(t), y(t), z(t)) \left(\frac{\partial}{\partial t} y(t)\right) + D_3(f)(x(t), y(t), z(t)) \left(\frac{\partial}{\partial t} z(t)\right)$

45. $\frac{\partial f}{\partial u} = D_1(f)(x(u, v), y(u, v)) \left(\frac{\partial}{\partial u} x(u, v)\right) + D_2(f)(x(u, v), y(u, v)) \left(\frac{\partial}{\partial u} y(u, v)\right)$ $\frac{\partial f}{\partial v} = D_1(f)(x(u, v), y(u, v)) \left(\frac{\partial}{\partial v} x(u, v)\right) +$
$D_2(f)(x(u, v), y(u, v)) \left(\frac{\partial}{\partial v} y(u, v)\right)$ For the special case: $\frac{\partial f}{\partial u} = e^{(u - 3uv + \sin(v))}(u + e^{u^2 v})$ $\frac{\partial f}{\partial v} = e^{(u - 3uv + \sin(v))}(u + e^{u^2 v})$

Exercise Set 14.3

1. $\frac{27}{5}$ **3.** $-\frac{9}{10\sqrt{2}}$ **5.** $-\frac{7}{5}$ **7.** $\frac{1}{\sqrt{5}}$ **9.** $-\frac{3}{2\sqrt{2}}$ **11.** $\frac{16}{3}$ **13.** $\frac{1}{\sqrt{2}}$ **15.** $\frac{82}{\sqrt{10}}$ **17.** $\mathbf{u} = \left\langle \frac{4}{\sqrt{41}}, -\frac{5}{\sqrt{41}} \right\rangle$; $\frac{\sqrt{41}}{27}$

19. $\mathbf{u} = \left\langle -\frac{1}{\sqrt{5}}, -\frac{2}{\sqrt{5}} \right\rangle$; $\sqrt{5}$ **21.** $\mathbf{u} = \langle 1, 0 \rangle$ or $\mathbf{u} = \left\langle -\frac{15}{17}, \frac{8}{17} \right\rangle$. f increases most rapidly if $\mathbf{u} = \left\langle \frac{1}{\sqrt{17}}, \frac{4}{\sqrt{17}} \right\rangle$. Maximum rate of change

$= 2\sqrt{17}$.   **23.** $-\frac{2}{5\sqrt{5}}$; $\left\langle \frac{4}{5}, \frac{3}{5} \right\rangle$; -1 **25.** $-\frac{14}{5\sqrt{13} \ln 4}$; $\mathbf{u} = \left\langle \frac{3}{\sqrt{10}}, \frac{1}{\sqrt{10}} \right\rangle$; $\frac{2\sqrt{10}}{5 \ln 4}$ **27.** $x^2 + y^2 = 4^c$, circles of radii between 1 and 2. y'

at $(x_0, y_0) = \frac{-x_0}{y_0}$, and a vector tangent to the curve is $\mathbf{u} = \left\langle 1, -\frac{x_0}{y_0} \right\rangle$. $\nabla V(x_0, y_0) = \frac{1}{\ln 4} \left\langle \frac{2x_0}{4^c}, \frac{2y_0}{4^c} \right\rangle$, and $\mathbf{u} \cdot \nabla V(x_0, y_0) = \frac{1}{\ln 4} \left(\frac{2x_0 - 2x_0}{4^c}\right) = 0$.
Hence, the gradient vector is orthogonal to the equipotential curve through (x_0, y_0).   **29.** $f(x, y) = \sin x - x \cos x + \sin y + 2$

31. $\nabla(cu) = \left\langle \frac{\partial}{\partial x} cu, \frac{\partial}{\partial y} cu \right\rangle = \left\langle c \frac{\partial u}{\partial x}, c \frac{\partial u}{\partial y} \right\rangle = c \left\langle \frac{\partial u}{\partial x}, \frac{\partial u}{\partial y} \right\rangle = c \nabla u$ **33.** $\nabla(uv) = \left\langle u \frac{\partial v}{\partial x} + v \frac{\partial u}{\partial x}, u \frac{\partial v}{\partial y} + v \frac{\partial u}{\partial y} \right\rangle = \left\langle u \frac{\partial v}{\partial x}, u \frac{\partial v}{\partial y} \right\rangle + \left\langle v \frac{\partial u}{\partial x}, v \frac{\partial u}{\partial y} \right\rangle =$
$u \left\langle \frac{\partial v}{\partial x}, \frac{\partial v}{\partial y} \right\rangle + v \left\langle \frac{\partial u}{\partial x}, \frac{\partial u}{\partial y} \right\rangle = u \nabla v + v \nabla u$ **35.** $\nabla u^\alpha = \left\langle \frac{\partial}{\partial x} u^\alpha, \frac{\partial}{\partial y} u^\alpha \right\rangle = \left\langle \alpha u^{\alpha-1} \frac{\partial u}{\partial x}, \alpha u^{\alpha-1} \frac{\partial u}{\partial y} \right\rangle = \alpha u^{\alpha-1} \nabla u$

37. $\nabla f = \langle 2, -2y \rangle$; the gradient at $(2, -1)$ is $\langle 2, 2 \rangle$.
The vector equation of the tangent line at $(1, -1)$
is $\mathbf{r}(t) = \left\langle 2 + \frac{1}{2} t \sqrt{2}, -1 - \frac{1}{2} t \sqrt{2} \right\rangle$

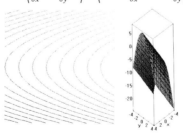

Exercise Set 14.4

1. Tangent plane: $\langle 6, -6, -5 \rangle \cdot \langle \mathbf{x} - (3, -2, 5) \rangle = 0$ or $6x - 5y - 5z = 5$; Normal line: $\mathbf{r}(t) = \langle 3 + 6t, -2 - 6t, 5 - 5t \rangle$.
3. Tangent Plane: $\langle 1, 0, -1 \rangle \cdot \langle \mathbf{x} - (1, 0, -2) \rangle = 0$, or $x - z = 3$; Normal Line: $\mathbf{r}(t) = \langle 1 + t, 0, -2 - t \rangle$
5. Tangent Plane: $\langle 3, 1, 4 \rangle \cdot \langle \mathbf{x} - (1, -1, 2) \rangle = 0$, or $3x + y + 4z = 10$; Normal Line: $\mathbf{r}(t) = \langle 1 + 3t, -1 + t, 2 + 4t \rangle$
7. Tangent Plane: $\langle -1, 0, 2 \rangle \cdot \langle \mathbf{x} - (3, -1, 1) \rangle = 0$, or $x - 2z = 1$; Normal Line: $\mathbf{r}(t) = \langle 3 - t, -1, 1 + 2t \rangle$
9. Tangent Plane: $\langle 10, -16, -1 \rangle \cdot \langle \mathbf{x} - (5, 4, -7) \rangle = 0$ or $10x - 16y - z + 7 = 0$; Normal Line: $\mathbf{r}(t) = \langle 5 + 10t, 4 - 16t, -7 - t \rangle$
11. Tangent Plane: $\langle 1, 1, -2 \rangle \cdot \left\langle \mathbf{x} - \left(1, -1, -\frac{\pi}{4}\right) \right\rangle = 0$ or $x + y - 2z = \frac{\pi}{2}$; Normal Line: $\mathbf{r}(t) = \left\langle 1 + t, -1 + t, -\frac{\pi}{4} - 2t \right\rangle$
13. Tangent Plane: $\langle 2, 0, 1 \rangle \cdot \left\langle \mathbf{x} - \left(0, \frac{\pi}{2}, -1\right) \right\rangle = 0$, or $2x + z + 1 = 0$; Normal Line: $\mathbf{r}(t) = \left\langle 2t, \frac{\pi}{2}, -1 + t \right\rangle$ **15.** $\left(\frac{1}{2}, -\frac{3}{4}, -\frac{19}{16}\right)$

17. (a) The surfaces have the same tangent plane at $P_0 = (x_0, y_0, z_0)$ if and only if their normal vectors are parallel at P_0, which is equivalent to $\nabla F(P_0) = k \nabla G(P_0)$. Note that neither vector is the zero vector, by assumption.   (b) $(2, 3, -1)$ and $(-2, -3, 1)$
19. $(0, 0, 0)$ and $(2, 1, 3)$ **21.** $\approx 6.34°; \approx 26.93°$ **23.** $-\frac{5}{6} + x + \frac{2}{3}y$ **25.** .2938577115 radians 16.83680664°

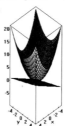

Exercise Set 14.5

1. Absolute minimum of -2 at $(-1, 2)$ **3.** Absolute minimum of -3 at $(0, 4)$ **5.** Local minimum at $(16, 4)$; saddle point at $(1, -1)$
7. Saddle point at $(0, 0)$; local minimum at $(3, 9)$ **9.** Saddle point at $(0, 0)$; local minimum at $(-1, -1)$; local minimum at $(1, 1)$
11. Saddle point at $(0, 0)$; local maximum at $(18, 6)$ **13.** Saddle point at $(0, 1/\sqrt{3})$; saddle point at $(0, -1/\sqrt{3})$; local minimum at $(-1, 1)$; local minimum at $(1, 1)$.   **15.** Local minimum at $\left(\frac{1}{\sqrt[3]{2}}, \sqrt[3]{4}\right)$ **17.** Local minimum at $\left(-\frac{1}{8}, -1\right)$; saddle point at $\left(-\frac{1}{8}, 1\right)$
19. No local extrema or saddle points   **21.** Saddle point at $(0, 0)$; local maximum at $(-2, 0)$ **23.** Saddle point at $(0, 0)$; local minima at $\left(\frac{\pi}{2}, -\frac{\pi}{2}\right), \left(-\frac{\pi}{2}, \frac{\pi}{2}\right)$; local maxima at $\left(\frac{\pi}{2}, \frac{\pi}{2}\right)$ and $\left(-\frac{\pi}{2}, -\frac{\pi}{2}\right)$ **25.** Local minimum at $(1, 0)$; saddle point at $\left(0, \frac{\pi}{2}\right)$; local maximum at $(-1, \pi)$; saddle point at $\left(0, \frac{3\pi}{2}\right)$ **27.** $\nabla f = \left\langle -\frac{2}{3}x^{-1/3} + \frac{2}{3}x^{-2/3}y^{1/3}, \frac{2}{3}x^{1/3}y^{-2/3} - \frac{2}{3}y^{-1/3} \right\rangle$ which does not exist at $(0, 0)$.
$f(0, 0) = 4$; for $(x, y) \neq (0, 0)$, $f(x, y) < 4$. $\therefore f$ has an absolute maximum (and local maximum) at $(0, 0)$.   **29.** Absolute maximum of 0; absolute minimum of $-\frac{5}{4}$.   **31.** $5 + 4\sqrt{2}$ = absolute maximum; $5 - 4\sqrt{2}$ = absolute minimum   **33.** $l = 8; w = 8; h = 4$
35. The hottest spots are $\left(\frac{1}{2}, \frac{1}{2}, \frac{1}{\sqrt{2}}\right)$ and $\left(-\frac{1}{2}, -\frac{1}{2}, \frac{1}{\sqrt{2}}\right)$, where the temperature is $50°$; the coldest spots are $\left(\frac{1}{2}, -\frac{1}{2}, \frac{1}{\sqrt{2}}\right)$ and $\left(-\frac{1}{2}, \frac{1}{2}, \frac{1}{\sqrt{2}}\right)$, where the temperature is $-50°$.   **37.** $l = 10$, $w = 8$, and $h = 4$ **39.** (a) $y = 1.051 + 0.00525x$ (b)

(c) ≈ 1.156 **41.** So the hottest point is $(-3/2, 0)$ at a temp of 12 and coldest is -2 at $(-1, 0)$.

Exercise Set 14.6

1. $f\left(-\frac{4}{3}, -\frac{8}{3}\right) = -\frac{16}{3}$ is a constrained local minimum.   **3.** $f(-1, -1) = 3$ is a constrained local maximum; $f\left(\frac{2}{3}, -\frac{14}{9}\right) = -\frac{44}{27}$ is a constrained local minimum.   **5.** $f(2, -2) = 20$ is a constrained local minimum.   **7.** $f(6, -3, -2) = 18$ is a constrained local minimum.   **9.** Let $g(x) = x^2 - \left(\frac{x-4}{2}\right)^2$; $f\left(-\frac{4}{3}, -\frac{8}{3}\right) = -\frac{16}{3}$, a constrained local minimum.   **11.** Let $g(x) = x^2 + 2x(x^2 - 2)$; $f\left(\frac{2}{3}, -\frac{14}{9}\right) = -\frac{44}{27}$, a constrained local minimum; $f(-1, -1) = 3$, a constrained local maximum.   **13.** Let $g(x) = x^3 + 3\left(\frac{16}{x^2}\right)$; $f(2, -2) = 20$, a constrained local minimum.   **15.** Let $g(x, y) = x - 2y - 3\left(\frac{36}{xy}\right)$; $f(6, -3, -2) = 18$, a constrained local minimum   **17.** $\frac{2\sqrt{5}}{5}$ **19.** $T(\pm\sqrt{3}, 1) = 80$ = maximum of T; $T(\pm\sqrt{3}, -1) = 20$ = minimum of T.
21. $x = 8, y = 8, z = 4$ **23.** Maximum of $50°$ at $\left(\frac{1}{2}, \frac{1}{2}, \frac{1}{\sqrt{2}}\right)$ and $\left(-\frac{1}{2}, -\frac{1}{2}, \frac{1}{\sqrt{2}}\right)$ and minimum of $-50°$ at $\left(-\frac{1}{2}, \frac{1}{2}, \frac{1}{\sqrt{2}}\right)$ and $\left(\frac{1}{2}, -\frac{1}{2}, \frac{1}{\sqrt{2}}\right)$.   **25.** $y = -x$: $2x^2 + 2z - 4 = 0$, $2x + 2z = 0$, Eliminate z to get $x^2 - x - 2 = (x - 2)(x + 1) = 0$. If $x = 2, y = -2, z = -2$; if $x = -1, y = 1, z = 1$. $\therefore (2, -2, -2)$ and $(-1, 1, 1)$ are critical points. $\lambda_1 = 1$: $2y - 2y = 0 = -\lambda_2 \rightarrow \lambda_2 = 0, z = \lambda_1 = 1, x^2 + y^2 - 2 = 0, x - y + 2 = 0 \rightarrow y = x + 2, x^2 + (x + 2)^2 - 2 = 2(x^2 + 2x + 1) = 0 \rightarrow x = -1 \therefore (-1, 1, 1)$ is a critical point.   **27.** $2\sqrt{2}$ by 4 by 2   **29.** $f\left(\pm 2\sqrt{3}, \frac{\pm 4}{\sqrt{3}}, \frac{7}{6}\right) = \frac{223}{12}$ are constrained local maxima. $-\frac{11}{4} = f\left(\pm 2, 0, \frac{-3}{2}\right)$ are constrained

local minima. **31.** $\left(0, \pm\frac{\sqrt{29}}{4}, \frac{3}{8}\right)$ = Closest point; $\left(\pm\frac{\sqrt{10}}{6}, \pm\frac{4}{3}, \frac{1}{6}\right)$ = Farthest point

Chapter Review

1. (a) $\frac{2}{x}dx + \frac{y}{1-y^2}dy$ **(b)** $\frac{\cos x}{\cosh y}dx - \frac{\sin x \sinh y}{\cosh^2 y}dy$ **3. (a)** $\Delta f \approx df = 0.0084$ **(b)** $\Delta f \approx df = 0.01$ **5.** $\Delta P \approx \$318$ increase in profit; $7,043   **7.** $f_x(0,0) = \lim_{h \to 0} \frac{f(h,0)-f(0,0)}{h} = \lim_{h \to 0} \frac{0}{h} = 0$; $f_y(0,0) = \lim_{k \to 0} \frac{f(0,k)-f(0,0)}{k} = \lim_{k \to 0} \frac{0}{k} = 0$. But f is not continuous at $(0, 0)$, since $\lim_{(x,y) \to (0,0)} f(x,y) = 0$ along the x-axis, but $\lim_{(x,y) \to (0,0)} f(x,y)$ along $y = x^2 = \lim_{x \to 0} \frac{x^4}{x^4 + x^4} = \frac{1}{2}$. Thus, f is not differentiable at $(0,0)$.
9. $\frac{dz}{dt}\Big|_{t=5\pi/6} = \sqrt{3} - \frac{1}{4}$ **11.** $\frac{dz}{dt} = -(\sinh t)\, e^{(1-\cosh t)}$ **13.** $\frac{\partial z}{\partial t}\Big|_{(1,1/2)} = \sqrt{2} - \frac{\pi}{2}(3 + \sqrt{2})$ **15.** $\frac{\partial z}{\partial x} = -\frac{z}{x}$; $\frac{\partial z}{\partial y} = \frac{-[\ln(\cos xz) + xz]}{xy(1 - \tan\, xz)}$
17. $\frac{dh}{dt} = -\frac{5}{18}$ ft/min $= -\frac{10}{3}$ in./min   **19.** -4 **21.** $y^2 - x^2 + \ln\frac{x}{y} + 1$ **23.** Tangent plane: $\langle 1, -1, -1\rangle \cdot \langle \mathbf{x} - (1,1,2)\rangle = 0$ or $x - y - z + 2 = 0$; Normal line: $\mathbf{r}(t) = \langle 1, 1, 2\rangle + t\langle 1, -1, -1\rangle$ **25.** 4.3°   **27.** Absolute maximum of $f = \frac{35}{2}$, at (4, 2); absolute minimum of $f = \frac{9}{\sqrt[3]{3}}$, at $\left(\sqrt[3]{3}, \frac{\sqrt[3]{3}}{2}\right)$ **29. (a)** $h = \frac{2a}{\sqrt{3}}, r = \frac{2a}{\sqrt{6}}$ **(b)** $r = \frac{a}{\sqrt{2}}, h = 2r = \sqrt{2}a$

CHAPTER 15

Exercise Set 15.1

1.

i	(x_i^*, y_i^*)	$f(x_i^*, y_i^*)$	$f(x_i^*, y_i^*)\Delta A_i$
1	(1/2, 1/4)	1	1/2
2	(3/2, 1/4)	2	1
3	(5/2, 1/4)	3	3/2
4	(7/2, 1/4)	4	2
5	(7/2, 3/4)	5	5/2
6	(5/2, 3/4)	4	2
7	(3/2, 3/4)	3	3/2
8	(1/2, 3/4)	2	1

$\sum_{i=1}^{8} f\left(x_i^*, y_i^*\right) \Delta A_i = 12$ **3.** $\sum_{i=1}^{8} f\left(x_i^*, y_i^*\right) \Delta A_i = 60$

5. $\sum_{i=1}^{9} f\left(x_i^*, y_i^*\right) \Delta A_i = 582$ **7.** $\sum_{i=1}^{6} f\left(x_i^*, y_i^*\right) \Delta A_i = 35 \approx$ volume   **9.** $4\sum_{i=1}^{6} f\left(x_i^*, y_i^*\right) \Delta A_i = \frac{69}{4} \approx$ volume

11. Since $g(x, y) \geq f(x, y), g(x, y) - f(x, y) \geq 0$. By Property 4, $\iint_R [g(x, y) - f(x, y)]dA \geq 0$. By Property 2,

$\iint_R g(x, y)dA - \iint_R f(x, y)dA \geq 0$ and $\iint_R f(x, y)dA \leq \iint_R g(x, y)dA$.   **13. (a)** 10.905   **(b)** 10.587   **(c)** 10.658   **(d)** 10.580

Exercise Set 15.2

1. $-\frac{7}{2}$ **3.** $\frac{15}{8}$ **5.** $-\frac{4352}{15}$ **7.** $\frac{5}{3}$ **9.** $4\ln\frac{6}{7}$

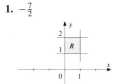
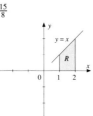
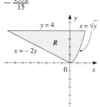
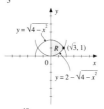
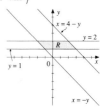

11. 40   **13.** $\frac{2}{3}\ln 2$ **15.** $4\sqrt{2} - \frac{8}{3}$ **17.** $\frac{49}{3}$ **19.** 4

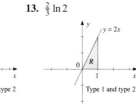
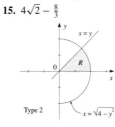
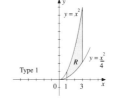
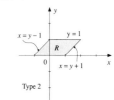

21. 2   **23.** $\frac{4}{3}$ **25.** $\frac{64}{3}$ **27.** $e - \frac{5}{3}$ **29.** $1 + \frac{3}{2}\ln\frac{4}{3}$

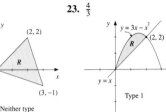   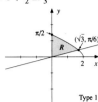

31. Either two type II regions or a type I and a type II region.

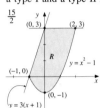   **33.** 4   **35.** $\frac{2\sqrt{2}}{3}$ **37.** $\frac{13}{2}$ **39.** 4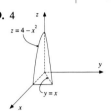

41. $\frac{16}{3}$ **43.** $\frac{e-1}{2e}$ **45.** $\frac{4}{27}$ **47.** $\int_0^2 \int_0^{y^2} f(x,y)\,dx\,dy$ **49.** $\int_2^4 \int_1^{x-1} f(x,y)\,dy\,dx$ **51.** $\int_0^1 \int_{1-\sqrt{1-y^2}}^{\sqrt{2y-y^2}} f(x,y)\,dx\,dy$

53. By Definition 15.2, the volume V under the surface $z = f(x,y)$ and above R is the double integral $V = \iint_R f(x,y)\,dA$. Let y_0 be an arbitrary value of y, such that $c \leq y_0 \leq d$. The plane $y = y_0$ intersects the given surface in a curve $z = f(x, y_0)$. Consider the portion from $x = a$ to $x = b$, and the area under its graph is $A(y_0) = \int_a^b f(x, y_0)\,dx$, the area of a typical cross-section of the volume taken perpendicular to the y-axis. Let y_0 now be variable; drop the subscript and write $V = \int_c^d A(y)\,dy = \int_c^d \left[\int_a^b f(x,y)\,dx\right]dy = \iint_R f(x,y)\,dA$.

55. $e + \frac{5}{e} - 4$ **57.** $\frac{82}{9} + \frac{16}{3}\ln 2$ **59.** $\frac{37}{12}$ **61.** $\frac{7}{9} - \ln 2$ **63.** $\frac{2\pi a^3}{35}$ **65.** 16   **67.** $\frac{16}{3}$ **69.** $a\pi b$

Exercise Set 15.3

1. $\frac{16}{3}\pi^2$ **3.** 99π **5.** $\frac{4}{5}$ **7.** $\frac{\pi}{8}$ **9.** 6π **11.** 2π

13. $\frac{10\pi}{3} + \frac{7\sqrt{3}}{2}$ **15.** $\sqrt{3} - \frac{\pi}{3}$ **17.** $\frac{19\pi}{3} - \frac{11\sqrt{3}}{2}$ **19.** 5π **21.** $\frac{\pi}{3}$ **23.** $\frac{16\sqrt{2}}{3}(1 - \sqrt{2})$

25. $\frac{\pi}{4} - \frac{2}{3}$ **27.** $\ln 4 - \frac{3}{4}$ **29.** $\frac{2\pi}{3} - \frac{10\sqrt{2}}{9}$ **31.** $\pi\left(1 - \frac{1}{e^{a^2}}\right)$ **33.** $\frac{8\pi}{3}$ **35.** $\frac{4\pi}{3}(a^2 - b^2)^{3/2}$ **37.** $\frac{32}{9}$ **39.** 8π

41. $\int_0^{\pi/3} \int_{2\sin\theta}^{4\sin\theta} (r^2 \cos\theta \sin\theta) r\,dr\,d\theta = \frac{135}{32}$ **43.** $6 - \frac{4\pi}{3} + \frac{\sqrt{3}}{3}$

Exercise Set 15.4

1. $\left(\frac{3}{8}, \frac{3}{4}\right)$ 3. $\left(0, \frac{4}{5}\right)$ 5. $\left(-\frac{1}{2}, \frac{2}{5}\right)$ 7. $\left(0, \frac{3}{\pi}\right)$ 9. $\left(0, \frac{3}{10}\right)$ 11. $\left(0, \frac{4a}{3\pi}\right)$ 13. $\left(\frac{16\sqrt{2}}{15}, \frac{176\sqrt{2}}{35}\right)$

15. $\left(\frac{e^2+1}{e^2-1}, \frac{e^2+1}{4}\right)$ 17. $I_x = \frac{2}{5}; I_y = \frac{2}{15}; I_0 = \frac{8}{15}; r_x = \frac{1}{\sqrt{5}}; r_y = \sqrt{\frac{3}{5}}$ 19. $I_x = \frac{1}{3}; I_y = \frac{1}{9}; I_0 = \frac{4}{9}; r_x = \frac{\sqrt{2}}{3}; r_y = \frac{\sqrt{6}}{3}$

21. $I_x = \frac{32}{15}; I_y = \frac{32}{15}; I_0 = \frac{64}{15}; r_x = \frac{2}{\sqrt{5}}; r_y = \frac{2}{\sqrt{5}}$ 23. $I_x = \frac{4\rho}{9}; I_y = \rho(\pi^2 - 4); I_0 = \rho\left(\pi^2 - \frac{32}{9}\right); r_x = \sqrt{\frac{\pi^2-4}{2}}; r_y = \frac{\sqrt{2}}{3}$

25. Let $R = \{(x, y) : 0 \le x \le b, 0 \le y \le h\}$; $I_x = \iint_R y^2 dm = \int_0^h \int_0^b y^2 \rho\, dx\, dy = \rho \int_0^h [y^2 x]_0^b dy = \rho b \int_0^h y^2 dy = \frac{\rho b h^3}{3}$

27. $I_{x=-2} = \rho \int_{-2}^{4} \int_{y^2/4}^{(y+4)/2} (x+2)^2 dx\, dy$; $I_{y=4} = \rho \int_{-2}^{4} \int_{y^2/4}^{(y+4)/2} (4-y)^2 dx\, dy$ 29. (a) $k = 4$ (b) $\mu_x = \frac{2}{3}; \mu_y = \frac{2}{3}$ (c) $\sigma_x^2 = \frac{1}{18}; \sigma_y^2 = \frac{1}{18}$ 31. (a) $k = 6$ (b) $\mu_x = \frac{1}{4}; \mu_y = \frac{1}{2}$ (c) $\sigma_x^2 = \frac{3}{80}; \sigma_y^2 = \frac{1}{20}$

33. (a) $I_{x=\bar{x}} = \iint_R (x - \bar{x})^2 dm = \iint_R x^2 dm - 2\bar{x} \iint_R x\, dm + \bar{x}^2 \iint_R dm = I_y - 2\bar{x}^2 m + \bar{x}^2 m = I_y - m\bar{x}^2$ (b) $I_{y=\bar{y}} = I_x - m\bar{y}^2$

Proof: $I_{y=\bar{y}} = \iint_R (y - \bar{y})^2 dm = \iint_R y^2 dm - 2\bar{y} \iint_R y\, dm + \bar{y}^2 \iint_R dm = I_x - 2\bar{y}^2 m + \bar{y}^2 m = I_x - m\bar{y}^2$ 35. Take the vertices as $(0,0)$,

$(b, 0)$, and (a, h). $I_b = \iint_R y^2 dm = \int_0^h \int_{ay/h}^{b-(b-a)y/h} y^2 \rho\, dx\, dy = \rho \int_0^h \left(b - \frac{by}{h}\right) y^2 dy = b\rho \left[\frac{y^3}{3} - \frac{y^4}{4h}\right]_0^h = \frac{bh^3 \rho}{12}$ 37. Orient the

coordinate axes so that l coincides with the y-axis. Then, we can compute the volume of the solid of revolution obtained by revolving R about the y-axis by using the method of cylindrical shells: $V = 2\pi \int_a^b x \int_{g_1(x)}^{g_2(x)} dy\, dx = 2\pi A\bar{x}$, where $\bar{x}$ is the x-coordinate of the centroid of R, and A is the area of R. $2\pi\bar{x}$ = distance traveled by the centroid of R. A similar formula can be derived to compute the volume of the solid of revolution about the x-axis. The result is $V = 2\pi A\bar{y}$, where $\bar{y}$ is the y-coordinate of the centroid of R.

Exercise Set 15.5

1. $2\sqrt{14}$ 3. $\frac{13}{6}$ 5. $\frac{13}{3}\pi$ 7. $4\pi\sqrt{2}$ 9. $\frac{\pi}{6}(17\sqrt{17} - 5\sqrt{5})$ 11. $\frac{1}{6}(13\sqrt{26} - 5\sqrt{10})$ 13. 2   15. $\frac{20}{9} - \frac{\pi}{3}$

17. $A(S) = \iint_R \frac{2\sqrt{x^2+y^2+z^2}}{2|z|} dA = 8 \int_0^{\pi/2} \int_0^a \frac{a}{\sqrt{a^2-r^2}} r\, dr\, d\theta = 4\pi a^2$ 19. At a point (x, y, z) on the surface, y and z form the

legs of a right triangle, and $f(x)$ is the hypotenuse. Hence, $y^2 + z^2 = [f(x)]^2$. By equation (15.26), with $y^2 + z^2 - [f(x)]^2 = F(x, y, z) = 0$; $F_x = -2f(x)f'(x)$, $F_y = -2y$, and $F_z = -2z$, $A(S) = \iint_R \frac{2\sqrt{[f(x)]^2[f'(x)]^2+y^2+z^2}}{2z} dA = \iint_R \frac{f(x)\sqrt{1+[f'(x)]^2}}{\sqrt{[f(x)]^2-y^2}} dy\, dx$.

Thus, $4\int_a^b \int_0^{f(x)} \frac{f(x)\sqrt{1+[f'(x)]^2}}{\sqrt{[f(x)]^2-y^2}} dy\, dx = 4 \int_a^b \left[f(x)\sqrt{1+f'(x)]^2}\, \sin^{-1} \frac{y}{f(x)}\right]_0^{f(x)} dx = 2\pi \int_a^b f(x)\sqrt{1+[f'(x)]^2}\, dx$

Exercise Set 15.6

1. 0   3. $\frac{1}{3}$ 5. $\frac{736}{3}$ 7. (a) $\int_1^3 \int_0^1 \int_{-1}^2 (2xy + 3yz^2) dy\, dx\, dz = \int_1^3 \int_0^1 \left(3x + \frac{9}{2}z^2\right) dx\, dz = \int_1^3 \left(\frac{3}{2} + \frac{9}{2}z^2\right) dz = 42$

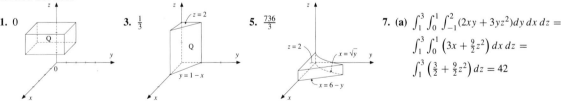

(b) $\int_{-1}^2 \int_1^3 \int_0^1 (2xy + 3yz^2) dx\, dz\, dy = \int_{-1}^2 \int_1^3 (y + 3yz^2) dz\, dy = \int_{-1}^2 y(3 + 27 - 1 - 1) dy = 42$

9. $\int_0^1 \int_0^{2\sqrt{1-x^2}} \int_0^{\sqrt{4-4x^2-y^2}} f(x,y,z) dz\, dy\, dx$ 11. $\int_1^2 \int_0^4 \int_{\sqrt{4-z^2}}^{\sqrt{3}} f(x,y,z) dx\, dy\, dz$ 13. $\frac{116}{15}$ 15. $\frac{2}{3}$ 17. $\frac{7}{6}$

19. $\int_0^4 \int_0^{\sqrt{4-z}} \int_{x^2}^{4-z} dy\, dx\, dz$; $\int_0^4 \int_0^{4-y} \int_0^{\sqrt{y}} dx\, dz\, dy$; $\int_0^4 \int_0^{\sqrt{y}} \int_0^{4-y} dz\, dx\, dy$; $\int_0^4 \int_{x^2}^4 \int_0^{4-y} dz\, dy\, dx$; and $\int_0^2 \int_0^{4-x^2} \int_{x^2}^{4-z} dy\, dz\, dx$

21. $m = \frac{1}{4}$; $\bar{x} = \frac{8}{15}$ 23. $\bar{z} = 4 \int_0^1 \int_0^{1-x} \int_0^{1-y} 2xz\, dz\, dy\, dx$; $I_y = \int_0^1 \int_0^{1-x} \int_0^{1-y} 2x(x^2+z^2) dz\, dy\, dx$

25. $m = 4\int_0^2 \int_0^{\sqrt{4-x^2}} \int_0^{\sqrt{16-x^2-y^2}} \sqrt{x^2+y^2}\, dz\, dy\, dx$; $\bar{z} = \frac{4}{m}\int_0^2 \int_0^{\sqrt{4-x^2}} \int_0^{\sqrt{16-x^2-y^2}} \sqrt{x^2+y^2}\, z\, dz\, dy\, dx$;

$I_z = 4\int_0^2 \int_0^{\sqrt{4-x^2}} \int_0^{\sqrt{16-x^2-y^2}} (x^2+y^2)^{3/2} dz\, dy\, dx$

27. $\bar{x} = \frac{1}{m}\int_0^1 \int_0^{\sqrt{1-x^2}} \int_0^{\sqrt{1-x^2}} x^2 yz\, dz\, dy\, dx$; $\bar{y} = \frac{1}{m}\int_0^1 \int_0^{\sqrt{1-x^2}} \int_0^{\sqrt{1-x^2}} xy^2 z\, dz\, dy\, dx$; $\bar{z} = \frac{1}{m}\int_0^1 \int_0^{\sqrt{1-x^2}} \int_0^{\sqrt{1-x^2}} x\, yz^2 dz\, dy\, dx$

29. $\bar{x} = \bar{y} = 0, \bar{z} = \frac{h}{4}$ **31.** $\bar{z} = \frac{3(4e-3)}{6e^2+6e-23} \approx 0.627$ **33.** Select the coordinate system so l' coincides with the z-axis and l passes through the point $(d, 0, 0)$. The distance from l to an element of mass located at (x, y, z) is $\sqrt{(d-x)^2 + y^2}$. Then
$I_l = \iiint_Q [(d-x)^2 + y^2] dm = \iiint_Q (x^2 + y^2) dm + \iiint_Q d^2 dm - 2d \iiint_Q x\, dm$. But, since l' goes through the center of mass, $\bar{x} = 0$, and $\iiint_Q x\, dm = m\bar{x}$, and so $\iiint_Q x\, dm = 0$. Thus, $I_l = \iiint_Q (x^2 + y^2) dm + d^2 \iiint_Q dm = I_{l'} + md^2$.

Exercise Set 15.7

1. 8 **3.** $\frac{\pi}{4}$ **5.** $\frac{2\pi}{3}$ **7.** 24π **9.** $\frac{2\pi\delta}{5}$ **11.** $\left(\frac{6}{5}, 0, \frac{27\pi}{128}\right)$ **13.** $\left(\frac{18\sqrt{3}}{7\pi}, \frac{18}{7\pi}, \frac{102}{35}\right)$ **15.** $\left(\frac{64a}{15\pi^2}, \frac{64a}{15\pi^2}, \frac{16a}{15\pi}\right)$

17. $m = 40\pi; I_z = \frac{1456\pi}{9}$ **19.** 3π **21.** $(2 - \sqrt{2})\pi$ **23.** $\frac{8}{9}(3\pi - 4)$ **25.** $\bar{z} = \frac{4}{15}\left[\frac{5a^2(h_2^3 - h_1^3) - 3(h_2^5 - h_1^5)}{2a^2(h_2^2 - h_1^2) - (h_2^4 - h_1^4)}\right], \bar{x} = \bar{y} = 0$

Exercise Set 15.8

1. R is bounded by a parallelogram with vertices $(2, 3), (4, 6), (-7, 9), (-5, 12)$. **3.** R is bounded by a triangle, with vertices $(0, 0), (0, 2),$ and $(2, 0)$. **5.** -10 **7.** e^u **9.** 0 **11.** $\frac{\partial(x, y, z)}{\partial(\rho, \theta, \phi)} = \begin{vmatrix} \cos\theta\sin\phi & -\rho\sin\theta\sin\phi & \rho\cos\theta\cos\phi \\ \sin\theta\sin\phi & \rho\cos\theta\sin\phi & \rho\sin\theta\cos\phi \\ \cos\phi & 0 & -\rho\sin\phi \end{vmatrix} =$
$-\rho^2 (\cos^2\theta \sin^3\phi + \sin^2\theta \sin\phi \cos^2\phi + \cos^2\theta \sin\phi \cos^2\phi + \sin^2\theta \sin^3\phi) = -\rho^2[\sin\phi(\sin^2\phi + \cos^2\phi)] = -\rho^2 \sin\phi$. If we take the absolute value, we get $\rho^2 \sin\phi$. **13.** $\frac{3}{2}$ **15.** $\frac{1}{2}(\sin 2 - 2) \approx -0.5454$ **17.** $V = 8\int_0^a \int_0^{b\sqrt{1-x^2/a^2}} \int_0^{c\sqrt{1-x^2/a^2-y^2/b^2}} dz\, dy\, dx; V = \frac{4\pi}{3}abc$
19. Let $u = x - y$ and $v = x + y$. $4\ln 2 \approx 2.773$ **21.** Let $x = u/v$ and $y = v$. $\ln 4$
23. Let $u = x + y$ and $v = x - y$. $\frac{1}{2}[\ln(\sec 1 + \tan 1) - 1] \approx 0.1131$

Chapter Review

1. $4 + \frac{2}{3}\ln 4$ **3.** $4\ln 2 - \frac{1}{6}$ **5.** $\frac{3-e}{2}$ **7.** $\frac{16}{3}$ **9.** $\frac{4}{3}$ **11.** $\frac{1}{16}(2 + \sin 2)$ **13.** $(\bar{x}, \bar{y}) = \left(\frac{34}{21}, \frac{110}{147}\right); (r_x, r_y) = \left(\frac{20\sqrt{3}}{21}, \frac{5\sqrt{14}}{21}\right)$
15. $(\bar{x}, \bar{y}) = \left(\frac{\pi}{2}, \frac{16}{9\pi}\right); I_x = \frac{3\pi}{32}; r_y = \frac{\sqrt{6}}{4}$ **17.** $\frac{14}{3}$ **19.** $\frac{321\sqrt{3}}{70} + 6\pi$ **21.** $\frac{4\pi}{3}$ **23.** (a) centroid: $\left(0, 0, \frac{5}{4}\right)$ (b) $I_z = \frac{3\pi}{10}$
25. $\frac{46\pi}{15}$ **27.** $\frac{4}{3}(4 - \sqrt{2})$ **29.** (a) $k = \frac{1}{8}$ (b) $1 - \frac{1}{4}\ln 3$ (c) $\mu_x = \frac{26}{9}, \mu_y = \frac{13}{6}$ (d) $\sigma_x = \frac{\sqrt{134}}{9}, \sigma_y = \frac{\sqrt{11}}{6}$
31. (a) $8uv + 1$ (b) $e^{u+v}(1 - uv)$ **33.** 0 **35.** $\frac{1}{6}(\ln 2)^3 \approx 0.05550$
37. Let $u = x - y$ and $v = x + y$; $-4\cosh 4 + \sinh 4 + 2\cosh 2 - \sinh 2 \approx -78.045$

CHAPTER 16

Exercise Set 16.1

1. **3.** **5.**

7. **9.** **11.** $\nabla f(x, y) = \langle 2x, -4y \rangle$ **13.** $\nabla f(x, y) = \left\langle \frac{x}{x^2+y^2}, \frac{y}{x^2+y^2} \right\rangle$

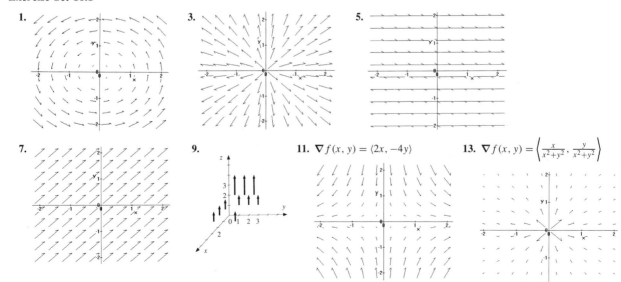

Answers 1311

15. $\nabla f(x, y) = \langle -e^{-x} \sin 2y, 2e^{-x} \cos 2y \rangle$ 17. $\nabla f(x, y, z) = \langle y - z, x + z, -x + y \rangle$ 19. $\nabla f(x, y, z) = \langle 2xy^3z, 3x^2y^2z, x^2y^3 \rangle$
21. $\nabla f(x, y, z) = \frac{-GMm}{(x^2+y^2+z^2)^{3/2}} \langle x, y, z \rangle$

Exercise Set 16.2

1. $\frac{27\sqrt{13}}{2}$ 3. $4(\sqrt{5} - 1)$ 5. $\pi - 2$ 7. 8π 9. $\frac{2}{3}(2\sqrt{2} - 1)$ 11. $A = \int_{-\infty}^{\ln 2} \frac{1}{2}\sqrt{e^{2t} + e^{4t}} \left(e^t \sqrt{1 + e^{2t}} \right) dt = 3$
13. $\frac{81}{64} - \ln \frac{5}{4}$ 15. $\frac{200}{3}$ 17. $-\frac{41}{6}$ 19. $-\frac{20}{3}$ 21. 3   23. $2 - \ln 4$ 25. $-\frac{7}{6}$ 27. 0   29. (a)

(b) $W = 0$ (c) $\mathbf{F}(x, y) = \left\langle -\frac{kx}{x^2+y^2}, -\frac{ky}{x^2+y^2} \right\rangle$ 31. $-\pi\sqrt{5}$ 33. 4   35. $\frac{3\pi}{16} + \frac{5}{12}$ 37. $-\frac{1}{2}$ 39. $\frac{3}{5}(1 - e^\pi)$

41. $\frac{49\pi}{2}$ 43. (a) Let C be partitioned into the subarcs $P_0P_1, P_1P_2, \cdots, P_{n-1}P_n$, and let Δs_i be the length of $P_{i-1}P_i$. If Δs_i is small, approximate $m_i \bar{x}_i$ for the subarc by $x_i^* \rho(x_i^*, y_i^*) \Delta s_i$, where (x_i^*, y_i^*) is some point on $P_{i-1}P_i$.

Let $\bar{x} = \dfrac{\lim\limits_{\|P\| \to 0} \sum_{i=1}^{n} x_i^* \rho(x_i^*, y_i^*) \Delta s_i}{\lim\limits_{\|P\| \to 0} \sum_{i=1}^{n} \rho(x_i^*, y_i^*) \Delta s_i} = \dfrac{\int_C x \rho(x, y) ds}{\int_C \rho(x, y) ds} = \dfrac{\int_C x \, dm}{m}$; Similarly, $\bar{y} = \dfrac{\lim\limits_{\|P\| \to 0} \sum_{i=1}^{n} y_i^* \rho(x_i^*, y_i^*) \Delta s_i}{\lim\limits_{\|P\| \to 0} \sum_{i=1}^{n} \rho(x_i^*, y_i^*) \Delta s_i} = \dfrac{\int_C y \, dm}{m}$ (b) Use the same partition of

C as in part (a). Then, by the definitions of I_x and I_y, $I_x = \lim\limits_{\|P\| \to 0} \sum_{i=1}^{n} (y_i^*)^2 \rho(x_i^*, y_i^*) \Delta s_i = \int_C y^2 \, dm$; $I_y = \lim\limits_{\|P\| \to 0} \sum_{i=1}^{n} (x_i^*)^2 \rho(x_i^*, y_i^*) \Delta s_i = \int_C x^2 \, dm$

45. Center of mass: $\left(-\frac{4}{3\pi}, \frac{4}{3\pi}, \frac{9\pi}{4} \right)$; $I_x = \frac{3\pi\sqrt{13}\rho}{8}(8 + 27\pi^2)$; $I_y = \frac{3\pi\sqrt{13}\rho}{8}(8 + 27\pi^2)$; $I_z = 6\pi\sqrt{13}\rho$ 47. $\int_C \mathbf{F} \cdot d\mathbf{r} = \int_a^b m\mathbf{r}'' \cdot \mathbf{r}' dt$ by Newton's Second Law of Motion, since $\mathbf{F} = m\mathbf{a} = m\mathbf{r}''(t)$. Next, we note that $\frac{d}{dt}(\mathbf{r}' \cdot \mathbf{r}') = \mathbf{r}'' \cdot \mathbf{r}' + \mathbf{r}' \cdot \mathbf{r}'' = 2\mathbf{r}'' \cdot \mathbf{r}'$, which allows us to rewrite the integral as $\frac{m}{2} \int_a^b \frac{d}{dt}(\mathbf{r}' \cdot \mathbf{r}') dt = \frac{m}{2} \int_a^b \frac{d}{dt} |\mathbf{r}'|^2 dt$.

Exercise Set 16.3

1. $\phi(x, y) = x^2 + 3xy - y^2 + y$ 3. not a gradient field   5. $\phi(x, y) = x^3y^2 + \frac{2}{3}x^3 - 2y$ 7. $\phi(x, y) = x^2 e^{xy}$
9. not a gradient field   11. $\phi(x, y) = y \sin xy - y^2$ 13. 33   15. $\frac{155}{3}$ 17. 15   19. $-\frac{3}{2}$
21. $\phi(x, y) = x^2 + y^3 - 2xy^2$ 23. $\phi(x, y) = x \sin y - y \cos x$ 25. $\frac{\partial f}{\partial y} = \frac{\partial g}{\partial x} = -2$; (a) $\int_C (2x - 2y)dx + (6y - 2x)dy = 16$
(b) $\int_C \mathbf{F} \cdot d\mathbf{r} = \int_0^1 \langle 2x - 2y, 6y - 2x \rangle \cdot \langle 4, -4 \rangle dt = 16$ 27. Let $\mathbf{F} = \langle f, g \rangle = \left\langle \frac{-y}{x^2+y^2}, \frac{x}{x^2+y^2} \right\rangle$. Then $\frac{\partial f}{\partial y} = \frac{\partial g}{\partial x} = \frac{y^2-x^2}{(x^2+y^2)^2}$;
(a) $\int_C \frac{-y\, dx + x\, dy}{x^2+y^2} = \frac{5\pi}{12}$ (b) $\int_C \mathbf{F} \cdot d\mathbf{r} = \int_{-\pi/4}^{\pi/6} \left[\frac{(-2\sqrt{3}\sin t)(-2\sqrt{3}\sin t)}{12} + \frac{(2\sqrt{3}\cos t)(2\sqrt{3}\cos t)}{12} \right] dt = \frac{5\pi}{12}$
29. (a) $\int_C \mathbf{F} \cdot d\mathbf{r} = \int_0^1 \langle 2(1 + 2t), 2(-2 + 8t) \rangle \cdot \langle 2, 8 \rangle dt = 40$ (b) $\int_C \mathbf{F} \cdot d\mathbf{r} = \int_0^1 \langle 2(1 + 2t), 2(-2) \rangle \cdot \langle 2, 0 \rangle dt + \int_0^1 \langle 2(3), 2(-2 + 8t) \rangle \cdot \langle 0, 8 \rangle dt = 40$ (c) $\int_C \mathbf{F} \cdot d\mathbf{r} = \int_1^3 \left\langle 2t, 4(t^2 - 2t) \right\rangle \cdot \langle 1, 4t - 4 \rangle dt = 40$ 31. $\frac{\partial f}{\partial y} = \frac{\partial g}{\partial x} = -3$; $W = 44$ 33. (a) $\frac{\sqrt{6}-2}{4}$ (b) 0
35. $\frac{\partial f}{\partial y} = \frac{\partial g}{\partial x} = y \cos xy + y \cos xy - xy^2 \sin xy$; $\phi(x, y) = x^2 + y \sin xy$ 37. $\frac{\partial f}{\partial y} = \frac{\partial g}{\partial x} = \tanh y$; $\phi(x, y) = x \ln \cosh y - y$
39. $\phi(x, y, z) = x^2 + y^2 + z^2$; 8   41. $\frac{\partial f}{\partial y} = \frac{\partial g}{\partial x} = 2xz$; $\frac{\partial f}{\partial z} = \frac{\partial h}{\partial x} = 2xy - 6z$; $\frac{\partial g}{\partial z} = \frac{\partial h}{\partial y} = x^2 + 24yz^2$; $\mathbf{F}$ is a gradient field
$\phi(x, y, z) = x^2yz - 3xz^2 + 4y^2z^3$

Exercise Set 16.4

1. $\frac{28}{3}$ 3. $-\frac{4}{\pi}$ 5. $\frac{3e^2-2e^3-1}{3}$ 7. 0   9. 0   11. -36 13. 0   15. 0   17. $\frac{3\pi}{2}$ 19. 5   21. $\frac{3}{2}$
23. $\frac{3}{8}\pi a^2$ 25. $3\pi a^2$ 27. Circulation of $\mathbf{F}$ around $C = \frac{4}{3}$. Flux of $\mathbf{F}$ through $C = \frac{4}{3}$ 29. Circulation of $\mathbf{F}$ around $C = 24\pi$. Flux of $\mathbf{F}$ through $C = 0$ 31. Circulation of $\mathbf{F}$ around $C = \frac{81\pi}{2}$.   33. Suppose $\mathbf{T}$ is the unit tangent vector at point $P(x, y)$ on the curve C and C is parameterized by arc length, then $\mathbf{T} = \left\langle \frac{dx}{ds}, \frac{dy}{ds} \right\rangle$. Let θ denote the smallest positive angle from the unit vector $\mathbf{i}$ to $\mathbf{T}$. Then $\overrightarrow{PA} = \frac{dx}{ds}$, $\overrightarrow{AC} = \frac{dy}{ds}$. Consider the outer normal $\mathbf{n}$. The angle between $\mathbf{T}$ and $\mathbf{n}$ is $\frac{\pi}{2}$, measured clockwise. Thus,

∠BPD = $\theta - \frac{\pi}{2}$, and $\mathbf{n} = \langle \cos(\theta - \frac{\pi}{2}), \sin(\theta - \frac{\pi}{2}) \rangle$. **35.** Suppose that C contains the origin. By Theorem 16.8, replace C by the circle $C_1 : \mathbf{r}(t) = \langle a\cos t, a\sin t\rangle, 0 \leq t \leq 2\pi$, where a is small enough that C_1 lies inside C. Then, $\int_C \mathbf{F} \cdot d\mathbf{r} = \int_{C_1} \mathbf{F} \cdot d\mathbf{r} = \int_0^{2\pi} \langle \cos t, \sin t\rangle \cdot \langle -a\sin t, a\cos t\rangle\, dt = \int_0^{2\pi}(-a\cos t \sin t + a\sin t \cos t)dt = 0$; If the origin is exterior to C, Green's theorem can be used: $\int_C \mathbf{F} \cdot d\mathbf{r} = \iint_R \left(\frac{\partial g}{\partial x} - \frac{\partial f}{\partial y}\right) dA = \iint_R \left[-\frac{xy}{(x^2+y^2)^{3/2}} + \frac{xy}{(x^2+y^2)^{3/2}}\right] dA = 0$ **37.** Assume that R can be divided by a vertical line segment into two simple regions R_1 and R_2, as shown. Then, $\iint_R \left(\frac{\partial g}{\partial x} - \frac{\partial f}{\partial y}\right)dA = \iint_{R_1}\left(\frac{\partial g}{\partial x} - \frac{\partial f}{\partial y}\right)dA + \iint_{R_2}\left(\frac{\partial g}{\partial x} - \frac{\partial f}{\partial y}\right)dA = \left(\int_{C_1}\mathbf{F}\cdot d\mathbf{r} + \int_{C_2}\mathbf{F}\cdot d\mathbf{r}\right) + \left(\int_{C_3}\mathbf{F}\cdot d\mathbf{r} + \int_{C_4}\mathbf{F}\cdot d\mathbf{r}\right) = \int_{C_1}\mathbf{F}\cdot d\mathbf{r} + \int_{C_4}\mathbf{F}\cdot d\mathbf{r} = \int_C \mathbf{F}\cdot d\mathbf{r}$, since $\int_{C_2}\mathbf{F}\cdot d\mathbf{r} = -\int_{C_3}\mathbf{F}\cdot d\mathbf{r}$. In a similar fashion, R_1 and R_2 can be subdivided into two regions each and the above proof used to show that Green's theorem applies to R. The process may be continued for a finite number of such divisions; each time the new integrals that are introduced along the common boundary of the subregions cancel.

Exercise Set 16.5

1. $\frac{8\sqrt{3}}{3}$ **3.** $\frac{81\sqrt{14}\pi}{2}$ **5.** $\frac{19\pi}{15}$ **7.** $\frac{32\pi}{3}$ **9.** $m = \frac{8\sqrt{21}}{3}$; $(\bar{x}, \bar{y}, \bar{z}) = (2, 1, 1)$ **11.** $m = \frac{13\pi k}{3}$; $(\bar{x}, \bar{y}, \bar{z}) = (0, 0, 2.854)$

13. $I_x = \frac{k}{3}\int_0^2\int_0^4\left(y^2 + \frac{x^4}{9}\right)\sqrt{9+4x^2}\,dy\,dx$; $I_y = \frac{k}{3}\int_0^2\int_0^4\left(x^2 + \frac{x^4}{9}\right)\sqrt{9+4x^2}\,dy\,dx$; $I_z = \frac{k}{3}\int_0^2\int_0^4(x^2+y^2)\sqrt{9+4x^2}\,dy\,dx$

15. (a) $\sqrt{6}\int_0^2\int_0^{4-2x}x^2y(2x+y)^3\,dy\,dx$ (b) $\sqrt{\frac{6}{8}}\int_0^4\int_0^z(z-y)^2yz^3\,dy\,dz$ (c) $\sqrt{6}\int_0^4\int_0^{z/2}x^2(z-2x)z^3\,dx\,dz$ **17.** Projection onto the xy-plane, $\int_1^{\sqrt{2}}\int_0^5 xy(2-x^2)\sqrt{1+4x^2}\,dy\,dx$; projection onto the yz-plane, $\frac{1}{2}\int_0^1\int_0^5 yz\sqrt{9-4z}\,dy\,dz$ **19.** Flux = 19

21. Flux = 56π **23.** Flux = $\frac{14\pi}{3}$ **25.** Flux = $\frac{8}{3}$ **27.** $m = 2\pi ka^2$; center of mass = $\left(0, 0, \frac{a}{2}\right)$

29. $\mathbf{F} = \frac{k\mathbf{u}}{|\mathbf{r}|^2} = \frac{k\mathbf{r}}{|\mathbf{r}|^3}$; $S : x^2 + y^2 + z^2 = a^2$, or $z = \sqrt{a^2 - x^2 - y^2}$; $\mathbf{n} = \frac{(2x, 2y, 2z)}{2\sqrt{x^2+y^2+z^2}} = \frac{\mathbf{r}}{a}$; $\mathbf{F} \cdot \mathbf{n}\,d\sigma = \left(\frac{k\mathbf{r}}{|\mathbf{r}|^3} \cdot \frac{\mathbf{r}}{a}\right) = \frac{k}{a^2}$; flux $= \iint_S \mathbf{F}\cdot\mathbf{n}\,d\sigma = \iint_S \frac{k}{a^2}\,d\sigma = 4\pi a^2\left(\frac{k}{a^2}\right) = 4\pi k$ **31.** Let S denote the surface of a lamina having continuous density $\rho(x, y, z)$. Partition S into finitely many small surface elements having areas $\Delta\sigma_1, \Delta\sigma_2, \ldots, \Delta\sigma_n$. Denote the partition by P and define its norm $||P||$ as the length of the longest diagonal of all the cells. Choose a point (x_i^*, y_i^*, z_i^*) arbitrarily in the ith element. The mass of the ith cell is approximately $\rho(x_i^*, y_i^*, z_i^*)\Delta\sigma_i$ and the mass of the entire lamina is $m = \lim_{||P||\to 0}\sum_{i=1}^n \rho(x_i^*, y_i^*, z_i^*)\Delta\sigma_i = \iint_S \rho(x, y, z)\,d\sigma$, provided the limit exists. The coordinates of the center of mass are approximated as follows: $\bar{x} \approx \frac{\sum_{i=1}^n (\Delta m_i)x_i^*}{\sum_{i=1}^n \Delta m_i}$, $\bar{y} = \frac{\sum_{i=1}^n (\Delta m_i)y_i^*}{\sum_{i=1}^n \Delta m_i}$, $\bar{z} = \frac{\sum_{i=1}^n (\Delta m_i)z_i^*}{\sum_{i=1}^n \Delta m_i}$.

Taking the limits as $||P|| \Rightarrow 0$, provided they exist, gives $\bar{x} = \frac{\iint_S x\,dm}{m}$, $\bar{y} = \frac{\iint_S y\,dm}{m}$, $\bar{z} = \frac{\iint_S z\,dm}{m}$, where $dm = \rho\,d\sigma$. Similarly, approximations for the moments of inertia are $I_x \approx \sum_{i=1}^n (\Delta m_i)(y_i^2 + z_i^2)$, $I_y \approx \sum_{i=1}^n (\Delta m_i)(x_i^2 + z_i^2)$, $I_z \approx \sum_{i=1}^n (\Delta m_i)(x_i^2 + y_i^2)$, where $(x_i, y_i, z_i) = (x_i^*, y_i^*, z_i^*)$ and $\Delta m_i = \rho(x_i^*, y_i^*, z_i^*)\Delta\sigma_i$. Taking the limits as $||P|| \Rightarrow 0$, provided they exist, gives $I_x = \iint_S (y^2 + z^2)\,dm$, $I_y = \iint_S (x^2 + z^2)\,dm$, $I_z = \iint_S (x^2 + y^2)\,dm$.

Exercise Set 16.6

1. 512 **3.** $\frac{52}{5}$ **5.** 24π **7.** Flux = $\frac{1792\pi}{3}$ **9.** Flux = $-\frac{59}{36}$ **11.** Flux = $\frac{7\pi}{6}$ **13.** $\iint_S \mathbf{F}\cdot\mathbf{n}\,d\sigma = \iiint_G \text{div }\mathbf{F}\,dV = 24\pi$

15. $\iint_S \mathbf{F}\cdot\mathbf{n}\,d\sigma = \iiint_G \text{div }\mathbf{F}\,dV = 4a^3$ **17.** $\mathbf{F} = x\mathbf{i} + y\mathbf{j} + z\mathbf{k}$; div $\mathbf{F} = 1 + 1 + 1 = 3$ $\iint_S \mathbf{F}\cdot\mathbf{n}\,d\sigma = \iiint_G \text{div }\mathbf{F}\,dV = 3\iiint_G dV = 3V$, if G and its boundary S satisfy the conditions of the Divergence Theorem. **19.** Equation (16.36): let G be a region that is yz-simple. Its boundary S can be decomposed into the surfaces S_1, S_2, and S_3 as in Figure 16.46 except that the projection of G is onto the yz-plane. If S_3 is nonempty, on it the outer unit normal vectors are parallel to the yz-plane, so that $\alpha = \pi/2$ and $\cos\alpha = 0$. Thus, $f\cos\alpha = 0$, and $\iint_S f\cos\alpha\,d\sigma = 0$. Whether or not S_3 is empty, $\iint_S f\cos\alpha\,d\sigma = \iint_{S_1} f\cos\alpha\,d\sigma + \iint_{S_2} f\cos\alpha\,d\sigma$. Let S_1 be the graph of $x = x_1(y, z)$ and S_2 the graph of $x = x_2(y, z)$. On S_2 the outer normal is directed in the positive x-direction, so that

α is acute. Thus, by equation (16.26) $\iint_{S_2} f \cos\alpha \, d\sigma = \iint_R f(x_2(y,z), y, z) dA$ On S_1 the outer normal is directed in the negative x-direction, so that α is obtuse, and $|\sec\alpha| = -\sec\alpha$. Equation (16.26) then gives $\iint_{S_1} f \cos\alpha \, d\sigma = -\iint_R f(x_1(y,z), y, z) dA$

Combining, we get $\iint_S f \cos\alpha \, d\sigma = \iint_R [f(x_2(y,z), y, z) - f(x_1(y,z), y, z)] dA = \iiint_G \frac{\partial f}{\partial x} dV = \iint_R \left[\int_{x_1(y,z)}^{x_2(y,z)} \left(\frac{\partial f}{\partial x}\right) dx \right] dA$; it follows that equation (16.36) is true. The proof of equation (16.37) is similar to the foregoing. The projection of G is onto the xz-plane, and the outward normals to S_3 are parallel to the xz-plane, so that $\beta = \pi/2$ and $\cos\beta = 0$. Thus, $g \cos\beta = 0$, and $\iint_{S_3} g \cos\beta \, d\sigma = 0$. Hence, $\iint_S g \cos\beta \, d\sigma = \iint_{S_1} g \cos\beta \, d\sigma + \iint_{S_2} g \cos\beta \, d\sigma$; Let S_1 be the graph of $y = y_1(x,z)$ and S_2 the graph of $y = y_2(x,z)$. Following the earlier steps, we get $\iint_S g \cos\beta \, d\sigma = \iint_R [g(x, y_2(x,z), z) - g(x, y_1(x,z), z)] dA$ and

$\iiint_G \frac{\partial g}{\partial y} dV = \iint_R \left[\int_{y_1(x,z)}^{y_2(x,z)} \left(\frac{\partial g}{\partial y}\right) dy \right] dA = \iint_R [g(x, y_2(x,z), z) - g(x, y_1(x,z), z)] dA$ from which it follows that equation (16.37) is true.

21. In spherical coordinates, $x^2 + y^2 + (z-a)^2 = a^2$ becomes $\rho = 2a\cos\phi$. $\iiint_G \text{div } \mathbf{F} \, dV$

$= 2\int_0^{2\pi} \int_0^{\pi/2} \int_0^{2a\cos\phi} [\rho \sin\phi(\cos\theta + \sin\theta) + \rho\cos\phi]\rho^2 \sin\phi \, d\rho \, d\phi \, d\theta = 8a^4 \int_0^{2\pi} \int_0^{\pi/2} [(\cos\theta + \sin\theta)\cos^4\phi \sin^2\phi + \cos^5\phi \sin\phi] d\phi \, d\theta$

Note that the terms $\int_0^{2\pi} (\cos\theta + \sin\theta) d\theta = 0$ (or we can reverse the order of integration). We now have $16\pi a^4 \int_0^{\pi/2} \cos^5\phi \sin\phi \, d\phi = \frac{8}{3}\pi a^4$

23. $\mathbf{F} = \left\langle u\frac{\partial v}{\partial x}, u\frac{\partial v}{\partial y}, u\frac{\partial v}{\partial z}\right\rangle$; div $\mathbf{F} = u\frac{\partial^2 v}{\partial x^2} + \frac{\partial u}{\partial x}\frac{\partial v}{\partial x} + u\frac{\partial^2 v}{\partial y^2} + \frac{\partial u}{\partial y}\frac{\partial v}{\partial y} + u\frac{\partial^2 v}{\partial z^2} + \frac{\partial u}{\partial z}\frac{\partial v}{\partial z} = u\nabla^2 v + \nabla u \cdot \nabla v$. $\therefore \iint_S \mathbf{F} \cdot \mathbf{n} \, d\sigma = \iint_S u\nabla v \cdot \mathbf{n} \, d\sigma = \iiint_G \text{div } \mathbf{F} \, dV = \iiint_G (u\nabla^2 v + \nabla u \cdot \nabla v) dV$ **25.** From Exercise 23, $\iiint_G (u\nabla^2 v + \nabla u \cdot \nabla v) dV = \iint_S u\nabla v \cdot \mathbf{n} \, d\sigma$. Replace v by u and u by $u(x,y,z) = 1$; $\iiint_G (\nabla^2 u + 0) dV = \iint_S \nabla u \cdot \mathbf{n} \, d\sigma$; $\therefore \iiint_G \nabla^2 u \, dV = \iint_S \nabla u \cdot \mathbf{n} \, d\sigma$

Exercise Set 16.7

1. curl $\mathbf{F} = \langle 1, 2xy, 1 - 2xz\rangle$ **3.** curl $\mathbf{F} = x(e^z - e^y)\mathbf{i} + y(e^x - e^z)\mathbf{j} + z(e^y - e^x)\mathbf{k}$ **5.** curl $\mathbf{F} = \langle 0, 0, y\cos xy - x\sin xy\rangle$ **7.** -35
9. -12π **11.** $\int_C \mathbf{F} \cdot d\mathbf{r} = \iint_S (\text{curl } \mathbf{F}) \cdot \mathbf{n} \, d\sigma = 0$ **13.** $\int_C \mathbf{F} \cdot d\mathbf{r} = \iint_S (\text{curl } \mathbf{F}) \cdot \mathbf{n} \, d\sigma = -\frac{41}{6}$ **15.** $\int_C \mathbf{F} \cdot d\mathbf{r} = \iint_S (\text{curl } \mathbf{F}) \cdot \mathbf{n} \, d\sigma = 0$
17. curl $\mathbf{F} = \langle 0,0,0\rangle$; $\phi(x,y,z) = x^2 yz$ **19.** curl $\mathbf{F} = \langle 0,0,0\rangle$; work $= 7$ **21.** curl $\mathbf{F} = \langle 0,0,0\rangle$; $\phi(x,y,z) = e^z \sin x \cos y$
23. div(curl $\mathbf{F}$) $= \frac{\partial}{\partial x}\left(\frac{\partial h}{\partial y} - \frac{\partial g}{\partial z}\right) + \frac{\partial}{\partial y}\left(\frac{\partial f}{\partial z} - \frac{\partial h}{\partial x}\right) + \frac{\partial}{\partial z}\left(\frac{\partial g}{\partial x} - \frac{\partial f}{\partial y}\right) = \frac{\partial^2 h}{\partial x\partial y} - \frac{\partial^2 g}{\partial x\partial z} + \frac{\partial^2 f}{\partial y\partial z} - \frac{\partial^2 h}{\partial y\partial x} + \frac{\partial^2 g}{\partial z\partial x} - \frac{\partial^2 f}{\partial z\partial y} = 0$
25. Let $\mathbf{H} = \text{curl } \mathbf{F} = \nabla \times \mathbf{F}$. By the divergence theorem, $\iint_S \mathbf{H} \cdot \mathbf{n} \, d\sigma = \iiint_G \text{div } \mathbf{H} \, dV = \iiint_G \nabla \cdot (\nabla \times \mathbf{F}) dV = 0$, from the result of Exercise 23.
$\therefore \iint_S (\text{curl } \mathbf{F}) \cdot \mathbf{n} \, d\sigma = 0$ **27. (a)** $\mathbf{a} \times \mathbf{r} = \langle a_2 z - a_3 y, a_3 x - a_1 z, a_1 y - a_2 x\rangle$; curl $(\mathbf{a} \times \mathbf{r}) = \langle a_1 + a_1, a_2 + a_2, a_3 + a_3\rangle = 2\mathbf{a}$
(b) $\int_C (\mathbf{a} \times \mathbf{r}) \cdot d\mathbf{r} = \iint_S \text{curl}(\mathbf{a} \times \mathbf{r}) \cdot \mathbf{n} \, d\sigma = 2\iint_S \mathbf{a} \cdot \mathbf{n} \, d\sigma$ **29. (a)** $\iint_S (\nabla u \times \nabla v) \cdot \mathbf{n} \, d\sigma = \iint_S \text{curl}(u\nabla v) \cdot \mathbf{n} \, d\sigma$ (by Exercise 28)
$= \int_C (u\nabla v) \cdot d\mathbf{r}$ (by Stokes's theorem) **(b)** $\int_C (u\nabla v - v\nabla u) \cdot d\mathbf{r} = \int_C u\nabla v \cdot d\mathbf{r} - \int_C v\nabla u \cdot d\mathbf{r} =$
$\iint_S (\nabla u \times \nabla v) \cdot \mathbf{n} \, d\sigma - \iint_S (\nabla v \times \nabla u) \cdot \mathbf{n} \, d\sigma$ (by part a) $= \iint_S (\nabla u \times \nabla v) \cdot \mathbf{n} \, d\sigma + \iint_S (\nabla u \times \nabla v) \cdot \mathbf{n} \, d\sigma = 2\iint_S (\nabla u \times \nabla v) \cdot \mathbf{n} \, d\sigma$

Chapter Review

1. $\frac{\pi}{2} + 1 + \frac{\sqrt{2}}{3}$ **3.** $\frac{e^{3/2}(1-e^3)}{3} + \frac{e^2+1}{4}$ **5. (a)** $\phi(x,y) = \frac{x}{\sqrt{x^2+y^2}} - 2x + 3y$ **(b)** $\phi(x,y) = \ln\frac{xy}{x+y}$ **(c)** not a gradient field
(d) $\phi(x,y) = x\tanh y - y^2 + \ln\cosh x^2$ **7.** 0 **9.** 3π **11.** $2\left(1 - \frac{\pi}{4}\right)$ **13.** π **15.** $\frac{4\pi}{3}(50 - 11\sqrt{5})$
17. xy-projection: $\int_0^2 \int_0^x x^2 y(4-x^2)^3\sqrt{1+4x^2} dy \, dx$; yz-projection: $\frac{1}{2}\int_0^2 \int_0^{4-y^2} yz^3\sqrt{(4-z)(17-4z)} dz \, dy$ **19.** Flux $= 6\pi$
21. $\frac{61\pi}{16}$ **23.** 0 **25.** Circulation $= -128\pi$ **27.** 2π **29.** $\int_C u\nabla v \, d\mathbf{r} = \iint_S (\nabla u \times \nabla v) \cdot \mathbf{n} \, d\sigma$ (from Exercise 29 of Exercise Set 16.7). Then, $\int_C (u\nabla v + v\nabla u) \cdot d\mathbf{r} = \iint_S (\nabla u \times \nabla v + \nabla v \times \nabla u) \cdot \mathbf{n} \, d\sigma = \iint_S (\nabla u \times \nabla v - \nabla u \times \nabla v) \cdot \mathbf{n} \, d\sigma = \iint_S (0) \cdot \mathbf{n} \, d\sigma = 0$

CHAPTER 17

Exercise Set 17.1

1. $y = \frac{x^2}{2} - 2x + \frac{11}{2}$ **3.** $y = \ln\sec x + 3$ **5.** $y = x^2 + x$ **7.** $C_1 e^{-x} + 4C_2 e^{-2x} + 3(-C_1 e^{-x} - 2C_2 e^{-2x}) + 2(C_1 e^{-x} + C_2 e^{-2x}) = 0$
9. $-C_1 a^2 \cos ax - C_2 a^2 \sin ax + a^2(C_1 \cos ax + C_2 \sin ax) = 0$ **11.** $e^{-t}(-2C_2 \cos t + 2C_1 \sin t) + 2e^{-t}[(C_2 - C_1)\cos t - (C_2 + C_1)\sin t] + 2e^{-t}(C_1 \cos t + C_2 \sin t) = 0$ **13.** $y = -2e^{-x} + 2e^{-2x}$ **15.** $y = 2\cos ax + \sin ax$
17. $x = e^{-t}\sin t$ **19.** Let $F(x,y) = 2x^3 y - 3xy^2 + y^4 - 7 = 0$; $\frac{dy}{dx} = -\frac{F_x}{F_y} = \frac{-(6x^2 y - 3y^2)}{2x^3 - 6xy + 4y^3} = \frac{3y(y-2x^2)}{2x^3 - 6xy + 4y^3}$ **21.** Let

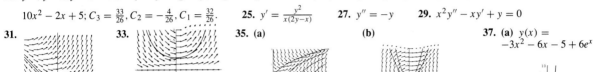

$F(x,y) = \tan^{-1}\frac{y}{x} + \ln\sqrt{x^2+y^2} - 1 = 0$; $\frac{dy}{dx} = -\frac{F_x}{F_y} = \frac{-\left(\frac{-y}{x^2+y^2} + \frac{x}{x^2+y^2}\right)}{\frac{x}{x^2+y^2} + \frac{y}{x^2+y^2}} = \frac{y-x}{y+x}$

23. $y''' + y' - 10y = e^{-x}\cos 2x(11C_1 - 2C_2 - C_1 + 2C_2 - 10C_1) + e^{-x}\sin 2x(2C_1 + 11C_2 - 2C_1 - C_2 - 10C_2) + e^{2x}(8C_3 + 2C_3 - 10C_3) - 2x + 10x^2 + 5 =$
$10x^2 - 2x + 5; C_3 = \frac{33}{26}, C_2 = -\frac{4}{26}, C_1 = \frac{32}{26}$. **25.** $y' = \frac{y^2}{x(2y-x)}$ **27.** $y'' = -y$ **29.** $x^2 y'' - xy' + y = 0$

31. 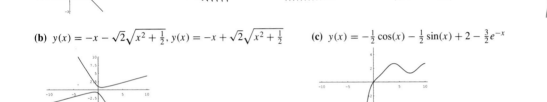 **33.** **35. (a)** **(b)** **37. (a)** $y(x) = -3x^2 - 6x - 5 + 6e^x$

(b) $y(x) = -x - \sqrt{2}\sqrt{x^2 + \frac{1}{2}}$, $y(x) = -x + \sqrt{2}\sqrt{x^2 + \frac{1}{2}}$ **(c)** $y(x) = -\frac{1}{2}\cos(x) - \frac{1}{2}\sin(x) + 2 - \frac{3}{2}e^{-x}$

Exercise Set 17.2

1. $y = C(x+1)$ **3.** $(y-2)^2 = (x+1)^2 + C$ **5.** $y = \tan(e^x + C)$ **7.** $Ce^{-x/y} = xy$
9. $(2y-x)^2(y+x) = C$ $(Cx > 0)$ **11.** $y = 2\sin\left(\frac{x^2}{2} + C\right)$ **13.** $y = x\sin(\ln Cx)$ $(Cx > 0)$
15. $2(x^2+y^2) = 5y$; $y = \frac{5 + \sqrt{25-16x^2}}{4}$ **17.** $e^{x/y} = \ln Cy$; $e^{x/y} = \ln y$ **19.** $e^x(y+1)^2 = Ce^y(x+1)$ **21.** $y' = F(x,y)$; let $y = vx$; $v + x\frac{dv}{dx} = F(1,v)$; $x\frac{dv}{dx} = F(1,v) - v$; $\frac{dv}{F(1,v)-v} = \frac{dx}{x}$; $\int(dv)/(F(1,v)-v) = \ln C_1|x|$; $C_1 x = e^{\int (dv)/(F(1,v)-v)}$; $x = Ce^{\int dv/F(1,v)-v}$
23. $\frac{(x-y)^2}{y} = Ce^{-(x+y)^2/(2y^2)}$ **25.** $y = 1 + 2\left(\frac{1-x}{2+x}\right)e^{-2x/(x-1)}$ **27.** If the differential equation is homogeneous, then F is homogeneous of degree 0, and so substituting $v = y/x$ gives $F(x,y) = F(x,vx) = x^0 F(1,v) = F(1,v)$. Define $G(v) = F(1,v)$, so that $F(x,y) = G(y/x)$. If, conversely, $F(x,y) = G(y/x)$, then $F(tx,ty) = G(ty/tx) = G(y/x)$. Thus, $F(tx,ty) = F(x,y) = t^0 F(x,y)$.

Exercise Set 17.3

1. $y = \frac{(x+1)^2}{2(x-2)} + \frac{C}{x-2}$ **3.** $x^3 - xy^2 = C$; $xy^2 - x^3 = 3$ **5.** $y\ln x + \frac{y^2}{2} = C$; $y\ln x + \frac{y^2}{2} = 8$ **7.** $y = x - 1 + Ce^{-x}$
9. $y = \frac{x^2}{3} + \frac{C}{x}$; $y = \frac{x^2}{3} - \frac{2}{3x}$ **11.** $\frac{y}{1-x} = \frac{(x+1)^2}{2} + C$; $y = \frac{2x-x^2-x^3}{2}$ **13.** exact; $x^2 - 4xy + 3y^2 = C$ **15.** exact; also variables separate; $x(y+1) - y = C$; $y = \frac{2+x}{1-x}$ **17.** linear; $\frac{y}{x^2+1} = \tan^{-1} x + C$; $y = (x^2+1)(\tan^{-1} x + \frac{\pi}{4})$ **19.** exact; $y\ln xy - e^x + 2(x-y) + y^3 = C$ **21.** exact; $\sqrt{x^2+y^2} + x - 2y = C$; $\sqrt{x^2+y^2} + x - 2y = 0$ **23.** exact; $x - y\ln x + \frac{y^2}{2} = C$
25. linear; $x = 1 + y^2 + C\sqrt{1+y^2}$ **27.** exact; $x\ln(x^2+y^2) - 2x - 2y\tan^{-1}\frac{y}{x} = C$ **29.** let $v = y^{-1} = \frac{1}{y}$; $y = \frac{2}{1+Ce^{x^2}}$
31. $f(x,y)dx + g(x,y)dy = 0$; multiply both sides by $\mu(x)$ and $[\mu(x)f(x,y)]dx + [\mu(x)g(x,y)]dy = 0$ is exact if and only if $\frac{\partial}{\partial y}[\mu(x)f(x,y)] = \frac{\partial}{\partial x}[\mu(x)g(x,y)]$, or $\mu f_y = \mu g_x + g\mu'$; $\frac{d\mu}{\mu} = \frac{f_y - g_x}{g}dx$. If $\frac{f_y - g_x}{g}$ is a function only of x, then $\ln\mu = \int \frac{f_y - g_x}{g}dx$, or $\mu(x) = e^{\int (f_y - g_x/g)dx}$ **33.** $\mu(x) = e^{2x}$; $e^{2x}\left(\frac{y^2}{2} + 2x - 1\right) = C$ **35.** If $\phi(x, y(x)) = C$, then $d\phi = \frac{\partial \phi}{\partial x} + \frac{\partial \phi}{\partial y}\frac{dy}{dx} = 0$. Conversely, if $d\phi(x,y) = 0$ and $y = y(x)$ solves $d\phi(x,y) = 0$, then $\phi(x, y(x)) = C$

Exercise Set 17.4

1. ≈ 3.42 hrs; ≈ 276 3. $A(30) \approx \$6049.65$; $t \approx 26.82$ yrs   5. $t \approx 3760$ yrs   7. $Q(3) \approx 33.7g$; $t \approx 6.35$ yrs
9. $T(30) \approx 6.17°C$; $T(t) \neq 5°$ C for any finite value of t, showing that the mathematical model is only an approximation of the physical system.
11. $Q(t) = 80 - 40e^{-t/20}$; concentration ≈ 0.355, or 35.5% 13. $\frac{dQ}{dt} = \alpha Q - \beta Q^2$, $(\alpha > 0, \beta > 0)$; separating variables, $\frac{dQ}{Q(\alpha - \beta Q)} = dt$; by partial fractions, $\frac{1}{Q(\alpha - \beta Q)} = \frac{1}{\alpha Q} + \frac{\beta}{\alpha(\alpha - \beta Q)}$; $\frac{1}{\alpha} \ln Q - \frac{1}{\alpha} \ln(\alpha - \beta Q) = t + \ln C_1$; $\frac{Q}{\alpha - \beta Q} = Ce^{\alpha t}$; $\frac{Q_0}{\alpha - \beta Q_0} = C$; $Q(t) = \frac{\alpha Q_0}{\beta Q_0 + (\alpha - \beta Q_0)e^{-\alpha t}}$
15. $v(t) = \sqrt{\frac{mg}{k}} \tanh \sqrt{\frac{kg}{m}} t$; $\lim_{t \to \infty} v(t) = \sqrt{\frac{mg}{k}}$ 17. $I(t) = 2(1 - e^{-6t})$ 19. $y(t) = \frac{a}{b}(1 - e^{-bt}) + y_0 e^{-bt}$
21. $\ln \frac{b(a-2x)}{a(b-x)} - \frac{2x(a-2b)}{a(a-2x)} = (a - 2b)^2 kt$ 23. concentration $\approx 0.602\%$; $t \approx 33.47$ min
25. $Q(t) = Q_0 e^{(\alpha/\beta) \sin \beta t}$ 27. $Q(t) = \frac{\alpha(Q_0 - m) + m(\alpha - \beta Q_0)e^{-(\alpha - m\beta)t}}{\beta(Q_0 - m) + (\alpha - \beta Q_0)e^{-(\alpha - m\beta)t}}$. If $Q_0 < m$, then $Q_0 - m < 0$ and $Q(t_1) = 0$ when $m(\alpha - \beta Q_0)e^{-(\alpha - m\beta)t_1} = \alpha(m - Q_0)$, or $e^{-(\alpha - m\beta)t_1} = \frac{\alpha(m - Q_0)}{m(\alpha - \beta Q_0)}$; $t_1 = \frac{1}{\alpha - m\beta} \ln \frac{m(\alpha - \beta Q_0)}{\alpha(m - Q_0)}$

Exercise Set 17.5

1. $y = C_1 e^{-x} + C_2 e^{-2x}$ 3. $y = C_1 e^{-4x} + C_2 x e^{-4x}$ 5. $y = C_1 \cos 3x + C_2 \sin 3x$ 7. $y = e^x(C_1 \cos 2x + C_2 \sin 2x)$
9. $y = C_1 e^{-x} + C_2 e^{5x/2}$ 11. $y = C_1 e^{4t} + C_2 e^{-t}$ 13. $u = C_1 + C_2 e^{3x/2}$ 15. $s = e^{3t/2}(C_1 + C_2 t)$
17. $y = \frac{3}{8} e^{-5x} + \frac{13}{8} e^{3x}$ 19. $y = 3e^{3x} - 1$ 21. $y = e^{2x}(2 - x)$ 23. $y = \sqrt{3} \sin x - \cos x$ 25. For real and unequal $m_1 = \frac{-b + \sqrt{b^2 - 4ac}}{2a}$, $m_2 = \frac{-b - \sqrt{b^2 - 4ac}}{2a}$, $y = C_1 e^{m_1 x} + C_2 e^{m_2 x}$. Let $C_1 = \frac{C_3 + C_4}{2}$, $C_2 = \frac{C_3 - C_4}{2}$, and then $y = \frac{C_3}{2}(e^{m_1 x} + e^{m_2 x}) + \frac{C_4}{2}(e^{m_1 x} - e^{m_2 x})$. Since $u = \frac{-b}{2a}$ and $v = \frac{\sqrt{b^2 - 4ac}}{2a}$, we have $m_1 = u + v$, $m_2 = u - v$.
$y = \frac{C_3}{2}(e^{(u+v)x} + e^{(u-v)x}) + \frac{C_4}{2}(e^{(u+v)x} - e^{(u-v)x}) = e^{ux}(C_3 \cosh vx + C_4 \sinh vx)$ (Now rename the constants.)
27. $t = \ln x \Rightarrow \frac{dt}{dx} = \frac{1}{x}$; $\frac{dy}{dx} = \frac{dy}{dt} \frac{dt}{dx} = \frac{1}{x} \frac{dy}{dt}$; $\frac{d^2 y}{dx^2} = -\frac{1}{x^2} \frac{dy}{dt} + \frac{1}{x^2} \frac{d^2 y}{dt^2}$. Substitute into the differential equation: $ax^2 \left(-\frac{1}{x^2} \frac{dy}{dt} + \frac{1}{x^2} \frac{d^2 y}{dt^2} \right) + bx \left(\frac{1}{x} \frac{dy}{dt} \right) + cy = 0$. Then, $a \frac{d^2 y}{dt^2} + (b - a) \frac{dy}{dt} + cy = 0$, which is linear with constant coefficients.
29. $s = C_1 \ln t + C_2$ 31. $y = C_1 + C_2 e^{3x} + C_3 e^{-x}$ 33. $y = C_1 \cos x + C_2 \sin x + C_3 \cos 2x + C_4 \sin 2x$
35. $y = C_1 e^x + C_2 \cos 2x + C_3 \sin 2x$; $y = \frac{2}{5} e^x - \frac{2}{5} \cos 2x + \frac{3}{10} \sin 2x$ 37. $L[ky] = aky'' + bky' + cky = k(ay'' + by' + cy) = kL[y]$; $L[y_1 + y_2] = a(y_1'' + y_2'') + b(y_1' + y_2') + c(y_1 + y_2) = (ay_1'' + by_1' + cy_1) + (ay_2'' + by_2' + cy_2) = L[y_1] + L[y_2]$ 39. If $am^2 + bm + c = 0$ has equal solutions, then they must be $-b/2a$, and $\sqrt{b^2 - 4ac}$ must be zero, $y = C_1 e^{-bx/2a} + C_2 x e^{-bx/2a}$, or $y_2 = xe^{-bx/2a}$. Using $y_2' = e^{-bx/2a} \left(1 - \frac{bx}{2a} \right)$ and $y_2'' = e^{-bx/2a} \left(-\frac{2b}{2a} + \frac{b^2 x}{4a^2} \right)$, we get $e^{-bx/2a} \left(-b + \frac{b^2 x}{4a} + b - \frac{b^2 x}{2a} + cx \right) = 0$.

Exercise Set 17.6

1. $y_p = 2x$; $y = C_1 \cos x + C_2 \sin x + 2x$ 3. $y_p = -\frac{3}{2}$; $y = C_1 e^{2x} + C_2 e^{-x} - \frac{3}{2}$ 5. $y_p = Ae^x = -\frac{3}{4} e^x$
7. $y_p = A \cos x + B \sin x = \frac{1}{3} \sin x$ 9. $y_p = Axe^{2x} = \frac{1}{3} xe^{2x}$ 11. $y_p = x(A \cos x + B \sin x) = \frac{3}{2} x \sin x$
13. $y_p = Ax^2 e^x = \frac{1}{2} x^2 e^x$ 15. $y_p = (Ax^2 + Bx)e^{2x} = \frac{1}{4} x(x - 1)e^{2x}$ 17. $y_p = e^{-x}(A \cos x + B \sin x) = -\frac{2}{3} e^{-x} \sin x$
19. $u_1' \cos x + u_2' \sin x = 0$ (1) $-u_1' \sin x + u_2' \cos x = \sec^3 x$ (2)   Multiply (1) by $\sin x$ and (2) by $\cos x$. Add the resulting equations to get $u_2'(\sin^2 x + \cos^2 x) = \sec^3 x \cos x = \sec^2 x$; $u_2' = \sec^2 x$. Substitute into (1) and solve for $u_1' = -\sec^2 x \tan x$.
21. $y = C_1 \cos x + C_2 \sin x + (x - \tan x) \cos x + (\sin x) \ln(\sec x)$
23. $x = C_1 \cos t + C_2 \sin t - \cos t \ln(\sec t) + t \sin t$ 25. $y = C_1 e^x + C_2 e^{-x} + \frac{x^2 e^x}{4} - \frac{1}{4} xe^x + \frac{1}{8} e^x$
27. $y = C_1 e^{-x} + C_2 e^{-2x} - e^{-2x} \cos(e^x)$ 29. $y = e^{-x}(C_1 \cos x + C_2 \sin x) + \frac{e^x}{8}(\sin x - \cos x)$ 31. $y_p = (Ax^4 + Bx^3 + Cx^2 + Dx)e^x$
33. $y_p = Ax^3 + Bx^2 + Cx + (Dx^2 + Ex)e^{2x}$ 35. $y_p = e^{3x}(Ax \cos 4x + Bx \sin 4x) + C$ 37. $y_p = e^{-x}[(Ax + B) \cos x + (Cx + D) \sin x]$
39. $y = \frac{11}{16} e^{2x} + \frac{1}{16} e^{-2x} - \frac{1}{4}(x + 3)$ 41. $y = \frac{7}{3} \cos x + \frac{8}{3} \sin x - \frac{1}{3} \cos 2x$ 43. $y = e^{2x} \left(-\frac{19}{17} \cos x + \frac{76}{17} \sin x \right) + \frac{2}{17} e^{-2x}$
45. $y = \frac{14}{9} e^{2x} + \frac{4}{9} e^{-x} + \frac{1}{3} xe^{2x}$ 47. $y = C_1 e^x + C_2 e^{-x} + \frac{x}{2} \sinh x$ 49. $y = C_1 \cos 2x + C_2 \sin 2x - \frac{1}{4} x \cos 2x$
51. $x(t) = C_1 e^{3t} + C_2 t e^{3t} + t^2 e^{3t} + \frac{1}{9} t^2 + \frac{4}{27} t + \frac{2}{27}$; $x(t) = \left(\frac{106}{27} - \frac{376}{27} t + t^2 \right) e^{3t} + \frac{1}{27}(3t^2 + 4t + 2)$
53. $y_p = -\frac{x^4}{36} - \frac{x^3}{27} - \frac{x^2}{27}$; $y = C_1 + C_2 x + C_3 e^{3x} - \frac{x^4}{36} - \frac{x^3}{27} - \frac{x^2}{27}$ 55. $y_p = \frac{1}{3} x^2 e^x + \frac{3}{2} x + \frac{9}{4}$; $y = C_1 e^x + C_2 x e^x + C_3 e^{-2x} + \frac{1}{3} x^2 e^x + \frac{3}{2} x + \frac{9}{4}$

Exercise Set 17.7

1. $y = \frac{1}{2}\cos 8t$ **3.** $y = \frac{2}{3}\cos 4t - 3\sin 4t$ **5.** $y = -10\cos 14t$ **7.** $y = 9e^{-7t} - 5e^{-14t}$; overdamped   **9.** $y = \frac{1}{2}\cos 4t$; underdamped   **11. (a)** $c > \sqrt{5}$; **(b)** $c = \sqrt{5}$; **(c)** $c < \sqrt{5}$ **13. (a)** $\frac{1}{2}$ radian; **(b)** $\frac{\pi}{2}$; **(c)** $\frac{5\pi}{24}$; **(d)** 2 ft/sec

15. $C_1 \cos\omega t + C_2 \sin\omega t = \sqrt{C_1^2+C_2^2}\left(\frac{C_1}{\sqrt{C_1^2+C_2^2}}\cos\omega t + \frac{C_2}{\sqrt{C_1^2+C_2^2}}\sin\omega t\right)$. Let α be such that $\sin\alpha = \frac{C_1}{\sqrt{C_1^2+C_2^2}}$ and $\cos\alpha = \frac{C_2}{\sqrt{C_1^2+C_2^2}}$. Let β be such that $\cos\beta = \frac{C_1}{\sqrt{C_1^2+C_2^2}}$ and $\sin\beta = \frac{C_2}{\sqrt{C_1^2+C_2^2}}$. Let $C = \sqrt{C_1^2+C_2^2}$. Then, **(a)** $C(\sin\alpha\cos\omega t + \cos\alpha\sin\omega t) = C\sin(\omega t + \alpha)$ **(b)** $C(\cos\beta\cos\omega t + \sin\beta\sin\omega t) = C\cos(\omega t - \beta)$ **17.** $I(t) = e^{-t}(C_1\cos 2t + C_2 \sin 2t) + 4\cos t + 2\sin t$; $\lim_{t\to\infty} I(t) = 4\cos t + 2\sin t$

Exercise Set 17.8

1. $y = a_0 \sum_{k=0}^{\infty} \frac{x^{3k}}{3^k k!} = Ce^{x^3/3}$ **3.** $y = a_0 \sum_{k=0}^{\infty} \frac{x^{2k}}{(2k)!} + a_1 \sum_{k=0}^{\infty} \frac{x^{2k+1}}{(2k+1)!} = a_0 \cosh x + a_1 \sinh x = C_1 e^x + C_2 e^{-x}$

5. $y = a_0\left[1 + \sum_{k=1}^{\infty} \frac{1\cdot 4\cdot 7\cdots(3k-2)}{(3k)!}x^{3k}\right] + a_1\left[x + \sum_{k=1}^{\infty} \frac{2\cdot 5\cdot 8\cdots(2k-1)}{(3k+1)!}x^{3k+1}\right]$ **7.** $y = a_0 \sum_{k=0}^{\infty} \frac{(-1)^k}{(2k)!}x^{2k} + a_1 \sum_{k=0}^{\infty} \frac{(-1)^k}{(2k+1)!}x^{2k+1}$; $y = \sum_{k=0}^{\infty} \frac{(-1)^k}{(2k)!}x^{2k} = \cos x$ **9.** $y = a_0\left(1 - \frac{x^4}{3\cdot 4} + \frac{x^8}{3\cdot 5\cdot 7\cdot 8} - \frac{x^{12}}{3\cdot 4\cdot 7\cdot 8\cdot 11\cdot 12} + \cdots\right) + a_1\left(x - \frac{x^5}{4\cdot 5} + \frac{x^9}{4\cdot 5\cdot 8\cdot 9} - \frac{x^{13}}{4\cdot 5\cdot 8\cdot 9\cdot 12\cdot 13} + \cdots\right)$

11. $y = a_0 \sum_{k=1}^{\infty} \frac{1\cdot 3\cdot 5\cdots(2k+1)}{2^k k!}x^{2k} + a_1 \sum_{k=1}^{\infty} \frac{2^k(k+1)!}{1\cdot 3\cdot 5\cdots(2k+1)}x^{2k+1}$ **13.** $y = 2 + x + \frac{x^2}{4} + \frac{7}{4}\sum_{k=1}^{\infty} \frac{(-1)^k x^{2k}}{1\cdot 3\cdot 5\cdots(2k-1)} + \sum_{k=1}^{\infty} \frac{(-1)^k x^{2k+1}}{2\cdot 4\cdot 6\cdots(2k)}$

15. $y_1 = 1 - \frac{x^2}{2!} + \frac{x^3}{3!} + \frac{x^4}{4!} - \frac{4x^5}{5!} + \frac{3x^6}{6!} + \frac{9x^7}{7!} - \frac{27x^8}{8!} + \frac{12x^9}{9!} + \cdots$; $y_2 = x - \frac{x^3}{3!} + \frac{2x^4}{4!} + \frac{x^5}{5!} - \frac{6x^6}{6!} + \frac{9x^7}{7!} + \frac{12x^8}{8!} - \frac{51x^9}{9!} + \frac{60x^{10}}{10!} + \cdots$; $y = C_1 y_1 + C_2 y_2$

17. $y(x) = 2 + x + \frac{2x^2}{2!} - \frac{x^3}{3!} - \frac{7x^5}{5!} + \frac{4x^6}{6!}x - \frac{7x^7}{7!} + \frac{46x^8}{8!} - \frac{35x^9}{9!} + \frac{102x^{10}}{10!}$

Chapter Review

1. $2x + 3y - e^{-x}\cos y = C$ **3.** $y = \sin\left(\frac{x^2}{2} + \frac{\pi}{2}\right)$ **5.** $Cx^2 = ye^{y/x}$ **7.** $\frac{3}{2}\tan^{-1}\left(\frac{y}{2x}\right) - \frac{1}{2}\ln\left[\left(\frac{y}{x}\right)^2 + 4\right] = \ln x + C$; $\tan^{-1}\frac{y}{2x} = \frac{2}{3}\ln\frac{\sqrt{y^2+4x^2}}{2}$ **9.** $y = -\sqrt{x(2-x)}\sin^{-1}(1-x) + C\sqrt{x(2-x)}$; $y = \sqrt{2x - x^2}[2 - \sin^{-1}(1-x)]$

11. $x\cos xy = C$ **13.** $ye^{x^2} = -\frac{1}{2}e^{-x^2} + C$; $y = -\frac{1}{2}e^{-2x^2} + \frac{3}{2}e^{-x^2}$ **15.** $y = C_1 e^{-3x} + C_2 x e^{-3x}$; $y = 3e^{-3x} + 5xe^{-3x}$

17. $s(t) = C_1 + C_2 e^{-2t}$; $s(t) = 2(1 - e^{-2t})$ **19.** $x_p = -\frac{1}{3}e^t$ **21.** $y_p = -\frac{x^2}{2} - x + 3xe^x$ **23.** $y_p = \frac{e^x}{53}(7\sin x - 2\cos x)$

25. $y_p = \frac{xe^x}{2} - \frac{1}{4}e^{-x}$ **27.** $y_p = x(Ax + B)e^{-x} + C\cos 2x + D\sin 2x$ **29.** ≈ 43.71 years   **31.** $Q(t) = e^{\alpha t}\left(Q_0 - \frac{H}{\alpha}\right) + \frac{H}{\alpha}$; (1) If $\frac{H}{\alpha} > Q_0$, then $Q(t) = 0$ when $\frac{H}{\alpha} = e^{\alpha t}\left(\frac{H}{\alpha} - Q_0\right)$ and $t = \frac{1}{\alpha}\ln\frac{H}{H - \alpha Q_0}$. Thus, the population dies out in a finite time. (2) If $\frac{H}{\alpha} < Q_0$, then the population grows without limit. (3) If $\frac{H}{\alpha} = Q_0$, then $Q(t) = \frac{H}{\alpha}$, a constant.   **33.** $y(t) = \frac{e^{-t}}{3}(2\sin 7t - \cos 7t)$

35. $y = a_0 \sum_{k=0}^{\infty} \frac{2^{2k}k!}{(2k)!}x^{2k} + a_1 \sum_{k=0}^{\infty} \frac{x^{2k+1}}{k!}$; series converges for $-\infty < x < \infty$.

Photo Credits

PROLOGUE: ix, top, Kathleen Olson; bottom, courtesy of Texas Instruments.

CHAPTER 10: 720, photo by Jay M. Pasachoff; **740**, left, © Herb Charles Ohlmeyer from Fran Heyl Associates; right, © Dwight Kuhn; **759**, Loren M. Winters; **765**, Larry Molmud/Mucking Otis Press; **769**, Loren M. Winters; **826**, Exercises 7 and 8, reprinted by permission of Dr. Judith H. Morrel, Department of Mathematics and Computer Science, Butler University.

CHAPTER 11: 829, © Diana Mara Henry, Carmel, CA; **884**, courtesy, McDonald Observatory.

CHAPTER 12: 886, photo by Jay M. Pasachoff; **890**, Dr. A. Lesk/Laboratory of Molecular Biology/Science Photo Library/Photo Researchers; **902**, © 1992, 1993, PhotoDisc, Inc.; **917**, © Mark Burnett/Photo Researchers; **919**, © Science Photo Library/Photo Researchers; **929**, top, Erich Lessing/Art Resource, NY; bottom, © 1990 Hansen Planetarium, Salt Lake City, Utah; reproduced with permission; **931**, photo by Jay M. Pasachoff.

CHAPTER 13: 938, photo by Jay M. Pasachoff; **940**, Science Source/Photo Researchers; **943**, Peter Arnold; **945**, Department of the Interior/U.S. Geological Survey.

CHAPTER 14: 1000, photo by Jay M. Pasachoff; **1007**, W. & D. McIntyre/Science Source/Photo Researchers; **1011**, courtesy, Central Scientific Company; **1022**, Superstock; **1038**, © Bill O'Connor/Peter Arnold, Inc.; **1063**, photo by Jay M. Pasachoff.

CHAPTER 16: 1143, photo by Jay M. Pasachoff; **1145**, visualization by Wayne Lytle, Cornell Theory Center/© 1991 Cornell University; **1148**, courtesy, Laboratory for Terrestrial Physics, NASA/Goddard Space Flight Center; **1146**, Andrew Sacks/Tony Stone Images; **1174**, courtesy of the Master and Fellows of Gonville and Caius College, University of Cambridge, England; **1189**, courtesy, Dr. Peter Cornillon/The University of Rhode Island; **1189**, Jet Propulsion Laboratory/NASA; **1210**, NASA.

CHAPTER 17: 1214, photo by Jay M. Pasachoff; **1240**, Los Alamos National Lab; **1243**, courtesy, Aquaculture Development Program; **1245**, top, Jet Propulsion Laboratory; bottom, © Richard Megna 1991/Fundamental Photographs; **1246**, courtesy, Central Scientific Company; **1250**, left, © Paul Silverman 1990/Fundamental Photographs; right, D. Nunuk/Science Source/Photo Researchers; **1251**, © Carl R. Sams II/Peter Arnold, Inc.; **1274**, F. B. Farquharson/courtesy, The University of Washington, Instructional Media Collection.

INDEX

Page numbers followed by a "d" refer to definitions on those pages.

Absolute convergence, 785–786, 785d
Absolute extrema, 1046
  of functions of two variables, 1038d
Acceleration, 917, 918d
  tangential components, 923–926
Addition of vectors, 830–831
Additive identity (vector), 833
Additive inverse (vector), 833
Air resistance, 1244–1245
Algebraic vectors, 831–832
Alternating harmonic series, 782, 784, 798
Alternating series, 781–785
Alternating Series Test, 782
Ampère, André Marie, 997
Ampère's Law, 997, 1211
Angular speed, 919
Approximate error, 1007
Arc length, 902–905
Arc length function, 903–904
Area,
  of a cylindrical surface, 1151
  using line integrals, 1177–1178
  of a parallelogram, 855–856
  of a plane region, 1071
  of a surface, 1103–1109
Ascent time, 1283
Auxiliary equation, 1253
Average velocity, 917

Basis vectors, 834–836
Bernoulli equation, 1239, 1241
Bernoulli family, 1241–1242
Binomial Formula, 807
Binomial series, 807–810
Binormal, 910
Bond angle, 883
Boundary, 974

Boundary point, 974
Boundary-value problem, 1218, 1257
Bounded sequence, 748d
Boundedness, 747–749
Brahe, Tycho, 929

C^1 transformation, 1129
Cauchy-Schwarz Inequality, 840
Center of curvature, 912
Center of mass,
  using a surface integral, 1188–1189
  using double integrals, 1092–1101
  using triple integrals, 1115
Central force field, 1210
  radially symmetric, 1210
Centripetal force, 919
Centroid,
  using double integrals, 1092–1101
  using triple integrals, 1115
Chain rule, 1012–1021
  generalized, 1016
  using Computer Algebra Systems, 1035–1036
Characteristic equation, 1253
Chemical reactions, 1245–1246
Circle of curvature, 912
Circular cone, 963
Circular helix, 889
Circulation, 1180–1183, 1203
Closed form of series, 757
Closed set, 994
Cobb-Douglas Production Function, 986, 1009
Common ratio, 756
Comparison test, 772–776
Complement set, 994
Complementary equation, 1259
Complementary solution, 1259
Completeness property, 748

Computer Algebra Systems,
  calculus of vector functions, 914–915
  computing Taylor polynomials, 732–734
  double integration, 1106–1109
  finding maximum values of functions
    of several variables, 1046–1049
  finding minimum values of functions
    of several variables, 1046–1049
  finding tangent planes, 1035–1036
  graphing vector functions, 891–892
  infinite series, 766–767
  investigating lines and planes in
    space, 876–879
  limits of functions of several
    variables, 987–990
  partial derivatives of functions of
    several variables, 987–990
  power series computations, 811
  sketching surfaces, 947–951
  sketching vector fields, 1158–1159
  solving differential equations, 1220–1222
  using chain rule, 1035–1036
  vector operations, 859–862
  work done by a force field, 1158–1159
Conditional convergence, 786–787, 786d
Connected region, 1164
  multiply, 1167
  simply, 1167
Conservation of energy, 1170–1171
Conservative force field, 1164
Conservative vector field, 1148
Constrained extremum problems, 1051–1059
  functions of more than two variables, 1056–1058
  two or more constraints, 1058–1059
Continuity,
  and differentiability, 1004
  of function of two or more variables, 974–975

Continuity, *continued*
  of function of two variables, 974d
Continuity equation, 1211
Contour map, 939, 945–947, 945d
Convergent series, 761–762, 770–776, 777–780
  properties, 763–765
Coordinate planes, 846
Coplanar vectors, 858–859
Coriolis acceleration, 935
Cormack, Allan,
  personal view, 885–886
Critical point, 1040d
Critically damped motion, 1272
Cross product, 853–859, 853d
  algebraic properties, 856–857
  geometric interpretation, 854–855
Curl, 1200d, 1203, 1204
curl **F**, 1200d, 1203, 1204
Curvature, 908–913
  in two dimensions, 912–913
Curves,
  length of, 903
  parameterized by arc length, 904–905
  positively oriented with respect to R, 1178
  positively oriented with respect to S, 1200
  rectifiable, 902
  smooth, 902
Cylindrical box, 1118
Cylindrical coordinates, 1117–1122
  changing from rectangular coordinates, 1121–1122
  in triple integrals, 1118–1121
Cylindrical surface, area, 1151

d'Alembert, Jean Le Rond, 984
d'Alembert's Solution, 984
Damped free vibrations, 1271–1273
Decreasing sequence, 747d
Degenerate sphere, 960
Derivative(s),
  directional, 1023–1030, 1024d
    extreme values, 1027–1028
    function of three variables, 1030
  implicit functions, 1019–1021
  partial, 977–990, 978d
  of a power series, 795
  second, 1017–1019
  of vector function, 894–895, 894d, 896–897
Descent time, 1283
Determinants, 853–854, 856
Differentiability,
  function of three or more variables, 1010
  function of two variables, 1001–1010, 1003d
  geometric interpretation, 1005
Differential,
  function of three or more variables, 1010
  function of two variables, 1006, 1006d
Differential equations (ordinary), 1215–1285
  applications, 1239–1249
  boundary values, 1218
  exact, 1232–1234, 1232d
  family of solutions, 1217
  general solution, 1217
  homogeneous, 1226–1231, 1227d
  implicit solution, 1218
  initial condition, 1217
  linear, 1234–1238, 1234d
    integrating factor, 1236
    procedure for solving, 1237
    standard form, 1235
  one-parameter family, 1217
  second-order homogeneous, 1251–1258
    general solution, 1258
  second-order nonhomogeneous, 1259–1266
  separable, 1224–1226, 1224d
  series solutions, 1275–1279
  solution, 1216
  solving using Computer Algebra Systems, 1220–1222
  vibrating spring, 1269
Differential equations (partial), 983–985, 1216
Differential of area in polar coordinates, 1087
Differential of volume, 1119
Differential of volume in spherical coordinates, 1124
Differentiation,
  power series, 794–800
Direction angles (three dimensions), 850
Direction cosines (three dimensions), 850
Direction field, *see* Slope fields
Direction numbers, 866
Direction vector, 864
Directional derivatives, 1001, 1023–1030, 1024d
  extreme values, 1027–1028
  function of three variables, 1030
Dirichlet, P. G. Lejeune, 818
Dirichlet's Theorem, 818
Distance,
  between plane and point, 875–876
Distance formula (three dimensions), 847–848
div **F**, 1181–1182, 1197–1198
Divergence, 1181–1182, 1197–1198
Divergence of a vector field, 1192d
Divergence Theorem, 1192–1198, 1205
Divergent series, 763–765, 770–776, 777–780
Dot product, 837–844, 838d
  properties, 838–839
  in three dimensions, 848
Double integral,
  centers of mass, 1092–1101
  centroids, 1092–1101
  evaluating by iterated integrals, 1073–1082
  geometric interpretation, 1070–1071
  moments, 1092–1101
  over general regions, 1070
  over rectangular regions, 1068–1069, 1068d
  in polar coordinates, 1085–1091
  properties, 1072
  volume defined by, 1071d
Double integration,
  using Computer Algebra Systems, 1106–1109
Doyle Log Rule, 995

Electrical circuits, 1246–1247
Ellipsoid, 961
Elliptic cone, 962
Elliptic hyperboloid, 961–962
Elliptic paraboloid, 944, 961
Elliptical cylinder, 958
Elson, Constance,
  personal view, 827–828
Energy,
  conservation of, 1170–1171
  kinetic, 1170
  potential, 1170
Equimarginal productivity, 1064
Equivalent vectors, 830
Error,
  in alternating series, 783–785
  approximate, 1007
  in Taylor polynomial approximation, 727–731
  percentage, 1007
  relative, 1007
Error function, 812
Euler, Leonhard, 719–720, 772
Euler's Formula, 1255
Exact differential, 1166
Extended Mean-Value Theorem, 728
Extreme values, 1037–1049

Falling body, 1244–1245
Faraday, Michael, 997
Faraday's Induction Law, 1211
Faraday's Law, 997
Fibonacci, Leonardo, 740
Fibonacci numbers, 740
Fibonacci sequence, 740
Final variable, 1012
Finite sequence, 737
First Fundamental Theorem of Calculus for Vector Functions, 898
First moment, 1096
Flux, 1180–1183, 1190, 1197–1198
Flux density, 1197
Force field, 1145
  conservative, 1164
Forced vibrations, 1273–1274
Forcing function, 1273
Fourier, Jean Baptiste Joseph, 813
Fourier series, 813–820, 816d
  convergence theorem, 817–818
  deriving, 815–817
  harmonics, 819–820
Fractals, 765–766
Free motion, 1269
Frenet Formulas, 917
Frequency, 1270
Fubini's Theorem, 1075–1082, 1119
  definition, 1075
  stronger form, 1079
  triple integrals, 1111
Function(s),
  approximation by polynomials, 721–734
  critical point, 1040d
  gradient, 1026–1027, 1026d, 1029–1030
  harmonic, 983, 991, 1208
  homogeneous, 1226d
  limit of, 967–974
  linearly dependent, 1252
  linearly independent, 1252
  of more than one variable, 939–997
  potential, 1148, 1161, 1167–1169
  sawtooth, 818
  scalar, 888

of several variables, 939–997
  finding maximum and minimum values
    using Computer Algebra Systems,
    1046–1049
  of three variables, 941, 947
  of two variables, 940–947, 940d
    absolute extrema, 1038d
    local extrema, 1038d
  vector, 887–935, 888d
Fundamental, 820
Fundamental Increment Formula, 1003
Fundamental Theorem for Line Integrals, 1162, 1205

G, 929
Gauss's Law, 1211
Gauss's Theorem, 1192–1198, 1205
General term of sequence, 737
Geoid, 1148
Geometric series, 756–760
Geometric vectors, 830
Gibbs, Josiah Willard, 829
Goldberg, Alfred L.,
  personal view, 1143–1144
Golden mean, 753
Gompertz model, 1251
Gradient, 1026–1027, 1026d, 1029–1030
Gradient field, 1148, 1161–1171
  testing for, 1167–1169
Graphs,
  of function of two variables, 942–944, 943d
Gravitational potential function, 984
Green, George, 1174
Green's Identities, 1199
Green's Theorem, 1173–1183, 1205
  and two-dimensional fluid flow, 1180–1183

Hamilton, William Rowan, 829
Harmonic conjugate, 991
Harmonic function, 983, 991, 1208
Harmonic series, 762–763
Harmonics, 819–820
Heading, 837
Heat equation, 983, 1216
Heaviside, Oliver, 829
Higher-order moments, 1096–1098
Homogeneous differential equations, 1226–1231
Homogeneous function, 1226d
Hyperbolic cylinder, 959
Hyperbolic paraboloid, 947, 963–966

i, 835
Ideal gas law, 954
Image, 1128
Implicit functions,
  derivative formulas, 1019–1021
Incompressible flow, 1190
Incompressible fluid, 1182
Increasing sequence, 747d
Independent variable, 1012
Infinite limits, 744
Infinite sequence, 737
Infinite series, 719–820, 754
  sum of, 756d
    using Computer Algebra Systems, 766–767
Infinite series of constants, 754–767
Initial point, 830

Initial-value problems, 1217, 1257
Integral,
  double, see Double integral
  line, see Line integral
  multiple, 1067–1141
  surface, see Surface integral
  triple, see Triple integral
  of vector function, 898–900
Integral test, 770–772
Integrating factor, 1235, 1236
Integration,
  interchanging order of, 1082
  of power series, 794–800
Interior point, 974
Intermediate variable, 1012
Interval of convergence, 791–793
Irrotational flow, 1204
Isoclines, 1223
Isolated point, 977
Isotherm, 939, 945

j, 835
Jacobi, Carl, 1130
Jacobian of a transformation, 1131d
Jacobi's identity, 881

k, 849
κ, 908–913
Kelvin, Lord, 1174
Kepler, Johannes, 928
Kepler's laws, 928–931
Kinetic energy, 1170
Kirchhoff's Second Law, 1246
Koch curve, 765
Koch island, 765
kth moment about the mean, 1099
kth moment about the origin, 1099

Lagrange, Joseph Louis, 727, 1051
Lagrange form of the remainder, 727
Lagrange Multiplier, 1052
  summary, 1056
Lagrange's Theorem, 1052
Laplace's Equation, 983, 1208
Laplacian of f, 1023, 1199
Lattice (crystal), 883
Law of Conservation of Energy, 1170
Law of Cosines, 839–840
Law of Logistic Growth, 1242
Least upper bound, 748
Length of curve, 903
Level curves, 945–947, 945d, 1029–1030
Level surfaces, 947
Lieberman, David,
  personal view, 999–1000
Limit,
  formal definition, 973d
  of function of more than two variables, 969
  of function of several variables,
    using Computer Algebra Systems, 987–990
  of function of two variables, 967–974, 969d
  infinite, 744
  properties, 972
  properties of sequence, 744–747
  of sequence, 741–744, 741d
  showing failure to exist, 970–972
Limit Comparison Test, 774

Line integral, 1149–1159, 1150d
  circulation, 1180–1183
  computational formula, 1151–1152
  finding area of a region, 1177–1178
  flux, 1180–1183
  Fundamental Theorem, 1162
  geometric interpretation, 1151
  independent of path, 1163
  in three dimensions, 1157–1158
Linear combination, 1252
Linear differential operator, 1252
Lines,
  angle between, 866–868
  investigating with Computer Algebra Systems, 876–879
  parametric equations of, 865
  skew, 866
  in space, 864–868
  symmetric equations, 869
Local extrema,
  functions of two variables, 1038d, 1041–1046
Logistic equation, 1242
Lotka, A. J., 1251
Lotka–Volterra model, 1251
Lower bound, 748

Maclaurin, Colin, 802
Maclaurin series, 802, 805–806
Malthus model, 1241
Mandelbrot, Benoit B., 765
  personal view, 717–718
Mapping, 1127
Marginal cost, 985
Marginal productivity, 987
Marginal propensity to consume, 824
Marginal propensity to save, 824
Mass,
  of a lamina, 1093
  using a surface integral, 1188–1189
  using triple integrals, 1115
Mass flux, 1190
Maxwell, James Clerk, 829, 1210
Maxwell's Equations, 1210
McGrath, Frank,
  personal view, 1213–1214
Mean, 941
Method of Lagrange Multipliers, 1051
Method of undetermined coefficients, 1261–1264
Mixing problems, 1243–1244
Möbius strip, 1192
Moment of inertia, 1096–1098
  using a surface integral, 1188–1189
  using triple integrals, 1115
Moments,
  higher-order, 1096–1098
  kth moment about the mean, 1099
  kth moment about the origin, 1099
  in probability theory, 1099–1101
  using double integrals, 1092–1101, 1094
  using triple integrals, 1115
Monotone Bounded Sequence Theorem, 748
Monotone sequence, 747d
Monotonicity, 747–751
  tests for, 751
Motion,
  along a curve, 917–923
  of a projectile, 921–923

Moving trihedral, 910
Multiple integral, 1067–1141
Multiplier effect, 824
Multiply-connected regions, 1167, 1178–1180

N, 907–913
Nappe, 962
Newhouse, Joseph,
   personal view, 937–938
Newton, Isaac, 929
Newton's Law of Cooling, 1249
Newton's Law of Gravitation, 929, 1063, 1147
Newton's Second Law of Motion, 929
Norm, 1068
Normal lines, 1032–1035, 1033d
Normal Probability Density Function, 1092
Normal vectors, 870, 871–872, 907–913
nth term of sequence, 737

Octants, 846
Open region, 1162
Open set, 994
Order (of a differential equation), 1216
Ordinary differential equations,
   see Differential equations
Orientable surface, 1192
Orthogonal trajectories, 1223, 1247–1249
Orthogonal vectors, 841
Orthonormal set, 882
Osculating circle, 912
Outer unit normal, 1190
Overdamped motion, 1272

p series, 772
Parabolic cylinder, 958
Parallel Axis Theorem, 1117
Parallel vectors, 836d
Parallelogram,
   area of, 855–856
Parallelogram law, 830
Parametric equations,
   of a curve, 889, 1154
   of a line, 865
Partial derivative(s), 977–990, 978d
   of functions of several variables,
      using CAS, 987–990
   higher-order, 981–983
   second-order, 981
   three or more variables, 980–981
Partial differential equations, 983–985, 1216
   applications in economics, 985–987
Partial integration, 1074
Partition, 1068
Path, 1163
Path independence, 1161–1171
   theorem, 1163
Percentage error, 1007
Period, 1270
Periodic extension of f, 817
Permeability, 997, 1210
Permittivity, 1210
Permittivity constant, 997
Personal view,
   Cormack, Allan, 885–886
   Elson, Constance, 827–828
   Goldberg, Alfred L., 1143–1144

   Lieberman, David, 999–1000
   Mandelbrot, Benoit B., 717–718
   McGrath, Frank, 1213–1214
   Newhouse, Joseph, 937–938
   Urquhart, N. Scott, 1065–1066
Planck's constant, 996
Plane curve, 890
Planes, 870–879
   angle between, 872–873
   distance to point, 875–876
   equations of, 870–871
   investigating with Computer
      Algebra Systems, 876–879
   line of intersection, 875
   parallel, 872–873
   perpendicular, 872–873
   sketching, 874–875
   traces, 874
Polar coordinates,
   changing from rectangular coordinates,
      1089–1091
   differential of area, 1087
   in double integrals, 1085–1091
Polar moment of inertia, 1097
Population growth, 1241–1243
Position vector, 831
Potential energy, 1170
Potential function, 1148, 1161
   finding, 1167–1169
Power series, 720, 789–793, 789d
   computations with Computer
      Algebra Systems, 811
   differentiation, 794–800
   integration, 794–800
   interval of convergence, 791–793
   radius of convergence, 791–793
Predator–prey problem, 1251
Principia Mathematica, 929
Prism, 1070
Probability density, 1099
Probability theory,
   moments in, 1099–1101
Projection (trace), 945
Projection (vectors), 843–844

Quadric surfaces, 960

Radially symmetric central force field, 1210
Radioactive decay, 1240–1241
Radius of convergence, 791–793
Radius of curvature, 912
Radius of gyration, 1096, 1097
Ratio Test, 777, 791–793
Rectangular coordinates,
   changing to cylindrical coordinates, 1121–1122
   changing to polar coordinates, 1089–1091
Rectifiable curve, 902
Recursion formula, 739, 1277
Relative error, 1007
Relative growth rate, 1063
Resonance, 1274
Resultant vector, 830
Riemann sum, 1068
Right-hand rule, 846, 855
RL series circuit, 1246
Root Test, 779

rot **F**, 1181–1182
Rotation of **F**, 1181–1182
Rotation of **F** around **n** at **P**, 1204

Saddle point, 963
Sawtooth function, 818
Scalar curl, 1181
Scalar function, 888
Scalar multiplication, 831
Scalar product, see Dot product
Scalars, 829
Schrödinger, Erwin, 996
Schrödinger's Equation, 996, 1141
Second derivative,
   by chain rule, 1017–1019
Second Fundamental Theorem of Calculus for
      Vector Functions, 898
Second-order reactions, 1245
Sectionally smooth curves, 1153
Self-similarity, 765
Separable differential equations, 1224–1226
Sequence, 736–751, 737d
   bounded, 748d
   boundedness, 747–749
   convergent, 741d
   decreasing, 747d
   divergent, 741d
   increasing, 747d
   limit properties, 744–747
   limits, 741–744, 741d
   monotone, 747d
   Monotone Bounded Theorem, 748
   monotonicity, 747–751
   of partial sums, 755
   recursive definition, 739–740
   Squeeze theorem, 746
Series,
   absolute convergence, 785–786, 785d
   alternating, 781–785
   alternating harmonic, 782, 784, 798
   binomial, 807–810
   conditional convergence, 786–787, 786d
   convergent, 719–720, 756d, 761–762, 763–765,
      770–776, 777–780
   divergent, 720, 756d, 761–762, 763–765,
      770–776, 777–780
   Fourier, 813–820
   geometric, 756–760
   harmonic, 762–763
   infinite, see Infinite series
   Maclaurin, 802, 805–806, 809–810
   power, 789–793
   Taylor, 802–810
   telescoping, 760
Series power, 789d
Simple harmonic motion, 1270
Simple region, 1173
Simple three-dimensional region, 1193
Simply connected region, 1167
Sink, 1182, 1190, 1197
Skew lines, 866
Slope fields, 1220
Smooth curve, 902
Solution (of a differential equation), 1216
Source, 1182, 1190, 1197
Space curve, 889
Speed, 918d

Speed, angular, 919
Spherical box, 1123
Spherical coordinates, 1122–1123
 triple integrals in, 1123–1125, 1123d
Squeeze theorem, 746
Steady-state flow, 1180
Stokes, George, 1199, 1201
Stokes's Theorem, 1199–1204, 1205
 application to fluid flow, 1203–1204
Streamlines, 955, 1146
Subrectangle, 1068
Subtraction of vectors, 831, 833d
Sum of infinite series, 756d
Surface(s), sketching, 943, 955–966
 cylinders, 957–959
 planes, 956–957
 quadric surfaces, 960–961
 spheres, 959–960
 using Computer Algebra Systems, 947–951
Surface area, 1103–1109
Surface integral, 1184–1190
Surface of revolution, 963

T, 907–913
Tangent lines, 895–896, 896d
Tangent planes, 1001, 1032–1036, 1033d
 finding using Computer Algebra Systems, 1035–1036
Tangent vectors, 895–896, 896d, 907–913
Taylor, Brook, 724
Taylor polynomials, 721–734, 724d
 computing using Computer Algebra Systems, 732–734
 error, 727–731
Taylor series, 731–732, 802–810
Taylor's formula with remainder, 727
Taylor's Theorem, 803
Telegraph equation, 983
Telescoping series, 760
Terminal point, 830
Theorem of Pappus, First, 1102
Thermal diffusivity, 983
Thompson, William, 1174
Three-dimensional fluid flow, 1189–1190
Topographical map, 945
Torsion, 917
Traces, 874, 945, 956
Transformation, 1127
Tree diagram, 1014
Triangle inequality, 834
Triple integral, 1110–1116, 1111d
 applications, 1114–1116
 changing variables in, 1135–1136
 in cylindrical coordinates, 1118–1121, 1119d
 evaluating by iterated integrals, 1111–1112
 over general regions, 1112–1114
 in spherical coordinates, 1123–1125, 1123d
 volume, 1114

Triple scalar product, 857
Twisted cubic, 890
Two-dimensional curve, 1181
Two-dimensional vector space, 832d
Type I regions, 1077
Type II regions, 1077

Undamped free vibrations, 1269–1271
Undamped motion, 1269
Underdamped motion, 1272
Unit normal vector, 907–913
Unit tangent vector, 907–913
Unit vectors, 834–836
Universal gravitational constant, 929
Upper bound, 748
Urquhart, N. Scott,
 personal view, 1065–1066

Variables,
 changing in double integrals, 1127–1135
 changing in multiple integrals, 1126–1136
 changing in triple integrals, 1135–1136
Variation of parameters, 1264–1266
Vector field, 1146–1148, 1146d
 central force field, 1210
 conservative, 1148
 curl, 1200d
 divergence of, 1192d
 Divergence Theorem, 1192–1198
 sketching using Computer Algebra Systems, 1158–1159
 three dimensional conservative, 1204
 in three dimensions, 1169–1170
Vector field theory, 1145–1211
Vector function(s), 887–935, 888d
 calculus of, 893–900
 calculus using Computer Algebra Systems, 914–915
 derivative of, 894–895, 894d, 896–897
 graph of, 889d
 graphing using Computer Algebra Systems, 891–892
 integrals of, 898–900
 limit of, 893d
Vector product, see Cross product
Vector space,
 three-dimensional, 848
 two-dimensional, 832d
Vector-valued function, see Vector function(s)
Vectors, 829–884
 addition, 830–831
 algebraic, 831–832
 angle between, 838–841, 839d
 angle between (three dimensions), 848
 basis, 834–836
 component along another vector, 841
 coplanar, 858–859
 cross product, 853–859, 853d

 algebraic properties, 856–857
 geometric interpretation, 854–855
 direction, 864
 distance formula (three dimensions), 847–848
 dot product (three dimensions), 848
 equations, 864–865
 equivalent, 830
 geometric, 830, 831
 horizontal component, 831
 magnitude, 833–834, 834d
 magnitude (three dimensions), 848
 negative of, 833d
 normal, 870, 871–872
 operations using Computer Algebra Systems, 859–862
 orthogonal, 841
 parallel, 836d
 position, 831
 projections, 843–844
 properties, 832–833
 resultant, 830
 scalar multiplication, 831
 subtraction, 831, 833d
 tangent, 895–896, 896d
 in three dimensions, 846–851
 triple scalar product, 857
 unit, 834–836
 unit normal, 907–913
 unit tangent, 907–913
 vertical component, 831
 zero vector, 830, 833d
Velocity, 917, 918d
 average, 917
Velocity field, 1145, 1146
Verhulst, Pierre-François, 1241
Verhulst's equation, 1241
Vertical Line Test, 955
Vibrating spring, 1268–1274
 damped free vibrations, 1271–1273
 forced vibrations, 1273–1274
 undamped free vibrations, 1269–1271
Volterra, Vito, 1251
Volume,
 defined by double integral, 1071d
 of parallelepiped, 858
 using triple integrals, 1114
von Koch, Helge, 765
Vortex, 1204

Wave equation, 983
Weierstrass, Karl, 813
Work, 842–843, 1149–1159
 using Computer Algebra Systems, 1158–1159
 by a variable force, 1154–1157, 1155d
Wronski, Josef, 1267
Wronskian, 1267

Zero vector, 830, 833d

Forms Containing $\sqrt{u^2 \pm a^2}$ (continued)

41. $\displaystyle\int \frac{\sqrt{u^2 - a^2}}{u} du = \sqrt{u^2 - a^2} - a\sec^{-1}\left|\frac{u}{a}\right| + C$

42. $\displaystyle\int \frac{\sqrt{u^2 \pm a^2}}{u^2} du = -\frac{\sqrt{u^2 \pm a^2}}{u} + \ln\left|u + \sqrt{u^2 \pm a^2}\right| + C$

43. $\displaystyle\int \frac{du}{\sqrt{u^2 \pm a^2}} = \ln\left|u + \sqrt{u^2 \pm a^2}\right| + C$

44. $\displaystyle\int \frac{u^2 du}{\sqrt{u^2 \pm a^2}} = \frac{u}{2}\sqrt{u^2 \pm a^2} \mp \frac{a^2}{2}\ln\left|u + \sqrt{u^2 \pm a^2}\right| + C$

45. $\displaystyle\int \frac{du}{u\sqrt{u^2 + a^2}} = -\frac{1}{a}\ln\left|\frac{a + \sqrt{u^2 + a^2}}{u}\right| + C$

46. $\displaystyle\int \frac{du}{u\sqrt{u^2 - a^2}} = \frac{1}{a}\sec^{-1}\left|\frac{u}{a}\right| + C$

47. $\displaystyle\int \frac{du}{u^2\sqrt{u^2 \pm a^2}} = \mp\frac{\sqrt{u^2 \pm a^2}}{a^2 u} + C$

48. $\displaystyle\int (u^2 \pm a^2)^{3/2} du = \frac{u}{8}(2u^2 \pm 5a^2)\sqrt{u^2 \pm a^2} + \frac{3a^4}{8}\ln\left|u + \sqrt{u^2 \pm a^2}\right| + C$

49. $\displaystyle\int \frac{du}{(u^2 \pm a^2)^{3/2}} = \pm\frac{u}{a^2\sqrt{u^2 \pm a^2}} + C$

Forms Containing $\sqrt{a^2 - u^2}$

50. $\displaystyle\int \sqrt{a^2 - u^2}\, du = \frac{u}{2}\sqrt{a^2 - u^2} + \frac{a^2}{2}\sin^{-1}\frac{u}{a} + C$

51. $\displaystyle\int u^2\sqrt{a^2 - u^2}\, du = \frac{u}{8}(2u^2 - a^2)\sqrt{a^2 - u^2} + \frac{a^4}{8}\sin^{-1}\left(\frac{u}{a}\right) + C$

52. $\displaystyle\int \frac{\sqrt{a^2 - u^2}}{u} du = \sqrt{a^2 - u^2} - a\ln\left|\frac{a + \sqrt{a^2 - u^2}}{u}\right| + C$

53. $\displaystyle\int \frac{\sqrt{a^2 - u^2}}{u^2} du = -\frac{\sqrt{a^2 - u^2}}{u} - \sin^{-1}\frac{u}{a} + C$

54. $\displaystyle\int \frac{u^2}{\sqrt{a^2 - u^2}} du = -\frac{u}{2}\sqrt{a^2 - u^2} + \frac{a^2}{2}\sin^{-1}\left(\frac{u}{a}\right) + C$

55. $\displaystyle\int \frac{du}{u\sqrt{a^2 - u^2}} = -\frac{1}{a}\ln\left|\frac{a + \sqrt{a^2 - u^2}}{u}\right| + C$

56. $\displaystyle\int \frac{du}{u^2\sqrt{a^2 - u^2}} = -\frac{\sqrt{a^2 - u^2}}{a^2 u} + C$

57. $\displaystyle\int (a^2 - u^2)^{3/2} du = \frac{u}{4}\left(a^2 - u^2\right)^{3/2} + \frac{3a^2 u}{8}\sqrt{a^2 - u^2} + \frac{3a^4}{8}\sin^{-1}\frac{u}{a} + C$

58. $\displaystyle\int \frac{du}{(a^2 - u^2)^{3/2}} = \frac{u}{a^2\sqrt{a^2 - u^2}} + C$

Forms Involving $\sqrt{2au - u^2}$

59. $\displaystyle\int \sqrt{2au - u^2}\, du = \frac{u - a}{2}\sqrt{2au - u^2} + \frac{a^2}{2}\cos^{-1}\left(\frac{a - u}{a}\right) + C$

60. $\displaystyle\int u\sqrt{2au - u^2}\, du = \frac{2u^2 - au - 3a^2}{6}\sqrt{2au - u^2} + \frac{a^3}{2}\cos^{-1}\left(\frac{a - u}{a}\right) + C$

61. $\displaystyle\int \frac{\sqrt{2au - u^2}}{u} du = \sqrt{2au - u^2} + a\cos^{-1}\left(\frac{a - u}{a}\right) + C$

62. $\displaystyle\int \frac{\sqrt{2au - u^2}}{u^2} du = -\frac{2\sqrt{2au - u^2}}{u} - \cos^{-1}\left(\frac{a - u}{a}\right) + C$

63. $\displaystyle\int \frac{du}{\sqrt{2au - u^2}} = \cos^{-1}\left(\frac{a - u}{a}\right) + C$

64. $\displaystyle\int \frac{u\, du}{\sqrt{2au - u^2}} = -\sqrt{2au - u^2} + a\cos^{-1}\left(\frac{a - u}{a}\right) + C$

65. $\displaystyle\int \frac{u^2\, du}{\sqrt{2au - u^2}} = -\frac{(u + 3a)}{2}\sqrt{2au - u^2} + \frac{3a^2}{2}\cos^{-1}\left(\frac{a - u}{a}\right) + C$

66. $\displaystyle\int \frac{du}{u\sqrt{2au - u^2}} = -\frac{\sqrt{2au - u^2}}{au} + C$

67. $\displaystyle\int \frac{\sqrt{2au - u^2}}{u^n} du = \frac{(2au - u^2)^{3/2}}{(3 - 2n)au^n} + \frac{n - 3}{(2n - 3)a}\int \frac{\sqrt{2au - u^2}}{u^{n-1}} du,\ n \neq \frac{3}{2}$

68. $\displaystyle\int \frac{u^n\, du}{\sqrt{2au - u^2}} = -\frac{u^{n-1}\sqrt{2au - u^2}}{n} + \frac{a(2n - 1)}{n}\int \frac{u^{n-1}}{\sqrt{2au - u^2}} du$

69. $\displaystyle\int \frac{du}{u^n\sqrt{2au - u^2}} = \frac{\sqrt{2au - u^2}}{a(1 - 2n)u^n} + \frac{n - 1}{(2n - 1)a}\int \frac{du}{u^{n-1}\sqrt{2au - u^2}}$

70. $\displaystyle\int \frac{du}{(2au - u^2)^{3/2}} = \frac{u - a}{a^2\sqrt{2au - u^2}} + C$

71. $\displaystyle\int \frac{u\, du}{(2au - u^2)^{3/2}} = \frac{u}{a\sqrt{2au - u^2}} + C$

Forms Containing Trigonometric Functions

72. $\displaystyle\int \sin^2 u\, du = \frac{u}{2} - \frac{\sin 2u}{4} + C$

73. $\displaystyle\int \cos^2 u\, du = \frac{u}{2} + \frac{\sin 2u}{4} + C$